PHYSICAL CHEMISTRY
Second Edition

Gilbert W. Castellan
University of Maryland

Addison-Wesley Publishing Company
Reading, Massachusetts
Menlo Park, California · London · Amsterdam · Don Mills, Ontario · Sydney

This book is in the
ADDISON-WESLEY SERIES IN CHEMISTRY

Consulting Editor
Francis T. Bonner

To Joan and our family

Preface
to the
Second Edition

The second edition differs from the first edition mainly by the addition of three chapters on quantum mechanics, and substantial revisions of several others. These chapters include a discussion of operator algebra and the mathematical underpinning of the quantum mechanics, a thorough discussion of simple problems, such as the free particle, the particle in a box and the harmonic oscillator. Multidimensional problems, such as a particle in a three-dimensional box, the rigid rotator, and the hydrogen atom, are introduced to illustrate both the separation technique and the properties of these systems. Some additional detail has been added to the discussion on the valence bond and molecular orbital methods.

The fundamental theory of spectroscopy has been presented in sufficient mathematical detail so that the student can understand how and why the transition moment integral enters the problem in such a crucial way. It is then an easy step to the understanding of selection rules and the role of symmetry. Symmetry is emphasized both in its relation to selection rules and in its relation to quantization.

At the same time, most of the original qualitative material has been retained for the benefit of those who do not desire so much mathematical detail.

Several other improvements in presentation have been made and problems have been added in a number of the chapters.

The empirical observation is that the second edition is longer than the first. The additional material should allow a wider range of choice among the various topics. Although the depth of the mathematical treatment may appear to go beyond that of some of the popular textbooks, it is not beyond the capability of the student who has a good grasp of the fundamentals of the calculus. The selection of the areas of physical chemistry which are to be given the greatest emphasis and the most thorough treatment is a judgmental matter which is best left to the individual teacher.

I wish to express my appreciation to all of the many students, teachers, and casual readers who have been kind enough to write letters with questions, criticisms, and suggestions. All of these have been very helpful. Particular thanks are due to Professor N. O. Smith of Fordham University, who was thoughtful enough to send me a very extensive list of errors, and to Reverend Dr. Robert J. Ratchford, S.J., of Loyola University (New Orleans), who has given me outstanding help by systematically discovering errors of various kinds. I can only hope that in this revision these and other errors have not crept in again!

Special thanks are due Mrs. Leone Havenner, who typed much of the new material, and Mrs. Louise Wells for her help with the index.

Finally, my best thanks go to my wife, Joan, and to Stephen, Billy, David, and Susan for their encouragement, patience, and understanding.

June, 1971   G. W. C.

Preface
to the
First Edition

Eighty years ago a physical chemist was a man whose principal interest was the study of electrolytic solutions. Today the physical chemist may be engaged in almost any scientific field. There are physical chemists whose work may take them from pure mathematics through theoretical and practical physics and chemistry, medicine, and biology, to botany. It is no longer practical to set limits "by definition" on physical chemistry. One might describe a physical chemist as a scientist whose first training was in chemistry.

In spite of the diversity in the ultimate activities of physical chemists, an introductory course must expose the fundamental principles which are applicable to all kinds of physico-chemical systems. Beyond the exposition of fundamentals, the first course in physical chemistry takes as many directions as there are teachers. In writing a book, an author is faced with the problem of selection among the many possible topics which can quite properly be considered under the title of physical chemistry. Rather than attempt an encyclopedic coverage of the subject and include a paragraph on almost everything, I have tried to cover fundamentals and some applications in depth. I hope that when a teacher finds that one of his favorite topics has been slighted or even omitted entirely that he will also find that the fundamental groundwork has been laid and that he has a good point of departure for the discussion of that topic. The primary aim has been to write a book which the student can, with effort, read and understand, to provide the beginner with a reliable and understandable guide for study when he is not in the teacher's presence. My hope is that this book may be readable enough so that the teacher may leave the side issues and the more elementary aspects for assigned reading while he uses the lectures to illuminate the more difficult points and to expand on some of the major points in the current of his own field and thought. Some of the material presented is intended exclusively for reading; Chapters 1, 5, 6, and most of Chapter 18 are designed in this way. These chapters provide background necessary for the material which follows.

There is a trend toward placing physical chemistry earlier in the curriculum than the third year. So long as the student has had introductory courses in chemistry, physics, and the calculus, this book should be suitable regardless of the academic year in which the course is taken.

Except where it would overburden the student needlessly, the subject is presented in a mathematically rigorous way. In spite of this, no mathematics beyond the elementary calculus is required. The justification for a rigorous treatment is peda-gogical; it makes the subject simpler. The beginner may find it difficult at first to follow a lengthy derivation, but he *can* follow it if it is rigorous and logical. Some of the "simplified" derivations are not difficult to follow, but impossible. These deriva-tions leave mysteries in their wake and give the student the feeling that some sixth sense is necessary, as indeed it is, to decide what to do next. Much of the beauty of an exact science lies in the unity of logical structure and in the precision with which the parts are joined. Good cabinet making does not permit the use of lumber which has not been trued, nor does good science permit the use of sloppy terminology and half-joined reasoning. To think always with complete precision and great clarity is im-possible for all of us, but the habit of doing it at least occasionally must begin some-time. At the very latest it should begin in the physical chemistry course. This is difficult in the beginning, but by the end of the course those who have begun to master the task know the feeling of accomplishment.

I have tried to give each of the broad areas, thermodynamics, structure, and kinetics, a unified treatment. For example, the condition of equilibrium is formulated in every instance in terms of the chemical potentials, rather than in terms of the statement that at equilibrium, $\Delta G = 0$. The repeated application of the requirement of equality of the chemical potentials to diverse equilibria has a unifying effect and a pedagogical value that, in my opinion, is absent in the statement that $\Delta G = 0$. Similarly, the Heisenberg principle has been emphasized in the discussion of electron clouds and of quantum numbers. For the same reason, the discussion of transport properties is presented first in a general way, then the principles are applied to various kinds of transport phenomena.

In the first parts of the book a persistent effort is made to read physical meaning into the equations; this is done in a less obvious way in the later parts since by then the student should be taking the initiative of doing the obvious part of the process himself. The earlier chapters are written with leading questions as a motif. This is to illustrate the mode of attack on an unsolved problem. This technique is suppressed later in the hope that by that time the questions to be asked will occur naturally to the reader's mind.

A word is in order about the notation for thermodynamic quantities. For energy, E is used since it suggests the word. For enthalpy, H, which is common usage, is used. For the Helmholtz function, A is used, and for the Gibbs function, G. In the text attention is called to the fact that the commonest tables of data use F for the

Gibbs free energy. I have avoided using F since it is used as commonly for the Helmholtz function as it is for the Gibbs function, but the trend is in the direction of using F for the Helmholtz function and G for the Gibbs function. I think that for the elementary student the simplest course at present is to avoid the use of F completely. The discussion of electrochemical cells has been done in terms of electrode potentials for which $\mathscr{V}$ is used, following the usage in Pitzer and Brewer's revision of Lewis and Randall's *Thermodynamics*. The letter $\mathscr{E}$ is used for the cell emf.

My thanks are due to earlier authors in physical chemistry who have shaped my thoughts on various topics. Most particular thanks are due to my first teachers in the subject, men such as Professors Karl F. Herzfeld, Walter J. Moore, and Francis O. Rice. In addition, I am deeply indebted to Professor James A. Beattie for his kind permission to reprint definitions from his book, *Lectures on Elementary Chemical Thermodynamics*. I believe that the influence of this remarkably clear exposition may be noticeable throughout the material on thermodynamics in this book. Chapter 8, the introduction to the second law, is particularly indebted to Professor Beattie's *Lectures*.

Special thanks are due Professor Francis T. Bonner of State University of New York at Stonybrook, for his careful review of the manuscript and helpful suggestions. Thanks are also due to the American Chemical Society for permission to reprint material; parts of Chapter 12 first appeared in the *Journal of Chemical Education*. The kind assistance of Dr. Paul L. Damour in the preparation of the manuscript is gratefully acknowledged. The work of proofreading was done during a sabbatical year as a National Science Foundation Postdoctoral Fellow at the Max Planck Institut für physikalische Chemie in Göttingen. I wish to express my appreciation to the National Science Foundation for the grant of the Fellowship and to the Max Planck Gesellschaft for their hospitality.

My thanks go to my colleagues who have withstood a barrage of questions. Finally, to my wife, Joan McDonald Castellan, and my family for their constant encouragement and patient endurance throughout the labor I am grateful in measure beyond words.

G. W. C.

Contents

Chapter 4   The Structure of Gases

Chapter 5   Some Properties of Liquids and Solids

Chapter 6   The Laws of Thermodynamics: Generalities and the Zeroth Law

Chapter 7   Energy and the First Law of Thermodynamics; Thermochemistry

Chapter 12   Phase Equilibrium in Simple Systems; The Phase Rule

Chapter 13   Solutions
I. The Ideal Solution and Colligative Properties

Chapter 14   Solutions
II. More than One Volatile Component; The Ideal Dilute Solution

Chapter 32   Chemical Kinetics
II. Theoretical Aspects

Chapter 33   Chemical Kinetics
III. Heterogeneous Reactions, Electrolysis, Photochemistry

Appendix 1

Foreword
to the
Student

On most campuses the course in physical chemistry has a certain reputation for difficulty. It is not, nor should it be, the easiest course available. To keep the matter in perspective it must be said that the IQ of a genius is not necessary for understanding the subject.

The greatest stumbling block that can be erected in the path of learning physical chemistry is the notion that memorizing equations is a sensible way to proceed. Memory should be reserved for the fundamentals and important definitions. Equations are meant to be understood, not to be memorized. In physics and chemistry an equation is not a jumbled mass of symbols, but is a statement of a relation between physical quantities. As you study keep a pencil and scratch paper handy. Play with the final equation from a derivation. If it expresses pressure as a function of temperature, turn it around and express the temperature as a function of pressure. Sketch the functions so that you can "see" the variation. How does the sketch look if one of the parameters is changed? Read physical meaning into the various terms and the algebraic signs which appear in the equation. If a simplifying assumption has been made in the derivation, go back and see what would happen if that assumption were omitted. Applying the derivation to a different special case. Invent problems of your own involving this equation and solve them. Juggle the equation back and forth until you understand its meaning.

In the first parts of the book much space is devoted to the meaning of equations; I hope that I have not been too long-winded about it, but it is important to be able to interpret the mathematical statement in terms of its physical content.

By all means try to keep a good grasp on the fundamental principles which are being applied; memorize *them* and above all *understand* them. Take the time to gain an appreciation of the methods which are used to attack a problem. The algebra and calculus are mechanical devices after the fundamental relations have been established, but they should be respected for the precision tools they are.

If problems baffle you, learn the technique of problem solving. The principles contained in G. Polya's book,* *How to Solve It*, have helped many of my students. It is available as a paperback and is well worth studying. Work as many problems as possible. Numerical answers to *all* problems can be found at the end of the book, p. 687 ff. Make up your own problems as often as possible. Watching your teacher perform will not make you into an actor; problem solving will.

Finally, don't let that reputation for difficulty fool you. Many people have learned physical chemistry and many have enjoyed it at the same time.

Chapter One

Some Fundamental Chemical Concepts

1–1 Introduction

We begin the study of physical chemistry with a brief statement of a few fundamental ideas and common usages in chemistry. These are very familiar things, but it is worth while recalling them to mind.

1–2 The Kinds of Matter

The various kinds of matter can be separated into two broad divisions: (1) substances and (2) mixtures.

Under a specified set of experimental conditions a substance exhibits a definite set of physical and chemical properties which do not depend on the previous history or on the method of preparation of the substance. For example, after appropriate purification, sodium chloride has the same properties whether it has been obtained from a salt mine or prepared in the laboratory by combining sodium hydroxide with hydrochloric acid.

On the other hand, mixtures may vary widely in chemical composition (a property, in the sense of the word in the preceding paragraph). Consequently their physical and chemical properties vary with composition, and may depend on the manner of preparation. By far the majority of naturally occurring materials are mixtures of substances. For example, a solution of salt in water, a handful of earth, or a splinter of wood are all mixtures.

1–3 The Kinds of Substances

Substances are of two kinds: elementary substances, or elements, and compound substances, or compounds. An element cannot be broken down into simpler

1

substances by ordinary chemical methods, but a compound can be. An ordinary chemical method is any method involving an energy of the order of 10 eV or less.†

For example, the element mercury cannot undergo any *chemical* decomposition of the type Hg → X + Y, in which X and Y individually have smaller masses than the original mass of mercury. For the purposes of this definition, both X and Y must have masses at least as large as that of the hydrogen atom, since the reaction Na → Na$^+$ + e$^-$ is a chemical reaction involving an energy of about 5 eV. In contrast, the compound methane can be decomposed chemically into simpler substances individually less massive than the original methane: CH_4 → C + $2H_2$.

All natural materials can be chemically broken down ultimately into 89 elements. In addition to these, 16 more elements have recently been made using the methods of nuclear physics (methods involving energies of the order of 10^6 eV or larger). Because of the great difference in the energies involved in chemical methods and nuclear methods, there is no likelihood of confusing the two. The nuclei of atoms endure through chemical reactions; only the outermost electrons of the atoms, the valence electrons, are affected.

1–4   Atoms, Molecules, Atomic and Molecular Weights

The atom is defined classically as the smallest part of an element which can exist in a chemical change. The atoms of a particular element have a definite mass, which is different for different elements. Since the discovery of *isotopes*, the mass of the atom has been replaced by *atomic number* (Z, the number of units of positive charge on the nucleus) as the characteristic property which differentiates one element from another. All atoms of hydrogen have one unit of positive charge on the nucleus, all atoms of helium have two units, and all atoms of oxygen have eight units, etc. Among the atoms of hydrogen, atoms of different mass may be distinguished; these are the isotopes of hydrogen. The isotope having one unit of mass is ordinary hydrogen, the isotope having two units of mass is deuterium, and the isotope having three units of mass is tritium. There are three naturally occurring isotopes of oxygen, which have masses of 16, 17, and 18 mass units. As they occur in nature the elements are mixtures of isotopes. For example, chlorine is a mixture of two isotopes:

$$75.4\% \text{ of the isotope of mass} = 34.98,$$

$$24.6\% \text{ of the isotope of mass} = 36.98,$$

$$\text{Average mass} = 0.754(34.98) + 0.246(36.98) = 35.47.$$

The table of atomic "weights" is a compilation of masses of the elements relative to the mass of one element (or one isotope) which is assigned an arbitrary, convenient value. Formerly, the chemist assigned to the average mass of the atoms in the naturally occurring isotopic mixture of oxygen atoms, the arbitrary value

---

† One electron-volt (1 eV) = 23.05 kcal/mole.

of exactly 16 units. The physicist found it more convenient to assign the mass of exactly 16 units to the oxygen isotope having 16 atomic mass units. Thus, two scales of atomic "weights" were in common use. In 1961, by international agreement, these two scales were abandoned and a unified scale came into use. The unified scale is based on the assignment of the value of exactly 12 units to the mass of the carbon isotope which has 12 atomic mass units. Atomic weights on the new scale are lower than those on the old chemical scale by 43 parts in a million; this change is so slight that the vast majority of the data in the chemical literature need not be recalculated.

Atoms of one element can combine chemically with atoms of another element to form the minute parts of the compound called molecules; e.g., four atoms of hydrogen can combine with one atom of carbon to form a molecule of methane, CH_4. Atoms of a single element can also combine with themselves to form molecules of the element, e.g., H_2, O_2, Cl_2, P_4, S_8.

The molecular weight of a molecule can be computed by adding the atomic weights of all the atoms in it. By adding the atomic weight of carbon, 12.011, to four times the atomic weight of hydrogen, 4(1.008), the molecular weight of CH_4, methane, 16.043, is obtained. This method of computing molecular weights assumes that there is no change in mass when the carbon atom combines with four hydrogen atoms to form methane. That is, in the reaction

$$C + 4H \rightarrow CH_4$$

the total mass on the left, 16.043 units, is equal to the total mass on the right, 16.043 units, if the molecular weight of CH_4 is computed by the rule given above.

The question of whether or not mass is conserved in chemical reactions has been the subject of very extensive and very precise experimental investigations, and in no case has any change in mass during a chemical reaction been demonstrated. The law of conservation of mass holds accurately for chemical reactions within the limits of precision of experiments conducted thus far. The expected change in mass accompanying any chemical reaction can be computed from the mass–energy equivalence law of relativity theory. If the energy involved in the chemical reaction is ΔE, and Δm is the associated change in mass, then $\Delta E = (\Delta m)c^2$, where c, the velocity of light, equals 3×10^{10} cm/sec. Computation shows that the change in mass is of the order of 10^{-10} gram per kilocalorie of energy involved in a reaction. This change in mass is too small to be detected by contemporary methods. Therefore, the law of conservation of mass may be considered exact in all chemical situations.

1–5 Gram-atomic and Gram-molecular Weights; the Avogadro Number

Atomic weights are dimensionless ratios which have no reference whatsoever to any particular unit of measurement. If the metric unit of mass, the gram, is attached to these numbers, then they are called the *gram-atomic weights* of the elements. A gram-molecular weight of any substance is called a *gram-mole*, or a mole, of the substance.

Since the mass of an individual atom is extremely small, the number of atoms in a gram-atomic weight of any element must be enormous. This number is the Avogadro number, N_0 ($= 6.023 \times 10^{23}$). One gram-atomic weight of any element contains an Avogadro number of atoms of that element. Similarly, one gram-molecular weight of any compound contains an Avogadro number of molecules of that compound.

1–6 Symbols; Formulas

Over the years a set of symbols for the elements has evolved. Depending on the context, the symbol for an element may stand for several different things: it may merely be an abbreviation of the name of the element; it may symbolize one atom of the element; most often it represents a gram-atomic weight of the element.

The formulas of compounds are interpreted in a variety of ways, but in every instance the formula describes the relative composition of the compound. In substances such as quartz and salt, discrete molecules are not present. Therefore, the formulas SiO_2 and $NaCl$ are given only empirical meaning; these formulas describe the relative numbers of atoms of the elements present in the compound and nothing more. Strictly, one should speak of gram-formula weights of such compounds rather than gram-molecular weights.

For substances which consist of discrete molecules, their formulas describe the relative numbers of the constituent atoms and the total number of atoms in a molecule; e.g., acetylene, C_2H_2; benzene, C_6H_6; sulfur hexafluoride, SF_6.

Structural formulas are used to describe the way in which the atoms are connected within a molecule. Within the limitations imposed by a two-dimensional diagram, they display the geometry of a molecule. Bonding within a molecule is illustrated by using conventional symbols for single and multiple bonds, electron pairs, and positive and negative centers of charge in the molecule. Structural formulas have their greatest utility in representing substances with discrete molecules. As yet, no satisfactory abbreviated way of representing the structural complexity of substances such as quartz and salt has been devised. In using any structural formula, a great deal must be mentally supplied to the diagram.

1–7 Chemical Equations

A chemical equation is a shorthand method for describing a chemical transformation. The substances on the left-hand side of the equation are called *reactants*, those on the right-hand side are called *products*. The equation

$$MnO_2 + HCl \rightarrow MnCl_2 + H_2O + Cl_2$$

expresses the fact that manganese dioxide will react with hydrogen chloride to form manganous chloride, water, and chlorine. As it is written, the equation does little besides record the fact of the reaction and the proper formulas for each substance.

If the equation is balanced,

$$MnO_2 + 4HCl \rightarrow MnCl_2 + 2H_2O + Cl_2$$

it expresses the fact that the number of atoms of a given kind must be the same on both sides of the equation. Most important, *the balanced chemical equation is an expression of the law of conservation of mass.* Chemical equations provide the relationship between the masses of the various reactants and products, which is ordinarily of utmost importance in chemical problems. The balanced equation above states that 86.93 gm of MnO_2 will react with 145.83 gm of HCl to form 125.84 gm of $MnCl_2$, 18.02 gm of H_2O, and 70.90 gm of Cl_2. Knowing this, one can easily calculate the weight relation for any specified weight of any of the substances involved.

Chapter Two

Empirical Properties
of Gases

2–1 Boyle's Law; Charles' Law

Of the three states of aggregation, only the gaseous state allows a comparatively simple quantitative description. For the present we shall restrict this description to the relations among such properties as mass, pressure, volume, and temperature. We shall assume that the system is in equilibrium so that the values of the properties do not change with time, so long as the external constraints on the system are not altered.

A system is in a definite state or condition when all of the properties of the system have definite values, which are determined by the state of the system. Thus the state of the system is described by specifying the values of some or all of its properties. The important question is whether it is necessary to give values of fifty different properties, or twenty, or five, to assure that the state of the system is completely described. The answer depends to a certain extent upon how accurate a description is required. If we were in the habit of measuring the values of properties to twenty significant figures, and thank Heaven we are not, then quite a long list of properties would be required. Fortunately, even in experiments of great refinement, only four properties, mass, volume, temperature, and pressure, are ordinarily required.

The *equation of state* of the system is the mathematical relationship between the values of these four properties. Only three of these must be specified to describe the state; the fourth can be calculated from the equation of state, which is obtained from knowledge of the experimental behavior of the system.

The first quantitative measurements of the pressure–volume behavior of gases were made by Robert Boyle in 1662. His data indicated that the volume is inversely proportional to the pressure: $V = C/p$, where p is the pressure, V is the volume, and

C is a constant. Figure 2–1 shows V as a function of p. Boyle's law may be written in the form

$$pV = C; \qquad (2\text{--}1)$$

it applies only to a fixed mass of gas at a constant temperature.

Later experiments of Charles showed that the constant C is a function of temperature. This is a rough statement of Charles' law.

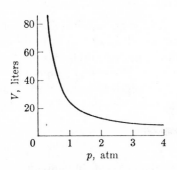

Fig. 2–1 Volume as a function of pressure, Boyle's law ($t = 25°C$).

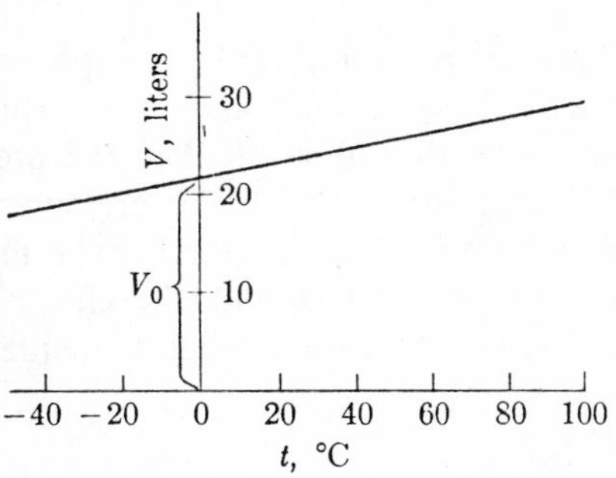

Fig. 2–2 Volume as a function of temperature, Charles' law ($p = 1$ atm).

Gay-Lussac made measurements of the volume of a fixed mass of gas under a fixed pressure and found that the volume was a linear function of the temperature. This is expressed by the equation

$$V = a + bt, \qquad (2\text{--}2)$$

where t is the temperature, and a and b are constants. A plot of volume as a function of temperature is shown in Fig. 2–2. The intercept on the vertical axis is $a = V_0$, the volume at 0°C. The slope of the curve is the derivative,[†] $b = (\partial V/\partial t)_p$. Thus, Eq. (2–2) can be written in the equivalent form

$$V = V_0 + \left(\frac{\partial V}{\partial t}\right)_p t. \qquad (2\text{--}3)$$

Charles' experiments showed that for a fixed mass of gas under a constant pressure, the *relative* increase in volume per degree increase in temperature was *the same for all gases* on which he made measurements. At a fixed pressure the increase in volume per degree is $(\partial V/\partial t)_p$; hence, the relative increase in volume per degree is $(1/V_0)(\partial V/\partial t)_p$. This quantity is the *coefficient of thermal expansion* at

---

[†]   The partial derivative is used rather than the ordinary derivative, since the volume depends on the pressure; a and b are constants only if the pressure is constant. The partial derivative $(\partial V/\partial t)_p$ is the rate of change of volume with temperature at constant pressure; this is the slope of the line under the conditions of the experiment.

$0°C$, for which we use the symbol α_0:

$$\alpha_0 = \frac{1}{V_0}\left(\frac{\partial V}{\partial t}\right)_p. \tag{2-4}$$

Then Eq. (2–3) may be written in terms of α_0:

$$V = V_0(1 + \alpha_0 t) = V_0\alpha_0\left(\frac{1}{\alpha_0} + t\right), \tag{2-5}$$

which is convenient because it expresses the volume of the gas in terms of the volume at zero degrees and a constant, α_0, which is the same for all gases and, as it turns out, is very nearly independent of the pressure at which the measurements are made. If we measure α_0 at various pressures we find that for all gases α_0 approaches the same limiting value at $p = 0$. The form of Eq. (2–5) suggests a transformation of coordinates which should be useful; namely, define a new temperature T in terms of the old temperature through the equation

$$T = 1/\alpha_0 + t. \tag{2-6}$$

Equation (2–6) defines a new temperature scale, called a *gas scale* of temperature, or, more exactly, an ideal gas scale of temperature. The importance of this scale lies in the fact that the limiting value of α_0, and consequently $1/\alpha_0$, has the same value for all gases. On the other hand, α_0 does depend on the scale of temperature used originally for t. In Eq. (2–6), if t is in degrees fahrenheit, then $1/\alpha_0 = 459.7°$ and the T scale is the rankine scale. If t is in degrees celsius (centigrade) then $1/\alpha_0 = 273.15°$ and the resulting T scale is the kelvin scale.† Temperatures on the kelvin scale are called *absolute temperatures*. Here we shall write temperatures either in degrees kelvin ($°K$) or degrees celsius ($°C$). According to Eq. (2–6)

$$T = 273.15 + t. \tag{2-7}$$

Equations (2–5) and (2–6) are combined to yield

$$V = \alpha_0 V_0 T, \tag{2-8}$$

which states that the volume of a gas under a fixed pressure is directly proportional to the kelvin temperature.

2–2 Molecular Weight of a Gas. Avogadro's Law; the Ideal Gas Law

So far, two relations between the four variables have been obtained: Boyle's law, Eq. (2–1) (fixed mass, constant temperature), and Gay-Lussac's, or Charles' law, Eq. (2–8) (fixed mass, constant pressure). These two equations may be combined

---

†   Strictly speaking, the kelvin scale is the thermodynamic temperature scale rather than the celsius gas scale. Since the scales are numerically identical, the distinction between them will be deferred for later discussion.

into one general equation by noting that V_0 is the volume at $0°C$, and so is related to the pressure by Boyle's law, $V_0 = C_0/p$, where C_0 is the value of the constant at $t = 0$. Then, Eq. (2–8) becomes

$$V = C_0\alpha_0 T/p \qquad \text{(fixed mass)}. \qquad (2\text{–}9)$$

The restriction of fixed mass is removed by realizing that if the temperature and pressure are kept constant and the mass of the gas is doubled, the volume will double. This means that the constant C_0 is proportional to the mass of gas; hence, we write $C_0 = Bw$, where B is a constant and w is the mass. Introducing this result into Eq. (2–9), we obtain

$$V = B\alpha_0 wT/p, \qquad (2\text{–}10)$$

which is the general relation between the four variables V, w, T, and p. Each gas has a different value of the constant B.

For Eq. (2–10) to be useful, one would have to have at hand a table of values for B for all the various gases. To avoid this, B is expressed in terms of a characteristic mass for each gas. Let M denote the mass of gas in the container under a set of standard conditions T_0, p_0, V_0. If different gases are confined in the standard volume V_0 under the standard temperature and pressure T_0 and p_0, then by Eq. (2–10), for each gas

$$M = \left(\frac{1}{B\alpha_0}\right)\left(\frac{p_0 V_0}{T_0}\right). \qquad (2\text{–}11)$$

Since the standard conditions are chosen to suit our convenience, the ratio $R = p_0 V_0/T_0$ has a fixed numerical value for any particular choice and has, of course, the same value for all the gases (R is called the *gas constant*). Equation (2–11) may then be written

$$M = R/B\alpha_0 \quad \text{or} \quad B = R/M\alpha_0.$$

Using this value for B in Eq. (2–10), we obtain

$$V = (w/M)RT/p. \qquad (2\text{–}12)$$

Let the number of characteristic masses of the gas contained in the mass w be $n = w/M$. Then $V = nRT/p$, or

$$pV = nRT. \qquad (2\text{–}13)$$

Equation (2–13), the *ideal gas law*, has great importance in the study of gases. It does not contain anything which is characteristic of an individual gas, but is a generalization applicable to all gases.

We now inquire about the significance of the characteristic mass M. Avogadro's law says that equal volumes of different gases under the same conditions of temperature and pressure contain equal numbers of molecules. We have compared equal volumes, V_0, under the same temperature and pressure, T_0 and p_0, to obtain the characteristic masses of the different gases. According to Avogadro's law these characteristic masses must contain the same number of molecules; let this number be N_0. Then the characteristic mass M is N_0 times the mass of the individual molecule m:

$$M = N_0 m. \tag{2–14}$$

Since M is proportional to the mass of the individual molecule, it is called the *molecular weight* of the gas, meaning the *mass* of N_0 molecules. This amount of the substance is called a *mole*. In Eq. (2–13) n is the number of moles of the gas present. Since the value of R is directly connected to the definition of molecular weight, we shall find that the gas constant appears in equations which describe molar properties of solids and liquids, as well as gases.

It is important to realize that M and N_0 depend upon the volume chosen originally to make the comparison. Before the adoption of the unified scale of atomic weights, it was the chemical convention to choose the volume, T_0, and p_0 so that the characteristic mass of gaseous oxygen was exactly 32.0000. This choice yields atomic weights which are all greater than unity. Then $V_0 = 22.414$ liters/mole, $T_0 = 273.15°$ K, $p_0 = 1$ atm, and $N_0 = 6.023 \times 10^{23}$ molecules/mole. The constant R is calculated using these values:

$$R = \frac{p_0 V_0}{T_0} = \frac{(1 \text{ atm})(22.414 \text{ liters/mole})}{(273.15 \text{ deg K})} = 0.08205 \text{ liter·atm·mole}^{-1}\text{·deg } K^{-1},$$

or, if V_0 had been expressed in milliliters, $R = 82.05$ ml·atm·mole^{-1}·deg K^{-1}. These two values of R are the most convenient ones in computations using the ideal gas law, since the volume is usually expressed either in liters or milliliters, the pressure in atmospheres, and the temperature in °K.

Values of R in other units are given in Table 2–1. Note that the pressure–volume product has the dimensions of energy. Writing physical magnitudes only, pressure = force/area = dyne/cm^2, while volume = cm^3. Then pressure $\times$ volume = (dyne/cm^2) $\times$ cm^3 = dyne·cm = erg. Thus, in the cgs system, R has the units ergs·mole^{-1} deg K^{-1}.

Any of the values in Table 2–1 may be used in any problem involving the gas constant R, so long as the units of the various quantities are carefully noted.

Table 2–1 Values of R in several units

Type of unit	Value	Units
Mechanical	0.082054	liter·atm·mole^{-1}·deg K^{-1}
Mechanical	82.054	ml·atm·mole^{-1}·deg K^{-1}
cgs	8.3144×10^7	ergs·mole^{-1}·deg K^{-1}
Electrical	8.3144	joules·mole^{-1}·deg K^{-1}
Thermal	1.9872	calories·mole^{-1}·deg K^{-1}

Example One mole of an ideal gas occupies 10 liters at 300° K. What is the pressure of the gas?

Solution.   The required relation between the data and the unknown is the ideal gas law, Eq. (2–13), written in the form

$$p = nRT/V.$$

To calculate the numerical answer, suppose that we choose the last entry in Table 2–1, $R = 1.987$ cal·mole^{-1}·deg K^{-1}. Then

$$p = (1 \text{ mole})(1.987 \text{ cal·mole}^{-1}\text{·deg K}^{-1})(300° \text{K})/10 \text{ liter} = 59.6 \text{ cal/liter}.$$

This value, 59.6 cal/liter, is a perfectly correct answer to the problem; the difficulty is that the unit *cal/liter* for pressure is so infrequently used that its use here would be considered eccentric. It would be better if the answer had come out in atmospheres. Examination of the other entries in Table 2–1 leads us to choose the first one. Then

$$p = (1 \text{ mole})(0.08205 \text{ liter·atm·mole}^{-1}\text{·deg K}^{-1})(300° \text{K})/10 \text{ liter} = 2.46 \text{ atm}.$$

Obviously, it would have been more sensible to use this value in the beginning. When difficulty is experienced in "getting the answer" to problems which involve the gas constant, check immediately to see if the value of R used was chosen in units compatible with the units of the other quantities involved.

2–3 The Equation of State; Extensive and Intensive Properties

The ideal gas law, $pV = nRT$, is a relation between the four variables which describe the state of any gas. As such, it is an *equation of state*. The variables in this equation fall into two classes: n and V are extensive variables (extensive properties), while p and T are intensive variables (intensive properties).

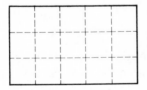

Fig. 2–3 Subdivision of the system.

The value of any extensive property is obtained by summing the values of that property in every part of the system. Suppose that the system is subdivided into many small parts, as in Fig. 2–3. Then the total volume of the system is obtained by adding together the volumes of each small part. Similarly, the total number of moles (or total mass) in the system is obtained by summing the number of moles in (or mass of) each part. By definition, such properties are extensive. It should be clear that the value obtained is independent of the way in which the system is subdivided.

Intensive properties are not obtained by such a process of summation but are measured at any point in the system, and each has a uniform value throughout a system at equilibrium; e.g., T and p.

Extensive variables are proportional to the mass of the system. For the ideal gas, as an example, $n = w/M$, and $V = wRT/Mp$. Both n and V are proportional to the mass of the system. Dividing V by n, we obtain $\bar{V}$, the volume per mole:

$$\bar{V} = V/n = RT/p. \tag{2–15}$$

The ratio of V to n is not proportional to the mass; since in forming the ratio the mass drops out, $\bar{V}$ is an intensive variable. The ratio of any two extensive variables is *always* an intensive variable.

If the ideal gas law is written in the form

$$p\bar{V} = RT, \tag{2–16}$$

it is a relation between *three intensive variables*, pressure, temperature, and molar volume. This is important because we can now discuss the properties of the ideal gas without continually worrying about whether we are dealing with ten moles or ten million moles. It should be clear that no fundamental property of the system depends on the accidental choice of 20 gm rather than 100 gm of material for study. In the atom bomb project, micro quantities of material were used in preliminary studies, and vast plants were built based on the properties determined on this tiny scale. If it were the case that fundamental properties depended on the amount of substance used, one could imagine the government giving research grants for the study of extremely large systems; enormous buildings might be required, depending upon the ambition of the investigators! For the discussion of principles, the intensive variables are the significant ones. In practical applications such as design of apparatus and engineering, the extensive properties are important as well, since they condition the size of apparatus, the horsepower of an engine, the production capacity of a plant in tons per day, etc.

2–4 Properties of the Ideal Gas

If arbitrary values are assigned to any two of the three variables p, $\bar{V}$, and T, the value of the third variable can be calculated from the ideal gas law. Hence, any set of two variables is a set of *independent* variables; the remaining variable is a *dependent* variable. The fact that the state of a gas is completely described if the values of any two intensive variables are specified allows a very neat geometric representation of the states of a system.

In Fig. 2–4, p and $\bar{V}$ have been chosen as independent variables. Any point, such as A, determines a pair of values of p and $\bar{V}$; this is sufficient to describe the state of the system. Therefore, every point in the p-$\bar{V}$ quadrant (both p and $\bar{V}$ must be positive to make physical sense) describes a different state of the gas. Furthermore, every state of the gas is represented by some point in the p-$\bar{V}$ diagram.

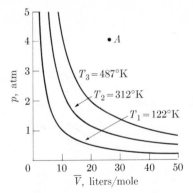

Fig. 2–4 Isotherms of the ideal gas.

It is frequently useful to pick out all of the points which correspond to a certain restriction on the state of the gas, as, for example, the points which correspond to the same temperature. In Fig. 2–4 the curves labeled T_1, T_2, and T_3 collect all the points which represent states of the ideal gas at the temperatures T_1, T_2, and T_3, respectively. The curves in Fig. 2–4 are called *isotherms*. The isotherms of the ideal gas are rectangular hyperbolas determined by the relation

$$p = RT/\overline{V}. \tag{2–17}$$

For each curve, T has a different constant value.

In Fig. 2–5 every point corresponds to a set of values for the coordinates $\overline{V}$ and T; again each point represents a state of the gas, just as in Fig. 2–4. In Fig. 2–5 points corresponding to the same pressure are collected on the lines, which are called *isobars*. The isobars of the ideal gas are described by the equation

$$\overline{V} = (R/p)T, \tag{2–18}$$

where the pressure is assigned various constant values.

As in the other figures, every point in Fig. 2–6 represents a state of the gas, since it determines values of p and T. The lines of constant molar volume, *isometrics*, are described by the equation

$$p = (R/\overline{V})T, \tag{2–19}$$

where $\overline{V}$ is assigned various constant values.

These diagrams derive their great utility from the fact that all the gaseous, liquid, and solid states of any pure substance can be represented on the same diagram. We will use this idea extensively, particularly in Chapter 12.

A careful examination of Figs. 2–4, 2–5, and 2–6 and of Eqs. (2–17), (2–18), and (2–19) leads to some rather bizarre conclusions about the ideal gas. For example, Fig. 2–5 and Eq. (2–18) say that the volume of an ideal gas confined under a constant pressure is zero at $T = 0°$. Similarly, Fig. 2–4 and Eq. (2–17) tell us that the volume of the ideal gas kept at a constant temperature approaches zero as the pressure

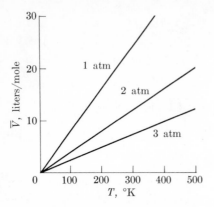

Fig. 2–5 Isobars of the ideal gas.

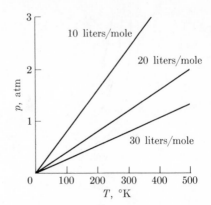

Fig. 2–6 Isometrics of the ideal gas.

becomes infinitely large. These predictions do not correspond to the observed behavior of real gases at low temperatures and high pressures. As a real gas under a constant pressure is cooled, we observe a decrease in volume, but at some definite temperature the gas liquefies; after liquefaction occurs, not much decrease in the volume is observed as the temperature is lowered. Similarly, isothermal compression of a real gas may produce liquefaction, and thereafter further increase in pressure produces little change in the volume. It is apparent from this that there is good reason for referring to the relation $p\bar{V} = RT$ as the *ideal* gas law. The above discussion shows that we may expect the ideal gas law to fail in predicting the properties of a real gas at low temperatures and at high pressures. Experiment shows that the behavior of all real gases approaches that of the ideal gas as the pressure approaches zero.

In Chapter 3 deviations from the ideal gas law are discussed in detail. For the moment, a few general remarks will suffice on the question of when the ideal gas law may reasonably be used for predicting properties of real gases. In practice, if only a rough approximation is required, the ideal gas law is used without hesitation. This rough approximation is in many cases quite good, within perhaps 5 %. For a rule of such broad scope, the ideal gas law is astonishingly accurate in many practical situations.

The ideal gas law is more accurate the higher the temperature is above the critical temperature of the substance, and the lower the pressure is below the critical pressure† of the substance. In precision work the ideal gas law is never used.

2–5 Determination of Molecular Weights of Gases and Volatile Substances

The ideal gas law is useful in determining the molecular weights of volatile substances. For this purpose a bulb of known volume is filled with the gas at a measured pressure

---

†   Above the critical temperature, and above the critical pressure, it is not possible to distinguish liquid and vapor as separate entities.

and temperature. The mass of the gas in the bulb is measured. These measurements suffice to determine the molecular weight of the substance. From Eq. (2–12) we have $pV = (w/M)RT$; then

$$M = (w/V)(RT/p) = (\rho/p)(RT),\tag{2–20}$$

where $\rho = w/V$; ρ is the density (mass/cm^3). All of the quantities on the right-hand side of Eq. (2–20) are known from the measurements; hence M can be calculated.

A rough value of the molecular weight is usually sufficient to determine the molecular formula of a substance. For example, if chemical analysis of a gas yields an empirical formula, $(CH_2)_n$, then the molecular weight must be some multiple of 14; the possibilities are 28, 42, 56, 70, etc. If a molecular weight determination using Eq. (2–20) yields a value of 54, then we may conclude that $n = 4$ and that the material is one of the butenes. The fact that the gas is not strictly ideal does not hinder us in this conclusion at all. In this example the possible values of M are well enough so that even if the ideal gas law were wrong by 5%, we would still have no difficulty in assigning the correct molecular formula to the gas. In this example it is unlikely that the ideal gas law would be in error by as much as 2% for a convenient choice of experimental conditions.

Since the determination of molecular weight together with chemical analysis establishes the molecular formula of the gaseous substance, the results are of great importance. For example, some very common substances exhibit *dimerization*, a doubling of a simple unit. Table 2–2 lists some of these substances, all of which are solids or liquids at room temperature. Measurements of molecular weight must be made at temperatures sufficiently high to vaporize the materials.

The fact that the behavior of a real gas approaches that of the ideal gas as the pressure is lowered is used as a basis for the precise determination of molecular weights of gases. According to Eq. (2–20), the ratio of density to pressure should be independent of pressure: $\rho/p = M/RT$. This is correct for an ideal gas, but if the density of a

Table 2–2 Dimerization

Compound	Empirical formula	Molecular formula in the vapor
Aluminum chloride	$AlCl_3$	Al_2Cl_6
Aluminum bromide	$AlBr_3$	Al_2Br_6
Formic acid	$HCOOH$	$(HCOOH)_2$
Acetic acid	CH_3COOH	$(CH_3COOH)_2$
Arsenic trioxide	As_2O_3	As_4O_6
Arsenic pentoxide	As_2O_5	As_4O_{10}
Phosphorus trioxide	P_2O_3	P_4O_6
Phosphorus pentoxide	P_2O_5	P_4O_{10}

real gas is measured at one temperature and at several different pressures, the ratio of density to pressure is found to depend slightly on the pressure. At sufficiently low pressures, ρ/p is a linear function of the pressure. The straight line can be extrapolated to yield a value of ρ/p at zero pressure $(\rho/p)_0$, which is appropriate to the ideal gas and can be used in Eq. (2–20) to give a precise value of M:

$$M = (\rho/p)_0 RT. \tag{2–21}$$

This procedure is illustrated for ammonia in Fig. 2–7.

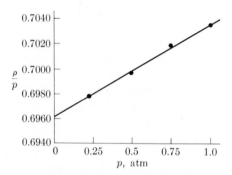

Fig. 2–7 Plot of ρ/p versus p for ammonia.

2–6 Mixtures; Composition Variables

The state or condition of a mixture of several gases depends not only on the pressure, volume, and temperature, but also on the composition of a mixture. Consequently, a method of specifying the composition must be devised. The simplest method would be to state the mole numbers $n_1, n_2, \ldots$ of the several substances in the mixture (the masses would also serve). This method has the disadvantage that the mole numbers are extensive variables. It is preferable to express the composition of a mixture in terms of a set of intensive variables.

It has been shown that the ratio of two extensive variables is an intensive variable. The mole numbers can be converted to intensive variables by dividing each one by some extensive variable. This can be done in several ways.

Molar concentrations, c_i (moles/liter), are obtained by dividing each of the mole numbers by the volume:

$$c_i = n_i/V. \tag{2–22}$$

Molar concentrations are satisfactory for describing the composition of liquid or solid mixtures because the volume is comparatively insensitive to changes in temperature and pressure. Since the volume of a gas depends markedly on temperature and pressure, molar concentrations are not usually convenient for describing the composition of gas mixtures.

Mole ratios, r_i, are obtained by choosing one of the mole numbers and dividing all the others by that one. Choosing n_1 as the divisor, we have

$$r_i = n_i/n_1. \tag{2-23}$$

A variant of the mole ratio description, the molal concentration m_i, is often used for liquid solutions. Let the solvent be component 1, with a molecular weight M_1. The molality of component i is the number of moles of i per kilogram of solvent. Then

$$m_i = \frac{1000 n_i}{n_1 M_1} = \frac{1000}{M_1} r_i. \tag{2-24}$$

The molality is the mole ratio multiplied by a constant, $1000/M_1$. Since the mole ratios and the molalities are completely independent of temperature and pressure, they are preferable to the molar concentrations for the physico-chemical description of mixtures of any kind.

Mole fractions, x_i, are obtained by dividing each of the mole numbers by the total number of moles of all the substances present, $n_t = n_1 + n_2 + \cdots$,

$$x_i = n_i/n_t. \tag{2-25}$$

The sum of the mole fractions of all the substances in a mixture must be unity:

$$x_1 + x_2 + x_3 + \cdots = 1. \tag{2-26}$$

Because of this relation, the composition of the mixture is described when the mole fractions of all but one of the substances are specified; the remaining mole fraction is computed using Eq. (2-26). Like molalities and mole ratios, mole fractions are independent of temperature and pressure, and thus are suitable for describing the composition of any mixture. Gas mixtures are commonly described by the mole fractions, since the p-V-T relations have a concise and symmetrical form in these terms.

2-7 Equations of State for a Gas Mixture; Dalton's Law

Experiment shows that for a mixture of gases, the ideal gas law is correct in the form

$$pV = n_t RT, \tag{2-27}$$

where n_t is the total number of moles of all the gases in the volume V. Equation (2-27) and the statement of the mole fractions of all but one of the constituents of the mixture constitute a complete description of the equilibrium state of the system.

It is desirable to relate the properties of complicated systems to those of simpler systems, so we attempt to describe the state of a gas mixture in terms of the states of pure unmixed gases. Consider a mixture of three gases described by the mole numbers n_1, n_2, n_3 in a container of volume V at a temperature T. If $n_t = n_1 + n_2 + n_3$, then the pressure exerted by this mixture is given by

$$p = n_t(RT/V). \tag{2-28}$$

We define the partial pressure of each gas in the mixture as the pressure the gas would exert if it were alone in the container of volume V at temperature T. Then the partial pressures p_1, p_2, p_3 are given by

$$p_1 = n_1 \frac{RT}{V}, \qquad p_2 = n_2 \frac{RT}{V}, \qquad p_3 = n_3 \frac{RT}{V}. \tag{2-29}$$

Adding these equations, we obtain

$$p_1 + p_2 + p_3 = (n_1 + n_2 + n_3)\frac{RT}{V} = n_t \frac{RT}{V}.$$

Comparison of this equation with Eq. (2–28) shows that

$$p = p_1 + p_2 + p_3. \tag{2-30}$$

This is Dalton's law of partial pressures, which states that at any specified temperature the total pressure exerted by a gas mixture is equal to the sum of the partial pressures of the constituent gases. The first gas is said to exert a partial pressure p_1, the second gas exerts a partial pressure p_2, and so on. Partial pressures are calculated using Eqs. (2–29).

Partial pressures are simply related to the mole fractions of the gases in the mixture. Dividing both sides of the first of Eqs. (2–29) by the total pressure p, we obtain

$$p_1/p = n_1(RT/pV); \tag{2-31}$$

but, by Eq. (2–28), $p = n_t RT/V$. Using this value for p on the right-hand side of Eq. (2–31), we have

$$p_1/p = n_1/n_t = x_1.$$

Thus,

$$p_1 = x_1 p, \qquad p_2 = x_2 p, \qquad p_3 = x_3 p.$$

These equations are conveniently abbreviated by writing

$$p_i = x_i p \qquad (i = 1, 2, 3, \ldots), \tag{2-32}$$

where p_i is the partial pressure of the gas which has a mole fraction x_i. Equation (2–32) allows the calculation of the partial pressure of any gas in a mixture from the mole fraction of that gas and the total pressure of the mixture.

Two things should be noted about Eq. (2–32): first, if either molar concentrations or mole ratios had been used, the final result would not be as simple an expression as Eq. (2–32); second, examination of the steps leading to Eq. (2–32) shows that it is not restricted to a mixture of three gases; it is correct for a mixture containing any number of gases.

2–8 The Partial-pressure Concept

The definition given in Eqs. (2–29) for the partial pressures of the gases in a mixture is a purely mathematical one; we now ask whether or not this mathematical concept of partial pressure has any physical significance. The results of two experiments, illustrated in Figs. 2–8 and 2–9, provide the answer to this question. First consider the experiment shown in Fig. 2–8. A container, Fig. 2–8(a), is partitioned into two compartments of equal volume V. The upper compartment contains hydrogen under a pressure of one atmosphere; the lower compartment is evacuated. One arm of a manometer is covered by a thin palladium foil and is connected to the hydrogen-filled compartment. The other arm of the manometer is open to a pressure of 1 atm, which is kept constant during the experiment as is the temperature. At the beginning of the experiment, the mercury levels in the two arms of the manometer stand at the same height. This is possible because the palladium membrane is permeable to hydrogen but not to other gases, and so the membrane does not block the entrance of hydrogen to the manometer arm.

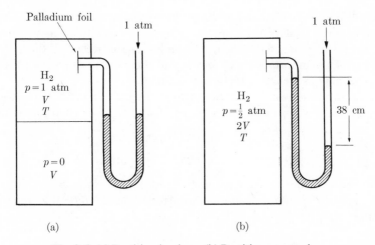

Fig. 2–8 (a) Partition in place. (b) Partition removed.

The partition is removed, and the hydrogen fills the entire vessel. After a period of time, the mercury levels rest in the final positions shown in Fig. 2–8(b). Since the volume available to the hydrogen has doubled, the pressure in the container has fallen to one-half its original value. (We neglect the volume of the manometer arm in this computation.)

A different experiment, Fig. 2–9, is performed. In this experiment, the lower compartment contains nitrogen (which cannot pass the palladium foil) under 1 atm pressure. At the beginning of the experiment, the mercury levels stand at the same height. The partition is removed and the gases mix throughout the container. After

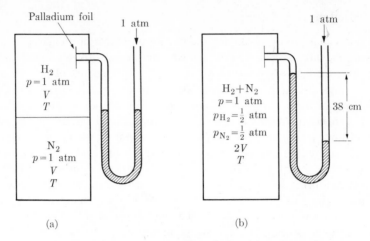

Fig. 2–9 (a) Partition in place. (b) Partition removed.

a period of time the levels stand at the positions shown in Fig. 2–9(b). The result of this experiment is *exactly* the same as in the first experiment in which the lower compartment was evacuated. The hydrogen behaves exactly as if the nitrogen were not present. This important result means that the concept of partial pressure has a physical meaning as well as a mathematical one.

The interpretation of each experiment is direct. In the first experiment, the manometer read the total pressure both before and after the partition was removed:

$$p_{\text{initial}} = n_{H_2}(RT/V) = 1 \text{ atm},$$

$$p_{\text{final}} = n_{H_2}(RT/2V) = \tfrac{1}{2} \text{ atm}.$$

In the second experiment, the manometer read *total pressure* before the membrane was removed, and *partial pressure of hydrogen in the mixture* after removal of the membrane:

$$p_{\text{initial}} = n_{H_2}(RT/V) = 1 \text{ atm},$$

$$p_{H_2 \text{(final)}} = n_{H_2}(RT/2V) = \tfrac{1}{2} \text{ atm},$$

$$p_{N_2 \text{(final)}} = n_{N_2}(RT/2V) = \tfrac{1}{2} \text{ atm},$$

$$p_{\text{total,final}} = p_{H_2} + p_{N_2} = \tfrac{1}{2} + \tfrac{1}{2} = 1 \text{ atm}.$$

Note that the total pressure in the container does not change upon removal of the partition.

It is possible to measure the partial pressure of any gas in a mixture directly if there is a membrane which is permeable to that gas alone; e.g., palladium is permeable to hydrogen and certain types of glass are permeable to helium. The fact that at present only a few such membranes are known does not destroy the physical reality of the concept of partial pressure. Later it will be shown that in chemical equilibria involving gases and in physical equilibria such as solubility of gases in liquids and solids, it is the partial pressures of the gases in the mixture which are significant (further confirmation of the physical content of the concept).

2–9   Partial Volumes; Amagat's Law

The partial volume of a gas in a mixture is defined as the volume the gas would occupy if it were by itself in a container at temperature T and pressure p. Thus,

$$V_1 = n_1(RT/p), \qquad V_2 = n_2(RT/p), \qquad V_3 = n_3(RT/p). \qquad (2\text{--}33)$$

Adding Eqs. (2–33), we obtain

$$V_1 + V_2 + V_3 = (n_1 + n_2 + n_3)(RT/p) = n_t(RT/p).$$

Since, by Eq. (2–27), $V = n_t RT/p$, we see that

$$V = V_1 + V_2 + V_3, \qquad (2\text{--}34)$$

which is Amagat's law of partial volumes. Dividing Eqs. (2–33) by $V = n_t RT/p$, we obtain

$$V_i = x_i V. \qquad (2\text{--}35)$$

Although for real gases Amagat's law is frequently more accurate than Dalton's law and consequently is convenient for some computations, the concept of partial volumes is a purely mathematical one and has no physical significance. It need not be discussed further here.

2–10   The Barometric Distribution Law

In the foregoing discussion of the behavior of ideal gases it has been tacitly assumed that the pressure of the gas has the same value everywhere in the container. Strictly speaking, this assumption is correct only in the absence of force fields. Since all measurements are made on laboratory systems which are always in the presence of a gravitational field, it is important to know what effect is produced by the influence of this field. It may be said that for gaseous systems of ordinary size, the influence of the gravity field is so slight as to be imperceptible even with extremely refined experimental methods. For a fluid of higher density such as a liquid, the effect is quite pronounced, and the pressure will be different at different vertical positions in a container.

    A column of fluid, Fig. 2–10, having a cross-sectional area A, at a uniform temperature T, is subjected to a gravitational field acting downward to give a particle an acceleration g. The vertical coordinate z is measured upward from ground level where $z = 0$. The pressure at any height z in the column is determined by the total weight of fluid in the column above that height. (Note that weight is a force;

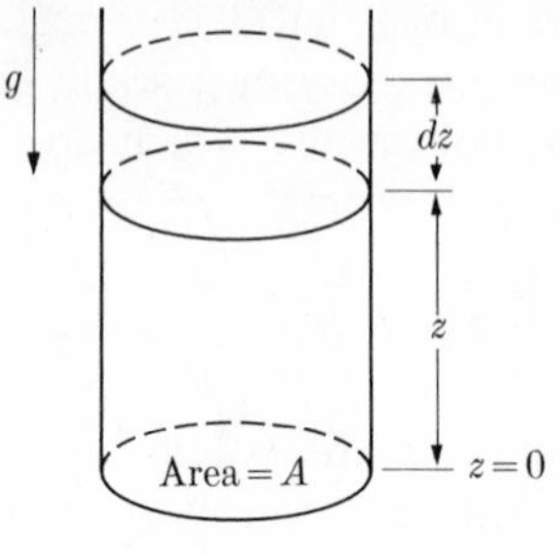

Fig. 2–10 Column of gas in a gravity field.

weight $= mg$. Hence, a weight per unit area is a pressure.) Let F_z equal the weight of fluid in the column above height z, F_{z+dz} equal the weight of fluid in the column above height $z + dz$, and dF equal the weight of fluid in the column between heights z and $z + dz$. Then

$$F_{z+dz} + dF = F_z. \tag{2–36}$$

If p is the pressure at height z, and $p + dp$ is the pressure at height $z + dz$, then

$$F_z = pA \qquad \text{and} \qquad F_{z+dz} = (p + dp)A.$$

Thus Eq. (2–36) becomes $(p + dp)A + dF = pA$, or,

$$A\,dp + dF = 0. \tag{2–37}$$

If ρ is the density of the fluid at height z, the mass of fluid in the column between heights z and $z + dz$ is $\rho A\,dz$; the weight between these heights is $\rho Ag\,dz = dF$. Equation (2–37) then becomes

$$dp = -\rho g\,dz. \tag{2–38}$$

The differential equation, Eq. (2–38), relates the change in pressure, dp, to the density of the fluid, the gravitational acceleration, and to the increment in height dz. The negative sign means that if the height increases (dz is $+$), the pressure of the fluid will decrease (dp is $-$). The effect of change in height on the pressure is proportional to the density of the fluid; thus the effect is important for liquids and negligible for gases.

If the density of a fluid is independent of pressure, as is the case for liquids, then Eq. (2–38) may be integrated immediately. Since ρ and g are constants, they are removed from the integral, and we obtain

$$\int_{p_0}^{p} dp = -\rho g \int_{0}^{z} dz,$$

which, after integrating, gives

$$p - p_0 = -\rho g z, \tag{2–39}$$

where p_0 is the pressure at the bottom of the column, and p is the pressure at the height

z above the bottom of the column. Equation (2–39) is the usual equation for hydrostatic pressure in a liquid. Since it is assumed that the reader is familiar with the variation of pressure with depth in a liquid from his study of physics, Eq. (2–39) will not be discussed further.

To apply Eq. (2–38) to a gas, it must be recognized that the density of the gas is a function of the pressure. If the gas is ideal, then from Eq. (2–20), $\rho = (M/RT)p$. Using this in Eq. (2–38), we have

$$dp = -(Mg/RT)p\,dz.$$

Separating variables yields

$$dp/p = -(Mg/RT)\,dz, \tag{2–40}$$

and integrating, we obtain

$$\ln p = -(Mgz/RT) + C. \tag{2–41}$$

The integration constant C is evaluated in terms of the pressure at ground level; when $z = 0$, $p = p_0$. Using these values in Eq. (2–41), we find that $\ln p_0 = C$. Substituting this value for C and rearranging reduces Eq. (2–41) to

$$\ln (p/p_0) = -(Mgz/RT), \tag{2–42}$$

or

$$p = p_0 e^{-Mgz/RT}. \tag{2–43}$$

Since the density is proportional to the pressure, and the number of particles per cubic centimeter is proportional to the pressure, Eq. (2–43) can be written in two other equivalent forms:

$$\rho = \rho_0 e^{-Mgz/RT} \quad \text{or} \quad N' = N'_0 e^{-Mgz/RT}, \tag{2–44}$$

where ρ and ρ_0 are the densities and N' and N'_0 are the number of particles per cubic centimeter at z and at ground level. Any of the equations (2–43), or (2–44) is called the barometric distribution law or the gravitational distribution law. The equation is a distribution law, since it describes the distribution of the gas in the column. In calculation with these equations, R, M, g, and z are all expressed in cgs units.

Equation (2–43) relates the pressure at any height z to the height, the temperature of the column, the molecular weight of the gas, and the acceleration produced by the gravity field. Figure 2–11 shows a plot of p/p_0 versus z for nitrogen at three temperatures, according to Eq. (2–43). Figure 2–11 shows that at the higher temperature, the distribution is smoother than at the lower temperature. The variation in pressure with height is less pronounced the higher the temperature; if the temperature were infinite, the pressure would be the same everywhere in the column.

It is advisable to look more closely at this exponential type of distribution law, since it occurs so frequently in physics and physical chemistry in a more general form as the *Boltzmann distribution*. Equation (2–40) is most informative in discussing the

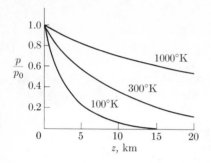

Fig. 2–11 Plot of p/p_0 versus z for nitrogen.

exponential distribution, and it can be written

$$(-dp/p) = (Mg/RT)\, dz \tag{2–45}$$

which says that the relative decrease† in pressure, $-dp/p$, is a constant, Mg/RT, multiplied by the increase in height, dz. It follows that this relative decrease is the same at all positions in the column; therefore, it cannot matter where the origin of z is chosen. For example, suppose that for a certain gas the pressure at ground level is 1 atm and the distribution shows that the pressure decreases to $\frac{1}{2}$ atm at a height of 10 km. Then for this same gas, the pressure at a height $z + 10$ km is one-half the value of the pressure at the height z. Thus at any height, the pressure is one-half the value it has at a height 10 km below. This aspect of the distribution is emphasized in Fig. 2–12.

The argument does not depend on the choice of one-half as the relative value. Suppose that for some gas the pressure at a height of 6.3 km is 0.88 of its value at ground level. Then in another interval of 6.3 km, the pressure will drop again by the factor 0.88. The pressure at $2(6.3) = 12.6$ km will then be $(0.88)(0.88) = 0.774$ of the ground level value (see Problem 2–22).

Another point to note about Eq. (2–45) is that the relative decrease in pressure is proportional to Mg/RT. Consequently, for any particular gas, the relative decrease is less at higher temperatures (see Fig. 2–11). At a specified temperature the relative decrease is larger for a gas having a high molecular weight than for a gas with a low molecular weight.

For a gas mixture in a gravity field, it can be shown that each of the gases obeys the distribution law independently of the others. For each gas

$$p_i = p_{io}e^{-M_i gz/RT}, \tag{2–46}$$

where p_i is the partial pressure of the ith gas in the mixture at the height z, p_{io} is the partial pressure of the gas at ground level, and M_i is the molecular weight of the gas. The interesting consequence of this law is that the partial pressures of very light gases

---

† Since dp is an increase, $-dp$ is a decrease.

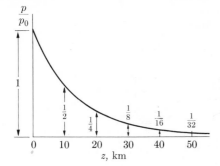

Fig. 2–12 The constant relative decrease in pressure with equal increments in height.

decrease less rapidly with height than do those of heavier gases. Thus, in the earth's atmosphere, the percentage composition at very great heights is quite different from that at ground level.† At a height of 100 km the light gases such as helium and neon form a higher percentage of the atmosphere than they do at ground level.

Problems

2–1. Five grams of ethane are confined in a bulb of one liter capacity. The bulb is so weak that it will burst if the pressure exceeds 10 atm. At what temperature will the pressure of the gas reach the bursting value?

2–2. A large cylinder for storing compressed gases has a volume of about 1.5 ft³. If the gas is stored under a pressure of 150 atm at 300°K, how many moles of gas are contained in the cylinder? What would be the weight of oxygen in such a cylinder?

2–3. Helium is contained at 30.2°C in the system illustrated in Fig. 2–13. The leveling bulb L can be raised so as to fill the lower bulb with mercury and force the gas into the upper part of the device. The volume of bulb 1 to the mark b is 100.5 cm³ and the volume of bulb 2 between the marks a and b is 110.0 cm³. The pressure exerted by the helium is measured by the difference between the mercury levels in the device and in the evacuated arm of the manometer. When the mercury level is at a, the pressure is 20.14 mm of Hg. What is the mass of helium in the container?

2–4. The same type of apparatus is used as in Problem 2–3. In this case the volume v_1 is not known; the volume of bulb 2, v_2, is 110.0 cm³. When the mercury level is at a the pressure is 15.42 mm of Hg. When the mercury level is raised to b, the gas pressure rises to 27.35 mm. The temperature is 30.2°C.

(a) What is the mass of helium in the system?

(b) What is the volume of bulb 1?

2–5. Suppose that in setting up the scales of atomic weights, the standard conditions had been chosen as $p_0 = 1$ atm, $V_0 = 30.000$ liters, and $T_0 = 300.00°$K. Compute the "gas

---

†   This does not mean to imply that the atmospheric composition at various levels can be calculated *accurately* by Eq. (2–46). The atmosphere is not isothermal, nor is it in equilibrium. Winds at low levels mix the gases more uniformly than would be predicted by the distribution law.

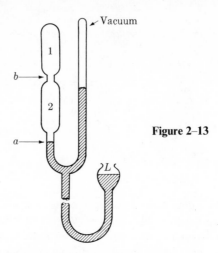

Figure 2–13

constant," the "Avogadro number," and the masses of an "Avogadro number" of hydrogen atoms and oxygen atoms.

2–6. The coefficient of thermal expansion α is defined by $\alpha = (1/V)(\partial V/\partial T)_p$. Using the equation of state, compute the value of α for an ideal gas.

2–7. The coefficient of compressibility β is defined by $\beta = -(1/V)(\partial V/\partial p)_T$. Compute the value of β for an ideal gas.

2–8. For an ideal gas, express the derivative $(\partial p/\partial T)_V$ in terms of α and β.

2–9. One gram of N_2 and 1 gm of O_2 are put in a 2-liter flask at 27°C. Compute the partial pressure of each gas, the total pressure, and the composition of the mixture in mole percent.

2–10. One gram of H_2 and 1 gm of O_2 are put in a 2-liter flask at 27°C. Compute the partial pressure of each gas, the total pressure, and the composition of the mixture in mole percent. Compare with the result of Problem 2–9.

2–11. A mixture of nitrogen and water vapor is admitted to a flask which contains a solid drying agent. Immediately after admission, the pressure in the flask is 760 mm. After standing some hours, the pressure reaches a steady value of 745 mm.

   a) Calculate the composition, in mole percent, of the original mixture.

   b) If the experiment is done at 20°C and the drying agent increases in weight by 0.150 gm, what is the volume of the flask? (The volume occupied by the drying agent may be ignored.)

2–12. A mixture of oxygen and hydrogen is analysed by passing it over hot copper oxide and through a drying tube. Hydrogen reduces the CuO according to the equation CuO + $H_2 \rightarrow$ Cu + H_2O; Oxygen then reoxidizes the copper formed: Cu + $\frac{1}{2}O_2 \rightarrow$ CuO. 100 cm^3 of the mixture measured at 25°C and 750 mm yields 84.5 cm^3 of dry oxygen measured at 25°C and 750 mm after passage over CuO and the drying agent. What is the original composition of the mixture?

2–13. Show that $x_i = (y_i/M_i)/[(y_1/M_1) + (y_2/M_2) + \cdots]$, in which x_i, y_i, and M_i are the mole fraction, the weight percent, and the molecular weight of component i, respectively.

2–14. A gas sample is known to be a mixture of ethane and butane. A bulb of 200 cm³ capacity is filled with the gas to a pressure of 750 mm at 20°C. If the weight of gas in the bulb is 0.3846 gm, what is the mole percent of butane in the mixture?

2–15. A bulb of 138.2 ml volume contains 0.6946 gm of gas at 756.2 mm and 100°C. What is the molecular weight of the gas?

2–16. Assuming that air has a mean molecular weight of 28.8, and that the atmosphere is isothermal at 25°C, compute the barometric pressure at Denver, Colo., which is one mile above sea level; compute the barometric pressure at the top of Mt. Evans, 14,260 ft above sea level. The pressure at sea level may be taken as 760 mm.

2–17. The approximate composition of the atmosphere at sea level is given in the table below.†

Gas	Mole percent
Nitrogen	78.09
Oxygen	20.93
Argon	0.93
Carbon dioxide	0.03
Neon	0.0018
Helium	0.0005
Krypton	0.0001
Hydrogen	5×10^{-5}
Xenon	8×10^{-6}
Ozone	5×10^{-5}

Ignoring the last four components, compute the partial pressures of the others, the total pressure, and the composition of the atmosphere in mole percent, at altitudes of 50 and 100 km. ($t = 25°C$.)

2–18. What would the molecular weight of a gas have to be if the pressure of the gas is to fall to one-half its value in a vertical distance of one meter? ($t = 25°C$.) What types of molecules have molecular weights of this magnitude?

2–19. Consider an "ideal potato gas" which has the following properties: it obeys the ideal gas law, the individual particles weigh 100 gm, but occupy no volume, i.e., they are point masses.

   a) At 25°C compute the height at which the number of potatoes per cubic centimeter falls to one-millionth of its ground-level value.

   b) Recognizing that real potatoes do occupy a volume, is there any correspondence between the result of the calculation in (a) and the observed spatial distribution of potatoes in a paper bag?

† By permission from *Scientific Encyclopedia*, 3rd ed., D. Van Nostrand Co., Inc., New York, 1958, p. 34.

2–20. A balloon having a capacity of $10,000 \, \text{m}^3$ is filled with helium at $20°C$ and 1 atm pressure. If the balloon is loaded with 80% of the load that it can lift at ground level, at what height will the balloon come to rest? Assume that the volume of the balloon is constant, the atmosphere isothermal, $20°C$, the molecular weight of air is 28.8, and the ground level pressure is 1 atm. The mass of the balloon is $1.3 \times 10^6 \, \text{gm}$.

2–21. Express the partial pressures in a mixture of gases (a) in terms of the molar concentrations c_i, and (b) in terms of the mole ratios r_i.

2–22. If, at a specified height Z, the pressure of a gas is p_Z, and that at $z = 0$ is p_0, show that at any height z, $p = p_0 f^{z/Z}$, where $f = p_Z/p_0$.

2–23. When Julius Caesar expired, his last exhalation had a volume of about $500 \, \text{cm}^3$. This expelled air was $1 \, \text{mole}\%$ argon. Assume that the temperature was $300°K$ and the ground-level pressure was 1 atm. Assume that the temperature and pressure are uniform over the earth's surface and still have the same values. If Caesar's argon molecules have all remained in the atmosphere and have been completely mixed throughout the atmosphere, how many inhalations, $500 \, \text{cm}^3$ each, must we make on average to inhale one of Caesar's argon molecules? (Watch out for units!)

2–24. a) Show that if we calculate the total number of molecules of a gas in the atmosphere using the barometric formula, we would get the same result if we assumed that the gas had the ground-level pressure up to a height, $z = RT/Mg$, and had zero pressure above that level.

b) Show that the total mass of the earth's atmosphere is given by Ap_0/g, where p_0 is the total ground-level pressure and A is the area of the earth's surface. Note that this result does not depend on the composition of the atmosphere. (Do this problem first by calculating the mass of each constituent, mole fraction $= x_i$, molecular weight $= M_i$, and summing. Then by examining the result, do it the easy way.)

c) If the mean radius of the earth is $6.37 \times 10^8 \, \text{cm}$, $g = 980 \, \text{cm/sec}^2$, and $p_0 = 1 \, \text{atm}$, calculate the mass of the atmosphere in grams. (Watch out for units!)

2–25. Since the gases in the atmosphere are distributed differently according to their molecular weights the *average* percentage of each gas is different from the percentage at ground level. The values, x_i^0, of mole fractions at ground level are given.

a) Derive a relation between the average mole fraction of the gas in the atmosphere and the mole fractions at ground level.

b) If the mole fractions of N_2, O_2, and A at ground level are 0.78, 0.21, and 0.01, respectively, compute the average mole fractions of N_2, O_2, and A in the atmosphere.

c) Show that the *average weight fraction* of any gas in the atmosphere is equal to its *mole fraction at ground level*.

2–26. Consider a column of gas in a gravity field. Calculate the height Z, determined by the condition that half the mass of the column lies below Z.

Chapter Three

Real Gases

3–1 Deviations from Ideal Behavior

In Chapter 2 we mentioned that the ideal gas law does not precisely represent the behavior of real gases. We will now attempt to formulate more realistic equations of state for gases and explore the implications of these equations.

If measurements of pressure, molar volume, and temperature of a gas do not confirm the relation $p\overline{V} = RT$, within the precision of the measurements, the gas is said to deviate from ideality or to exhibit nonideal behavior. Deviations from ideality can be broadly classified into two types: "apparent" deviations and real deviations.

3–2 "Apparent" Deviations

"Apparent" deviations from the ideal gas law occur when the total number of molecules in the system depends on the pressure. This happens in gas mixtures in which a chemical equilibrium is established if the chemical reaction produces a change in the number of molecules in the system. Then, by the principle of LeChatelier, the number of moles of gas present is a function of the pressure. If we use the law $pV = n_t RT$, allowance must be made for this dependence of n_t on the pressure.

Consider a vessel containing n moles of N_2O_4 at constant temperature. At ordinary temperatures this substance contains a certain amount of NO_2 because of the establishment of the equilibrium $N_2O_4 \rightleftharpoons 2NO_2$. If α is the degree of dissociation, then at equilibrium the mixture contains $n(1 - \alpha)$ moles of N_2O_4, and $2\alpha n$ moles of NO_2. The total number of moles of gases present is

$$n_t = n(1 - \alpha) + 2\alpha n = n(1 + \alpha).$$

Then, for the mixture,

$$pV = n_t RT = n(1 + \alpha)RT. \qquad (3\text{–}1)$$

From the measured values of n and of the pV product, the value of α can be calculated using Eq. (3–1). From the dependence of α on pressure, the equilibrium constant for the chemical reaction can be obtained. This is a useful experimental method for determining the equilibrium constant of a chemical reaction.

3–3 Real Deviations

Real deviations are small compared with "apparent" deviations; they are observed in pure gases and in nonreacting gas mixtures. Real deviations from ideality are the ones of interest in the study of the properties of gases. Figures 3–1 and 3–2 show schematically the types of deviation which occur; the curves are for room temperature. In both Figs. 3–1 and 3–2 the curve for the real gas is sensibly coincident with that of the ideal gas at low pressures. On the other hand, at high pressures, or low volumes, the curves diverge. Figure 3–1 shows that nitrogen at moderate pressures has a smaller molar volume than the ideal gas and at very high pressures has a larger volume than the ideal gas. Hydrogen, Fig. 3–2, at all pressures has a larger molar volume than that of the ideal gas. It is clear from the illustrations that the divergence becomes more marked at higher pressures; in Figs. 3–1 and 3–2 the magnitude of the deviations has been exaggerated to show the effect.

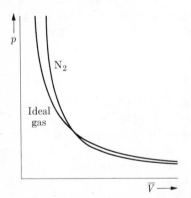

Fig. 3–1 Isotherm for N_2 compared with the ideal gas.

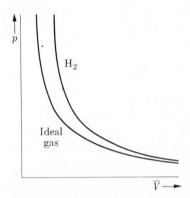

Fig. 3–2 Isotherm for H_2 compared with the ideal gas.

To display the deviations more clearly, the ratio of the observed molar volume $\bar{V}$ to the ideal molar volume $\bar{V}_{id}$ ($= RT/p$) is plotted as a function of pressure at constant temperature. This ratio is called the *compressibility factor*† Z. Then,

$$Z = \frac{\overline{V}}{\overline{V}_{id}} = \frac{p\overline{V}}{RT}. \tag{3–2}$$

For the ideal gas, $Z = 1$ and is independent of pressure and temperature. For real gases, $Z = Z(T, p)$, a function of both temperature and pressure.

    Figure 3–4 shows a plor of Z versus p for several gases at 0°C. Note that for those the ideal gas. For hydrogen, Z is greater than unity (the ideal value) at all pressures. For nitrogen, Z is less than unity in the lower part of the pressure range, but is greater than unity at very high pressures. Note that the pressure range in Fig. 3–3 is very large; near one atmosphere both of these gases behave nearly ideally.

    Figure 3–4 shows a plot of Z versus p for several gases at 0°C. Note that for those gases which are easily liquefied, Z dips sharply below the ideal line in the low-pressure region.

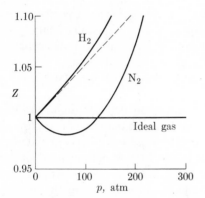

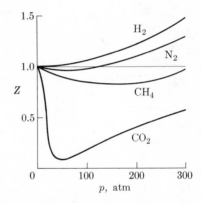

Fig. 3–3 Plot of Z versus p for H_2, N_2, and ideal gas at 0°C.

Fig. 3–4 Plot of Z versus p for several gases at 0°C.

3–4   Modifying the Ideal Gas Equation; the van der Waals Equation

How can the ideal gas law be modified to yield an equation which will represent the experimental data more precisely? We begin by correcting an obvious defect in the ideal gas law, namely the prediction that under any finite pressure the volume of the gas is zero at the absolute zero of temperature: $\overline{V} = RT/p$. On cooling, real gases liquefy and ultimately solidify; after liquefaction the volume does not change very much. We can arrange the new equation so that it predicts a finite, positive

---

†   Not to be confused with the *compressibility*, which is defined in Section 5–2.

volume for the gas at 0°K by adding a positive constant b to the ideal volume:

$$\overline{V} = b + RT/p. \tag{3-3}$$

According to Eq. (3–3) the molar volume at 0°K is b, and we expect that b will be roughly comparable with the molar volume of the liquid or solid.

Equation (3–3) also predicts that as the pressure becomes infinite the molar volume approaches the limiting value b. This prediction is more in accord with experience than the prediction of the ideal gas law that the molar volume approaches zero at very high pressures.

Now it would be interesting to see how well Eq. (3–3) predicts the curves in Figs. 3–3 and 3–4. Since by definition $Z = p\overline{V}/RT$, multiplication of Eq. (3–3) by p/RT yields

$$Z = 1 + \frac{b}{RT}p. \tag{3-4}$$

Since Eq. (3–4) requires Z to be a linear function of pressure with a positive slope b/RT, it cannot possibly fit the curve for nitrogen in Fig. 3–3 which starts from the origin with a negative slope. However, Eq. (3–4) can represent the behavior of hydrogen. In Fig. 3–3 the dashed line is a plot of Eq. (3–4) fitted at the origin to the curve for hydrogen. In the low-pressure range the dashed line represents the data very well.

We can conclude from Eq. (3–4) that the assumption that the molecules of a gas have finite size is sufficient to explain values of Z greater than unity. Apparently this size effect is the dominating one in producing deviations from ideality in hydrogen at 0°C. It is also clear that some other effect must produce the deviations from ideality in gases such as nitrogen and methane, since the size effect cannot explain their behavior in the low-pressure range. This other effect must now be sought.

We have already noted that the worst offenders in the matter of having values of Z less than unity are methane and carbon dioxide, which are easily liquefied. Thus, we begin to suspect a connection between ease of liquefaction and the compressibility factor, and to ask why a gas liquefies. First of all, energy, the heat of vaporization, must be supplied to take a molecule out of the liquid and put it into the vapor. This energy is required because of the forces of attraction acting between the molecule and its neighbors in the liquid. The force of attraction is strong if the molecules are close together, as they are in a liquid, and very weak if the molecules are far apart, as they are in a gas. The problem is to find an appropriate way of modifying the gas equation to take account of the effect of these weak attractive forces.

The pressure exerted by a gas on the walls of a container acts in an outward direction. Attractive forces between the molecules tend to pull them together, thus diminishing the outward thrust against the wall and reducing the pressure below that exerted by the ideal gas. This reduction in pressure should be proportional to the force of attraction between the molecules of the gas.

Consider two small volume elements v_1 and v_2 in a container of gas (Fig. 3–5). Suppose that each volume element contains one molecule and that the attractive force between the two volume elements is some small value f. If another molecule is added to v_2, keeping one molecule in v_1, the force acting between the two elements should be $2f$; addition of a third molecule to v_2 should increase the force to $3f$, and so on. The force of attraction between the two volume elements is therefore proportional to c_2, the concentration of molecules in v_2. If at any point in the argument, the number of molecules in v_2 is kept constant and molecules are added to v_1, then the force should double and triple, etc. The force is therefore proportional to c_1, the concentration of molecules in v_1. Thus, the force acting between the two elements can be written as force $\propto c_1 c_2$. Since the concentration in a gas is everywhere the same, $c_1 = c_2 = c$, and so force $\propto c^2$. But $c = n/V = 1/\overline{V}$; consequently force $\propto 1/\overline{V}^2$.

Fig. 3–5 Volume elements in a gas.

We rewrite Eq. (3–3) in the form

$$p = RT/(\overline{V} - b). \tag{3–5}$$

Because of the attractive forces between the molecules, the pressure is less than that given by Eq. (3–5) by an amount proportional to $1/\overline{V}^2$, so a term is subtracted from the right-hand side to yield

$$p = \frac{RT}{\overline{V} - b} - \frac{a}{\overline{V}^2}, \tag{3–6}$$

where a is a positive constant roughly proportional to the energy of vaporization of the liquid. Two things should be noted about the introduction of the $a/\overline{V}^2$ term. First, forces acting on any volume element in the interior of the gas balance out to zero, only those elements of volume near the wall of the container experience an unbalanced force which tends to pull them toward the center. Thus the effect of the attractive forces is felt only at the walls of the vessel; this is as it should be. Second, the derivation assumed an effective range of action of the attractive forces of the order of centimeters; in fact the range of these forces is of the order of angstroms. The derivation can be done without this assumption to yield the same result.

Equation (3–6) is the *van der Waals equation*, proposed by van der Waals, who was the first to recognize the influence of molecular size and intermolecular forces on the pressure of a gas. These weak forces of attraction are called van der Waals

Table 3-1 van der Waals constants†

Gas	a, liter^2·atm·moles^{-2}	b, liter/mole
He	0.0340	0.0234
H_2	0.244	0.0266
O_2	1.36	0.0318
CO_2	3.61	0.0429
H_2O	5.72	0.0319
Hg	2.88	0.0055

† Francis Weston Sears, *An Introduction to Thermodynamics, the Kinetic Theory of Gases, and Statistical Mechanics.* Addison-Wesley Publishing Co., Inc., Reading, Mass., 1950.

forces. The van der Waals constants a and b for a few gases are given in Table 3-1. The van der Waals equation is frequently written in the equivalent but less instructive forms

$$\left(p + \frac{a}{\overline{V}^2}\right)(\overline{V} - b) = RT$$

or

$$\left(p + \frac{n^2 a}{V^2}\right)(V - nb) = nRT, \tag{3-7}$$

where $V = n\overline{V}$ has been used in the second writing.

3-5 Implications of the van der Waals Equation

The van der Waals equation takes two effects into account: first, the effect of molecular size, Eq. (3-3),

$$p = RT/(\overline{V} - b).$$

Since the denominator in the above equation is smaller than the denominator in the ideal gas equation, the size effect by itself increases the pressure above the ideal value. According to this equation it is the empty space between the molecules, the "free" volume, which follows the ideal gas law. Second, the effect of intermolecular forces, Eq. (3-6),

$$p = \frac{RT}{\overline{V} - b} - \frac{a}{\overline{V}^2},$$

is taken into account. The effect of attractive forces by itself reduces the pressure below the ideal value and is taken into account by subtracting a term from the pressure.

To calculate Z for the van der Waals gas we multiply Eq. (3–6) by $\overline{V}$ and divide by RT; this yields

$$Z = \frac{p\overline{V}}{RT} = \frac{\overline{V}}{\overline{V} - b} - \frac{a}{RT\overline{V}}.$$

The numerator and denominator of the first term on the right-hand side are divided by $\overline{V}$:

$$Z = \frac{1}{1 - b/\overline{V}} - \frac{a}{RT\overline{V}}.$$

At low pressures $b/\overline{V}$ is small compared with unity, so the first term on the right may be developed into a power series in $1/\overline{V}$ by division; thus $1/(1 - b/\overline{V}) = 1 + (b/\overline{V}) + (b/\overline{V})^2 + \cdots$. Using this result in the preceding equation for Z and collecting terms, we have

$$Z = 1 + \left(b - \frac{a}{RT}\right)\frac{1}{\overline{V}} + \left(\frac{b}{\overline{V}}\right)^2 + \left(\frac{b}{\overline{V}}\right)^3 + \cdots, \tag{3–8}$$

which expresses Z as a function of temperature and molar volume. It would be preferable to have Z as a function of temperature and pressure; however, this would entail solving the van der Waals equation for $\overline{V}$ as a function of T and p, then multiplying the result by p/RT to obtain Z as a function of T and p. Since the van der Waals equation is a cubic equation in $\overline{V}$, the solutions are too complicated to be particularly informative. We content ourselves with an approximate expression for $Z(T, p)$ which we obtain from Eq. (3–8) by observing that as $p \to 0$, $(1/\overline{V}) \to 0$, and $Z = 1$. Presumably at low pressures we can expand Z as a power series in the pressure;

$$Z = 1 + A_1 p + A_2 p^2 + A_3 p^3 + \cdots$$

in which the coefficients A_1, A_2, $A_3, \ldots$, are functions of temperature only. To determine these coefficients, we use the definition of Z in Eq. (3–2) to write $(1/\overline{V}) = p/RTZ$. Using this value of $(1/\overline{V})$ in Eq. (3–8) brings it to the form

$$1 + A_1 p + A_2 p^2 + A_3 p^3 + \cdots$$

$$= 1 + \left(b - \frac{a}{RT}\right)\frac{p}{RTZ} + \left(\frac{b}{RT}\right)^2 \frac{p^2}{Z^2} + \left(\frac{b}{RT}\right)^3 \frac{p^2}{Z^3} + \cdots.$$

We subtract 1 from each side of this equation and divide the result by p to obtain

$$A_1 + A_2 p + A_3 p^2 + \cdots$$

$$= \frac{1}{RT}\left(b - \frac{a}{RT}\right)\frac{1}{Z} + \left(\frac{b}{RT}\right)^2 \frac{p}{Z^2} + \left(\frac{b}{RT}\right)^3 \frac{p^2}{Z^3} + \cdots. \tag{3–8a}$$

In the limit of zero pressure, $Z = 1$, and this equation becomes

$$A_1 = \frac{1}{RT}\left(b - \frac{a}{RT}\right)$$

which is the required value of A_1. Using this value of A_1 in Eq. (3–8a) brings it to the form,

$$A_1 + A_2 p + A_3 p^2 + \cdots = A_1\left(\frac{1}{Z}\right) + \left(\frac{b}{RT}\right)^2 \frac{p}{Z^2} + \left(\frac{b}{RT}\right)^3 \frac{p^2}{Z^3} + \cdots.$$

We repeat the procedure by subtracting A_1 from both sides of this equation, dividing by p and taking the limiting value at zero pressure. Note that $(Z - 1)/p = A_1$ at zero pressure. Then,

$$A_2 = \left(\frac{b}{RT}\right)^2 - A_1^2 = \frac{a}{(RT)^3}\left(2b - \frac{a}{RT}\right)$$

Thus, the expansion of Z, correct to the term in p^2 is

$$Z = 1 + \frac{1}{RT}\left(b - \frac{a}{RT}\right)p + \frac{a}{(RT)^3}\left(2b - \frac{a}{RT}\right)p^2 + \cdots. \tag{3–9}$$

The correct coefficient for p could have been obtained by simply replacing $1/\overline{V}$ by the ideal value in Eq. (3–8); however, this would yield incorrect values of the coefficients of the higher powers of pressure.

The second term on the right of Eq. (3–9) should be compared with the second term on the right of Eq. (3–4) which considered only the effect of finite molecular volume. The slope of the Z versus p curve is obtained by differentiating Eq. (3–9) with respect to pressure, keeping the temperature constant:

$$\left(\frac{\partial Z}{\partial p}\right)_T = \frac{1}{RT}\left(b - \frac{a}{RT}\right) + \frac{2a}{(RT)^3}\left(2b - \frac{a}{RT}\right)p + \cdots.$$

At $p = 0$, all of the higher terms drop out and this derivative reduces simply to

$$\left(\frac{\partial Z}{\partial p}\right)_T = \frac{1}{RT}\left(b - \frac{a}{RT}\right), \qquad p = 0, \tag{3–10}$$

where the derivative is the *initial slope* of the Z versus p curve. If $b > a/RT$, the slope is positive; the size effect dominates the behavior of the gas. On the other hand, if $b < a/RT$, then the initial slope is negative; the effect of the attractive forces dominates the behavior of the gas. Thus, the van der Waals equation, which includes both the effects of size and of the intermolecular forces, can interpret either positive or negative slopes of the Z versus p curve. In interpreting Fig. 3–4, we can say that at 0°C the effect of the attractive forces dominates the behavior of methane and carbon dioxide, while the molecular size effect dominates the behavior of hydrogen.

Having examined the Z versus p curves for several gases at one temperature, we focus attention on the Z versus p curves for a *single* gas at different temperatures. Equation (3–10) shows that if the temperature is low enough, the term a/RT will be larger than b and so the initial slope of Z versus p will be negative. As the temperature increases, a/RT becomes smaller and smaller; if the temperature is high enough, a/RT becomes less than b, and the initial slope of the Z versus p curve becomes positive. Finally, if the temperature is extremely high, Eq. (3–10) shows that the slope of Z versus p must approach zero. This behavior is shown in Fig. 3–6.

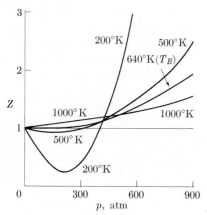

Fig. 3–6 Plot of Z versus p for methane at several temperatures (T_B = Boyle temperature).

At some intermediate temperature T_B, the Boyle temperature, the initial slope must be zero. The condition for this is given by Eq. (3–10) as $b - a/RT_B = 0$. This yields

$$T_B = a/Rb. \qquad (3\text{–}11)$$

At the Boyle temperature, the Z versus p curve is tangent to the curve for the ideal gas at $p = 0$ and rises above the ideal gas curve only very slowly. In Eq. (3–9) the second term drops out at T_B, and the remaining terms are small until the pressure becomes very high. Thus at the Boyle temperature the real gas behaves ideally over a wide range of pressures, because the effects of size and of intermolecular forces roughly compensate. This is also shown in Fig. 3–6. The Boyle temperatures for several different gases are given in Table 3–2.

Table 3–2 Boyle temperatures for several gases

Gas	H_2	He	N_2	CH_4	NH_3
t_B, °C	-156	-249	59	224	587

From the data in Table 3–2, the curves in Fig. 3–4 are comprehensible. All of them are drawn at 0°C. Thus hydrogen is above its Boyle temperature and so always has Z-values greater than unity. The other gases are below their Boyle temperature and so have Z-values less than unity in the low-pressure range.

By multiplying out the parentheses in Eq. (3–7), dividing by RT, and rearranging, we can write the van der Waals equation in the form

$$\frac{p\overline{V}}{RT} = 1 + \frac{bp}{RT} - \frac{a}{RT\overline{V}} + \frac{ab}{RT\overline{V}^2}. \tag{3–12}$$

The last three terms on the right-hand side are those which account for the deviations from ideality. Since as the pressure approaches zero as a limit the volume approaches infinity, all of these terms vanish, and the gas behaves ideally in the limit of zero pressure. Similarly, as the temperature approaches infinity, all of these terms vanish and the gas behaves ideally. Consequently, as a general rule, real gases are more nearly ideal the lower the pressure and the higher the temperature.

The van der Waals equation is a distinct improvement over the ideal gas law in that it gives qualitative reasons for the deviations from ideal behavior. This improvement is gained at considerable sacrifice, however. The ideal gas law contains nothing which depends on the individual gas; the constant R is a universal constant. The van der Waals equation contains two constants which are different for every gas. In this sense a different van der Waals equation must be used for each different gas. In Section 3–9 it will be seen that this loss in generality can be remedied for the van der Waals equation and for certain other equations of state.

3–6   The Isotherms of a Real Gas

If the pressure–volume relations for a real gas are measured at various temperatures, a set of isotherms such as are shown in Fig. 3–7 are obtained. At high temperatures the isotherms look much like those of an ideal gas, while at low temperatures the curves

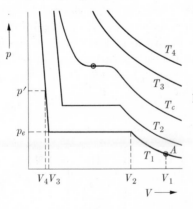

Fig. 3–7 Isotherms of a real gas.

have quite a different appearance. The horizontal portion of the low-temperature curves is particularly striking. Consider a container of gas in a state described by point A in Fig. 3–7. Imagine one wall of the container to be movable (a piston); keeping the temperature at T_1, we slowly push in this wall thus decreasing the volume. As the volume becomes smaller, the pressure rises slowly along the curve until the volume V_2 is reached. Reduction of the volume beyond V_2 produces no change in pressure until V_3 is reached. The small reduction in volume from V_3 to V_4 produces a large increase in pressure from p_e to p'. This is a rather remarkable sequence of events; particularly the decrease in volume over a wide range in which the pressure remains at the constant value p_e.

If we look into the container while all this is going on, we observe that at V_2 the first drops of liquid appear. As the volume goes from V_2 to V_3 more and more liquid forms; the constant pressure p_e is the equilibrium vapor pressure of the liquid at the temperature T_1. At V_3 the last trace of gas disappears. Further reduction of the volume simply compresses the liquid; the pressure rises very steeply, since the liquid is almost incompressible. The steep lines at the left of the diagram are therefore isotherms of the liquid. At a somewhat higher temperature the behavior is qualitatively the same, but the range of volume over which condensation occurs is smaller and the vapor pressure is larger. In going to still higher temperatures, the plateau finally shrinks to a point at a temperature T_c, the critical temperature. As the temperature is increased above T_c, the isotherms approach more and more closely those of the ideal gas; no plateau appears above T_c.

3–7 Continuity of States

In Fig. 3–8 the endpoints of the plateaus in Fig. 3–7 have been connected with a dashed line. Just as in any p–V diagram every point in Fig. 3–8 represents a state of the system. From the discussion in the preceding paragraph it can be seen that a point, such as A, on the extreme left of the diagram represents a liquid state of the substance. A point, such as C, on the right side of the diagram represents a gaseous

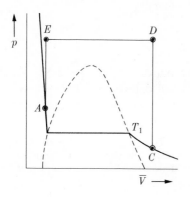

Fig. 3–8 Two-phase region and continuity of states.

state of the substance. Points under the "dome" formed by the dashed line represent states of the system in which liquid and vapor coexist in equilibrium. It is always possible to make a sharp distinction between states of the system in which one *phase* is present and states in which two phases† coexist in equilibrium, that is, between those points on and under the "dome" and those outside the "dome". However, it should be noted that there is no dividing line between the liquid states and the gaseous states. The fact that it is not always possible to distinguish between a liquid and a gas is the *principle of continuity of states*.

In Fig. 3–8 points A and C lie on the same isotherm, T_1. Point C clearly represents a gaseous state, and point A clearly represents the liquid obtained by compressing the gas isothermally. However, suppose that we begin at C and increase the temperature of the gas, keeping the volume constant. The pressure rises along the line CD. Having arrived at point D, the pressure is kept constant and the gas is cooled; this decreases the volume along the line DE. Having arrived at point E, the volume is again kept constant and the gas is cooled; this decreases the pressure until the point A is reached. At no time in this series of changes did the state point pass through the two-phase region. Condensation in the usual sense of the term did not occur. Point A could reasonably be said to represent a highly compressed gaseous state of the substance. The statement that point A *clearly* represented a liquid state must be modified. The distinction between liquid and gas is not always clear at all. As this demonstration shows, these two states of matter can be transformed into one another continuously. Whether we refer to states in the region of point A as liquid states or as highly compressed gaseous states depends purely upon which viewpoint happens to be convenient at the moment.

If the state point of the system lies under the dome, the liquid and gas can be distinguished, since both are present in equilibrium and there is a surface of discontinuity separating them. In the absence of this surface of discontinuity there is no fundamental way of distinguishing between liquid and gas.

3–8 The Isotherms of the van der Waals Equation

Consider the van der Waals equation in the form

$$p = \frac{RT}{\overline{V} - b} - \frac{a}{\overline{V}^2}.$$ (3–13)

When $\overline{V}$ is very large this equation approximates the ideal gas law, since $\overline{V}$ is very large compared with b and $a/\overline{V}^2$ is very small compared with the first term. This

---

† A *phase* is a region of uniformity in a system. This means a region of uniform chemical composition and uniform physical properties. Thus, a system containing liquid and vapor has two regions of uniformity. In the vapor phase, the density is uniform throughout. In the liquid phase, the density is uniform throughout but has a value different from that in the vapor phase.

is true at all temperatures. At high temperatures, the term $a/\overline{V}^2$ can be ignored, since it is small compared with $RT(\overline{V} - b)$. A plot of the isotherms, p versus $\overline{V}$, calculated from the van der Waals equation, is shown in Fig. 3–9. It is apparent from the figure that in the high-volume region the isotherms look much like the isotherms for the ideal gas, as does the isotherm at high temperature T_3.

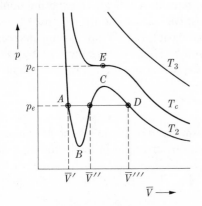

Fig. 3–9 Isotherms of the van der Waals gas.

At lower temperatures and smaller volumes, none of the terms in the equation may be neglected. The result is rather curious. At the temperature T_c the isotherm develops a point of inflection, point E. At still lower temperatures, the isotherms exhibit a maximum and a minimum.

Comparison of the van der Waals isotherms with those of a real gas shows similarity in certain respects. The curve at T_c in Fig. 3–9 resembles the curve at the critical temperature in Fig. 3–7. The curve at T_2 in Fig. 3–9 predicts three values of the volume, $\overline{V}'$, $\overline{V}''$, and $\overline{V}'''$, at the pressure p_e. The corresponding plateau in Fig. 3–7 predicts infinitely many volumes of the system at the pressure p_e. It is worthwhile to realize that even if a very complicated function had been written down, it would not exhibit a plateau such as appears in Fig. 3–7. The oscillation of the van der Waals equation in this region is as much as can be expected of a simple continuous function.

The sections AB and DC of the van der Waals curve at T_2 can be realized experimentally. If the volume of a gas at temperature T_2 is gradually reduced, the pressure rises along the isotherm until the point D, at pressure p_e, is reached. At this point condensation should occur; however, it may happen that liquid does not form, so that further reduction in volume produces an increase in pressure along the line DC. In this region (DC) the pressure of the gas exceeds the equilibrium vapor pressure of the liquid, p_e, at the temperature T_2; these points are therefore state points of a supersaturated (or supercooled) vapor. Similarly, if the volume of a liquid at temperature T_2 is increased, the pressure falls until point A, at pressure p_e, is reached. At this point vapor should form; however, it may happen that vapor does not form, so that further increase in the volume produces a reduction of pressure along the line

AB. Along the line *AB* the liquid exists under pressures which correspond to equilibrium vapor pressures of the liquid at temperatures below T_2. The liquid is *at* T_2 and so these points are state points of a superheated liquid. The states of the superheated liquid and those of the supercooled vapor are *metastable* states; they are unstable in the sense that slight disturbances are sufficient to cause the system to revert spontaneously into the stable state with the two phases present in equilibrium.

The section *BC* of the van der Waals isotherm cannot be realized experimentally. In this region the slope of the p-$\overline{V}$ curve is positive; increasing the volume of such a system would increase the pressure, and decreasing the volume would decrease the pressure! States in the region *BC* are *unstable*; slight disturbances of a system in such states as *B* to *C* would produce either explosion or collapse of the system.

3–9   The Critical State

If the van der Waals equation is taken in the form given by Eq. (3–7), the parentheses multiplied out, and the result divided by p, it can be arranged in the form

$$\overline{V}^3 - \left(b + \frac{RT}{p}\right)\overline{V}^2 + \frac{a}{p}\overline{V} - \frac{ab}{p} = 0. \tag{3–14}$$

Since Eq. (3–14) is a cubic equation, it may have three real roots for certain values of pressure and temperature. In Fig. 3–9 these three roots for T_2 and p_e are the intersections of the horizontal line at p_e with the isotherm at T_2. All three roots lie on the boundary of or within the two-phase region. As we have seen in both Figs. 3–8 and 3–9 the two-phase region narrows and finally closes at the top. This means that there is a certain maximum pressure p_c and a certain maximum temperature T_c at which liquid and vapor can coexist. This condition of temperature and pressure is the critical point and the corresponding volume is the critical volume $\overline{V}_c$. As the two-phase region narrows, the three roots of the van der Waals equation approach one another, since they must lie on the boundary or in the region. At the critical point the three roots are all equal to $\overline{V}_c$. The cubic equation can be written in terms of its roots $\overline{V}'$, $\overline{V}''$, $\overline{V}'''$:

$$(\overline{V} - \overline{V}')(\overline{V} - \overline{V}'')(\overline{V} - \overline{V}''') = 0.$$

At the critical point $\overline{V}' = \overline{V}'' = \overline{V}''' = \overline{V}_c$, so that the equation becomes $(\overline{V} - \overline{V}_c)^3 = 0$. Expanding, we obtain

$$\overline{V}^3 - 3\overline{V}_c\overline{V}^2 + 3\overline{V}_c^2\overline{V} - \overline{V}_c^3 = 0. \tag{3–15}$$

Under these same conditions, Eq. (3–14) becomes

$$\overline{V}^3 - \left(b + \frac{RT_c}{p_c}\right)\overline{V}^2 + \frac{a}{p_c}\overline{V} - \frac{ab}{p_c} = 0. \tag{3–16}$$

Since Eqs. (3–15) and (3–16) are simply different ways of writing the same equation, the coefficients of the individual powers of $\overline{V}$ must be the same in both equations.

Setting the coefficients equal, we obtain three equations:†

$$3\overline{V}_c = b + \frac{RT_c}{p_c}, \qquad 3\overline{V}_c^2 = \frac{a}{p_c}, \qquad \overline{V}_c^3 = \frac{ab}{p_c}. \tag{3-17}$$

Equations (3–17) may be looked at in two ways. First, the set of equations can be solved for $\overline{V}_c$, p_c, and T_c in terms of a, b, and R; thus,

$$\overline{V}_c = 3b, \qquad p_c = \frac{a}{27b^2}, \qquad T_c = \frac{8a}{27Rb}. \tag{3-18}$$

If the values of a and b are known, Eqs. (3–18) can be used to calculate $\overline{V}_c$, p_c, and T_c.

Taking the second point of view, we solve the equations for a, b, and R in terms of p_c, $\overline{V}_c$, and T_c. Then

$$b = \frac{\overline{V}_c}{3}, \qquad a = 3p_c\overline{V}_c^2, \qquad R = \frac{8p_c\overline{V}_c}{3T_c}. \tag{3-19}$$

Using Eqs. (3–19), we can calculate values of the constants a, b, and R from the critical data. However, the value of R so obtained does not agree well at all with the known value of R, and some difficulty arises.

Since experimentally it is hard to determine $\overline{V}_c$ accurately, it would be better if a and b could be obtained from p_c and T_c only. This is done by taking the third member of Eqs. (3–19) and solving it for $\overline{V}_c$. This yields

$$\overline{V}_c = \frac{3RT_c}{8p_c}.$$

---

† An equivalent method of obtaining these relations is to use the fact that the point of inflection on the p versus $\overline{V}$ curve occurs at the critical point p_c, T_c, $\overline{V}_c$. The conditions for the point of inflection are

$$(\partial p/\partial \overline{V})_T = 0, \qquad (\partial^2 p/\partial \overline{V}^2)_T = 0.$$

From the van der Waals equation,

$$\left(\frac{\partial p}{\partial \overline{V}}\right)_T = \frac{-RT}{(\overline{V} - b)^2} + \frac{2a}{\overline{V}^3},$$

$$\left(\frac{\partial^2 p}{\partial \overline{V}^2}\right)_T = \frac{2RT}{(\overline{V} - b)^3} - \frac{6a}{\overline{V}^4}.$$

Hence, at the critical point,

$$0 = -RT_c/(\overline{V}_c - b)^2 + 2a/\overline{V}_c^3, \qquad 0 = 2RT_c/(\overline{V}_c - b)^3 - 6a/\overline{V}_c^4.$$

These two equations, together with the van der Waals equation itself,

$$p_c = RT_c/(\overline{V}_c - b) - a/\overline{V}_c^2,$$

are equivalent to Eqs. (3–17).

This value of $\overline{V}_c$ is put in the first two of Eqs. (3–19) to yield

$$b = \frac{RT_c}{8p_c}, \qquad a = \frac{27(RT_c)^2}{64p_c}. \qquad (3\text{–}20)$$

Using Eqs. (3–20) and the ordinary value of R, we can calculate a and b from p_c and T_c only. This is the more usual procedure. However, to be honest we should compare the value, $\overline{V}_c = 3RT_c/8p_c$, with the measured value of $\overline{V}_c$. The result is again very bad. The observed and calculated values of $\overline{V}_c$ disagree by more than can be accounted for by the experimental difficulties.

The whole trouble lies in the fact that the van der Waals equation is not very accurate at all near the critical state. This fact, together with the fact that the tabulated values of these constants are nearly always calculated, one way or another, from the critical data, means that the van der Waals equation, although an improvement over the ideal gas law, cannot be used for a precision calculation of the gas properties. The great virtue of the van der Waals equation lies in the fact that the study of its predictions gives an excellent insight into the behavior of gases and to their relation to liquids and the phenomenon of liquefaction. The important thing is that the equation does predict a critical state; it is too bad that it does not describe its properties to six significant figures, but that is of secondary importance. Other equations are available which are very precise. Critical data for a few gases are given in Table 3–3.

Table 3–3 Critical constants for gases[†]

Gas	p_c, atm	$\overline{V}_c$, liter/mole	T_c, °K
He	2.26	0.062	5.25
H_2	12.8	0.065	33.2
N_2	33.6	0.090	126
O_2	50.3	0.075	154
CO_2	73.0	0.095	304
H_2O	218	0.057	647
Hg	3550	0.040	1900

† Francis Weston Sears, *An Introduction to Thermodynamics, the Kinetic Theory of Gases, and Statistical Mechanics.* Addison-Wesley Publishing Co., Inc., Reading, Mass., 1950.

3–10   The Law of Corresponding States

Using the values of a, b, and R given by Eqs. (3–19), we can write the van der Waals equation in the equivalent form

$$p = \frac{8p_c\overline{V}_c T}{3T_c(\overline{V} - \overline{V}_c/3)} - \frac{3p_c\overline{V}_c^2}{\overline{V}^2},$$

which can be rearranged to the form

$$\frac{p}{p_c} = \frac{8(T/T_c)}{3(\overline{V}/\overline{V}_c) - 1} - \frac{3}{(\overline{V}/\overline{V}_c)^2}. \tag{3-21}$$

Equation (3–21) involves only the ratios p/p_c, T/T_c, and $\overline{V}/\overline{V}_c$. This suggests that these ratios, rather than p, T, and $\overline{V}$, are the significant variables for the characterization of the gas. These ratios are called the *reduced variables* of state, π, τ, and φ:

$$\pi = p/p_c, \qquad \tau = T/T_c, \qquad \varphi = \overline{V}/\overline{V}_c.$$

Written in terms of these variables, the van der Waals equation becomes

$$\pi = \frac{8\tau}{3\varphi - 1} - \frac{3}{\varphi^2}. \tag{3-22}$$

The important thing about Eq. (3–22) is that it does not contain any constants which are peculiar to the individual gas; therefore it should be capable of describing all gases. In this way, the loss in generality entailed by the van der Waals equation compared with the ideal gas equation is regained. Equations, such as Eq. (3–22), which express one of the reduced variables as a function of the other two reduced variables are expressions of the *law of corresponding states*.

Two gases at the same reduced temperature and under the same reduced pressure are in corresponding states. By the law of corresponding states, they should both occupy the same reduced volume. For example, argon at 302°K and under 16 atm pressure, and ethane at 381°K and under 18 atm are in corresponding states, since each has $\tau = 2$ and $\pi = \frac{1}{3}$.

Any equation of state which involves only two constants in addition to R can be written in terms of the reduced variables only. For this reason equations which involved more than two constants were, at one time, frowned upon as contradicting the law of corresponding states. At the same time, hopes were high that an accurate two-constant equation could be devised to represent the experimental data. These hopes have been abandoned, since it is now recognized that the experimental data do not support the law of corresponding states as a law of great accuracy over all ranges of pressure and temperature. Although the law is not exact, it has a good deal of importance in engineering practice; in the range of industrial pressures and temperatures, the law often holds with accuracy sufficient for engineering calculations. Plots of Z versus p/p_c at various reduced temperatures are ordinarily used rather than an equation.

3–11 Other Equations of State

The van der Waals equation is only one of many equations which have been proposed over the years to account for the observed p-V-T data for gases. Several of these equations are listed in Table 3–4, together with the expression for the law of corresponding states for the two-constant equations, and the predicted value of the critical

Table 3–4

van der Waals equation:	$p = RT/(\bar{V} - b) - a/\bar{V}^2$
	$\pi = 8\tau/(3\varphi - 1) - 3/\varphi^2$
	$RT_c/p_c\bar{V}_c = 8/3 = 2.67$
Dieterici equation:	$p = RT e^{-a/\bar{V}RT}/(\bar{V} - b)$
	$\pi = \dfrac{\tau}{2\varphi - 1} e^{2 - 2/\tau\varphi}$
	$RT_c/p_c\bar{V}_c = e^2/2 = 3.69$
Berthelot equation:	$p = RT/(\bar{V} - b) - a/T\bar{V}^2$
	$\pi = 8\tau/(3\varphi - 1) - 3/\tau\varphi^2$
	$RT_c/p_c\bar{V}_c = 8/3 = 2.67$
Modified Berthelot equation:	$p = (RT/\bar{V})[1 + (9/128\tau - 27/64\tau^3)\pi]$
	$\pi = \dfrac{128\tau}{9(4\varphi - 1)} - \dfrac{16}{3\tau\varphi^2}$
	$RT_c/p_c\bar{V}_c = 32/9 = 3.56$
General virial equation:	$p\bar{V} = RT(1 + B/\bar{V} + C/\bar{V}^2 + D/\bar{V}^3 + \cdots).$

B, C, D, . . . are called the second, third, fourth, . . . virial coefficients.
They are functions of temperature.

Series expansion in terms of pressure:

$$p\bar{V} = RT(1 + B'p + C'p^2 + D'p^3 + \cdots)$$

B', C', etc., are functions of temperature.

Beattie-Bridgeman equation:

(1) Virial form　$p\bar{V} = RT + \dfrac{\beta}{\bar{V}} + \dfrac{\gamma}{\bar{V}^2} + \dfrac{\delta}{\bar{V}^3}$

(2) Form explicit in the volume:　$\bar{V} = \dfrac{RT}{p} + \dfrac{\beta}{RT} + \gamma'p + \delta'p^2 + \cdots$

$\beta = RTB_0 - A_0 - Rc/T^2$ $\qquad\qquad \gamma' = \dfrac{1}{(RT)^2}\left(\gamma - \dfrac{\beta^2}{RT}\right)$

$\gamma = -RTB_0b + A_0a - RB_0c/T^2$

$\delta = RB_0bc/T^2$ $\qquad\qquad\qquad \delta' = \dfrac{1}{(RT)^3}\left[\delta - \dfrac{3\beta\gamma}{RT} + \dfrac{2\beta^3}{(RT)^2}\right]$

ratio $RT_c/p_c\bar{V}_c$. Of these equations in Table 3–4, the Beattie-Bridgeman equation or the virial equation are best suited for precise work. The Beattie-Bridgeman equation involves five constants in addition to R: A_0, a, B_0, b, and c. The values of the Beattie-Bridgeman constants for a few gases are given in Table 3–5.

Table 3–5† Constants for the Beattie-Bridgeman equation‡

Gas	A_0	a	B_0	b	$c \times 10^{-4}$
He	0.0216	0.05984	0.01400	0.0	0.0040
H_2	0.1975	−0.00506	0.02096	−0.04359	0.0504
O_2	1.4911	+0.02562	0.04624	+0.004208	4.80
CO_2	5.0065	0.07132	0.10476	+0.07235	66.00
NH_3	2.3930	0.17031	0.03415	0.19112	476.87

† Francis Weston Sears, *An Introduction to Thermodynamics, the Kinetic Theory of Gases, and Statistical Mechanics.* Addison-Wesley Publishing Co., Inc., Reading, Mass., 1950.

‡ Units: p in atm, $\overline{V}$ in liters/mole, $R = 0.08206$, T in °K.

It is interesting to examine the values of the critical ratio $RT_c/p_c\overline{V}_c$ predicted by the various equations in Table 3–4. The average value of the critical ratio for a large number of nonpolar gases, H_2 and He excepted, is 3.65. Clearly then, the van der Waals equation will be useless at temperatures and pressures near the critical values; see Section 3–8. The Dieterici equation is much better near the critical point; however, it is little used because of the transcendental function involved. Of the two-constant equations, the modified Berthelot equation is most frequently used for estimates of volumes which are better than the ideal gas estimate. The critical temperature and pressure of the gas must be known to use this equation.

Finally, it should be pointed out that all of the equations of state which are proposed for gases are based on the two fundamental ideas first suggested by van der Waals: (1) molecules have size, and (2) forces act between molecules. The more modern equations include the dependence of the intermolecular forces on the distance of separation of the molecules.

Problems

3–1. For the dissociation $N_2O_4 \rightleftharpoons 2NO_2$, the equilibrium constant at 25°C is $K = 0.115$; it is related to the degree of dissociation α and the pressure in atm by $K = 4\alpha^2 p/(1 - \alpha^2)$. If n is the number of moles of N_2O_4 which would be present if no dissociation occurred, calculate V/n at $p = 2$ atm, 1 atm, and 0.5 atm, assuming that the equilibrium mixture behaves ideally. Compare the results with the volumes if dissociation did not occur.

3–2. For the mixture described in Problem 3–1, show that as p approaches zero, the compressibility factor $Z = pV/nRT$ approaches 2 instead of the usual value of unity. Why does this happen?

3–3. A certain gas at 0°C and 1 atm pressure has $Z = 1.00054$. Estimate the value of b for this gas.

3–4. If $Z = 1.00054$ at 0°C and 1 atm and the Boyle temperature of the gas is 107°K, estimate the values of a and of b. (Only the first two terms in the expression of Z are needed.)

3–5. The critical constants for water are 374°C, 218 atm, and 0.0566 liter/mole. Calculate values of a, b, and R; compare the value of R with the correct value and notice the discrepancy. Compute the constants a and b from p_c and T_c only. Using these values and the correct value of R, calculate the critical volume and compare with the correct value.

3–6. Find the relation of the constants a and b of the Berthelot equation to the critical constants.

3–7. Find the relation of the constants a and b of the Dieterici equation to the critical constants. (Note that this cannot be done by setting three roots of the equation equal to one another.)

3–8. The critical temperature of ethane is 32.3°C, the critical pressure is 48.2 atm. Compute the critical volume using

a) the ideal gas law,

b) the van der Waals equation, realizing that for a van der Waals gas $p_c \overline{V}_c / R T_c = \frac{3}{8}$,

c) the modified Berthelot equation,

d) Compare the results with the experimental value, 0.139 liter/mole.

3–9. The vapor pressure of liquid water at 25°C is 23.8 mm and at 100°C it is 760 mm. Using the van der Waals equation in one form or another as a guide, show that saturated water vapor behaves more nearly as an ideal gas at 25°C than it does at 100°C.

3–10. The compressibility factor for methane is given by $Z = 1 + Bp + Cp^2 + Dp^3$. If p is in atm., the values of the constants are as follows:

T, °K	B	C	D
200	-5.74×10^{-3}	6.86×10^{-6}	18.0×10^{-9}
1000	$+0.189 \times 10^{-6}$	0.275×10^{-6}	0.144×10^{-9}

Plot the values of Z as a function of p at these two temperatures in the range from 0 to 1000 atm.

3–11. If the compressibility factor of a gas is $Z(p, T)$, the equation of state may be written $p\overline{V}/RT = Z$. Show how this affects the equation for the distribution of the gas in a gravity field. From the differential equation for the distribution, show that if Z is greater than unity, the distribution is broader for the real gas than for the ideal gas and that the converse is true if Z is less than unity. If $Z = 1 + Bp$, where B is a function of temperature, integrate the equation and evaluate the constant of integration to obtain the explicit form of the distribution function.

3–12. At high pressures (small volumes), the van der Waals equation, Eq. (3–14), can be rearranged to the form

$$\overline{V} = b + \frac{p}{a}\left(b + \frac{RT}{p}\right)\overline{V}^2 - \left(\frac{p}{a}\right)\overline{V}^3.$$

If the quadratic and cubic terms are dropped, then we obtain $\overline{V}_0 = b$ as a first approximation to the smallest root of the equation. This would represent the volume of the liquid. Using this approximate value of $\overline{V}$ in the higher terms, show that the next approximation for the volume of the liquid is $\overline{V} = b + b^2RT/a$. From this expression show that the first approximation for the coefficient of thermal expansion of a van der Waals liquid is $\alpha = bR/a$.

3–13. Using the same technique as that used to obtain Eq. (3–9), prove the relation given in Table 3–4 between γ and γ' for the Beattie-Bridgeman equation; namely, $\gamma'(RT)^2 = \gamma - \beta^2/RT$.

3–14. At what temperature does the slope of the Z versus p curve (at $p = 0$) have a maximum value for the van der Waals gas? What is the value of the maximum slope?

Chapter Four

The Structure of Gases

4–1 Introduction

The aim of physics and chemistry is to interpret quantitatively the observed properties of macroscopic systems in terms of the kinds and arrangement of atoms or molecules which make up these systems. We seek an interpretation of behavior in terms of the structure of a system. Having studied the properties of a system, we construct in our mind's eye a model of the system, built of atoms and molecules, and forces of interaction between them. The laws of mechanics and of statistics are applied to this model to predict the properties of such an ideal system. If many of the predicted properties are in agreement with the observed properties, the model is a good one. If none, or only a few, of the predicted properties are in agreement with the observed properties, the model is poor. This ideal model of the system may be altered or replaced by a different model until its predictions are satisfactory.

Structurally, gases are nature's simplest substances, so that a simple model and elementary calculation yield results in excellent agreement with experiment. The kinetic theory of gases provides a beautiful and important illustration of the relation of theory to experiment in physics, as well as of the techniques which are used commonly in relating structure to properties.

4–2 Kinetic Theory of Gases; Fundamental Assumptions

The model used in the kinetic theory of gases may be described by three fundamental assumptions about the structure of gases.

1. A gas is composed of a very large number of minute particles (atoms or molecules).

2. In the absence of a force field, these particles move in straight lines. (Newton's first law of motion is obeyed.)

3. These particles interact (i.e., collide) with one another only infrequently.

In addition to these assumptions we impose the condition that in any collision, the total kinetic energy of the two molecules is the same before and after the collision. This kind of collision is an *elastic collision*.

If the gas consists of a very large number of moving particles, then the motion of the particles must be completely random or chaotic. The particles move in all directions with a variety of speeds, some moving quickly, others slowly. If the motion were orderly (let us say that all the particles in a rectangular box were moving in precisely parallel paths), such a condition could not persist. Any slight irregularity in the wall of the box would deflect some particle out of its path; collision of this deflected particle with another particle would deflect the second one, and so on. Clearly, the motion would soon be chaotic.

4–3 Calculation of the Pressure of a Gas

If a particle collides with a wall and rebounds, a force is exerted on the wall at the moment of collision. This force divided by the area of the wall would be the momentary pressure exerted on the wall by the impact and rebound of the particle. By calculating the force exerted on the wall by the impacts of many molecules, we can evaluate the pressure exerted by the gas.

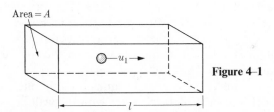

Figure 4–1

Consider a rectangular box of length l and cross-sectional area A (Fig. 4–1). In the box there is one particle of mass m traveling with a velocity u_1 in a direction parallel to the length of the box. When the particle hits the right-hand end of the box it is reflected and travels in the opposite direction with a velocity $-u_1$. After a period of time it returns to the right-hand wall, the collision is repeated, and so again and again. If a pressure gauge, sufficiently sensitive to respond to the impact of this single particle, were attached to the wall, the gauge reading as a function of time would be as shown in Fig. 4–2(a). The time interval between the peaks is the time required for the particle to traverse the length of the box and back again, and thus is the distance traveled divided by the speed, $2l/u_1$. If a second particle of the same mass and traveling in a parallel path with a higher velocity is put in the box, the gauge reading will be as shown in Fig. 4–2(b).

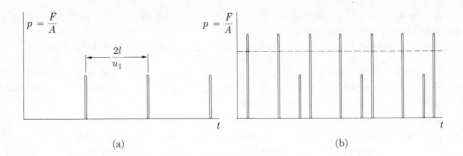

Fig. 4–2 Force resulting from a collision of particles with the wall.

In fact a pressure gauge which responds to the impact of individual molecules does not exist. In any laboratory situation, a pressure gauge reads a steady, *average value* of the force per unit area exerted by the impacts of an enormous number of molecules; this is indicated by the dashed line in Fig. 4–2(b).

To compute the average value of the pressure we begin with Newton's second law of motion:

$$F = ma = m\frac{du}{dt} = \frac{d(mu)}{dt}, \qquad (4\text{--}1)$$

where F is the force acting on the particle of mass m, a is the acceleration, and u is the velocity of the particle. According to Eq. (4–1) the force acting on the particle is equal to the change of momentum per unit time. The force acting on the wall is equal and opposite in sign to this. For the particle in Fig. 4–1, the momentum before collision is mu_1, while the momentum after collision is $-mu_1$. Then the change in momentum in collision is equal to the difference of the final momentum minus the initial momentum. Thus we have $(-mu_1) - mu_1 = -2mu_1$. The change in momentum in unit time is the change in momentum in one collision multiplied by the number of collisions the particle makes with the wall per second. Since the time between collisions is equal to the time to travel distance $2l$, $t = 2l/u_1$. Then the number of collisions per second is $u_1/2l$. Therefore the change in momentum per second equals $-2mu_1(u_1/2l)$. Thus the force acting on the particle is given by $F = -mu_1^2/l$, and the force acting on the wall by $F_w = +mu_1^2/l$. But the pressure p' is F_w/A; therefore

$$p' = mu_1^2/Al = mu_1^2/V, \qquad (4\text{--}2)$$

in which $Al = V$, the volume of the box.

Equation (4–2) gives the pressure p', exerted by one particle only; if more particles are added, each traveling parallel to the length of the box with speeds $u_2, u_3, \ldots$, the total force, and so the total pressure p, will be the sum of the forces exerted by each particle:

$$p = (m/V)(u_1^2 + u_2^2 + u_3^2 + \cdots). \qquad (4\text{--}3)$$

The average of the squares of the velocities, $\overline{u^2}$, is defined by

$$\overline{u^2} = (u_1^2 + u_2^2 + u_3^2 + \cdots)/N, \tag{4–4}$$

where N is the number of particles in the box. It is this average of the squares of the velocities which appears in Eq. (4–3). Using Eq. (4–4) in Eq. (4–3), we obtain

$$p = Nm\overline{u^2}/V, \tag{4–5}$$

the final equation for the pressure of a *one-dimensional gas*.† Before using Eq. (4–5), we must examine the derivation to see what effects collisions and the varied directions of motion will have on the result.

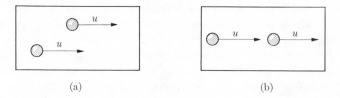

(a)   (b)

Figure 4–3

The effect of collisions is readily determined. It was assumed that all of the particles were traveling in parallel paths. This situation is illustrated for two particles, having the same velocity u, in Fig. 4–3(a). If the two particles travel on the same path, we have the situation shown in Fig. 4–3(b). In this latter case, the molecules collide with one another and each is reflected. One of the molecules never hits the right-hand wall and so cannot transfer momentum to it. However, the other molecule hits the right-hand wall twice as often as in the parallel-path case, and so the momentum transferred to the wall in a given time does not depend on whether the particles travel on parallel paths or on the same path. We conclude that collisions in the gas do not affect the result in Eq. (4–5). The same is true if the two molecules move with different velocities. An analogy may be helpful; a bucket brigade carries water to a fire. If the brigade consists of two men, the same amount of water will arrive in unit time whether one man relays the bucket to the other at the midpoint between the well and the fire, or both men run the entire distance to the well.

The fact that the molecules are traveling in different directions rather than all in the same direction as we originally assumed has an important effect on the result. As a first guess we might say that, on the average, only one-third of the molecules are moving in each of the three directions, so that the factor N in Eq. (4–5) should be replaced by $N/3$. This alteration gives

$$p = Nm\overline{u^2}/3V. \tag{4–6}$$

---

† A one-dimensional gas is a gas in which all the molecules are imagined to be moving in one direction (or its reverse) only.

This simple guess gives the correct result, but the reason is more complex than the one on which the guess was based. To gain a better insight into the effect of directions, Eq. (4–6) will be derived in a different way.

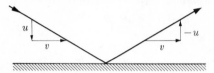

Fig. 4–4 Reversal of the normal component of velocity at the wall.

The velocity vector c of the particle can be resolved into one component normal to the wall, u, and two tangential components, v and w. Consider a particle which hits the wall at an arbitrary angle and is reflected (Fig. 4–4). The only component of the velocity which is reversed on collision is the *normal* component u. The tangential component v has the same direction and magnitude before and after the collision. This is true also of the second tangential component w, which is not shown in Fig. 4–4. Since it is only the reversal of the normal component which matters, the change in momentum per collision with the wall is $-2mu$; the number of impacts per second is equal to $u/2l$. Thus Eq. (4–5) should read

$$p = N m \overline{u^2} / V, \qquad (4–7)$$

where $\overline{u^2}$ is the average value of the square of the normal component of the velocity. If the components are taken along the three axes x, y, z, as in Fig. 4–5, then the square of the velocity vector is related to the squares of the components by

$$c^2 = u^2 + v^2 + w^2. \qquad (4–8)$$

For any individual molecule, the components of velocity are all different, and so each term on the right-hand side of Eq. (4–8) has a different value. However, if Eq. (4–8) is averaged over all the molecules, we obtain

$$\overline{c^2} = \overline{u^2} + \overline{v^2} + \overline{w^2}. \qquad (4–9)$$

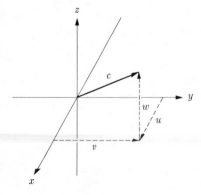

Fig. 4–5 Components of the velocity vector.

There is no reason to expect that any one of the three directions is preferred after averaging over all the molecules. Thus we expect that $\overline{u^2} = \overline{v^2} = \overline{w^2}$. Using this result in Eq. (4–9), we obtain

$$\overline{u^2} = \overline{c^2}/3. \tag{4–10}$$

The x-direction is taken as the direction normal to the wall; thus, putting $\overline{u^2}$ from Eq. (4–10) into (4–7), we obtain the exact equation for the pressure:

$$p = N m \overline{c^2}/3V, \tag{4–11}$$

the same as Eq. (4–6) obtained by the guess. Realize that in Eq. (4–6), $u = c$ since v and w were zero in the derivation of Eq. (4–6).

Let the kinetic energy of any molecule be $\epsilon = \frac{1}{2}mc^2$. If both sides of this equation are averaged over all the molecules, then $\bar{\epsilon} = \frac{1}{2}m\overline{c^2}$. Using this result in Eq. (4–11), yields $p = 2N\bar{\epsilon}/3V$, or

$$pV = \tfrac{2}{3}N\bar{\epsilon}. \tag{4–12}$$

It is encouraging to note that Eq. (4–12) bears a marked resemblance to the ideal gas law. Consequently, we examine the reason for the form in which the volume appears in Eq. (4–12). If the container in Fig. 4–1 is lengthened slightly, the volume increases by a small amount. If the velocities of the particles are the same, a longer time is required for a particle to travel between the walls and so it makes fewer collisions with the wall per second, reducing the pressure on the wall. Thus, an increase in volume reduces the pressure simply because there are fewer collisions with the wall in any given time interval.

We now compare Eq. (4–12) with the ideal gas law,

$$pV = nRT.$$

If Eq. (4–12) describes the ideal gas, then it must be that

$$nRT = \tfrac{2}{3}N\bar{\epsilon}.$$

Now n and N are related by $n = N/N_0$, where N_0 is the Avogadro number. Thus,

$$RT = \tfrac{2}{3}N_0\bar{\epsilon}. \tag{4–13}$$

Let E be the total kinetic energy associated with the random motion of the molecules in one mole of gas. Then $E = N_0\bar{\epsilon}$, and

$$E = \tfrac{3}{2}RT. \tag{4–14}$$

Equation (4–14) is one of the most fascinating results of the kinetic theory, for it provides us with an interpretation of temperature. It says that the kinetic energy of the random motion is proportional to the absolute temperature. For this reason, the random or chaotic motion is often called the *thermal motion* of the molecules. At the absolute zero of temperature, this thermal motion ceases entirely. Thus, temperature is a measure of the average kinetic energy of the chaotic motion. It is important to realize that temperature is *not* associated with the kinetic energy of one molecule, but with the *average* kinetic energy of an enormous number of molecules, i.e., it is a statistical concept. It is $\bar{\epsilon}$ and not ϵ which appears in Eq. (4–13). A system composed of one molecule or even of a few molecules would not have a temperature, properly speaking.†

The fact that the ideal gas law does not contain anything which is characteristic of a particular gas implies that at a specified temperature all gases have the same average kinetic energy. Applying Eq. (4–13) to two different gases, we have $\frac{3}{2}RT = N_0\bar{\epsilon}_1$, and $\frac{3}{2}RT = N_0\bar{\epsilon}_2$, when $\bar{\epsilon}_1 = \bar{\epsilon}_2$, or $\frac{1}{2}m_1\overline{c_1^2} = \frac{1}{2}m_2\overline{c_2^2}$. The root-mean-square speed, c_{rms}, is defined by

$$c_{rms} = \sqrt{\overline{c^2}}. \tag{4–15}$$

The ratio of the root-mean-square speeds of two molecules of different masses is equal to the square root of the inverse ratio of the masses:

$$\frac{(c_{rms})_1}{(c_{rms})_2} = \sqrt{\frac{m_2}{m_1}} = \sqrt{\frac{M_2}{M_1}}, \tag{4–16}$$

where $M = N_0 m$ is the molecular weight. The heavier gas has the smaller rms speed. Thus, if we compare hydrogen, $M_1 = 2$, and oxygen, $M_2 = 32$, we have

$$(c_{rms})_{H_2} = (c_{rms})_{O_2}\sqrt{(32/2)} = 4(c_{rms})_{O_2}.$$

At every temperature, hydrogen has an rms speed four times as great as that of oxygen, while their average kinetic energies are the same.

The numerical value of the rms velocity of any gas is calculated by combining Eqs. (4–13) and $\bar{\epsilon} = \frac{1}{2}m\overline{c^2}$; thus, $RT = \frac{2}{3}N_0\frac{1}{2}m\overline{c^2}$, or $\overline{c^2} = 3RT/M$, and

$$c_{rms} = \sqrt{3RT/M}. \tag{4–17}$$

Example For oxygen at room temperature, 27°C, $T = 300°K$, $M = 32$, and in *cgs* units $R = 8.314 \times 10^7$ ergs·mole^{-1}·deg^{-1}. Then

$$c_{rms} = \sqrt{\frac{3(8.314)(10^7)(300)}{32}} = \sqrt{23.4 \times 10^8}$$

$$= 4.84 \times 10^4 \text{ cm/sec} = 484 \text{ m/sec} = 1080 \text{ miles/hour}.$$

The last figure is more dramatic in bringing home the magnitude of these molecular speeds.

At room temperature, the usual range of molecular speeds is 300 to 500 m/sec. Hydrogen is unusual because of its low mass; its rms speed is about 1900 m/sec.

---

† It is sometimes convenient to define the "temperature" of a molecule in terms of the energy of the molecule. In such cases care must be taken not to confuse this "temperature" with the ordinary temperature of the system.

4–4 Dalton's Law of Partial Pressures

In a mixture of gases, the total pressure is the sum of the forces per unit area produced by the impacts of each kind of molecule on a wall of a container. Each kind of molecule contributes a term of the type in Eq. (4–11) to the pressure. For a mixture of gases we have

$$p = \frac{N_1 m_1 \overline{c_1^2}}{3V} + \frac{N_2 m_2 \overline{c_2^2}}{3V} + \frac{N_3 m_3 \overline{c_3^2}}{3V} + \cdots \tag{4–18}$$

or

$$p = p_1 + p_2 + p_3 + \cdots, \tag{4–19}$$

where $p_1 = N_1 m_1 \overline{c_1^2}/3V$, $p_2 = N_2 m_2 \overline{c_2^2}/3V, \ldots$. Dalton's law is thus an immediate consequence of the kinetic theory of gases.

4–5 Distributions and Distribution Functions

The distribution of molecules in a gravity field has been discussed. It was shown that the pressure decreased regularly with increase in height, which implies that the molecules are distributed in such a way that there are fewer per cubic centimeter at the upper levels than at lower levels. The analytical expression which describes this situation is the *distribution function*. A distribution over a space coordinate is a *spatial* distribution. In the kinetic theory of gases it is important to know the velocity distribution, i.e., how many molecules have velocities in a given range. The task of the following sections is to derive the velocity distribution function. Before proceeding to that problem, it is helpful to mention a few important ideas about distributions.

First of all, a distribution is the division of a group of things into classes. If we have a thousand ball bearings and five boxes, and place the ball bearings in the boxes in any particular way we please, the result is a distribution. If we divide the population of the U.S. into classes according to age, the result is an age distribution. Such a distribution shows how many people are between the ages of 0 to 20 years, between 20 to 40 years, 40 to 60 years, etc. The population could be divided into classes according to the amount of money in individual savings accounts, or according to the amount of money owed to pawnbrokers. Each of these classifications constitutes a distribution of greater or less importance.

The distribution is used to compute average values. From the distributions mentioned, we could compute the average age of persons in the U.S., the average amount per person in savings accounts, and the average amount per person owed to pawnbrokers. For these averages to have a reasonable accuracy, some attention must be given to the choice of the width of the classification interval. Without going into the details which enter into the choice of the interval width, it suffices to say that it must be small, but not too small. Consider an age distribution;

clearly it is senseless to choose 100 years as the width of the interval; essentially all of any group falls in that one interval and the group would not be divided into classes at all. So the interval width must be smaller. On the other hand, if we choose a very small interval width, e.g., one day, then in any small group, say of 10 people, we will find that one person falls in each of ten intervals and zeros fall in all the others. For any large group, the time required just to write down such a detailed distribution would be enormous. Furthermore, if the information were gathered on a different day, the entire distribution would be shifted. Consequently, in constructing a distribution, the interval width chosen must be wide enough to smooth out details of no interest and narrow enough so that significant aspects of the distribution are displayed and meaningful averages can be calculated.

4–6 The Maxwell Distribution

In a container of gas, the individual molecules are traveling in various directions with different speeds. We assume that the motions of the molecules are completely random. Then we set the following problem. What is the probability of finding a molecule with a speed between the values c and $c + dc$, regardless of the direction in which the molecule is traveling?

   This problem can be broken down into simpler parts; the solution of the problem is achieved by combining the solutions of the simpler problems. Let u, v, and w denote the components of velocity in the x-, y-, and z-directions, respectively. Let dn_u be the number of molecules which have an x-component of velocity with a value in the range between u and $u + du$. Then the probability of finding such a molecule is by definition dn_u/N, where N is the number of molecules in the container. If the interval width, du, is small, it is reasonable to expect that doubling the width will double the number of molecules in the interval. Thus dn_u/N is proportional to du. Also the probability dn_u/N will depend on the velocity component u. Thus, we write

$$dn_u/N = f(u^2)\,du, \qquad\qquad (4\text{–}20)$$

where the mathematical form of the function $f(u^2)$ is yet to be determined.†

   At this point it must be made clear why the function depends on u^2 and not simply on u. Because of the random nature of the molecular motion, the probability of finding a molecule with an x-component in the range u to $u + du$ must be the same as the probability of finding one with an x-component in the range $-u$ to $-(u + du)$. In other words, the molecule has the same chance of going east with a certain speed as it has of going west with the same speed. If the direction mattered, the motion would not be random and the entire mass of gas would have a net velocity in the preferred direction. The required symmetry in the function

---

† In writing Eq. (4–20) in this way, we assume implicitly that the probability dn_u/N is not in any way dependent on the values of the y- or z-components, v and w. This assumption is valid but will not be justified here.

is assured if we write $f(u^2)$ rather than $f(u)$. In the same way, if the number of molecules having a y-component of velocity between v and $v + dv$ is dn_v, the probability of finding a molecule whose y-component lies in the range v to $v + dv$ is given by

$$dn_v/N = f(v^2)\,dv, \qquad (4\text{-}21)$$

where the function $f(v^2)$ *must have exactly the same form* as the function $f(u^2)$ in Eq. (4–20). These functions must have the same form, since the randomness of the distribution does not allow one direction to be different from another.† With an analogous notation we have for the z-component,

$$dn_w/N = f(w^2)\,dw. \qquad (4\text{-}22)$$

We now ask a more involved question. What is the probability of finding a molecule which has simultaneously an x-component in the range u to $u + du$ and a y-component in the range v to $v + dv$? Let the number of molecules which satisfy this condition be dn_{uv}; then the probability of finding such a molecule is by definition dn_{nv}/N, the product of the probabilities of finding molecules which satisfy the conditions separately. That is, $dn_{uv}/N = (dn_u/N)(dn_v/N)$, or

$$dn_{uv}/N = f(u^2)f(v^2)\,du\,dv. \qquad (4\text{-}23)$$

Figure 4–6 illustrates the meaning of Eqs. (4–20), (4–21), and (4–23). The values of u and v for each molecule determine a representative point, marked with a dot, in the u-v coordinate system of Fig. 4–6. The representative points for two different molecules might conceivably coincide; this does not matter. The important thing is that every molecule is so represented. The total number of representative points is N, the total number of molecules in the container. Then the number of molecules having an x-component of velocity between the values u and $u + du$ is the number of representative points in the vertical strip at position u and of width du. This number is dn_u, and, by Eq. (4–20), is equal to $Nf(u^2)\,du$. Similarly, the number of representative points in the horizontal strip at the position v and of width dv is the number of molecules having a y-component of velocity between v and $v + dv$. The number of molecules which satisfy both conditions simultaneously is the number of representative points in the little rectangle formed by the intersection of the vertical and horizontal strips. By Eq. (4–23) this number of molecules is $dn_{uv} = Nf(u^2)f(v^2)\,du\,dv$. The *density* of representative points at the position (u, v) is the number dn_{uv} divided by the area of the little rectangle $du\,dv$:

$$\text{Point density at } (u, v) = dn_{uv}/du\,dv = Nf(u^2)f(v^2). \qquad (4\text{-}24)$$

To derive the form of the function $f(u^2)$, a new set of coordinate axes u' and v' is introduced in the position shown in Fig. 4–7. The velocity ranges in the new coordinate system are du' and dv'. The number of representative points in the area

---

†   It is assumed here that there is no force field, such as a gravity field, acting in a particular direction.

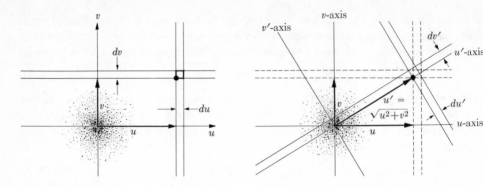

Fig. 4-6 Two-dimensional velocity space.

Fig. 4-7 Two-dimensional velocity space with different coordinate system.

$du'\, dv'$ is given by $dn_{u'v'} = Nf(u'^2)f(v'^2)\,du'\,dv'$. The density of representative points at the position (u', v') is

$$\text{Point density at } (u', v') = dn_{u'v'}/du'\,dv' = Nf(u'^2)f(v'^2). \qquad (4\text{-}25)$$

However, the position (u', v') is the same as the position (u, v), so the density of representative points must be the same regardless of which coordinate system is used to describe it. From Eqs. (4-24) and (4-25),

$$Nf(u'^2)f(v'^2) = Nf(u^2)f(v^2). \qquad (4\text{-}26)$$

The position (u, v) in the first coordinate system corresponds to the position $u' = (u^2 + v^2)^{1/2}, v' = 0$, in the second coordinate system. Using this relation in Eq. (4-26), we obtain

$$f(u^2 + v^2)f(0) = f(u^2)f(v^2).$$

Since $f(0)$ is a constant, set $f(0) = A$. Then

$$Af(u^2 + v^2) = f(u^2)f(v^2). \qquad (4\text{-}27)$$

It is shown in Appendix I that the only functions which satisfy Eq. (4-27) are

$$f(u^2) = Ae^{\beta u^2} \quad \text{and} \quad f(u^2) = Ae^{-\beta u^2},$$

where β is a positive constant. The physical situation forces us to choose the negative sign in the exponential, that is

$$f(u^2) = Ae^{-\beta u^2}, \qquad f(v^2) = Ae^{-\beta v^2}. \qquad (4\text{-}28)$$

Equation (4-20) becomes

$$\frac{dn_u}{N} = Ae^{-\beta u^2}\,du. \qquad (4\text{-}29)$$

If the positive sign in the exponential had been chosen, Eq. (4–29) would predict that as the velocity component u becomes infinite, the probability of finding such molecules becomes infinite. This would require infinite kinetic energy for the system and consequently is an impossible case. As it stands, with the negative exponential, Eq. (4–29) predicts that the probability of finding a molecule with infinite x-component of velocity is zero; this makes physical sense.

Although the original problem has not been solved, we have made considerable progress. It is well at this point to look back at what has been accomplished. First of all, we assumed that the probability of finding a molecule with an x-component of velocity in the range u to $u + du$ depended only on the value of u and the width of the range du. This was expressed, Eq. (4–20), as $dn_u/N = f(u^2)\,du$. A rather lengthy argument on the basis of probabilities led us finally to the functional form of $f(u^2) = A\exp(-\beta u^2)$. The important point about the whole matter is the use of the notion of randomness. The argument is almost completely mathematical. Only two specifically physical assumptions are involved: randomness in the motion and a finite value of $f(u^2)$ as $u \to \infty$. The form of the distribution function is completely determined by these two assumptions. The success of the treatment will give us confidence in the picture of a gas as a collection of molecules whose motions are completely random. Randomness prompts the use of probability theory, and the distribution function $A\exp(-\beta u^2)$ which appears is a famous one in probability theory, being the normal probability curve, or the Gaussian curve of error. This function is the governing rule in any completely random distribution; for example, it expresses the distribution of random errors in all types of experimental measurements.

We are now in a position to solve the original problem, namely to find the distribution of molecular speeds and to evaluate the constants A and β which appear in the distribution function.

The probability dn_{uvw}/N of finding a molecule with velocity components simultaneously in the ranges u to $u + du$, v to $v + dv$, w to $w + dw$ is given by the product of the individual probabilities: $dn_{uvw}/N = (dn_u/N)(dn_v/N)(dn_w/N)$, or

$$dn_{uvw}/N = f(u^2)f(v^2)f(w^2)\,du\,dv\,dw.$$

According to Eqs. (4–28),

$$dn_{uvw}/N = A^3 e^{-\beta(u^2 + v^2 + w^2)}\,du\,dv\,dw. \tag{4.30}$$

A three-dimensional velocity space† is constructed in Fig. 4–8. In this space a molecule is represented by a point determined by the values of the three components of velocity u, v, w. The total number of representative points in the parallelepiped at (u, v, w) is dn_{uvw}. The density of points in this parallelepiped is

$$\text{Point density at } (u, v, w) = dn_{uvw}/du\,dv\,dw = NA^3 e^{-\beta(u^2 + v^2 + w^2)}, \tag{4–31}$$

---

† Figures 4–6 and 4–7 are examples of a two-dimensional velocity space.

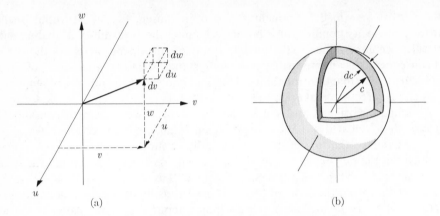

Fig. 4–8 (a) Three-dimensional velocity space. (b) Spherical shell.

where Eq. (4–30) has been used to obtain the last member of Eq. (4–31). Since $c^2 = u^2 + v^2 + w^2$ [see Eq. (4–8) and Fig. 4–5], we have

$$\text{Point density at } (u, v, w) = NA^3 e^{-\beta c^2}. \tag{4–32}$$

The right-hand side of Eq. (4–32) depends only on the constants N, A, β and on c^2; consequently, it does not depend in any way on the particular direction of the velocity vector but only on the length of the vector, i.e., on the speed. The density of representative points then has the same value everywhere on the sphere of radius c in the velocity space, Fig. 4–8(b).

We now pose the question: How many points lie in the shell between spheres of radii c and $c + dc$? This number of points, dn_c, will be equal to the number of molecules having speeds between c and $c + dc$, without regard for the different directions in which the molecules are traveling. The number of points dn_c in the shell is the density of points on the sphere of radius c multiplied by the volume of the shell, that is,

$$dn_c = \text{point density on sphere} \times \text{volume of shell}. \tag{4–33}$$

The volume of the shell, dV_{shell}, is the difference in volume between the outer and the inner sphere:

$$dV_{\text{shell}} = \frac{4\pi}{3}(c + dc)^3 - \frac{4\pi}{3}c^3 = \frac{4\pi}{3}[3c^2\, dc + 3c\,(dc)^2 + (dc)^3].$$

The terms on the right which involve $(dc)^2$ and $(dc)^3$ are infinitesimals of higher order which vanish more rapidly than dc in the limit as $dc \to 0$; these higher terms are dropped out and we obtain $dV_{\text{shell}} = 4\pi c^2\, dc$. Using this result and Eq. (4–32) in Eq. (4–33), we have

$$dn_c = 4\pi NA^3 e^{-\beta c^2} c^2\, dc, \tag{4–34}$$

which relates dn_c, the number of molecules with speeds between c and $c + dc$, to N, c, dc, and the constants A and β. Equation (4–34) is one form of the Maxwell distribution and is the solution of the problem posed at the beginning of this section. Before Eq. (4–34) can be useful, the constants A and β must be evaluated.

4–7   Mathematical Interlude

In the kinetic theory of gases we deal with integrals of the general type:

$$I_n(\beta) = \int_0^\infty x^{2n+1} e^{-\beta x^2} \, dx \qquad (\beta > 0; n > -1). \qquad (4\text{–}35)$$

If we make the substitution, $y = \beta x^2$, the integral reduces to the form

$$I_n(\beta) = \tfrac{1}{2}\beta^{-(n+1)} \int_0^\infty y^n e^{-y} \, dy.$$

However, the factorial function, $n!$, is defined by

$$n! = \int_0^\infty y^n e^{-y} \, dy \qquad (n > -1), \qquad (4\text{–}36)$$

so that

$$I_n(\beta) = \int_0^\infty x^{2n+1} e^{-\beta x^2} \, dx = \tfrac{1}{2}(n!)\beta^{-(n+1)}. \qquad (4\text{–}37)$$

The higher-order integrals can be obtained from those of lower order by differentiation; differentiating Eq. (4–37) with respect to β yields

$$\frac{dI_n(\beta)}{d\beta} = -\int_0^\infty x^{2n+3} e^{-\beta x^2} \, dx = -\tfrac{1}{2}[(n+1)!]\beta^{-(n+2)} = -I_{n+1}(\beta),$$

or

$$I_{n+1}(\beta) = -\, dI_n(\beta)/d\beta. \qquad (4\text{–}38)$$

Two cases commonly arise.

Case I. $n = 0$ or a positive integer.

In this case we apply Eq. (4–37) directly and no difficulty ensues. The lowest member is

$$I_0(\beta) = \tfrac{1}{2}\beta^{-1}$$

All other members can be obtained from Eq. (4–37) or by differentiating $I_0(\beta)$ and using Eq. (4–38).

Case II. $n = -\tfrac{1}{2}, \tfrac{1}{2}, \tfrac{3}{2}$ or $n = m - \tfrac{1}{2}$ where $m = 0$ or a positive integer.

In this case we may also use Eq. (4–37) directly, but unless we know the value of the factorial function for half-integral values of the argument we will be in trouble. If $n = m - \frac{1}{2}$, the function takes the form

$$I_{m-\frac{1}{2}}(\beta) = \int_0^\infty x^{2m} e^{-\beta x^2}\, dx = \tfrac{1}{2}[(m - \tfrac{1}{2})!]\beta^{-(m+\frac{1}{2})}. \tag{4–39}$$

When $m = 0$, we have

$$I_{-\frac{1}{2}}(\beta) = \int_0^\infty e^{-\beta x^2}\, dx = \beta^{-\frac{1}{2}} \int_0^\infty e^{-y^2}\, dy = \beta^{-\frac{1}{2}} I_{-\frac{1}{2}}(1), \tag{4–40}$$

where in the second writing, $x = \beta^{-\frac{1}{2}} y$, has been used. Comparing this result with the last member of Eq. (4–39) we find that

$$I_{-\frac{1}{2}}(1) = \int_0^\infty e^{-y^2}\, dy = \tfrac{1}{2}(-\tfrac{1}{2})!. \tag{4–41}$$

The integral, $I_{-\frac{1}{2}}(1)$, cannot be evaluated by elementary methods. We proceed by writing the integral in two ways,

$$I_{-\frac{1}{2}}(1) = \int_0^\infty e^{-x^2}\, dx \quad \text{and} \quad I_{-\frac{1}{2}}(1) = \int_0^\infty e^{-y^2}\, dy,$$

then multiply them together to obtain,

$$I^2_{-\frac{1}{2}}(1) = \int_0^\infty \int_0^\infty e^{-(x^2+y^2)}\, dx\, dy.$$

The integration is over the area of the first quadrant; we change variables to $r^2 = x^2 + y^2$ and replace $dx\, dy$ by the element of area in polar coordinates, $r\, d\varphi\, dr$. To cover the first quadrant we integrate φ from zero to $\pi/2$ and r from 0 to ∞: the integral becomes

$$I^2_{-\frac{1}{2}}(1) = \int_0^{\pi/2} d\varphi \int_0^\infty e^{-r^2} r\, dr = (\pi/2)(\tfrac{1}{2}) \int_0^\infty e^{-r^2}\, d(r^2) = (\pi/4) \int_0^\infty e^{-y}\, dy.$$

The last integral is equal to unity, so we have, taking the square root of both sides,

$$I_{-\frac{1}{2}}(1) = \tfrac{1}{2}\sqrt{\pi}. \tag{4–42}$$

Comparing Eqs. (4–41) and (4–42), it follows that $(-\tfrac{1}{2})! = \sqrt{\pi}$; now from Eqs. (4–40) and (4–42),

$$I_{-\frac{1}{2}}(\beta) = \tfrac{1}{2}\sqrt{\pi}\beta^{-\frac{1}{2}}.$$

By differentiation, and by using Eq. (4–38) we obtain

$$I_{\frac{1}{2}}(\beta) = -dI_{-\frac{1}{2}}/d\beta = \tfrac{1}{2}\sqrt{\pi}(\tfrac{1}{2}\beta^{-3/2}),$$

$$I_{\frac{3}{2}}(\beta) = -dI_{\frac{1}{2}}/d\beta = \tfrac{1}{2}\sqrt{\pi}(\tfrac{1}{2} \cdot \tfrac{3}{2}\beta^{-5/2}).$$

Repetition of this procedure yields ultimately

$$I_{m-\frac{1}{2}}(\beta) = \int_0^\infty x^{2m} e^{-\beta x^2}\, dx = \tfrac{1}{2}\sqrt{\pi}\,\frac{(2m)!}{2^{2m}m!}\,\beta^{-(m+\frac{1}{2})}. \tag{4–43}$$

By comparing this result with Eq. (4–39) we obtain the interesting result for half-integral factorials

$$(m - \tfrac{1}{2})! = \sqrt{\pi}\,\frac{(2m)!}{2^{2m}m!}. \tag{4–44}$$

Table 4–1 collects the most commonly used formulas.

Table 4–1 Integrals which occur in the kinetic theory of gases

(1)	$\displaystyle\int_{-\infty}^\infty x^{2n} e^{-\beta x^2}\, dx = 2\int_0^\infty x^{2n} e^{-\beta x^2}\, dx$	(6)	$\displaystyle\int_{-\infty}^\infty x^{2n+1} e^{-\beta x^2}\, dx = 0$
(2)	$\displaystyle\int_0^\infty e^{-\beta x^2}\, dx = \tfrac{1}{2}\sqrt{\pi}\,\beta^{-1/2}$	(7)	$\displaystyle\int_0^\infty x e^{-\beta x^2}\, dx = \tfrac{1}{2}\beta^{-1}$
(3)	$\displaystyle\int_0^\infty x^2 e^{-\beta x^2}\, dx = \tfrac{1}{2}\sqrt{\pi}\,\tfrac{1}{2}\beta^{-3/2}$	(8)	$\displaystyle\int_0^\infty x^3 e^{-\beta x^2}\, dx = \tfrac{1}{2}\beta^{-2}$
(4)	$\displaystyle\int_0^\infty x^4 e^{-\beta x^2}\, dx = \tfrac{1}{2}\sqrt{\pi}\,\tfrac{3}{4}\beta^{-5/2}$	(9)	$\displaystyle\int_0^\infty x^5 e^{-\beta x^2}\, dx = \beta^{-3}$
(5)	$\displaystyle\int_0^\infty x^{2n} e^{-\beta x^2}\, dx = \tfrac{1}{2}\sqrt{\pi}\,\frac{(2n)!\,\beta^{-(n+\frac{1}{2})}}{2^{2n}n!}$	(10)	$\displaystyle\int_0^\infty x^{2n+1} e^{-\beta x^2}\, dx = \tfrac{1}{2}(n!)\beta^{-(n+1)}$

The error function

We frequently have occasion to use integrals of the type of Case II above in which the upper limit is not extended to infinity but only to some finite value. These integrals are related to the error function (erf). We define

$$\text{erf}\,(x) = \frac{2}{\sqrt{\pi}}\int_0^x e^{-u^2}\, du. \tag{4–45}$$

If the upper limit is extended to $x \to \infty$, the integral is $\tfrac{1}{2}\sqrt{\pi}$ so that

$$\text{erf}\,(\infty) = 1.$$

Thus as x varies from zero to infinity, erf (x) varies from zero to unity. If to both sides of the definition we add the integral from x to ∞ multiplied by $2/\sqrt{\pi}$ we obtain

$$\text{erf}\,(x) + \frac{2}{\sqrt{\pi}}\int_x^\infty e^{-u^2}\, du = \frac{2}{\sqrt{\pi}}\left[\int_0^x e^{-u^2}\, du + \int_x^\infty e^{-u^2}\, du\right] = \frac{2}{\sqrt{\pi}}\int_0^\infty e^{-u^2}\, du = 1.$$

Therefore,

$$\frac{2}{\sqrt{\pi}} \int_x^\infty e^{-u^2} \, du = 1 - \mathrm{erf}(x).$$

We define the co-error function, $\mathrm{erfc}(x)$, by

$$\mathrm{erfc}(x) = 1 - \mathrm{erf}(x). \tag{4-46}$$

Thus

$$\int_x^\infty e^{-u^2} \, du = \frac{\sqrt{\pi}}{2} \mathrm{erfc}(x) \tag{4-47}$$

Some values of the error function are given in Table 4-2.

Table 4-2 The error function: $\mathrm{erf}(x) = \dfrac{2}{\sqrt{\pi}} \displaystyle\int_0^x e^{-u^2} \, du$

x	$\mathrm{erf}(x)$	x	$\mathrm{erf}(x)$	x	$\mathrm{erf}(x)$
0.00	0.000	0.80	0.742	1.60	0.976
0.10	0.112	0.90	0.797	1.70	0.984
0.20	0.223	1.00	0.843	1.80	0.989
0.30	0.329	1.10	0.880	1.90	0.993
0.40	0.428	1.20	0.910	2.00	0.995
0.50	0.521	1.30	0.934	2.20	0.998
0.60	0.604	1.40	0.952	2.40	0.9993
0.70	0.678	1.50	0.966	2.50	0.9996

4-8 Evaluation of A and β

The constants A and β are determined by requiring that the distribution yield correct values of the total number of molecules and the average kinetic energy. The total number of molecules is obtained by summing dn_c over all possible values of c between zero and infinity:

$$N = \int_{c=0}^{c=\infty} dn_c. \tag{4-48}$$

The average kinetic energy is calculated by multiplying the kinetic energy, $\frac{1}{2}mc^2$, by the number which have that kinetic energy, dn_c, summing over all values of c, and dividing by N, the total number of molecules:

$$\bar{\epsilon} = \frac{\displaystyle\int_{c=0}^{c=\infty} \frac{1}{2}mc^2 \, dn_c}{N}. \tag{4-49}$$

Equations (4-48) and (4-49) determine A and β.

Replacing dn_c in Eq. (4–48) by the value given by Eq. (4–34), we have

$$N = \int_0^\infty 4\pi N A^3 e^{-\beta c^2} c^2 \, dc.$$

Dividing through by N and removing constants from under the integral sign yields

$$1 = 4\pi A^3 \int_0^\infty c^2 e^{-\beta c^2} \, dc.$$

From Table 4–1 we have $\int_0^\infty c^2 e^{-\beta c^2} \, dc = \pi^{1/2}/4\beta^{3/2}$. Hence, $1 = 4\pi A^3 \pi^{1/2}/4\beta^{3/2}$. So finally

$$A^3 = (\beta/\pi)^{3/2}, \tag{4–50}$$

which gives the value of A^3 in terms of β.

In the second condition, Eq. (4–49), we use the value for dn_c from Eq. (4–34):

$$\bar{\epsilon} = \frac{\int_0^\infty \frac{1}{2}mc^2 4\pi N A^3 e^{-\beta c^2} c^2 \, dc}{N}.$$

Using Eq. (4–50), we have

$$\bar{\epsilon} = 2\pi m \left(\frac{\beta}{\pi}\right)^{3/2} \int_0^\infty c^4 e^{-\beta c^2} \, dc.$$

From Table 4–1, we have $\int_0^\infty c^4 e^{-\beta c^2} \, dc = 3\pi^{1/2}/8\beta^{5/2}$. So $\bar{\epsilon}$ becomes $\bar{\epsilon} = 3m/4\beta$, and, therefore,

$$\beta = 3m/4\bar{\epsilon}, \tag{4–51}$$

which expresses β in terms of the average energy per molecule $\bar{\epsilon}$. However, Eq. (4–13) relates the average energy per molecule to the temperature:

$$\bar{\epsilon} = \frac{3}{2}\left(\frac{R}{N_0}\right)T = \tfrac{3}{2}\mathbf{k}T. \tag{4–13a}$$

The gas constant per molecule is the Boltzmann constant, $\mathbf{k} = R/N_0 = 1.380 \times 10^{-16}$ erg/deg. Using this relation in Eq. (4–51) gives β explicitly in terms of m and T:

$$\beta = \frac{m}{2\mathbf{k}T}. \tag{4–52}$$

Using Eq. (4–52) in Eq. (4–50), we obtain

$$A^3 = \left(\frac{m}{2\pi\mathbf{k}T}\right)^{3/2}, \qquad A = \left(\frac{m}{2\pi\mathbf{k}T}\right)^{1/2}. \tag{4–53}$$

Using Eqs. (4–52) and (4–53) for β and A^3 in Eq. (4–34), we obtain the Maxwell distribution in explicit form:

$$dn_c = 4\pi N \left(\frac{m}{2\pi kT}\right)^{3/2} c^2 e^{-mc^2/2kT} \, dc. \qquad (4\text{-}54)$$

The Maxwell distribution expresses the number of molecules having speeds between c and $c + dc$ in terms of the total number present, the mass of the molecules, the temperature, and the speed. (To simplify computations with the Maxwell distribution, note that the ratio $m/k = M/R$, where M is the mass of one mole.) The Maxwell distribution is customarily plotted with the function $(1/N)(dn_c/dc)$ as the ordinate and c as the abscissa. The fraction of the molecules in the speed range c to $c + dc$ is dn_c/N; dividing this by dc gives the fraction of the molecules in this speed range per unit width of the interval:

$$\frac{1}{N}\frac{dn_c}{dc} = 4\pi \left(\frac{m}{2\pi kT}\right)^{3/2} c^2 e^{-mc^2/2kT}. \qquad (4\text{-}55)$$

The plot of the function at two temperatures is shown in Fig. 4–9.

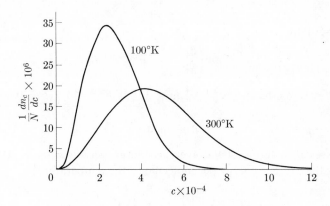

Fig. 4–9 Maxwell distribution for nitrogen at two temperatures.

The function shown in Fig. 4–9 is the probability of finding a molecule having a speed between c and $c + dc$, divided by the width dc of the range. Roughly speaking, the ordinate is the probability of finding a molecule with a speed between c and $(c + 1)$ cm/sec. The curve is parabolic near the origin, since the factor c^2 is dominant in this region, the exponential function being about equal to unity; at high values of c, the exponential factor dominates the behavior of the function, causing it to decrease rapidly in value. As a consequence of the contrasting behavior of the two factors, the product function has a maximum at a speed c_{mp}. This speed is called the *most probable speed*, since it corresponds to the maximum in the probability curve; c_{mp} can be calculated by differentiating the function on the right of Eq. (4–55) and setting the derivative equal to zero to find the location of the horizontal tangents. This procedure yields

$$c e^{-mc^2/2kT}(2 - mc^2/kT) = 0.$$

The curve has three horizontal tangents: at $c = 0$; at $c \to \infty$, when $\exp\left(-\frac{1}{2}mc^2/kT\right) = 0$; and when $2 - mc^2/kT = 0$. This last condition determines c_{mp}:

$$c_{mp} = \sqrt{2kT/m} = \sqrt{2RT/M}. \tag{4–56}$$

Figure 4–9 shows that the chance of finding molecules with very low speeds or with very high speeds is nearly zero. The majority of molecules have speeds which cluster around c_{mp} in the middle of the range of c.

Figure 4–9 also shows that an increase in temperature broadens the speed distribution and shifts the maximum to higher values of c. The area under the two curves in Fig. 4–9 must be the same, since it is equal to unity in both cases. This requires the curve to broaden as the temperature is increased. The speed distribution also depends on the mass of the molecule. At the same temperature a heavy gas has a narrower distribution of speeds than a light gas.

The appearance of the temperature as the parameter of the distribution yields another interpretation of the, as yet mysterious, notion of temperature. Roughly, the temperature is a measure of the broadness of the speed distribution. If by any means we manage to narrow the distribution, we will discover that the temperature of the system has dropped. At the absolute zero of temperature, the distribution becomes infinitely narrow; all of the molecules have the same kinetic energy, zero.

4–9   Calculation of Average Values Using the Maxwell Distribution

From the Maxwell distribution, the average value of any quantity which depends upon the speed can be calculated. If we wish to calculate the average value $\bar{g}$ of some function of speed $g(c)$, we multiply the function $g(c)$ by dn_c, the number of molecules which have the speed c; then sum over all values of c from zero to infinity and divide by the total number of molecules in the gas:

$$\bar{g} = \frac{\displaystyle\int_{c=0}^{c=\infty} g(c)\, dn_c}{N}. \tag{4–57}$$

As an example of the use of Eq. (4–57), we can calculate the average kinetic energy of the gas molecules; for this case, $g(c) = \epsilon = \frac{1}{2}mc^2$. Thus Eq. (4–57) becomes

$$\bar{\epsilon} = \frac{\displaystyle\int_{c=0}^{c=\infty} \frac{1}{2}mc^2\, dn_c}{N},$$

which is identical with Eq. (4–49). If we put in the value of dn_c and integrate, we would, of course, find that $\bar{\epsilon} = \frac{3}{2}kT$, since we used this relation to determine the constant β in the distribution function.

Another average value of importance is the average speed $\bar{c}$. Using Eq. (4–57), we have

$$\bar{c} = \frac{\int_{c=0}^{c=\infty} c \, dn_c}{N}.$$

Using the value of dn_c from Eq. (4–54), we obtain

$$\bar{c} = 4\pi \left(\frac{m}{2\pi kT}\right)^{3/2} \int_0^\infty c^3 e^{-mc^2/2kT} \, dc.$$

The integral can be obtained from Table 4–1, or can be evaluated by elementary methods through the change in variable: $x = \frac{1}{2}mc^2/kT$. This substitution yields

$$\bar{c} = \sqrt{\frac{8kT}{\pi m}} \int_0^\infty x e^{-x} \, dx.$$

But $\int_0^\infty x e^{-x} \, dx = 1$, therefore

$$\bar{c} = \sqrt{8kT/\pi m}. \tag{4–58}$$

It should be noted that the average speed is not equal to the rms speed, $c_{rms} = (3kT/m)^{1/2}$, but is somewhat smaller. The most probable speed, $c_{mp} = (2kT/m)^{1/2}$, is smaller yet. The average speed and the rms speed occur most frequently in physico-chemical calculations.

Since the speeds of the molecules are distributed, we can talk about the deviation of the speed of a molecule from the mean value, $\delta = c - \bar{c}$. The average deviation from the mean value is zero, of course. However, the square of the deviations from the mean, $\delta^2 = (c - \bar{c})^2$, has an average value different from zero. This quantity gives us a measure of the breadth of the distribution. Calculation of this kind of average value (Problems 4–9 and 4–10) gives us an important insight into the meaning of temperature, particularly in the case of the energy distribution.

4-10 The Maxwell Distribution as an Energy Distribution

The speed distribution, Eq. (4–54), can be converted to an energy distribution. The kinetic energy of a molecule is $\epsilon = \frac{1}{2}mc^2$. Then $c = (2/m)^{1/2}\epsilon^{1/2}$. Differentiating, we obtain $dc = (1/2m)^{1/2}\epsilon^{-1/2} \, d\epsilon$. The energy range $d\epsilon$ corresponds to the speed range dc, and so the number of particles dn_c in the speed range corresponds to the number of particles dn_ϵ in the energy range. By replacing c and dc in the velocity distribution by their equivalents according to these equations, we obtain the energy distribution

$$dn_\epsilon = 2\pi N \left(\frac{1}{\pi \mathbf{k} T} \right)^{3/2} \epsilon^{1/2} e^{-\epsilon/\mathbf{k}T} \, d\epsilon, \qquad (4\text{–}59)$$

where dn_ϵ is the number of molecules having kinetic energies between ϵ and $\epsilon + d\epsilon$. This form of the distribution function is plotted as a function of ϵ in Fig. 4–10(a). Attention should be given to the distinctly different shape of this curve compared with that of the speed distribution. In particular, the energy distribution has a vertical tangent at the origin and thus it rises much more quickly than the velocity distribution which starts with a horizontal tangent. After passing the maximum, the energy distribution falls off more gently than does the velocity distribution. As usual, the distribution is broadened at higher temperatures, a greater proportion of the

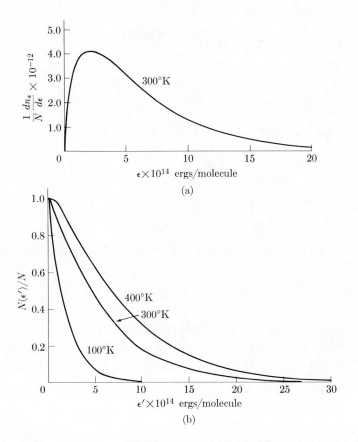

Fig. 4–10 (a) Energy distribution at 300°K. (b) Fraction of molecules having energies in excess of ϵ'.

molecules having higher energies. As before, the areas under the curves for different temperatures must be the same.

It is frequently important to know what fraction of the molecules in a gas have kinetic energies exceeding a specified value ϵ'. This quantity can be calculated from the distribution function. Let $N(\epsilon')$ be the number of molecules with energies greater than ϵ'. Then $N(\epsilon')$ is the sum of the number of molecules in the energy range above ϵ':

$$N(\epsilon') = \int_{\epsilon'}^{\infty} dn_\epsilon. \tag{4-60}$$

The fraction of the molecules with energies above ϵ' is $N(\epsilon')/N$; using the expression in Eq. (4–59) for the integrand in Eq. (4–60), this fraction becomes

$$\frac{N(\epsilon')}{N} = 2\pi \left(\frac{1}{\pi \mathbf{k} T}\right)^{3/2} \int_{\epsilon'}^{\infty} \epsilon^{1/2} e^{-\epsilon/kT} \, d\epsilon. \tag{4-61}$$

The substitutions

$$\epsilon = \mathbf{k} T x^2, \qquad d\epsilon = \mathbf{k} T \, d(x^2), \qquad \epsilon^{1/2} = (\mathbf{k} T)^{1/2} x,$$

reduce Eq. (4–61) to

$$\frac{N(\epsilon')}{N} = \frac{2}{\sqrt{\pi}} \int_{\sqrt{\epsilon'/kT}}^{\infty} x e^{-x^2} \, d(x^2) = -\frac{2}{\sqrt{\pi}} \int_{\sqrt{\epsilon'/kT}}^{\infty} x \, d(e^{-x^2}).$$

Integrating by parts, we have

$$\frac{N(\epsilon')}{N} = -\frac{2}{\sqrt{\pi}} \left[x e^{-x^2} \Big|_{\sqrt{\epsilon'/kT}}^{\infty} - \int_{\sqrt{\epsilon'/kT}}^{\infty} e^{-x^2} \, dx \right],$$

$$\frac{N(\epsilon')}{N} = 2\left(\frac{\epsilon'}{\pi \mathbf{k} T}\right)^{1/2} e^{-\epsilon'/kT} + \frac{2}{\sqrt{\pi}} \int_{\sqrt{\epsilon'/kT}}^{\infty} e^{-x^2} \, dx. \tag{4-62}$$

The integral in Eq. (4–62) can be expressed in terms of the co-error function defined in Eq. (4–47).

$$\frac{N(\epsilon')}{N} = 2\left(\frac{\epsilon'}{\pi \mathbf{k} T}\right)^{1/2} e^{-\epsilon'/kT} + \operatorname{erfc}\left(\sqrt{\epsilon'/\mathbf{k} T}\right). \tag{4-63}$$

However, if the energy ϵ' is very much larger than $\mathbf{k} T$, the value of the integral in Eq. (4–62) is approximately zero (since the area under the curve of the integrand is very small from a large value of the lower limit to infinity). In this important case, Eq. (4–62) becomes

$$\frac{N(\epsilon')}{N} = 2\left(\frac{\epsilon'}{\pi \mathbf{k} T}\right)^{1/2} e^{-\epsilon'/kT}, \qquad \epsilon' \gg \mathbf{k} T. \tag{4-64}$$

Equation (4–64) has the property that the right-hand side varies quite rapidly with temperature, particularly at low temperatures. Figure 4–10(b) shows the variation of $N(\epsilon')/N$ with ϵ' at three temperatures, calculated from Eq. (4–62). Also, Fig. 4–10(b) shows graphically that the fraction of molecules having energies greater than ϵ' increases markedly with temperature, particularly if ϵ' is in the high-energy range. This property of gases, and of liquids and solids as well, has fundamental significance in connection with the increase in the rates of chemical reactions with temperature. Since only molecules which have more than a certain minimum energy can react chemically, and since the fraction of molecules which have energies exceeding this minimum value increases with temperature according to Eq. (4–62), the rate of a chemical reaction increases with temperature.†

4-11   Average Values of Individual Components; Equipartition of Energy

It is instructive to compute the average values of the individual components of the velocity. To this end, the Maxwell distribution is most conveniently used in the form of Eq. (4–30). The average value of u is then given by an equation analogous to Eq. (4–57):

$$\bar{u} = \frac{\int_{-\infty}^{\infty} \int_{-\infty}^{\infty} \int_{-\infty}^{\infty} u \, dn_{uvw}}{N}.$$

The integration is taken over all possible values of all three components; note that any component may have any value from minus infinity to plus infinity. Using dn_{uvw} from Eq. (4–30), we obtain

$$\bar{u} = A^3 \int_{-\infty}^{\infty} \int_{-\infty}^{\infty} \int_{-\infty}^{\infty} ue^{-\beta(u^2 + v^2 + w^2)} \, du \, dv \, dw$$

$$= A^3 \int_{-\infty}^{\infty} ue^{-\beta u^2} \, du \int_{-\infty}^{\infty} e^{-\beta v^2} \, dv \int_{-\infty}^{\infty} e^{-\beta w^2} \, dw. \tag{4–65}$$

By Formula (6) in Table 4–1, the first integral on the right-hand side of Eq. (4–65) is zero; thus $\bar{u} = 0$. The same result is obtained for the average value of the other components:

$$\bar{u} = \bar{v} = \bar{w} = 0. \tag{4–66}$$

The reason the average value of any individual component must be zero is physically obvious. If the average value of any one of the components had a value other than zero, this would correspond to a net motion of the entire mass of gas in that particular direction; the present discussion applies only to gases at rest.

---

† Other conditions being comparable, the rate of chemical reaction depends on temperature through the factor $Ae^{-\epsilon_a/kT}$, where A is a constant and ϵ_a is a characteristic energy. Note the similarity in form to the right-hand side of Eq. (4–64).

The distribution function for the x-component can be written [see Eqs. (4–20), (4–29), (4–52), (4–53)] as

$$\frac{1}{N}\frac{dn_u}{du} = f(u^2) = \left(\frac{m}{2\pi\mathbf{k}T}\right)^{1/2} e^{-mu^2/2\mathbf{k}T}, \tag{4–67}$$

which is plotted in Fig. 4–11. It is the symmetry of the function with respect to the origin of u which leads to the vanishing value of $\bar{u}$. The interpretation of temperature as a measure of the width of the distribution is quite clearly illustrated by the two curves in Fig. 4–11. The area under each curve must have the same value, unity. The probability of finding a molecule of velocity u is the same as finding one with a velocity $-u$; this was assured in our original choice of the function as one which depended only on u^2.

Although the average value of the velocity component in any one direction is zero, because equal numbers of molecules have components u and $-u$, the average value of kinetic energy associated with a particular component has a positive value. The molecules with velocity component u contribute $\frac{1}{2}mu^2$ to the average and those with component $-u$ contribute $\frac{1}{2}m(-u)^2 = \frac{1}{2}mu^2$. The contributions of particles moving in opposite directions add up in averaging the energy, while in averaging the velocity component the contributions of particles moving in opposite directions exactly cancel one another. To calculate the average value of $\epsilon_x = \frac{1}{2}mu^2$, we use the Maxwell distribution in the same way as before:

$$\bar{\epsilon}_x = \frac{\displaystyle\int_{-\infty}^{\infty}\int_{-\infty}^{\infty}\int_{-\infty}^{\infty} \frac{1}{2}mu^2\,dn_{uvw}}{N}.$$

Using Eq. (4–30), we obtain

$$\epsilon_x = \frac{1}{2}mA^3 \int_{-\infty}^{\infty}\int_{-\infty}^{\infty}\int_{-\infty}^{\infty} u^2 e^{-\beta(u^2+v^2+w^2)}\,du\,dv\,dw$$

$$= \frac{1}{2}mA^3 \int_{-\infty}^{\infty} u^2 e^{-\beta u^2}\,du \int_{-\infty}^{\infty} e^{-\beta v^2}\,dv \int_{-\infty}^{\infty} e^{-\beta w^2}\,dw.$$

Using Formulas (1) and (2), Table 4–1, we have

$$\int_{-\infty}^{\infty} e^{-\beta v^2}\,dv = \int_{-\infty}^{\infty} e^{-\beta w^2}\,dw = \left(\frac{\pi}{\beta}\right)^{1/2},$$

and, by Formulas (1) and (3), Table 4–1,

$$\int_{-\infty}^{\infty} u^2 e^{-\beta u^2}\,du = \frac{\pi^{1/2}}{2\beta^{3/2}}.$$

Introducing these values for the integrals leads to

$$\bar{\epsilon}_x = \frac{1}{2}mA^3 \left(\frac{\pi^{1/2}}{2\beta^{3/2}}\right)\left(\frac{\pi}{\beta}\right)^{1/2}\left(\frac{\pi}{\beta}\right)^{1/2} = \frac{mA^3\pi^{3/2}}{4\beta^{5/2}}.$$

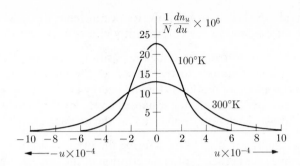

Fig. 4–11 x-component distribution in nitrogen.

Using the value of A^3 from Eq. (4–53) and the value of β from Eq. (4–52), we obtain finally

$$\bar{\epsilon}_x = \tfrac{1}{2}\mathbf{k}T.$$

The same result can be obtained for $\bar{\epsilon}_y$ and $\bar{\epsilon}_z$; therefore

$$\bar{\epsilon}_x = \bar{\epsilon}_y = \bar{\epsilon}_z = \tfrac{1}{2}\mathbf{k}T. \tag{4–68}$$

Since the average total kinetic energy is the sum of the three terms, its value is $\tfrac{3}{2}\mathbf{k}T$, the value given by Eq. (4–13a):

$$\bar{\epsilon} = \bar{\epsilon}_x + \bar{\epsilon}_y + \bar{\epsilon}_z = \tfrac{1}{2}\mathbf{k}T + \tfrac{1}{2}\mathbf{k}T + \tfrac{1}{2}\mathbf{k}T = \tfrac{3}{2}\mathbf{k}T. \tag{4–69}$$

Equation (4–68) expresses the important law of *equipartition of energy*. It says that the average total energy is equally divided among the three independent components of the motion, which are called *degrees of freedom*. The molecule has three translational degrees of freedom. The equipartition law may be stated in the following way. If the energy of the *individual* molecule can be written in the form of a sum of terms, each of which is proportional to the square of either a velocity component or as a coordinate, then each of these square terms contributes $\tfrac{1}{2}\mathbf{k}T$ to the *average* energy. As an example, in the gas the translational energy of the individual molecule is

$$\epsilon = \tfrac{1}{2}mu^2 + \tfrac{1}{2}mv^2 + \tfrac{1}{2}mw^2. \tag{4–70}$$

Since each term is proportional to the square of a velocity component, each contributes $\tfrac{1}{2}\mathbf{k}T$ to the average energy; thus we may write

$$\bar{\epsilon} = \tfrac{1}{2}\mathbf{k}T + \tfrac{1}{2}\mathbf{k}T + \tfrac{1}{2}\mathbf{k}T = \tfrac{3}{2}\mathbf{k}T. \tag{4–71}$$

4–12 Equipartition of Energy and Quantization

A mechanical system consisting of N particles is described by specifying three co-ordinates for each particle or a total of $3N$ coordinates. Thus, there are $3N$ independent components of the motion, or degrees of freedom, in such a system. If the N

particles are bound together to form a polyatomic molecule, then the $3N$ coordinates and components of the motion are conveniently chosen as follows:

Translational.   Three coordinates describe the position of the center of mass; motion in these coordinates corresponds to translation of the molecule as a whole. The energy stored in this mode of motion is kinetic energy only, $\epsilon_{\text{trans}} = \frac{1}{2}mu^2 + \frac{1}{2}mv^2 + \frac{1}{2}mw^2$. Each of these terms contains the square of a velocity component and therefore, as we have seen above, each contributes $\frac{1}{2}kT$ to the average energy.

Rotational.   Two angles are needed to describe the orientation of a linear molecule in space; three angles are needed for the description of the orientation of a nonlinear molecule. Motion in these coordinates corresponds to rotation about two axes (linear) or three axes (nonlinear) in space. The equation for the energy of rotation has the form

$$\epsilon_{\text{rot}} = \frac{1}{2}I\omega_x^2 + \frac{1}{2}I\omega_y^2 \qquad \text{(linear molecule)}$$

$$\epsilon_{\text{rot}} = \frac{1}{2}I_x\omega_x^2 + \frac{1}{2}I_y\omega_y^2 + \frac{1}{2}I_z\omega_z^2 \qquad \text{(nonlinear molecule)}$$

in which ω_x, ω_y, ω_z are angular velocities and I_x, I_y, I_z are moments of inertia about the x-, y-, and z-axes, respectively. (In the linear case, $I_x = I_y = I$.) Since each term in the energy expression is proportional to the square of a velocity component, each term on the average should have its equal share, $\frac{1}{2}kT$, of energy. Thus, the average rotational energy of linear molecules is $\frac{2}{2}kT$, while that of nonlinear molecules is $\frac{3}{2}kT$. The rotational modes of a diatomic molecule are illustrated in Fig. 4–12.

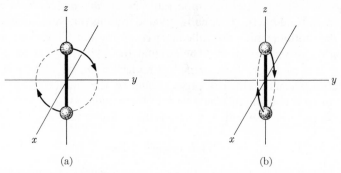

(a)   (b)

Fig. 4–12 Rotational modes of a diatomic molecule. (a) Rotation about the x-axis. (b) Rotation about the y-axis.

Vibrational.   There remain $3N - 5$ coordinates for linear molecules, $3N - 6$ coordinates for nonlinear molecules. These coordinates describe the bond distances and bond angles within the molecule. Motion in these coordinates corresponds to the vibrations (stretching or bending) of the molecule. Thus, linear molecules have $3N - 5$ vibrational modes; nonlinear molecules have $3N - 6$ vibrational modes. Assuming that the vibrations are harmonic, the energy of each vibrational mode can be written in the form

$$\epsilon_{\text{vib}} = \tfrac{1}{2}\mu\left(\frac{dr}{dt}\right)^2 + \tfrac{1}{2}k(r - r_0)^2,$$

in which μ is an appropriate mass, k is the force constant, r_0 is the equilibrium value of the coordinate r, and dr/dt is the velocity. The first term in this expression is the kinetic energy, the second term is the potential energy. By the equipartition law, the first term should contribute $\tfrac{1}{2}kT$ to the average energy, since it contains a velocity squared. The second term, since it contains the square of the coordinate $r - r_0$, should also contribute $\tfrac{1}{2}kT$ to the average energy. Each vibrational mode should contribute $\tfrac{1}{2}kT + \tfrac{1}{2}kT = kT$ to the average energy of the system. Thus, the average energy of the vibrations should be $(3N - 5)kT$ for linear molecules or $(3N - 6)kT$ for nonlinear molecules. The total average energy per molecule should be

$$\bar{\epsilon}_t = \tfrac{3}{2}kT + \tfrac{2}{2}kT + (3N - 5)kT \qquad \text{(linear molecules)}$$

$$\bar{\epsilon}_t = \tfrac{3}{2}kT + \tfrac{3}{2}kT + (3N - 6)kT \qquad \text{(nonlinear molecules)}.$$

If we multiply these values by N_0, the Avogadro number, to convert to average energies per mole, we obtain

Monatomic gases:

$$E = \tfrac{3}{2}RT \tag{4–72}$$

Polyatomic gases:

$$E = \tfrac{3}{2}RT + \tfrac{2}{2}RT + (3N - 5)RT \qquad \text{(linear)} \tag{4–73}$$

$$E = \tfrac{3}{2}RT + \tfrac{3}{2}RT + (3N - 6)RT \qquad \text{(nonlinear)}. \tag{4–74}$$

If heat flows into a gas kept at constant volume, the energy of the gas is increased by the amount of energy transferred by the heat flow. The ratio of the increase in energy to the increase in temperature of the system is the constant volume heat capacity, C_v. Thus, by definition,

$$C_v \equiv \left(\frac{\partial E}{\partial T}\right)_V. \tag{4–75}$$

By differentiating the molar energies with respect to temperature, we obtain the molar heat capacities, $\bar{C}_v$, predicted by the equipartition law:

Monatomic gases:

$$\bar{C}_v = \tfrac{3}{2}R. \tag{4–76}$$

Polyatomic gases:

$$\bar{C}_v = \tfrac{3}{2}R + \tfrac{2}{2}R + (3N - 5)R \qquad \text{(linear)} \tag{4–77}$$

$$\bar{C}_v = \tfrac{3}{2}R + \tfrac{3}{2}R + (3N - 6)R \qquad \text{(nonlinear)}. \tag{4–78}$$

If we examine the values of the heat capacities we find for monatomic gases, $\bar{C}_v/R = 1.5000$, with a high degree of accuracy. This value is independent of temperature over a very wide range.

If we examine the heat capacities of polyatomic gases, Table 4-3, we find two points of disagreement between the data and the equipartition law prediction. The observed heat capacities (1) are always substantially lower than the predicted values, and (2) depend noticeably on temperature. The equipartition principle is a law of classical physics, and these discrepancies were one of the first indications that classical

Table 4-3. Heat capacities of gases at 298.15°K

Monatomic

Species	$\bar{C}_v/R$
He, Ne, A, Kr, Xe	1.5000

Diatomic

Species	$\bar{C}_v/R$	Species	$\bar{C}_v/R$
H_2	2.468	F_2	2.78
N_2, HF, HBr, HCl	2.50	Cl_2	3.08
CO	2.505	ICl	3.26
HI	2.51	Br_2	3.33
O_2	2.531	IBr	3.37
NO	2.591	I_2	3.43

Triatomic

Linear	$\bar{C}_v/R$	Nonlinear	$\bar{C}_v/R$
CO_2	3.466	H_2O	3.038
N_2O	3.655	H_2S	3.09
COS	3.99	NO_2	3.56
CS_2	4.490	SO_2	3.79

Tetratomic

Linear	$\bar{C}_v/R$	Nonlinear	$\bar{C}_v/R$
C_2H_2	4.283	H_2CO	3.25
		NH_3	3.289
		HN_3	4.042
		P_4	7.05

mechanics was not adequate to describe molecular properties. To illustrate the difficulty we choose the case of diatomic molecules, which are of necessity linear. For diatomic molecules, $N = 2$, and we obtain from the equipartition law:

$$\bar{C}_v/R = \tfrac{3}{2} + \tfrac{2}{2} + 1 = \tfrac{7}{2} = 3.5.$$

With the exception of H_2, the observed values for diatomic molecules at ordinary temperature fall between 2.5 and 3.5, a number of them being very close to 2.50. Since the translational value 1.5 is observed so accurately in monatomic molecules we suspect the difficulty lies in either the rotational or vibrational motion. When we note that nonlinear molecules have $\bar{C}_v/R > 3.0$ we can narrow the difficulty to the vibrational motion.

The explanation of the observed behavior lies in the fact that the vibrational motion is quantized. The energy of an oscillator is restricted to certain discrete values and no others. This is in contrast to the classical oscillator which could have any energy value whatsoever. Now instead of the energies of the various oscillators being distributed continuously over the entire range of energies, the oscillators are distributed in the various quantum states (energy levels). The permissible values of the energy of a harmonic oscillator are given by the expression

$$\epsilon_s = (s + \tfrac{1}{2})h\nu \qquad (s = 0, 1, 2, \ldots), \tag{4–79}$$

in which s, the quantum number, is zero or a positive integer, h is Planck's constant, $h = 6.625 \times 10^{-27}$ erg sec, and ν is the classical frequency of the oscillator, $\nu = (1/2\pi)\sqrt{k/\mu}$. The distribution of the oscillators is governed by an exponential law (for proof see Section 28–1)

$$n_s = (N/Q)e^{-\epsilon_s/kT} \tag{4–80}$$

where n_s is the number of oscillators having the energy ϵ_s. The *partition function*, Q, is determined by the condition that the sum of the number of oscillators in all the energy levels will yield the total number of oscillators, N. That is,

$$\sum_{s=0}^{\infty} n_s = N. \tag{4–81}$$

Hence, substituting from Eq. (4–80) for n_s

$$\sum_{s=0}^{\infty} \left(\frac{N}{Q}\right) e^{-\alpha\epsilon_s} = N.$$

Thus,

$$Q = \sum_{s=0}^{\infty} e^{-\alpha\epsilon_s}, \tag{4–82}$$

where we have set $\alpha = 1/kT$ for momentary mathematical convenience. If we differentiate Q with respect to α, we obtain

$$\frac{dQ}{d\alpha} = -\sum_{s=0}^{\infty} \epsilon_s e^{-\alpha \epsilon_s}. \tag{4–83}$$

The average energy is obtained by multiplying the energy of each level by the number in that level, summing over all the levels and dividing by the total number of molecules;

$$\bar{\epsilon} = \frac{\sum_{s=0}^{\infty} \epsilon_s n_s}{N}.$$

Replacing n_s/N by its value from Eq. (4–80) we obtain

$$\bar{\epsilon} = \sum_{s=0}^{\infty} \epsilon_s \left(\frac{1}{Q}\right) e^{-\alpha \epsilon_s} = \frac{1}{Q} \sum_{s=0}^{\infty} \epsilon_s e^{-\alpha \epsilon_s}.$$

Comparing this equation with Eq. (4–83), we find that

$$\bar{\epsilon} = -\frac{1}{Q} \frac{dQ}{d\alpha} = -\frac{d \ln Q}{d\alpha}. \tag{4–84}$$

To evaluate Q, we insert $\epsilon_s = (s + \frac{1}{2})h\nu$ in the expression for Q:

$$Q = \sum_{s=0}^{\infty} e^{-\alpha h\nu(s + \frac{1}{2})} = e^{-\alpha h\nu/2} \sum_{s=0}^{\infty} e^{-\alpha h\nu s} = e^{-\alpha h\nu/2} \sum_{s=0}^{\infty} x^s;$$

in the right-hand side we have set $x = e^{-\alpha h\nu}$. But $\sum_{s=0}^{\infty} x^s = 1 + x + x^2 + \cdots$, and this series is the expansion of $1/(1 - x)$; consequently, we have

$$Q = \frac{e^{-\alpha h\nu/2}}{1 - x} = \frac{e^{-\alpha h\nu/2}}{1 - e^{-\alpha h\nu}},$$

whence

$$\ln Q = -\frac{\alpha h\nu}{2} - \ln (1 - e^{-\alpha h\nu}).$$

Differentiating, we obtain

$$\frac{d \ln Q}{d\alpha} = -\frac{h\nu}{2} - \frac{h\nu e^{-\alpha h\nu}}{1 - e^{-\alpha h\nu}} = -\frac{h\nu}{2} - \frac{h\nu}{e^{\alpha h\nu} - 1}.$$

Using this expression in Eq. (4–84) for $\bar{\epsilon}$, we obtain, after setting $\alpha = 1/kT$,

$$\bar{\epsilon} = \frac{h\nu}{2} + \frac{h\nu}{e^{h\nu/kT} - 1}. \tag{4–85}$$

We observe that the average energy is made up of the zero point energy $\frac{1}{2}h\nu$ which is the lowest energy possible for the quantum oscillator, plus a term which depends on temperature. At very low temperature, $hv/kT \gg 1$, hence $\exp(hv/kT) \gg 1$ so that the second term is very small, and

$$\bar{\epsilon} = \tfrac{1}{2}hv.$$

Effectively, all of the oscillators are in lowest quantum state with $s = 0$. At very high temperatures such that $hv/kT \ll 1$, we may expand the exponential function: $e^{hv/kT} \approx 1 + hv/kT$; then $e^{hv/kT} - 1 \approx hv/kT$, and we have

$$\bar{\epsilon} = \frac{hv}{2} + \frac{hv}{hv/kT} = \frac{hv}{2} + kT.$$

For one mole,

$$E = N_0 hv/2 + RT$$

and

$$\bar{C}_v = R.$$

Thus it is only at high temperatures that the vibrational heat capacity attains the classical value, R. It is customary to define a characteristic temperature $\theta = hv/k$ for each oscillator. Then

$$\bar{\epsilon} = \frac{hv}{2} + \frac{hv}{e^{\theta/T} - 1} = \frac{hv}{2} + \frac{k\theta}{e^{\theta/T} - 1}, \tag{4-86}$$

$$E = N_0\bar{\epsilon} = N_0\frac{hv}{2} + \frac{R\theta}{e^{\theta/T} - 1} \tag{4-87}$$

$$\bar{C}_v(\text{vib})/R = \left(\frac{\theta}{T}\right)^2 \frac{e^{\theta/T}}{(e^{\theta/T} - 1)^2}. \tag{4-88}$$

The function on the right-hand side of Eq. (4-88) is called an Einstein function. The Einstein function is shown as a function of (T/θ) in Fig. 4-13. Thus, for a diatomic molecule we have for the heat capacity

$$\bar{C}_v/R = \frac{5}{2} + \left(\frac{\theta}{T}\right)^2 \frac{e^{\theta/T}}{(e^{\theta/T} - 1)^2}.$$

In the case of polyatomic molecules which have more than one vibration, e.g. H_2O which has three vibrations, there are three distinct frequencies, therefore three distinct characteristic temperatures, so that the heat capacity contains three distinct Einstein functions,

$$\bar{C}_v/R = 3.0 + \left(\frac{\theta_1}{T}\right)^2 \frac{e^{\theta_1/T}}{(e^{\theta_1/T} - 1)^2} + \left(\frac{\theta_2}{T}\right)^2 \frac{e^{\theta_2/T}}{(e^{\theta_2/T} - 1)^2} + \left(\frac{\theta_3}{T}\right)^2 \frac{e^{\theta_3/T}}{(e^{\theta_3/T} - 1)^2}.$$

Values of the characteristic temperatures for a number of molecules are given in Table 4-4.

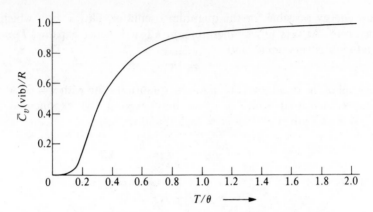

Fig. 4–13 Einstein function for $\bar{C}_v/R$ versus T/θ.

Table 4–4. Values of θ for various gases, in $^\circ K$†

H$_2$	6210°	Br$_2$	470°
N$_2$	3340°	I$_2$	310°
O$_2$	2230°	CO$_2$:	$\theta_1 = 1890°$
CO	3070°		$\theta_2 = 3360°$
NO	2690°		$\theta_3 = \theta_4 = 954°$
HCl	4140°		
HBr	3700°	H$_2$O:	$\theta_1 = 5410°$
HI	3200°		$\theta_2 = 5250°$
Cl$_2$	810°		$\theta_3 = 2290°$

† Terrell L. Hill, *Introduction to Statistical Thermodynamics.* Addison-Wesley Publishing Co., Inc. , Reading, Mass., 1960.

4–13 The Maxwell-Boltzmann Distribution Law

Two types of distribution functions have been discussed so far: the spatial distribution of molecules in a gravitational field, the Boltzmann distribution, and the speed distribution in a gas (the Maxwell distribution). These can be written in a combined form, the Maxwell-Boltzmann distribution law.

The barometric formula governs the spatial distribution of the molecules in a gravity field according to the equation

$$n = n_0 e^{-Mgz/RT}, \tag{4–89}$$

where n and n_0 are the number of particles per cubic centimeter at the heights z

and zero, respectively. The Boltzmann distribution law governs the spatial distribution of any system in which the particles have a potential energy which depends on the position. For any potential field, the Boltzmann distribution may be written in the form

$$n = n_0 e^{-\epsilon_p/\mathbf{k}T}, \qquad (4\text{--}90)$$

where ϵ_p is the potential energy of the particle at the point (x, y, z), and n is the number of particles per cubic centimeter at this position.

For the special case of the gravity field, $\epsilon_p = mgz$. This value of ϵ_p in Eq. (4–90) reduces it to Eq. (4–89), since $m/\mathbf{k} = M/R$.

The combined velocity and space distribution is written

$$\frac{dn^*}{n_0} = 4\pi \left(\frac{m}{2\pi\mathbf{k}T} \right)^{3/2} c^2 e^{-(mc^2/2 + \epsilon_p)/\mathbf{k}T} \, dc, \qquad (4\text{--}91)$$

where dn^* is the number of particles per cubic centimeter at the position (x, y, z) which have speeds in the range c to $c + dc$. Equation (4–91) is the Maxwell–Boltzmann distribution, which is similar to the Maxwell distribution, except that the exponential factor contains the total energy, kinetic plus potential, of the molecule instead of only the kinetic energy.

At any specified position in space, ϵ_p has a definite constant value, and so $\exp(-\epsilon_p/\mathbf{k}T)$ is a constant. Then the right-hand side of Eq. (4–91) is simply the Maxwell distribution multiplied by a constant. This means that at any position the distribution of speeds is Maxwellian regardless of the value of the potential energy at that point. For a gas in a gravity field, this means that although there are fewer molecules per cubic centimeter at 50 km height than at ground level, the fraction of them that have speeds in a given range is the same at both levels.

4–14 Experimental Verification of the Maxwell Distribution Law

The amount of indirect evidence for the correctness of the Maxwell distribution law is overwhelming. The relationship of the distribution law to the rate of a chemical reaction has already been mentioned briefly (Section 4–10). We shall see later that the functional form of the experimentally determined temperature dependence of the rate constant agrees with the dependence we expect from the Maxwell distribution. This agreement may be regarded as indirect evidence for the correctness of both the Maxwell distribution and our ideas concerning reaction rates.

For further illustration, suppose for the sake of argument that speeds were not distributed and that all the molecules moved with the same velocity. Now consider the effect of a gravity field on such a gas. If at ground level all the molecules had the same vertical component of velocity W then all would have a kinetic energy $\frac{1}{2}mW^2$. The maximum height any molecule could reach is that at which all of the ground-level kinetic energy is converted into potential energy; this height H is determined by the equality $mgH = \frac{1}{2}mW^2$, or $H = W^2/2g$. No molecule could reach a height greater than H, and if this situation prevailed, the atmosphere would have a sharp upper boundary. Furthermore the density of the atmosphere would increase with the height above ground-level, since the molecules at the higher levels are moving slowly and thus would spend a larger part of the time at these high levels. None of these predictions is confirmed by observation. The Maxwell distribution, however, says that some molecules have large kinetic energies and so can attain great heights; but the proportion of the molecules with these high energies is small. The Maxwell distribution predicts that the atmospheric density will decrease with increase in height, and that there will be no sharp upper boundary.

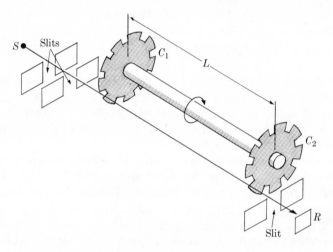

Fig. 4-14. Experiment to verify the Maxwell distribution. (Redrawn by permission from K. F. Herzfeld and H. Smallwood, *A Treatise on Physical Chemistry*, H. S. Taylor and S. Glasstone, eds., Vol. II, 3rd ed. New York: D. Van Nostrand Co. Inc., 1951, p. 37.)

A number of direct experimental determinations of the velocity distribution have been made, all of which have verified the Maxwell law within the experimental error. Because of the difficulty of the experiments, the experimental errors are rather large. A sketch of the apparatus used in one method is shown in Fig. 4-14. The apparatus is entirely enclosed in a highly evacuated chamber. The molecules escape through a pinhole in the source S, are collimated by the slits, and then pass through one of the openings between the cogs in the cogwheel C_1. The cogwheels C_1 and C_2 are mounted on the same axle, which is rotated rapidly. Only those molecules which have a speed such that they travel the length L in the time required for the cogwheel to be displaced by the width of one opening can get to R, where they deflect a radiometer. The radiometer deflection is measured to determine the number of molecules which have the requisite speed. By changing the velocity of rotation of the cogwheels, molecules having different speeds can be admitted to R. The resemblance of this method to that of Fizeau for measuring the velocity of light should be noted.†

Problems

4-1. Compute the root-mean-square speed, the average speed, and the most probable speed of an oxygen molecule at 300°K and at 500°K. Compare with the values for hydrogen.

4-2. a) Compare the average speed of an oxygen molecule with that of a molecule of carbon tetrachloride.
b) Compare their average kinetic energies.

4-3. a) Compute the kinetic energy in calories of a mole of a gas at 300°K and 500°K.
b) Compute the average kinetic energy of a molecule at 300°K in calories and in ergs.

4-4. Suppose that at some initial time all the molecules in a container have the same translational energy, 2.0×10^{-14} ergs. As time passes, the motion becomes chaotic and the energies finally are distributed in a Maxwellian way.
a) Compute the final temperature of the system.
b) What fraction of the molecules finally have energies in the range between 1.98×10^{-14} and 2.02×10^{-14} ergs? [*Hint:* Since the range of energies in part (b) is small, the differential form of the Maxwell distribution may be used.]

4-5. What fraction of molecules have energies within the range $\bar{\epsilon} - \frac{1}{2}kT$ to $\bar{\epsilon} + \frac{1}{2}kT$?

4-6. Compute the energy corresponding to the maximum of the energy distribution curve.

4-7. What fraction of molecules have energies greater than kT? $2kT$? $10kT$?

4-8. Kinetic theory was once criticized on the grounds that it should apply even to potatoes. Compute the average thermal speed at 25°C of a potato weighing 100 gm. Assuming that the earth's gravity field were turned off, how long would it take a potato to traverse 1 cm? (After working the problem, compare with the result of Problem 2-19.)

4-9. The quantity $(c - \bar{c})^2 = c^2 - 2c\bar{c} + \bar{c}^2$ is the square of the deviation of the speed of a molecule from the average speed. Compute the average value of this quantity using the Maxwell distribution, then take the square root of the result to obtain the root-mean-square deviation of the distribution. Note the way in which this last quantity depends on temperature and on the mass of the molecule:

4-10. The quantity $(\epsilon - \bar{\epsilon})^2 = \epsilon^2 - 2\epsilon\bar{\epsilon} + \bar{\epsilon}^2$ is the square of the deviation of the energy of the molecule from the average energy. Compute the average value of this quantity using the Maxwell distribution. The square root of this quantity is the root-mean-square deviation of the distribution. Note its dependence on temperature and the mass of the molecule.

4-11. The velocity of escape from a planet's surface is given by $v_e = \sqrt{2gR}$. On earth the gravitational acceleration is $g = 980 \text{ cm/sec}^2$, the earth's radius is $R = 6.37 \times 10^8$ cm.

---

† For a description of several methods of direct determination of the velocity distribution see K. F. Herzfeld and H. Smallwood in *A Treatise on Physical Chemistry*, H. S. Taylor and S. Glasstone, eds., Vol. II, 3rd ed. D. Van Nostrand Co., Inc., New York, 1951, p. 35ff.

At 300°K what fraction of

a) hydrogen molecules have velocities exceeding the escape velocity?
b) nitrogen molecules have velocities exceeding the escape velocity?

On the moon, $g = 167\,\text{cm/sec}^2$, $R = 1.74 \times 10^8$ cm. Assuming a temperature of 300°K, what fraction of

c) hydrogen molecules have velocities exceeding the escape velocity?
d) nitrogen molecules have velocities in excess of the escape velocity?

4–12. What fraction of Cl_2 molecules ($\theta = 810$°K) are in excited vibrational states at 300°K? at 500°K? at 700°K?

4–13. The characteristic vibrational temperature for chlorine is 810°K. Calculate the heat capacity of chlorine at 298°K; at 500°K; at 700°K.

4–14. The vibrational frequencies in CO_2 are 7.002×10^{13}, 3.939×10^{13}, 1.988×10^{13}, and $1.988 \times 10^{13}\,\text{sec}^{-1}$. Calculate the corresponding characteristic temperatures and the contributions of each to the heat capacity at 298°K.

4–15. The heat capacity of F_2 at 298°K is $\bar{C}_v/R = 2.78$. Calculate the characteristic vibrational frequency.

4–16. What is the contribution to $\bar{C}_v(\text{vib})/R$ at $T = 0.1\theta$, 0.2θ, 0.5θ, θ, 1.5θ?

4–17. The water molecule has three vibrational frequencies: $11.27 \times 10^{13}\,\text{sec}^{-1}$, $10.94 \times 10^{13}\,\text{sec}^{-1}$, $4.767 \times 10^{13}\,\text{sec}^{-1}$. Which of these contributes significantly to the vibrational heat capacity at 298°K? What is the total heat capacity at 298°K?

Chapter Five

Some Properties of
Liquids and Solids

5–1 Condensed Phases

Solids and liquids are referred to collectively as *condensed phases*. This name emphasizes the high density of the liquid or solid as compared with the low density of gases. This difference in density is one of the most striking differences between gases and solids or liquids. The weight of air in a room of moderate size would not exceed two hundred pounds; the weight of liquid required to fill the same room would be some hundreds of tons. Conversely, the volume per mole is very large for gases and very small for liquids and solids. At STP a gas occupies 22,400 cm^3/mole, while the majority of liquids and solids occupy between 10 and 100 cm^3/mole. Under these conditions the molar volume of a gas is 500 to 1000 times larger than that of a liquid or solid.

If the ratio of the gas volume to liquid volume is 1000, then the ratio of distances between the molecules in the gas as compared with the liquid is the cube root of this, ten. The molecules of the gas are ten times farther apart on the average than are those of the liquid. The distance between molecules in the liquid is roughly equal to the molecular diameter; hence, in the gas the molecules are separated by a distance which is, on the average, ten times their diameter. This large spacing in the gas as compared with the liquid results in the characteristic properties of the gas and the contrast of these properties with those of the liquid. This comes about simply because of the short-range nature of the intermolecular forces, the van der Waals forces. The effect of these forces decreases very sharply with increase in distance between the molecules and falls to an almost negligible value at distances of four to five times the molecular diameter. If we measure the forces by the magnitude of the term $a/\overline{V}^2$ in the van der Waals equation, then an increase in volume by a factor of 1000 in going from liquid to gas decreases the term by a factor of 10^6. Conversely, in the liquid the effect of the van der Waals forces is a million times larger than it is in the gas.

In gases the volume occupied by the molecules is small compared with the total volume, and the effect of the intermolecular forces is very small. In the first approximation these effects are ignored and any gas is described by the ideal gas law, which is strictly correct only at $p = 0$. This condition implies an infinite separation of the molecules; the intermolecular forces would be exactly zero, and the molecular volume would be completely negligible.

Is it possible to find an equation of state for solids or for liquids which has the same generality as the ideal gas law? On the basis of what has been said, the answer must be in the negative. The distances between molecules in liquids and solids are so small, and the effect of the intermolecular forces is correspondingly so large, that the properties of the condensed phases depend on the details of the forces acting between the molecules. Therefore, we must expect that the equation of state will be different for each different solid or liquid. If the force law acting between the molecules were a particularly simple one and had the same analytical form for all molecules, one could expect that the law of corresponding states would have such a universal validity. In fact, the intermolecular forces do not follow such a simple law with precision, so that the law of corresponding states must also be expected to fall short of complete generality. It remains a convenient approximation in many practical situations.

5–2   Coefficients of Thermal Expansion and Compressibility

The dependence of the volume of a solid or liquid on temperature at constant pressure can be expressed by the equation

$$V = V_0(1 + \alpha t), \tag{5–1}$$

where t is the celsius temperature, V_0 is the volume of the solid or liquid at 0°C, and α is the coefficient of thermal expansion. Equation (5–1) is formally the same as Eq. (2–5), which relates the volume of a gas to the temperature. The important difference between the two equations is that the value of α is approximately the same for all gases, while each liquid or solid has its own particular value of α. Any particular substance has different values of α in the solid and in the liquid state. The value of α is constant over limited ranges of temperature. If the data are to be represented with precision over a wide range of temperature, it is necessary to use an equation with higher powers of t:

$$V = V_0(1 + at + bt^2 + \cdots), \tag{5–2}$$

where a and b are constants. For gases and solids, α is *always* positive, while for liquids α is *usually* positive. There are a few liquids for which α is negative over a small range of temperature. For example, between 0 and 4°C water has a negative value of α. In this small temperature interval, the specific volume of water becomes smaller as the temperature increases.

In Eq. (5–1), V_0 is a function of pressure. Experimentally, it is found that the relation between volume and pressure is given by

$$V_0 = V_0^0[1 - \beta(p - 1)], \tag{5–3}$$

where V_0^0 is the volume at 0°C under one atmosphere pressure, p is the pressure in atmospheres, and β is the *coefficient of compressibility*, which is a constant for a particular substance over fairly wide ranges of pressure. The value of β is different for each substance and for the solid and liquid states of the same substance. It is shown in Section 9–2 that the necessary condition for mechanical stability of a substance is that β must be positive.

According to Eq. (5–3) the volume of a solid or liquid decreases linearly with pressure. This behavior is in sharp contrast to the behavior of gases in which the volume is inversely proportional to the pressure. Furthermore, the values of β for liquids and solids are extremely small, being of the order of 10^{-6} to 10^{-5} atm^{-1}. If we take $\beta = 10^{-5}$, then for a pressure of two atmospheres, the volume of the condensed phase is, by Eq. (5–3), $V = V_0^0[1 - 10^{-5}(1)]$. The decrease in volume in going from 1 atm to 2 atm pressure is 0.001%. If a gas were subjected to the same change in pressure, the volume would be halved. Because moderate changes in pressure produce only very tiny changes in the volume of liquids and solids, it is often convenient to consider them to be *incompressible* ($\beta = 0$) in the first approximation.

The coefficients α and β are usually given more general definitions than are implied by Eqs. (5–1) and (5–3). The general definitions are

$$\alpha = \frac{1}{V}\left(\frac{\partial V}{\partial t}\right)_p, \qquad \beta = -\frac{1}{V}\left(\frac{\partial V}{\partial p}\right)_t. \tag{5–4}$$

According to Eq. (5–4), α is the relative increase $(\partial V/V)$ in volume per unit increase in temperature at constant pressure. Similarly, β is the relative decrease in volume $(-\partial V/V)$ per unit increase in pressure at constant temperature.

If the temperature increment is small, the general definition of α yields the result in Eq. (5–1). Rearranging Eq. (5–4), we have

$$\frac{dV}{V} = \alpha \, dt. \tag{5–5}$$

If the temperature is changed from 0° to t°, then the volume changes from V_0 to V. Integration assuming α is constant yields $\ln(V/V_0) = \alpha t$, or $V = V_0 e^{\alpha t}$. If $\alpha t \ll 1$, we can expand the exponential in series to obtain $V = V_0(1 + \alpha t)$, which is identical with Eq. (5–1). By a similar argument, the definition of β can be reduced for a small increment in pressure to Eq. (5–3).

Combining Eqs. (5–1) and (5–3) by eliminating V_0 yields an equation of state for the condensed phase:

$$V = V_0^0(1 + \alpha t)[1 - \beta(p - 1)]. \tag{5–6}$$

To use the equation for any particular solid or liquid, the values of α and β for that substance must be known. Values of α and β for a few common solids and liquids are given in Table 5–1.

Table 5–1. Coefficients of thermal expansion and compressibility at 20°C.

Coefficients	Solids					
	Copper	Graphite	Platinum	Quartz	Silver	NaCl
$\alpha \times 10^4$, deg^{-1}	0.492	0.24	0.265	0.15	0.583	1.21
$\beta \times 10^6$, atm^{-1}	0.78	3.0	0.38	2.8	1.0	4.2

Coefficients	Liquids					
	C_6H_6	CCl_4	C_2H_5OH	CH_3OH	H_2O	Hg
$\alpha \times 10^4$, deg^{-1}	12.4	12.4	11.2	12.0	2.07	1.81
$\beta \times 10^6$, atm^{-1}	94	103	110	120	45.3	3.85

5–3 Heats of Fusion; Vaporization; Sublimation

The absorption or release of heat without any accompanying temperature change is characteristic of a change in the state of aggregation of a substance. The quantity of heat absorbed in the transformation of solid to liquid is the *heat of fusion*. The quantity of heat absorbed in the transformation of liquid to vapor is the *heat of vaporization*. The direct transformation of a solid to vapor is called *sublimation*. The quantity of heat absorbed is the *heat of sublimation*, which is approximately equal to the sum of the heats of fusion and vaporization.

An obvious but important fact about condensed phases is that the intermolecular forces hold the molecules together. The vaporization of a liquid requires the molecules to be pulled apart against the intermolecular forces. The energy required is measured quantitatively by the heat of vaporization. Similarly, energy is required to pull the molecules out of the ordered arrangement in the crystal to the disordered arrangement, usually at a slightly larger distance of separation, existing in the liquid. This energy is measured by the heat of fusion.

Liquids composed of molecules which have comparatively strong forces acting between them have high heats of vaporization, while those composed of weakly interacting molecules have low heats of vaporization. The van der Waals a is a measure of the strength of the attractive forces; we expect the heats of vaporization of substances to fall in the same order as the values of a. This is in fact correct; it can be shown that for a van der Waals fluid the heat of vaporization per mole, Q_{vap}, is equal to a/b.

5–4 Vapor Pressure

If a quantity of a pure liquid is placed in an evacuated container which has a volume greater than that of the liquid, a portion of the liquid will evaporate so as to fill the remaining volume of the container with vapor. Provided that some liquid remains after the equilibrium is established, the pressure of the vapor in the container is a function only of the temperature of the system. The pressure developed is the *vapor pressure* of the liquid, which is a characteristic property of a liquid; it increases rapidly with temperature. The temperature at which the vapor pressure is equal to 1 atm is the *normal boiling point* of the liquid, T_b. Some solids are sufficiently volatile to produce a measurable vapor pressure even at ordinary temperatures; if it should happen that the vapor pressure of the solid reaches 1 atm at a temperature below the melting point of the solid, the solid sublimes. This temperature is called the normal sublimation point, T_s. The boiling point and sublimation point depend upon the pressure imposed upon the substance.

The existence of a vapor pressure and its increase with temperature are consequences of the Maxwell–Boltzmann energy distribution. Even at low temperatures a fraction of the molecules in the liquid have, because of the energy distribution, energies in excess of the cohesive energy of the liquid. As shown in Section 4–10, this fraction increases rapidly with increase in temperature. The result is a rapid increase in the vapor pressure with increase in temperature. The same is true of volatile solids.

The argument implies that at a specified temperature a liquid with a large cohesive energy (i.e., a large molar heat of vaporization Q_{vap}) will have a smaller vapor pressure than one with a small cohesive energy. At 20°C the heat of vaporization of water is 10.53 kcal/mole, while that of carbon tetrachloride is 7.7 kcal/mole; correspondingly, the vapor pressures at this temperature are 17.5 mm for water and 91.0 mm for carbon tetrachloride.

From the general Boltzmann distribution, the relation between the vapor pressure and heat of vaporization can be made plausible. A system containing liquid and vapor in equilibrium has two regions in which the potential energy of a molecule has different values. The strong effect of the intermolecular forces makes the potential energy low in the liquid; $W = 0$. Comparatively, in the gas the potential energy is high, W. By the Boltzmann law, Eq. (4–90), the number of moles of gas per cubic centimeter, $n/V = A \exp(-W/RT)$, where A is a constant. In the gas, the number of moles per cubic centimeter is proportional to the vapor pressure, so we have $p = B \exp(-W/RT)$, where B is another constant. The energy required to take a molecule from the liquid and put it in the vapor is W, the energy of vaporization. As we shall see later, the heat of vaporization is related to W by: $Q_{\text{vap}} = W + RT$. Putting this value of W in the expression for p, we obtain

$$p = p_\infty e^{-Q_{\text{vap}}/RT}, \tag{5-7}$$

where p_∞ is also a constant. Equation (5–7) relates the vapor pressure, temperature, and the heat of vaporization; it is one form of the Clausius-Clapeyron equation for which we will give a more rigorous derivation in Section 12–9. The constant p_∞ has the same units as p, and can be evaluated in terms of Q_{vap} and the normal boiling point T_b. At T_b the vapor pressure is 1 atmosphere, so that 1 atm $= p_\infty \, e^{-Q_{\text{vap}}/RT_b}$. Then

$$p_\infty = (1 \text{ atm})e^{+Q_{\text{vap}}/RT_b}. \tag{5-8}$$

The auxiliary equation (5–8) suffices to evaluate the constant p_∞.

Taking logarithms, Eq. (5–7) becomes

$$\ln p = -\frac{Q_{\text{vap}}}{RT} + \ln p_\infty, \tag{5-9}$$

which is useful for the graphical representation of the variation of vapor pressure with temperature. The function $\ln p$ is plotted against the function $1/T$. Equation (5–9) is then the equation of a straight line, whose slope is $-Q_{\text{vap}}/R$. (If common logarithms are used, the slope is $-Q_{\text{vap}}/2.303R$.) The intercept on the vertical axis is $\ln p_\infty$ (or $\log_{10} p_\infty$). Figure 5–1 is a typical plot of this kind; the vapor-pressure data are for benzene.

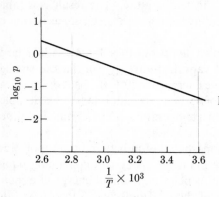

Fig. 5–1 Plot of $\log_{10}p$ versus $1/T$ for benzene.

A convenient method for determining the heat of vaporization of a liquid is to measure its vapor pressure at several temperatures. After the experimental data are plotted in the manner of Fig. 5–1, the slope of the line is measured and from this the value of Q_{vap} is calculated. If only simple apparatus is used, this method is capable of yielding results of higher accuracy than would a calorimetric determination of Q_{vap} using simple apparatus.

5–5   Other Properties of Liquids

The viscosity, or more precisely, the coefficient of viscosity, of a liquid measures the resistance to flow under stress. Because the molecules of liquid are very close to one another, a liquid is much more viscous than a gas. The close spacing and the intermolecular forces both contribute to this resistance to flow. Viscosity is discussed in somewhat greater detail in Chapter 29.

A molecule in the bulk of a liquid is attracted by its neighbors about equally, and over a long time interval does not experience an unbalanced force in any particular direction. A molecule in the surface layer of a liquid is attracted by its neighbors, but since it only has neighbors below it in the liquid, it is attracted toward the body of the liquid. Since the molecules on the surface are bound only to the molecules on one side, they do not have as low an energy as do those in the body of the liquid. To move a molecule from the body of the liquid to the surface requires the addition of energy. Since the presence of another molecule in the surface increases the surface area, it follows that energy must be supplied to increase the area of the liquid surface. The energy required to effect an increase of 1 cm² is called the *surface tension* of the liquid. Surface tension is dealt with in more detail in the chapter on surface properties. For the moment we simply note that the intermolecular forces are responsible for this phenomenon.

5–6   Review of Structural Differences between Solids, Liquids, and Gases

We have described the structure of a gas simply in terms of the chaotic motion of molecules (thermal motion), which are separated from one another by distances which are very large compared with their own diameter. The influence of intermolecular forces and finite molecular size is very small and vanishes in the limit of zero pressure.

Since in a liquid the molecules are separated by a distance of the same magnitude as the molecular diameter, the volume occupied by a liquid is about the same as the volume of the molecules themselves. At these close distances the effect of the intermolecular forces is very large, with the result that each molecule has a low potential energy compared with its energy in the gas. The difference in potential energy between gas and liquid is the energy which must be supplied to vaporize the liquid. The motion of the molecules in the liquid is still chaotic, but since the liquid occupies a much smaller volume, there is less randomness in the space distribution of the molecules. The liquid has a very low compressibility simply because there is very little empty space left between the molecules. The liquid is capable of flow under stress because the molecule does have freedom to move anywhere within the volume; it must, however, push other molecules aside to do so, and as a consequence the resistance to flow is greater than for the gas.

The molecules in a solid are locked in a regular pattern; the spatial arrangement is not random as in the gas or liquid, but completely ordered. The solid does not flow under the application of a small stress, as do liquids or gases, but deforms slightly, snapping back when the stress is removed. This highly ordered arrangement is always accompanied by a lower potential energy, so that energy is required to convert the solid to a liquid. The ordered arrangement usually has a smaller volume than the liquid volume, but not greatly smaller (perhaps 5 to 10%). The solid has a coefficient of compressibility which is about the same magnitude as that of the liquid.

The distribution of energies in solids and liquids is essentially the same as in the gas, and, so long as the temperature is sufficiently high, is described by the Maxwell–Boltzmann distribution function. The motion in gases is characterized by kinetic energy only; in solids and liquids there is a potential energy as well. The motion in solids consists purely of vibration. In liquids, some of the molecules are moving through the liquid while others are momentarily caged by their neighbors and are vibrating in the cage. The motion in the liquid has some of the characteristics of the unhampered motion of molecules in the gas and some of the characteristics of the vibration of molecules in the solid. When all is said and done the liquid bears a closer resemblance to a solid than to a gas.

Problems

5–1. At 25°C a sealed, rigid container is completely filled with liquid water. If the temperature is raised by 10 deg, what pressure will develop in the container? For water, $\alpha = 2.07 \times 10^{-4} \deg^{-1}$; $\beta = 4.50 \times 10^{-5} \text{atm}^{-1}$.

5–2. The coefficient of linear expansion is defined by $a = (1/l)(dl/dt)$. If a is very small and has the same value in any direction for a solid, show that the volume expansion coefficient α is approximately equal to $3a$.

5–3. The correction term to the pressure in the van der Waals equation, $a/\bar{V}^2$, has the dimensions of molar energy per liter; therefore, $a/\bar{V}$ is an energy per mole. Suppose that the energy per mole of a van der Waals fluid has the form $E = f(T) - a/\bar{V}$. At a given temperature find the difference between the energy of water as a gas and the energy of liquid water, assuming that $\bar{V}_{gas} = 24$ liters/mole and $\bar{V}_{liq} = 18 \text{ cm}^3$/mole. For water, $a = 5.72 \text{ liter}^2 \cdot \text{atm} \cdot \text{mole}^{-2}$. Convert this difference to calories and compare with the heat of vaporization, 9820 cal/mole.

5–4. The heat of vaporization of water is 9820 cal/mole. The normal (1 atm) boiling point is 100°C. Compute the value of the constant p_∞ in Eq. (5–7) and the vapor pressure of water at 25°C.

5–5. The Clausius-Clapeyron equation relates the equilibrium vapor pressure p to the temperature T. This implies that the liquid boils at the temperature T if it is subjected to a pressure p. Use this idea together with the Boltzmann distribution to derive a relation between the boiling point of a liquid T, the boiling point under 1 atm pressure T_0, and the height above sea level z. Assume that the pressure at sea level is $p_0 = 1$ atm. The temperature of the atmosphere is T_a. If the atmosphere is at 27°C compute the boiling point of water at 2 km above sea level; $Q_{vap} = 9820$ cal/mole; $T_0 = 373$°K. (Watch out for units!)

5–6. If $\alpha = (1/V)(\partial V/\partial T)_p$, show that $\alpha = -(1/\rho)(\partial \rho/\partial T)_p$, where ρ is the density.

5–7. Show that $(d\rho/\rho) = -\alpha \, dt + \beta \, dp$, where ρ is the density, $\rho = w/V$, where the mass, w, is constant, and V is the volume.

5–8. Since in forming second derivatives of a function of two variables, the order of differentiation does not matter, we have $(\partial^2 V/\partial t \, \partial p) = (\partial^2 V/\partial p \, \partial t)$. Use this relation to show that $(\partial \alpha/\partial p)_t = -(\partial \beta/\partial t)_p$.

5–9. The following vapor pressure data are available for liquid metallic zinc.

p (mm)	10	40	100	400
t (°C)	593	673	736	844

From an appropriate plot of the data, determine the heat of vaporization of zinc and the normal boiling point.

5–10. From the general definition of α, we find $V = V_0 \exp(\int_0^t \alpha \, dt)$. If α has the form: $\alpha = \alpha_0 + \alpha' t + (\alpha''/2)t^2$, where α_0, α', and α'' are constants, find the relation between α_0, α', and α'' and the constants a, b, and c in the empirical equation

$$V = V_0(1 + at + bt^2 + ct^3).$$

Chapter Six

The Laws of
Thermodynamics:
Generalities and the
Zeroth Law

6–1 Kinds of Energy and the First Law of Thermodynamics

Since a physical system may possess energy in a variety of ways, we speak of various kinds of energy.

1. Kinetic energy: energy possessed by a body in virtue of its motion.
2. Potential energy: energy possessed by a body in virtue of its position in a force field; e.g., a mass in a gravity field, a charged particle in an electrical field.
3. Thermal energy: energy possessed by a body in virtue of its temperature.
4. Energy possessed by a substance in virtue of its constitution; e.g., a compound has "chemical" energy, nuclei have "nuclear" energy.
5. Energy possessed by a body in virtue of its mass; the relativistic mass–energy equivalence.
6. A generator "produces" electrical energy.
7. A motor "produces" mechanical energy.

Many other examples could be mentioned: magnetic energy, strain energy, surface energy, etc. The object of thermodynamics is to seek out logically the relations between kinds of energy and their diverse manifestations. The laws of thermodynamics govern the transformation of one kind of energy into another.

In the last two examples a "production" of energy is mentioned. The electrical energy "produced" by the generator did not come from nothing. Some mechanical device, such as a turbine, was needed to run the generator. Mechanical energy disappeared and electrical energy appeared. The quantity of electrical energy "produced" by the generator, plus any frictional losses, is exactly equal to the quantity of mechanical energy "lost" by the turbine. Similarly, in the last example, the mechanical energy produced by the motor, plus the frictional losses, is exactly equal to the electrical energy supplied to the motor from the power lines. The validity of this conservation law has been established by many most careful and most painstaking direct experimental tests and by hundreds of thousands of experiments which confirm it indirectly.

The first law of thermodynamics is the most general statement of this law of conservation of energy; no exception to this law is known. The law of conservation of energy is a generalization from experience and is not derivable from any other principle.

6–2   Restrictions on the Conversion of Energy from One Form to Another

The first law of thermodynamics does not place any restriction on the conversion of energy from one form to another; it simply requires that the total quantity of energy be the same before and after the conversion.

It is always possible to convert any kind of energy into an equal quantity of thermal energy. For example, the output of the generator can be used to operate a toaster immersed in a tub of water. The thermal energy of the water and the toaster is increased by just the amount of electrical energy expended. The electric motor can turn a paddle wheel in the tub of water (as in Joule's experiments), the mechanical energy being converted to an increase in the thermal energy of the water, which is manifested by an increase in the temperature of the water. All kinds of energy can be completely transformed into thermal energy manifested by an increase in temperature of some sample of matter, usually water. The quantity of energy involved can be measured by measuring the temperature rise of a specified mass of water.

Energy may also be classified according to its ability to increase the potential energy of a mass by lifting it against the force of gravity. Only a limited number of the kinds of energy can be completely converted into the lifting of a mass against gravity, e.g., the mechanical energy produced by the electric motor. The thermal energy of a steam boiler or the chemical energy of a compound can be only partly converted into the lifting of a mass. The limitations on the conversion of energy from one form to another lead us to the second law of thermodynamics.

6–3   The Second Law of Thermodynamics

Imagine the following situation. A hard steel ball is suspended at a height h above a hard steel plate. Upon release the ball travels downward losing its potential energy and simultaneously increasing its velocity and hence its kinetic energy. The ball hits the plate and rebounds. We assume that the collision with the plate is elastic; no energy is lost to the plate in the collision. On the rebound the ball travels upward, gaining in potential energy and losing kinetic energy until it returns to the original height h. At this point the ball has its original energy, mgh, and its original kinetic energy, zero. We can either stop the motion at this point or let the ball repeat the motion as often as we please. The first law of thermodynamics in this case is simply the law of conservation of mechanical energy. The sum of the potential energy and the kinetic energy must be a constant throughout the course of the motion. The first law is not in the least concerned with how much of the energy is potential or how much is kinetic, but requires only that the sum remain constant.

Now imagine a somewhat different situation. The ball is poised above a beaker of water. Upon release, it loses potential energy, gains kinetic energy, then enters the water and comes to rest at the bottom of the beaker. Strictly from the stand-point of mechanics, it seems that some energy has been destroyed, since in the final state the ball has neither potential energy nor kinetic energy, while initially it possessed potential energy. Mechanics makes no prediction of the fate of this energy which has "disappeared." However, careful examination of the system reveals that the temperature of the water is slightly higher after the ball has entered and come to rest than before. The potential energy of the ball has been converted to thermal energy of the ball and the water. The first law of thermodynamics requires that both the ball and the beaker of water be included in the system, and that the total of potential energy, kinetic energy, and thermal energy of both ball and water be constant throughout the motion. Using E_b and E_w for the energies of the ball and water, the requirement can be expressed as

$$E_{b(kin)} + E_{w(kin)} + E_{b(pot)} + E_{w(pot)} + E_{b(therm)} + E_{w(therm)} = \text{constant}.$$

Just as in the case of the ball and the plate, the first law is not concerned with how the constant amount of energy is distributed among the various forms.

There is an important difference between the case of the steel plate and that of the beaker of water. The ball can bounce up and down on the plate for an indefinite period, but it falls only once into the beaker of water. Fortunately, we never observe a ball bearing in a glass of water suddenly leaping out of the glass, leaving the water slightly cooler than it was. It is important to realize, however, that the first law of thermodynamics does not rule out this disconcerting event.

The behavior of the ball and the beaker of water is typical of all real processes in one respect. Every real process has a sequence which we recognize as natural; the opposite sequence is unnatural. We recognize the falling of the ball and its coming to rest in the water as a natural sequence. If the ball were at rest in the beaker and then hopped out of the water, we would admit that this is not a natural sequence of events.

The second law of thermodynamics is concerned with the direction of natural processes. In combination with the first law, it enables us to predict the natural direction of any process, and as a result to predict the equilibrium situation. To choose a complicated example, if the system consists of a gasoline tank and a motor mounted on wheels, the second law allows us to predict that the natural sequence of events is: consumption of gasoline, the production of carbon dioxide and water, and the forward motion of the whole device. From the second law, the maximum possible efficiency of the conversion of the chemical energy of the gasoline into mechanical energy can be calculated. The second law also predicts that one cannot manufacture gasoline by feeding carbon dioxide and water into the exhaust and pushing this contraption along the highway; not even if we push it *backwards* along the highway!

Obviously, if thermodynamics can predict results of this type, it must be of enormous importance. In addition to having far-reaching theoretical consequences, thermodynamics is an immensely practical science. A simpler example of the importance of the second law to the chemist is that it allows the calculation of the equilibrium position of any chemical reaction, and it defines the parameters which characterize the equilibrium, e.g., the equilibrium constant.

We shall not deal with the third law of thermodynamics at this point. The principal utility of the third law to the chemist is that it enables him to calculate equilibrium constants from calorimetric data (thermal data) exclusively.

6–4 The Zeroth Law of Thermodynamics

The law of thermal equilibrium, the zeroth law of thermodynamics, is another important principle. The importance of this law to the temperature concept was not fully realized until after the other parts of thermodynamics had reached a rather advanced state of development, hence the unusual name, zeroth law.

To illustrate the zeroth law we consider two samples of gas.† One sample is confined in a volume V_1, the other in a volume V_2. The pressures are p_1 and p_2, respectively. At the beginning the two systems are isolated from each other and are in complete equilibrium. The volume of each container is fixed, and we imagine that each has a pressure gauge, as shown in Fig. 6–1(a).

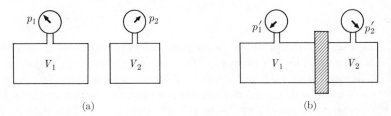

(a)   (b)

Fig. 6–1 (a) Systems isolated. (b) Systems in thermal contact.

The two systems are brought in contact through a wall. Two possibilities exist: when in contact through the wall either the systems influence each other or they do not. If the systems do not influence each other, the wall is an *insulating*, or *adiabatic*, wall; of course, in this situation the pressures of the two systems are not

---

† The argument does not depend in the least upon whether gases, real or ideal, or liquids or solids were chosen.

affected by putting the systems in contact. If the systems do influence one another after putting them in contact, we will observe that the readings of the pressure gauges change with time, finally reaching two new values p'_1 and p'_2 which no longer change with time, Fig. 6–1(b). In this situation the wall is a *thermally conducting* wall; the systems are in *thermal contact*. After the properties of two systems in thermal contact settle down to values which no longer change with time, the two systems are in *thermal equilibrium*. These two systems then have a property in common, the property of being in thermal equilibrium with each other.

Consider three systems *A*, *B*, and *C* arranged as in Fig. 6–2(a). Systems *A* and *B* are in thermal contact, and systems *B* and *C* are in thermal contact. This composite system is allowed sufficient time to come to thermal equilibrium. Then *A* is in thermal equilibrium with *B*, and *C* is in thermal equilibrium with *B*. Now we remove *A* and *C* from their contact with *B* and place them in thermal contact with each other; Fig. 6–2(b). We then observe that no changes in the properties of *A* and *C* occur with time. Therefore *A* and *C* are in thermal equilibrium with each other. This experience is summed up in the zeroth law of thermodynamics: Two systems which are both in thermal equilibrium with a third system are in thermal equilibrium with each other.

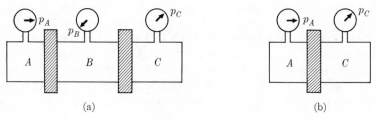

(a)   (b)

Fig. 6–2 The zeroth law.

The temperature concept can be made precise by the statements: (1) Systems in thermal equilibrium with each other have the same temperature. (2) Systems not in thermal equilibrium with each other have different temperatures. The zeroth law, therefore, gives us an operational definition of temperature which does not depend on the physiological sensation of "hotness" or "coldness." This definition is in agreement with the physiological one, since two bodies in thermal equilibrium feel the same as far as hotness is concerned. The zeroth law is based on the experience that systems in thermal contact are not in complete equilibrium with one another until they have the same degree of hotness, i.e., the same temperature.

6–5 Thermometry

The zeroth law suggests a method for measuring the temperature of any system. We choose a system, the thermometer, having some property *y* which is conveniently measurable and which varies reasonably rapidly with temperature. The thermometer is allowed to come to thermal equilibrium with a system whose temperature is reproducible, e.g., melting ice. The value of y is measured. For the thermometer, suppose that we choose a small quantity of gas confined in a box of constant volume which is fitted with a pressure gauge. After this thermometer comes to thermal equilibrium with the melting ice, the needle of the pressure gauge will stand in a definite position. This position we can mark with any number we please; let us follow Celsius and mark it zero. The thermometer is next allowed to come to thermal equilibrium with another system having a reproducible temperature, water boiling under 1 atm pressure. The needle stands at some new position which we can mark with any arbitrary number; again following Celsius we mark the new position with 100. Between the 0 and the 100 we place 99 evenly spaced marks; the dial above 100 and below 0 is divided into intervals of the same width. The thermometer is ready. To measure the temperature of any body, the thermometer is allowed to come to thermal equilibrium with the body; the position of the needle indicates the temperature of the body in degrees. One caution is in order here; the property chosen as the thermometric property must continually increase or continually decrease in value as the temperature rises in the range of application of the thermometer. The thermometric property may not have a maximum or minimum or stationary value in the temperature range in which the thermometer is to be used.

The procedure is easily reduced to a formula by which the temperature can be calculated from the measured value of the thermometric property y. Let y_0 be the value at the ice point and y_{100} be the value at the steam point. These points are separated by 100 degrees. Then

$$\frac{dy}{dt} = \frac{y_{100} - y_0}{100 - 0} = \frac{y_{100} - y_0}{100}.$$

The right-hand side of this equation is a constant; multiplying through by dt and integrating, we obtain

$$y = \frac{y_{100} - y_0}{100} t + C, \tag{6-1}$$

where C is an integration constant. But at $t = 0$, $y = y_0$; using these values, Eq. (6–1) becomes $y_0 = C$. Using this value for C, Eq. (6–1) reduces to

$$y = \frac{y_{100} - y_0}{100} t + y_0.$$

Solving for t, we obtain

$$t = \frac{y - y_0}{y_{100} - y_0} 100, \tag{6-2}$$

which is the thermometric equation. From the measured value of the thermometric property y the temperature on this particular scale can be calculated.

An objection may be raised against this procedure on the grounds that it seems to require that the thermometric property be a linear function of the temperature. The objection is without substance, since we have no way of knowing whether or not a property is linear with temperature until we have chosen some method of measuring temperature. In fact, the method of operation by its very nature automatically makes the thermometric property a linear function of the temperature measured on that particular scale. This reveals a very real difficulty associated with thermometry. A different scale of temperature is obtained for every different property which is chosen as the thermometric property. Even with one substance, different scales of temperature will be obtained depending on which property is chosen as the thermometric property. Truly, this is an outrageous turn of events; imagine the consequences if a similar state of affairs existed in the measurement of length. The size of the centimeter would be different depending upon whether the meter stick was made of metal or wood or paper.

We can attempt to save the situation by searching for a class of substances all of which have some property which behaves in much the same way with temperature. Gases come to mind. For a given change in temperature, the relative change in pressure at constant volume (or relative change in volume under constant pressure) has nearly the same value for all real gases. The behavior of gases can be generalized in the limit of zero pressure to that of the ideal gas. So we might use an ideal gas in the thermometer and define an *ideal gas scale* of temperature. This procedure is quite useful as we have seen in Chapter 2. In spite of its utility, the ideal gas scale of temperature does not solve the difficulty. In the first situation different substances yielded different temperature scales, but at least each of the scales depended upon some property of a *real* substance. The ideal gas scale is a generalization to be sure, but the scale depends upon the properties of a *hypothetical* substance!

Fortunately, there is a way out of this predicament. Using the second law of thermodynamics it is possible to establish a temperature scale which is independent of the particular properties of any substance, real or hypothetical. This scale is the *absolute*, or the *thermodynamic*, temperature scale, also called the kelvin scale after Lord Kelvin, who first demonstrated the possibility of establishing such a scale. By choosing 100 degrees between the ice point and the steam point and with the usual definition of the mole of substance, the kelvin scale and the ideal gas scale become numerically identical. The fact of this identity does not destroy the more general character of the kelvin scale. We establish this identity because of the convenience of the ideal gas scale compared with other possible scales of temperature.

Having overcome the fundamental difficulties, we use all sorts of thermometers with confidence, requiring only that if the temperatures of two bodies A and B are measured with different thermometers, the thermometers must agree that $t_A > t_B$ or that $t_A = t_B$ or that $t_A < t_B$. The different thermometers need not agree on the numerical value of either t_A or t_B. If it is necessary, the reading of each thermometer can be translated into the temperature in degrees kelvin; then the numerical values must agree.

Problems

Conversion factors:

1 joule = 1 watt-second = 10^7 ergs

1 calorie = 4.184 joules

6–1. An electric motor requires 1 kilowatt-hour to run for a specified period of time. In this same period it produces 3200 kilojoules of mechanical work. How much energy, in calories, is dissipated in friction and in the windings of the motor?

6–2. A ball bearing having a mass of 10 gm falls through a distance of 1 meter and comes to rest. How much energy, in calories, is dissipated into thermal energy?

6–3. A bullet, mass = 30 gm, leaves the muzzle of a rifle with a velocity of 900 m/sec. How much energy, in calories, is dissipated in bringing the bullet to rest?

6–4. Suppose that we use the equilibrium vapor pressure of water as a thermometric property in constructing a scale of temperature, t'. In terms of the ordinary celsius temperature, t, the vapor pressure is (to the nearest mm)

t, °C	0	25	50	75	100
p, mm	5	24	93	289	760

If the fixed points, ice point and steam point, are separated by 100° on the t' scale, what will be the temperatures t' corresponding to $t = 0°, 25°, 50°, 75°$, and 100°? Plot t' versus t.

6–5. The length of a metal rod is given in terms of the ordinary celsius temperature t by

$$l = l_0(1 + at + bt^2),$$

where a and b are constants. A temperature scale, t', is defined in terms of the length of the metal rod, taking 100° between the ice point and the steam point. Find the relation between t' and t.

6–6. With the present scale of absolute temperature T there are 273.15° between the ice point and the absolute zero of temperature. Suppose we were to define an absolute temperature scale, T', such that 300° separate the ice point and the absolute zero. What would be the boiling point of water on this scale?

Chapter Seven

Energy and the First
Law of Thermodynamics;
Thermochemistry

7–1 Thermodynamic Terms: Definitions

In beginning the study of thermodynamics it is important to understand the precise thermodynamic sense of the terms which are employed. The following definitions have been given succinct expression by J. A. Beattie.†

"*System, Boundary, Surroundings.* A thermodynamic *system* is that part of the physical universe the properties of which are under investigation . . . .

"The system is confined to a definite place in space by the *boundary* which separates it from the rest of the universe, the *surroundings* . . . .

"A system is *isolated* when the boundary prevents any interaction with the surroundings. An isolated system produces no observable effect or disturbance in its surroundings . . . .

"A system is called *open* when mass passes across the boundary, *closed* when no mass passes the boundary . . . .

"*Properties of a System.* The properties of a system are those physical attributes that are perceived by the senses, or are made perceptible by certain experimental methods of investigation. Properties fall into two classes: (1) non-measurable, as the kinds of substances composing a system and the states of aggregation of its parts; and (2) measurable, as pressure and volume, to which a numerical value can be assigned by a direct or indirect comparison with a standard.

"*State of a System.* A system is in a definite state when each of its properties has a definite value. We must know, from an experimental study of a system or from experience with similar systems, what properties must be taken into consideration in order that the state of a system be defined with sufficient precision for the purpose at hand . . . .

---

† J. A. Beattie, *Lectures on Elementary Chemical Thermodynamics.* Printed by permission from the author.

"*Change in State, Path, Cycle, Process.* Let a system undergo a change in its state from a specified initial to a specified final state.

"The *change in state* is completely defined when the initial and the final states are specified.

"The *path* of the change in state is defined by giving the initial state, the sequence of intermediate states arranged in the order traversed by the system, and the final state.

"A *process* is the method of operation by means of which a change in state is effected. The description of a process consists in stating some or all of the following: (1) the boundary; (2) the change in state, the path, or the effects produced in the system during each stage of the process; and (3) the effects produced in the surroundings during each stage of the process.

"Suppose that a system having undergone a change in state returns to its initial state. The path of this cyclical transformation is called a *cycle*, and the process by means of which the transformation is effected is called a cyclical process.

"*State Variable, ...* A state variable is one that has a definite value when the state of a system is specified . . . ."

The reader should not be misled by the simplicity and clarity of these definitions. The meanings, while apparently "obvious," are precise. These definitions should be thoroughly digested so that when a term appears, it will be immediately recognized as one which has a precise meaning. In the illustrations which follow, the mental questions should be posed: What is the system? Where is the boundary? What is the initial state? What is the final state? What is the path of the transformation? Asking such questions, and other pertinent ones, will help a great deal in clarifying the discussion and is absolutely indispensable before beginning to work any problem.

A system ordinarily must be in a container so that *usually* the boundary is located at the inner surface of the container. As we have seen in Chapter 2, the state of a system is described by giving the values of a sufficient number of state variables; in the case of pure substances, two intensive variables such as T and p are ordinarily sufficient.

7–2 Work and Heat

The concepts of work and of heat are of fundamental importance in thermodynamics, and their definitions must be thoroughly understood, since the use of the term *work* in thermodynamics is much more restricted than its use in physics generally, and the use of the term *heat* is quite different from the everyday meaning of the word. Again, the definitions are those given by J. A. Beattie.†

"*Work.* In thermodynamics work is defined as any quantity that flows across the boundary of a system during a change in its state and is completely convertible into the lifting of a weight in the surroundings."

Several things should be noted in this definition of work.

1. Work appears only at the boundary of a system.

---

† J. A. Beattie, *op. cit.*

2. Work appears only *during* a change in state.

3. Work is manifested by an effect in the *surroundings*.

4. The quantity of work is equal to *mgh*, where *m* is the mass lifted, *g* is the acceleration due to gravity, *h* is the height through which the weight has been raised.

5. Work is an algebraic quantity; it is positive if the weight is lifted (*h* is +), in which case we say that work has been produced in the surroundings or has flowed *to* the surroundings; it is negative if the weight is lowered (*h* is −), in which case we say that work has been destroyed in the surroundings or has flowed *from* the surroundings.†

"*Heat*. We explain the attainment of thermal equilibrium of two systems by asserting that a quantity of heat *Q* has flowed from the system of higher temperature to the system of lower temperature.

"In thermodynamics heat is defined as a quantity that flows across the boundary of a system during a change in its state in virtue of a difference in temperature between the system and its surroundings and flows from a point of higher to a point of lower temperature."‡

Again several things must be emphasized.

1. Heat appears only at the boundary of the system.

2. Heat appears only *during* a change in state.

3. Heat is manifested by an effect in the *surroundings*.

4. The quantity of heat is equal to the number of grams of water in the surroundings which are increased by one degree in temperature starting at a specified temperature under a specified pressure. (We must agree to use one particular thermometer.)

5. Heat is an algebraic quantity; it is positive if a mass of water in the surroundings is cooled, in which case we say that heat has flowed *from* the surroundings; it is negative if a mass of water in the surroundings is warmed, in which case we say that heat has flowed *to* the surroundings.‡

In these definitions of work and heat, it is of utmost importance that the judgment as to whether or not a heat flow or a work flow has occurred in a transformation is based on observation of *effects produced in the surroundings*, not upon what happens within the system. The following example clarifies this point, as well as the distinction between work and heat.

Consider a system consisting of 10 gm of liquid water contained in an open beaker under constant pressure of 1 atm. Initially the water is at 25°C, so that we describe the initial state by $p = 1$ atm, $t = 25°C$. The system is now immersed in, let us say, 100 gm of water at a high temperature, 90°C. The system is kept in contact with this 100 gm of water until the temperature of the 100 gm has fallen to 89°C,

---

†   Parts of this paragraph follow Beattie's discussion closely. By permission from the author.
‡   J. A. Beattie, *op. cit.*

whereupon the system is removed. We say that 100 units of heat has flowed from the surroundings, since the 100 gm of water in the surroundings dropped 1 deg C in temperature. The final state of the system is described by $p = 1$ atm, $t = 35°C$.

Now consider the same system, 10 gm of water, $p = 1$ atm, $t = 25°C$, and immerse a stirring paddle driven by a falling weight (Fig. 7–1). By properly adjusting the mass of the falling weight and the height h through which it falls, the experiment can be arranged so that after the weight falls once, the temperature of the system rises to 35°C. Then the final state is $p = 1$ atm, $t = 35°C$. In this experiment the *change in state* of the system is exactly the same as in the previous experiment. There has been no heat flow, but a flow of work instead. A mass is lower in the surroundings.

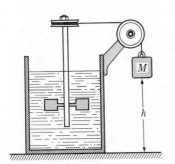

Figure 7–1

If we turned our backs on the experimenter while the change in state had been effected, but had *observed the system* before and after the change in state, we could deduce nothing whatsoever about the heat flow or work flow involved. We could conclude only that the temperature of the system was higher afterward than before; as we shall see later, this implies that the *energy* of the system increased. On the other hand, if we observed the surroundings before and after, we would find cooler bodies of water and/or weights at lower elevations. From these observations on the surroundings, we could immediately deduce the quantities of heat and work which flowed in the transformation.†

It should be clear that the fact that a system is hotter, that is, has a higher temperature, after some transformation does not mean that it has more "heat"; it could equally well have more "work." The system has neither "heat" nor "work"; this use of these terms is to be avoided at all costs. It seems to arise because of a confusion between the concepts of heat and temperature.

The experiment in Fig. 7–1 is Joule's classic experiment on "the mechanical equivalent of heat." This experiment together with earlier ones of Rumford were instrumental in demolishing the caloric theory of heat and establishing that "heat" is equivalent in a certain sense to ordinary mechanical energy. Even today this

---

† The work of expansion accompanying the temperature increase is negligibly small and has been ignored to avoid obscuring the argument.

experiment is described in the words "work is converted into 'heat'." In the modern definition of the word, there is no heat involved in the Joule experiment. Today Joule's observation is described by saying that the destruction of work in the surroundings produces an increase in temperature of the system. Or, less rigidly, work in the surroundings is converted into thermal energy of the system.

The two experiments, immersion of the system in hot water and rotating a paddle in the system, involved the same *change in state* but different heat and work effects. The quantities of heat and work which flow depend upon the process and therefore on the *path* connecting the initial and final states. Heat and work are called *path functions*.

7–3   Expansion Work

If a system alters its volume against an opposing pressure, a work effect is produced in the surroundings. This expansion work appears in most practical situations. The system is a quantity of a gas contained in a cylinder fitted with a piston D; Fig. 7–2(a). The piston is assumed to be weightless and to move without friction. The cylinder is immersed in a thermostat so that the temperature of the system is constant throughout the change in state. Unless a specific statement to the contrary is made, in all of these experiments with cylinders it is understood that the space above the piston is evacuated so that no air pressure is pushing down on the piston.

In the initial state the piston D is held against a set of stops S by the pressure of the gas. A second set of stops S' is provided to arrest the piston after the first set is pulled out. The initial state of the system is described by T, p_1, V_1. We place a small mass M on the piston; this mass must be small enough so that when the stops S are pulled out, the piston will rise and be forced against the stops S'. The final state of the system is T, p_2, V_2; Fig. 7–2(b). The boundary is the inner walls of the

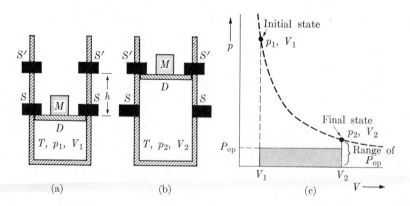

Fig. 7–2 Single-stage expansion.   (a) Initial state.   (b) Final state.
(c) Work produced in a single-stage expansion, $W = P_{op}(V_2 - V_1)$.

cylinder and the piston; in the change the boundary expands to enclose a larger volume V_2. Work is produced in this transformation, since a mass M *in the surroundings*, has been lifted a vertical distance h against the force of gravity Mg. The quantity of work produced is

$$W = Mgh. \tag{7–1}$$

If the area of the piston is A, then the downward pressure acting on the piston is $Mg/A = P_{op}$, the pressure which *opposes* the motion of the piston. Thus $Mg = P_{op}A$. Using this value in Eq. (7–1), we obtain

$$W = P_{op}Ah.$$

However, the product Ah is simply the additional volume enclosed by the boundary in the change of state. Thus, $Ah = V_2 - V_1 = \Delta V$, and we have†

$$W = P_{op}(V_2 - V_1). \tag{7–2}$$

The work produced in the change in state, Eq. (7–2), is represented graphically by the shaded area in the p-V diagram of Fig. 7–2(c). The dashed curve is the isotherm of the gas, on which the initial and final states have been indicated. It is evident that M can have any arbitrary value from zero to some definite upper limit and still permit the piston to rise to the stops S'. It follows that P_{op} can have any value in the range $0 \le P_{op} \le p_2$, and so the quantity of work produced may have any value between zero and some upper limit. *Work is a function of the path.* It must be kept in mind that P_{op} is arbitrary and is not related to the pressure of the system.

The sign of W is determined by the sign of ΔV, since $P_{op} = Mg/A$ is always positive. In expansion, $\Delta V = +$, and $W = +$; the weight rises. In compression, $\Delta V = -$, $W = -$; the weight falls.

As it stands, Eq. (7–2) is correct only if P_{op} is constant throughout the change in state. It is easy to imagine more complicated ways of performing the expansion. Suppose that a large weight were placed on the piston during the first part of the expansion from V_1 to some intermediate volume V'; then a smaller weight replaced the large one in the expansion from V' to V_2. In such a two-stage expansion, we apply Eq. (7–2) to each stage of the expansion, using different values of P_{op} in each stage. Then the total work produced is the sum of the amounts produced in each stage:

$$W = W_{\text{first stage}} + W_{\text{second stage}} = P'_{op}(V' - V_1) + P''_{op}(V_2 - V').$$

The quantity of work produced in the two-stage expansion is represented by the shaded areas in Fig. 7–3 for the special case $P''_{op} = p_2$.

---

†   Differences between the values of a state function in the final and initial states occur so frequently in thermodynamics that a special short-hand notation is used. The Greek capital delta, Δ, is prefixed to the symbol of the state function. The symbol ΔV is read "delta vee" or "the increase in volume" or "the difference in volume." The symbol Δ always signifies a *difference* of two values, which is *always* taken in the order, final value minus initial value.

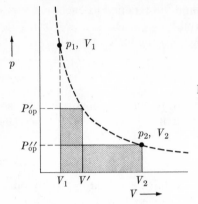

Fig. 7–3 Work produced in a two-stage expansion, $W = P'_{op}(V'' - V_1) + P''_{op}(V_2 - V')$.

Comparison of Figs. 7–2(c) and 7–3 shows that for the same change in state the two-stage expansion produces more work than the single-state expansion could possibly produce. If the heats had been measured, we would have found different quantities of heat associated with the two paths also.

In a multistage expansion the work produced is the sum of the small amounts of work produced in each stage. If P_{op} is constant as the volume increases by an infinitesimal amount dV, then the small quantity of work dW is given by

$$dW = P_{op}\, dV. \tag{7–3}$$

The total work produced in the expansion from V_1 to V_2 is the integral

$$W = \int_1^2 dW = \int_{V_1}^{V_2} P_{op}\, dV, \tag{7–4}$$

which is the general expression for the work of expansion of any system whatsoever. Once P_{op} is known as a function of the volume, the integral is evaluated by the usual methods.

Observe that the differential dW does not integrate in the ordinary way. The integral of an ordinary differential dx between limits yields a finite difference,

$$\int_{x_1}^{x_2} dx = x_2 - x_1 = \Delta x,$$

but the integral of dW is the sum of small quantities of work produced along each element of the path,

$$\int_1^2 dW = W,$$

where W is the total amount of work produced. This explains the use of d instead of the ordinary d. The differential dW is an *inexact differential*, dx is an *exact differential*. More about that later.

7-4 Work of Compression

The work destroyed in compression is computed using the same equation that is used to compute the work produced in expansion. In compression the final volume is less than the initial volume, so in every stage ΔV is negative; therefore the total work destroyed is negative. The sign is automatically taken care of by the integration process if the volume of the final state is the upper limit and the volume of the initial state the lower limit in the integral of Eq. (7–4). However, in comparing work of compression with work of expansion, more than a sign change is involved; to compress the gas we need larger weights on the piston than those that were lifted in the expansion. Thus, more work is destroyed in the compression of a gas than is produced in the corresponding expansion. The single-stage compression of a gas illustrates this point.

The system is the same as before; a gas, kept at a constant temperature T; but now the initial state is the expanded state T, p_2, V_2, while the final state is the compressed state T, p_1, V_1. The positions of the stops are arranged so that the piston rests on top of them. Figure 7–4(a, b) shows that if the gas is to be compressed to the final volume V_1 in one stage, we must choose a mass large enough to produce an opposing pressure P_{op} which is *at least* as great as the final pressure p_1. The mass may be larger than this but not smaller. If we choose the mass M to be equivalent to $P_{op} = p_1$, then the work destroyed is equal to the area of the shaded rectangle in Fig. 7–4(c) with, of course, a negative sign:

$$W = P_{op}(V_1 - V_2).$$

The work destroyed in the single-stage compression is very much greater than the work produced in the single-stage expansion; Fig. 7–2(c). We could destroy any greater amount of work in this compression by using larger masses.

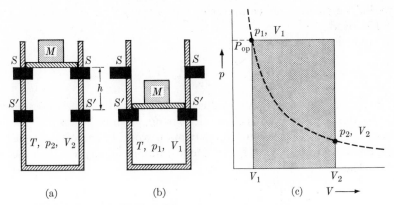

Fig. 7-4 Single-stage compression. (a) Initial state. (b) Final state. (c) Work destroyed in a single-stage compression, $W = P_{op}(V_1 - V_2)$.

If the compression is done in two stages, compressing first with a lighter weight to an intermediate volume and then with the heavy weight to the final volume, less work is destroyed; the work destroyed is the area of the shaded rectangles in Fig. 7–5.

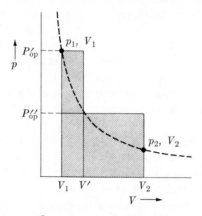

Fig. 7–5 Work destroyed in a two-stage compression, $W = P''_{op}(V' - V_2) + P'_{op}(V_1 - V')$.

7–5 Maximum and Minimum Quantities of Work

In the two-stage expansion more work was produced than in the single-stage expansion. It seems reasonable that if the expansion were done in many stages using a large mass in the beginning and making it smaller as the expansion proceeded, even more work should be produced. This is correct, but there is a limitation to the procedure. The weights that we use must not be so large as to compress the system instead of allowing it to expand. By doing the expansion in a progressively larger number of stages, the work produced can be increased up to a definite maximum value.†
Correspondingly, the work destroyed in the two-stage compression is less than that destroyed in the single-stage compression. In a multistage compression, even less work is destroyed.

The expansion work is given by

$$W = \int_{V_i}^{V_f} P_{op}\, dV.$$

For the integral to have a maximum value, P_{op} must have the largest possible value at each stage of the process. But if the gas is to expand, P_{op} must be less than the pressure p of the gas. Therefore, to obtain the maximum work, at each stage we adjust the opposing pressure to $P_{op} = p - dp$, a value just infinitesimally less than the pressure of the gas. Then

$$W_m = \int_{V_i}^{V_f} (p - dp)\, dV = \int_{V_i}^{V_f} (p\, dV - dp\, dV),$$

† This is true only if the temperature is constant along the path of the change in state. If the temperature is allowed to vary along the path, there is no upper limit on the work produced.

where V_i and V_f are the initial and final volumes. The second term in the integral is an infinitesimal of higher order than the first and so has a limit of zero. Thus, for the maximum work in expansion

$$W_m = \int_{V_i}^{V_f} p \, dV. \tag{7–5}$$

Similarly, we find the minimum work required for compression by setting the value of P_{op} at each stage just infinitesimally greater than the pressure of the gas; $P_{op} = p + dp$. The argument will obviously yield Eq. (7–5) for the minimum work required for compression if V_i and V_f are the initial and final volumes in the compression. Equation (7–5) is, of course, general and not restricted to gases.

For the ideal gas, the maximum quantity of work produced in the expansion or the minimum destroyed in the compression is equal to the shaded area under the isotherm, Fig. 7–6. The maximum or minimum work in an isothermal change in state is easily evaluated, since $p = nRT/V$. Using this value for the pressure in Eq. (7–5), we obtain

$$W_{\text{max, min}} = \int_{V_i}^{V_f} \frac{nRT}{V} \, dV = nRT \int_{V_i}^{V_f} \frac{dV}{V} = nRT \ln \frac{V_f}{V_i}. \tag{7–6}$$

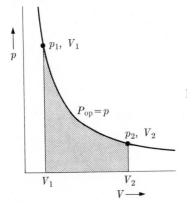

Fig. 7–6 W_{max} or W_{min}.

Under the conditions described, n and T are constant throughout the change and so can be removed from under the integral sign. Note that in expansion $V_f > V_i$, so the logarithm of the ratio is positive; in compression, $V_f < V_i$, the ratio is less than unity so the logarithm is negative. In this way the sign of W takes care of itself.

7–6 Reversible and Irreversible Transformations

Consider the same system as before: a quantity of gas, confined in a cylinder at a constant temperature T. We expand the gas from the state T, p_1, V_1 to the state T, p_2, V_2, then we compress the gas to the original state. The gas has been subjected to a *cyclical* transformation returning at the end to its initial state. Suppose that we perform this cycle by two different processes and calculate the net work effect W_{cy} for each process.

Process I. Single-stage expansion with $P_{op} = p_2$; then single-stage compression with $P_{op} = p_1$.

The work produced in the expansion is, by Eq. (7–4),

$$W_{exp} = p_2(V_2 - V_1),$$

while the work produced in the compression is

$$W_{comp} = p_1(V_1 - V_2).$$

The net work effect in the cycle is the sum of these two:

$$W_{cy} = p_2(V_2 - V_1) + p_1(V_1 - V_2) = (p_2 - p_1)(V_2 - V_1).$$

Since $V_2 = V_1$ is positive, and $p_2 - p_1$ is negative, W_{cy} is negative. Net work has been destroyed in this cycle. The system has been restored to its initial state, but the surroundings have not been restored; weights are lower in the surroundings after the cycle.

Process II. The limiting multistage expansion with $P_{op} = p$; then the limiting multistage compression with $P_{op} = p$.

By Eq. (7–5), the work produced in expansion is

$$W_{exp} = \int_{V_1}^{V_2} p\, dV,$$

while the work produced in compression is, Eq. (7–5),

$$W_{comp} = \int_{V_2}^{V_1} p\, dV.$$

The net work effect in the cycle is

$$W_{cy} = \int_{V_1}^{V_2} p\, dV + \int_{V_2}^{V_1} p\, dV = \int_{V_1}^{V_2} p\, dV - \int_{V_1}^{V_2} p\, dV = 0.$$

(The change in sign of the second integral is effected by interchanging the limits of integration.) If the transformation is conducted by this second method, the system is restored to its initial state, and *the surroundings are also restored to their initial condition*, since no net work effect is produced.

Suppose that a system undergoes a change in state through a specified sequence of intermediate states and then is restored to its original state by traversing the same sequence of states in reverse order. Then if the surroundings are also brought to their original state, the transformation in either direction is *reversible*. The corresponding process is a reversible process. If the surroundings are not restored to their original state after the cycle, the transformation and the process are *irreversible*.

Clearly, the second process described above is a reversible process, while the first is irreversible. There is another important characteristic of reversible and irreversible processes. In the irreversible process described above, a single weight is placed on the piston, the stops are released, and the piston shoots up and settles in the final position. As this occurs the internal equilibrium of the gas is completely upset, convection currents are set up, the temperature fluctuates. A finite length of time is required for the gas to equilibrate under the new set of conditions. A similar situation prevails in the irreversible compression. This behavior contrasts with the reversible expansion in which at each stage the opposing pressure differs only infinitesimally from the equilibrium pressure in the system, and the volume increases only infinitesimally. In the reversible process the internal equilibrium of the gas is disturbed only infinitesimally and in the limit not at all. Therefore, at any stage in a reversible transformation, the system does not depart from equilibrium by more than an infinitesimal amount.

Obviously, we cannot actually conduct a transformation reversibly. An infinite length of time would be required if the volume increment in each stage were truly infinitesimal. Reversible processes are, therefore, not real processes, but *ideal ones*. *Real processes are always irreversible*. With patience and skill the goal of reversibility can be very closely approached, but not attained. Reversible processes are important because the work effects associated with them represent maximum or minimum values. Thus, limits are set on the ability of a specified transformation to produce work; in actuality we will get less, but we must not expect to get more.

In the isothermal cycle described above, the net work produced in the irreversible cycle was negative, that is, net work was destroyed. This is a fundamental characteristic of every *irreversible* and therefore every *real* isothermal cyclical transformation. If any system is kept at a constant temperature and subjected to a cyclical transformation by irreversible processes (real processes), a net amount of work is destroyed in the surroundings. This is in fact a statement of the second law of thermodynamics. The greatest work effect will be produced in a reversible isothermal cycle, and this, as we have seen, is $W_{cy} = 0$. Therefore, we cannot expect to get a positive amount of work in the surroundings from the cyclical transformation of a system kept at a constant temperature.

Examination of the arguments presented above shows that the general conclusions reached do not depend on the fact that the system chosen for illustration consisted of a gas; the conclusions are valid regardless of how the system is constituted. Therefore to calculate the expansion work produced in the transformation of any system whatsoever we use Eq. (7–4), and to calculate the work produced in the reversible transformation, we set $P_{op} = p$ and use Eq. (7–5).

By appropriate modification of the argument, the general conclusions reached could be shown to be correct for any kind of work whatsoever; electrical work, work done against a magnetic field, etc. To calculate the quantities of these other kinds of work, we would not, of course, use the integral of pressure over volume, but the integral of the appropriate force over the corresponding displacement.

7–7   Energy and the First Law of Thermodynamics

The work produced in a cyclic transformation is the sum of the small quantities of work dW produced at each stage of the cycle. Similarly, the heat withdrawn from the surroundings in a cyclic transformation is the sum of the small quantities of heat dQ withdrawn at each stage of the cycle. These sums are symbolized by the *cyclic integrals* of dW and dQ:

$$W_{cy} = \oint dW, \qquad Q_{cy} = \oint dQ.$$

In general, W_{cy} and Q_{cy} are not zero; this is characteristic of path functions.

In contrast, note that if we sum the differential of any *state property* of the system over any cycle, the total difference, the cyclic integral, must be zero. Since in any cycle the system returns at the end to its initial state, the total difference in value of any state property must be zero. Conversely, if we find a differential quantity dy such that

$$\oint dy = 0 \quad \text{(all cycles)}, \tag{7–7}$$

then dy is the differential of some property of the state of the system. This is a purely mathematical theorem, stated here in physical language. Using this theorem and the first law of thermodynamics, we discover the existence of a property of the state of the system, the *energy*.

The first law of thermodynamics is a statement of the following universal experience. *If a system is subjected to any cyclic transformation, the work produced in the surroundings is equal to the heat withdrawn from the surroundings.* In mathematical terms, the first law states that

$$\oint dW = \oint dQ \quad \text{(all cycles)}. \tag{7–8}$$

The system suffers no net change in the cycle, but the condition of the surroundings changes. If weights in the surroundings are higher after the cycle than before, then some bodies in the surroundings must be colder. If weights are lower, then some bodies must be hotter.

Rearranging Eq. (7–8), we have

$$\oint (dQ - dW) = 0 \quad \text{(all cycles)}. \tag{7–9}$$

But if Eq. (7–9) is true, then the mathematical theorem requires that the quantity under the integral sign must be the differential of some property of the state of the system. This property of the state is called the *energy*[†] E of the system, and the differential is dE, defined by

$$dE \equiv dQ - dW; \tag{7–10}$$

[†]   This is sometimes called the "internal" energy, but the adjective is redundant.

then, of course,

$$\oint dE = 0 \quad \text{(all cycles)}. \tag{7–11}$$

Thus, from the first law which relates the heat and work effects observed in the surroundings in a cyclic transformation, we deduce the existence of a property of the state of the system, the energy. Equation (7–10) is an equivalent statement of the first law.

Equation (7–10) shows that when small amounts of heat and work dQ and dW appear at the boundary, the energy of the system suffers a change dE. For a finite change in state, we integrate Eq. (7–10):

$$\int_i^f dE = \int_i^f dQ - \int_i^f dW,$$

$$\Delta E = Q - W, \tag{7–12}$$

where $\Delta E = E_{\text{final}} - E_{\text{initial}}$. Note that only a difference in energy dE or ΔE has been defined, so we can calculate the difference in energy in a change in state, but we cannot assign an absolute value to the energy of the system in any particular state.

We can show that energy is conserved in any change in state. Consider two systems A and B in contact with each other but with the composite system $A + B$ isolated from the rest of the universe. In the composite system we can choose A as our system and B as its surroundings. If A undergoes a change in its state, a quantity of heat Q will flow from B and a quantity of work W will flow to B. According to Eq. (7–12) for the system A

$$E_{A(\text{final})} = E_{A(\text{initial})} + Q - W.$$

However, if we had chosen B as the system and A as the surroundings, then a quantity of heat $-Q$ has flowed from A and a quantity of work $-W$ has flowed to A. Then for the system B

$$E_{B(\text{final})} = E_{B(\text{initial})} + (-Q) - (-W).$$

Adding these two equations, we obtain

$$(E_A + E_B)_{\text{final}} = (E_A + E_B)_{\text{initial}},$$

or

$$\Delta E_A = -\Delta E_B.$$

The total energy of the two systems is the same after the change in state as it was before; alternatively, the increase in energy of A, ΔE_A, is just balanced by the decrease in energy of B, $-\Delta E_B$. The energy of the system plus the energy of the surroundings is conserved in the change in state. If we imagine that B is so large as to encompass the entire universe excluding only system A, then A and B together compose the universe. Thus we have the famous statement of the first law of thermodynamics due to Clausius: "The energy of the universe is a constant."

7–8 Properties of the Energy

For a specified change in state, the increase in energy ΔE of the system depends only on the initial and final states of the system and not upon the path connecting those states. Both Q and W depend upon the path, but their difference $Q - W = \Delta E$ is independent of the path. This is equivalent to the statement that dQ and dW are inexact differentials, while dE is an exact differential.

The energy is an *extensive* state property of the system; under the same conditions of T and p, ten moles of the substance composing the system has ten times the energy of one mole. The energy per mole is an intensive state property of the system.

Energy is conserved in all transformations. A perpetual motion machine of the first kind is a machine which by its action creates energy by some transformation of a system. The first law of thermodynamics asserts that it is impossible to construct such a machine; not that people have not tried! No one has ever succeeded, but there have been some famous frauds in this field.

7–9 Mathematical Interlude; Exact and Inexact Differentials

An exact differential integrates to a finite difference, $\int_1^2 dE = E_2 - E_1$, which is independent of the path of integration. An inexact differential integrates to a total quantity, $\int_1^2 dQ = Q$, which depends on the path of integration. The cyclic integral of an exact differential is zero for any cycle, Eq. (7–7). The cyclic integral of an inexact differential is usually not zero.

Note that the symbolism ΔQ and ΔW is *meaningless*. If ΔW meant anything, it would mean $W_2 - W_1$; but the system in either the initial state or the final state does not have any work W_1 or W_2, nor does it have any heat Q_1 or Q_2. Work and heat appear *during a change* in state; they are not properties of the state, but properties of the path.

Properties of the state of a system, such as T, p, V, E, have differentials which are exact. Differentials of properties of the path, such as Q and W, are inexact.

7–10 Changes in Energy in Relation to Changes in Properties of the System

Using the first law in the form

$$\Delta E = Q - W,$$

we can calculate ΔE for the change in state from the measured values of Q and W, effects in the surroundings. However, a change in state of the system implies changes in properties of the system, such as T and V. These properties of the system are readily measurable in the initial and final states, and it is useful to relate the change in energy of the system to, let us say, changes in its temperature and volume. It is to this problem that we now direct our attention.

Choosing a system of fixed mass, we can describe the state by T and V. Then $E = E(T, V)$, and the change in energy dE is related to the changes in temperature dT and in volume dV through the total differential expression†

$$dE = \left(\frac{\partial E}{\partial T}\right)_V dT + \left(\frac{\partial E}{\partial V}\right)_T dV. \tag{7–13}$$

Expressions of this sort are used so often that it is essential to understand their physical and mathematical meaning. Equation (7–13) says that if the temperature of the system increases by an amount dT and the volume increases by an amount dV, then the total increase in energy dE is the sum of two contributions: the first term, $(\partial E/\partial T)_V \, dT$, is the increase in energy resulting from the temperature increase alone; the second term, $(\partial E/\partial V)_T \, dV$, is the increase in energy resulting from the volume increase alone. The first term is the rate of increase of energy with temperature at constant volume, $(\partial E/\partial T)_V$, multiplied by the increase in temperature dT. The second term is interpreted in an analogous way. Each time an expression of this kind appears, the effort should be made to give this interpretation to each term until it becomes a habit. The habit of reading a physical meaning into an equation will help enormously in clarifying the derivations which follow.

Since the energy is an important property of the system, the partial derivatives $(\partial E/\partial T)_V$ and $(\partial E/\partial V)_T$ are also important properties of the system. These derivatives tell us the rate of change of energy with temperature at constant volume, or with volume at constant temperature. If the values of these derivatives are known, we can integrate Eq. (7–13) and obtain the change in energy from the change of temperature and volume of the system. Therefore, we must express these derivatives in terms of measurable quantities.

We begin by combining Eqs. (7–10) and (7–13) to obtain

$$dQ - P_{op} \, dV = \left(\frac{\partial E}{\partial T}\right)_V dT + \left(\frac{\partial E}{\partial V}\right)_T dV, \tag{7–14}$$

where $P_{op} \, dV$ has replaced dW, and work other than expansion work has been ignored. (If other kinds of work must be included, we set $dW = P_{op} \, dV + dU$, where dU represents the small amounts of other kinds of work.) Next we apply Eq. (7–14) to various changes in state.

---

†   Appendix I. The differential of any state property, any exact differential, can be written in the form of Eq. (7–13).

7–11 Changes in State at Constant Volume

If the volume of a system is constant in the change in state, then $dV = 0$, and the first law, Eq. (7–10), becomes

$$dE = dQ_V, \tag{7–15}$$

where the subscript indicates the restriction to constant volume. But at constant volume, Eq. (7–14) becomes

$$dQ_V = \left(\frac{\partial E}{\partial T}\right)_V dT, \tag{7–16}$$

which relates the heat withdrawn from the surroundings, dQ_V, to the increase in temperature dT of the system at constant volume. Both dQ_V and dT are easily measurable; the ratio, dQ_V/dT, of the heat withdrawn from the surroundings to the temperature increase of the system is C_v, the heat capacity of the system at constant volume. Thus, dividing Eq. (7–16) by dT, we obtain

$$C_v \equiv \frac{dQ_V}{dT} = \left(\frac{\partial E}{\partial T}\right)_V. \tag{7–17}$$

Either member of Eq. (7–17) is an equivalent definition of C_v. The important point about Eq. (7–17) is that it identifies the partial derivative $(\partial E/\partial T)_V$ with an easily measurable quantity C_v. Using C_v for the derivative in Eq. (7–13), we obtain, since $dV = 0$,

$$dE = C_v\, dT \quad \text{(infinitesimal change)}, \tag{7–18}$$

or, integrating, we have

$$\Delta E = \int_{T_1}^{T_2} C_v\, dT \quad \text{(finite change)}. \tag{7–19}$$

Using Eq. (7–19) we can calculate ΔE exclusively from properties of the system. Integrating Eq. (7–15), we obtain the additional relation

$$\Delta E = Q_V \quad \text{(finite change)}. \tag{7–20}$$

Both Eqs. (7–19) and (7–20) express the energy change in a transformation at constant volume in terms of measurable quantities. These equations apply to any system whatsoever: solids, liquids, gases, mixtures, old razor blades, etc.

Note in Eq. (7–20) that ΔE and Q_V have the same sign. According to the convention for Q, if heat flows *from* the surroundings, $Q_V > 0$, and so $\Delta E > 0$; the energy of the system increases. If heat flows *to* the surroundings, both Q_V and ΔE are negative; the energy of the system decreases. Furthermore, since C_v is always positive, Eq. (7–18) shows that if the temperature increases, $dT > 0$, the energy of the system increases; conversely, a decrease in temperature, $dT < 0$, means a decrease in the energy of the system, $\Delta E < 0$. For a system maintained at a constant volume, the temperature is a direct reflection of the energy of the system.

Since the energy of the system is an extensive state property, the heat capacity is also. The heat capacity per mole $\bar{C}$, an intensive property, is the quantity which is found in tables of data. If the heat capacity of the system is a constant in the range of temperature of interest, then Eq. (7–19) reduces to the special form

$$\Delta E = C_v \Delta T. \tag{7–21}$$

This equation is quite useful, particularly if the temperature range ΔT is not very large. Over short ranges of temperature the heat capacity of most substances does not change very much.

Although Eqs. (7–19) and (7–20) are completely general for a constant-volume process, a practical difficulty arises if the system consists entirely of solids or liquids. If a liquid or a solid is confined in a container of fixed volume and the temperature is increased by a small amount, the pressure goes up to fantastic values, because of the very small compressibility of the liquid. Any ordinary container will be deformed and increase in volume or it will burst. From the experimental standpoint, constant volume processes are practical only for those systems which are, at least partly, gaseous.

7–12 Measurement of $(\partial E/\partial V)_T$; Joule's Experiment

The identification of the differential coefficient $(\partial E/\partial V)_T$ with readily measurable quantities is not so easily managed. For gases it can be done, in principle at least, by an experiment devised by Joule. Two containers A and B are connected through a stopcock. In the initial state, A is filled with a gas at a pressure p, while B is evacuated. The apparatus is immersed in a large vat of water and is allowed to equilibrate with the water at the temperature T, which is read on the thermometer, Fig. 7–7. The water is stirred vigorously to hasten the attainment of thermal equilibrium. The stopcock is opened and the gas expands to fill the containers A and B uniformly. After allowing time for the system to come to thermal equilibrium with the water in the vat, the temperature of the water is read again. Joule observed no temperature difference in the water before and after opening the stopcock.

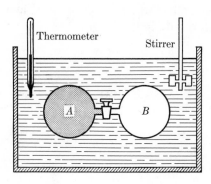

Fig. 7–7 Joule expansion experiment.

The interpretation of this experiment is as follows. To begin with, no work is produced in the surroundings. The boundary, which is initially along the interior walls of the vessel A, moves in such a way that it always encloses the entire mass of gas; the boundary therefore expands against zero opposing pressure so no work is produced. This is called a *free expansion* of the gas. Setting $dW = 0$, we see that the first law becomes $dE = dQ$. Since the temperature of the surroundings (the water) is unchanged, it follows that $dQ = 0$. Hence, $dE = 0$. Since the system and the water are in thermal equilibrium, the temperature of the system is also unchanged, $dT = 0$. In this situation, Eq. (7–13) becomes

$$dE = \left(\frac{\partial E}{\partial V}\right)_T dV = 0.$$

Since $dV \neq 0$, it follows that

$$\left(\frac{\partial E}{\partial V}\right)_T = 0. \tag{7–22}$$

If the derivative of energy with respect to volume is zero, the energy is independent of the volume. This means that the energy of the gas is a function only of temperature. This rule of behavior is *Joule's law*, which may be expressed either by Eq. (7–22) or by $E = E(T)$.

Later experiments, notably the Joule-Thomson experiment, have shown that Joule's law is not precisely correct for real gases. In Joule's apparatus the large heat capacity of the vat of water and the small heat capacity of the gas reduced the magnitude of the effect below the limits of observation. For real gases, the derivative $(\partial E/\partial V)_T$ is a very small quantity, usually positive. The ideal gas obeys Joule's law exactly.

Until we have the equations from the second law of thermodynamics, the problem of identifying the derivative $(\partial E/\partial V)_T$ with readily measurable quantities is a clumsy procedure at best. The Joule experiment, which does not work very well with gases, is completely unsuitable for liquids and solids. A fortunate circumstance intervenes to simplify matters for liquids and solids. Very great pressures are required to effect even a small change in volume of a liquid or solid kept at a constant temperature. The energy change accompanying an isothermal change in volume of a liquid or solid is, by integrating Eq. (7–13) with $dT = 0$,

$$\Delta E = \int_{V_1}^{V_2} \left(\frac{\partial E}{\partial V}\right)_T dV.$$

The initial and final volumes V_1 and V_2 are so nearly equal that the derivative is constant over this small range of volume; removing it from under the integral sign and integrating dV, the equation becomes

$$\Delta E = \left(\frac{\partial E}{\partial V}\right)_T \Delta V. \tag{7–23}$$

Even though for liquids and solids the value of the derivative is very large, the value of ΔV is so small that the product in Eq. (7–23) is very nearly zero. Consequently, to a good approximation the energy of all substances can be considered to be a function of temperature only. The statement is precisely true only for the ideal gas. To avoid errors in derivations the derivative will be carried along. Having identified $(\partial E/\partial T)_V$ with C_v, we shall, from now on, write the total derivative of E, Eq. (7–13), in the form

$$ dE = C_v\, dT + \left(\frac{\partial E}{\partial V}\right)_T dV. \tag{7–24}$$

7–13   Changes in State at Constant Pressure

In laboratory practice most changes in state are carried out under a constant atmospheric pressure, which is equal to the pressure of the system. The change in state at constant pressure can be envisioned by confining the system to a cylinder closed by a weighted piston which floats freely (Fig. 7–8), rather than being held in some position by a set of stops. Since the piston floats freely, its equilibrium position is determined by the balance of the opposing pressure developed by the mass M and the pressure in the system. No matter what we do to the system, the piston will move until the condition $p = P_{op}$ is fulfilled. The pressure p in the system may be adjusted to any constant value by appropriately adjusting the mass M. Under ordinary laboratory conditions the mass of the column of air above the system floats on top of the system and maintains the pressure at the constant value p.

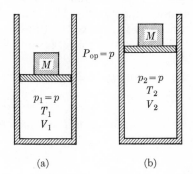

Fig. 7–8 Change in state at constant pressure. (a) Initial state. (b) Final state.

Since $P_{op} = p$, for a change in state at constant pressure the first-law statement becomes

$$ dE = dQ_p - p\, dV. \tag{7–25}$$

Since p is constant, this integrates at once to yield

$$ \int_1^2 dE = \int_1^2 dQ_p - \int_{V_1}^{V_2} p\, dV, $$

$$ E_2 - E_1 = Q_p - p(V_2 - V_1). $$

Rearranging, we obtain

$$(E_2 + pV_2) - (E_1 + pV_1) = Q_p. \tag{7-26}$$

Since $p_1 = p_2 = p$, in Eq. (7–26), the first p can be replaced by p_2, the second by p_1:

$$(E_2 + p_2 V_2) - (E_1 + p_1 V_1) = Q_p. \tag{7-27}$$

Since the pressure and the volume of the system depend only on the state, the product pV depends only on the state of the system. The function $E + pV$, being a combination of state variables, is itself a state variable H. We define

$$H \equiv E + pV; \tag{7-28}$$

H is called the *enthalpy* of the system,† an extensive state property.

Using the definition of H, we can rewrite Eq. (7–27) as $H_2 - H_1 = Q_p$, or

$$\Delta H = Q_p, \tag{7-29}$$

which shows that in a constant pressure process the heat withdrawn from the surroundings is equal to the increase in enthalpy of the system. Ordinarily, heat effects are measured at constant pressure; therefore, these heat effects indicate changes in enthalpy of the system, not changes in its energy. To compute the change in energy in a constant pressure process, Eq. (7–26) is written as

$$\Delta E + p\Delta V = Q_p. \tag{7-30}$$

Knowing Q_p and the change in volume ΔV, we can calculate the value of ΔE.

Equation (7–29) finds immediate application to the vaporization of a liquid under a constant pressure and at a constant temperature. The heat withdrawn from the surroundings is the heat of vaporization Q_{vap} since the transformation is done at constant pressure, $Q_{vap} = \Delta H_{vap}$. Similarly, the heat of fusion of a solid is the enthalpy increase in fusion: $Q_{fus} = \Delta H_{fus}$.

For an infinitesimal change in state of a system, Eq. (7–29) takes the form

$$dH = dQ_p. \tag{7-31}$$

Since H is a state function, dH is an exact differential; choosing T and p as convenient variables for H, we can write the total differential as

$$dH = \left(\frac{\partial H}{\partial T}\right)_p dT + \left(\frac{\partial H}{\partial p}\right)_T dp. \tag{7-32}$$

For a transformation at constant pressure, $dp = 0$, and Eq. (7–32) becomes $dH = (\partial H/\partial T)_p \, dT$. Combining this with Eq. (7–31) yields

$$dQ_p = \left(\frac{\partial H}{\partial T}\right)_p dT,$$

---

† It is worth noting that the appearance of the product pV in the definition of the enthalpy results from the algebraic form for the expansion work; it has nothing to do with the presence of the pV product in the ideal gas law!

which relates the heat withdrawn from the surroundings to the temperature increment of the system. The ratio, dQ_p/dT, is C_p, the heat capacity of the system at
constant pressure. Hence, we have

$$C_p \equiv \frac{dQ_p}{dT} = \left(\frac{\partial H}{\partial T}\right)_p, \qquad (7\text{--}33)$$

which identifies the important partial derivative $(\partial H/\partial T)_p$ with the measurable
quantity C_p. Henceforward, the total differential in Eq. (7–32) will be written in the
form

$$dH = C_p\, dT + \left(\frac{\partial H}{\partial p}\right)_T dp. \qquad (7\text{--}34)$$

For any constant pressure transformation, since $dp = 0$, Eq. (7–34) reduces to

$$dH = C_p\, dT, \qquad (7\text{--}35)$$

or for a finite change in state from T_1 to T_2,

$$\Delta H = \int_{T_1}^{T_2} C_p\, dT. \qquad (7\text{--}36)$$

If C_p is constant in the temperature range of interest, Eq. (7–36) becomes

$$\Delta H = C_p \Delta T. \qquad (7\text{--}37)$$

The equations in this section are quite general and are applicable to any transformation at constant pressure of any system of fixed mass, provided no phase changes
or chemical reactions occur.

7–14   The Relation between C_p and C_v

For a specified change in state of a system which has a definite temperature change
dT associated with it, the heat withdrawn from the surroundings may have different
values, since it depends upon the path of the change in state. Therefore, it is not surprising that a system has more than one value of heat capacity. In fact, the heat
capacity of a system may have any value from minus infinity to plus infinity. Only
two values, C_p and C_v, have major importance, however. Since they are not equal, it is
important to find the relation between them.

We attack this problem by calculating the heat withdrawn at constant pressure
using Eq. (7–14) in the form

$$dQ = C_v\, dT + \left(\frac{\partial E}{\partial V}\right)_T dV + P_{op}\, dV.$$

For a change at constant pressure with $P_{op} = p$, this equation becomes

$$dQ_p = C_v(\partial T)_p + \left[p + \left(\frac{\partial E}{\partial V}\right)_T\right](\partial V)_p.$$

Since $C_p = dQ_p/(\partial T)_p$, we divide by $(\partial T)_p$ and obtain

$$C_p = C_v + \left[p + \left(\frac{\partial E}{\partial V} \right)_T \right] \left(\frac{\partial V}{\partial T} \right)_p, \tag{7–38}$$

which is the required relation between C_p and C_v. It is usually written in the form

$$C_p - C_v = \left[p + \left(\frac{\partial E}{\partial V} \right)_T \right] \left(\frac{\partial V}{\partial T} \right)_p. \tag{7–39}$$

This equation is a general relation between C_p and C_v. It will be shown later that the quantity on the right-hand side is always positive; thus C_p is always larger than C_v for any substance whatsoever. The excess of C_p over C_v is made up of a sum of two terms. The first term,

$$p \left(\frac{\partial V}{\partial T} \right)_p,$$

is the work produced $p\,dV$ per unit increase in temperature in the constant pressure process. The second term,

$$\left(\frac{\partial E}{\partial V} \right)_T \left(\frac{\partial V}{\partial T} \right)_p,$$

is the energy required to pull the molecules farther apart against the attractive intermolecular forces.

If a gas is expanded, the average distance between the molecules increases. A small amount of energy must be supplied to the gas to pull the molecules to this greater separation against the attractive forces; the energy required per unit increase in volume is given by the derivative $(\partial E/\partial V)_T$. In a constant volume process, no work is produced, and the average distance between the molecules remains the same. Therefore, the heat capacity is small; all of the heat withdrawn goes into the chaotic motion and is reflected by a temperature increase. In a constant pressure process, the system expands against the resisting pressure and produces work in the surroundings; the heat withdrawn from the surroundings is divided into three portions. The first portion produces work in the surroundings; the second portion provides the energy necessary to separate the molecules to a larger distance; the third portion goes into increasing the energy of the chaotic motion. Only this last portion is reflected by a temperature increase. To produce a temperature increment of one degree, more heat must be withdrawn in the constant pressure process than is withdrawn in the constant volume process. Thus, C_p is greater than C_v.

Another useful quantity is the heat capacity ratio, γ, defined by

$$\gamma \equiv \frac{C_p}{C_v}. \tag{7–40}$$

From what has been said, it is clear that γ is always greater than unity.

The heat capacity difference for the ideal gas has a particularly simple form because $(\partial E/\partial V)_T = 0$; Joule's law. Then Eq. (7–39) is

$$C_p - C_v = p\left(\frac{\partial V}{\partial T}\right)_p. \tag{7–41}$$

If we are speaking of molar heat capacities, then the volume in the derivative is the molar volume; from the equation of state, $\overline{V} = RT/p$. Differentiating with respect to temperature, keeping the pressure constant, yields $(\partial \overline{V}/\partial T)_p = R/p$. Putting this value in Eq. (7–41) reduces it to the simple result

$$\overline{C}_p - \overline{C}_v = R. \tag{7–42}$$

Although Eq. (7–42) is precisely correct only for the ideal gas, it is a useful approximation for real gases. In Eq. (7–42), R is in cal·deg^{-1}·mole^{-1}; $R = 1.987$ cal.deg^{-1}·mole$^{-1} \approx 2.0$ cal·deg^{-1}·mole^{-1}.

The heat capacity difference for liquids and solids is usually rather small, and except in work of great accuracy it is sufficient to take

$$C_p = C_v, \tag{7–43}$$

although there are some notable exceptions to this rule. The physical reason for the approximate equality of C_p and C_v is plain enough. The thermal expansion coefficients of liquids and solids are very small, so that the volume change on increasing the temperature by one degree is very small; correspondingly, the work produced by the expansion is small and little energy is required for the small increase in the spacing of the molecules. Almost all of the heat withdrawn from the surroundings goes into increasing the energy of the chaotic motion, and so is reflected in a temperature increase which is nearly as large as that in a constant volume process. For the reasons mentioned at the end of Section 7–11, it is impractical to measure the C_v of a liquid or solid directly; C_p is readily measurable. The tabulated values of heat capacities of liquids and solids are values of C_p.

7–15 The Measurement of $(\partial H/\partial p)_T$; Joule-Thomson Experiment

The identification of the partial derivative $(\partial H/\partial p)_T$ with quantities which are readily accessible to experiment is beset with the same difficulties we experienced with $(\partial E/\partial V)_T$ in Section 7–12. These two derivatives are related. By differentiating the definition $H = E + pV$, we obtain

$$dH = dE + p\,dV + V\,dp.$$

Introducing the values of dH and dE from Eqs. (7–24) and (7–34), we have

$$C_p\,dT + \left(\frac{\partial H}{\partial p}\right)_T dp = C_v\,dT + \left[\left(\frac{\partial E}{\partial V}\right)_T + p\right]dV + V\,dp. \tag{7–44}$$

Restricting this formidable equation to constant temperature, $dT = 0$, and dividing by $(\partial p)_T$, we can simplify it to

$$\left(\frac{\partial H}{\partial p}\right)_T = \left[p + \left(\frac{\partial E}{\partial V}\right)_T\right]\left(\frac{\partial V}{\partial p}\right)_T + V, \tag{7–45}$$

which is at best a clumsy equation.

For liquids and solids the first term on the right-hand side of Eq. (7–45) is ordinarily very much smaller than the second term, so that a good approximation is

$$\left(\frac{\partial H}{\partial p}\right)_T = V \quad \text{(solids and liquids)}. \tag{7–46}$$

Since the molar volume of liquids and solids is very small, unless the change in pressure is enormous, the variation of the enthalpy with pressure can be ignored.

For the ideal gas,

$$\left(\frac{\partial H}{\partial p}\right)_T = 0. \tag{7–47}$$

This result is most easily obtained from the definition $H = E + pV$. For the ideal gas, $p\overline{V} = RT$, so that

$$\overline{H} = \overline{E} + RT. \tag{7–48}$$

Since the energy of the ideal gas is a function of temperature only, by Eq. (7–48) the enthalpy is a function of temperature only, and is independent of pressure. The result in Eq. (7–47) could also be obtained from Eq. (7–45) and Joule's law.

The derivative $(\partial H/\partial p)_T$ is very small for real gases, but can be measured. The Joule experiment, in which the gas expanded freely, failed to show a measurable difference in temperature between the initial and final states. Later, Joule and Thomson performed a different experiment, the Joule-Thomson experiment (Fig. 7–9).

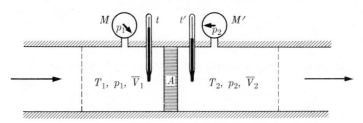

Fig. 7–9 The Joule-Thomson experiment.

A steady flow of gas passes through an insulated pipe in the direction of the arrows; at the position A there is an obstruction, which may be a porous disc or a diaphragm with a small hole in it or, as in the original experiment, a silk handkerchief. Because of the obstruction there is a drop in pressure, measured by the gauges

M and M', in passing from the left to the right side. Any drop in temperature is measured by the thermometers t and t'. The boundary of the system moves with the gas, always enclosing the same mass. Consider the passage of one mole of gas through the obstruction. The volume on the left decreases by the molar volume $\bar{V}_1$; since the gas is pushed by the gas behind it, which exerts a pressure p_1, work in the amount $p_1\bar{V}_1$ is destroyed. The volume on the right increases by the molar volume $\bar{V}_2$; the gas coming through must push the gas ahead of it, which exerts an opposing pressure p_2. On the right, work in the amount $p_2\bar{V}_2$ is produced. The net work produced is

$$W = p_2\bar{V}_2 - p_1\bar{V}_1.$$

Since the pipe is insulated, $Q = 0$, and we have the first-law statement

$$\bar{E}_2 - \bar{E}_1 = Q - W = -(p_2\bar{V}_2 - p_1\bar{V}_1).$$

Rearranging, we have

$$\bar{E}_2 + p_2\bar{V}_2 = \bar{E}_1 + p_1\bar{V}_1, \qquad \bar{H}_2 = \bar{H}_1.$$

The enthalpy of the gas is a constant in the Joule-Thomson expansion. The measured decrease in temperature $-\Delta T$ and the measured decrease in pressure $-\Delta p$ are combined in the ratio

$$\left(\frac{-\Delta T}{-\Delta p}\right)_H = \left(\frac{\Delta T}{\Delta p}\right)_H.$$

The Joule-Thomson coefficient μ_{JT} is defined as the limiting value of this ratio as Δp approaches zero:

$$\mu_{\mathrm{JT}} = \left(\frac{\partial T}{\partial p}\right)_H. \tag{7–49}$$

The drop in temperature (Joule-Thomson effect) is easily measurable in this experiment, particularly if the pressure difference is large. A noisy but dramatic demonstration of this effect can be made by partially opening the valve on a tank of compressed nitrogen; after a few minutes the valve is cold enough to form a coating of snow by condensing moisture from the air. (This should not be done with hydrogen or oxygen because of the explosion or fire hazard!) If the tank of gas is nearly full, the driving pressure is about 150 atm, the exit pressure is 1 atm. With this pressure drop, the temperature drop is quite large.

   The relation between μ_{JT} and the derivative $(\partial H/\partial p)_T$ is simple. The total differential of H,

$$dH = C_p\, dT + \left(\frac{\partial H}{\partial p}\right)_T dp,$$

expresses the change in H in terms of the changes in T and in p. Is it possible to change T and p in such a way that H remains unchanged? This imposes the condition

$dH = 0$. Under this condition the relation becomes

$$0 = C_p(\partial T)_H + \left(\frac{\partial H}{\partial p}\right)_T (\partial p)_H.$$

Dividing by $(\partial p)_H$, we obtain

$$0 = C_p\left(\frac{\partial T}{\partial p}\right)_H + \left(\frac{\partial H}{\partial p}\right)_T.$$

Using the definition of μ_{JT} and rearranging, we have

$$\left(\frac{\partial H}{\partial p}\right)_T = -C_p\mu_{JT}. \tag{7-50}$$

Thus, if we measure C_p and μ_{JT}, the value of $(\partial H/\partial p)_T$ can be calculated from Eq. (7–50). Note that by combining Eqs. (7–50) and (7–45), a value of $(\partial E/\partial V)_T$ can be obtained in terms of measurable quantities.

The Joule-Thomson coefficient is positive at and below room temperature for all gases, except hydrogen and helium, which have negative Joule-Thomson coefficients. These two gases will be hotter after undergoing this kind of expansion. Every gas has a characteristic temperature above which the Joule-Thomson coefficient is negative, the Joule-Thomson inversion temperature. The inversion temperature for hydrogen is about $-80°C$; below this temperature hydrogen will cool in a Joule-Thomson expansion. The inversion temperatures of most gases are very much higher than room temperature.

The Joule-Thomson effect can be used as the basis for a refrigerating device. The cooled gas on the low-pressure side is passed back over the high-pressure line to reduce the temperature of the gas before it is expanded; repetition of this can reduce the temperature on the high-pressure side to quite low values. If the temperature is low enough, then on expansion the temperature falls below the boiling point of the liquid and drops of liquid are produced. This procedure is essentially what is done in the Linde method for producing liquid air. The ordinary household refrigerator has a high- and a low-pressure side separated by an expansion valve, but the cooling results from the evaporation of a liquid refrigerant on the low-pressure side; the refrigerant is liquefied by compression on the high-pressure side.

7–16 Adiabatic Changes in State

If no heat flows during a change in state, $dQ = 0$, and the change in state is *adiabatic*. Experimentally we approximate this condition by wrapping the system in a layer of insulating material or by using a vacuum bottle to contain the system. For an adiabatic change in state, since $dQ = 0$, the first law statement is

$$dE = -dW, \tag{7-51}$$

or, for a finite change in state,

$$\Delta E = -W. \tag{7-52}$$

Turning Eq. (7–52) around, we find that $W = -\Delta E$, which means that the work produced, W, is at the expense of a decrease in energy of the system, $-\Delta E$. A decrease in energy in a system is evidenced almost entirely by a decrease in temperature of the system; hence, if work is produced in an adiabatic change in state, the temperature of the system falls. If work is destroyed in the adiabatic change, W is $-$, and so ΔE is $+$; the work destroyed increases the energy and the temperature of the system.

If only pressure–volume work is involved, Eq. (7–51) becomes

$$dE = -P_{op}\, dV, \tag{7-53}$$

from which it is clear that in expansion dV is $+$ and dE is $-$; the energy decreases and so does the temperature. If the system is compressed adiabatically, dV is $-$, and dE is $+$; the energy and the temperature both increase.

Special case: adiabatic changes in state in the ideal gas

Because of Joule's law we have for the ideal gas $dE = C_v\, dT$. Using this relation in Eq. (7–53), we obtain

$$C_v\, dT = -P_{op}\, dV, \tag{7-54}$$

which shows immediately that dT and dV have opposite signs. The drop in temperature is proportional to P_{op}, and for a specified increase in volume will have a maximum value when P_{op} has its maximum value; that is, when $P_{op} = p$. Consequently, for a fixed change in volume, the *reversible adiabatic* expansion will produce the greatest drop in temperature; conversely, a reversible adiabatic compression between two specified volumes produces the least increase in temperature.

For a reversible adiabatic change in state of the ideal gas, $P_{op} = p$, and Eq. (7–54) becomes

$$C_v\, dT = -p\, dV. \tag{7-55}$$

To integrate this equation, C_v and p must be expressed as functions of the variables of integration T and V. Since E is a function only of temperature, C_v is also a function only of temperature; from the ideal gas law, $p = nRT/V$. Equation (7–55) becomes

$$C_v\, dT = -nRT\frac{dV}{V}.$$

Dividing by T to separate variables, and using $\bar{C}_v = C_v/n$, we have

$$\bar{C}_v\frac{dT}{T} = -R\frac{dV}{V}.$$

Describing the initial state by T_1, V_1, the final state by T_2, V_2, and integrating, we have

$$\int_{T_1}^{T_2} \bar{C}_v \frac{dT}{T} = -R \int_{V_1}^{V_2} \frac{dV}{V}.$$

If $\bar{C}_v$ is a constant, it can be removed from the integral. Integration yields

$$\bar{C}_v \ln \left(\frac{T_2}{T_1} \right) = -R \ln \frac{V_2}{V_1}. \tag{7–56}$$

Since $R = \bar{C}_p - \bar{C}_v$, then $R/\bar{C}_v = (\bar{C}_p/\bar{C}_v) - 1 = \gamma - 1$. This value of $R/\bar{C}_v$ reduces Eq. (7–56) to

$$\ln \left(\frac{T_2}{T_1} \right) = -(\gamma - 1) \ln \left(\frac{V_2}{V_1} \right),$$

which can be written

$$\frac{T_2}{T_1} = \left(\frac{V_1}{V_2} \right)^{\gamma - 1}$$

or

$$T_1 V_1^{\gamma - 1} = T_2 V_2^{\gamma - 1}. \tag{7–57}$$

Using the ideal gas law, we can transform this equation to the equivalent forms

$$T_1^\gamma p_1^{1-\gamma} = T_2^\gamma p_2^{1-\gamma}, \tag{7–58}$$

$$p_1 V_1^\gamma = p_2 V_2^\gamma. \tag{7–59}$$

Equation (7–59), for example, says that any two states of an ideal gas which can be connected by a reversible adiabatic process fulfill the condition that $pV^\gamma = $ constant. Equations (7–57) and (7–58) can be given analogous interpretations. Although these equations are rather specialized, occasional use will be made of them.

7–17 A Note about Problem Working

So far we have more than fifty equations. Working a problem would be a terrible task if it were done by searching through such a bewildering array of equations in the hope of quickly finding the right one. Only the fundamental equations should be used in application to any problem. The conditions set in the problem immediately limit these fundamental equations to simple forms from which it should be clear how to calculate the "unknowns" in the problem. So far we have only seven fundamental equations:

1. The formula for expansion work: $dW = P_{op}\, dV$.
2. The definition of energy: $dE = dQ - dW$.
3. The definition of enthalpy: $H = E + pV$.

4. The definition of the heat capacities:

$$C_v = \frac{dQ_V}{dT} = \left(\frac{\partial E}{\partial T}\right)_V, \qquad C_p = \frac{dQ_p}{dT} = \left(\frac{\partial H}{\partial T}\right)_p.$$

5. Two purely mathematical consequences:

$$dE = C_v\, dT + \left(\frac{\partial E}{\partial V}\right)_T dV, \qquad dH = C_p\, dT + \left(\frac{\partial H}{\partial p}\right)_T dp.$$

Of course, it is essential to understand the meaning of these equations and the meaning of such terms as isothermal, adiabatic, reversible, etc. These terms have definite mathematical consequences to the equations. For problems involving the ideal gas, the equation of state should be known, the mathematical consequences of Joule's law, and the relation between the heat capacities. The equations which solve each problem must be derived from these few fundamental equations. Other methods of attack, such as attempting to memorize as many equations as possible, result in drowning in vast seas of ΔH's and ΔE's.

7–18 Application of the First Law of Thermodynamics to Chemical Reactions. The Heat of Reaction

If a chemical reaction takes place in a system, the temperature of the system immediately after the reaction generally is different from the temperature immediately before the reaction. To restore the system to its initial temperature, heat must flow either to or from the surroundings. If the system is hotter after the reaction than before, heat must flow *to* the surroundings to restore the system to the initial temperature. In this event the reaction is *exothermic*; by the convention for heat flow, the heat of the reaction is negative. If the system is colder after the reaction than before, heat must flow *from* the surroundings to restore the system to the initial temperature. In this event the reaction is *endothermic*, and the heat of the reaction is positive. The *heat of a reaction* is the heat withdrawn from the surroundings in the transformation of reactants at T and p to products at the *same T and p*.

In the laboratory the majority of chemical reactions are performed under a constant pressure; therefore the heat withdrawn from the surroundings is equal to the change in enthalpy of the system. To avoid mixing the enthalpy change associated with the chemical reaction and that associated with a temperature or pressure change in the system, the initial and final states of the system must have the same temperature and pressure.

For example, in the reaction

$$\mathrm{Fe_2O_3(s) + 3H_2(g) \rightarrow 2Fe(s) + 3H_2O(l)},$$

the initial and final states are:

Initial state	Final state
T, p	T, p
1 mole solid Fe_2O_3	2 moles solid Fe
3 moles gaseous H_2	3 moles liquid H_2O

Since the state of aggregation of each substance must be specified, the letters s, l, g appear in parentheses after the formulas of the substances. Suppose that we think of the change in state as occurring in two distinct steps. In the first step, reactants at T and p are transformed adiabatically to products at T' and p.

Step 1:

$$\underbrace{Fe_2O_3(s) + 3H_2(g)}_{T, p} \rightarrow \underbrace{2Fe(s) + 3H_2O(l)}_{T', p} .$$

At constant pressure, $\Delta H = Q_p$; but, since this first step is adiabatic, $(Q_p)_1 = 0$ and $\Delta H_1 = 0$. In the second step, the system is placed in a heat reservoir at the initial temperature T. Heat flows into or out of the reservoir as the products of the reaction come to the initial temperature.

Step 2:

$$\underbrace{2Fe(s) + 3H_2O(l)}_{T', p} \rightarrow \underbrace{2Fe(s) + 3H_2O(l)}_{T, p},$$

for which $\Delta H_2 = Q_p$. The sum of the two steps is the overall change in state

$$Fe_2O_3(s) + 3H_2(g) \rightarrow 2Fe(s) + 3H_2O(l),$$

and the ΔH for the overall reaction is the sum of the enthalpy changes in the two steps: $\Delta H = \Delta H_1 + \Delta H_2 = 0 + Q_p$,

$$\Delta H = Q_p, \tag{7-60}$$

where Q_p is the heat of the reaction, the increase in enthalpy of the system resulting from the chemical reaction.

The increase in enthalpy in a chemical reaction can be viewed in a different way. At a specified temperature and pressure, the molar enthalpy $\bar{H}$ of each substance has a definite value. For any reaction, we can write

$$\Delta H = H_{\text{final}} - H_{\text{initial}}. \tag{7-61}$$

But the enthalpy of the initial or the final state is the sum of the enthalpies of the substances present initially or finally. Therefore, for the example,

$$H_{\text{final}} = 2\bar{H}_{\text{Fe(s)}} + 3\bar{H}_{\text{H}_2\text{O(l)}},$$

$$H_{\text{initial}} = \bar{H}_{\text{Fe}_2\text{O}_3\text{(s)}} + 3\bar{H}_{\text{H}_2\text{(g)}},$$

and Eq. (7–61) becomes

$$\Delta H = [2\bar{H}_{\text{Fe(s)}} + 3\bar{H}_{\text{H}_2\text{O(l)}}] - [\bar{H}_{\text{Fe}_2\text{O}_3\text{(s)}} + 3\bar{H}_{\text{H}_2\text{(g)}}]. \tag{7–62}$$

It seems reasonable that measuring ΔH could lead ultimately to the evaluation of the four molar enthalpies in Eq. (7–62). However, there are four "unknowns" and only one equation. We could measure the heats of several different reactions, but this will introduce more "unknowns". Since it is impossible to determine "absolute" values of enthalpy, it is customary to make a conventional assignment of the values of the enthalpy of elements.

7–19 Conventional Values of $\bar{H}$; Formation Reactions

The molar enthalpy $\bar{H}$ of any substance is a function of T and p; $\bar{H} = \bar{H}(T, p)$. Choosing $p = 1$ atm as the *standard* pressure, we define the *standard* molar enthalpy $\bar{H}^0$ of a substance by

$$\bar{H}^0 = \bar{H}(T, 1\text{ atm}). \tag{7–63}$$

From this it is clear that $\bar{H}^0$ is a function only of temperature. The zero superscript on any thermodynamic quantity indicates the value of that quantity at the standard pressure. (Because the dependence of enthalpy on pressure is very slight, Section 7–15, we will often use standard enthalpies at pressures other than one atm; the error will not be serious unless the pressure is very large, for example, 1000 atm.)

Consider the solid element zinc; its standard molar enthalpy is a function only of temperature, so at 25°C (298.15°K) it has a definite value. Since we have no way of discovering what this value is, we arbitrarily assign a convenient, or a conventional, value to it. The most convenient value is zero; we do the same for all the other elements. The enthalpy of every *element* in its stable state of aggregation at 1 atm pressure and at 298.15°K is assigned the value zero. For example, at 1 atm and 298.15°K the stable state of aggregation of bromine is the liquid state. Hence, liquid bromine, gaseous hydrogen, solid zinc, solid (rhombic) sulfur, and solid (graphite) carbon all have $\bar{H}^0_{298.15} = 0$.

For elementary solids which exist in more than one crystalline form, the modification which is stable at 25°C and 1 atm is assigned $\bar{H} = 0$; e.g., the zero assignment goes to rhombic sulfur rather than to monoclinic sulfur, and to graphite rather than to diamond. In cases where more than one molecular species is possible (e.g., oxygen atoms, O; diatomic oxygen, O_2; ozone, O_3) the stable form at 25°C and 1 atm pressure is assigned the zero enthalpy value. After the value of the standard enthalpy of elements at 298.15°K has been assigned, the value at any other temperature can be computed. Since at constant pressure, $d\bar{H}^0 = \bar{C}_p^0 \, dT$, then

$$\int_{298.15}^{T} d\bar{H}^0 = \int_{298.15}^{T} \bar{C}_p^0 \, dT, \qquad \bar{H}_T^0 - \bar{H}_{298.15}^0 = \int_{298.15}^{T} \bar{C}_p^0 \, dT,$$

$$\bar{H}_T^0 = \bar{H}_{298.15}^0 + \int_{298.15}^{T} \bar{C}_p^0 \, dT, \tag{7-64}$$

which is correct for both elements and compounds; for elements, the first term on the right-hand side is zero.

The usefulness of this conventional assignment of enthalpies will be apparent if we consider the *formation reaction* of a compound. The formation reaction of a compound has *one mole* of the compound and *nothing else* on the product side; only elements in their stable states of aggregation appear on the reactant side. The increase in enthalpy at 1 atm pressure and at 25°C in such a reaction is the standard heat of formation of the compound, ΔH_f^0. For example,

$$\tfrac{1}{2}H_2(g) + \tfrac{1}{2}Br_2(l) \rightarrow HBr(g),$$

$$H_2(g) + \tfrac{1}{2}O_2(g) \rightarrow H_2O(l) ,$$

$$2Fe(s) + \tfrac{3}{2}O_2(g) \rightarrow Fe_2O_3(s),$$

$$\tfrac{1}{2}N_2(g) + 2H_2(g) + \tfrac{1}{2}Cl_2(g) \rightarrow NH_4Cl(s).$$

If the ΔH of these reactions is written in terms of the molar enthalpies, we obtain, using the first reaction as an example,

$$\Delta H_f^0 = \bar{H}_{HBr(g)}^0 - \left[\tfrac{1}{2}\bar{H}_{H_2(g)}^0 + \tfrac{1}{2}\bar{H}_{Br_2(l)}^0\right].$$

By the conventional assignment, each of the enthalpies in the bracket is zero. Then $\Delta H_f^0 = \bar{H}_{HBr(g)}^0$. From the character of the formation reaction, it is clear that this relation is correct for any compound; that is

$$\bar{H}^0 = \Delta H_f^0. \tag{7-65}$$

The standard heat of formation ΔH_f^0 is the molar enthalpy of the compound relative to the elements which compose it. Accordingly, if the heats of formation ΔH_f^0 of all the compounds in a chemical reaction are known, the heat of the reaction can be calculated from equations formulated in the manner of Eq. (7–62). Table 7–1 lists values of standard heats of formation at 25°C for a number of compounds.

Returning to the example of the reduction of ferric oxide with hydrogen, setting the enthalpies of the elements equal to zero in Eq. (7–62) yields

$$\Delta H = 3\bar{H}_{H_2O(l)} - \bar{H}_{Fe_2O_3(s)}.$$

In view of Eq. (7–65) this can be written

$$\Delta H = 3\Delta H_{f(H_2O,l)}^0 - \Delta H_{f(Fe_2O_3,s)}^0.$$

Table 7–1.† Standard heats of formation at 25°C.

Compound	ΔH_f^0, kcal/mole	Compound	ΔH_f^0, kcal/mole
$O_3(g)$	34.0	$ZnO(s)$	−83.17
$H_2O(g)$	−57.7979	$HgO(s, red)$	−21.68
$H_2O(l)$	−68.3174	$CuO(s)$	−37.1
$HF(g)$	−64.2	$Cu_2O(s)$	−39.84
$HCl(g)$	−22.063	$Ag_2O(s)$	−7.306
$Br_2(g)$	7.34	$AgCl(s)$	−30.362
$HBr(g)$	−8.66	$Ag_2S(s)$	−7.60
$HI(g)$	6.20	$FeO(s, wustite)$	−63.7
$SO_2(g)$	−70.96	$Fe_2O_3(s)$	−196.5
$SO_3(g)$	−94.45	$Fe_3O_4(s)$	−267.0
$H_2S(g)$	−4.815	$FeS(s, \alpha)$	−22.72
$H_2SO_4(l)$	−193.91	$FeS_2(s)$	−42.52
$NO(g)$	21.600	$TiO_2(s, rutile)$	−218.0
$NO_2(g)$	8.091	$TiCl_4(l)$	−179.3
$N_2O(g)$	19.49	$Al_2O_3(s, \alpha)$	−399.09
$N_2O_3(g)$	20.0	$MgO(s)$	−143.84
$N_2O_4(g)$	2.309	$MgCO_3(s)$	−266
$N_2O_5(g)$	3.6	$CaO(s)$	−151.9
$NH_3(g)$	−11.04	$CaC_2(s)$	−15.0
$HNO_3(l)$	−41.404	$Ca(OH)_2(s)$	−235.80
$NOCl(g)$	12.57	$CaCO_3(s, calcite)$	−288.45
$NH_4Cl(s)$	−75.38	$SrO(s)$	−141.1
$PCl_3(g)$	−73.22	$SrCO_3(s)$	−291.2
$PCl_5(g)$	−95.35	$BaO(s)$	−133.4
$C(s, diamond)$	0.4532	$BaCO_3(s)$	−291.3
$CO(g)$	−26.4157	$Na_2O(s)$	−99.4
$CO_2(g)$	−94.0518	$NaOH(s)$	−101.99
$CH_4(g)$	−17.889	$NaF(s)$	−136.0
$HCHO(g)$	−27.7	$NaCl(s)$	−98.232
$CH_3OH(l)$	−57.02	$NaBr(s)$	−86.030
$C_2H_2(g)$	54.194	$NaI(s)$	−68.84
$C_2H_4(g)$	12.496	$Na_2SO_4(s)$	−330.90
$C_2H_6(g)$	−20.236	$Na_2SO_4 \cdot 10H_2O(s)$	−1033.48
$C_2H_5OH(l)$	−66.356	$NaNO_3(s)$	−111.54
$C_6H_6(g)$	19.820	$KF(s)$	−134.46
$SiO_2(s, quartz)$	−205.4	$KCl(s)$	−104.175
$SiH_4(g)$	−14.8	$KClO_3(s)$	−93.50
$SiF_4(g)$	−370	$KClO_4(s)$	−103.6
$PbO(s, red)$	−52.40	$KBr(s)$	−93.73
$PbS(s)$	−22.4	$KI(s)$	−78.31

†   NBS Circular 500. U.S. G.P.O., Washington, D.C., 1952.

From Table 7–1 we find

$$\Delta H^0_{f(\text{H}_2\text{O},\text{l})} = -68.3 \text{ kcal/mole,}$$

and

$$\Delta H^0_{f(\text{Fe}_2\text{O}_3,\text{s})} = -196.5 \text{ kcal/mole}$$

so that

$$\Delta H = 3(-68.3) - (-196.5) = -204.9 + 196.5 = -8.4 \text{ kcal.}$$

The negative sign indicates that the reaction is exothermic.

7–20   The Determination of Heats of Formation

In some cases it is possible to determine the heat of formation of a compound directly by carrying out the formation reaction in a calorimeter and measuring the heat effect produced. Two important examples are

$$\text{C(graphite)} + \text{O}_2(\text{g}) \rightarrow \text{CO}_2(\text{g}), \qquad \Delta H^0_f = -94.0518 \text{ kcal/mole.}$$

$$\text{H}_2(\text{g}) + \tfrac{1}{2}\text{O}_2(\text{g}) \rightarrow \text{H}_2\text{O(l)}, \qquad \Delta H^0_f = -68.3174 \text{ kcal/mole.}$$

These reactions can be conducted easily in a calorimeter; the reactions go to completion, and conditions can easily be arranged so that only one product is formed. Because of the importance of these two reactions, the values have been determined quite accurately.

The majority of formation reactions are unsuitable for calorimetric measurement; these heats of formation must be determined by indirect methods. For example,

$$\text{C(graphite)} + 2\text{H}_2(\text{g}) \rightarrow \text{CH}_4(\text{g}).$$

This reaction has three strikes against it as far as its use in calorimetry is concerned. The combination of graphite with hydrogen does not occur readily; if we did manage to get these materials to react in a calorimeter, the product would not be pure methane, but an exceedingly complex mixture of hydrocarbons. Even if we succeeded in analysing the product mixture, the result of such an experiment would be impossible to interpret.

There is one method which is generally applicable if the compound burns easily to form definite products. The heat of formation of a compound can be calculated from the measured value of the heat of combustion of the compound. Using methane as an illustration, the combustion reaction is

$$\text{CH}_4(\text{g}) + 2\text{O}_2(\text{g}) \rightarrow \text{CO}_2(\text{g}) + 2\text{H}_2\text{O(l)}.$$

The measured heat of combustion is $\Delta H^0_{\text{comb}} = -212.798 \text{ kcal/mole}$. In terms of the enthalpies of the individual substances,

$$\Delta H^0_{\text{comb}} = \overline{H}^0_{\text{CO}_2(\text{g})} + 2\overline{H}^0_{\text{H}_2\text{O(l)}} - \overline{H}^0_{\text{CH}_4(\text{g})}.$$

Therefore

$$\bar{H}^0_{CH_4(g)} = \bar{H}^0_{CO_2(g)} + 2\bar{H}^0_{H_2O(l)} - \Delta H^0_{comb}. \tag{7-66}$$

The molar enthalpies of CO_2 and H_2O are known to a high accuracy; from this knowledge and the measured value of the heat of combustion, the molar enthalpy of methane (the heat of formation) can be calculated from Eq. (7-66):

$$\bar{H}^0_{CH_4}(g) = -94.0518 + 2(-68.3174) - (-212.798)$$

$$= -230.687 + 212.798$$

$$= -17.889 \text{ kcal/mole}.$$

The measurement of the heat of combustion is used to determine the heats of formation of all organic compounds which contain only carbon, hydrogen, and oxygen. These compounds burn completely to carbon dioxide and water in the calorimeter. The combustion method is used also for organic compounds containing sulfur and nitrogen; however, in these cases the reaction products are not so definite. The sulfur may end up as sulfurous acid or sulfuric acid, the nitrogen may end up in the elementary form or as a mixture of oxy-acids. In these cases considerable skill and ingenuity are required in the determination of the conditions for the reaction and in the analysis of the reaction products. The accuracy of the values obtained for this latter class of compounds is very much less than that obtained for the compounds containing only carbon, hydrogen, and oxygen.

The problem of determining the heat of formation of any compound resolves into that of finding some chemical reaction involving the compound which is suitable for calorimetric measurement, then measuring the heat of this reaction. If the heats of formation of all the other substances involved in this reaction are known, then the problem is solved. If the heat of formation of one of the other substances in the reaction is not known, then we must find a calorimetric reaction for that substance; and so on.

Devising a series of reactions from which an accurate value of the heat of formation of a particular compound can be obtained can be a challenging problem. A calorimetric reaction must take place quickly, being completed within a few minutes at most, with as few side reactions as possible, preferably none at all. Very few chemical reactions take place without concomitant side reactions, but their effect can be minimized by controlling the reaction conditions so as to favor the main reaction as much as possible. The final product mixture must be carefully analysed, and the thermal effect of the side reactions must be subtracted from the measured value. Precision calorimetry is at best exacting work.

7-21 Sequences of Reactions; Hess's Law

The change in state of a system produced by a specified chemical reaction is definite. The corresponding enthalpy change is definite, since the enthalpy is a function of the state. Thus, if we transform a specified set of reactants to a specified set of products by more than one sequence of reactions, the total enthalpy change must be the same for every sequence. This rule, which is a consequence of the first law of thermodynamics, was originally known as Hess's law of constant heat summation. Suppose that we compare two different methods of synthesizing sodium chloride from sodium and chlorine.

Method 1:

$$Na(s) + H_2O(l) \rightarrow NaOH(s) + \tfrac{1}{2}H_2(g), \qquad \Delta H = -33.67,$$
$$\tfrac{1}{2}H_2(g) + \tfrac{1}{2}Cl_2(g) \rightarrow HCl(g), \qquad \Delta H = -22.06,$$
$$HCl(g) + NaOH(s) \rightarrow NaCl(s) + H_2O(l), \qquad \Delta H = -42.50.$$

Net change: $Na(s) + \tfrac{1}{2}Cl_2(g) \rightarrow NaCl(s),$ $\qquad \Delta H_{net} = -98.23.$

Method 2:

$$\tfrac{1}{2}H_2(g) + \tfrac{1}{2}Cl_2(g) \rightarrow HCl(g), \qquad \Delta H = -22.06,$$
$$Na(s) + HCl(g) \rightarrow NaCl(s) + \tfrac{1}{2}H_2(g), \qquad \Delta H = -76.17.$$

Net change: $Na(s) + \tfrac{1}{2}Cl_2(g) \rightarrow NaCl(s),$ $\qquad \Delta H_{net} = -98.23.$

The net chemical change is obtained by adding together all the reactions in the sequence; the net enthalpy change is obtained by adding together all the enthalpy changes in the sequence. The net enthalpy change must be the same for every sequence which has the same net chemical change. Any number of reactions can be added or subtracted to yield the desired chemical reaction; the enthalpy changes of the reactions are added or subtracted algebraically in the corresponding way.

If a certain chemical reaction is combined in a sequence with the reverse of the same reaction, there is no net chemical effect, and $\Delta H = 0$ for the combination. It follows immediately that the ΔH of the reverse reaction is equal in magnitude but opposite in sign to that of the forward reaction.

The utility of this property of sequences, which is really nothing more than the fact that the enthalpy change in a system is independent of the path, is illustrated by the sequence

1) $\qquad\qquad C(graphite) + \tfrac{1}{2}O_2(g) \rightarrow CO(g), \qquad \Delta H_1,$

2) $\qquad\qquad CO(g) + \tfrac{1}{2}O_2(g) \rightarrow CO_2(g), \qquad \Delta H_2.$

The net change in the sequence is

3) $\qquad\qquad C(graphite) + O_2(g) \rightarrow CO_2(g), \qquad \Delta H_3.$

Therefore, $\Delta H_3 = \Delta H_1 + \Delta H_2$. In this particular instance, ΔH_2 and ΔH_3 are readily measurable in the calorimeter, while ΔH_1 is not. Since the value of ΔH_1 is computed from the other two values, there is no need to measure it.

Similarly, by subtracting reaction (2) from reaction (1) we obtain

4) $C(graphite) + CO_2(g) \rightarrow 2CO(g), \quad \Delta H_4 = \Delta H_1 - \Delta H_2,$

and the heat of this reaction can also be obtained from the measured values.

7–22 Heats of Solution and Dilution

The *heat of solution* is the enthalpy change associated with the addition of a specified amount of solute to a specified amount of solvent at constant temperature and pressure. For convenience we shall use water as the solvent in the illustrations; the argument can be applied to any solvent with slight modification. The change in state is represented by

$$X + nAq \rightarrow X \cdot nAq, \quad \Delta H_S.$$

One mole of solute X is added to n moles of water; the water is given the symbol Aq in this equation, since it is convenient to assign a conventional enthalpy of zero to the water in these solution reactions.

Consider the examples

$$
\begin{aligned}
HCl(g) + 10Aq &\rightarrow HCl \cdot 10Aq, & \Delta H_1 &= -16.608 \text{ kcal,} \\
HCl(g) + 25Aq &\rightarrow HCl \cdot 25Aq, & \Delta H_2 &= -17.272 \text{ kcal,} \\
HCl(g) + 40Aq &\rightarrow HCl \cdot 40Aq, & \Delta H_3 &= -17.453 \text{ kcal,} \\
HCl(g) + 200Aq &\rightarrow HCl \cdot 200Aq, & \Delta H_4 &= -17.735 \text{ kcal,} \\
HCl(g) + \infty Aq &\rightarrow HCl \cdot \infty Aq, & \Delta H_5 &= -17.960 \text{ kcal.}
\end{aligned}
$$

The values of ΔH show the general dependence of the heat of solution on the amount of solvent. As more and more solvent is used, the heat of solution approaches a limiting value, the value in the "infinitely dilute" solution; for HCl this limiting value is given by ΔH_5.

If we subtract the first equation from the second in the above set, we obtain

$$HCl \cdot 10Aq + 15Aq \rightarrow HCl \cdot 25Aq, \quad \Delta H = \Delta H_2 - \Delta H_1 = -0.664 \text{ kcal.}$$

This value of ΔH is a heat of dilution, the heat withdrawn from the surroundings when additional solvent is added to a solution. The heat of dilution of a solution is dependent on the original concentration of the solution and on the amount of solvent added.

The heat of formation of a solution is the enthalpy associated with the reaction (using hydrochloric acid as the example):

$$\tfrac{1}{2}H_2(g) + \tfrac{1}{2}Cl_2(g) + nAq \rightarrow HCl \cdot nAq, \quad \Delta H_f^0,$$

where the solvent Aq is counted as having zero enthalpy.

The heat of solution defined above is the *integral* heat of solution. This distinguishes it from the *differential* heat of solution which is defined in Section 11–23.

7–23 Heats of Reaction at Constant Volume

If any of the reactants or products of the calorimetric reaction are gaseous, it is necessary to conduct the reaction in a sealed bomb. Under this condition the system is initially and finally in a constant volume rather than being under a constant pressure. The measured heat of reaction at constant volume is equal to an energy increment, rather than an enthalpy increment:

$$Q_V = \Delta E. \qquad (7\text{–}67)$$

The corresponding change in state is

$$R(T, V, p) \rightarrow P(T, V, p'),$$

where $R(T, V, p)$ represents the reactants in the initial condition T, V, p; and $P(T, V, p')$ represents the products in the final condition T, V, p'. The temperature and volume remain constant, but the pressure may change from p to p' in the transformation.

To relate the ΔE in Eq. (7–67) to the corresponding ΔH, we apply the defining equation for H to the initial and final states:

$$H_{final} = E_{final} + p'V, \qquad H_{initial} = E_{initial} + pV.$$

Subtracting the second equation from the first, we have

$$\Delta H = \Delta E + (p' - p)V. \qquad (7\text{–}68)$$

The initial and final pressures in the bomb are determined by the number of moles of gases present initially and finally; assuming that the gases behave ideally, we have

$$p = n_R(RT/V), \qquad p' = n_P(RT/V),$$

where n_R and n_P are the total number of moles of *gaseous* reactants and *gaseous* products in the reaction. Equation (7–68) becomes

$$\Delta H = \Delta E + (n_P - n_R)RT,$$

$$\Delta H = \Delta E + \Delta nRT. \qquad (7\text{–}69)$$

Strictly speaking, the ΔH in Eq. (7–69) is the ΔH for the constant-volume transformation. To convert it to the appropriate value for the constant-pressure transformation, we must add to it the enthalpy change of the process:

$$P(T, V, p') \rightarrow P(T, V', p).$$

For this change in pressure at constant temperature, the enthalpy change is very nearly zero (Section 7–15) and exactly zero if only ideal gases are involved. Thus, for practical purposes the ΔH in Eq. (7–69) is equal to the ΔH for the constant-pressure process, while the ΔE refers to the constant-volume transformation. To a good approximation, Eq. (7–69) can be interpreted as

$$Q_p = Q_V + \Delta n R T. \tag{7-70}$$

It is through Eqs. (7–69) or (7–70) that measurements in the bomb calorimeter, $Q_V = \Delta E$, are converted to values of $Q_p = \Delta H$. In precision measurements it may be necessary to include the effect of gas imperfection or of the change in enthalpy of the products with pressure; this would depend on the conditions employed in the experiment.

Example. Consider the combustion of benzoic acid in the bomb calorimeter:

$$C_6H_5COOH(s) + \tfrac{15}{2}O_2(g) \rightarrow 7CO_2(g) + 3H_2O(l).$$

In this reaction, $n_P = 7$, while $n_R = \tfrac{15}{2}$; thus, $\Delta n = 7 - \tfrac{15}{2} = -\tfrac{1}{2}$. Using $R = 1.9872$ cal·deg^{-1}·mole^{-1}, and $T = 298.15°K$, we have

$$Q_p = Q_V - \tfrac{1}{2}(1.9872)(298.15), \qquad Q_p = Q_V - 296.1 \text{ cal}.$$

Note that only the number of moles of *gases* are counted in computing Δn.

7–24 Dependence of the Heat of Reaction on Temperature

If we know the value of ΔH^0 for a reaction at a particular temperature, let us say at 25°C, then we can calculate the heat of reaction at any other temperature if the heat capacities of all the substances taking part in the reaction are known. The ΔH^0 of any reaction is

$$\Delta H^0 = H^0(\text{products}) - H^0(\text{reactants}).$$

To find the dependence of this quantity on temperature we differentiate with respect to temperature:

$$\frac{d\,\Delta H^0}{dT} = \frac{dH^0}{dT}(\text{products}) - \frac{dH^0}{dT}(\text{reactants}).$$

But, by definition, $dH^0/dT = C_p^0$. Hence,

$$\frac{d\,\Delta H^0}{dT} = C_p^0\,(\text{products}) - C_p^0\,(\text{reactants})$$

$$\frac{d\,\Delta H^0}{dT} = \Delta C_p^0. \tag{7-71}$$

Note that since H^0 and ΔH^0 are functions only of temperature (Section 7–19), these are ordinary derivatives rather than partial derivatives.

The value of ΔC_p^0 is calculated from the individual heat capacities in the same way as ΔH^0 is calculated from the individual values of the molar enthalpies. We multiply the molar heat capacity of each product by the number of moles of that product involved in the reaction; the sum of these quantities for every product yields the heat capacity of the products. A similar procedure yields the heat capacity of the reactants. The difference in value of the heat capacity of products less that of reactants is ΔC_p.

Writing Eq. (7–71) in differential form, we have

$$d\,\Delta H^0 = \Delta C_p^0\,dT.$$

Integrating between a fixed temperature T_0 and any other temperature T, we obtain

$$\int_{T_0}^{T} d\,\Delta H^0 = \int_{T_0}^{T} \Delta C_p^0\,dT.$$

The integral of the differential on the left is simply ΔH^0, which, when evaluated between the limits, becomes

$$\Delta H_T^0 - \Delta H_{T_0}^0 = \int_{T_0}^{T} \Delta C_p^0\,dT.$$

Rearranging, we have

$$\Delta H_T^0 = \Delta H_{T_0}^0 + \int_{T_0}^{T} \Delta C_p^0\,dT. \tag{7–72}$$

Knowing the value of the enthalpy increase at the fixed temperature T_0, we can calculate the value at any other temperature T, using Eq. (7–72). If any of the substances change their state of aggregation in the temperature interval, the corresponding enthalpy change must be included.

If the range of temperature covered in the integration of Eq. (7–72) is short, the heat capacities of all the substances involved may be considered constant. If the temperature interval is very large, the heat capacities must be taken as functions of temperature. For many substances this function has the form

$$C_p = a + bT + cT^2 + dT^3 + \cdots, \tag{7–73}$$

where $a, b, c, d, \ldots$ are constants for a specified material. In Table 7–2 values of these constants are given for a number of substances.

Example 1. Calculate the ΔH^0 at 85°C for the reaction

$$Fe_2O_3(s) + 3H_2(g) \rightarrow 2Fe(s) + 3H_2O(l)$$

from the data $\Delta H_{298}^0 = -8.4\,\text{kcal}$.

Substance	$Fe_2O_3(s)$	$Fe(s)$	$H_2O(l)$	$H_2(g)$
$\bar{C}_p^0$, cal·deg^{-1}·mole^{-1}	25.0	6.1	18.0	6.9

First compute ΔC_p^0:

$$\Delta C_p^0 = 2\bar{C}_{p(Fe)}^0 + 3\bar{C}_{p(H_2O)}^0 - [\bar{C}_{p(Fe_2O_3)}^0 + 3\bar{C}_{p(H_2)}^0]$$

$$= 2(6.1) + 3(18.0) - 25.0 - 3(6.9) = 66.2 - 45.7 = 20.5 \text{ cal/deg}.$$

Table 7–2†   Heat capacity of gases as a function of temperature
$\bar{C}_p$, cal·deg^{-1}·mole$^{-1} = a + bT + cT^2 + dT^3$
Range: 300 to 1500°K

Gas	a	$b \times 10^3$	$c \times 10^7$	$d \times 10^9$
H_2	6.9469	−0.1999	4.808	
O_2	6.0954	+3.2533	−10.171	
Cl_2	7.5755	2.4244	−9.650	
Br_2	8.4228	0.9739	−3.555	
N_2	6.4492	1.4125	−0.807	
CO	6.3424	1.8363	−2.801	
HCl	6.7319	0.4325	+3.697	
HBr	6.5776	0.9549	1.581	
NO	7.020	−0.370	25.46	−1.087
CO_2	6.369	+10.100	−34.05	
H_2O	7.219	2.374	2.67	
NH_3	6.189	7.887	−7.28	
H_2S	6.385	5.704	−12.10	
SO_2	6.147	13.844	−91.03	+2.057
CH_4	3.381	18.044	−43.00	
C_2H_6	2.247	38.201	−110.49	
C_2H_4	2.830	28.601	−87.26	
C_2H_2	7.331	12.622	−38.89	
C_3H_8	2.410	57.195	−175.33	
C_3H_6	3.253	45.116	−137.40	
C_3H_4	6.334	30.990	−94.57	
C_6H_6	−0.409	77.621	−264.29	
$C_6H_5CH_3$	0.576	93.493	−312.27	
C (graphite)	−1.265	14.008	−103.31	2.751

†   From the compilations of H. M. Spencer and J. L. Justice, *J. Am. Chem. Soc.*, **56.** 2311 (1934); H. M. Spencer and G. N. Flanagan, *J. Am. Chem. Soc.*, **64.** 2511 (1942); H. M. Spencer, *Ind. Eng. Chem.*, **40.**, 2152 (1948).

Since 85°C = 358°K, we have

$$\Delta H^0_{358} = \Delta H^0_{298} + \int_{298}^{358} 20.5 \, dT$$

$$= -8.4 \, \text{kcal} + 20.5(358 - 298) \, \text{cal}$$

$$= -8.4 \, \text{kcal} + 20.5(60) \, \text{cal} = -8.4 \, \text{kcal} + 1230 \, \text{cal}$$

$$= -8.4 \, \text{kcal} + 1.2 \, \text{kcal} = -7.2 \, \text{kcal}.$$

Note that care must be taken to express both terms in kcal or both in calories before adding them together!

Example 2. Compute the heat of reaction at $1000°C = 1273°K$ for

$$\tfrac{1}{2}H_2(g) + \tfrac{1}{2}Cl_2(g) \rightarrow HCl(g)$$

if $\Delta H^0_{298} = -22.063$ kcal and

$$\bar{C}^0_{p(H_2)} = 6.9469 - 0.1999(10^{-3})T + 4.808(10^{-7})T^2 \text{ cal·deg}^{-1}\text{·mole}^{-1},$$

$$\bar{C}^0_{p(Cl_2)} = 7.5755 + 2.4244(10^{-3})T - 9.650(10^{-7})T^2 \text{ cal·deg}^{-1}\text{·mole}^{-1},$$

$$\bar{C}^0_{p(HCl)} = 6.7319 + 0.4325(10^{-3})T + 3.697(10^{-7})T^2 \text{ cal·deg}^{-1}\text{·mole}^{-1}.$$

We begin by computing the value of ΔC^0_p for the integral in Eq. (7–72). It is best to arrange the work in columns:

$$
\begin{aligned}
\Delta C^0_p = \quad & 6.7319 + 0.4325(10^{-3})T + 3.697(10^{-7})T^2 \\
& -\tfrac{1}{2}[6.9469 - 0.1999(10^{-3})T + 4.808(10^{-7})T^2] \\
& -\tfrac{1}{2}[7.5755 + 2.4244(10^{-3})T - 9.650(10^{-7})T^2] \\
= \quad & -0.5293 - 0.6797(10^{-3})T + 6.118(10^{-7})T^2
\end{aligned}
$$

$$
\begin{aligned}
\int_{298}^{1273} \Delta C^0_p \, dT = & -0.5293 \int_{298}^{1273} dT - 0.6797(10^{-3}) \int_{298}^{1273} T \, dT \\
& + 6.118(10^{-7}) \int_{298}^{1273} T^2 \, dT \\
= & -0.5293(1273 - 298) - \tfrac{1}{2}(0.6797)(10^{-3})(1273^2 - 298^2) \\
& + \tfrac{1}{3}(6.118)(10^{-7})(1273^3 - 298^3) \\
= & -516 - 521 + 415 \\
= & -622 \text{ cal} = -0.622 \text{ kcal.}
\end{aligned}
$$

Then

$$\Delta H^0_{1273} = \Delta H^0_{298} + \int_{298}^{1273} \Delta C^0_p \, dT$$

$$= -22.063 \text{ kcal} - 0.622 \text{ kcal} = -22.685 \text{ kcal.}$$

Note that the heat capacities of *all* the substances taking part in the reaction must be included; the elements cannot be omitted as they were in calculating enthalpy differences.

7–25 Adiabatic Flame Temperature

It is occasionally useful to characterize compounds by the hypothetical temperature which would be attained if the compound were burned adiabatically, the *adiabatic flame temperature.* As a typical example, we consider the combustion of methane:

$$CH_4(g) + 2O_2(g) \rightarrow CO_2(g) + 2H_2O(g).$$

If this reaction occurs adiabatically, then the products in the final state are at a very high temperature. For this reason the gaseous state of water is chosen on the product side; this simplifies the computation.

We suppose that the reaction occurs in two steps, both at constant pressure; then
Step 1:

$$\text{Reactants}\,(T_0, p) \rightarrow \text{Products}\,(T_0, p) \qquad \Delta H_{T_0}.$$

Step 2:

$$\text{Products}\,(T_0, p) \rightarrow \text{Products}\,(T_f, p) \qquad \Delta H_2 = \int_{T_0}^{T_f} C_p(\text{products})\, dT.$$

The overall transformation is the sum of the two steps:

$$\text{Reactants}\,(T_0, p) \rightarrow \text{Products}\,(T_f, p) \qquad 0 = \Delta H_{T_0} + \int_{T_0}^{T_f} C_p(\text{products})\, dT.$$

Since the transformation is adiabatic, $\Delta H = 0$ for the overall transformation. This yields the result

$$\int_{T_0}^{T_f} C_p(\text{products})\, dT = -\Delta H_{T_0}.$$

Since heats of combustion are of the order of kcal/mole while heat capacities of gases are of the order of cal/deg-mole, it follows that flame temperatures are typically of the order of thousands of degrees.

In the case of methane, at 298°K,

$$\Delta H_{298} = -94.0518 - 2(-57.7979) - (-17.889) \approx -192 \text{ kcal},$$

$$C_p(CO_2 + 2H_2O)_{298} = 8.874 + 2(8.025) \approx 25 \text{ cal/deg}.$$

Making the totally unwarranted assumption that the heat capacity remains constant over the temperature range, yields a first estimate of T_f,

$$25(T_f - 298) = 192{,}000 \qquad \text{or} \qquad T_f \approx 8000°K.$$

If the compound were to be burned in air, then each mole of oxygen is associated with four moles of nitrogen. The nitrogen must also be raised to the final temperature so that the temperature reached is significantly lower. In the case of methane which requires two moles of oxygen, eight moles of nitrogen must be raised to the final temperature; the heat capacity of eight moles of nitrogen is about 56 cal/deg. Hence the total heat capacity of the products plus nitrogen is 81 cal/deg, and

$$81(T_f - 298) = 192{,}000 \qquad \text{or} \qquad T_f \approx 2700°K.$$

It should be apparent that these temperatures represent hypothetical values which would be achieved provided the molecules did not dissociate under such extreme conditions. For comparative purposes the numbers have some utility.

7–26 Bond Energies

In a simple case, if we consider the atomization of a gaseous diatomic molecule, e.g.,

$$O_2(g) \rightarrow 2O(g) \qquad \Delta H^{\circ}_{298} = 118.318 \text{ kcal},$$

the value 118.318 kcal is called the bond energy of the oxygen molecule. Similarly we can write

$$H_2O(g) \rightarrow 2H(g) + O(g) \qquad \Delta H^{\circ}_{298} = 221.14 \text{ kcal}$$

and call $\frac{1}{2}(221.14) = 110.57$ kcal the average bond energy of the O—H bond in water. So long as one deals only with molecules in which the bonds are equivalent, such as H_2O, NH_3, CH_4, the procedure is straightforward. On the other hand, even in dealing with a molecule such as H_2O_2 in which two different bonds exist, some assumption must be introduced. The assumption is usually made that the average OH bond strength in H_2O_2 is the same as that in water. The enthalpy of atomization of H_2O_2 is

$$H_2O_2(g) \rightarrow 2H(g) + 2O(g) \qquad \Delta H^{\circ}_{298} = 254.33 \text{ kcal}.$$

If we subtract the energy of two OH bonds in water we obtain: 254.33–221.14 = 33.19 kcal as the strength of the O—O single bond. Clearly the method does not warrant keeping the fraction and we would say that the oxygen–oxygen single bond has a strength of about 33 kcal.

The heats of formation of the atoms must be known to compute the bond strengths. Some values are given in Table 7–3. Although the value for carbon has been the subject of considerable controversy, that given is probably close to the correct one.

Table 7–3 Heats of formation of atoms at 25°C (kcal/mole)

H	52.089	P	75.18	F	18.3
C	171.698	O	59.159	Cl	29.012
Si	88.04	S	53.25	Br	26.71
N	85.565	Se	48.37	I	25.482

7–27 Calorimetric measurements

It is worthwhile to describe how the heat of a reaction is calculated from the quantities which are actually measured in a calorimetric experiment. It is not possible in a brief space to describe all the types of calorimeters or all of the variations and refinements of technique which are necessary in individual cases and in precision work. A highly idealized situation will be described to illustrate the methods involved.

The situation is simplest if the calorimeter is an *adiabatic* calorimeter. In the laboratory, this device is quite elaborate; on paper we shall simply say that the vessel containing the system is perfectly insulated so that no heat flows into or out of the system. Under constant pressure, the first law for any transformation within the calorimeter is

$$\Delta H = Q_p = 0. \tag{7–74}$$

The change in state can be represented by

$$K(T_1) + R(T_1) \rightarrow K(T_2) + P(T_2) \qquad (p = \text{constant}),$$

where K symbolizes the calorimeter, R the reactants, and P the products. Since the system is insulated, the final temperature T_2 differs from the initial temperature T_1; both temperatures are measured as accurately as possible with a sensitive thermometer.

The change in state can be supposed to occur in two steps:

1) $R(T_1) \rightarrow P(T_1),$ $\Delta H_{T_1},$

2) $K(T_1) + P(T_1) \rightarrow K(T_2) + P(T_2),$ $\Delta H_2.$

By Eq. (7–74) the overall $\Delta H = 0$, so that $\Delta H_{T_1} + \Delta H_2 = 0$, or $\Delta H_{T_1} = -\Delta H_2$. The second step is simply a temperature change of the calorimeter, and the reaction products, so

$$\Delta H_2 = \int_{T_1}^{T_2} [C_p(K) + C_p(P)] \, dT,$$

and we obtain for the heat of the reaction at T_1

$$\Delta H_{T_1} = - \int_{T_1}^{T_2} [C_p(K) + C_p(P)] \, dT. \tag{7–75}$$

If the heat capacities of the calorimeter and the products of the reaction are known, the heat of reaction at T_1 can be calculated from the measured temperatures T_1 and T_2.

An alternative scheme can be imagined for the steps in the reaction:

3) $K(T_1) + R(T_1) \rightarrow K(T_2) + R(T_2),$ $\Delta H_3,$

4) $R(T_2) \rightarrow P(T_2),$ $\Delta H_{T_2}.$

Again, the overall $\Delta H = 0$, so that $\Delta H_3 + \Delta H_{T_2} = 0$, or

$$\Delta H_{T_2} = -\Delta H_3 = - \int_{T_1}^{T_2} [C_p(K) + C_p(R)] \, dT. \tag{7–76}$$

If the heat capacities of the calorimeter and the reactants are known, the heat of reaction at T_2 can be calculated from Eq. (7–76).

If the required heat capacities are not known, the value of ΔH_2 can be measured as follows. Cool the calorimeter and the products to the initial temperature T_1. (This assumes that T_2 is greater than T_1.) The calorimeter and the products are then taken from T_1 to T_2 by allowing an electrical current to flow in a resistor immersed in the calorimeter; the change in enthalpy in this step is equal to ΔH_2. This can be related to the electrical work expended in the resistance wire, which can be measured quite accurately, being the product of the current, the potential difference across the resistance, and the time.

If we include electrical work, dU, at constant pressure, the first law becomes

$$dE = dQ - p\,dV - dU. \tag{7–77}$$

Differentiating $H = E + pV$ under the constant pressure condition, we get $dH = dE + p\,dV$. Adding this equation to Eq. (7–77) yields

$$dH = dQ - dU. \tag{7–78}$$

For an adiabatic process, $dQ = 0$, and Eq. (7–78) integrates to

$$\Delta H = -U. \tag{7–79}$$

Applying Eq. (7–79) to the electrical method of carrying the products and calorimeter from the initial to the final temperature, we have $\Delta H_2 = -U$, and so, since $\Delta H_{T_1} + \Delta H_2 = 0$, we obtain

$$\Delta H_{T_1} = U. \tag{7–80}$$

Since work was destroyed in the surroundings, U, and therefore ΔH_{T_1}, are negative. The reaction is exothermic; a result of the assumption that T_2 is greater than T_1. For endothermic reactions the procedure is modified in an obvious way.

If one deals with a bomb calorimeter so that the volume is constant rather than the pressure, the argument is unchanged. In all of the equations ΔH's are simply replaced by ΔE's, and C_p's by C_v's.

Problems

(Before working problems, see Section 7–17.)

7–1. a) An ideal gas undergoes a single-stage expansion against a constant opposing pressure from T, p_1, V_1 to T, p_2, V_2. What is the largest mass M which can be lifted through a height h in this expansion?

b) The system in (a) is restored to its initial state by a single-stage compression. What is the smallest mass M' which must fall through the height h to restore the system?

c) What is the net mass lowered through the height h in the cyclic transformation in (a) and (b)?

d) If $h = 10$ cm, $p_1 = 10$ atm, $p_2 = 5$ atm, $T = 300°$K, and one mole of gas is involved, calculate the numerical values of the quantities required in (a) and (b).

7–2. One mole of an ideal gas is expanded from T, p_1, V_1 to T, p_2, V_2 in two stages:

	Opposing pressure	Volume change
First stage	P' (constant)	V_1 to V'
Second stage	p_2 (constant)	V' to V_2

We specify that the point P', V' lies on the isotherm at the temperature T.

a) Formulate the expression for the work produced in this expansion in terms of T, p_1, p_2, and P'.

b) For what value of P' does the work in this two-stage expansion have a maximum value? What is the maximum value of the work produced?

7–3. Three moles of an ideal gas expand isothermally against a constant opposing pressure of 1 atm from 20 liters to 60 liters. Compute W, Q, ΔE, and ΔH.

7–4. Three moles of an ideal gas at 27°C expand isothermally and reversibly from 20 liters to 60 liters. Compute W, Q, ΔE, and ΔH.

7–5. One mole of a van der Waals gas at 27°C expands isothermally and reversibly from 10 liters to 30 liters. Compute the work produced; $a = 5.49$ liters^2·atm·moles^{-2}, $b = 0.064$ liter/mole.

7–6. One mole of the ideal gas is confined under a constant pressure, $P_{op} = p = 2$ atm. The temperature is changed from 100°C to 25°C.

a) What is the value of W?

b) If $\bar{C}_v = 3$ cal·deg^{-1}·mole^{-1}, calculate Q, ΔE, and ΔH.

7–7. If an ideal gas undergoes a reversible polytropic expansion, the relation $pV^n = C$ holds; C and n are constants, $n > 1$.

a) Calculate W for such an expansion if one mole of gas expands from V_1 to V_2 and if $T_1 = 300°$C, $T_2 = 200°$C, and $n = 2$.

b) If $\bar{C}_v = 5$ cal·deg^{-1}·mole^{-1}, calculate Q, ΔE, and ΔH.

7–8. a) The coefficient of thermal expansion of liquid water is 2.1×10^{-4} deg^{-1} and the density is 1 gm/cm^3. If 200 cm^3 of water are warmed from 25°C to 50°C under a constant pressure of 1 atm, calculate W.

b) If $\bar{C}_p = 18$ cal·deg^{-1}·mole^{-1}, calculate Q and ΔH.

7–9. One mole of an ideal gas is compressed adiabatically in a single stage with a constant opposing pressure equal to 10 atm. Initially the gas is at 27°C and 1 atm pressure; the final pressure is 10 atm. Calculate the final temperature of the gas, W, Q, ΔE, and ΔH. Do this for two cases: *Case I.* Monatomic gas, $\bar{C}_v = \frac{3}{2}R$. *Case II.* Diatomic gas, $\bar{C}_v = \frac{5}{2}R$. How would the various quantities be affected if n moles were used instead of one mole?

7-10. One mole of an ideal gas at 27°C and 1 atm is compressed adiabatically and reversibly to a final pressure of 10 atm. Compute the final temperature, Q, W, ΔE, and ΔH for the same two cases as in Problem 7–9.

7-11. One mole of an ideal gas at 27°C and 10 atm pressure is expanded adiabatically to a final pressure of 1 atm against a constant opposing pressure of 1 atm. Calculate the final temperature, Q, W, ΔE, and ΔH for the two cases, $\bar{C}_v = \frac{3}{2}R$, and $\bar{C}_v = \frac{5}{2}R$.

7-12. Repeat Problem 7–11 assuming that the expansion is reversible.

7-13. The Joule-Thomson coefficient for a van der Waals gas is given by

$$\mu_{JT} = [(2a/RT) - b]/C_p.$$

Calculate the value of ΔH (cal) for the isothermal, 300°K, compression of 1 mole of nitrogen from 1 to 500 atm: $a = 1.34$ liters^2·atm·moles^{-2}, $b = 0.039$ liter/mole.

7-14. The boiling point of nitrogen is -196°C, and $\bar{C}_p = 5.0$ cal·deg^{-1}·mole^{-1}. The van der Waals constant and μ_{JT} are given in Problem 7–13. What must the initial pressure be if nitrogen in a single-stage Joule–Thomson expansion is to drop in temperature from 25°C to the boiling point? (The final pressure is to be 1 atm.)

7-15. Repeat the calculation in Problem 7–14 for ammonia: b.p. $= -34$°C, $\bar{C}_p = 8.5$ cal·deg^{-1}·mole^{-1}, $a = 4.17$ liter^2·atm·moles^{-2}, $b = 0.037$ liter/mole.

7-16. From the value of $\bar{C}_p$ given in Table 7–2 for oxygen calculate Q, W, ΔE, and ΔH for the change in state:

a) $p = $ constant, 100°C to 300°C;

b) $V = $ constant, 100°C to 300°C.

7-17. From the data in Table 7–1, compute the values of ΔH^0_{298} for the following reactions.

a) $2O_3(g) \rightarrow 3O_2(g)$

b) $H_2S(g) + \frac{3}{2}O_2(g) \rightarrow H_2O(l) + SO_2(g)$

c) $TiO_2(s) + 2Cl_2(g) \rightarrow TiCl_4(l) + O_2(g)$

d) $C(graphite) + CO_2(g) \rightarrow 2CO(g)$

e) $CO(g) + 2H_2(g) \rightarrow CH_3OH(l)$

f) $Fe_2O_3(s) + 2Al(s) \rightarrow Al_2O_3(s) + 2Fe(s)$

g) $NaOH(s) + HCl(g) \rightarrow NaCl(s) + H_2O(l)$

h) $CaC_2(s) + 2H_2O(l) \rightarrow Ca(OH)_2(s) + C_2H_2(g)$

i) $CaCO_3(s) \rightarrow CaO(s) + CO_2(g)$

7-18. Assuming the gases are ideal, calculate the value of ΔE_{298} for each of the reactions in Problem 7–17.

7-19. For the reaction

$$C(graphite) + H_2O(g) \rightarrow CO(g) + H_2(g),$$

$\Delta H^0_{298} = 31.3822$ kcal. The values of $\bar{C}_p$ (cal·deg^{-1}·mole^{-1}) are: graphite, 2.066; $H_2O(g)$, 8.025; $CO(g)$, 6.965; and $H_2(g)$, 6.892. Calculate the value of ΔH^0 at 125°C.

7–20. From the data at 25°C:

$$Fe_2O_3(s) + 3C(graphite) \rightarrow 2Fe(s) + 3CO(g), \qquad \Delta H^0 = 117.3 \text{ kcal,}$$
$$FeO(s) + C(graphite) \rightarrow Fe(s) + CO(g), \qquad \Delta H^0 = 37.3 \text{ kcal,}$$
$$C(graphite) + O_2(g) \rightarrow CO_2(g), \qquad \Delta H^0 = -94.05 \text{ kcal,}$$
$$CO(g) + \tfrac{1}{2}O_2(g) \rightarrow CO_2(g), \qquad \Delta H^0 = -67.63 \text{ kcal,}$$

compute the standard heat of formation of FeO(s) and of $Fe_2O_3(s)$.

7–21. From the data at 25°C:

$$O_2(g) \rightarrow 2O(g), \qquad \Delta H^0 = 118.318 \text{ kcal,}$$
$$Fe(s) \rightarrow Fe(g), \qquad \Delta H^0 = 96.68 \text{ kcal.}$$

The heat of formation of FeO(s) is -63.7 kcal/mole.

a) Compute the ΔH^0 at 25°C for the reaction

$$Fe(g) + O(g) \rightarrow FeO(s).$$

b) Assuming that the gases are ideal, calculate ΔE for this reaction. (The negative of this quantity, $+218.4$ kcal, is the cohesive energy of the crystal.)

7–22. From the data at 25°C:

$$\tfrac{1}{2}H_2(g) + \tfrac{1}{2}O_2(g) \rightarrow OH(g), \qquad \Delta H^0 = 10.06 \text{ kcal,}$$
$$H_2(g) + \tfrac{1}{2}O_2(g) \rightarrow H_2O(g), \qquad \Delta H^0 = -57.80 \text{ kcal,}$$
$$H_2(g) \rightarrow 2H(g), \qquad \Delta H^0 = 104.178 \text{ kcal,}$$
$$O_2(g) \rightarrow 2O(g), \qquad \Delta H^0 = 118.318 \text{ kcal,}$$

compute ΔH^0 for

a) $OH(g) \rightarrow H(g) + O(g)$,

b) $H_2O(g) \rightarrow 2H(g) + O(g)$,

c) $H_2O(g) \rightarrow H(g) + OH(g)$.

d) Assuming the gases are ideal, compute the values of ΔE for these three reactions.

[*Note:* The enthalpy change for (a) is called the bond energy of the OH radical; one-half the enthalpy change in (b) is the average O—H bond energy in H_2O. The enthalpy change for (c) is the bond-dissociation energy of the O—H bond in H_2O.]

7–23. a) From the data in Table 7–1 compute the heat of vaporization of water at 25°C.

b) Compute the work produced in the vaporization of one mole of water at 25°C under a constant pressure of 1 atm.

c) Compute the ΔE of vaporization of water at 25°C.

d) The values of $\bar{C}_p$ (cal·deg^{-1}·mole^{-1}) are: water vapor, 8.025; liquid water, 17.996. Calculate the heat of vaporization at 100°C.

7–24. From the values of $\bar{C}_p$ as a function of temperature given in Table 7–2 and from the data:

$$\tfrac{1}{2}H_2(g) + \tfrac{1}{2}Br_2(l) \rightarrow HBr(g), \qquad \Delta H^0_{298} = -8.66 \text{ kcal,}$$
$$Br_2(l) \rightarrow Br_2(g), \qquad \Delta H^0_{298} = 7.34 \text{ kcal.}$$

Calculate the ΔH^0_{1000} for the reaction

$$\tfrac{1}{2}H_2(g) + \tfrac{1}{2}Br_2(g) \rightarrow HBr(g).$$

7–25. From the data in Tables 7–1 and 7–2, calculate the ΔH^0_{1000} for the reaction

$$2C(graphite) + O_2(g) \rightarrow 2CO(g).$$

7–26. A sample of sucrose, $C_{12}H_{22}O_{11}$, weighing 0.1265 gm is burned in a bomb calorimeter. After the reaction is over, it is found that to produce an equal temperature increment electrically, 2082.3 joules must be expended.

a) Calculate the heat of combustion of sucrose.

b) From the heat of combustion and appropriate data in Table 7–1 calculate the heat of formation of sucrose.

c) If the temperature increment in the experiment is 1.743°C, what is the heat capacity of the calorimeter and contents?

7–27. From the heats of solution at 25°C:

$$HCl(g) + 100Aq \rightarrow HCl \cdot 100Aq, \qquad \Delta H = -17.650 \, kcal,$$
$$NaOH(s) + 100Aq \rightarrow NaOH \cdot 100Aq, \qquad \Delta H = -10.12 \, kcal,$$
$$NaCl(s) + 200Aq \rightarrow NaCl \cdot 200Aq, \qquad \Delta H = +1.016 \, kcal,$$

and the heats of formation of HCl(g), NaOH(s), NaCl(s), and $H_2O(l)$ in Table 7–1, calculate ΔH for the reaction

$$HCl \cdot 100Aq + NaOH \cdot 100Aq \rightarrow NaCl \cdot 200Aq + H_2O(l).$$

7–28. From the heats of formation at 25°C:

Solution	$H_2SO_4 \cdot 600Aq$	$KOH \cdot 200Aq$	$KHSO_4 \cdot 800Aq$	$K_2SO_4 \cdot 1000Aq$
ΔH, kcal	− 212.35	− 114.82	− 274.3	− 336.75

calculate the ΔH for the reactions

$$H_2SO_4 \cdot 600Aq + KOH \cdot 200Aq \rightarrow KHSO_4 \cdot 800Aq + H_2O(l),$$
$$KHSO_4 \cdot 800Aq + KOH \cdot 200Aq \rightarrow K_2SO_4 \cdot 1000Aq + H_2O(l).$$

[Use Table 7–1 for the heat of formation of $H_2O(l)$.]

7–29. From the heats of formation at 25°C

Solution	$H_2SO_4(l)$	$H_2SO_4 \cdot 1Aq$	$H_2SO_4 \cdot 2Aq$	$H_2SO_4 \cdot 4Aq$
ΔH, kcal	− 193.91	− 200.62	− 203.93	− 206.83

Solution	$H_2SO_4 \cdot 10Aq$	$H_2SO_4 \cdot 20Aq$	$H_2SO_4 \cdot 100Aq$	$H_2SO_4 \cdot \infty Aq$
ΔH, kcal	− 209.93	− 211.00	− 211.59	− 216.90

Calculate the heat of solution of sulfuric acid for these various solutions, and plot ΔH_S against the mole fraction of water in the solution.

7–30. Using Eq. (7–45) and Joule's law show that for the ideal gas, $(\partial H/\partial p)_T = 0$.

7–31. From the ideal gas law and Eq. (7–57), derive Eqs. (7–58) and (7–59).

7–32. By applying Eq. (7–44) to a constant volume transformation, show that

$$C_p - C_v = [V - (\partial H/\partial p)_T](\partial p/\partial T)_v.$$

7–33. Given the data from Table 7–3 and the heats of formation (kcal/mole) at 25°C of the gaseous compounds:

Compound	SiF$_4$	SiCl$_4$	CF$_4$	NF$_3$	OF$_2$	HF
ΔH_f^0	-370	-145.7	-162.5	-27.2	5.5	-64.2

calculate the single bond energies: Si—F; Si—Cl; C—F; N—F; O—F; H—F.

7–34. Given the data in Table 7–3 and the heats of formation (kcal/mole) at 25°C of the gaseous compounds:

Compound	CH$_4$	C$_2$H$_6$	C$_2$H$_4$	C$_2$H$_2$
ΔH_f^0	-17.889	-20.236	12.496	54.194

calculate the bond energy of

a) the C—C single bond in C$_2$H$_6$;

b) the C=C double bond in C$_2$H$_4$;

c) the C≡C triple bond in C$_2$H$_2$.

Chapter Eight

Introduction to the Second
Law of Thermodynamics

8–1 General Remarks

In Chapter 6 we mentioned the fact that all real changes have a direction which we consider natural. The transformation in the opposite sense would be unnatural; it would be unreal. In nature, rivers run from the mountains to the sea, never in the opposite way. A tree blossoms, bears fruit, and later sheds its leaves. The thought of dry leaves rising, attaching themselves to the tree, and later shrinking into buds is grotesque. An isolated metal rod initially hot at one end and cold at the other comes to a uniform temperature; such a metal rod initially at a uniform temperature never develops a hot and a cold end spontaneously.

Yet the first law of thermodynamics tells us nothing of this preference of one direction over the opposite one. The first law requires only that the energy of the universe remain the same before and after the change takes place. In the changes described above, the energy of the universe is not one whit altered; the transformation may go in either direction and satisfy the first law.

It would be helpful if a system possessed one or more properties which always change in one direction when the system undergoes a natural change, and change in the opposite direction if we imagine the system to undergo an "unnatural change." Fortunately, there exists such a property of a system, the entropy, as well as several others derived from it. To prepare a foundation for the mathematical definition of the entropy, we must divert our attention for a little while to the study of the characteristics of cyclic transformations. Having done that, we will return to chemical systems and the chemical implications of the second law.

8-2 The Carnot Cycle

In 1824 a French engineer, Sadi Carnot, investigated the principles governing the transformation of thermal energy, "heat," into mechanical energy, work. He based his discussion on a cyclical transformation of a system which is now called the Carnot cycle. The Carnot cycle consists of four reversible steps, and therefore is a reversible cycle. A system is subjected consecutively to the *reversible* changes in state:

Step 1. Isothermal expansion.   *Step 2.* Adiabatic expansion.

Step 3. Isothermal compression.   *Step 4.* Adiabatic compression.

Since the mass of the system is fixed, the state can be described by any two of the three variables T, p, V. A system of this sort which produces only heat and work effects in the surroundings is called a *heat engine*.

Imagine that the material composing the system, the "working" substance, is confined in a cylinder fitted with a piston. In Step 1, the cylinder is immersed in a heat reservoir at a temperature T_1, and is expanded isothermally from the initial volume V_1 to a volume V_2. The cylinder is now taken out of the reservoir, insulated, and, in Step 2, is expanded adiabatically from V_2 to V_3; in this step the temperature of the system drops from T_1 to a lower temperature T_2. The insulation is removed and the cylinder placed in a heat reservoir at T_2. In Step 3 the system is compressed isothermally from V_3 to V_4. The cylinder is removed from the reservoir and insulated again. In Step 4 the system is compressed adiabatically from V_4 to V_1, the original volume. In this adiabatic compression, the temperature rises from T_2 to T_1, the original temperature. Thus, as it must be in a cycle, the system is restored to its initial state.

Before continuing, the term *heat reservoir* must be clearly understood. A heat reservoir is a system which has the same temperature everywhere within it; this temperature is unaffected by the transfer of any desired quantity of heat into or out of the reservoir.

The initial and final states and the application of the first law to each step in the Carnot cycle are described in Table 8-1. For the cycle, $\Delta E = 0 = Q_{cy} - W_{cy}$, or

$$W_{cy} = Q_{cy}. \qquad (8-1)$$

Table 8-1

Step	Initial state	Final state	First-law statement
1	T_1, p_1, V_1	T_1, p_2, V_2	$\Delta E_1 = Q_1 - W_1$
2	T_1, p_2, V_2	T_2, p_3, V_3	$\Delta E_2 = -W_2$
3	T_2, p_3, V_3	T_2, p_4, V_4	$\Delta E_3 = Q_2 - W_3$
4	T_2, p_4, V_4	T_1, p_1, V_1	$\Delta E_4 = -W_4$

Summing the first law statements for the four steps yields

$$W_{cy} = W_1 + W_2 + W_3 + W_4, \tag{8-2}$$

$$Q_{cy} = Q_1 + Q_2. \tag{8-3}$$

Combining Eqs. (8-1) and (8-3), we have

$$W_{cy} = Q_1 + Q_2. \tag{8-4}$$

(Note that the subscripts on the Q's have been chosen to correspond with those on the T's.) If W_{cy} is positive, then work has been produced at the expense of the thermal energy of the surroundings. The system suffers no net change in the cycle.

8-3   The Second Law of Thermodynamics

The important thing about Eq. (8-4) is that W_{cy} is the sum of *two* terms, each associated with a *different* temperature. We might imagine a complicated cyclical process involving many heat reservoirs at different temperatures; for such a case

$$W_{cy} = Q_1 + Q_2 + Q_3 + Q_4 + \cdots,$$

where Q_1 is the heat withdrawn from the reservoir at T_1, and so on. Some of the Q's will have positive signs, some will have negative signs; the net work effect in the cycle is the algebraic sum of all the values of Q.

It is possible to devise a cyclical process so that W_{cy} is positive; that is, such that after the cycle, weights are truly higher in the surroundings. It can be done in complicated ways using reservoirs at many different temperatures, or it can be done using only two reservoirs at different temperatures, as in the Carnot cycle. However, experience has shown that it is not possible to build such an engine using only one heat reservoir. Thus, if

$$W_{cy} = Q_1,$$

where Q_1 is the heat withdrawn from a single heat reservoir at a uniform temperature, then W_{cy} must be negative or, at best, zero; that is,

$$W_{cy} \leq 0.$$

This experience is embodied in the second law of thermodynamics. *It is impossible for a system operating in a cycle and connected to a single heat reservoir to produce a positive amount of work in the surroundings.* This statement is equivalent to that proposed by Kelvin in about 1850.

8-4   Characteristics of a Reversible Cycle

According to the second law, the simplest cyclical process capable of producing a positive amount of work in the surroundings must involve at least two heat reservoirs at different temperatures. The Carnot engine operates in such a cycle; because of its simplicity it has come to be the prototype of cyclical heat engines. An important property of the Carnot cycle is the fact that it is reversible. In a cyclical transformation, reversibility requires, after the complete cycle has been traversed once in the forward sense and once in the opposite sense, that the surroundings be restored to their original condition. This means that the reservoirs and weights must be restored to their initial condition, which can only be accomplished if reversing the cycle reverses the sign of W and of Q_1 and Q_2, individually. The magnitudes of W and of the individual values of Q are not changed by running a reversible engine backwards, only the signs. Hence for a reversible engine we have:

$$\text{Forward cycle:} \quad W_{cy}, \quad Q_1, \quad Q_2, \quad W_{cy} = Q_1 + Q_2;$$
$$\text{Reverse cycle:} \quad -W_{cy}, \quad -Q_1, \quad -Q_2, \quad -W_{cy} = -Q_1 + (-Q_2).$$

8–5 A Perpetual-motion Machine of the Second Kind

A Carnot engine and two heat reservoirs are usually represented schematically by a drawing such as that in Fig. 8–1. The work W produced in the surroundings by the reversible engine E_r is indicated by the arrow directed away from the system. The quantities of heat Q_1 and Q_2 withdrawn from the reservoirs are indicated by arrows directed toward the system. In all the discussions which follow, we choose T_1 as the higher temperature.

The second law has the immediate consequence that Q_1 and Q_2 cannot have the same algebraic sign. We prove this by making the contrary assumption. Assume that both Q_1 and Q_2 are positive; then W, being the sum of Q_1 and Q_2, is also positive. If Q_2 is positive, then heat flows out of the reservoir at T_2 as indicated by the arrow in Fig. 8–1. Suppose that we restore this quantity of heat Q_2 to the reservoir at T_2 by connecting the two reservoirs with a metal rod, so that heat may flow directly from the high-temperature reservoir to the low-temperature reservoir (Fig. 8–2). By making the rod in the proper shape and size, matters can be arranged so that the time needed for the engine to traverse one cycle in which it extracts Q_2 from the reservoir, an equal quantity of heat Q_2 flows to the reservoir through the rod. Therefore, after a cycle, the reservoir at T_2 is restored to its initial state; thus, the engine and the reservoir at T_2 form a composite cyclic engine, enclosed by the light line in Fig. 8–2. This composite cyclic engine is connected to a single heat reservoir at T_1 and produces a positive quantity of work. The second law asserts that such an engine is an impossibility. Our assumption that Q_1 and Q_2 are both positive has led to a contradiction of the second law. If we assume that Q_1 and Q_2 are both negative, then W is negative. We reverse the engine; then Q_1 and Q_2 and W are all positive, and the proof goes as before. We conclude that Q_1 and Q_2 must differ in sign; otherwise we could build this impossible engine.

In imagination suppose that we install the impossible engine shown in Fig. 8–2 in the parlor. The room itself can serve as the heat reservoir. We set the machine in motion; note that we do not have to plug it in! The machine is now extracting

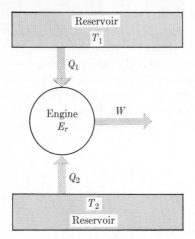

Fig. 8–1 Schematic representation of the Carnot engine.

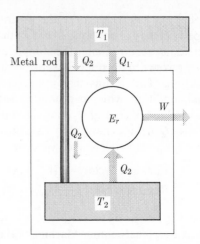

Fig. 8–2 An impossible engine.

heat from the room and producing mechanical work. Anyone with an ounce of frugality in his nature would use this work to run an electrical generator. As we gaily pen a note to the local power company stating that we no longer have need of their services, we note that the room is getting chilly. Air-conditioning, very nice in the summer, but definitely a nuisance in the winter. In winter we can put the machine out of doors. The heat is then extracted from the atmosphere; that engine will run for a long time before the atmosphere cools by so much as one degree; meanwhile we aren't paying any light bills. The delightful thing about this engine is that of course the atmosphere will never go permanently cold on us. As we use the stored electrical energy, it is returned to the universe mainly in the form of "heat." This wonderful machine is not available on the market. The truth of the matter is that experience has shown that it is not possible to build such a machine. It is a perpetual-motion machine of the second kind.

8–6 The Efficiency of Heat Engines

Experience shows that if a heat engine operates between two temperature reservoirs so that a positive amount of work is produced, then Q_1, the heat withdrawn from the high-temperature reservoir, is positive, and Q_2, the heat withdrawn from the low-temperature reservoir, is negative. The negative value of Q_2 means that the heat flows to the low-temperature reservoir. The engine in producing work extracts an amount of heat Q_1 from the high-temperature reservoir and rejects an amount $-Q_2$ to the low-temperature reservoir. The arrow between reservoir T_2 and the engine in Figs. 8–1 and 8–2 thus seems deceptive. However, we will retain the direction of the arrow and keep in mind that in every case one of the Q's is negative. This preserves our original sign convention for Q: positive when it flows from the surroundings, and the signs will care for themselves without our juggling them back and forth.

The efficiency ϵ of a heat engine is defined as the ratio of the work produced to the quantity of heat extracted from the high-temperature reservoir:

$$\epsilon \equiv W/Q_1. \tag{8–5}$$

But, since $W = Q_1 + Q_2$,

$$\epsilon = 1 + Q_2/Q_1. \tag{8–6}$$

Since Q_1 and Q_2 differ in sign, the second term in Eq. (8–6) is negative; consequently, the efficiency is less than unity. The efficiency is the fraction of the heat withdrawn from the high-temperature reservoir which is converted into work in the cyclic process.

8–7   Another Impossible Engine

We consider two engines E_r and E' both operating in a cycle between the same two heat reservoirs. Is it possible that the efficiencies of these two engines are different? The engines may be designed differently and may use different working substances. Let E_r be a reversible engine, and E' any engine at all, reversible or not. The reservoirs are at T_1 and T_2; $T_1 > T_2$. For the engine E_r we may write

$$W = Q_1 + Q_2, \qquad \text{(forward cycle)},$$
$$-W = -Q_1 + (-Q_2), \qquad \text{(reverse cycle)}.$$

For the engine E'

$$W' = Q_1' + Q_2', \qquad \text{(forward cycle)}.$$

Suppose that we run engine E_r in its reverse cycle and couple it to engine E' running in its forward cycle. This gives us a composite cyclic engine which produces heat and work effects which are simply the sum of the individual effects of the appropriate cycles:

$$-W + W' = -Q_1 + (-Q_2) + Q_1' + Q_2'. \tag{8–7}$$

By making the engine E_r of the proper size, matters are arranged so that the composite engine produces no work effect in the surroundings; that is, we adjust E_r until $-W + W' = 0$, or

$$W = W'. \tag{8–8}$$

Equation (8–7) can then be rearranged to the form

$$Q_1 - Q_1' = -(Q_2 - Q_2'). \tag{8–9}$$

We now examine these heat effects in the reservoirs under the assumption that the efficiency of E' is greater than that of E_r, that is,

$$\epsilon' > \epsilon.$$

By the definition of the efficiency, this implies that

$$W'/Q'_1 > W/Q_1.$$

Since by Eq. (8–8), $W = W'$, the inequality becomes

$$\frac{1}{Q'_1} > \frac{1}{Q_1},$$

which is equivalent to $Q_1 > Q'_1$, or

$$Q'_1 - Q_1 < 0, \qquad \text{(a negative quantity)}.$$

The heat withdrawn from reservoir at T_1 is Q'_1 by E' running forward, and $-Q_1$ by E_r running in reverse. The total heat withdrawn from T_1 is the sum of these two, $Q'_1 - Q_1$, and this, by our argument, has a negative value. If the heat withdrawn from the reservoir is negative, then the heat actually flows *to* that reservoir. Thus this engine pours heat into the reservoir at T_1. The heat withdrawn from the reservoir at T_2 is, by the same token, $Q'_2 - Q_2$. Our argument, together with Eq. (8–9), would show that this quantity of heat is positive. Heat is extracted from the reservoir at T_2. The various quantities are tabulated in Table 8–2. The quantities for the composite engine are the sums of the quantities for the separate engines, the sum of the preceding two columns in Table 8–2.

Table 8–2

	E_r forward	E_r reverse	E' forward	Composite engine E_r(reverse) + E'(forward)
Work produced	W	$-W$	W	0
Heat out of T_1	Q_1	$-Q_1$	Q'_1	$Q'_1 - Q_1 = -$
Heat out of T_2	Q_2	$-Q_2$	Q'_2	$Q'_2 - Q_2 = +$
First law	$W = Q_1 + Q_2$	$-W = -Q_1 - Q_2$	$W = Q'_1 + Q'_2$	$0 = (Q'_1 - Q_1) + (Q'_2 - Q_2)$

The fifth column shows that the composite engine takes a negative quantity of heat, $Q'_1 - Q_1$, out of the reservoir at T_1. Consequently, the engine puts a positive quantity of heat into the high-temperature reservoir and extracts an equal quantity of heat from the low-temperature reservoir. The remarkable aspect of this engine is that it produces no work, nor does it require work to operate it.

Again in imagination we install this engine in the living room. We pour a bucket of hot water into one end and a bucket of cold water into the other end, then set the machine in motion. It commences to pump heat from the cold end to the hot end. Before long the water in the hot end is boiling while that in the cold end is freezing. If the designer has been provident enough to make the cold end in the shape of an insulated box, we can keep the beer in that end and boil the coffee on the hot end. Any thrifty hostess or housewife would be delighted with this gadget. What a kitchen appliance! A combination stove-refrigerator! And again, no bill from the power company. Experience shows that it is not possible to build this engine; this is another example of perpetual motion of the second kind.

The argument which led to this impossible engine was based only on the first law and an assumption. The assumption that the efficiency of E' is greater than E_r is in error. We therefore conclude that the efficiency of any engine E' must be less than or equal to the efficiency of a reversible engine E_r, both engines operating between the same two temperature reservoirs:

$$\epsilon' \leq \epsilon. \qquad (8\text{–}10)$$

The relation in Eq. (8–10) is another important consequence of the second law. The engine E' is any engine whatsoever; the engine E_r is any *reversible* engine. Consider two *reversible* engines, with efficiencies ϵ' and ϵ''. Since the second one is reversible, the efficiency of the first must be less than or equal to that of the second, by Eq. (8–10):

$$\epsilon' \leq \epsilon''. \qquad (8\text{–}11)$$

But the first engine is reversible, therefore, by Eq. (8–10) the efficiency of the second must be less than or equal to that of the first:

$$\epsilon'' \leq \epsilon'. \qquad (8\text{–}12)$$

The only way that both (8–11) and (8–12) can be satisfied simultaneously is if

$$\epsilon' = \epsilon''. \qquad (8\text{–}13)$$

Equation (8–13), which results from the second law, means that *all reversible engines operating between the same two temperature reservoirs have the same efficiency.*

According to Eq. (8–13) the efficiency does not depend on the engine, so it cannot depend on the design of the engine or on the working substance used in the engine. only the reservoirs are left; the only specifications placed on the reservoirs were the temperatures. Hence, the efficiency is a function only of the temperatures of the reservoirs:

$$\epsilon = f(T_1, T_2). \qquad (8\text{–}14)$$

Since from Eq. (8–6), $\epsilon = 1 + Q_2/Q_1$, the ratio Q_2/Q_1 must be a function only of the temperatures:

$$Q_2/Q_1 = g(T_1, T_2). \qquad (8\text{–}15)$$

From the concept of reversibility it follows that an irreversible engine will produce heat and work effects in the surroundings which are different from those produced by a reversible engine. Therefore, the efficiency of the irreversible engine is different from that of a reversible one; the efficiency cannot be greater, so it must be less.

8–8 The Thermodynamic Temperature Scale

For a reversible engine, both the efficiency and the ratio Q_2/Q_1 can be calculated directly from the measurable quantities of work and heat flowing to the surroundings. Therefore, we have measurable properties which depend on temperatures only and are independent of the properties of any special kind of substance. Consequently, it is possible to establish a scale of temperature, independent of the properties of any individual substance. This overcomes the difficulty associated with empirical scales of temperature described in Section 6–5. This scale is the absolute, or the thermodynamic, scale of temperature.

We operate a reversible heat engine in the following way. The low-temperature reservoir is at some fixed low temperature t_0. The t_0 is the temperature on any empirical scale. The heat withdrawn from this reservoir is Q_0. If we run the engine with the high-temperature reservoir at t, an amount of heat Q will flow from this reservoir, and a positive amount of work is produced. Keeping t_0 and Q_0 constant, we increase the temperature of the other reservoir to some higher temperature t'. Experimentally we find that more heat Q' is withdrawn from the reservoir at t'. Thus, the heat withdrawn from the high-temperature reservoir increases with increase in temperature. For this reason we choose the heat withdrawn from the high-temperature reservoir as the thermometric property. We can define the thermodynamic temperature θ by

$$Q = a\theta, \tag{8–16}$$

where a is a constant, and Q is the heat withdrawn from the reservoir. Writing Eq. (8–15) in the notation for this situation, it becomes $Q_0/Q = g(t, t_0)$. From this it is clear that if Q_0 and t_0 are constant, then Q is a function of t only. In Eq. (8–16) we have arbitrarily chosen Q as a simple and reasonable function of the absolute temperature.

The work produced in the cycle is $W = Q + Q_0$, which, using Eq. (8–16), becomes

$$W = a\theta + Q_0. \tag{8–17}$$

Now if the high-temperature reservoir is cooled until it reaches θ_0, the temperature of the cold reservoir, then the cycle becomes an isothermal cycle, and no work can be produced. Since it is a reversible cycle, $W = 0$, and so $0 = a\theta_0 + Q_0$; hence, $Q_0 = -a\theta_0$. Then Eq. (8–17) becomes

$$W = a(\theta - \theta_0). \tag{8–18}$$

For the efficiency we obtain

$$\epsilon = \frac{\theta - \theta_0}{\theta}.$$
(8–19)

Since there is nothing special about the temperature of the cold reservoir, except that $\theta > \theta_0$, Eqs. (8–18) and (8–19) apply to any reversible heat engine operating between any two thermodynamic temperatures θ and θ_0. Equation (8–18) shows that the work produced in a reversible heat engine is directly proportional to the difference in temperatures on the thermodynamic scale, while the efficiency is equal to the ratio of the difference in temperature to the temperature of the hot reservoir. The Carnot formula, Eq. (8–19), which relates the efficiency of a reversible engine to the temperatures of the reservoirs is probably the most celebrated formula in all of thermodynamics.

Lord Kelvin was the first to define the absolute temperature scale, named in his honor, from the properties of reversible engines. If we fix the size of the degree on the kelvin scale so that one-hundred degrees lie between the ice point and the steam point, and adjust the proportionality constant a in Eq. (8–16) to conform with the ordinary definition of one mole of an ideal gas, then the ideal gas scale of temperature and the kelvin scale become *numerically* identical. However, the kelvin scale is the fundamental one. From now on we will use T for the thermodynamic temperature, $\theta = T$, except where the use of θ can supply needed emphasis.

Once one value of the thermodynamic temperature has been assigned a positive value, all other temperatures must be positive, otherwise in some circumstances the Q's for the two reservoirs would have the same sign. This would result, as we have seen, in perpetual motion.

8–9 Retrospection

From the characteristics of a particularly simple kind of heat engine, the Carnot engine, and from the postulate that certain kinds of engine cannot be constructed, we concluded that the efficiencies of all reversible heat engines operating between the same two reservoirs is the same, and depends only on the temperature of the reservoirs. Thus, it was possible to establish the thermodynamic scale of temperature, which is independent of the properties of any individual substance, and to relate the efficiency of the engine to the temperatures on this scale:

$$\epsilon = \frac{\theta_1 - \theta_2}{\theta_1} = \frac{T_1 - T_2}{T_1},$$

where $\theta_1 = T_1$ is the temperature of the hot reservoir.

The second law has been stated in the sense that it is impossible for an engine operating in a cycle and connected to a reservoir at only one temperature to produce a positive amount of work in the surroundings. This is equivalent to the Kelvin-Planck statement of the second law. The possibility of another kind of engine is also denied. It is impossible for an engine operating in a cycle to have as its *only* effect the transfer of a quantity of heat from a reservoir of low temperature to a reservoir at a higher temperature. This is the content of the Clausius statement of the second law. Both engines are perpetual motion machines of the second kind. If it were possible to build one of them, the other could be built. (The proof of the equivalence is left as an exercise, Problem 8–1.) The Kelvin-Planck statement and the Clausius statement of the second law of thermodynamics are, of course, completely equivalent.

In this study of thermodynamic engines, our hope has been to arrive at the definition of some state property the variation of which in a given change in state would yield a clue as to whether the change in state was a real or natural change. We have arrived at the brink of that definition, but we will first look at the Carnot cycle using an ideal gas as the working substance, and also describe the operation of the Carnot refrigerator.

8–10   Carnot Cycle with an Ideal Gas

If an ideal gas is used as the working substance in a Carnot engine, the application of the first law to each of the steps in the cycle can be written in the scheme of Table 8–3. The values of W_1 and W_3, which are quantities of work produced in an isothermal reversible expansion of an ideal gas, are obtained from Eq. (7–6). The values of ΔE are computed by integrating the equation $dE = C_v\, dT$. The total work produced in the cycle is the sum of the individual quantities:

$$W = RT_1 \ln\left(\frac{V_2}{V_1}\right) - \int_{T_1}^{T_2} C_v\, dT + RT_2 \ln\left(\frac{V_4}{V_3}\right) - \int_{T_2}^{T_1} C_v\, dT.$$

The two integrals sum to zero, as can be shown by interchanging the limits and thus changing the sign of either of them. Hence

$$W = RT_1 \ln\left(\frac{V_2}{V_1}\right) - RT_2 \ln\left(\frac{V_3}{V_4}\right), \qquad (8\text{–}20)$$

Table 8–3

Step no.	General case	Ideal gas
1	$\Delta E_1 = Q_1 - W_1$	$0 = Q_1 - RT_1 \ln (V_2/V_1)$
2	$\Delta E_2 = -W_2$	$\int_{T_1}^{T_2} C_v\, dT = -W_2$
3	$\Delta E_3 = Q_2 - W_3$	$0 = Q_2 - RT_2 \ln (V_4/V_3)$
4	$\Delta E_4 = -W_4$	$\int_{T_2}^{T_1} C_v\, dT = -W_4$

where the sign of the second term has been changed by inverting the argument of the logarithm.

Equation (8–20) can be simplified by realizing that the volumes V_2 and V_3 are connected by an adiabatic reversible transformation; the same is true for V_4 and V_1. By Eq. (7–57),

$$T_1 V_2^{\gamma-1} = T_2 V_3^{\gamma-1}, \qquad T_1 V_1^{\gamma-1} = T_2 V_4^{\gamma-1}.$$

By dividing the first equation by the second, we obtain

$$\left(\frac{V_2}{V_1}\right)^{\gamma-1} = \left(\frac{V_3}{V_4}\right)^{\gamma-1} \quad \text{or} \quad \frac{V_2}{V_1} = \frac{V_3}{V_4}.$$

Putting this result in Eq. (8–20), we obtain

$$W = R(T_1 - T_2)\ln\left(\frac{V_2}{V_1}\right). \tag{8–21}$$

From the equation for the first step in the cycle, we have

$$Q_1 = RT_1 \ln\left(\frac{V_2}{V_1}\right),$$

and the efficiency is given by

$$\epsilon = \frac{W}{Q_1} = \frac{T_1 - T_2}{T_1} = 1 - \frac{T_2}{T_1}. \tag{8–22}$$

Equation (8–21) shows that the total work produced depends on the difference in temperature between the two reservoirs [compare Eq. (8–18)] and the volume ratio V_2/V_1 (the compression ratio). The efficiency is a function only of the two temperatures [compare Eq. (8–19)]. It is apparent from Eq. (8–22) that if the efficiency is to be unity, either the cold reservoir must be at $T_2 = 0$, or the hot reservoir must have T_1 equal to infinity. Neither situation is physically realizable.

8–11 The Carnot Refrigerator

If a reversible heat engine operates so as to produce a positive amount of work in the surroundings, then a positive amount of heat is extracted from the hot reservoir and heat is rejected to the cold reservoir. Suppose we call this the forward cycle of the engine. If the engine is reversed, the signs of all the quantities of heat and work are reversed. Work is destroyed, $W < 0$; heat is withdrawn from the cold reservoir and rejected to the hot reservoir. In this reverse cycle, by destroying work, heat is pumped from a cold reservoir to a hot reservoir; the machine is a refrigerator. Note that the refrigerator is quite different from our impossible engine which pumped heat from a cold end to a hot end of the machine. The impossible engine did not destroy work in the process, as a proper refrigerator would. The signs of the

Table 8–4

Cycle	Q_1	Q_2	W
Forward	+	−	+
Reverse	−	+	−

quantities of work and heat in the two modes of operation are shown in Table 8–4 (T_1 is the higher temperature).

The "coefficient of performance" η of a refrigerator is the ratio of the heat extracted from the low-temperature reservoir to the work destroyed:

$$\eta = \frac{Q_2}{-W} = \frac{Q_2}{-(Q_1 + Q_2)}, \tag{8–23}$$

since $W = Q_1 + Q_2$. Also, since $(Q_2/Q_1) = -(T_2/T_1)$, we obtain

$$\eta = \frac{T_2}{T_1 - T_2}. \tag{8–24}$$

The coefficient of performance is the heat extracted from the cold box for each unit of work expended. From Eq. (8–24) it is apparent that as T_2, the temperature inside the cold box, gets smaller, the coefficient of performance drops off very rapidly; this happens because the numerator in Eq. (8–24) decreases and the denominator increases. The amount of work which must be expended to maintain a cold temperature against a specified heat leak into the box goes up very rapidly as the temperature of the box goes down.

8–12 Definition of the Entropy

Just as the first law led to the definition of the energy, so also the second law leads to a definition of a state property of the system, the entropy. It is characteristic of a state property that the sum of the changes of that property in a cycle is zero. For example, the sum of changes in energy of a system in a cycle is given by $\oint dE = 0$. From the second law we must find some new quantity whose changes sum to zero in a cycle.

We begin by comparing two expressions for the efficiency of a simple reversible heat engine which operates between the two reservoirs at the thermodynamic temperatures θ_1 and θ_2. We have seen that

$$\epsilon = 1 + \frac{Q_2}{Q_1} \quad \text{and} \quad \epsilon = 1 - \frac{\theta_2}{\theta_1}.$$

Subtracting these two expressions yields the result

$$\frac{Q_2}{Q_1} + \frac{\theta_2}{\theta_1} = 0,$$

which can be rearranged to the form

$$\frac{Q_1}{\theta_1} + \frac{Q_2}{\theta_2} = 0. \tag{8–25}$$

The left-hand side of Eq. (8–25) is simply the sum over the cycle of the quantity Q/θ. It could be written as the cyclic integral of the differential quantity dQ/θ:

$$\oint \frac{dQ}{\theta} = 0 \quad \text{(reversible cycles).} \tag{8–26}$$

Since the sum over the cycle of the quantity dQ/θ is zero, this quantity is the differential of some property of state; this property is called the *entropy* of the system and is given the symbol S. The defining equation for the entropy is then

$$dS \equiv \frac{dQ_{rev}}{T}, \tag{8–27}$$

where the subscript "rev" has been used to indicate the restriction to reversible cycles. The symbol θ for the thermodynamic temperature has been replaced by the more usual symbol T. Note that while dQ_{rev} is not the differential of a state property, dQ_{rev}/T is; dQ_{rev}/T is an *exact* differential.

8–13 General Proof

We have shown that dQ_{rev}/T has a cyclic integral equal to zero only for cycles which involve only two temperatures. The result can be generalized to any cycle whatsoever.

Consider a Carnot engine. Then in a cycle

$$W = \oint dQ, \tag{8–28}$$

and we have shown for the Carnot engine that

$$\oint \frac{dQ}{T} = 0. \tag{8–29}$$

(By the definition of the Carnot cycle, the Q is a reversible Q.) Consider another engine E'. Then in a cycle, by the first law,

$$W' = \oint dQ'; \tag{8–30}$$

but let us assume that for this engine,

$$\oint \frac{dQ'}{T} > 0. \tag{8–31}$$

This second engine may execute as complicated a cycle as we please; it may have many temperature reservoirs; it may use any working substance.

The two engines are coupled together to make a composite cyclic engine. The work produced by the composite engine in its cycle is $W_c = W + W'$, which, by Eqs. (8–28) and (8–30), is equal to

$$W_c = \oint (dQ + dQ') = \oint dQ_c \tag{8–32}$$

where $dQ_c = dQ + dQ'$.

If we add Eqs. (8–29) and (8–31), we obtain

$$\oint \frac{(dQ + dQ')}{T} > 0,$$

$$\oint \frac{dQ_c}{T} > 0. \tag{8–33}$$

We now adjust the direction of operation and the size of the Carnot engine so that the composite engine produces no work; the work required to operate E' is supplied by the Carnot engine, or vice versa. Then, $W_c = 0$, and Eq. (8–32) becomes

$$\oint dQ_c = 0. \tag{8–34}$$

Under what condition will the relations Eqs. (8–33) and (8–34) be compatible?

Since each of the cyclic integrals can be considered as a sum of terms, we write Eqs. (8–34) and (8–33) in the forms

$$Q_1 + Q_2 + Q_3 + Q_4 + \cdots = 0, \tag{8–35}$$

$$\frac{Q_1}{T_1} + \frac{Q_2}{T_2} + \frac{Q_3}{T_3} + \frac{Q_4}{T_4} + \cdots > 0. \tag{8–36}$$

The sum on the left-hand side of Eq. (8–35) consists of a number of terms, some positive and some negative. But the positive ones just balance the negative ones, and the sum is zero. We have to find numbers (temperatures) such that by dividing each term in Eq. (8–35) by a proper number we can obtain a sum in which the positive terms predominate, and thus fulfill the requirement of the inequality (8–36). We can make the positive terms predominate if we divide the positive terms in Eq. (8–35) by small numbers and the negative terms by larger numbers. However, this means that we are associating positive values of Q with low temperatures and negative values with high temperatures. This implies that heat is extracted from reservoirs at low temperatures and rejected to reservoirs at higher temperatures in the operation of the composite engine. The composite engine is consequently an impossible engine, and our assumption, Eq. (8–31), must be incorrect. It follows that for any engine E',

$$\oint \frac{dQ'}{T} \leqslant 0. \tag{8–37}$$

We distinguish two cases:

Case I. The engine E' is reversible.
    We have excluded the possibility expressed by (8–31). If we assume that for E'

$$\oint \frac{dQ'}{T} < 0,$$

then we can reverse this engine, which changes all the signs but not the magnitudes of the Q's. Then we have

$$\oint \frac{dQ'}{T} > 0,$$

and the proof is the same as before. This forces us to the conclusion that for any system whatsoever,

$$\oint \frac{dQ_{\text{rev}}}{T} = 0 \quad \text{(all reversible cycles).} \tag{8–38}$$

Therefore, every system has a state property S, the entropy, such that

$$dS = \frac{dQ_{\text{rev}}}{T}. \tag{8–39}$$

The study of the properties of the entropy will be undertaken in the next chapter.

Case II. The engine E' is not reversible.
    For any engine we have only the possibilities expressed by (8–37). We have shown that the equality holds for the reversible engine. Since the heat and work effects associated with an irreversible cycle are different from those associated with a reversible cycle, this implies that the value of $\oint dQ/T$ for an irreversible cycle is different from the value, zero, associated with the reversible cycle. We have shown that for any engine the value cannot be greater than zero; consequently, it must be less than zero. Therefore, for irreversible cycles we must have

$$\oint \frac{dQ}{T} < 0 \quad \text{(all irreversible cycles).} \tag{8–40}$$

8–14 The Clausius Inequality

Consider the following cycle: a system is transformed irreversibly from state 1 to state 2, then restored reversibly from state 2 to state 1. The cyclic integral is

$$\oint \frac{dQ}{T} = \int_1^2 \frac{dQ_{\text{irr}}}{T} + \int_2^1 \frac{dQ_{\text{rev}}}{T} < 0,$$

and it is less than zero, by (8–40), since it is an irreversible cycle. Using the definition of dS, this relation becomes

$$\int_1^2 \frac{dQ_{\text{irr}}}{T} + \int_2^1 dS < 0.$$

The limits can be changed on the second integral (but not on the first!) by changing the sign. Thus we have

$$\int_1^2 \frac{dQ_{\text{irr}}}{T} - \int_1^2 dS < 0,$$

or, by rearranging, we have

$$\int_1^2 dS > \int_1^2 \frac{dQ_{\text{irr}}}{T}. \tag{8–41}$$

If the change in state from state 1 to state 2 is an infinitesimal one, we have

$$dS > \frac{dQ_{\text{irr}}}{T}, \tag{8–42}$$

the Clausius inequality, which is a fundamental requirement for a real transformation. The inequality (8–42) enables us to decide whether or not some proposed transformation will occur in nature. We will not ordinarily use (8–42) just as it stands but will manipulate it to express the inequality in terms of properties of the state of a system, rather than in terms of a path property such as dQ_{irr}.

The Clausius inequality can be applied directly to changes in an isolated system. For any change in state in an isolated system, $dQ_{\text{irr}} = 0$. The inequality then becomes

$$dS > 0. \tag{8–43}$$

The requirement for a real transformation in an isolated system is that dS be positive; the entropy must increase. Any natural change occurring within an isolated system is attended by an increase in entropy of the system. The entropy of an isolated system continues to increase so long as changes occur within it. When the changes cease, the system is in equilibrium and the entropy has reached a maximum value. Therefore, the condition of equilibrium *in an isolated system* is that the entropy have a maximum value.

These, then, are also fundamental properties of the entropy: (1) the entropy of an isolated system is increased by any natural change which occurs within it; and

(2) the entropy of an isolated system has a maximum value at equilibrium. Changes in a nonisolated system produce effects in the system and in the immediate surroundings. The system and its immediate surroundings constitute a composite isolated system in which the entropy increases as natural changes occur within it. Thus, in the universe the entropy increases continually as natural changes occur within it.

Clausius expressed the two laws of thermodynamics in the famous aphorism: "The energy of the universe is constant; the entropy strives to reach a maximum."

8-15 Conclusion

By what may seem a rather long route, the existence of a property of a system, the entropy, has been demonstrated. The existence of this property is a consequence of the second law of thermodynamics. The zeroth law defined the temperature of a system; the first law, the energy; and the second law, the entropy. Our interest in the second law stems from the fact that this law has something to say about the natural direction of a transformation. It denounces a machine that causes heat to flow from a cold to a hot reservoir without any other effect. In the same way, the second law will denounce the "unnatural" direction of a chemical reaction. In some situations, the second law declares that neither direction of the chemical reaction is natural; the reaction must then be at equilibrium. The application of the second law to chemical reactions is the most fruitful approach to the subject of chemical equilibrium. Fortunately, this application can be made quite easily and is done without interminable combinations of cyclic engines.

Problems

8-1. a) Consider the impossible engine which is connected to only one heat reservoir and produces net work in the surroundings. Couple this impossible engine to an ordinary Carnot engine in such a way that the composite engine is the "stove–refrigerator."

   b) Couple the "stove–refrigerator" to an ordinary Carnot engine in such a way that the composite engine produces work in an isothermal cycle.

8-2. What is the maximum possible efficiency of a heat engine which has a hot reservoir of water boiling under pressure at 125°C and a cold reservoir at 25°C?

8-3. A refrigerator is operated by a $\frac{1}{4}$-hp motor (1 hp = 10.688 kcal/min). If the interior of the box is to be maintained at $-20°C$ against a maximum exterior temperature of 35°C, what is the maximum heat leak into the box (cal/min) which can be tolerated if the motor runs continuously. Assume that the coefficient of performance is 75% of the value for a reversible engine.

8-4. a) Suppose we choose the efficiency of a reversible engine as the thermometric property for a thermodynamic temperature scale. Let the cold reservoir have a fixed temperature. Measure the efficiency of the engine with the hot reservoir at the ice point, 0°, and with the hot reservoir at the steam point, 100°. What is the relation between temperatures, t on this scale and the usual thermodynamic temperatures T?

b) Suppose the hot reservoir has a fixed temperature and we define the temperature scale by measuring efficiency with the cold reservoir at the steam point and at the ice point. Find the relation between t and T for this case. (Choose $100°$ between the ice point and the steam point.)

8-5. a) Liquid helium boils at about $4°K$, and liquid hydrogen boils at about $20°K$. What is the efficiency of a reversible engine operating between heat reservoirs at these temperatures?

b) If we wanted the same efficiency as in (a) for an engine with a cold reservoir at ordinary temperature, $300°K$; what must the temperature of the hot reservoir be?

8-6. Consider the following cycle using 1 mole of an ideal gas, initially at $25°C$ and 1 atm pressure.

Step 1. Isothermal expansion against zero pressure to double the volume (Joule expansion).

Step 2. Isothermal, reversible compression from $\frac{1}{2}$ atm to 1 atm.

a) Calculate the value of $\oint dQ/T$; note that the sign conforms with (8–40).

b) Calculate ΔS for Step 2.

c) Realizing that for the cycle, $\Delta S_{cycle} = 0$, find ΔS for Step 1.

d) Show that ΔS for Step 1 is *not* equal to the Q for Step 1 divided by T.

Chapter Nine

Properties of the Entropy and the Third Law of Thermodynamics

9–1 The Properties of Entropy

Each year the question, "What is entropy?", echoes plaintively in physical chemistry classrooms. The questioner rarely regards the answer given as a satisfactory one. The question springs from a strange feeling most people have that entropy is something they can see or feel or put in a bottle, if only they could squint at the system from the proper angle. The difficulty arises for two reasons. Firstly, it must be admitted that entropy is a more impalpable thing than a quantity of heat or work. Secondly, the question itself is vague; unintentionally, of course. Sleepless nights can be saved if, at least for the present, we simply ignore the vague question, "What is entropy?", and consider precise questions and statements about entropy. How does the entropy change with temperature under constant pressure? How does the entropy change with volume at constant temperature? If we know how the entropy behaves in various circumstances, we will know a great deal about what it "is." Later, the entropy will be related to "randomness" in a spatial or energy distribution of the constituent particles. However, this relation to "randomness" depends upon the assumption of a structural model for a system, while the purely thermodynamic definition is independent of any structural model and, in fact, does not require such a model. The entropy is defined by the differential equation

$$dS = dQ_{rev}/T, \tag{9-1}$$

from which it follows that the entropy is a *single-valued*, *extensive* state property of the system. The differential dS is an *exact* differential. For a finite change in state from state 1 to state 2, we have from Eq. (9–1)

$$\Delta S = S_2 - S_1 = \int_1^2 dQ_{rev}/T. \tag{9–2}$$

Since the values of S_2 and S_1 depend only on the states 1 and 2, it does not matter in the least whether the change in state is *effected* by a reversible process or an irreversible process; ΔS is the same regardless. However, if we use Eq. (9–2) to *calculate* ΔS, we must use the heat withdrawn along any *reversible* path connecting the two states.

9–2 Conditions of Thermal and Mechanical Stability of a System

Before beginning a detailed discussion of the properties of the entropy, two facts must be established. The first is that the heat capacity at constant volume C_v is always positive for a pure substance in a single state of aggregation; the second is that the coefficient of compressibility β is always positive for such a substance. Although each of these statements is capable of elegant mathematical proof from the second law, a simple physical argument will be convincing enough for our purposes.

Suppose that for the system specified, C_v is negative, and that the system is kept at constant volume. If a warm draft strikes the system, an amount of heat, $dQ_V = +$, flows from the surroundings; by definition, $dQ_V = C_v dT$. Since dQ_V is positive, and by supposition C_v is negative, dT would have to be negative to fulfill this relation. Thus, the flow of heat into this system lowers its temperature, which causes more heat to flow in, and the system cools even more. Ultimately, the system would get very cold for no reason but that an accidental draft struck it. By the same argument, an accidental cold draft would result in the system getting extremely hot. It would be too distressing to have objects in a room glowing red hot and freezing up just because of drafts. Therefore C_v must be positive to ensure the thermal stability of a system against chance variations in external temperature.

The coefficient of compressibility has been defined, Eq. (5–4), as

$$\beta = -\frac{1}{V}\left(\frac{\partial V}{\partial p}\right)_T; \tag{9–3}$$

thus at constant temperature $dp = -(dV/V\beta)$. Suppose that at constant temperature the system is accidentally pushed in a little bit, dV is then negative. If β is negative, to fulfill the relation dp must be negative. The pressure in the system goes down, which allows the external pressure to push the system in a little more, which lowers the pressure further. The system would collapse. If the volume of the system were accidentally increased, the system would explode. We conclude that β must be positive if the system is to be mechanically stable against accidental variations in its volume.

9–3 Entropy Changes in Isothermal Transformations

For any isothermal change in state, T, being constant, can be removed from the integral in Eq. (9–2), which then reduces immediately to

$$\Delta S = Q_{\text{rev}}/T. \qquad (9-4)$$

The entropy change for the transformation can be calculated by evaluating the quantities of heat required to conduct the change in state reversibly. Since quantities of heat are usually expressed in calories, the entropy unit (eu) is calories per degree (1 eu = 1 cal/degree).

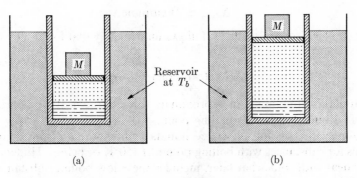

Fig. 9–1 Reversible vaporization of a liquid.

Equation (9–4) is used to calculate the entropy change associated with a change in state of aggregation at the equilibrium temperature. Consider a liquid in equilibrium with its vapor under a pressure of 1 atm. The temperature is the equilibrium temperature, the normal boiling point of the liquid. Imagine that the system is confined in a cylinder by a floating piston carrying a weight equivalent to the 1 atm pressure; Fig. 9–1(a). The cylinder is immersed in a temperature reservoir at the equilibrium temperature T_b. If the temperature of the reservoir is raised infinitesimally, a small quantity of heat flows from the reservoir to the system, some liquid vaporizes, and the mass M rises; Fig. 9–1(b). If the temperature of the reservoir is lowered infinitesimally, the same quantity of heat flows back to the reservoir. The vapor formed originally condenses, and the mass falls to its original position. Both the system and the reservoir are restored to their initial condition in this small cycle, and the transformation is reversible; the quantity of heat required is a Q_{rev}. The pressure is constant, so that $Q_p = \Delta H$; hence for the vaporization of a liquid at the boiling point, Eq. (9–4) becomes

$$\Delta S_{\text{vap}} = \Delta H_{\text{vap}}/T_b. \qquad (9-5)$$

By the same argument, the entropy of fusion at the melting point is given by

$$\Delta S_{\text{fus}} = \Delta H_{\text{fus}}/T_m, \qquad (9-6)$$

where ΔH_{fus} is the heat of fusion at the melting point T_m. For any change of phase at the equilibrium temperature T_e, the entropy of transition is given by

$$\Delta S = \Delta H / T_e, \tag{9-7}$$

where ΔH is the heat of transition at T_e.

9–4 Trouton's Rule

For many liquids, the entropy of vaporization at the normal boiling point has approximately the same value:

$$\Delta S_{vap} \approx 21 \text{ eu/mole.} \tag{9-8}$$

Equation (9–8) is Trouton's rule. It follows immediately that for liquids which obey this rule,

$$\Delta H_{vap} \text{ (cal/mole)} = 21 T_b, \tag{9-9}$$

which is useful for obtaining an approximate value of the heat of vaporization of a liquid from a knowledge of its boiling point.

Trouton's rule fails for associated liquids such as water, alcohols, and amines. It also fails for substances with boiling points of 150°K or below. Hildebrand's rule, which we deal with somewhat later, includes these low-boiling substances, but not associated liquids.

There is no equally general rule for entropies of fusion at the melting point. For most substances the entropy of fusion is much less than the entropy of vaporization, lying usually in the range from 2 eu/mole to at most 9 eu/mole. If the particles composing the substances are atoms, such as in the metals, the entropy of fusion is about 2 eu/mole. If the molecule composing the substance is quite large, a long chain hydrocarbon for example, the entropy of fusion may be as high as 30 eu/mole.

9–5 Relation of the Entropy Changes to Changes in the Other Properties of the System

The defining equation for the entropy,

$$dS = dQ_{rev}/T, \tag{9-10}$$

relates the change in entropy to an effect, dQ_{rev}, in the surroundings. It would be useful to transform this equation so as to relate the change in entropy to changes in value of properties of the system. This is quite easily done.

If only pressure–volume work is done, then in a reversible transformation, we have $P_{op} = p$, the pressure of the system, so that the first law becomes

$$dQ_{rev} = dE + p \, dV. \tag{9-11}$$

Dividing Eq. (9–11) by T and using the definition of dS, we obtain

$$dS = \frac{1}{T} dE + \frac{p}{T} dV, \tag{9–12}$$

which relates the change in entropy dS to changes in energy and volume, dE and dV, and to the pressure and temperature of the system. Equation (9–12), a combination of the first and second laws of thermodynamics, is the fundamental equation of thermodynamics; all our discussions of the equilibrium properties of a system will begin from this equation or equations immediately related to it.

For the present, it is sufficient to remark that both of the differential coefficients $1/T$ and p/T are always positive. According to Eq. (9–12) there are two independent ways of varying the entropy of a system: by varying the energy or the volume. Note carefully that if the volume is constant ($dV = 0$), an increase in energy (dE is +) implies an increase in entropy. Also, if the energy is constant ($dE = 0$), an increase in volume (dV is +) implies an increase in entropy. This behavior is a fundamental characteristic of the entropy. At constant volume, the entropy goes up as the energy goes up. At constant energy, the entropy goes up as the volume goes up.

In the laboratory we do not ordinarily exercise control of the energy of the system directly. Since we can conveniently control the temperature and volume, or the temperature and pressure, it is useful to transform Eq. (9–12) to the more convenient sets of variables, T and V_g or T and p.

9–6 Mathematical Interlude. More Properties of Exact Differentials. The Cyclic Rule

The total differential of a function of two variables $f(x, y)$ is written in the form

$$df = \frac{\partial f}{\partial x} dx + \frac{\partial f}{\partial y} dy. \tag{9–13}$$

Since the differential coefficients $(\partial f/\partial x)$ and $(\partial f/\partial y)$ are functions of x and y, we may write

$$M(x, y) = \frac{\partial f}{\partial x}, \qquad N(x, y) = \frac{\partial f}{\partial y}, \tag{9–14}$$

and Eq. (9–13) becomes

$$df = M(x, y)\, dx + N(x, y)\, dy. \tag{9–15}$$

If we form second derivatives of the function $f(x, y)$, there are several possibilities: $(\partial f/\partial x)$ can be differentiated with respect to either x or y, and the same is true of $(\partial f/\partial y)$. We get

$$\frac{\partial^2 f}{\partial x^2}, \qquad \frac{\partial^2 f}{\partial y\, \partial x}, \qquad \frac{\partial^2 f}{\partial x\, \partial y}, \qquad \frac{\partial^2 f}{\partial y^2}.$$

Of these four, only three are distinct. It can be shown that for a function of several variables, the order of differentiation with respect to two variables such as x and y does not matter and the mixed derivatives are equal; that is,

$$\frac{\partial^2 f}{\partial y\, \partial x} = \frac{\partial^2 f}{\partial x\, \partial y}. \tag{9–16}$$

Differentiating the first of Eqs. (9–14) with respect to y, and the second with respect to x, we obtain

$$\frac{\partial M}{\partial y} = \frac{\partial^2 f}{\partial y\, \partial x}, \qquad \frac{\partial N}{\partial x} = \frac{\partial^2 f}{\partial x\, \partial y}.$$

These two equations in the light of Eq. (9–16) yield

$$\frac{\partial M}{\partial y} = \frac{\partial N}{\partial x}. \tag{9–17}$$

The derivatives in Eq. (9–17) are sometimes called "cross-derivatives" because of their relation to the total differential, Eq. (9–15):

$$df = M\, dx + N\, dy.$$

(In all the above equations, the subscript on the derivatives denoting constancy of x or y has been dropped to simplify the writing.)

Applying the cross-derivative rule in Eq. (9–17) to the fundamental equation (9–12), in the form

$$dE = T\, dS - p\, dV,$$

we obtain

$$\left(\frac{\partial T}{\partial V}\right)_S = \left[\frac{\partial(-p)}{\partial S}\right]_V,$$

$$\left(\frac{\partial T}{\partial V}\right)_S = -\left(\frac{\partial p}{\partial S}\right)_V. \tag{9–18}$$

Equation (9–18) is one of an important group of equations called the Maxwell relations; its meaning will be discussed later along with that of the other members of the group. The equality of the cross-derivatives is used frequently in later arguments.

The rule in Eq. (9–17) follows from the fact that the differential expression $M\, dx + N\, dy$ is the total differential of some function $f(x, y)$; that is, $M\, dx + N\, dy$ is an exact differential expression. The converse is also true. For example, suppose that we have an expression of the form

$$R(x, y)\, dx + Q(x, y)\, dy. \tag{9–19}$$

This is an exact differential expression *if and only if*

$$\frac{\partial R}{\partial y} = \frac{\partial Q}{\partial x}. \tag{9–20}$$

If Eq. (9–20) is satisfied, then there exists some function of x and y, $g(x, y)$, for which

$$dg = R\,dx + Q\,dy.$$

If Eq. (9–20) is not satisfied, then no such function $g(x, y)$ exists, and the differential expression (9–19) is an inexact differential.

The cyclic rule

Another useful relation between partial derivatives is the cyclic rule. The total differential of a function $z(x, y)$ is written

$$dz = \left(\frac{\partial z}{\partial x}\right)_y dx + \left(\frac{\partial z}{\partial y}\right)_x dy. \qquad (9\text{–}21)$$

We now restrict Eq. (9–21) to variations of x and y which leave the value of z unchanged; $dz = 0$:

$$0 = \left(\frac{\partial z}{\partial x}\right)_y (\partial x)_z + \left(\frac{\partial z}{\partial y}\right)_x (\partial y)_z.$$

Dividing by $(\partial y)_z$, we have

$$0 = \left(\frac{\partial z}{\partial x}\right)_y \left(\frac{\partial x}{\partial y}\right)_z + \left(\frac{\partial z}{\partial y}\right)_x.$$

Multiplying by the reciprocal of the second term, $(\partial y/\partial z)_x$, we obtain

$$0 = \left(\frac{\partial z}{\partial x}\right)_y \left(\frac{\partial x}{\partial y}\right)_z \left(\frac{\partial y}{\partial z}\right)_x + 1.$$

A slight rearrangement brings this to

$$\left(\frac{\partial x}{\partial y}\right)_z \left(\frac{\partial y}{\partial z}\right)_x \left(\frac{\partial z}{\partial x}\right)_y = -1, \qquad (9\text{–}22)$$

which is the cyclic rule. The variables x, y, z in the numerators are related to y, z, x in the denominators and to the subscripts z, x, y by a cyclic permutation. If any three variables are connected by a functional relationship, then the three partial derivatives satisfy a relation of the type of Eq. (9–22). Since in many thermodynamic situations, the variables of state are functions of two other variables, Eq. (9–22) has frequent application. The lovely part of an equation such as Eq. (9–22) is that we do not have to memorize it. Write the three variables down in any order, x, y, z, then underneath them write the variables again in any order so that the vertical columns do not match; there are only two possibilities:

$$xyz, \qquad xyz,$$

$$yzx, \qquad zxy.$$

The first row yields the numerators of the derivatives, the second row the denominators; the subscripts are easily obtained, since in any derivative the same symbol does not occur twice. From the diagrams we write

$$\left(\frac{\partial x}{\partial y}\right)_z \left(\frac{\partial y}{\partial z}\right)_x \left(\frac{\partial z}{\partial x}\right)_y = -1 \quad \text{and} \quad \left(\frac{\partial x}{\partial z}\right)_y \left(\frac{\partial y}{\partial x}\right)_z \left(\frac{\partial z}{\partial y}\right)_x = -1.$$

The first expression is Eq. (9–22), the second is the reciprocal of Eq. (9–22); since the reciprocal of -1 is also -1, it is almost impossible to write this equation incorrectly.

Suppose that the three variables are pressure, temperature, and volume. We write the cyclic rule using the variables p, T, V:

$$\left(\frac{\partial p}{\partial T}\right)_V \left(\frac{\partial T}{\partial V}\right)_p \left(\frac{\partial V}{\partial p}\right)_T = -1.$$

From the definitions of the coefficient of thermal expansion and the coefficient of compressibility, we have

$$\left(\frac{\partial V}{\partial T}\right)_p = V\alpha \quad \text{and} \quad \left(\frac{\partial V}{\partial p}\right)_T = -V\beta.$$

Using the definitions of α and β, the cyclic rule becomes

$$\left(\frac{\partial p}{\partial T}\right)_V \frac{1}{V\alpha}(-V\beta) = -1,$$

so that

$$\left(\frac{\partial p}{\partial T}\right)_V = \frac{\alpha}{\beta}. \tag{9–23}$$

With the cross-derivative rule and the cyclic reaction at our disposal, we are ready to manipulate the equations of thermodynamics into useful forms.

9–7 Entropy as a Function of Temperature and Volume

Considering the entropy as a function of T and V, we have $S = S(T, V)$; the total differential is written as

$$dS = \left(\frac{\partial S}{\partial T}\right)_V dT + \left(\frac{\partial S}{\partial V}\right)_T dV. \tag{9–24}$$

Equation (9–12) can be brought into the form of Eq. (9–24) if we express dE in terms of dT and dV. In these variables,

$$dE = C_v\, dT + \left(\frac{\partial E}{\partial V}\right)_T dV. \tag{9–25}$$

Using this value of dE in Eq. (9–12), we have

$$dS = \frac{C_v}{T} dT + \frac{1}{T}\left[p + \left(\frac{\partial E}{\partial V}\right)_T \right] dV. \tag{9–26}$$

Since Eq. (9–15) expresses the change in entropy in terms of changes in T and V, it must be identical with Eq. (9–24), which does the same thing. In view of this identity, we may write

$$\left(\frac{\partial S}{\partial T}\right)_V = \frac{C_v}{T}, \tag{9–27}$$

$$\left(\frac{\partial S}{\partial V}\right)_T = \frac{1}{T}\left[p + \left(\frac{\partial E}{\partial V}\right)_T \right]. \tag{9–28}$$

Since C_v/T is always positive (Section 9–2), Eq. (9–27) expresses the important fact that at constant volume the entropy increases with increase in temperature. Note that the dependence of entropy on temperature is simple, the differential coefficient being the appropriate heat capacity divided by the temperature. For a finite change in temperature at constant volume

$$\Delta S = \int_{T_1}^{T_2} \frac{C_v}{T} dT. \tag{9–29}$$

Example. One mole of argon is heated at a constant volume from 300°K to 500°K; $C_v = 3 \text{ cal·deg}^{-1}\text{·mole}^{-1}$. Compute the change in entropy for this change in state.

Solution.

$$\Delta S = \int_{300}^{500} \frac{3}{T} dT = 3 \ln \frac{500}{300} = 3(0.510) \text{ eu/mole} = 1.53 \text{ eu/mole}.$$

Note that if two moles were used, C_v would be doubled and so the entropy change would be doubled.

In contrast to the simplicity of the temperature dependence, the volume dependence *at constant temperature* given by Eq. (9–28) is quite complicated. Remember that the volume dependence *at constant energy*, Eq. (9–12), was very simple. We can obtain a simpler expression for the isothermal volume dependence of the entropy by the following device. We differentiate Eq. (9–27) with respect to volume, keeping temperature constant; this yields

$$\frac{\partial^2 S}{\partial V \partial T} = \frac{1}{T}\frac{\partial C_v}{\partial V} = \frac{1}{T}\frac{\partial^2 E}{\partial V \partial T}.$$

In the right-hand side we have replaced C_v by $(\partial E/\partial T)_V$. Similarly, we differentiate Eq. (9–28) with respect to temperature keeping volume constant, to obtain

$$\frac{\partial^2 S}{\partial T \partial V} = \frac{1}{T}\left[\left(\frac{\partial p}{\partial T}\right)_V + \frac{\partial^2 E}{\partial T \partial V} \right] - \frac{1}{T^2}\left[p + \left(\frac{\partial E}{\partial V}\right)_T \right].$$

However, since S is a function of T and V (dS is an exact differential) the mixed second derivatives must be equal; hence we have

$$\frac{\partial^2 S}{\partial V \partial T} = \frac{\partial^2 S}{\partial T \partial V},$$

or

$$\frac{1}{T}\left(\frac{\partial^2 E}{\partial V \partial T}\right) = \frac{1}{T}\left(\frac{\partial p}{\partial T}\right)_V + \frac{1}{T}\left(\frac{\partial^2 E}{\partial T \partial V}\right) - \frac{1}{T^2}\left[p + \left(\frac{\partial E}{\partial V}\right)_T\right].$$

Now the same consideration applies to E: the mixed second derivatives are equal. This reduces the preceding equation to

$$p + \left(\frac{\partial E}{\partial V}\right)_T = T\left(\frac{\partial p}{\partial T}\right)_V. \tag{9-30}$$

Comparing Eqs. (9–30) and (9–28) we obtain

$$\left(\frac{\partial S}{\partial V}\right)_T = \left(\frac{\partial p}{\partial T}\right)_V. \tag{9-31}$$

Equation (9–31) is a relatively simple expression for the isothermal volume dependence of the entropy in terms of a derivative, $(\partial p/\partial T)_V$, which is readily measurable for any system. From Eq. (9–23), the cyclic rule, we have $(\partial p/\partial T)_V = \alpha/\beta$. Using this result, we obtain

$$\left(\frac{\partial S}{\partial V}\right)_T = \frac{\alpha}{\beta}. \tag{9-32}$$

Since β is positive, the sign of this derivative depends on the sign of α; for the vast majority of substances the volume increases with temperature so that α is positive. According to Eq. (9–32) then, for the majority of substances the entropy will increase with increase in volume. Water between 0°C and 4°C has a negative value of α and so is an exception to the rule.

The equations written in this section are applicable to any substance whatsoever. Thus for any substance we can write the total differential of the entropy in terms of T and V in the form

$$dS = \frac{C_v}{T}dT + \frac{\alpha}{\beta}dV. \tag{9-33}$$

Except for gases, the dependence of entropy on volume at constant temperature is negligibly small in most practical situations.

9–8 Entropy as a Function of Temperature and Pressure

If the entropy is considered as a function of temperature and pressure, $S = S(T, p)$, the total differential is written

$$dS = \left(\frac{\partial S}{\partial T}\right)_p dT + \left(\frac{\partial S}{\partial p}\right)_T dp. \qquad (9\text{–}34)$$

To bring Eq. (9–12) into this form, we introduce the relation between energy and enthalpy in the form $E = H - pV$; differentiating yields

$$dE = dH - p\, dV - V dp.$$

Using this value for dE in Eq. (9–12), we have

$$dS = \frac{1}{T} dH - \frac{V}{T} dp, \qquad (9\text{–}35)$$

which is another version of the fundamental equation (9–12); it relates dS to changes in enthalpy and pressure. We can express dH in terms of dT and dp, as we have seen before:

$$dH = C_p\, dT + \left(\frac{\partial H}{\partial p}\right)_T dp. \qquad (9\text{–}36)$$

Using this value of dH in Eq. (9–35), we obtain

$$dS = \frac{C_p}{T} dT + \frac{1}{T}\left[\left(\frac{\partial H}{\partial p}\right)_T - V\right] dp. \qquad (9\text{–}37)$$

Since Eqs. (9–34) and (9–37) both express dS in terms of dT and dp, they must be identical. Comparison of the two equations shows that

$$\left(\frac{\partial S}{\partial T}\right)_p = \frac{C_p}{T}, \qquad (9\text{–}38)$$

$$\left(\frac{\partial S}{\partial p}\right)_T = \frac{1}{T}\left[\left(\frac{\partial H}{\partial p}\right)_T - V\right]. \qquad (9\text{–}39)$$

For any substance, the ratio C_p/T is always positive. Therefore, Eq. (9–38) says that at constant pressure, the entropy always increases with temperature. Here again, the dependence of entropy on temperature is simple, the derivative being the ratio of the appropriate heat capacity to the temperature.

In Eq. (9–39) we have a rather messy expression for the pressure dependence of the entropy at constant temperature. To simplify matters, we again form the mixed second derivatives and set them equal. Differentiation of Eq. (9–38) with respect to pressure at constant temperature yields

$$\frac{\partial^2 S}{\partial p\, \partial T} = \frac{1}{T}\left(\frac{\partial C_p}{\partial p}\right)_T = \frac{1}{T}\frac{\partial^2 H}{\partial p\, \partial T}.$$

To obtain right-hand equality we have set $C_p = (\partial H/\partial T)_p$. Similarly, differentiation of Eq. (9–39) with respect to temperature yields

$$\frac{\partial^2 S}{\partial T \, \partial p} = \frac{1}{T}\left[\frac{\partial^2 H}{\partial T \, \partial p} - \left(\frac{\partial V}{\partial T}\right)_p\right] - \frac{1}{T^2}\left[\left(\frac{\partial H}{\partial p}\right)_T - V\right].$$

Setting the mixed derivatives equal yields

$$\frac{1}{T}\frac{\partial^2 H}{\partial p \, \partial T} = \frac{1}{T}\frac{\partial^2 H}{\partial T \, \partial p} - \frac{1}{T}\left(\frac{\partial V}{\partial T}\right)_p - \frac{1}{T^2}\left[\left(\frac{\partial H}{\partial p}\right)_T - V\right].$$

Since the mixed second derivatives of H are also equal this equation reduces to

$$\left(\frac{\partial H}{\partial p}\right)_T - V = -T\left(\frac{\partial V}{\partial T}\right)_p. \tag{9–40}$$

Combining this result with Eq. (9–39) we have

$$\left(\frac{\partial S}{\partial p}\right)_T = -\left(\frac{\partial V}{\partial T}\right)_p = -V\alpha. \tag{9–41}$$

To obtain the right-hand equality the definition of α has been used. In Eq. (9–41) we have an expression for the isothermal pressure dependence of the entropy in terms of the quantities V and α which are easily measurable for any system. The entropy can be written in terms of the temperature and pressure in the form

$$dS = \frac{C_p}{T}\,dT - V\alpha\,dp. \tag{9–42}$$

The value of α for solids is generally of the order of 10^{-4} deg^{-1} or less; for liquids it is 10^{-3} deg^{-1} or less. Suppose the liquid has a molar volume of $100\ cm^3/$mole; using Eq. (9–42) we can calculate the change in entropy at constant temperature $dT = 0$, for an increase in pressure of 1 atm. Since V and α are constants, they can be removed from the integral and we obtain

$$\Delta S = -V\alpha\,\Delta p$$
$$= -(100\ cm^3/mole)(10^{-3}\ deg^{-1})(1\ atm) = -0.1\ cm^3\cdot atm\cdot deg^{-1}\cdot mole^{-1}$$
$$= -0.002\ cal\cdot deg^{-1}\cdot mole^{-1} = -0.002\ eu/mole.$$

To produce a decrease in entropy of 1 eu a pressure of at least 1000 atm must be applied to the liquid. Since the variation of entropy of a liquid or a solid with pressure is so small, we will usually ignore it completely. If the pressure on a gas were increased from 1 atm to 2 atm, the corresponding change in entropy would be $\Delta S = -1.4$ eu/mole; the decrease is large simply because the volume has decreased greatly. We cannot ignore the entropy change of a gas accompanying a change in pressure.

9–9   The Temperature Dependence of the Entropy

Attention has been directed to the simplicity of the dependence of entropy on temperature both at constant volume and constant pressure. This simplicity results from the fundamental definition of the entropy. If the state of the system is described in terms of the temperature and any other independent variable x, then the heat capacity of the system in a reversible transformation at constant x is by definition $C_x = (dQ_{rev})_x/dT$. Combining this equation with the definition of dS, we obtain at constant x

$$dS = \frac{C_x}{T}\,dT \quad \text{or} \quad \left(\frac{\partial S}{\partial T}\right)_x = \frac{C_x}{T}. \tag{9–43}$$

Thus, under any constraint, the dependence of the entropy on temperature is simple; the differential coefficient is always the appropriate heat capacity divided by the temperature. In the majority of practical applications, x is either V or p. Thus, we may take as equivalent definitions of the heat capacities

$$C_v = T\left(\frac{\partial S}{\partial T}\right)_V \quad \text{or} \quad C_p = T\left(\frac{\partial S}{\partial T}\right)_p. \tag{9–44}$$

9–10   Entropy Changes in the Ideal Gas

The relations derived in the preceding sections are applicable to any system at all. They have a particularly simple form when applied to the ideal gas, which is the result of the fact that in the ideal gas the energy and the temperature are equivalent variables: $dE = C_v\,dT$. Using this value of dE in Eq. (9–12), we obtain immediately

$$dS = \frac{C_v}{T}\,dT + \frac{p}{T}\,dV. \tag{9–45}$$

The same result could be obtained by using Joule's law, $(\partial E/\partial V)_T = 0$, in Eq. (9–26). To use Eq. (9–45), all of the quantities must be expressed as functions of the two variables T and V. Hence, we replace the pressure by $p = RT/\overline{V}$; and the equation becomes, for 1 mole,

$$d\overline{S} = \frac{\overline{C}_v}{T}\,dT + \frac{R}{\overline{V}}\,d\overline{V}. \tag{9–46}$$

By comparing Eq. (9–46) with (9–24), we see that

$$\left(\frac{\partial \overline{S}}{\partial \overline{V}}\right)_T = \frac{R}{\overline{V}}. \tag{9–47}$$

This derivative is always positive; in an isothermal transformation, the entropy of the ideal gas increases with increase in volume. The rate of increase is less at large volumes, since V appears in the denominator.

For a finite change in state, we integrate Eq. (9–46) to

$$\Delta \bar{S} = \int_{T_1}^{T_2} \frac{\bar{C}_v}{T} dT + R \int_{\bar{V}_1}^{\bar{V}_2} \frac{d\bar{V}}{\bar{V}}.$$

If $\bar{C}_v$ is a constant, this integrates immediately to

$$\Delta \bar{S} = \bar{C}_v \ln \left(\frac{T_2}{T_1} \right) + R \ln \left(\frac{\bar{V}_2}{\bar{V}_1} \right). \tag{9–48}$$

The entropy of the ideal gas is expressed as a function of T and p by using the property of the ideal gas, $dH = C_p \, dT$, in Eq. (9–35) which reduces to

$$dS = \frac{C_p}{T} dT - \frac{V}{T} dp.$$

To express everything in terms of T and p, we use $\bar{V} = RT/p$, so that, for 1 mole,

$$d\bar{S} = \frac{\bar{C}_p}{T} dT - \frac{R}{p} dp. \tag{9–49}$$

Comparing Eq. (9–49) with Eq. (9–34), we have

$$\left(\frac{\partial \bar{S}}{\partial p} \right)_T = -\frac{R}{p}, \tag{9–50}$$

which shows that the entropy decreases with isothermal increase in pressure, a result that would be expected from the volume dependence of the entropy. For a finite change in state, Eq. (9–49) integrates to

$$\Delta \bar{S} = \bar{C}_p \ln \left(\frac{T_2}{T_1} \right) - R \ln \left(\frac{p_2}{p_1} \right), \tag{9–51}$$

where $\bar{C}_p$ has been taken as a constant in the integration.

9–11 Standard State for the Entropy of an Ideal Gas

For a change in state at constant temperature, Eq. (9–50) can be written

$$d\bar{S} = -\frac{R}{p} dp.$$

Suppose that we integrate this equation from $p = 1$ atm to any pressure p. Then

$$\bar{S} - \bar{S}^0 = -R \ln \left(\frac{p}{1 \text{ atm}} \right), \tag{9–52}$$

where $\bar{S}^0$ is the value of the molar entropy under 1 atm pressure; it is the standard entropy at the temperature in question.

To calculate a numerical value of the logarithm on the right-hand side of Eq. (9–52), it is essential that the pressure be expressed in atmospheres. Then the ratio $(p/1 \text{ atm})$ will be a pure number, and the operation of taking the logarithm is possible. (Note that it is not possible to take the logarithm of five oranges.) It is customary to abbreviate Eq. (9–52) to the simple form

$$\bar{S} - \bar{S}^0 = R \ln p. \tag{9–53}$$

It must be clearly understood that in Eq. (9–53) the value of p is a pure number, the number obtained by dividing the pressure in atm by 1 atm.

The quantity $\bar{S} - \bar{S}^0$ is the molar entropy at the pressure p relative to that at 1 atm pressure. A plot of $\bar{S} - \bar{S}^0$ for the ideal gas is shown as a function of pressure in Fig. 9–2(a). The rate of decrease of the entropy with pressure is rapid at low pressures and becomes less rapid at higher pressures. There is an evident advantage in using a plot of $\bar{S} - \bar{S}^0$ against $\ln p$ in this situation; Fig. 9–2(b). The plot is linear and a wider range of pressures can be represented on a scale of reasonable length.

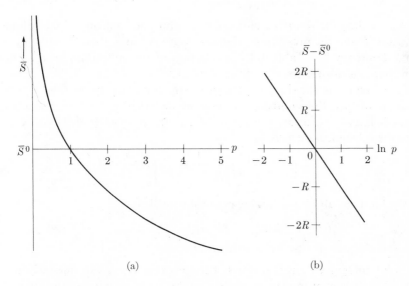

Fig. 9–2 a) Entropy of the ideal gas as a function of pressure. b) Entropy of the ideal gas versus $\ln p$.

9–12 The Third Law of Thermodynamics

Consider the constant-pressure transformation of a solid from the absolute zero of temperature to some temperature T below its melting point:

$$\text{Solid } (0°\text{K}, p) \rightarrow \text{Solid } (T, p).$$

The entropy change is given by Eq. (9–38),

$$\Delta S = S_T - S_{0^\circ} = \int_0^T \frac{C_p}{T} \, dT,$$

$$S_T = S_{0^\circ} + \int_0^T \frac{C_p}{T} \, dT. \tag{9–54}$$

The integral in Eq. (9–54) is positive, since the entropy can only increase with temperature. Thus, at 0°K the entropy has its smallest possible algebraic value S_{0°; the entropy at any higher temperature is greater than S_{0°. In 1913, M. Planck suggested that the value of S_{0° is zero for every pure, perfectly crystalline substance. This is the third law of thermodynamics: *The entropy of a pure, perfectly crystalline substance is zero at the absolute zero of temperature.*

Applying the third law of thermodynamics to Eq. (9–54), it reduces to

$$S_T = \int_0^T \frac{C_p}{T} \, dT, \tag{9–55}$$

where S_T is called the third-law entropy,† or simply the entropy of the solid at the temperature T and the pressure p. If the pressure is 1 atm, then the entropy is also a standard entropy S_T^0. Table 9–1 is a selection of entropy values for a number of different types of substances.

Since a change in the state of aggregation (melting or vaporization) involves an increase in entropy, this contribution must be included in the computation of the entropy of a liquid or of a gas. For the standard entropy of a liquid above the melting point of the substance, we have

$$S_T^0 = \int_0^{T_m} \frac{C_p^0(s)}{T} \, dT + \frac{\Delta H_{fus}^0}{T_m} + \int_{T_m}^T \frac{C_p^0(l)}{T} \, dT. \tag{9–56}$$

Similarly, for a gas above the boiling point of the substance

$$S_T^0 = \int_0^{T_m} \frac{C_p^0(s)}{T} \, dT + \frac{\Delta H_{fus}^0}{T_m} + \int_{T_m}^{T_b} \frac{C_p^0(l)}{T} \, dT + \frac{\Delta H_{vap}^0}{T_b} + \int_{T_b}^T \frac{C_p^0(g)}{T} \, dT. \tag{9–57}$$

If the solid undergoes any transition between one crystalline modification and another, the entropy of transition at the equilibrium temperature must be included also. To calculate the entropy, the heat capacity of the substance in its various states of aggregation must be measured accurately over the range of temperature from the absolute zero to the temperature of interest. The values of the heats of transition and the transition temperatures must also be measured. All of these measurements can be made calorimetrically.

---

†   The term "absolute entropy" is also commonly used for the third-law entropy. Since the values obtained by application of the third law are not really "absolute" values at all, the term is unfortunate.

Measurements of the heat capacity of some solids have been made at temperatures as low as a few hundredths of a degree above the absolute zero. However, this is unusual. Ordinarily, measurements of heat capacity are made down to a low temperature T', which frequently lies in the range from 10° to 15°K. At such low temperatures, the heat capacity of solids follows the Debye "T-cubed" law accurately; that is

$$C_v = aT^3, \tag{9–58}$$

where a is a constant for each substance. At these temperatures C_p and C_v are indistinguishable, so the Debye law is used to evaluate the integral of C_p/T over the interval from 0°K to the lowest temperature of measurement T'. The constant a is determined from the value of $C_p(= C_v)$ measured at T'. From the Debye law, $a = (C_p)_{T'}/T'^3$.

In the range of temperature above T', the integral

$$\int_{T'}^{T} \frac{C_p}{T}\, dT = \int_{T'}^{T} C_p\, d(\ln T) = 2.303 \int_{T'}^{T} C_p\, d(\log_{10} T)$$

is evaluated graphically by plotting either C_p/T versus T, or C_p versus $\log_{10} T$. The area under the curve is the value of the integral. Figure 9–3 shows the plot of C_p versus $\log_{10} T$ for a solid from 12°K to 298°K. The total area under the curve when multiplied by 2.303 yields a value of $S^0_{298.15}$ of 7.8 eu.

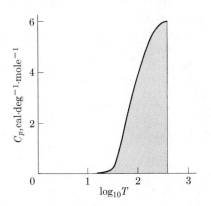

Fig. 9–3 Plot of C_p versus $\log_{10} T$.

In conclusion, we should note that the first statement of the third law of thermodynamics was made by Nernst in 1906, *the Nernst heat theorem*, which states that in any chemical reaction involving only pure, crystalline solids the change in entropy is zero. This form is less restrictive than the statement of Planck.

The third law of thermodynamics lacks the generality of the other laws, since it applies only to a special class of substances, namely pure, crystalline substances, and not to all substances. In spite of this restriction the third law is extremely useful.

Table 9–1† Standard entropies at 25°C (cal·deg^{-1}·mole^{-1})

Solids				
Solids		SrO	13.0	
Single unit, simple		BaO	16.8	
C (diamond)	0.5829	NaF	14.0	
Si	4.47	NaCl	17.30	
Sn (white)	12.3	KF	15.91	
Pb	15.51	KCl	19.76	
Zn	9.95	KBr	23.05	
Cu	7.96	KI	24.94	
Ag	10.206	*Two unit, complex*		
Ni	7.20	FeS$_2$	12.7	
Co	6.8	CaC$_2$	16.8	
Fe	6.49	NH$_4$Cl	22.6	
Mn(α)	7.59	MgCO$_3$	15.7	
Ti	7.24	CaCO$_3$(calcite)	22.2	
Al	6.769	SrCO$_3$	23.2	
Mg	7.77	BaCO$_3$	26.8	
Ca	9.95	NaNO$_3$	27.8	
Sr	13.0	KClO$_3$	34.17	
Ba	16.0	KClO$_4$	36.1	
Na	12.2	*Three units, simple*		
K	15.2	SiO$_2$(quartz, II)	10.0	
Single unit, complex		Cu$_2$O	24.1	
I$_2$	27.9	Cu$_2$S	28.9	
P$_4$	42.4	Ag$_2$O	29.09	
S$_8$(rhombic)	60.96	Ag$_2$S(rhombic)	34.8	
C (graphite)	1.3609	MnO$_2$	12.7	
Two unit, simple		TiO$_2$(rutile)	12.01	
SnO	13.5	Na$_2$O	17.4	
PbO(red)	16.2	*Three units, complex*		
PbS	21.8	Na$_2$SO$_4$	35.73	
ZnO	10.5	*Five units*		
HgO(red)	17.2	Al$_2$O$_3$(α)	12.186	
CuO	10.4	Fe$_2$O$_3$	21.5	
CuS	15.9	*Seven units*		
AgCl	22.97	Fe$_3$O$_4$	35.0	
NiO	9.22			
CoO	10.5	**Liquids**		
FeO(wusite)	12.9	Hg	18.5	
FeS(α)	16.1	Br$_2$	36.4	
MnO	14.4	H$_2$O	16.716	
MgO	6.4	HNO$_3$	37.19	
CaO	9.5	TiCl$_4$	60.4	

†   NBS Circular 500. U.S. G.P.O., Washington, D.C., 1952.

Table 9–1 (*cont.*)

Liquids (*cont.*)		Triatomic	
CH_3OH	30.3	H_2O	45.106
C_2H_5OH	38.4	O_3	56.8
		H_2S	49.15
		SO_2	59.40
		NO_2	57.47
Gases		N_2O	52.58
Monatomic		NOCl	63.0
He	30.126	CO_2	51.061
Ne	34.948	*Tetratomic*	
Ar	36.983	SO_3	61.24
Kr	39.19	NH_3	46.01
Xe	40.53	P_4	66.90
Diatomic		PF_3	64.12
H_2	31.211	PCl_3	74.79
HF	41.47	HCHO	52.26
HCl	44.617	C_2H_2	47.997
HBr	47.437	*Pentatomic*	
HI	49.314	CH_4	44.50
F_2	48.6	SiH_4	48.7
Cl_2	53.286	SiF_4	68.0
Br_2	58.639	*Hexatomic*	
I_2	62.280	N_2O_4	72.73
O_2	49.003	PCl_5	84.3
N_2	45.767	C_2H_4	52.45
NO	50.339	*Octatomic*	
CO	47.301	C_2H_6	54.85

The reasons for exceptions to the law can be better understood after we have discussed the statistical interpretation of the entropy; the entire matter of exceptions to the third law will be deferred until then.

The following general comments may be made about the entropy values which appear in Table 9.1.

1. Entropies of gases are larger than those of liquids which are larger than those of solids. This is an immediate consequence of Eq. (9–7).

2. The entropy of gases increases logarithmically with the mass; this is illustrated by the monatomic gases, or the series of diatomics, HF, HCl, HBr, HI.

3. Comparing gases having the same mass; Ne, HF, H_2O, we see the effect of the rotational heat capacity. Two degrees of rotational freedom add 6.52 eu in passing from Ne to HF; one additional rotation in H_2O compared to HF adds 3.64 eu. Similarly, H_2O and NH_3 have nearly the same entropy. (Both have 3 rotational degrees of freedom.) Molecules with the same mass and same heat capacities but

different shapes: the more symmetrical molecule has the lower entropy; clear-cut examples are few, but compare N_2 with CO and NH_3 with CH_4.

4. In the case of solids consisting structurally of a single simple unit, the heat capacity is exclusively vibrational. A tightly bound solid (high cohesive energy) has high characteristic frequencies (in the sense of Sec. 4–12), hence a lower heat capacity and a low entropy; e.g., diamond has very high cohesive energy, very low entropy; silicon has lower cohesive energy (also lower vibrational frequencies due to higher mass), hence a higher entropy.

5. Solids made up of two, three, ..., simple units have entropies which are roughly two, three, ..., times greater than those composed of one simple unit. The entropy per particle is roughly the same throughout.

6. Where there is a single complex unit, van der Waals forces (very low cohesive forces) bind the solid. The entropy is correspondingly high. Note the masses are quite large in the examples given in the table.

7. Where complex units occur in the crystal, the entropy is correspondingly greater since the heat capacity is greater due to the additional degrees of freedom associated with these units.

9–13 Entropy Changes in Chemical Reactions

The standard entropy change in a chemical reaction is computed from tabulated data in much the same way as the standard change in enthalpy. However, there is one important difference; the standard entropy of elements is *not* assigned a conventional value of zero. The characteristic value of the entropy of each element at 25°C and 1 atm pressure is known from the third law. As an example, in the reaction

$$Fe_2O_3(s) + 3H_2(g) \rightarrow 2Fe(s) + 3H_2O(l),$$

the standard entropy change is given by

$$\Delta S^0 = S^0_{(final)} - S^0_{(initial)}. \tag{9–59}$$

Since

$$S^0_{(final)} = 2\bar{S}^0_{Fe(s)} + 3\bar{S}^0_{H_2O(l)},$$

$$S^0_{(initial)} = \bar{S}^0_{Fe_2O_3(s)} + 3\bar{S}^0_{H_2(g)}.$$

Thus Eq. (9–59) becomes

$$\Delta S^0 = 2\bar{S}^0_{Fe(s)} + 3\bar{S}^0_{H_2O(l)} - [\bar{S}^0_{FeO_3(s)} + 3\bar{S}^0_{H_2(g)}]. \tag{9–60}$$

From the values in Table 9–1, we find for this reaction at 25°C

$$\Delta S^0 = 2(6.49) + 3(16.716) - [21.5 + 3(31.211)]$$

$$= 13.0 + 50.1 - 21.5 - 93.6 = -52.0 \text{ eu.}$$

Since the entropy of gases is much larger than the entropy of condensed phases, there is a large decrease in entropy in this reaction, since a gas, hydrogen, is consumed to form condensed materials. Conversely, in reactions in which a gas is formed at the expense of condensed materials, the entropy will increase markedly; e.g. (at 25°C),

$$Cu_2O(s) + C(s) \rightarrow 2Cu(s) + CO(g) \qquad \Delta S^0 = +37.8 \text{ eu}.$$

From the value of ΔS^0 for a reaction at any particular temperature T_0, the value at any other temperature is easily obtained by applying Eq. (9–38):

$$\Delta S^0 = S^0(\text{products}) - S^0(\text{reactants}).$$

Differentiating this equation with respect to temperature at constant pressure, we have

$$\left(\frac{\partial \Delta S^0}{\partial T} \right)_p = \left(\frac{\partial S^0 (\text{products})}{\partial T} \right)_p - \left(\frac{\partial S^0 (\text{reactants})}{\partial T} \right)_p$$

$$= \frac{C_p (\text{products})}{T} - \frac{C_p (\text{reactants})}{T} = \frac{\Delta C_p}{T}. \qquad (9\text{–}61)$$

Writing Eq. (9–61) in differential form and integrating between the reference temperature T_0, and any other temperature T, we obtain

$$\int_{T_0}^{T} d(\Delta S^0) = \int_{T_0}^{T} \frac{\Delta C_p}{T} dT$$

$$\Delta S_T^0 = \Delta S_{T_0}^0 + \int_{T_0}^{T} \frac{\Delta C_p}{T} dT, \qquad (9\text{–}62)$$

which is applicable to any chemical reaction so long as none of the reactants or products undergoes a change in its state of aggregation in the temperature interval T_0 to T.

9–14 Entropy and Probability

The entropy of a system in a definite state can be related to what is called the probability of that state of the system. To make this relation, or even to define what is meant by the probability of the state, it is necessary to have some structural model of the system. In contrast, the definition of the entropy from the second law does not require a structural model; the definition does not depend in the least on whether we suppose that the system is composed of atoms and molecules or imagine that it is built with waste paper and baseball bats. For simplicity we will postulate that the system is composed of a very large number of small particles, or molecules.

Imagine the following situation. A large room is sealed and completely evacuated. In one corner of the room there is a small box which confines a gas under atmospheric pressure. The sides of the box are now taken away so that the molecules of gas are free to move into the room. After a period of time we observe that the gas is distributed uniformly throughout the room. At the time the box was opened each gas molecule had a definite position and velocity, if we take a classical view of the matter. At some instant after the gas has filled the room, the position and velocity of each molecule have values which are related in a complicated way to the values of the positions and velocities of all the molecules at the instant of opening the box. At this later time, imagine that each velocity component of every molecule is exactly reversed. Then the molecules will just reverse their original motion; after a period of time the gas will collect itself in the corner of the room where it was originally sealed in the box.

The strange thing about the matter is this. There is no reason to suppose that the one particular motion which led to the uniform filling of the room is any more probable than the same motion reversed which leads to the collection of the gas in one corner of the room. If this is so, why is it that we never observe the air in a room collecting in one particular portion of the room? The fact that we never observe some motions of a system which are inherently just as probable as those we do observe is called the *Boltzmann paradox*.

This paradox is resolved in the following way. It is true that any exactly specified motion of the molecules has the same probability as any other exactly specified motion. But it is also true that of all the possible exactly specified motions of a group of molecules, the total number of these motions which lead to the uniform filling of the available space is enormously greater than the number of these motions which leads to the occupation of only a small part of the available space. And so, although each individual motion of the system has the same probability, the probability of observing the available space filled uniformly is proportional to the total number of motions which would result in this observation; consequently, the probability of observing the uniform filling is overwhelmingly large compared with the probability of any other observation.

It is difficult to imagine the detailed motion of even one particle, much less that of many particles. Fortunately, for the calculation we do not have to deal with the motions of the particles, but only with the number of ways of distributing the particles in a given volume. A simple illustration suffices to show how the probability of the uniform distribution compares with that of the nonuniform one.

Suppose we have a set of four cells each of which can contain one ball. The set of four cells is then divided in half, each half has two cells, as in Fig. 9–4(a). We place two balls in the cells; the arrangements in Fig. 9–4(b) are possible ($\bigcirc$ indicates an empty cell, $\otimes$ indicates an occupied cell). Of these six arrangements, four correspond to uniform filling; that is, one ball in each half of the box. The probability of uniform filling is therefore $\frac{4}{6} = \frac{2}{3}$, while the probability of finding both balls on one side of the box is $\frac{2}{6} = \frac{1}{3}$. The probability of any *particular* arrangement is $\frac{1}{6}$. But four particular arrangements lead to uniform filling, only two particular arrangements lead to nonuniform filling.

Suppose that there are eight cells and two balls; then the total number of arrangements is 28. Of the 28 arrangements, 16 of them correspond to one ball in each half of the box. The probability of the uniform distribution is therefore $\frac{16}{28} = \frac{4}{7}$. It is easy to show that as the number of cells increases without limit, the probability

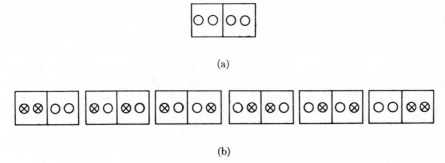

(a)

(b)

Figure 9–4

of finding one ball in one half of the box and the other in the other half of the box approaches the value $\frac{1}{2}$.

At this point it seems reasonable to ask what all this has to do with entropy. The entropy of a system in a specified state can be defined in terms of the number of possible arrangements of the particles composing the system which are consonant with the state of the system. Each such possible arrangement is called a *complexion* of the system. Following Boltzmann, we define the entropy by the equation

$$S = \mathbf{k} \ln \Omega, \tag{9–63}$$

where $\mathbf{k}$ is the Boltzmann constant, $\mathbf{k} = R/N_0$, and Ω is the number of complexions of the system which are consonant with the specified state of the system. Since the probability of a specified state of a system is proportional to the number of complexions which make up that state, it is clear from Eq. (9–63) that the entropy depends on the logarithm of the probability of the state.

Suppose we calculate the entropy for two situations in the foregoing example.

Situation 1. The two balls are confined to the left half of the box. There is only one arrangement (complexion) which produces this situation; hence, $\Omega = 1$, and

$$S_1 = \mathbf{k} \ln (1) = 0.$$

The entropy of this state is zero.

Situation 2. The two balls may be anywhere in the box. As we have seen, there are six complexions corresponding to this situation; hence, $\Omega = 6$, and

$$S_2 = \mathbf{k} \ln (6).$$

The entropy increase associated with the expansion of the system from 2 cells to 4 cells is then

$$\Delta S = S_2 - S_1 = \mathbf{k} \ln 6 \qquad \text{for 2 balls}$$

$$= (\mathbf{k}/2) \ln 6 \quad \text{for 1 ball.}$$

This result is readily generalized to apply to a box having N cells. How many arrangements are possible for two balls in N cells? There are N choices for the placement of the first ball; for each choice of cell for the first ball there are $N - 1$ choices for the second ball. The total number of arrangements of 2 balls in N cells is apparently $N(N - 1)$. However, since we cannot distinguish between ball 1 in position x, ball 2 in position y, and the arrangement ball 1 in y, ball 2 in x, this number must be divided by 2 to obtain the number of distinct arrangements; hence,

$$\Omega_1 = \frac{N(N - 1)}{2}.$$

The entropy of this system is, by Eq. (9–63),

$$S_1 = \mathbf{k} \ln \left[\tfrac{1}{2} N(N - 1) \right].$$

If we increase the number of cells available to N', then $\Omega_2 = \tfrac{1}{2} N'(N' - 1)$, and

$$S_2 = \mathbf{k} \ln \left[\tfrac{1}{2} N'(N' - 1) \right].$$

The increase in entropy associated with increasing the number of cells from N to N' is

$$\Delta S = S_2 - S_1 = \mathbf{k} \ln \left[\frac{N'(N' - 1)}{N(N - 1)} \right]. \tag{9–64}$$

If $N' = 4$ and $N = 2$, this yields the result obtained originally for the expansion from 2 to 4 cells.

A more instructive application of Eq. (9–64) is obtained if we suppose that both N and N' are very large, so large that $N - 1$ can be replaced by N and $N' - 1$ by N'. Then Eq. (9–64) becomes

$$\Delta S = \mathbf{k} \ln \left(\frac{N'}{N} \right)^2 = 2\mathbf{k} \ln \left(\frac{N'}{N} \right). \tag{9–65}$$

If we ask to what physical situation this random placing of balls in cells might be applied, the ideal gas comes to mind. In the ideal gas the position of a molecule at any time is a result of pure chance. The proximity of the other molecules does not affect the chance of the molecule being where it is. If we apply Eq. (9–65) to an ideal gas, the balls become molecules and the number of cells is proportional to the volume occupied by the gas; thus, $N'/N = V'/V$, and Eq. (9–65) becomes

$$\Delta S \text{ (two molecules)} = 2\mathbf{k} \ln (V'/V), \qquad \Delta S \text{ (one molecule)} = \mathbf{k} \ln (V'/V).$$

Since $N_0 \mathbf{k} = R$, the gas constant, for one mole, we have

$$\Delta S \text{ (one mole)} = R \ln (V'/V), \tag{9–66}$$

which is identical with the second term of Eq. (9–48), the expression for the increase in entropy accompanying the isothermal expansion of one mole of an ideal gas from volume V to volume V'.

From the standpoint of this structural and statistical definition of entropy, isothermal expansion of a gas increases the entropy because there are more ways of arranging a given number of molecules in a larger volume than in a small volume. Since the probability of a given state is proportional to the number of ways of arranging the molecules in that state, the gas confined in a large volume is in a more probable state than if it is confined in a small volume. If we assume that the equilibrium state of the gas is the state of highest probability, then it is understandable why the gas in a room never collects in a small corner. The gas achieves its most probable state by occupying as much volume as is available to it. The equilibrium state has the maximum probability consistent with the constraints on the system and so has a maximum entropy.

9–15 General Form for Omega

To calculate the number of arrangements of three particles in N cells, we proceed in the same way as before. There are N choices for placing the first particle, $N - 1$ choices for the second, and $N - 2$ choices for the third. This would seem to make a total of $N(N - 1)(N - 2)$ arrangements; but again we cannot distinguish between arrangements which are only permutations of the three particles between the cells x, y, z. There are $3!$ such permutations: $xyz, xzy, yxz, yzx, zxy, zyx$. Hence, for three particles in N cells the number of complexions is

$$\Omega = \frac{N(N - 1)(N - 2)}{3!}. \qquad (9\text{–}67)$$

Again if the number of cells N is much larger than the number of particles, this reduces for three particles to

$$\Omega = \frac{N^3}{3!}.$$

From this approximate form we can immediately jump to the conclusion that for N_a particles, if N is much larger than N_a, then, approximately,

$$\Omega = \frac{N^{N_a}}{N_a!}. \qquad (9\text{–}68)$$

On the other hand, if we need the exact form for Ω, Eq. (9–67) can be generalized for N_a particles to

$$\Omega = \frac{N(N - 1)(N - 2)(N - 3) \cdots (N - N_a + 1)}{N_a!}.$$

If we multiply this last equation by $(N - N_a)!$ in both numerator and denominator, it reduces to

$$\Omega = \frac{N!}{N_a!(N - N_a)!}. \qquad (9\text{–}69)$$

The entropy attending the expansion from N to N' cells is easily calculated using Eq. (9–68). For N cells,

$$S = \mathbf{k}[\ln N^{N_a} - \ln (N_a!)],$$

while for N' cells,

$$S' = \mathbf{k}[\ln N'^{N_a} - \ln (N_a!)].$$

The value of ΔS is

$$\Delta S = S' - S = N_a \mathbf{k} \ln \left(\frac{N'}{N}\right).$$

As before, we take the ratio $N'/N = V'/V$; then if $N_a = N_0$, the equation becomes

$$\Delta S = R \ln \left(\frac{V'}{V}\right),$$

which is identical with Eq. (9–66).

9–16 The Energy Distribution

It is rather easy to make the translation from arrangements of balls in cells to the physical situation of arrangement of molecules in small elements of volume. By arranging molecules in the elements of volume we obtain a space distribution of the molecules. The problem in the space distribution was simplified considerably by the implicit assumption that there is at most one molecule in a given volume element.

The problem of translating arrangements of balls in cells to an energy distribution is only slightly more difficult. We assume that any molecule can have an energy value between zero and infinity. We partition this entire range of energy into small compartments of width $d\epsilon$; the compartments are labeled, beginning with the one of lowest energy, by $\epsilon_1, \epsilon_2, \epsilon_3, \ldots$, as in Fig. 9–5. The energy distribution is described by specifying the number of molecules n_1 having energies lying in the first compartment, the number n_2 in the second compartment, and so on.

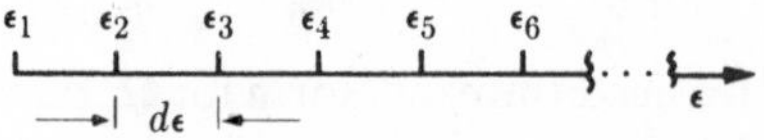

Fig. 9–5 Division of the energy range into compartments.

Consider a collection of N molecules for which the energy distribution is described by the numbers $n_1, n_2, n_3, n_4, n_5, \ldots$. In how many ways can this particular distribution be achieved? We begin by supposing that there are three molecules in ϵ_1; there are N ways of choosing the first molecule, $(N - 1)$ of choosing the second, and $(N - 2)$

ways of choosing the third. Thus, there appear to be $N(N - 1)(N - 2)$ ways of selecting three molecules from N molecules; however, the order of choice does not matter, the same distribution is obtained with molecules 1, 2, and 3 whether they are chosen in the order 123, 132, 213, 231, 312, or 321. We must divide the total number of ways of choosing by 3! to get the number of distinguishable ways of choosing, which is therefore

$$\frac{N(N - 1)(N - 2)}{3!}.$$

Suppose that there are two molecules in the second compartment; these must be chosen from the $N - 3$ molecules remaining; the first may be chosen in $N - 3$ ways, the second in $N - 4$ ways. Again the order does not matter, so we divide by 2!. The two molecules in the second compartment can be chosen in

$$\frac{(N - 3)(N - 4)}{2!}$$

different ways. The total number of ways of choosing three molecules in the first compartment and two molecules in the second compartment is the product of these expressions:

$$\frac{N(N - 1)(N - 2)(N - 3)(N - 4)}{3!2!}.$$

We then find how many ways there are of choosing the number of molecules in compartment three from the remaining $N - 5$ molecules, and so on. Repetition of this procedure yields the final result for Ω, the total number of ways of placing n_1 molecules in compartment 1, n_2 molecules in compartment 2,... :

$$\Omega = \frac{N!}{n_1!n_2!n_3!n_4!\dots}. \tag{9–70}$$

The value of Ω, the number of complexions for a particular distribution, given by Eq. (9–70) seems rather forbidding. However, we do not need to do very much with it to get the information that we need. As usual, the entropy resulting from the distribution of molecules over a range of energies is related to the number of complexions by $S = k \ln \Omega$. If Ω is very large, the entropy will be large. It is clear from Eq. (9–70) that the smaller the populations of the compartments, $n_1, n_2, n_3, \dots$, the larger will be the value of Ω. For example, if every compartment was either empty or contained only 1 molecule, all of the factors in the denominator would be either 0! or 1!; the denominator would then be unity and $\Omega = N!$. This would be the largest possible value of Ω, and would correspond to the largest possible value of the entropy. Note that in this situation the molecules are spread out very widely over the energy range; thus, a broad energy distribution means a high entropy.

In contrast, consider the situation where all the molecules but one are crowded into the first level; then

$$\Omega = \frac{N!}{(N-1)!1!0!0!\cdots} = N.$$

If N is large, then N is very much smaller than $N!$; the entropy in this case is very much smaller than that for the broad distribution.

To achieve a high entropy, the molecules will therefore try to spread out into as broad an energy distribution as possible, just as gas molecules fill as much space as is available. The space distribution is limited by the walls of the container. The energy distribution is subject to an analogous limitation. In a specified state, a system has a fixed value of its total energy; from the distribution this value is

$$E = n_1\epsilon_1 + n_2\epsilon_2 + n_3\epsilon_3 + n_4\epsilon_4 + \cdots.$$

It is clear that the system may not have many molecules in the high-energy compartments; if it did, the distribution would yield a value of energy above the fixed value in the particular state. This restriction limits the number of complexions of a system quite severely. The value of Ω nonetheless reaches a maximum consistent with the restriction that the energy must sum up to the fixed value E. The molecules spread themselves over as broad a range of energy as is consistent with the fixed total energy of the system.

If the energy of the system is increased, the distribution can be broader; the number of complexions and the entropy of the system goes up. This is a statistical interpretation of the fact illustrated by the fundamental equation (9–12):

$$dS = \frac{1}{T}dE + \frac{p}{T}dV,$$

from which we obtain the differential coefficient

$$\left(\frac{\partial S}{\partial E}\right)_V = \frac{1}{T}.$$

We noted in Section 9–5 that this coefficient was always positive. For the present we simply note the agreement in the sign of this coefficient with the statistical argument that increase in energy increases the number of complexions and the entropy.

The two fundamental ways of varying the entropy of a system expressed by the fundamental equation are interpreted as the two ways of achieving a broader distribution. By increasing the volume, the spatial distribution broadens; by increasing the energy, the energy distribution broadens. The broader distribution is the more probable one, since it can be made up in a greater number of ways.

It is easy now to understand why the entropies of liquids and solids are nearly unchanged by a change in pressure. The volume of condensed materials is altered so little by a change in pressure that the breadth of the space distribution remains about the same. The entropy therefore remains at very nearly the same value.

We can also understand the phenomena in the adiabatic reversible expansion of a gas; in such an expansion, $dQ_{rev} = 0$, so that $dS = 0$. Since the volume goes up, the distribution over space broadens, and this part of the entropy increases. If the total entropy change is to be zero, the distribution over energies must get narrower; this corresponds to a decrease in energy which is reflected in a decrease in the temperature of the gas. The work produced in such an adiabatic expansion of a gas is produced at the expense of the decrease in energy of the system.

In Chapter 4 the Maxwell distribution of kinetic energies in a gas was discussed in detail. There we found that the average energy was given by $\frac{3}{2}RT$. Thus, an increase in temperature corresponds to an increase in the energy of the gas; it should also correspond to a broadening of the energy distribution. This broadening of the energy distribution with increase in temperature was emphasized at that time.

From what has been said, it seems reasonable to expect the direction of natural changes to correspond to the direction which increases the probability of the system. Thus in natural transformations we might expect the entropy of the system to increase. This is not quite true. In a natural change both the system and the surroundings are involved. Therefore, in any natural change, it is required that the universe reach a state of higher probability and thus of higher entropy. In a natural transformation, the entropy of the system may decrease if there is an increase in entropy in the surroundings which more than compensates the decrease in the system. The entropy change in a transformation is a powerful clue to the natural direction of the transformation.

9–17 Entropy of Mixing and Exceptions to the Third Law of Thermodynamics

The third law of thermodynamics is applicable only to those substances which attain a completely ordered configuration at the absolute zero of temperature. In a pure crystal, for example, the atoms are located in an exact pattern of lattice sites. If we calculate the number of complexions of N atoms arranged on N sites, we find that although there are $N!$ ways of arranging the atoms, since the atoms are identical, these arrangements differ only in the order of choosing the atoms. Since the arrangements are not distinguishable, we must divide by $N!$, and we obtain $\Omega = 1$ for the perfectly ordered crystal. The entropy is therefore

$$S = k \ln (1) = 0.$$

Suppose that we arrange different kinds of atoms A and B on the N sites of the crystal. If N_a is the number of A atoms, and N_b is the number of B atoms, then $N_a + N_b = N$, the total number of sites. In how many distinguishable ways can we select N_a sites for the A atoms and N_b sites for the B atoms? This number is given by Eq. (9–70):

$$\Omega = \frac{N!}{N_a! N_b!}. \tag{9–71}$$

The entropy of the mixed crystal is given by

$$S = k \ln \frac{N!}{N_a! N_b!}. \tag{9-72}$$

To evaluate this expression we take advantage of the Stirling approximation: when N is very large, then

$$\ln N! = N \ln N - N. \tag{9-73}$$

The expression for the entropy becomes

$$S = k(N \ln N - N - N_a \ln N_a + N_a - N_b \ln N_b + N_b).$$

Since $N = N_a + N_b$, this becomes

$$S = -k(N_a \ln N_a + N_b \ln N_b - N \ln N).$$

But, $N_a = x_a N$, and $N_b = x_b N$, where x_a is the mole fraction of A and x_b is the mole fraction of B. The expression for the entropy reduces to

$$S_{\mathrm{mix}} = -Nk(x_a \ln x_a + x_b \ln x_b). \tag{9-74}$$

Since the terms in the parentheses in Eq. (9–74) are negative (the logarithm of a fraction is negative), the entropy of the mixed crystal is positive. If we imagine the mixed crystal to be formed from a pure crystal of A and a pure crystal of B, then for the mixing process

$$\text{pure } A + \text{pure } B \to \text{mixed crystal}.$$

The entropy change is

$$\Delta S_{\mathrm{mix}} = S \, (\text{mixed crystal}) - S \, (\text{pure } A) - S \, (\text{pure } B).$$

The entropies of the pure crystals are zero, so the ΔS of mixing is simply

$$\Delta S_{\mathrm{mix}} = -Nk(x_a \ln x_a + x_b \ln x_b), \tag{9-75}$$

and is a positive quantity.

Since any impure crystal has at least the entropy of mixing at the absolute zero, its entropy cannot be zero; such a substance does not follow the third law of thermodynamics. Some substances which are pure from the chemical standpoint do not fulfill the requirement that the crystal be perfectly ordered at the absolute zero of temperature. Carbon monoxide, CO, and nitric oxide, NO, are classic examples. In the crystals of CO and NO, some molecules are oriented differently than others. In a perfect crystal of CO, all the molecules should be lined up with the oxygen pointing north and the carbon pointing south, for example. In the actual crystal, the two ends of the molecule are oriented randomly; it is as if two kinds of carbon monoxide were mixed, half and half. The entropy of mixing would be

$$\Delta S = -N_0 k(\tfrac{1}{2} \ln \tfrac{1}{2} + \tfrac{1}{2} \ln \tfrac{1}{2}) = N_0 k \ln 2$$

$$= R \ln 2 = 1.377 \text{ eu}.$$

The actual value for the residual entropy of crystalline carbon monoxide is 1.1 eu; the mixing is apparently not quite half and half. In the case of NO, the residual entropy is 0.66 eu, which is about one-half of 1.38; this has been explained by the observation that the molecules in the crystal of NO are dimers, $(NO)_2$. Thus, one mole of NO contains only $\frac{1}{2}N_0$ double molecules; this reduces the residual entropy by a factor of two.

In ice, a residual entropy remains at the absolute zero because of randomness in the hydrogen bonding of the water molecules in the crystal. The magnitude of residual entropy has been computed and is in agreement with that observed.

It has been found that crystalline hydrogen has a residual entropy of 1.49 eu at the absolute zero of temperature. This entropy is not the result of disorder in the crystal, but of a distribution over several quantum states. Ordinary hydrogen is a mixture of ortho- and para-hydrogen, which have different values of the total nuclear spin angular momentum. As a consequence of this difference, the rotational energy of ortho-hydrogen at low temperatures does not approach zero as does that of para-hydrogen, but achieves a finite value. Ortho-hydrogen can be in any one of nine states, all having the same energy, while the para-hydrogen exists in a single state. As a result of the mixing of the two kinds of hydrogen and the distribution of the ortho-hydrogen in nine different energy states, the system has a randomness and therefore a residual entropy. Pure para-hydrogen, since it exists in a single state at low temperature, would have no residual entropy and would follow the third law. Pure ortho-hydrogen would be distributed over nine states at the absolute zero and would have a residual entropy.

From what has been said it is clear that glassy or amorphous substances will have a random arrangement of the constituent particles and so will possess a residual entropy at the absolute zero. The third law is therefore restricted to pure crystalline substances. A final restriction should be made in the application of the third law; the substance must be in a single quantum state. This last requirement would take care of the difficulty which arises in the case of hydrogen.

Problems

9–1. What is the entropy change if the temperature of one mole of an ideal gas is increased from 100°K to 300°K, $\bar{C}_v = \frac{3}{2}R$,

a) if the volume is constant,

b) if the pressure is constant?

c) What would the change in entropy be if three moles were used instead of one mole?

9–2. A monatomic solid has a heat capacity, $\bar{C}_p = 6.2$ cal·deg^{-1}·mole^{-1}. Calculate the increase in entropy of one mole of this solid if the temperature is increased from 300°K to 500°K at constant pressure.

9–3. One mole of an ideal gas, $\bar{C}_v = \frac{3}{2}R$, is transformed from 0°C and 2 atm to -40°C and 0.4 atm. Calculate ΔS for this change in state.

9–4. One mole of an ideal gas is expanded isothermally to twice its initial volume.

a) Calculate ΔS.

b) What would be the value of ΔS if five moles of an ideal gas were doubled in volume isothermally?

9–5. a) What is the entropy change if one mole of water is warmed from 0°C to 100°C under constant pressure; $\bar{C}_p = 18.0$ cal·deg^{-1}·mole^{-1}.

b) The melting point is 0°C and the heat of fusion is 1.4363 kcal/mole. The boiling point is 100°C and the heat of vaporization is 9.7171 kcal/mole. Calculate ΔS for the transformation

$$\text{ice (0°C, 1 atm)} \rightarrow \text{steam (100°C, 1 atm)}.$$

9–6. At the boiling point, 35°C, the heat of vaporization of MoF_6 is 6.0 kcal/mole. Calculate ΔS_{vap}.

9–7. a) At the transition temperature, 95.4°C, the heat of transition from rhombic to monoclinic sulfur is 0.09 kcal/mole. Calculate the entropy of transition.

b) At the melting point, 119°C, the heat of fusion of monoclinic sulfur is 0.293 kcal/mole. Calculate the entropy of fusion.

c) The values given in (a) and (b) are for one mole of S, that is for 32 gm; however, in crystalline and liquid sulfur the molecule is S_8. Convert the values in parts (a) and (b) to ones appropriate to S_8. These converted values are more representative of the usual magnitudes of entropies of fusion and transition.

9–8. One mole of gaseous hydrogen is heated at constant pressure from 300°K to 500°K. Calculate the entropy change for this transformation using the heat capacity data in Table 7–2.

9–9. One mole of an ideal gas, initially at 25°C, is expanded

a) isothermally and reversibly from 20 to 40 liters/mole, and

b) isothermally and irreversibly against zero opposing pressure (Joule expansion) from 20 to 40 liters/mole.

Calculate ΔE, ΔS, Q, and W for both (a) and (b). Note the relation between ΔS and Q in (a) and in (b).

9–10. a) One mole of an ideal gas, $\bar{C}_v = \frac{3}{2}R$, is expanded adiabatically and reversibly: Initial state: 300°K, 1 atm; final state: 0.5 atm. Calculate Q, W, ΔE, and ΔS.

b) The same gas, initially at 300°K and 1 atm, is expanded adiabatically against a constant opposing pressure equal to the final pressure, 0.5 atm. Calculate Q, W, ΔE, and ΔS.

9–11. From the data for graphite: $\bar{S}^0_{298.15} = 1.3609$ eu/mole

$$\bar{C}_p(\text{cal·mole}^{-1}\text{·deg}^{-1}) = -1.265 + 14.008 \times 10^{-3}T - 103.31 \times 10^{-7}T^2$$
$$+ 2.751 \times 10^{-9}T^3.$$

Calculate the molar entropy of graphite at 1500°K.

9–12. In the limit, $T = 0$°K, it is known empirically that the value of the coefficient of thermal expansion of solids, $\alpha \rightarrow 0$. Show that as a consequence, the entropy is independent of pressure at 0°K so that no specification of pressure is necessary in the third law statement.

9-13. In a Dewar flask (an adiabatic enclosure) 20 gm of ice at $-5°C$ are added to 30 gm of water at $+25°C$. If the heat capacities are $C_p(\text{liquid}) = 1.0 \text{ gbs/gm}$ and $C_p(\text{ice}) = 0.5 \text{ gbs/gm}$, what is the final state of the system; the pressure is constant. $\Delta H_{\text{fusion}} = 80 \text{ cal/gm}$; 1 gbs = 1 cal/deg. Calculate ΔS and ΔH for the transformation.

9-14. How many gms of water at $25°C$ are required in the Dewar flask in Problem 9-13 to satisfy the following conditions? Compute the entropy change in each case.

a) The final temperature is $-20°C$; all the water freezes.

b) The final temperature is $0°C$; half the water freezes.

c) The final temperature is $0°C$; half the ice melts.

d) The final temperature is $10°C$; all the ice melts.

Predict the sign of ΔS in each case before doing the calculation.

9-15. For liquid water at $25°C$, $\alpha = 2.0 \times 10^{-4} \text{ deg}^{-1}$; the density may be taken as 1.0 gm/cm^3. One mole of liquid water at $25°C$ is compressed isothermally from 1 atm to 1000 atm. Calculate ΔS.

a) supposing that water is incompressible; i.e., $\beta = 0$.

b) supposing that $\beta = 4.53 \times 10^{-5} \text{ atm}^{-1}$.

9-16. For copper, at $25°C$, $\alpha = 0.492 \times 10^{-4} \text{ deg}^{-1}$ and $\beta = 0.78 \times 10^{-6} \text{ atm}^{-1}$; the density is 8.92 gm/cm^3. Calculate ΔS for the isothermal compression of one mole of copper from 1 atm to 1000 atm, under the same two conditions as in Problem 9-15.

9-17. Show that $(\partial \alpha / \partial p)_T = -(\partial \beta / \partial T)_p$.

9-18. Consider the expression:

$$dS = \frac{C_p}{T}dT - V\alpha\, dp$$

Suppose that water has $\beta = 4.53 \times 10^{-5} \text{ atm}^{-1}$, $\bar{V} = 18 \text{ cm}^3/\text{mole}$, $\bar{C}_p = 18 \text{ cal/deg·mole}$, and $\alpha = 2.0 \times 10^{-4} \text{ deg}^{-1}$. Compute the decrease in temperature which occurs if water at $25°C$ and 1000 atm pressure is brought reversibly and adiabatically to 1 atm pressure.

9-19. For metallic zinc the values of $\bar{C}_p$ as a function of temperature are given. Calculate S^0 for zinc at $100°K$.

T (°K)	$\bar{C}_p$ (cal/deg·mole)	T (°K)	$\bar{C}_p$ (cal/deg·mole)	T (°K)	$\bar{C}_p$ (cal/deg·mole)
1	0.000172	10	0.0391	50	2.671
2	0.000437	15	0.172	60	3.250
3	0.000906	20	0.406	70	3.687
4	0.00172	25	0.766	80	4.031
6	0.00453	30	1.187	90	4.328
8	0.0150	40	1.953	100	4.578

9-20. Sketch the possible indistinguishable arrangements of

a) two balls in six cells;

b) four balls in six cells.

c) What is the probability of the uniform distribution in each case?

9–21. Suppose that three indistinguishable molecules are distributed among three energy levels. The energies of the levels are: 0, 1, 2 units.

a) How many complexions are possible if there is no restriction on the energy of the three molecules?

b) How many complexions are possible if the total energy of the three molecules is fixed at one unit?

c) Find the number of complexions if the total energy is two units, and calculate the increase in entropy accompanying the energy increase from one to two units.

9–22. Pure ortho-hydrogen can exist in any of nine quantum states at the absolute zero. Calculate the entropy of this mixture of nine "kinds" of ortho-hydrogen; each has a mole fraction of $\frac{1}{9}$.

9–23. The entropy of a binary mixture relative to its pure components is given by Eq. (9–74). Since $x_a + x_b = 1$, write the entropy of the mixture in terms of x_a or x_b only, and show that the entropy is a maximum when $x_a = x_b = \frac{1}{2}$. Calculate values of S_{mix} for $x_a = 0, 0.2, 0.4,$ 0.5, 0.6, 0.8, and 1. Plot these values of S_{mix} as a function of x_a.

Chapter Ten

Spontaneity and Equilibrium

10–1 The General Conditions for Equilibrium and for Spontaneity

Our aim now is to find out what characteristics distinguish irreversible (real) transformations from reversible (ideal) transformations. We begin by asking what relation exists between the entropy change in a transformation and the irreversible heat flow which accompanies it. At every stage of a reversible transformation, the system departs from equilibrium only infinitesimally. The system is transformed, yet remains effectively at equilibrium throughout a reversible change in state. The condition for reversibility is therefore a condition of equilibrium; from the defining equation for dS, the condition of reversibility is that

$$T \, dS = dQ_{rev}. \tag{10–1}$$

Therefore, Eq. (10–1) is the condition of equilibrium.

The condition placed on an irreversible change in state is the Clausius inequality, (8–42), which we write in the form

$$T \, dS > dQ. \tag{10–2}$$

Irreversible changes are real changes, or natural changes, or spontaneous changes. We shall refer to changes in the natural direction as spontaneous changes, and the inequality (10–2) as the condition of spontaneity. The two relations Eq. (10–1) and (10–2) can be combined in the single one

$$T \, dS \geq dQ, \tag{10–3}$$

where it is understood that the equality sign implies a reversible value of dQ.

By using the first law in the form $dQ = dE + dW$, the relation in (10–3) can be written

$$T\,dS \geq dE + dW,$$

or

$$-dE - dW + T\,dS \geq 0. \tag{10–4}$$

The work includes all kinds; $dW = P_{op}\,dV + dU$. This value for dW brings (10–4) to the form

$$-dE - P_{op}\,dV - dU + T\,dS \geq 0. \tag{10–5}$$

Both (10–4) and (10–5) express the condition of equilibrium (=) and of spontaneity (>) for a transformation in terms of changes in properties of the system dE, dV, dS, and the amount of work dW or dU associated with the transformation.

10–2 Conditions for Equilibrium and Spontaneity under Constraints

Under the combinations of restraints usually imposed in the laboratory, (10–4) and (10–5) can be expressed in simple and convenient terms. We consider each set of constraints separately.

1. Transformations in an isolated system

For an isolated system, $dE = 0$, $dW = 0$, $dQ = 0$; thus, (10–4) becomes immediately

$$dS \geq 0. \tag{10–6}$$

This requirement for an isolated system was discussed in detail in Section 8–14, where it was shown that in an isolated system the entropy can only increase and reaches a maximum at equilibrium.

2. Transformations at constant temperature

If a system undergoes an isothermal change in state, then $T\,dS = d(TS)$, and the relation (10–4) can be written

$$-dE + d(TS) \geq dW,$$

$$-d(E - TS) \geq dW. \tag{10–7}$$

The combination of variables $E - TS$ appears so frequently that it is given a special symbol, A. By definition,

$$A \equiv E - TS. \tag{10–8}$$

Being a combination of other functions of the state, A is a function of the state of the system; A is called the *work function* of the system.† The relation (10–7) reduces to the form

$$-dA \geq dW, \qquad (10\text{–}9)$$

or, by integrating,

$$-\Delta A \geq W. \qquad (10\text{–}10)$$

The significance of A is given by relation (10–10); the work produced in an isothermal transformation is less than or equal to the decrease in the work function. The equality sign applies to the reversible transformation, so the maximum work obtainable in an isothermal change in state is equal to the decrease in the work function. This maximum work includes all the kinds of work produced in the transformation.

3. Transformations at constant temperature and under constant pressure

The system is confined under a constant pressure, $P_{op} = p$, the equilibrium pressure of the system. Since p is a constant, $p\,dV = d(pV)$. The temperature is constant so that $T\,dS = d(TS)$. The relation (10–5) becomes

$$-[dE + d(pV) - d(TS)] \geq dU,$$

$$-d(E + pV - TS) \geq dU. \qquad (10\text{–}11)$$

The combination of variables $E + pV - TS$ is a frequently occurring one and is given a special symbol, G. By definition,

$$G \equiv E + pV - TS = H - TS = A + pV. \qquad (10\text{–}12)$$

Being a composite of properties of the state of a system, G is a property of the state; G is called the *free energy* of the system.‡

Using Eq. (10–12), relation (10–11) becomes

$$-dG \geq dU, \qquad (10\text{–}13)$$

or, by integrating,

$$-\Delta G \geq U. \qquad (10\text{–}14)$$

Fixing our attention on the equality sign in (10–14), we have

$$-\Delta G = U_{rev}, \qquad (10\text{–}15)$$

---

† The quantity A has a number of names: *maximum work function, Helmholtz function, Helmholtz free energy*, and simply *free energy*. The simple name *work function* will be used exclusively for A in this book. The name *free energy* will be used exclusively to refer to the Gibbs function G, which we will define shortly.

‡ Other names for G are: *Gibbs free energy*, or *Gibbs function*. A good deal of confusion exists over the letter which should be used to symbolize the free energy; F is frequently used. The commonest tables of data use F, so the symbol G should be so translated in using those tables. In using any table of data it is best to make sure just what the symbols stand for.

which reveals an important property of the free energy; the decrease in free energy $(-\Delta G)$ associated with a change in state at constant T and p is equal to the maximum work U_{rev} over and above expansion work, which is obtainable in the transformation. By (10–14), in any real transformation the work obtained over and above expansion work is less than the decrease in free energy which accompanies the change in state at constant T and p.

If the work U is to show up in the laboratory, the transformation must be conducted in a device which enables the work to be produced; the most usual chemical example of such a device is the electrochemical cell. If granulated zinc is dropped into a solution of copper sulfate, metallic copper precipitates and the zinc dissolves according to the reaction

$$Zn + Cu^{2+} \rightarrow Cu + Zn^{2+}.$$

Quite obviously the only work produced in this mode of performing the reaction is expansion work, and there is very little of that. On the other hand, this same chemical reaction can be carried out in such a way as to produce a quantity of electrical work U. In the Daniell cell shown in Fig. 10–1, a zinc electrode is immersed in a solution of zinc sulfate and a copper electrode is immersed in a solution of copper sulfate; the solutions are in electrical contact through a porous partition which prevents the solutions mixing. The Daniell cell can produce electrical work U, which is related to the decrease in free energy $-\Delta G$ of the chemical reaction by (10–14). If the cell operates reversibly, then the electrical work produced is equal to the decrease in free energy. The performance of the electrochemical cell is discussed in detail in Chapter 17.

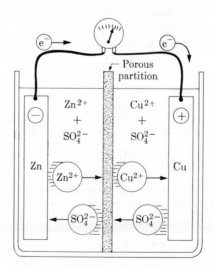

Fig. 10–1 Daniell cell.

Any spontaneous transformation *may* be harnessed to do some kind of work in addition to expansion work, but it need not necessarily be so harnessed. For the present, our interest is in those transformations which are not harnessed to do special kinds of work; for these cases, $dU = 0$, and the condition of equilibrium and spontaneity for a transformation at constant p and T, (10–14), becomes

$$-dG \geq 0, \tag{10–16}$$

or, for a finite change,

$$-\Delta G \geq 0. \tag{10–17}$$

Both (10–16) and (10–17) require the free energy to decrease in any real transformation at constant T and p; if the free energy decreases, ΔG is negative and $-\Delta G$ is positive. Spontaneous changes can continue to occur in such a system as long as the free energy of the system can decrease, that is, *until the free energy of the system reaches a minimum value*. The system at equilibrium has a minimum value of the free energy; this equilibrium condition is expressed by the equality sign in (10–16): $dG = 0$, the usual mathematical condition for a minimum value.

Of the several criteria for equilibrium and spontaneity, we shall have the most use for the one involving dG or ΔG, simply because most chemical reactions and phase transformations are subject to the conditions, constant T and p. If we know how to compute the change in free energy for any transformation, the algebraic sign of ΔG tells us if this transformation can occur in the direction in which we imagine it. There are three possibilities:

1. ΔG is $-$; the transformation can occur spontaneously, or naturally;

2. $\Delta G = 0$; the system is at equilibrium with respect to this transformation;

3. ΔG is $+$; the natural direction is opposite to the direction we have envisioned for the transformation (the transformation is nonspontaneous).

The third case is best illustrated by an example. Suppose we ask if water can flow uphill. The transformation can be written

$$H_2O \text{ (low level)} \rightarrow H_2O \text{ (high level)} \qquad (T \text{ and } p \text{ constant}).$$

The value of ΔG for this transformation is calculated and found to be positive. We conclude that the direction of this transformation as it is written is not the natural direction, and that the natural, spontaneous direction is opposite to the way we have written it. In the absence of artificial restraints, water at a high level will flow to a lower level; the ΔG for water flowing down hill is equal in magnitude but opposite in sign to that for water flowing uphill. Transformations with positive values for ΔG include such outlandish things as water flowing uphill, a ball bearing jumping out of the glass of water, the automobile which manufactures gasoline from water and carbon dioxide as it is pushed down the street. "Nonspontaneous" seems too mild a word to describe them; "fantastic" might be better.

10–3 Recollection

By comparing real transformations with reversible ones we arrived at the Clausius inequality, $dS > dQ/T$, which gives us a criterion for a real, or spontaneous, transformation. By algebraic manipulation this criterion was given simple expression in terms of the entropy change, or changes in value of two new functions A and G. By examining the algebraic sign of ΔS, or ΔA, or ΔG for the transformation in question, we can decide whether it can occur spontaneously or not. At the same time we obtain the condition of equilibrium for the transformation. These conditions of spontaneity and equilibrium are summarized in Table 10–1. Of all the conditions in Table 10–1, we shall make the greatest use of those on the last line, since the constraints $U = 0$, T and p constant are those most frequently used in the laboratory.

Table 10–1

Constraint	Condition for spontaneity		Equilibrium condition	
None	$-(dE + p\,dV - T\,dS) - dU = +$		$-(dE + p\,dV - T\,dS) - dU = 0$	
	Infinitesimal change	Finite change	Infinitesimal change	Finite change
Isolated system	$dS = +$	$\Delta S = +$	$dS = 0$	$\Delta S = 0$
T constant	$dA + dW = -$	$\Delta A + W = -$	$dA + dW = 0$	$\Delta A + W = 0$
T, p constant	$dG + dU = -$	$\Delta G + U = -$	$dG + dU = 0$	$\Delta G + U = 0$
$U = 0; T, V$ constant	$dA = -$	$\Delta A = -$	$dA = 0$	$\Delta A = 0$
$U = 0; T, p$ constant	$dG = -$	$\Delta G = -$	$dG = 0$	$\Delta G = 0$

The word "spontaneous" applied to changes in state in a thermodynamic sense must not be given too broad a meaning. It means only that the change in state is *possible*. Thermodynamics cannot provide any information about how long a time is required for the change in state. For example, thermodynamics predicts that at 25°C and 1 atm pressure the reaction between hydrogen and oxygen to form water is a spontaneous reaction. However, in the absence of a catalyst or an initiating event, such as a spark, they do not react to form water in any measurable length of time. The length of time required for a spontaneous transformation to come to equilibrium is proper subject matter for kinetics, not thermodynamics. Thermodynamics tells us what *can* happen; kinetics tells us whether it will take a thousand years or a millionth of a second. Once we know that a certain reaction *can* happen, it may be worth our while to search for a catalyst which will shorten the time interval required for the reaction to reach equilibrium. It is futile to seek a catalyst for a reaction which is thermodynamically impossible.

What can be done about those transformations which have ΔG positive and which are thermodynamically impossible, or nonspontaneous? Human nature being what it is, it does not submit lightly to the judgment that a certain change is "impossible." The "impossible" flow of water uphill can be made "possible," not through the agency of a catalyst which is unchanged in the transformation, but by *coupling* the nonspontaneous uphill flow of a certain mass of water with the spontaneous downhill flow of a greater mass of some substance. A weight cannot by itself jump three feet up from the floor, but if it is coupled through a pulley to a heavier weight which falls three feet, it will jump up. The composite change, light weight up, heavier weight down, is accompanied by a decrease in free energy and thus is a "possible" one. As we shall see later, coupling one change in state with another can be turned to great advantage in dealing with chemical reactions.

10–4   Driving Forces for Natural Changes

In a natural change at constant temperature and pressure, ΔG must be negative. By definition, $G = H - TS$, so that at constant temperature

$$\Delta G = \Delta H - T\,\Delta S. \tag{10–18}$$

Two contributions to the value of ΔG can be distinguished in Eq. (10–18); an energetic one, ΔH, and an entropic one, $T\,\Delta S$. From Eq. (10–18) it is clear that to make ΔG negative, it is best if ΔH is negative (exothermic transformation) and if ΔS is positive. In a natural change, the system attempts to achieve the lowest enthalpy (roughly, the lowest energy) and the highest entropy. It is also clear from Eq. (10–18) that a system can tolerate a decrease in entropy in the change in state which makes the second term positive, if the first term is negative enough to over-balance the positive second term. Similarly, an increase in enthalpy, ΔH positive, can be tolerated if ΔS is sufficiently positive so that the second term over-balances the first. In such instances the compromise between low enthalpy and high entropy is reached in such a way as to minimize the free energy at equilibrium. The majority of common chemical reactions are exothermic in their natural direction, often so highly exothermic that the term $T\,\Delta S$ has little influence in determining the equilibrium position. In the case of reactions which are endothermic in their natural direction, the term $T\,\Delta S$ is all-important in determining the equilibrium position.

10–5   The Fundamental Equations of Thermodynamics

In addition to the mechanical properties p and V, a system has three fundamental properties T, E, and S, defined by the laws of thermodynamics, and three composite properties H, A, and G, which are important. We are now in a position to develop the important differential equations which relate these properties to one another.

For the present we restrict the discussion to systems which produce only expansion work so that $dU = 0$. With this restriction, the general condition of equilibrium is

$$dE = T\,dS - p\,dV. \tag{10-19}$$

This combination of the first and second laws of thermodynamics is the fundamental equation of thermodynamics. Using the definitions of the composite functions,

$$H = E + pV, \quad A = E - TS, \quad G = E + pV - TS,$$

and differentiating each one, we obtain

$$dH = dE + p\,dV + V\,dp,$$

$$dA = dE - T\,dS - S\,dT,$$

$$dG = dE + p\,dV + V\,dp - T\,dS - S\,dT.$$

In each of these three equations, dE is replaced by its value from Eq. (10–19); after collecting terms, the equations become [Eq. (10–19) is repeated first]

$$dE = T\,dS - p\,dV,$$

$$dH = T\,dS + V\,dp, \tag{10-20}$$

$$dA = -S\,dT - p\,dV, \tag{10-21}$$

$$dG = -S\,dT + V\,dp. \tag{10-22}$$

These four equations are sometimes called the four fundamental equations of thermodynamics; in fact, they are simply four different ways of looking at the one fundamental equation, Eq. (10–19).

Equation (10–19) relates the change in energy to the changes in entropy and volume. Equation (10–20) relates the change in enthalpy to changes in entropy and pressure. Equation (10–21) relates the change in the work function dA to changes in temperature and volume. Equation (10–22) relates the change in free energy to changes in temperature and pressure. Because of the simplicity of these equations, S and V are called the "natural" variables for the energy; S and p are the natural variables for the enthalpy; T and V are the natural variables for the work function; and T and p are the natural variables for the free energy.

Since each of the expressions on the right-hand side of these equations is an exact differential expression, it follows that the cross-derivatives are equal. From this, we immediately obtain the four Maxwell relations:

$$\left(\frac{\partial T}{\partial V}\right)_S = -\left(\frac{\partial p}{\partial S}\right)_V, \tag{10-23}$$

$$\left(\frac{\partial T}{\partial p}\right)_S = \left(\frac{\partial V}{\partial S}\right)_p, \tag{10-24}$$

$$\left(\frac{\partial S}{\partial V}\right)_T = \left(\frac{\partial p}{\partial T}\right)_V, \tag{10–25}$$

$$-\left(\frac{\partial S}{\partial p}\right)_T = \left(\frac{\partial V}{\partial T}\right)_p. \tag{10–26}$$

The first two of these equations relate to changes in state at constant entropy, that is, adiabatic, reversible changes in state. The derivative $(\partial T/\partial V)_S$ represents the rate of change of temperature with volume in a reversible adiabatic transformation. We shall not be much concerned with Eqs. (10–23) and (10–24).

Equations (10–25) and (10–26) are of great importance because they relate the isothermal volume dependence of the entropy and the isothermal pressure dependence of the entropy to easily measured quantities. We obtained these relations earlier, Eqs. (9–31) and (9–41), by utilizing the fact that dS is an exact differential. Here we obtain them with much less algebraic labor from the facts that dA and dG are exact differentials. The two derivations are clearly equivalent since A and G are functions of the state only if S is a function of the state.

10–6   The "Thermodynamic" Equation of State

The equations of state discussed so far, the ideal gas law, the van der Waals equation, etc., were relations between p, V, and T obtained from empirical data on the behavior of gases or from speculation about the effects of molecular size and attractive forces on the behavior of the gas. The equation of state for a liquid or solid was simply expressed in terms of the experimentally determined coefficients of thermal expansion and compressibility. These relations applied to systems at equilibrium, but there is a more general condition of equilibrium. The second law of thermodynamics requires the relation, Eq. (10–19),

$$dE = T\,dS - p\,dV$$

as an equilibrium condition. From this we should be able to derive an equation of state for any system whatsoever. Let the changes in E, S, and V of Eq. (10–19) be changes at constant T:

$$(\partial E)_T = T(\partial S)_T - p(\partial V)_T.$$

Now dividing by $(\partial V)_T$, we have

$$\left(\frac{\partial E}{\partial V}\right)_T = T\left(\frac{\partial S}{\partial V}\right)_T - p, \tag{10–27}$$

where, from the writing of the derivatives, E and S are considered as functions of T and V. Therefore, the partial derivatives in Eq. (10–27) are functions of T and V. This equation relates the pressure to functions of T and V; it is an equation of state. Using the value for $(\partial S/\partial V)_T$ from Eq. (10–25) and rearranging, Eq. (10–27) becomes

$$p = T\left(\frac{\partial p}{\partial T}\right)_V - \left(\frac{\partial E}{\partial V}\right)_T, \tag{10–28}$$

which is perhaps a neater form for the equation.

By restricting the second fundamental equation, Eq. (10–20), to constant temperature and dividing by $(\partial p)_T$, we obtain

$$\left(\frac{\partial H}{\partial p}\right)_T = T\left(\frac{\partial S}{\partial p}\right)_T + V. \tag{10–29}$$

Using Eq. (10–26) and rearranging, this equation becomes

$$V = T\left(\frac{\partial V}{\partial T}\right)_p + \left(\frac{\partial H}{\partial p}\right)_T, \tag{10–30}$$

which is a general equation of state expressing the volume as a function of temperature and pressure. These equations of state are applicable to any substance whatsoever. Eqs. (10–28) and (10–30) were obtained earlier, Eqs. (9–30) and (9–40), but were not discussed at that point.

If we knew the value of either $(\partial E/\partial V)_T$ or $(\partial H/\partial p)_T$ for a substance, we would know its equation of state immediately from Eqs. (10–28) or (10–30). More commonly we do not know the values of these derivatives, so we arrange Eq. (10–28) in the form

$$\left(\frac{\partial E}{\partial V}\right)_T = T\left(\frac{\partial p}{\partial T}\right)_V - p. \tag{10–31}$$

From the empirical equation of state, the right-hand side of Eq. (10–31) can be evaluated to yield a value of the derivative $(\partial E/\partial V)_T$. For example, for the ideal gas, $p = RT/\overline{V}$, so $(\partial p/\partial T)_V = R/\overline{V}$. Using these values in Eq. (10–31), we obtain $(\partial E/\partial V)_T = RT/\overline{V} - p = p - p = 0$. We have used this result, Joule's law, before; this demonstration proves its validity for the ideal gas.

Since, from Eq. (9–23), $(\partial p/\partial T)_V = \alpha/\beta$, Eq. (10–31) is often written in the form

$$\left(\frac{\partial E}{\partial V}\right)_T = T\frac{\alpha}{\beta} - p = \frac{\alpha T - \beta p}{\beta}, \tag{10–32}$$

and Eq. (10–30) in the form

$$\left(\frac{\partial H}{\partial p}\right)_T = V(1 - \alpha T). \tag{10–33}$$

It is now possible, using Eqs. (10–32) and (10–33), to write the total differentials of E and H in a form containing only quantities which are easily measurable:

$$dE = C_v\,dT + \frac{(\alpha T - \beta p)}{\beta}dV, \tag{10–34}$$

$$dH = C_p\,dT + V(1 - \alpha T)\,dp. \tag{10–35}$$

These equations together with the two equations for dS, Eqs. (9–33), and (9–42), are helpful in deriving others.

Using Eq. (10–32), we can obtain a simple expression for $C_p - C_v$. From Eq. (7–39), we have

$$C_p - C_v = \left[p + \left(\frac{\partial E}{\partial V} \right)_T \right] V\alpha.$$

Using the value of $(\partial E/\partial V)_T$ from Eq. (10–32), we obtain

$$C_p - C_v = \frac{TV\alpha^2}{\beta}, \qquad (10\text{–}36)$$

which permits the evaluation of $C_p - C_v$ from quantities which are readily measurable for any substance. Since T, V, β, and α^2 must all be positive, C_p is always greater than C_v.

For the Joule-Thomson coefficient we have from Eq. (7–50),

$$C_p\mu_{JT} = -(\partial H/\partial p)_T.$$

Using Eq. (10–33), we obtain for μ_{JT},

$$C_p\mu_{JT} = V(\alpha T - 1). \qquad (10\text{–}37)$$

Thus, if we know C_p, V, and α for the gas, we can calculate μ_{JT}.

These quantities are much more easily measured than is μ_{JT} itself. At the Joule-Thomson inversion temperature, μ_{JT} changes sign, that is, $\mu_{JT} = 0$; using this condition in Eq. (10–37), we find at the inversion temperature, $T\alpha - 1 = 0$.

10–7 The Properties of A

The properties of the work function A are expressed by the fundamental equation (10–21),

$$dA = -S\,dT - p\,dV.$$

This equation views A as a function of T and V, and we have the identical equation

$$dA = \left(\frac{\partial A}{\partial T} \right)_V dT + \left(\frac{\partial A}{\partial V} \right)_T dV.$$

Comparing these two equations shows that

$$\left(\frac{\partial A}{\partial T} \right)_V = -S, \qquad (10\text{–}38)$$

$$\left(\frac{\partial A}{\partial V} \right)_T = -p. \qquad (10\text{–}39)$$

Since the entropy of any substance is positive, Eq. (10–38) shows that the work function of any substance decreases (minus sign) with an increase in temperature. The rate of this decrease is greater the greater the entropy of the substance. For gases which have high entropies, the rate of decrease of A with temperature is larger than for liquids and solids which have comparatively small entropies.

Similarly, the minus sign in Eq. (10–39) shows that an increase in volume decreases the work function; the rate of decrease is greater the higher the pressure.

10–8 The Properties of G

The fundamental equation (10–22),

$$dG = -S\,dT + V\,dp,$$

views the free energy as a function of temperature and pressure; the equivalent expression is, therefore,

$$dG = \left(\frac{\partial G}{\partial T}\right)_p dT + \left(\frac{\partial G}{\partial p}\right)_T dp. \qquad (10\text{–}40)$$

Comparing these two equations shows that

$$\left(\frac{\partial G}{\partial T}\right)_p = -S, \qquad (10\text{–}41)$$

$$\left(\frac{\partial G}{\partial p}\right)_T = V. \qquad (10\text{–}42)$$

Because of the importance of the free energy, Eqs. (10–41) and (10–42) contain two of the most important pieces of information in thermodynamics. Again, since the entropy of any substance is positive, the minus sign in Eq. (10–41) shows that increase in temperature decreases the free energy if the pressure is constant. The rate of decrease is greater for gases which have large entropies than for liquids or solids which have small entropies. The fact that V is always positive means through Eq. (10–42) that increase in pressure increases the free energy at constant temperature. The larger the volume of the system the greater is the increase in free energy for a given increase in pressure. The comparatively large volume of a gas implies that the free energy of a gas increases much more rapidly with pressure than would that of a liquid or a solid.

The free energy of any pure material is conveniently expressed by integrating Eq. (10–22) at constant temperature from the standard pressure, 1 atm, to any other pressure p:

$$\int_1^p dG = \int_1^p V\,dp, \qquad G - G^0 = \int_1^p V\,dp,$$

$$G = G^0(T) + \int_1^p V\,dp, \qquad (10\text{–}43)$$

where $G^0(T)$ is the free energy of the substance under 1 atm pressure, the *standard* free energy, which is a function of temperature.

If the substance in question is either a liquid or a solid, then the volume is nearly independent of the pressure and can be removed from under the integral sign; then

$$G(T, p) = G^0(T) + V(p - 1) \qquad \text{(liquids and solids).} \qquad (10\text{–}44)$$

Since the volume of liquids and solids is small, unless the pressure is enormous, the second term on the right of Eq. (10–44) is negligibly small; ordinarily for condensed phases we will write simply

$$G = G^0(T) \qquad (10\text{–}45)$$

and ignore the pressure dependence of G.

The volume of gases is very much larger than that of liquids or solids and depends sharply on pressure; applying Eq. (10–43) to the ideal gas, it becomes

$$G = G^0(T) + \int_1^p \frac{nRT}{p}\,dp, \qquad \frac{G}{n} = \frac{G^0(T)}{n} + RT\ln\left(\frac{p(\text{atm})}{1\ \text{atm}}\right).$$

It is customary to use a special symbol, μ, for the free energy per mole; we define

$$\mu = G/n. \qquad (10\text{–}46)$$

Thus for the molar free energy of the ideal gas, we have

$$\mu = \mu^0(T) + RT\ln p. \qquad (10\text{–}47)$$

As in Section 9–11, the symbol p in Eq. (10–47) represents a pure number, the number which when multiplied by 1 atm yields the value of the pressure in atmospheres.

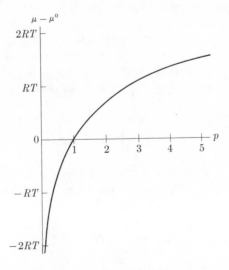

Fig. 10–2 Free energy of ideal gas as a function of pressure.

The logarithmic term in Eq. (10–47) is quite large in most circumstances and cannot be ignored. From this equation it is clear that at a specified temperature, the pressure describes the free energy of the ideal gas; the higher the pressure the greater the free energy (Fig. 10–2).

It is worth emphasizing that if we know the functional form of $G(T, p)$, then we can obtain all the other thermodynamic functions by differentiation, using Eqs. (10–41) and (10–42), and combining with definitions. (See Problem 10–19.)

10–9   The Free Energy of Real Gases

The functional form of Eq. (10–47) is particularly simple and convenient. It would be helpful if the molar free energy of real gases could be expressed in the same mathematical form. We therefore "invent" a function of the state which will express the molar free energy of a real gas by the equation

$$\mu = \mu^0(T) + RT \ln f. \tag{10–48}$$

The function f is called the *fugacity* of the gas. Clearly, the fugacity measures the free energy of a real gas in the same way as the pressure measures the free energy of an ideal gas.

An invented function such as the fugacity has little use unless it can be related to measurable properties of the gas. Dividing the fundamental equation (10–22) by n, the number of moles of gas, and restricting to constant temperature, $dT = 0$, we obtain for the real gas $d\mu = \overline{V} \, dp$, while for the ideal gas $d\mu_{id} = \overline{V}_{id} \, dp$, where $\overline{V}$ and $\overline{V}_{id}$ are the molar volumes of the real and ideal gases, respectively. Subtracting these two equations, we obtain $d(\mu - \mu_{id}) = (\overline{V} - \overline{V}_{id}) \, dp$.

Integrating between the limits p^* and p yields

$$(\mu - \mu_{id}) - (\mu^* - \mu_{id}^*) = \int_{p^*}^{p} (\overline{V} - \overline{V}_{id}) \, dp.$$

We now let $p^* \to 0$; the properties of any real gas approach their ideal values as the pressure of the gas approaches zero. Therefore, as $p^* \to 0$, $\mu^* \to \mu_{id}^*$. The equation becomes

$$\mu - \mu_{id} = \int_{0}^{p} (\overline{V} - \overline{V}_{id}) \, dp. \tag{10–49}$$

But by Eq. (10–47), $\mu_{id} = \mu^0(T) + RT \ln p$, and by the definition of f, Eq. (10–48), $\mu = \mu^0(T) + RT \ln f$. Using these values for μ and μ_{id}, Eq. (10–49) becomes

$$RT(\ln f - \ln p) = \int_{0}^{p} (\overline{V} - \overline{V}_{id}) \, dp,$$

$$\ln f = \ln p + \frac{1}{RT} \int_{0}^{p} (\overline{V} - \overline{V}_{id}) \, dp. \tag{10–50}$$

The integral in Eq. (10–50) can be evaluated graphically; knowing $\bar{V}$ as a function of pressure, we plot the quantity $(\bar{V} - \bar{V}_{id})/RT$ as a function of pressure. The area under the curve from $p = 0$ to p is the value of the second term on the right of Eq. (10–50). Or, if $\bar{V}$ can be expressed as a function of pressure by an equation of state, the integral can be evaluated analytically, since $\bar{V}_{id} = RT/p$. The integral can be expressed neatly in terms of the compressibility factor Z; by definition, $\bar{V} = Z\bar{V}_{id}$. Using this value for $\bar{V}$, and $\bar{V}_{id} = RT/p$, in the integral of Eq. (10–50), it reduces to

$$\ln f = \ln p + \int_0^p \frac{(Z - 1)}{p} dp. \qquad (10\text{–}51)$$

The integral in Eq. (10–51) is evaluated graphically by plotting $(Z - 1)/p$ against p and measuring the area under the curve. For gases below their Boyle temperatures, $Z - 1$ is negative at moderate pressures, and the fugacity, by Eq. (10–51), will be less than the pressure. For gases above their Boyle temperatures, the fugacity is greater than the pressure.

The free energy of gases will usually be discussed as if the gases were ideal and Eq. (10–47) will be used. The algebra will be exactly the same for real gases, we need only replace the pressure in the final equations by the fugacity, keeping in mind that the fugacity depends on temperature as well as pressure.

10–10 Temperature Dependence of the Free Energy

The dependence of the free energy on temperature is expressed in several different ways for convenience in different problems. Rewriting Eq. (10–41), we have

$$\left(\frac{\partial G}{\partial T}\right)_p = -S. \qquad (10\text{–}52)$$

From the definition $G = H - TS$, we obtain $-S = (G - H)/T$, and Eq. (10–52) becomes

$$\left(\frac{\partial G}{\partial T}\right)_p = \frac{G - H}{T}, \qquad (10\text{–}53)$$

a form which is sometimes useful.

It is frequently important to know how the function G/T depends on temperature. By the ordinary rule of differentiation, we obtain

$$\left(\frac{\partial (G/T)}{\partial T}\right)_p = \frac{1}{T}\left(\frac{\partial G}{\partial T}\right)_p - \frac{1}{T^2}G.$$

Using Eq. (10–52) this becomes

$$\left(\frac{\partial (G/T)}{\partial T}\right)_p = -\frac{TS + G}{T^2},$$

which reduces to

$$\left(\frac{\partial(G/T)}{\partial T}\right)_p = -\frac{H}{T^2}, \tag{10-54}$$

the Gibbs–Helmholtz equation, which we use frequently.

Since $d(1/T) = -(1/T^2)\, dT$, we can replace ∂T in the derivative in Eq. (10–54) by $-T^2\, \partial(1/T)$; this reduces it to

$$\left(\frac{\partial(G/T)}{\partial(1/T)}\right)_p = H, \tag{10-55}$$

which is another frequently used relation.

Any of Eqs. (10–52), (10–53), (10–54), (10–55) are simply different versions of the fundamental equation, Eq. (10–52). We will refer to them as the first, second, third, and fourth forms of the Gibbs–Helmholtz equation.

Problems

10–1. Using the van der Waals equation together with the thermodynamic equation of state, evaluate $(\partial E/\partial V)_T$ for the van der Waals gas.

10–2. From the purely mathematical properties of the exact differential

$$dE = C_v\, dT + \left(\frac{\partial E}{\partial V}\right)_T dV,$$

show that if $(\partial E/\partial V)_T$ is a function only of volume, then C_v is a function only of temperature.

10–3. By integrating the total differential dE for a van der Waals gas, show that if C_v is a constant, $E = E' + C_v T - a/V$, where E' is a constant of integration. (The answer to Problem 10–1 is needed for this problem.)

10–4. Calculate ΔE for the isothermal expansion of one mole of a van der Waals gas from 20 liters/mole to 80 liters/mole; if $a = 1.39$ liter^2·atm·moles^{-2} (nitrogen) and if $a = 31.5$ liter^2·atm·moles^{-2} (heptane).

10–5. By integrating Eq. (10–39) derive an expression for the work function of

   a) the ideal gas,

   b) the van der Waals gas. (Do not forget the "constant" of integration!)

10–6. a) Find the value of $(\partial S/\partial V)_T$ for the van der Waals gas.

   b) Derive an expression for the change in entropy for the isothermal expansion of one mole of the van der Waals gas from V_1 to V_2.

   c) Compare the result in (b) with the expression for the ideal gas. For the same increase in volume, will the entropy increase be greater for the van der Waals gas or for the ideal gas?

10–7. By taking the reciprocal of both sides of Eq. (10–23) we obtain $(\partial S/\partial p)_V = -(\partial V/\partial T)_S$. Using this equation and the cyclic relation between V, T, and S, show that $(\partial S/\partial p)_V = \beta C_v/\alpha T$.

10-8. a) Write the thermodynamic equation of state for a substance which follows Joule's law.

b) By integrating the differential equation obtained in (a), show that at constant volume the pressure is proportional to the absolute temperature for such a substance.

10-9. To a first approximation the compressibility factor of the van der Waals gas is given by

$$p\bar{V}/RT = 1 + [b - (a/RT)](p/RT).$$

Calculate the fugacity of the van der Waals gas.

10-10. From the definition of the fugacity and the Gibbs–Helmholtz equation, show that the molar enthalpy $\bar{H}$ of a real gas is related to the molar enthalpy of the ideal gas $\bar{H}^0$ by

$$\bar{H} = \bar{H}^0 - RT^2\left(\frac{\partial \ln f}{\partial T}\right)_p.$$

10-11. Combining the results of Problems 10-9 and 10-10 show that the enthalpy of the van der Waals gas is

$$\bar{H} = \bar{H}^0 + \left(b - \frac{2a}{RT}\right)p.$$

10-12. a) Show that Eq. (10-28) can be written in the form

$$\left(\frac{\partial E}{\partial V}\right)_T = T^2\left[\frac{\partial(p/T)}{\partial T}\right]_V = -\left[\frac{\partial(p/T)}{\partial(1/T)}\right]_V.$$

b) Show that Eq. (10-30) can be written in the form

$$\left(\frac{\partial H}{\partial p}\right)_T = -T^2\left[\frac{\partial(V/T)}{\partial T}\right]_p = \left[\frac{\partial(V/T)}{\partial(1/T)}\right]_p.$$

10-13. Show that for real gas; $C_p\mu_{JT} = (RT^2/p)\,(\partial Z/\partial T)_p$ where μ_{JT} is the Joule-Thomson coefficient, $Z = p\bar{V}/RT$ is the compressibility factor of the gas. [Compare Eq. (7–50).]

10-14. Using the value of Z for the van der Waals gas given in Problem 10–9, calculate the value μ_{JT}. Show that μ_{JT} changes sign at the inversion temperature, $T_{inv} = 2a/Rb$.

10-15. Given $dE = C_v\,dT + [(\alpha T - \beta p)/\beta]\,dV$, show that $dE = [C_v + (TV\alpha^2/\beta) - pV\alpha]\,dT + V(p\beta - T\alpha)\,dp$. [Hint: Expand dV in terms of dT and dp.]

10-16. Using the result in Problem 10–15 and the data for carbon tetrachloride at 20°C: $\alpha = 12.4 \times 10^{-4}\,\mathrm{deg}^{-1}$; $\beta = 103 \times 10^{-6}\,\mathrm{atm}^{-1}$;   density $= 1.5942\,\mathrm{gm/cm^3}$ and $M = 153.8$, show that near 1 atm pressure, $(\partial E/\partial p)_T \approx -VT\alpha$. Calculate the change in molar energy per atm at 20°C in cal/atm.

10-17. Using the approximate value of the compressibility factor given in Problem 10–9, show that for the van der Waals gas:

a) $C_p - C_v = R + 2ap/RT^2$.

b) $(\partial E/\partial p)_T = -a/RT$. [Hint: Problem 10–15.]

c) $(\partial E/\partial T)_p = C_v + ap/RT^2$.

10-18. Knowing that $dS = (C_p/T)\,dT - V\alpha\,dp$, show that

a) $(\partial S/\partial p)_V = \beta C_v/T\alpha$.

b) $(\partial S/\partial V)_p = C_p/TV\alpha$.

c) $-(1/V)(\partial V/\partial p)_S = \beta/\gamma$, where $\gamma = C_p/C_v$.

10–19. By using the fundamental differential equations and the definitions of the functions, determine the functional form of $\bar{S}, \bar{V}, \bar{H}, \bar{E}$ for

 a) the ideal gas, given that $\mu = \mu^0(T) + RT \ln p$.

 b) the van der Waals gas, given that

$$\mu = \mu^0(T) + RT \ln p + (b - a/RT)p.$$

10–20. Show that if $Z = 1 + B(T)p$, then $f = pe^{Z-1}$ and this implies that at low to moderate pressures $f \approx pZ$ and that $p = \sqrt{fp_{ideal}}$. This last relation says that the pressure is the geometric mean of the ideal pressure and the fugacity.

Chapter Eleven

Systems of Variable Composition; Chemical Equilibrium

11–1 The Fundamental Equation

In our study so far we have assumed implicitly that the system is composed of a pure substance, or, if it was composed of a mixture, that the composition of the mixture was unaltered in the change of state. As a chemical reaction proceeds, the composition of the system changes; correspondingly, the thermodynamic properties change. Consequently, we must introduce the dependence on composition into the thermodynamic equations. We do this first only for the free energy G, since it is the most immediately useful.

For a pure substance or for a mixture of fixed composition the fundamental equation for the free energy is

$$dG = -S\,dT + V\,dp. \tag{11-1}$$

If the mole numbers, $n_1, n_2, \ldots,$ of the substances present vary, then $G = G(T, p, n_1, n_2, \ldots)$, and the total differential is

$$dG = \left(\frac{\partial G}{\partial T}\right)_{p,n_i} dT + \left(\frac{\partial G}{\partial p}\right)_{T,n_i} dp + \left(\frac{\partial G}{\partial n_1}\right)_{T,p,n_j} dn_1 + \left(\frac{\partial G}{\partial n_2}\right)_{T,p,n_j} dn_2 + \cdots, \tag{11-2}$$

where the subscript n_i on the partial derivative means that *all* the mole numbers are constant in the differentiation, and the subscript n_j on the partial derivative means that all the mole numbers except the one in that derivative are constant in the differentiation. For example, $(\partial G/\partial n_2)_{T,p,n_j}$ means that T, p, and all the mole numbers except n_2 are constant in the differentiation.

If the system does not suffer any change in composition, then

$$dn_1 = 0, \qquad dn_2 = 0,$$

and so on, and Eq. (11–2) reduces to

$$dG = \left(\frac{\partial G}{\partial T}\right)_{p,n_i} dT + \left(\frac{\partial G}{\partial p}\right)_{T,n_i} dp. \qquad (11\text{–}3)$$

Comparison of Eq. (11–3) with Eq. (11–1), shows that

$$\left(\frac{\partial G}{\partial T}\right)_{p,n_i} = -S \qquad (11\text{–}4a)$$

and

$$\left(\frac{\partial G}{\partial p}\right)_{T,n_i} = V. \qquad (11\text{–}4b)$$

To simplify writing, we define

$$\mu_i = \left(\frac{\partial G}{\partial n_i}\right)_{T,p,n_j}. \qquad (11\text{–}5)$$

In view of Eqs. (11–4) and (11–5), the total differential of G in Eq. (11–2) becomes

$$dG = -S\,dT + V\,dp + \mu_1\,dn_1 + \mu_2\,dn_2 + \cdots. \qquad (11\text{–}6)$$

Equation (11–6) relates the change in free energy to changes in the temperature, pressure, and the mole numbers; it is usually written in the more compact form

$$dG = -S\,dT + V\,dp + \sum_i \mu_i\,dn_i, \qquad (11\text{–}7)$$

where the sum includes all the constituents of the mixture.

11–2 The Properties of μ_i

If a small amount of substance i, dn_i moles, is added to a system, keeping T, p, and all the other mole numbers constant, then the increase in free energy is given by Eq. (11–7), which reduces to $dG = \mu_i\,dn_i$. The increase in free energy *per mole* of the substance added is, therefore,

$$(\partial G/\partial n_i)_{T,p,n_j} = \mu_i.$$

This equation expresses the immediate significance of μ_i, and is simply the content of the definition of μ_i in Eq. (11–5). For any substance i in a mixture, the value of μ_i is the increase in free energy which attends the addition of an infinitesimal number of moles of that substance to the mixture *per mole* of the substance added. (The amount added is restricted to an infinitesimal quantity so that the composition of the mixture, and therefore the value of μ_i, does not change.)

An alternative approach involves an extremely large system, let us say a roomful of a water solution of sugar. If one mole of water is added to such a large system, the composition of the system remains the same for all practical purposes, and therefore the μ_{H_2O} of the water is constant. The increase in free energy attending the addition of one mole of water to the roomful of solution is the value of μ_{H_2O} in the solution.

Since μ_i is the derivative of one extensive variable by another, it is an *intensive* property of the system and must have the same value everywhere within a system at equilibrium.

Suppose that μ_i had different values, μ_i^A and μ_i^B, in two regions of the system, A and B. Then keeping T, p, and all the other mole numbers constant, suppose that we transfer dn_i moles of i from region A to region B. For the increase in free energy in the two regions, we have from Eq. (11–7), $dG^A = \mu_i^A(-dn_i)$, and $dG^B = \mu_i^B\, dn_i$, since $+dn_i$ moles go into B and $-dn_i$ moles go into A. The total change in free energy of the system is the sum $dG = dG^A + dG^B$, or

$$dG = (\mu_i^B - \mu_i^A)\, dn_i.$$

Now if μ_i^B is less than μ_i^A, then dG is negative, and this transfer of matter decreases the free energy of the system; the transfer therefore occurs spontaneously. Thus, substance i flows spontaneously from a region of high μ_i to a region of low μ_i; this flow continues until the value of μ_i is uniform throughout the system, that is, until the system is in equilibrium. The fact that μ_i must have the same value throughout the system is an important equilibrium condition which we will use again and again.

The property μ_i is called the *chemical potential*† of the substance i. Matter flows spontaneously from a region of high chemical potential to a region of low chemical potential just as electric current flows spontaneously from a region of high electrical potential to one of lower electrical potential, or as mass flows spontaneously from a position of high gravitational potential to one of low gravitational potential. Another name frequently given to μ_i is the *escaping tendency* of i. If the chemical potential of a component in a system is high, that component has a large escaping tendency, while if the chemical potential is low, the component has a small escaping tendency.

11–3 The Free Energy of a Mixture

The fact that the μ_i are intensive properties implies that they can depend only on other intensive properties such as temperature, pressure, and intensive composition variables such as the mole ratios, or the mole fractions. Since the μ_i depend on the mole numbers only through intensive composition variables, an important relation is easily derived.

---

† Strictly, the chemical potential is $(\partial G/\partial m_i)_{T,p,m_j}$ where m_i is the *mass* of the ith substance. No difficulty arises with the definition in Eq. (11–5) so long as a change in molecular weight is treated as a chemical transformation.

Consider the following transformation:

	Initial state					Final state			
	T, p					T, p			
Substance:	1	2	3	...	Substance:	1	2	3	...
Mole number: 0	0	0	...	Mole number: n_1	n_2	n_3	...	
Free energy:	$G = 0$				Free energy:	G			

We achieve this transformation by considering a large quantity of a mixture of uniform composition, in equilibrium at constant temperature and constant pressure. Imagine a small, closed mathematical surface such as a sphere which lies completely in the interior of this mixture and forms the boundary which encloses our thermodynamic system. We denote the free energy of this system by G^* and the number of moles of the ith species in the system by n_i^*. We now ask by how much the free energy of the system increases if we enlarge this mathematical surface so that it encloses a greater quantity of the mixture. We may imagine that the final boundary enlarges and deforms in such a way as to enclose any desired amount of mixture in a vessel of any shape. Let the free energy of the enlarged system be G and the mole numbers be n_i. We obtain this change in free energy by integrating Eq. (11–7) at constant T and p; that is,

$$\int_{G^*}^{G} dG = \sum_i \mu_i \int_{n_i^*}^{n_i} dn_i$$

$$G - G^* = \sum_i \mu_i (n_i - n_i^*). \tag{11–8}$$

The μ_i were taken out of the integrals since, as we have shown above, each μ_i must have the same value everywhere throughout a system at equilibrium. Now we allow our initial small boundary to shrink to the limit of enclosing zero volume; then $n_i^* = 0$, and $G^* = 0$. This reduces Eq. (11–8) to

$$G = \sum_i n_i \mu_i. \tag{11–9}$$

The addition rule in Eq. (11–9) is a very important property of chemical potentials. Knowing the chemical potential and the number of moles of each constituent of a mixture, we can compute, using Eq. (11–9), the total free energy G of the mixture at the specified temperature and pressure. If the system contains only one substance, then Eq. (11–9) reduces to $G = n\mu$, or

$$\mu = G/n. \tag{11–10}$$

By Eq. (11–10), the μ of a pure substance is simply the *molar free energy*; for this reason the symbol μ was introduced for molar free energy in Section 10–8. In mixtures, μ_i is the *partial molar free energy* of the substance i.

11–4 The Chemical Potential of a Pure Ideal Gas

The chemical potential of a pure ideal gas is given explicitly by Eq. (10–47):

$$\mu = \mu^0(T) + RT\ln p. \tag{11–11}$$

This equation shows that at a given temperature, the pressure is a measure of the chemical potential of the gas. If inequalities in pressure exist in a container of a gas, then matter will flow from the high-pressure regions (high chemical potential) to those of lower pressure (lower chemical potential) until the pressure is equalized throughout the vessel. The equilibrium condition, equality of the chemical potential everywhere, requires that the pressure be uniform throughout the vessel. For nonideal gases it is the fugacity that must be uniform throughout the vessel; however, since the fugacity is a function of temperature and pressure, at a given temperature equal values of fugacity imply equal values of pressure.

11–5 Chemical Potential of an Ideal Gas in a Mixture of Ideal Gases

Consider the system shown in Fig. 11–1. The right-hand compartment contains a mixture of hydrogen under a partial pressure p_{H_2} and nitrogen under a partial pressure p_{N_2}, the total pressure being $p = p_{H_2} + p_{N_2}$. The mixture is separated from the left-hand side by a palladium membrane. Since hydrogen can pass freely through the membrane, the left-hand side contains pure hydrogen. When equilibrium is attained, the pressure of the pure hydrogen on the left-hand side is equal by definition to the partial pressure of the hydrogen in the mixture; see Section 2–8. The equilibrium condition requires that the chemical potential of the hydrogen must have the same value in both sides of the vessel:

$$\mu_{H_2(pure)} = \mu_{H_2(mix)}.$$

The chemical potential of pure hydrogen under a pressure p_{H_2} is, by Eq. (11–11),

$$\mu_{H_2(pure)} = \mu^0_{H_2}(T) + RT\ln p_{H_2}.$$

Therefore, in the mixture it must be that

$$\mu_{H_2(mix)} = \mu^0_{H_2}(T) + RT\ln p_{H_2}.$$

Palladium membrane

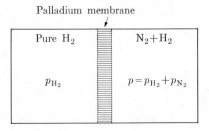

Fig. 11–1 Chemical potential of a gas in a mixture.

This equation shows that the chemical potential of hydrogen in a mixture is a logarithmic function of the *partial pressure* of hydrogen in the mixture. By repeating the argument using a mixture of any number of ideal gases, and a membrane† permeable only to substance i, it may be shown that the chemical potential of substance i in the mixture is given by

$$\mu_i = \mu_i^0(T) + RT \ln p_i, \tag{11–12}$$

where p_i is the partial pressure of substance i in the mixture. The $\mu_i^0(T)$ has the same significance as for a pure gas; it is the chemical potential of the pure gas under 1 atm pressure at the temperature T.

By using the relation $p_i = x_i p$, where x_i is the mole fraction of substance i in the mixture and p is the total pressure, for p_i in Eq. (11–12), and expanding the logarithm, we obtain

$$\mu_i = \mu_i^0(T) + RT \ln p + RT \ln x_i. \tag{11–13}$$

By Eq. (11–11), the first two terms in Eq. (11–13) are the μ for pure i under the pressure p, so Eq. (11–13) reduces to

$$\mu_i = \mu_{i(\text{pure})}(T, p) + RT \ln x_i. \tag{11–14}$$

Since x_i is a fraction and its logarithm is negative, Eq. (11–14) shows that the chemical potential of any gas in a mixture is always less than the chemical potential of the pure gas under the same total pressure. If a pure gas under a pressure p is placed in contact with a mixture under the same total pressure, the pure gas will spontaneously flow into the mixture. This is the thermodynamic interpretation of the fact that gases, and for that matter liquids and solids as well, diffuse into one another.

The form of Eq. (11–14) suggests a generalization. Suppose we define an *ideal mixture*, or *ideal solution*, in any state of aggregation, solid, liquid, or gaseous, as one in which the chemical potential of every species is given by the expression

$$\mu_i = \mu_i^0(T, p) + RT \ln x_i \tag{11–14a}$$

In Eq. (11–14a) we interpret $\mu_i^0(T, p)$ as the chemical potential of the *pure* species i in the *same state of aggregation as the mixture*; i.e., in a liquid mixture, $\mu_i^0(T, p)$ is the chemical potential, or molar free energy, of *pure liquid i* at temperature T and pressure p, and x_i is the mole fraction of i in the liquid mixture. We will introduce particular empirical evidence to justify this generalization in Chapter 13.

11–6   Free Energy and Entropy of Mixing

Since the formation of a mixture from the pure constituents always occurs spontaneously, this process must be attended by a decrease in free energy. Our object now is to calculate the free energy of mixing. The initial state is shown in Fig. 11–2(a).

---

† The fact that such membranes are known for only a few gases does not impair the argument.

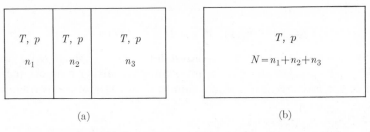

Fig. 11–2 Free energy of mixing. (a) Initial state. (b) Final state.

Each of the compartments contains a pure substance under a pressure p. The partitions separating the substances are pulled out and the final state, shown in Fig. 11–2(b), is the mixture under the same pressure p. The temperature is the same initially and finally. For the pure substances, the free energies are

$$G_1 = n_1\mu_1^0, \qquad G_2 = n_2\mu_2^0, \qquad G_3 = n_3\mu_3^0.$$

The free energy of the initial state is simply the sum

$$G_{\text{initial}} = G_1 + G_2 + G_3 = n_1\mu_1^0 + n_2\mu_2^0 + n_3\mu_3^0 = \sum_i n_i\mu_i^0.$$

The free energy in the final state is given by the addition rule, Eq. (11–9):

$$G_{\text{final}} = n_1\mu_1 + n_2\mu_2 + n_3\mu_3 = \sum_i n_i\mu_i.$$

The free energy of mixing, $\Delta G_{\text{mix}} = G_{\text{final}} - G_{\text{initial}}$, on inserting the values of G_{final} and G_{initial}, becomes

$$\Delta G_{\text{mix}} = n_1(\mu_1 - \mu_1^0) + n_2(\mu_2 - \mu_2^0) + n_3(\mu_3 - \mu_3^0) = \sum_i n_i(\mu_i - \mu_i^0)$$

Using the value of $\mu_i - \mu_i^0$ from Eq. (11–14a), we obtain

$$\Delta G_{\text{mix}} = RT(n_1 \ln x_1 + n_2 \ln x_2 + n_3 \ln x_3) = RT \sum_i n_i \ln x_i,$$

which can be put in a slightly more convenient form by the substitution $n_i = x_i N$, where N is the total number of moles in the mixture, and x_i is the mole fraction of i. Then

$$\Delta G_{\text{mix}} = NRT(x_1 \ln x_1 + x_2 \ln x_2 + x_3 \ln x_3), \tag{11–15}$$

which is the final expression for the free energy of mixing in terms of the mole fractions of the constituents of the mixture. Every term on the right-hand side is negative, and so the sum is always negative. From the derivation, it can be seen that in forming an ideal mixture of any number of species the free energy of mixing will be

$$\Delta G_{\text{mix}} = NRT \sum_i x_i \ln x_i. \tag{11–16}$$

If there are only two substances in the mixture, then if $x_1 = x, x_2 = 1 - x$, Eq. (11–16) becomes

$$\Delta G_{\text{mix}} = NRT[x \ln x + (1 - x) \ln (1 - x)]. \tag{11–17}$$

A plot of the function in Eq. (11–17) is shown in Fig. 11–3. The curve is symmetrical about $x = \frac{1}{2}$. The greatest decrease in free energy on mixing is associated with the formation of the mixture having equal numbers of moles of the two constituents.

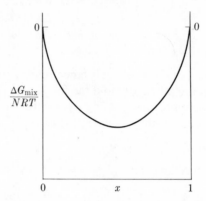

Fig. 11–3 $\Delta G_{\text{mix}}/NRT$ for a binary ideal mixture.

In a ternary system, the greatest decrease in free energy on mixing occurs if the mole fraction of each substance is equal to $\frac{1}{3}$, and so on.

Differentiation of $\Delta G_{\text{mix}} = G_{\text{final}} - G_{\text{initial}}$, with respect to temperature, yields ΔS_{mix} directly, through Eq. (11–4a):

$$\left(\frac{\partial \Delta G_{\text{mix}}}{\partial T}\right)_{p,n_i} = \left(\frac{\partial G_{\text{final}}}{\partial T}\right)_{p,n_i} - \left(\frac{\partial G_{\text{initial}}}{\partial T}\right)_{p,n_i} = -(S_{\text{final}} - S_{\text{initial}}),$$

$$\left(\frac{\partial \Delta G_{\text{mix}}}{\partial T}\right)_{p,n_i} = -\Delta S_{\text{mix}}. \tag{11–18}$$

Differentiating both sides of Eq. (11–16) with respect to temperature, we have

$$\left(\frac{\partial \Delta G_{\text{mix}}}{\partial T}\right)_{p,n_i} = NR \sum_i x_i \ln x_i,$$

so that Eq. (11–18) becomes

$$\Delta S_{\text{mix}} = -NR \sum_i x_i \ln x_i. \tag{11–19}$$

The functional form of the entropy of mixing is the same as for the free energy of mixing except that T does not appear as a factor and a minus sign occurs in the expression for the entropy of mixing. The minus sign means that the entropy of mixing is always positive, while the free energy of mixing is always negative. The positive entropy of mixing corresponds to the increase in randomness which occurs in mixing the molecules of several kinds. The expression for the entropy of mixing in Eq. (11–19) should be compared with that in Eq. (9–75), which was obtained from the statistical argument. Note that N in Eq. (9–75) is the number of molecules, while in Eq. (11–19) N is the number of moles; therefore, different constants, R and $\mathbf{k}$, appear in the two equations.

A plot of the entropy of mixing of a binary mixture according to the equation

$$\Delta S_{\text{mix}} = -NR[x \ln x + (1 - x) \ln (1 - x)] \tag{11–20}$$

is shown in Fig. 11–4. The entropy of mixing has a maximum value when $x = \frac{1}{2}$. Using $x = \frac{1}{2}$ in Eq. (11–20), we obtain for the entropy of mixing per mole of mixture

$$\Delta S_{\text{mix}}/N = -1.987(\tfrac{1}{2} \ln \tfrac{1}{2} + \tfrac{1}{2} \ln \tfrac{1}{2}) = -1.987 \ln \tfrac{1}{2} = +1.38 \text{ eu/mole}.$$

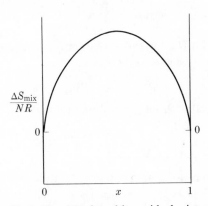

Fig. 11–4 $\Delta S_{\text{mix}}/NR$ for a binary ideal mixture.

In a mixture containing only two substances, the entropy of mixing per mole of the final mixture varies between 0 and 1.38 eu, depending on the composition.

The heat of mixing can be calculated by the equation

$$\Delta G_{\text{mix}} = \Delta H_{\text{mix}} - T \Delta S_{\text{mix}}, \tag{11–21}$$

using the values of the free energy and entropy of mixing from Eqs. (11–16) and (11–19). This reduces Eq. (11–21) to

$$NRT \sum_i x_i \ln x_i = \Delta H_{\text{mix}} + NRT \sum_i x_i \ln x_i,$$

which becomes

$$\Delta H_{\text{mix}} = 0 . \tag{11–22}$$

There is no heat effect associated with the formation of an ideal mixture.

Using the result that $\Delta H_{\text{mix}} = 0$, Eq. (11–21) becomes

$$-\Delta G_{\text{mix}} = T \Delta S_{\text{mix}}. \tag{11–23}$$

Equation (11–23) shows that the driving force, $-\Delta G_{mix}$, which produces the mixing is entirely an entropy effect. The mixed state is a more random state, and therefore is a more probable state. If the value of 1.38 eu is used for the entropy of mixing, then at $T = 300°K$, $\Delta G_{mix} = -(300)(1.38) = -414$ cal/mole. Thus, the free energy of mixing of an ideal binary mixture ranges from 0 to -414 cal/mole. Since -414 cal is not large, in nonideal mixtures for which the heat of mixing is not zero, the heat of mixing must either be negative or only slightly positive if the substances are to mix spontaneously. If the heat of mixing is more positive than about 300 or 400 cal/mole of mixture, then ΔG_{mix} is positive, and the liquids are not miscible but remain in two distinct layers.

The volume of mixing is obtained by differentiating the free energy of mixing with respect to pressure, the temperature and composition being constant,

$$\Delta V_{mix} = \left(\frac{\partial \Delta G_{mix}}{\partial p} \right)_{T, n_i}.$$

However, inspection of Eq. (11–16) shows that the free energy of mixing is independent of pressure, so the derivative is zero; hence,

$$\Delta V_{mix} = 0. \tag{11–24}$$

Ideal mixtures are formed without any volume change.

11–7 Chemical Equilibrium in a Mixture

Consider a closed system at a constant temperature and under a constant total pressure. The system consists of a mixture of several chemical species which can react according to the equation

$$\alpha A + \beta B \rightarrow \gamma C + \delta D, \tag{11–25}$$

where A, B, C, and D represent the chemical formulas of the substances, while α, β, γ, and δ represent the stoichiometric coefficients. We now inquire whether the free energy of the mixture will increase or decrease if the reaction advances in the direction indicated by the arrow. If the free energy decreases as the reaction advances, then the reaction goes spontaneously in the direction of the arrow; the advance of the reaction and the decrease in free energy continue until the free energy of the system reaches a minimum value. When the free energy of the system is a minimum, the reaction is at equilibrium. If the free energy of the system increases as the reaction advances in the direction of the arrow, then the reaction will go spontaneously, with a decrease in free energy, in the opposite direction; again the mixture will reach a minimum value of free energy at the equilibrium position.

Since T and p are constant, as the reaction advances the change in free energy of the system is given by Eq. (11–7), which becomes

$$dG = \mu_A \, dn_A + \mu_B \, dn_B + \mu_C \, dn_C + \mu_D \, dn_D, \tag{11–26}$$

where the changes in the mole numbers, dn_A, dn_B, ..., are those resulting from the chemical reaction. These changes are not independent because the substances react in the stoichiometric ratios. After the reaction takes place once as written, that is after α moles of A and β moles of B have been consumed to form γ moles of C and δ moles of D, we say that one unit of reaction has occurred. Let the reaction advance by ξ units; then the number of moles of each of the substances present is

$$n_C = n_C^0 + \gamma\xi, \qquad n_D = n_D^0 + \delta\xi,$$

$$n_A = n_A^0 - \alpha\xi, \qquad n_B = n_B^0 - \beta\xi, \tag{11–27}$$

where the n^0's are the number of moles of the substances present before the reaction advanced by ξ units. Since the reactants are consumed and products are produced, the plus and minus signs appear in Eq. (11–27) as shown. Since the n^0's are constant, by differentiating the equations in (11–27) we obtain

$$dn_C = \gamma \, d\xi, \qquad dn_D = \delta \, d\xi,$$

$$dn_A = -\alpha \, d\xi, \qquad dn_B = -\beta \, d\xi. \tag{11–28}$$

The variable ξ was first introduced by DeDonder and called the degree of advancement of the reaction. Here we will use the term *advancement* for ξ; then $d\xi$ is the differential advancement.

Using the relations in (11–28) in Eq. (11–26), we obtain

$$dG = [\gamma\mu_C + \delta\mu_D - (\alpha\mu_A + \beta\mu_B)] \, d\xi,$$

which becomes

$$\left(\frac{\partial G}{\partial \xi}\right)_{T,p} = \gamma\mu_C + \delta\mu_D - (\alpha\mu_A + \beta\mu_B). \tag{11–29}$$

The derivative, $(\partial G/\partial \xi)_{T,p}$, is the rate of increase of the free energy of the mixture with the advancement ξ of the reaction. If this derivative is negative, the free energy of the mixture decreases as the reaction progresses in the direction indicated by the arrow, which implies that the reaction is spontaneous. If this derivative is positive, progress of the reaction in the forward direction would lead to an increase in free energy of the system; since this is not possible, the reverse reaction would go spontaneously. If $(\partial G/\partial \xi)_{T,p}$ is zero, then the free energy has a minimum value and the reaction is at equilibrium. The equilibrium condition for the chemical reaction is then

$$\left(\frac{\partial G}{\partial \xi}\right)_{T,p,\text{eq}} = 0 \tag{11–30}$$

$$[\gamma\mu_C + \delta\mu_D - (\alpha\mu_A + \beta\mu_B)]_{\text{eq}} = 0. \tag{11–31}$$

The derivative in Eq. (11–29) has the form of an increase of free energy, ΔG, since it is the sum of the free energies of the products of the reaction less the sum of the free energies of the reactants. Consequently we will write ΔG for $(\partial G/\partial \xi)_{T,p}$ and call ΔG the *reaction free energy*. From the above derivation it is clear that for any chemical reaction whatever

$$\Delta G = \sum_P v_P \mu_P - \sum_R v_R \mu_R, \tag{11-32}$$

where v_P is the stoichiometric coefficient of the product P, and μ_P is its chemical potential, while v_R and μ_R are the stoichiometric coefficient and chemical potential of the reactant R. The first sum is taken over all the products of the reaction; the second sum is over all the reactants. The equilibrium condition for *any chemical reaction whatsoever* is

$$\left(\sum_P v_P \mu_P - \sum_R v_R \mu_R \right)_{eq} = 0. \tag{11-33}$$

The subscript eq is placed on the quantities in Eqs. (11–31) and (11–33) to emphasize the fact that at equilibrium the values of the μ's are related in the special way indicated by these equations.

The result in Eq. (11–33) may be obtained somewhat more elegantly as follows. We note that each of the mole numbers in Eq. (11–27) has the form

$$n_i = n_i^0 + v_i \xi, \tag{11-27a}$$

where v_i is the stoichiometric coefficient of the substance i in the chemical reaction, if we agree that v_i will be given a negative sign for reactants and a positive sign for products. Then corresponding to Eq. (11–28), we have

$$dn_i = v_i \, d\xi. \tag{11-28a}$$

Since $dG = \sum_i \mu_i \, dn_i$ at constant T and p, this becomes

$$dG = \left(\sum_i v_i \mu_i \right) d\xi,$$

or

$$\left(\frac{\partial G}{\partial \xi} \right)_{T,p} = \sum_i v_i \mu_i. \tag{11-29a}$$

At equilibrium $(\partial G/\partial \xi)_{T,p} = 0$, hence the equilibrium condition for any chemical reaction whatsoever is

$$\left(\sum_i v_i \mu_i \right)_{eq} = 0. \tag{11-33a}$$

Since each μ_i is $\mu_i(T, p, n_1^0, n_2^0, \dots, \xi)$ the equilibrium condition determines ξ_e as a function of T, p, and the specified values of the initial mole numbers.

11–8   The General Behavior of G as a Function of ξ

Figure 11–5a shows the general behavior of G as a function of ξ. The advancement, ξ, has a limited range of variation between a least value, ξ_1, and a greatest value, ξ_g. At ξ_1, one or more of the products has been exhausted, while at ξ_g one or more of the reactants has been exhausted. At some intermediate value, ξ_e, G passes through a minimum. The value ξ_e is the equilibrium value of the advancement. To the left of the minimum, $\partial G/\partial \xi$ is negative, indicating spontaneity in the forward direction, while to the right of the minimum, $\partial G/\partial \xi$ is positive, indicating spontaneity in the reverse direction. Note that even though in the case illustrated the products have an intrinsically higher free energy than the reactants, nonetheless the reaction does form some products. This is a consequence of the contribution of the free energy of mixing.

At any composition the free energy of the mixture has the form

$$G = \sum_i n_i \mu_i.$$

If we add and subtract $\mu_i^0(T, p)$, the chemical potential of the *pure* species i in each term of the sum, we obtain

$$G = \sum_i n_i(\mu_i^0 + \mu_i - \mu_i^0) = \sum_i n_i\mu_i^0(T, p) + \sum_i n_i(\mu_i - \mu_i^0).$$

The first sum is the total free energy of the pure gases separately, G_{pure}; the last sum is the free energy of mixing, ΔG_{mix}. The free energy of the system is given by

$$G = G_{\text{pure}} + \Delta G_{\text{mix}}. \tag{11–34}$$

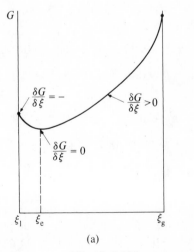 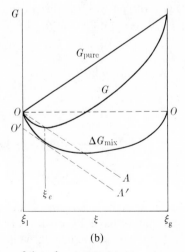

Fig. 11–5   Free energy as a function of the advancement.

The plot of G_{pure}, ΔG_{mix}, and G as a function of the advancement is shown in Fig. 11–5b. Since G_{pure} depends on ξ only through the n_i, each of which is a linear function of ξ, we see that G_{pure} is a linear function of ξ. The minimum in G occurs at the point where ΔG_{mix} decreases as rapidly as G_{pure} increases; by differentiating,

$$\left(\frac{\partial G}{\partial \xi}\right)_{T,p} = \left(\frac{\partial G_{pure}}{\partial \xi}\right)_{T,p} + \left(\frac{\partial \Delta G_{mix}}{\partial \xi}\right)_{T,p}.$$

At equilibrium

$$\left(\frac{\partial G_{pure}}{\partial \xi}\right)_{eq} = -\left(\frac{\partial \Delta G_{mix}}{\partial \xi}\right)_{eq}.$$

This condition can be established geometrically by reflecting the line for G_{pure} in the horizontal line OO, to yield the line OA; the point of tangency of the line $O'A'$, parallel to OA, with the curve for ΔG_{mix} yields the value of the advancement at equilibrium. Equation (11–34) is correct for any equilibrium in a true mixture of substances.

Equation (11–34) shows that a system approaches the equilibrium state of minimum free energy by forming substances of intrinsically lower free energy; this makes G_{pure} small. It also lowers its free energy by mixing the reactants and products. A compromise is reached between a pure material having a low intrinsic free energy and the highly mixed state.

11–9 Chemical Equilibrium in a Mixture of Ideal Gases

It has been shown, Eq. (11–12), that the μ of an ideal gas in a gas mixture is given by

$$\mu_i = \mu_i^0 + RT \ln p_i, \tag{11–35}$$

where p_i is the partial pressure of the gas in the mixture. We use this value of μ_i in Eq. (11–29) to compute the ΔG for the reaction (11–25):

$$\Delta G = \gamma \mu_C^0 + \gamma RT \ln p_C + \delta \mu_D^0 + \delta RT \ln p_D - \alpha \mu_A^0 - \alpha RT \ln p_A - \beta \mu_B^0 - \beta RT \ln p_B,$$

$$\Delta G = \gamma \mu_C^0 + \delta \mu_D^0 - (\alpha \mu_A^0 + \beta \mu_B^0) + RT[\gamma \ln p_C + \delta \ln p_D - (\alpha \ln p_A + \beta \ln p_B)].$$

Let

$$\Delta G^0 = \gamma \mu_C^0 + \delta \mu_D^0 - (\alpha \mu_A^0 + \beta \mu_B^0); \tag{11–36}$$

ΔG^0 is the *standard* reaction free energy. Then, combining the logarithmic terms,

$$\Delta G = \Delta G^0 + RT \ln \frac{p_C^\gamma p_D^\delta}{p_A^\alpha p_B^\beta}. \tag{11–37}$$

The argument of the logarithm is called the *proper quotient of pressures*; the numerator is the product of partial pressures of the chemical products each raised to the power of its stoichiometric coefficient, while the denominator is the product of the partial pressures of the reactants each raised to the power of its stoichiometric coefficient.

Ordinarily the quotient is abbreviated by the symbol Q_p:

$$Q_p = \frac{p_C^\gamma p_D^\delta}{p_A^\alpha p_B^\beta}. \tag{11–38}$$

This reduces Eq. (11–37) to

$$\Delta G = \Delta G^0 + RT \ln Q_p. \tag{11–39}$$

The sign of ΔG is determined by the sign and magnitude of $\ln Q_p$, since at a given temperature ΔG^0 is a constant characteristic of the reaction. If, for example, we compose the mixture so that the partial pressures of the reactants are very large, while those of the products are very small, then Q_p will have a small fractional value, and $\ln Q_p$ will be a large negative number. This in turn will make ΔG more negative and increase the tendency for products to form.

At equilibrium, $\Delta G = 0$, and Eq. (11–37) becomes

$$0 = \Delta G^0 + RT \ln \frac{(p_C)_e^\gamma (p_D)_e^\delta}{(p_A)_e^\alpha (p_B)_e^\beta}, \tag{11–40}$$

where the subscript e indicates that these are *equilibrium* partial pressures. The quotient of *equilibrium* partial pressures, is the pressure equilibrium constant K_p:

$$K_p = \frac{(p_C)_e^\gamma (p_D)_e^\delta}{(p_A)_e^\alpha (p_B)_e^\beta}. \tag{11–41}$$

Equation (11–40) becomes

$$\Delta G^0 = -RT \ln K_p. \tag{11–42}$$

The quantity ΔG^0 is a combination of μ^0's each of which is a function only of temperature; therefore, ΔG^0 is a function only of temperature, and so K_p is a function only of temperature. From a measurement of the equilibrium constant of the reaction ΔG^0 can be calculated using Eq. (11–42). This is the way in which the value of ΔG^0 for any reaction is obtained.

Example. For the reaction

$$\tfrac{1}{2}N_2(g) + \tfrac{3}{2}H_2(g) \rightarrow NH_3(g),$$

the equilibrium constant is 6.59×10^{-3} at 450°C. Compute the standard reaction free energy at 450°C.

Solution. $\Delta G^0 = -(1.99)(723) \ln (6.59 \times 10^{-3}) = -1440(-5.02) = 7230 \, \text{cal.}$

11–10 Chemical Equilibrium in a Mixture of Real Gases

If the corresponding algebra were carried out for real gases using Eq. (10–48), the equation equivalent to Eq. (11–41) is

$$K_f = \frac{(f_C)_e^\gamma (f_D)_e^\delta}{(f_A)_e^\alpha (f_B)_e^\beta}, \tag{11–43}$$

and, corresponding to Eq. (11–42),

$$\Delta G^0 = -RT \ln K_f. \tag{11–44}$$

For real gases, it is K_f rather than K_p which is a function of temperature only.

11–11   The Equilibrium Constants, K_x and K_c

It is sometimes advantageous to express the equilibrium constant for gaseous systems in terms of either mole fractions, x_i, or molar concentrations, c_i, rather than partial pressures. The partial pressure, p_i, the mole fraction, and the total pressure, p, are related by $p_i = x_i p$. Using this relation for each of the partial pressures in the equilibrium constant, we obtain from Eq. (11–41)

$$K_p = \frac{(p_C)_e^\gamma (p_D)_e^\delta}{(p_A)_e^\alpha (p_B)_e^\beta} = \frac{(x_C p)_e^\gamma (x_D p)_e^\delta}{(x_A p)_e^\alpha (x_B p)_e^\beta} = \frac{(x_C)_e^\gamma (x_D)_e^\delta}{(x_A)_e^\alpha (x_B)_e^\beta} p^{\gamma + \delta - \alpha - \beta}.$$

The mole fraction equilibrium constant is defined by

$$K_x = \frac{(x_C)_e^\gamma (x_D)_e^\delta}{(x_A)_e^\alpha (x_B)_e^\beta}. \tag{11–45}$$

Then

$$K_p = K_x p^{\Delta \nu}, \tag{11–46}$$

where $\Delta \nu$ is the sum of stoichiometric coefficients on the right-hand side of the chemical equation minus the sum of the coefficients on the left-hand side. Rearranging Eq. (11–46), we obtain $K_x = K_p p^{-\Delta \nu}$. Since K_p is independent of pressure, K_x will depend on pressure unless $\Delta \nu$ is zero.

In a similar way, since the partial pressure of a gas is given by $p_i = n_i (RT/V)$ and the concentration in moles per liter is $c_i = n_i/V$, we obtain $p_i = c_i RT$. Repetition of the argument above yields

$$K_p = K_c (RT)^{\Delta \nu}, \tag{11–47}$$

where K_c is the quotient of equilibrium concentrations; K_c is a function of temperature only.

11–12   Standard Free Energies of Formation

Having obtained values of ΔG^0 from measurements of equilibrium constants, it is possible to calculate conventional values of the standard molar free energy μ^0 of individual compounds. Just as in the case of the standard enthalpies of substances, we are at liberty to assign a value of zero to the free energy of the elements in their stable state of aggregation at 25°C and 1 atm pressure. For example, at 25°C

$$\mu^0_{H_2}(g) = 0, \qquad \mu^0_{Br_2}(l) = 0, \qquad \mu^0_S(s, \text{rhombic}) = 0.$$

Table 11–1†   Standard free energies of formation of compounds at 25°C

Compound	ΔG_f^0, kcal/mole	Compound	ΔG_f^0, kcal/mole
$O_3(g)$	39.06	$ZnO(s)$	-76.05
$H_2O(g)$	-54.6357	$HgO(s, red)$	-13.990
$H_2O(l)$	-56.6902	$CuO(s)$	-30.4
$HF(g)$	-64.7	$Cu_2O(s)$	-34.98
$HCl(g)$	-22.769	$Ag_2O(s)$	-2.586
$Br_2(g)$	0.751	$AgCl(s)$	-26.224
$HBr(g)$	-12.72	$Ag_2S(s, rhombic)$	-9.62
$HI(g)$	0.31	$FeO(s, wustite)$	-58.4
$SO_2(g)$	-71.79	$Fe_2O_3(s)$	-177.1
$SO_3(g)$	-88.52	$Fe_3O_4(s)$	-242.4
$H_2S(g)$	-7.892	$FeS(s, \alpha)$	-23.32
$NO(g)$	20.719	$FeS_2(s)$	-39.84
$NO_2(g)$	12.390	$TiO_2(s, rutile)$	-203.8
$N_2O(g)$	24.76	$TiCl_4(l)$	-161.2
$N_2O_4(g)$	23.491	$Al_2O_3(s, \alpha)$	-376.77
$NH_3(g)$	-3.976	$MgO(s)$	-136.13
$HNO_3(l)$	-19.100	$MgCO_3(s)$	-246
$NOCl(g)$	15.86	$CaO(s)$	-144.4
$NH_4Cl(s)$	-48.73	$Ca(OH)_2(s)$	-214.33
$PCl_3(g)$	-68.42	$CaC_2(s)$	-16.2
$PCl_5(g)$	-77.59	$CaCO_3(s, calcite)$	-269.78
$C(s, diamond)$	0.6850	$SrO(s)$	-133.8
$CO(g)$	-32.8079	$SrCO_3(s)$	-271.9
$CO_2(g)$	-94.2598	$BaO(s)$	-126.3
$CH_4(g)$	-12.140	$BaCO_3(s)$	-272.2
$HCHO(g)$	-26.3	$Na_2O(s)$	-90.0
$CH_3OH(l)$	-39.73	$NaF(s)$	-129.3
$C_2H_2(g)$	50.000	$NaCl(s)$	-91.785
$C_2H_4(g)$	16.282	$Na_2SO_4(s)$	-302.78
$C_2H_6(g)$	-7.860	$Na_2SO_4 \cdot 10H_2O$	-870.93
$C_2H_5OH(l)$	-41.77	$NaNO_3(s)$	-87.45
$C_6H_6(g)$	30.989	$KF(s)$	-127.42
$SiO_2(s, quartz II)$	-192.4	$KCl(s)$	-97.592
$SiH_4(g)$	-9.4	$KClO_3(s)$	-69.29
$SiF_4(g)$	-360	$KClO_4(s)$	-72.7
$PbO(s, red)$	-45.25	$KBr(s)$	-90.63
$PbS(s)$	-22.15	$KI(s)$	-77.03

†   NBS Circular 500. U.S. G.P.O., Washington, D.C., 1952. Note that this publication uses F rather than G for the Gibbs free energy.

For the formation reaction of a compound such as CO, we have

$$C(\text{graphite}) + \tfrac{1}{2}O_2(g) \rightarrow CO(g),$$

$$\Delta G_f^0 = \mu_{CO}^0 - (\mu_C^0 + \tfrac{1}{2}\mu_{O_2}^0).$$

Since $\mu_C^0 = 0$ and $\mu_{O_2}^0 = 0$ by convention, we have

$$\Delta G_f^0 = \mu_{CO}^0. \tag{11–48}$$

Consequently, the standard free energy of formation of any compound is equal to the conventional standard molar free energy of that compound. Some values of · the standard free energy of formation at 25°C are given in Table 11–1.

11–13 Examples

1. The dissociation of nitrogen tetroxide,

$$N_2O_4(g) \rightleftharpoons 2NO_2(g),$$

is an equilibrium which can be easily studied in the laboratory through a measurement of the vapor density of the equilibrium mixture. Suppose a system contains n moles of N_2O_4 and at equilibrium a fraction α_e is dissociated. Then in the equilibrium mixture, we have the conditions shown in Table 11–2. If N_e is the total number of moles in the mixture, then $N_e = n(1 + \alpha_e)$. The equilibrium constant has the form

$$K_p = \frac{(p_{NO_2})_e^2}{(p_{N_2O_4})_e}.$$

Writing the partial pressures in terms of the degree of dissociation and the total pressure as in the last column of Table 11–2, we have

$$K_p = \frac{\left[\left(\dfrac{2\alpha_e}{1 + \alpha_e}\right)p\right]^2}{\left(\dfrac{1 - \alpha_e}{1 + \alpha_e}\right)p} = \frac{4\alpha_e^2 p}{1 - \alpha_e^2}. \tag{11–49}$$

Table 11–2

Substance	Mole number	Mole fraction	Partial pressure
N_2O_4	$n(1 - \alpha_e)$	$\dfrac{1 - \alpha_e}{1 + \alpha_e}$	$\dfrac{1 - \alpha_e}{1 + \alpha_e}p$
NO_2	$2n\alpha_e$	$\dfrac{2\alpha_e}{1 + \alpha_e}$	$\dfrac{2\alpha_e}{1 + \alpha_e}p$

A measurement of α_e at any pressure p suffices to determine K_p. From K_p, ΔG^0 can be calculated. The dependence of α_e on the pressure can be obtained explicitly by solving Eq. (11–49) for α_e:

$$\alpha_e = \sqrt{\frac{K_p}{K_p + 4p}}.$$

It is clear that as $p \to 0$, $\alpha_e \to 1$, while as $p \to \infty$, $\alpha_e \to 0$. This is what would be expected from the LeChatelier principle. At moderately high pressures, $K_p \ll 4p$ and $\alpha_e = \frac{1}{2}K_p^{1/2}/p^{1/2}$, approximately.

    2. The equilibrium constant for the reaction

$$\tfrac{1}{2}N_2 + \tfrac{3}{2}H_2 \rightleftharpoons NH_3$$

is

$$K_p = \frac{(p_{NH_3})_e}{(p_{N_2})_e^{1/2}(p_{H_2})_e^{3/2}}.$$

The total pressure is $p = p_{NH_3} + p_{N_2} + p_{H_2}$. If nitrogen and hydrogen are mixed in the stoichiometric ratio, and brought to equilibrium, then at all times $p_{H_2} = 3p_{N_2}$. Let $p_{H_2} + p_{N_2} = p - p_{NH_3}$. Then $p_{N_2} = \frac{1}{4}(p - p_{NH_3})$, and $p_{H_2} = \frac{3}{4}(p - p_{NH_3})$, and K_p becomes

$$K_p = \frac{p_{NH_3}}{[\frac{1}{4}(p - p_{NH_3})]^{1/2}[\frac{3}{4}(p - p_{NH_3})]^{3/2}} = \frac{16 p_{NH_3}}{3^{3/2}(p - p_{NH_3})^2},$$

$$0.325 K_p = \frac{p_{NH_3}}{(p - p_{NH_3})^2}.$$

From this relation, the partial pressure of NH_3 can be calculated at any total pressure. If the conversion to NH_3 is low, then $p - p_{NH_3} \approx p$, and $p_{NH_3} = 0.325 K_p p^2$, so that the partial pressure of ammonia is approximately proportional to the square of the pressure. If the reactants are not mixed originally in the stoichiometric ratio, the expression is more complex.

    A measurement of the equilibrium partial pressure of NH_3 at a given temperature and pressure yields a value of ΔG^0 for this reaction, which is the conventional standard molar free energy of NH_3 at this temperature.

11–14   The Temperature Dependence of the Equilibrium Constant

The equilibrium constant can be written as

$$\ln K_p = -\frac{\Delta G^0}{RT}. \tag{11–50}$$

Differentiating, we obtain

$$\frac{d \ln K_p}{dT} = -\frac{1}{R}\frac{d(\Delta G^0/T)}{dT}. \tag{11–51}$$

Dividing Eq. (11–32) by T, we obtain

$$\frac{\Delta G^0}{T} = \sum_P \nu_P \left(\frac{\mu_P^0}{T}\right) - \sum_R \nu_R \left(\frac{\mu_R^0}{T}\right).$$

Differentiating, we have

$$\frac{d(\Delta G^0/T)}{dT} = \sum_P \nu_P \frac{d(\mu_P^0/T)}{dT} - \sum_R \nu_R \frac{d(\mu_R^0/T)}{dT}, \tag{11–52}$$

where the μ^0's are standard molar free energies of pure substances. Using molar values in the Gibbs-Helmholtz equation, Eq. (10–54), we have $d(\mu^0/T)/dT = -\bar{H}^0/T^2$. This relation reduces Eq. (11–52) to

$$\frac{d(\Delta G^0/T)}{dT} = -\frac{1}{T^2}\left[\sum_P \nu_P \bar{H}_P^0 - \sum_R \nu_R \bar{H}_R^0\right] = -\frac{\Delta H^0}{T^2}, \tag{11–53}$$

since the quantity in the bracket is the standard enthalpy increase for the reaction, ΔH^0. Equation (11–53) reduces Eq. (11–51) to

$$\frac{d \ln K_p}{dT} = \frac{\Delta H^0}{RT^2}, \quad \text{or} \quad \frac{d \log_{10} K_p}{dT} = \frac{\Delta H^0}{2.303\, RT^2}. \tag{11–54}$$

Equation (11–54) is also called the Gibbs–Helmholtz equation.

If the reaction is endothermic, ΔH^0 is positive; then Eq. (11–54) shows that $\ln K_p$, and so K_p itself, increases with increase in temperature. If the reaction is exothermic, ΔH^0 is negative, and the equilibrium constant decreases with increase in temperature. Since an increase in the equilibrium constant implies an increase in the yield of products, Eq. (11–54) is the mathematical expression of one aspect of the LeChatelier principle.

Equation (11–54) can be expressed readily in a form convenient for plotting:

$$d \ln K_p = \frac{\Delta H^0}{R} \frac{dT}{T^2} = -\frac{\Delta H^0}{R} d\left(\frac{1}{T}\right),$$

$$\frac{d \ln K_p}{d(1/T)} = -\frac{\Delta H^0}{R}, \qquad \frac{d \log_{10} K_p}{d(1/T)} = -\frac{\Delta H^0}{2.303\, R}. \tag{11–55}$$

Equation (11–55) shows that a plot of $\ln K_p$ versus $1/T$ has a slope equal to $-\Delta H^0/R$. Since ΔH^0 is almost constant, at least over moderate ranges of temperature, the plot is often linear.

If K_p is measured at several temperatures and the data plotted as $\ln K_p$ versus $1/T$, the slope of the line yields a value of ΔH^0 for the reaction through Eq. (11–55). Consequently it is possible to determine heats of reaction by measuring equilibrium constants over a range of temperature. The values of the heats of reaction obtained by this method are usually not so precise as those obtained by precision calorimetric methods. However, the equilibrium method can be used for reactions which are not suited to direct calorimetric measurement. Later we will find that certain equilibrium constants can be calculated from calorimetrically measured quantities only.

Having obtained values of ΔG^0 at several temperatures and a value of ΔH^0 from the plot of Eq. (11–55), we can calculate the values of ΔS^0 at each temperature from the equation

$$\Delta G^0 = \Delta H^0 - T \Delta S^0. \tag{11–56}$$

The equilibrium constant can be written as an explicit function of temperature by integrating Eq. (11–54). Suppose that at some temperature T_0, the value of the equilibrium constant is $(K_p)_0$ and at any other temperature T the value is K_p:

$$\int_{\ln (K_p)_0}^{\ln K_p} d(\ln K_p) = \int_{T_0}^{T} \frac{\Delta H^0}{RT^2} dT, \qquad \ln K_p - \ln (K_p)_0 = \int_{T_0}^{T} \frac{\Delta H^0}{RT^2} dT,$$

$$\ln K_p = \ln (K_p)_0 + \int_{T_0}^{T} \frac{\Delta H^0}{RT^2} dT. \tag{11–57}$$

If ΔH^0 is a constant, then by integrating, we have

$$\ln K_p = \ln (K_p)_0 + \frac{\Delta H^0}{R} \left(\frac{1}{T_0} - \frac{1}{T} \right). \tag{11–58}$$

From the knowledge of ΔH_0 and a value of $(K_p)_0$ at any temperature T_0, we can calculate K_p at any other temperature.

If ΔH^0 is not a constant, it can ordinarily be expressed (Section 7–24) as a power series in T:

$$\Delta H^0 = \Delta H_0^0 + A'T + B'T^2 + C'T^3 + \cdots.$$

Using this value for ΔH^0 in Eq. (11–57) and integrating, we obtain

$$\ln K_p = \ln (K_p)_0 - \frac{\Delta H_0^0}{R} \left(\frac{1}{T} - \frac{1}{T_0} \right) + \frac{A'}{R} \ln \left(\frac{T}{T_0} \right) + \frac{B'}{R} (T - T_0)$$

$$+ \frac{C'}{2R} (T^2 - T_0^2) + \cdots, \tag{11–59}$$

which has the general functional form

$$\ln K_p = A/T + B + C \ln T + DT + ET^2 + \cdots, \tag{11–60}$$

in which A, B, C, D, and E are constants. Equations having the general form of Eq. (11–60) are often used to calculate an equilibrium constant at 25°C (so that it can be tabulated) from a measurement at some other (usually higher) temperature. To evaluate the constants, the values of ΔH^0 and the heat capacities of all the reactants and products must be known.

11–15   Equilibria Between Ideal Gases and Pure Condensed Phases

If the substances participating in the chemical equilibrium are in more than one phase, the equilibrium is *heterogeneous*. If the substances are all present in a single phase, the equilibrium is *homogeneous*. We have dealt so far only with homogeneous equilibria in gases. If, in addition to gases, a chemical reaction involves one or more *pure* liquids or solids, the expression for the equilibrium constant is slightly different.

Consider the reaction

$$CaCO_3(s) \rightleftharpoons CaO(s) + CO_2(g).$$

The equilibrium condition is

$$(\mu_{CaO} + \mu_{CO_2} - \mu_{CaCO_3})_{eq} = 0.$$

For each gas present, e.g., CO_2, $(\mu_{CO_2})_{eq} = \mu_{CO_2}^0 + RT \ln (p_{CO_2})_e$. While for the pure solids (and for pure liquids if they appear), because of the insensitivity of the free energy of a condensed phase to change in pressure, we have

$$\mu_{CaCO_3} = \mu_{CaCO_3}^0, \qquad \mu_{CaO} = \mu_{CaO}^0.$$

The equilibrium condition becomes

$$0 = \mu_{CaO}^0 + \mu_{CO_2}^0 - \mu_{CaCO_3}^0 + RT \ln (p_{CO_2})_e,$$

$$0 = \Delta G^0 + RT \ln (p_{CO_2})_e. \tag{11–61}$$

In this case, the equilibrium constant is simply

$$K_p = (p_{CO_2})_e.$$

The equilibrium constant contains only the pressure of the gas; however, the ΔG^0 contains the standard free energies of *all* the reactants and products.

From Table 11–1, we have at 25°C

$$\mu_{CaCO_3}^0(s) = -269.8 \text{ kcal/mole}, \qquad \mu_{CaO}^0(s) = -144.4 \text{ kcal/mole},$$

$$\mu_{CO_2}^0(g) = -94.3 \text{ kcal/mole},$$

$$\Delta G_{298}^0 = -144.4 - 94.3 - (-269.8) = 31.1 \text{ kcal}.$$

The equilibrium pressure is calculated from Eq. (11–61);

$$\log_{10} (p_{CO_2})_e = -\frac{31{,}100}{2.303(1.987)(298)} = -22.8$$

$$(p_{CO_2})_e = 1.6 \times 10^{-23} \text{ atm at } 25°C.$$

Consider the reaction

$$HgO(s) \rightleftharpoons Hg(l) + \tfrac{1}{2}O_2(g).$$

The equilibrium constant is $K_p = (p_{O_2})_e^{1/2}$. Also

$$\Delta G^0 = \mu_{Hg(l)}^0 + \tfrac{1}{2}\mu_{O_2}^0 - \mu_{HgO}^0 = -\mu_{HgO}^0.$$

Liquid–vapor equilibrium

An important example of equilibrium between ideal gases and pure condensed phases is the equilibrium between a pure liquid and its vapor:

$$A(l) \rightleftharpoons A(g).$$

Let p be the equilibrium vapor pressure. Then

$$K_p = p \quad \text{and} \quad \Delta G^0 = \mu_{gas}^0 - \mu_{liq}^0.$$

Using the Gibbs-Helmholtz equation, Eq. (11–54), we have

$$\frac{d \ln p}{dT} = \frac{\Delta H_{vap}^0}{RT^2}, \tag{11–62}$$

which is the Clausius-Clapeyron equation; it relates the temperature dependence of the vapor pressure of a liquid to the heat of vaporization. A similar expression holds for the sublimation of a solid. Consider the reaction

$$A(s) \rightleftharpoons A(g); \quad K_p = p,$$

where p is the equilibrium vapor pressure of the solid. By the same argument as above

$$\frac{d \ln p}{dT} = \frac{\Delta H_{sub}^0}{RT^2}, \tag{11–63}$$

where ΔH_{sub}^0 is the heat of sublimation of the solid. In either case, a plot of $\ln p$ versus $1/T$ is linear with a slope equal to $-\Delta H^0/R$.

11–16   The LeChatelier Principle

It is fairly easy to show how a change in temperature or pressure affects the equilibrium value of the advancement ξ_e of a reaction. We need only determine the sign of the derivatives $(\partial \xi_e / \partial T)_p$ and $(\partial \xi_e / \partial p)_T$. We begin by writing the identity

$$\left(\frac{\partial G}{\partial \xi} \right)_{T,p} = \Delta G. \tag{11–64}$$

Since $(\partial G / \partial \xi)_{T,p}$ is itself a function of T, p, and ξ we may write the total differential expression,

$$d \left(\frac{\partial G}{\partial \xi} \right) = \frac{\partial}{\partial T} \left(\frac{\partial G}{\partial \xi} \right) dT + \frac{\partial}{\partial p} \left(\frac{\partial G}{\partial \xi} \right) dp + \frac{\partial}{\partial \xi} \left(\frac{\partial G}{\partial \xi} \right) d\xi. \tag{11–65}$$

Using Eq. (11–64) and setting $(\partial^2 G / \partial \xi^2) = G''$, Eq. (11–65) becomes

$$d \left(\frac{\partial G}{\partial \xi} \right) = \frac{\partial \Delta G}{\partial T} dT + \frac{\partial \Delta G}{\partial p} dp + G'' d\xi.$$

From the fundamental equation; $(\partial \Delta G/\partial T) = -\Delta S$ and $(\partial \Delta G/\partial p) = \Delta V$, in which ΔS is the entropy change for the reaction and ΔV is the volume change for the reaction. Thus,

$$d\left(\frac{\partial G}{\partial \xi}\right) = -\Delta S\, dT + \Delta V\, dp + G''\, d\xi.$$

If we insist that these variations in temperature, pressure and advancement occur all the while keeping the reaction at equilibrium, then $\partial G/\partial \xi = 0$ and hence also $d(\partial G/\partial \xi) = 0$. At equilibrium, $\Delta S = \Delta H/T$ so the equation becomes

$$0 = -\left(\frac{\Delta H}{T}\right)(\partial T)_{eq} + \Delta V(\partial p)_{eq} + G''_e(\partial \xi_e). \tag{11-66}$$

At equilibrium G is a minimum, therefore G''_e must be positive.

At constant pressure, $dp = 0$, and Eq. (11–66) becomes

$$\left(\frac{\partial \xi_e}{\partial T}\right)_p = \frac{\Delta H}{TG''_e}. \tag{11-67}$$

At constant temperature, $dT = 0$, and Eq. (11–66) becomes

$$\left(\frac{\partial \xi_e}{\partial p}\right)_T = -\frac{\Delta V}{G''_e}. \tag{11-68}$$

Equations (11–67) and (11–68) are quantitative statements of the principle of LeChatelier; they describe the dependence of the advancement of the reaction at equilibrium on temperature and on pressure. Since G''_e is positive, the sign of $(\partial \xi_e/\partial T)_p$ depends on the sign of ΔH. If ΔH is $+$, an endothermic reaction, then $(\partial \xi_e/\partial T)_p$ is $+$, and an increase in temperature increases the advancement at equilibrium. For an exothermic reaction, ΔH is $-$, so $(\partial \xi_e/\partial T)_p$ is $-$; increase in temperature will decrease the equilibrium advancement of the reaction.

Similarly, the sign of $(\partial \xi_e/\partial p)_T$ depends on ΔV. If ΔV is $-$, the product volume is less than the reactant volume, $(\partial \xi_e/\partial p)_T$ is positive; increase in pressure increases the equilibrium advancement. Conversely, if ΔV is $+$, then $(\partial \xi_e/\partial p)_T$ is $-$; increase in pressure decreases the equilibrium advancement.

The net effect of these relations is that an increase in pressure shifts the equilibrium to the low-volume side of the reaction while a decrease in pressure shifts the equilibrium to the high-volume side. Similarly an increment in temperature shifts the equilibrium to the high-enthalpy side, while a decrease in temperature shifts it to the low-enthalpy side.

We may state the principle of LeChatelier in the following way. If the external constraints under which an equilibrium is established are changed, the equilibrium will shift in such a way as to moderate the effect of the change.

For example, if the volume of a nonreactive system is decreased by a specified amount, the pressure rises correspondingly. In a reactive system, the equilibrium shifts to the low-volume side (if $\Delta V \neq 0$), so the pressure increment is less than in the nonreactive case. The response of the system is moderated by the shift in equilibrium position. This implies that the compressibility of a reactive system is much greater than that of a nonreactive one.

Similarly, if we extract a fixed quantity of heat from a nonreactive system, the temperature decreases by a definite amount. In a reactive system, withdrawing the same amount of heat will not produce as large a decrease in temperature since the equilibrium shifts to the low-enthalpy side (if $\Delta H \neq 0$). This implies that the heat capacity of a reactive system is much larger than that of a nonreactive one. This is useful if the system can be used as a heat-transfer medium.

It must be noted here that there are certain types of systems which do not obey the LeChatelier principle in all circumstances, e.g., open systems. A very general validity has been claimed for the LeChatelier principle. However, if the principle does have such broad application, the statement of the principle must be very much more complex than that given here or in other elementary discussions.

11–17 Equilibrium Constants from Calorimetric Measurements. The Third Law in its Historical Context

Using the Gibbs-Helmholtz equation, we can calculate the equilibrium constant of a reaction at any temperature T from a knowledge of the equilibrium constant at one temperature T_0 and the ΔH^0 of the reaction. For convenience we rewrite Eq. (11–57):

$$\ln K_p = \ln (K_p)_0 + \int_{T_0}^{T} \frac{\Delta H^0}{RT^2} dT.$$

The ΔH^0 for any reaction and its temperature dependence can be determined by purely thermal (i.e., calorimetric) measurements. Thus, according to Eq. (11–57), a measurement of the equilibrium constant at only *one* temperature together with the thermal measurements of ΔH^0 and ΔC_p suffice to determine the value of K_p at any other temperature.

The question naturally arises whether or not it is possible to calculate the equilibrium constant exclusively from quantities which have been determined calorimetrically. In view of the relation $\Delta G^0 = -RT \ln K_p$, the equilibrium constant can be calculated if ΔG^0 is known. At any temperature T, by definition,

$$\Delta G^0 = \Delta H^0 - T \Delta S^0. \tag{11–69}$$

Since ΔH^0 can be obtained from thermal measurements, the problem resolves itself into the question of whether or not ΔS^0 can be obtained solely from thermal measurements.

For any single substance

$$S_T^0 = S_0^0 + S_{0 \to T}^0, \tag{11–70}$$

where S_T^0 is the entropy of the substance at temperature T; S_0^0, the entropy at $0°K$, and

$S_{0 \to T}^0$ is the entropy increase if the substance is taken from $0°K$ to the temperature T. The $S_{0 \to T}^0$ can be measured calorimetrically. For a chemical reaction, using Eq. (11–70) for each substance,

$$\Delta S^0 = \Delta S_0^0 + \Delta S_{0 \to T}^0.$$

Putting this result into Eq. (11–69), we obtain

$$\Delta G^0 = \Delta H^0 - T\,\Delta S_0^0 - T\,\Delta S_{0 \to T}^0.$$

Therefore,

$$\ln K = \frac{\Delta S_0^0}{R} + \frac{\Delta S_{0 \to T}^0}{R} - \frac{\Delta H^0}{RT}. \tag{11–71}$$

Since the last two terms in Eq. (11–71) can be calculated from heat capacities and heats of reaction, the only unknown quantity is ΔS_0^0, the change in entropy of the reaction at $0°K$. In 1906, Nernst suggested that for all chemical reactions involving pure crystalline solids, ΔS_0^0 is zero at the absolute zero; the *Nernst heat theorem*. In 1913, Planck suggested that the reason that ΔS_0^0 is zero is that the entropy of each individual substance taking part in such a reaction is zero. It is clear that Planck's statement includes the Nernst theorem. However, either one is sufficient for the solution of the problem of determining the equilibrium constant from thermal measurements. Setting $\Delta S_0^0 = 0$ in Eq. (11–71), we obtain

$$\ln K = \frac{\Delta S^0}{R} - \frac{\Delta H^0}{RT}, \tag{11–72}$$

where ΔS^0 is the difference, at temperature T, in the third-law entropies of the substances involved in the reaction. Thus it is possible to calculate equilibrium constants from calorimetric data exclusively, provided that every substance in the reaction follows the third law.

Nernst based the heat theorem on the evidence from several chemical reactions. The data showed that, at least for these reactions, ΔG^0 approached ΔH^0 as the temperature decreased; from Eq. (11–69)

$$\Delta G^0 - \Delta H^0 = -T\,\Delta S^0.$$

If ΔG^0 and ΔH^0 approach each other in value, it follows that the product $T\,\Delta S^0 \to 0$ as the temperature decreases. This could be because T is getting smaller; however, the result was observed when the value of T was still of the order of $250°K$. This strongly suggests that $\Delta S^0 \to 0$ as $T \to 0$, which is the Nernst heat theorem.

The validity of the third law is tested by comparing the change in entropy of a reaction computed from the third-law entropies, with the entropy change computed from equilibrium measurement. Discrepancies appear whenever one of the substances in the reaction does not follow the third law. A few of these exceptions to the third law were described in Section 9–17.

11–18   Chemical Reactions and the Entropy of the Universe

A chemical reaction proceeds from some arbitrary initial state to the equilibrium state. If the initial state has the properties T, p, G_1, H_1, and S_1, and the equilibrium state has the properties T, p, G_e, H_e, S_e, then the free-energy change in the reaction is $\Delta G = G_e - G_1$; the enthalpy change is $\Delta H = H_e - H_1$, and the entropy change of the *system* is $\Delta S = S_e - S_1$. Since the temperature is constant, we have

$$\Delta G = \Delta H - T \Delta S,$$

and since the pressure is constant, $Q_p = \Delta H$. The heat which flows to the surroundings is $Q_s = -Q_p = -\Delta H$. If we suppose that Q_s is transferred reversibly to the immediate surroundings at the temperature T, then the entropy increase of the surroundings is $\Delta S_s = Q_s/T = -\Delta H/T$; or $\Delta H = -T \Delta S_s$. In view of this relation we have

$$\Delta G = -T(\Delta S_s + \Delta S).$$

The sum of the entropy changes in the system and the immediate surroundings is the entropy change in the universe; we have the relation

$$\Delta G = -T \Delta S_{\text{universe}}.$$

In this equation we see the equivalence of the two criteria for spontaneity: the free energy decrease of the system and the increase in entropy of the universe. If $\Delta S_{\text{universe}}$ is positive, then ΔG is negative. Note that it is not necessary for spontaneity that the entropy *of the system* increase and in many spontaneous reactions the entropy of the system decreases, e.g., $Na + \frac{1}{2}Cl_2 \rightarrow NaCl$. The entropy *of the universe* must increase in any spontaneous transformation.

11–19   Coupled Reactions

It often happens that a reaction which would be useful to produce a desirable product has a positive value of ΔG. For example, the reaction

$$TiO_2(s) + 2Cl_2(g) \rightarrow TiCl_4(l) + O_2(g), \qquad \Delta G^0_{298} = +38.7 \text{ kcal},$$

would be highly desirable for producing titanium tetrachloride from the common ore TiO_2. The high positive value of ΔG^0 indicates that at equilibrium only traces of $TiCl_4$ and O_2 are present. Increasing the temperature will improve the yield of $TiCl_4$ but not enough to make the reaction useful. However, if this reaction is coupled with another reaction which involves a ΔG more negative than -38.7 kcal, then the composite reaction can go spontaneously. If we are to pull the first reaction along, the second reaction must consume one of the products; since $TiCl_4$ is the desired product, the second reaction must consume oxygen. A likely prospect for the second reaction is

$$C(s) + O_2(g) \rightarrow CO_2(g), \qquad \Delta G^0_{298} = -94.3 \text{ kcal}.$$

The reaction scheme is

coupled reactions $\begin{cases} TiO_2(s) + 2Cl_2(g) \rightarrow TiCl_4(l) + O_2(g), & \Delta G^0_{298} = +38.7 \text{ kcal}, \\ C(s) + O_2(g) \rightarrow CO_2(g), & \Delta G^0_{298} = -94.3 \text{ kcal}, \end{cases}$

overall reaction:

$$C(s) + TiO_2(s) + 2Cl_2(g) \rightarrow TiCl_4(l) + CO_2(g), \qquad \Delta G^0_{298} = -55.6 \text{ kcal}.$$

Since the overall reaction has a highly negative ΔG^0, it is spontaneous. As a general rule metal oxides cannot be converted to chlorides by simple replacement; in the presence of carbon, the chlorination proceeds easily.

Coupled reactions have great importance in biological systems. Vital functions in an organism often depend on reactions which by themselves involve a positive ΔG; these reactions are coupled with the metabolic reactions which have highly negative values of ΔG. As a trivial example, the lifting of a weight by Mr. Universe is a nonspontaneous event involving a large increase in free energy. The weight goes up only because that event is coupled with the metabolic processes in the body which involve decreases in free energy sufficient to over-compensate the increase associated with the lifting of the weight.

11–20 Dependence of the Other Thermodynamic Functions on Composition

Having established the relation between the free energy and the composition, we can readily obtain the relation of the other functions to the composition. Considering the fundamental equation, Eq. (11–7),

$$dG = -S \, dT + V \, dp + \sum_i \mu_i \, dn_i.$$

We write the definitions of the other functions in terms of G:

$$E = G - pV + TS,$$

$$H = G + TS,$$

$$A = G - pV.$$

Differentiating each of these definitions, we have

$$dE = dG - p \, dV - V \, dp + T \, dS + S \, dT,$$

$$dH = dG + T \, dS + S \, dT,$$

$$dA = dG - p \, dV - V \, dp.$$

Replacing dG by its value in Eq. (11–7), we obtain

$$dE = T \, dS - p \, dV + \sum_i \mu_i \, dn_i, \qquad (11–73)$$

$$dH = T\,dS + V\,dp + \sum_i \mu_i\,dn_i, \tag{11–74}$$

$$dA = -S\,dT - p\,dV + \sum_i \mu_i\,dn_i, \tag{11–75}$$

$$dG = -S\,dT + V\,dp + \sum_i \mu_i\,dn_i. \tag{11–76}$$

Equations (11–73), (11–74), (11–75), and (11–76) are the fundamental equations for systems of variable composition, and they imply that μ_i may be interpreted in four different ways:

$$\mu_i = \left(\frac{\partial E}{\partial n_i}\right)_{S,V,n_j} = \left(\frac{\partial H}{\partial n_i}\right)_{S,p,n_j} = \left(\frac{\partial A}{\partial n_i}\right)_{T,V,n_j} = \left(\frac{\partial G}{\partial n_i}\right)_{T,p,n_j} \tag{11–77}$$

The last equality in Eq. (11–77), namely

$$\mu_i = \left(\frac{\partial G}{\partial n_i}\right)_{T,p,n_j}, \tag{11–78}$$

is the one we have used previously.

11–21 Partial Molar Quantities and Additivity Rules

Any extensive property of a mixture can be considered as a function of $T, p, n_1, n_2, \ldots$. Therefore, corresponding to any extensive property E, V, S, H, A, G, there are partial molar properties, $\bar{E}_i, \bar{V}_i, \bar{S}_i, \bar{H}_i, \bar{A}_i, \bar{G}_i$. The partial molar quantities are defined by

$$\bar{E}_i = \left(\frac{\partial E}{\partial n_i}\right)_{T,p,n_j}, \quad \bar{H}_i = \left(\frac{\partial H}{\partial n_i}\right)_{T,p,n_j}, \quad \bar{S}_i = \left(\frac{\partial S}{\partial n_i}\right)_{T,p,n_j},$$

$$\bar{V}_i = \left(\frac{\partial V}{\partial n_i}\right)_{T,p,n_j}, \quad \bar{A}_i = \left(\frac{\partial A}{\partial n_i}\right)_{T,p,n_j}, \quad \bar{G}_i = \mu_i = \left(\frac{\partial G}{\partial n_i}\right)_{T,p,n_j}. \tag{11–79}$$

If we differentiate the defining equations for H, A, and G with respect to n_i, keeping T, p, n_j constant, and use the definitions in Eqs. (11–79), we obtain

$$\bar{H}_i = \bar{E}_i + p\bar{V}_i, \qquad \bar{A}_i = \bar{E}_i - T\bar{S}_i, \qquad \mu_i = \bar{H}_i - T\bar{S}_i. \tag{11–80}$$

Equations (11–80) show that the partial molar quantities are related to each other in the same way as the total quantities. (The use of μ_i rather than $\bar{G}_i$ for the partial molar free energy is common.)

The total differential of any extensive property then takes a form analogous to Eq. (11–7). Choosing S, V, and H as examples,

$$dS = \left(\frac{\partial S}{\partial T}\right)_{p,n_i} dT + \left(\frac{\partial S}{\partial p}\right)_{T,n_i} dp + \sum_i \bar{S}_i\,dn_i \tag{11–81}$$

$$dV = \left(\frac{\partial V}{\partial T}\right)_{p,n_i} dT + \left(\frac{\partial V}{\partial p}\right)_{T,n_i} dp + \sum_i \bar{V}_i\,dn_i \tag{11–82}$$

$$dH = \left(\frac{\partial H}{\partial T}\right)_{p,n_i} dT + \left(\frac{\partial H}{\partial p}\right)_{T,n_i} dp + \sum_i \bar{H}_i \, dn_i. \tag{11-83}$$

Since $\bar{S}_i$, $\bar{V}_i$, and $\bar{H}_i$ are *intensive* properties they must have the same value everywhere in a system at equilibrium. Consequently, we could use precisely the same argument as was used for G in Section 11–3 to arrive at the additivity rules, namely,

$$S = \sum_i n_i \bar{S}_i, \qquad V = \sum_i n_i \bar{V}_i, \qquad H = \sum_i n_i \bar{H}_i. \tag{11-84}$$

However, by proceeding differently we gain some additional insights.

The free energy of a mixture is given by Eq. (11–9), $G = \sum_i n_i \mu_i$. If we differentiate this with respect to temperature (p and n_i are constant), we obtain

$$\left(\frac{\partial G}{\partial T}\right)_{p,n_i} = \sum_i n_i \left(\frac{\partial \mu_i}{\partial T}\right)_{p,n_i}. \tag{11-85}$$

By Eq. (11–76), the derivative on the left of Eq. (11–85) is equal to $-S$. The derivative on the right is evaluated by differentiating Eq. (11–78) with respect to T (suppressing subscripts to simplify writing):

$$\left(\frac{\partial \mu_i}{\partial T}\right)_{p,n_i} = \frac{\partial}{\partial T}\left(\frac{\partial G}{\partial n_i}\right) = \frac{\partial}{\partial n_i}\left(\frac{\partial G}{\partial T}\right) = -\left(\frac{\partial S}{\partial n_i}\right)_{T,p,n/j} = -\bar{S}_i.$$

The second equality is correct since the order of differentiation does not matter (Section 9–6); the third since $\partial G/\partial T = -S$. This reduces Eq. (11–85) to

$$S = \sum_i n_i \bar{S}_i, \tag{11-86}$$

which is the additivity rule for the entropy.

By differentiating Eq. (11–9) with respect to p, keeping T and n_i constant, we obtain

$$\left(\frac{\partial G}{\partial p}\right)_{T,n_i} = \sum_i n_i \left(\frac{\partial \mu_i}{\partial p}\right)_{T,n_i}. \tag{11-87}$$

Differentiating Eq. (11–78) with respect to p, we obtain

$$\left(\frac{\partial \mu_i}{\partial p}\right)_{T,n_i} = \frac{\partial}{\partial p}\left(\frac{\partial G}{\partial n_i}\right) = \frac{\partial}{\partial n_i}\left(\frac{\partial G}{\partial p}\right) = \left(\frac{\partial V}{\partial n_i}\right)_{T,p,n_j} = \bar{V}_i,$$

since $(\partial G/\partial p)_{T,n_i} = V$. Equation (11–87) then reduces to

$$V = \sum_i n_i \bar{V}_i, \tag{11-88}$$

which is the additivity rule for the volume. The other additivity rules can be established from these by taking the appropriate equation from the set (11–80). For example, multiply the last equation in the set by n_i and sum:

$$\sum_i n_i \mu_i = \sum_i n_i \bar{H}_i - T \sum_i n_i \bar{S}_i.$$

In view of Eqs. (11–9) and (11–86) this becomes

$$G = \sum_i n_i \bar{H}_i - TS,$$

but, by definition, $G = H - TS$; therefore,

$$H = \sum_i n_i \bar{H}_i. \tag{11–89}$$

In the same way, the additivity rules for E and A can be derived.

Any extensive property J of a system follows the additivity rule

$$J = \sum_i n_i \bar{J}_i, \tag{11–90}$$

where $\bar{J}_i$ is the partial molar quantity

$$\bar{J}_i = \left(\frac{\partial J}{\partial n_i}\right)_{T,p,n_j}. \tag{11–91}$$

This is true also for the total number of moles, $N = \sum_i n_i$; or the total mass, $M = \sum_i n_i M_i$. The partial molar mole numbers are all equal to unity. The partial molar mass of a substance is its molecular weight.

11–22 The Gibbs-Duhem Equation

An additional relation between the μ_i can be obtained by differentiating Eq. (11–9):

$$dG = \sum_i (n_i \, d\mu_i + \mu_i \, dn_i),$$

but, by the fundamental equation,

$$dG = -S \, dT + V \, dp + \sum_i \mu_i \, dn_i.$$

Subtracting, the two equations yield

$$\sum_i n_i \, d\mu_i = -S \, dT + V \, dp, \tag{11–92}$$

which is the Gibbs-Duhem equation. An important special case arises if the temperature and pressure are constant and only variations in composition occur; Eq. (11–92) becomes

$$\sum_i n_i \, d\mu_i = 0 \qquad (T, p \text{ constant}). \tag{11–93}$$

Equation (11–93) shows that if the composition varies, the chemical potentials do not change independently but in a related way. For example, in a system of two constituents, Eq. (11–93) becomes

$$n_1 \, d\mu_1 + n_2 \, d\mu_2 = 0 \qquad (T, p \text{ constant}).$$

Rearranging, we have

$$d\mu_2 = -(n_1/n_2)\,d\mu_1. \tag{11–94}$$

If a given variation in composition produces a change $d\mu_1$ in the chemical potential of the first component, then the concomitant change in the chemical potential of the second component $d\mu_2$ is given by Eq. (11–94).

By a similar argument it can be shown that the variations with composition of any of the partial molar quantities are related by the equation

$$\sum_i n_i\,d\bar{J}_i = 0 \qquad (T, p \text{ constant}), \tag{11–95}$$

where $\bar{J}_i$ is any partial molar quantity.

11–23 Partial Molar Quantities in Mixtures of Ideal Gases

The various partial molar quantities for the ideal gas are obtained from μ_i. From Eq. (11–13),

$$\mu_i = \mu_i^0(T) + RT\ln p + RT\ln x_i = \mu_{i(\text{pure})} + RT\ln x_i.$$

Differentiating, we have

$$\left(\frac{\partial\mu_i}{\partial T}\right)_{p,n_i} = \left(\frac{\partial\mu_i^0}{\partial T}\right)_{p,n_i} + R\ln p + R\ln x_i.$$

But $(\partial\mu_i/\partial T)_{p,n_i} = -\bar{S}_i$, so that

$$\bar{S}_i = \bar{S}_i^0 - R\ln p - R\ln x_i = \bar{S}_{i(\text{pure})} - R\ln x_i. \tag{11–96}$$

Similarly, differentiation of μ_i with respect to pressure, keeping T and all n_i constant, yields

$$\left(\frac{\partial\mu_i}{\partial p}\right)_{T,n_i} = \frac{RT}{p}.$$

Since $(\partial\mu_i/\partial p)_{T,n_i} = \bar{V}_i$, we obtain

$$\bar{V}_i = RT/p. \tag{11–97}$$

For an ideal gas mixture we have $V = NRT/p$, where N is the total number of moles of all the gases in the mixture. Therefore,

$$\bar{V}_i = V/N, \tag{11–98}$$

which shows that in a mixture of ideal gases, the partial molar volume is simply the average molar volume, and that the partial molar volume of all the gases in the mixture has the same value. This result is the reason for the statement made in Section 2–9, to the effect that the concept of "partial volume" developed there had little fundamental significance. Note that the partial molar volume and the "partial volume" are quite distinct entities; the partial molar volume has fundamental significance, the "partial volume" does not.

From Eqs. (11–13), (11–80), (11–96), and (11–97) it is easy to show that $\bar{H}_i = \mu_i^0 + T\bar{S}_i^0 = \bar{H}_i^0$, and that $\bar{E}_i = \bar{H}_i^0 - RT = \bar{E}_i^0$.

11–24 Differential Heat of Solution

If dn moles of pure solid i, with molar enthalpy $\bar{H}_i^0$, are added at constant T and p to a solution in which the partial molar enthalpy is $\bar{H}_i$, then the heat absorbed is $dq = dH = (\bar{H}_i - \bar{H}_i^0)\, dn$. (The system contains both solid and solution). The *differential heat of solution* is defined as dq/dn:

$$dq/dn = \bar{H}_i - \bar{H}_i^0. \tag{11–99}$$

The differential heat of solution is a more generally useful quantity than the integral heat of solution defined in Section 7–22.

Problems

In all the following problems, the gases are assumed to be ideal.

11–1. Plot the value of $(\mu - \mu^0)/RT$ for an ideal gas as a function of pressure.

11–2. a) Calculate the entropy of mixing 3 moles of hydrogen with 1 mole of nitrogen.

b) Calculate the free energy of mixing at 25°C.

11–3. At 25°C, calculate the free energy of mixing $1 - y$ moles of nitrogen, $3(1 - y)$ moles of hydrogen, and $2y$ moles of ammonia as a function of y. Plot the values from $y = 0$ to $y = 1$ at intervals of 0.2.

11–4. Consider two pure cases A and B, each at 25°C and 1 atm pressure. Calculate the free energy relative to the unmixed gases of

a) a mixture of 10 moles of A and 10 moles of B;

b) a mixture of 10 moles of A and 20 moles of B.

c) Calculate the change in free energy if 10 moles of B is added to the mixture of 10 moles of A with 10 moles of B.

11–5. The conventional standard free energy of ammonia at 25°C is -3976 cal/mole. Calculate the value of the molar free energy at $\frac{1}{2}$, 2, 10, and 100 atm.

11–6. For ozone at 25°C, $\Delta G_f^0 = 39.06$ kcal/mole. Compute the equilibrium constant K_p for the reaction

$$3O_2(g) \rightleftharpoons 2O_3(g)$$

at 25°C. Assuming that the advancement at equilibrium, ξ_e, is very much less than unity, show that $\xi_e = \frac{3}{2}\sqrt{pK_p}$. (Let the original number of moles of O_2 be three, and of O_3 be zero.)

11–7. Equimolar amounts of H_2 and CO are mixed. Using data from Table 11–1 calculate the equilibrium mole fraction of formaldehyde, HCHO(g), at 25°C as a function of the total pressure; evaluate this mole fraction for a total pressure of 1 atm and for 10 atm.

11–8. Consider the equilibrium: $2NO(g) + Cl_2(g) \rightleftharpoons 2NOCl(g)$. At 25°C for NOCl(g), $\Delta G_f^0 = 15.86$ kcal/mole. (For the value for NO(g) see Table 11–1.) If NO and Cl_2 are mixed in the molar ratio 2:1, show that at equilibrium, $x_{NO} = (2/pK_p)^{1/3}$ and $x_{NOCl} = 1 - \frac{3}{2}(2/pK_p)^{1/3}$. (Assume that $x_{NOCl} \approx 1$.) Note how each one of these quantities depends on pressure. Evaluate x_{NO} at 1 atm and at 10 atm.

11–9. Consider the following equilibrium at 25°C:

$$PCl_5(g) \rightleftharpoons PCl_3(g) + Cl_2(g).$$

a) From the data in Tables 7–1 and 11–1 compute ΔG^0 and ΔH^0 at 25°C.

b) Calculate the value of K_p at 600°K.

c) At 600°K calculate the degree of dissociation α at 1 atm and at 5 atm total pressure.

11–10. Consider the reaction $H_2(g) + I_2(g) \rightarrow 2HI(g)$.

a) If there are one mole of H_2, one mole of I_2, and zero moles of HI present before the reaction advances, express the free energy of the reaction mixture in terms of the advancement ξ.

b) What form would the expression for G have if the iodine were present as the solid?

11–11. From the data in Table 11–1 compute K_p at 25°C for the reaction $H_2(g) + S(rhombic) \rightleftharpoons H_2S(g)$. What is the mole fraction of H_2 present in the gas phase at equilibrium?

11–12. Consider the reaction $Ag_2O(s) \rightleftharpoons 2Ag(s) + \frac{1}{2}O_2(g)$, for which $\Delta G^0 = 7740 + 4.14T \log_{10} T - 27.84T$ cal.

a) At what temperature will the equilibrium pressure of oxygen be one atmosphere?

b) Express $\log_{10} K_p$, ΔH^0, and ΔS^0 as functions of temperature.

11–13. From the data in Tables 7–1 and 11–1 find the values of ΔG^0 and ΔH^0 for the reactions

$$MCO_3(s) \rightleftharpoons MO(s) + CO_2(g), \qquad (M = Mg, Ca, Sr, Ba).$$

Under the rash assumption that ΔH^0 for these reactions does not depend on temperature, calculate the temperatures at which the equilibrium pressure of CO_2 in these carbonate–oxide systems reaches 1 atm. (This is the decomposition temperature of the carbonate.)

11–14. Liquid bromine boils at 58.2°C; the vapor pressure at 9.3°C is 100 mm. Calculate the standard free energy of $Br_2(g)$ at 25°C.

11–15. Solid white phosphorus has a conventional standard free energy of zero at 25°C. The melting point is 44.2°C; and $\Delta H_{fus}^0 = 150$ cal/mole. The vapor pressure of white phosphorus has the values

p, mm	1	10	100
t, °C	76.6	128.0	197·3

a) Calculate ΔH_{vap}^0 of liquid phosphorus.

b) Calculate the boiling point of the liquid.

c) Calculate the vapor pressure at the melting point.

d) Assuming that solid, liquid, and gaseous phosphorus are in equilibrium at the melting point, calculate the vapor pressure of *solid* white phosphorus at 25°C.

e) Calculate the standard free energy of *gaseous* phosphorus at 25°C.

11–16. The values of ΔG^0 and ΔH^0 for the reactions

$$C(\text{graphite}) + \tfrac{1}{2}O_2(g) \rightleftharpoons CO(g), \qquad CO(g) + \tfrac{1}{2}O_2(g) \rightleftharpoons CO_2(g)$$

can be obtained from the data in Tables 7–1 and 11–1.

a) Assuming the values of ΔH^0 do not vary with temperature, compute the composition (mole percent) of the gas in equilibrium with solid graphite at 600°K and 1000°K if the total pressure is 1 atm. Qualitatively, how would the composition change if the pressure were increased?

b) Using the heat capacity data in Table 7–2, compute the composition at 600°K and 1000°K (1 atm) and compare the results with those in (a).

c) Using the equilibrium constants from (b) compute the composition at 1000°K and 10 atm pressure.

11–17. Show that in an ideal ternary mixture, the minimum free energy is obtained if $x_1 = x_2 = x_3 = \tfrac{1}{3}$.

11–18. Show that if two or more equilibria are established in a system, each one is governed by an equilibrium condition of the type in Eq. (11–33a).

11–19. Consider the dissociation of nitrogen tetroxide: $N_2O_4(g) \rightleftharpoons 2NO_2(g)$ at 25°C. Suppose one mole of N_2O_4 is confined in a vessel under 1 atm pressure. Using data from Table 11–1,
a) calculate the degree of dissociation.

b) If 5 moles of argon are introduced and the mixture confined under 1 atm total pressure, what is the degree of dissociation?

c) If the volume of the vessel, determined by the conditions specified in (a), is kept constant and 5 moles of argon are introduced, what will be the degree of dissociation?

11–20. The degree of dissociation of N_2O_4 is a function of the pressure. Show that if the mixture remains in equilibrium as the pressure is changed, the apparent compressibility $(-1/V)(\partial V/\partial p)_T = (1/p)[1 + \tfrac{1}{2}\alpha(1 - \alpha)]$. Show that this quantity has a maximum value at $p = \tfrac{1}{4}K_p$.

11–21. At 25°C, the various isomers of C_5H_{10} in the gas phase have the free energies and enthalpies of formation as follows (all values in kcal/mole):

	A	B	C	D
ΔH_f°	− 5.000	− 6.710	− 7.590	− 8.680
ΔG_f°	18.787	17.173	16.575	15.509
$\log_{10} K_f$	− 13.7704	− 12.5874	− 12.1495	− 11.3680

	E	F	G
ΔH_f°	-6.920	-10.170	-18.46
ΔG_f°	17.874	14.267	9.23
$\log_{10} K_f$	-13.1017	-10.4572	-6.7643

A = 1-pentene;   B = cis-2-pentene;
C = trans-2-pentene;   D = 2-methyl-1-butene;
E = 3-methyl-1-butene;   F = 2-methyl-2-butene;
G = cyclopentane.

Consider the equilibria:

$$A \rightleftharpoons B \rightleftharpoons C \rightleftharpoons D \rightleftharpoons E \rightleftharpoons F \rightleftharpoons G$$

which might be established using a suitable catalyst.

a) Calculate the mole ratios: (A/G); (B/G), ..., (F/G) present at equilibrium at 25°C.

b) Do these ratios depend on the total pressure?

c) Calculate the mole percent of the various species in the equilibrium mixture.

d) Calculate the composition of the equilibrium mixture at 500°K.

11–22. Consider the synthesis of formaldehyde:

$$CO(g) + H_2(g) \rightleftharpoons CH_2O(g).$$

At 25°C, $\Delta G^0 = 6.5$ kcal and $\Delta H^0 = -1.3$ kcal. For $CH_2O(g)$ we have: $C_p = 4.498 + 13.953(10^{-3})T - 3.730(10^{-6})T^2$ cal/deg-mole. The heat capacities of $H_2(g)$ and $CO(g)$ are given in Table 7–2.

a) Calculate the value of K_p at 1000°K assuming ΔH^0 is independent of temperature.

b) Calculate the value of K_p at 1000°K taking into account the variation of ΔH^0 with temperature, and compare the result with that in (a).

c) At 1000°K compare the value of K_x at 1 atm pressure with that at 5 atm pressure.

11–23. In a gravity field the chemical potential of a species is increased by the potential energy required to raise one mole of the material from ground level to the height z. Then $\mu_i(T, p, z) = \mu_i(T, p) + M_i g z$, in which $\mu_i(T, p)$ is the value of μ_i at ground level, M_i is the molecular weight, and g is the gravitational acceleration.

a) Show that if we require the chemical potential to be everywhere the same in an iso-thermal column of an ideal gas, this form of the chemical potential yields the baro-metric distribution law: $p_i = p_{i0} \exp(-M_i g z / RT)$.

b) Show that the condition of chemical equilibrium is independent of the presence or absence of a gravity field.

c) Derive expressions for the entropy and enthalpy as functions of z. (*Hint*: Write the differential of μ_i in terms of dT, dp, and dz.)

11–24. For the reaction at 25°C,

$$Zn(s) + Cl_2(g) \rightleftharpoons ZnCl_2(s),$$

$$\Delta G^0 = -88.255 \text{ kcal} \quad \text{and} \quad \Delta H^0 = -99.40 \text{ kcal}.$$

Sketch the ΔG^0 as a function of temperature in the range from 298°K to 1500°K for this reaction in the situation that all the substances are in their stable states of aggregation at every temperature. The data are

	Melting point, °K	ΔH_{fus}, kcal/mole	Boiling point, °K	ΔH_{vap} kcal/mole
Zn	692.7	1.595	1180	27.43
$ZnCl_2$	548	5.5	1029	30.9

11.25. The following data are given at 25°C.

Compound	$CuO(s)$	$Cu_2O(s)$	$Cu(s)$	$O_2(g)$
ΔH_f^0 (kcal/mole)	−37.1	−39.84	—	—
ΔG_f^0 (kcal/mole)	−30.4	−34.98	—	—
C_p^0 (cal/deg-mole)	10.6	16.7	5.8	7.0

a) Calculate the equilibrium pressure of oxygen over copper and cupric oxide at 900°K and at 1200°K; that is, the equilibrium constant for the reaction : $2CuO(s) \rightleftharpoons 2Cu(s) + O_2(g)$.

b) Calculate the equilibrium pressure of oxygen over Cu_2O and Cu, at 900°K and 1200°K.

c) At what temperature and pressure do Cu, CuO, Cu_2O, and O_2 coexist in equilibrium?

11–26. The standard state of zero free energy for phosphorus is solid white phosphorus, $P_4(s)$. At 25°C,

$$P_4(s) \rightleftharpoons P_4(g) \quad \Delta H^0 = 13.12 \text{ kcal} \quad \Delta G^0 = 5.82 \text{ kcal},$$

$$\tfrac{1}{4}P_4(s) \rightleftharpoons P(g) \quad \Delta H^0 = 75.18 \text{ kcal} \quad \Delta G^0 = 66.71 \text{ kcal},$$

$$\tfrac{1}{2}P_4(s) \rightleftharpoons P_2(g) \quad \Delta H^0 = 33.82 \text{ kcal} \quad \Delta G^0 = 24.60 \text{ kcal},$$

a) The P_4 molecule consists of four phosphorus atoms at the corners of a tetrahedron. Calculate the bond strength of the P—P bond in the tetrahedral molecule. Calculate the bond strength in P_2.

b) Calculate the mole fractions of P, P_2, and P_4 in the vapor at 900°K and 1200°K.

Chapter Twelve

Phase Equilibrium in
Simple Systems; The Phase Rule

12–1 The Equilibrium Condition

For a system in equilibrium the chemical potential of each constituent must be the same everywhere in the system. If there are several phases present, the chemical potential of each substance must have the same value in every phase in which that substance appears.

For a system of one component, $\mu = G/n$; dividing the fundamental equation by n, we obtain

$$d\mu = -\bar{S}\,dT + \bar{V}dp, \tag{12–1}$$

where $\bar{S}$ and $\bar{V}$ are the molar entropy and molar volume. Then

$$\left(\frac{d\mu}{\partial T}\right)_p = -\bar{S} \tag{12–2a}$$

and

$$\left(\frac{\partial\mu}{\partial p}\right)_T = \bar{V}. \tag{12–2b}$$

The derivatives in Eqs. (12–2a, b) are the slopes of the curves μ versus T and μ versus p, respectively.

12–2 Stability of the Phases of a Pure Substance

By the third law of thermodynamics, the entropy of a substance is always positive. This fact combined with Eq. (12–2a) shows that $(\partial\mu/\partial T)_p$ is always negative. Consequently, the plot of μ versus T at constant pressure is a curve with a negative slope.

For the three phases of a single substance we have

$$\left(\frac{\partial \mu_{\text{solid}}}{\partial T}\right)_p = -\bar{S}_{\text{solid}} \qquad \left(\frac{\partial \mu_{\text{liq}}}{\partial T}\right)_p = -\bar{S}_{\text{liq}} \qquad \left(\frac{\partial \mu_{\text{gas}}}{\partial T}\right)_p = -\bar{S}_{\text{gas}}. \qquad (12\text{–}3)$$

At any temperature, $\bar{S}_{\text{gas}} \gg \bar{S}_{\text{liq}} > \bar{S}_{\text{solid}}$. The entropy of the solid is small so that in Fig. 12–1 the μ versus T curve for the solid, curve S, has a slight negative slope. The μ versus T curve for the liquid has a slope which is slightly more negative than that of the solid, curve L. The entropy of the gas is very much larger than that of the liquid, so the slope of curve G has a large negative value.[†]

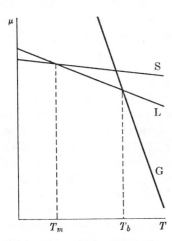

Fig. 12–1 μ versus T at constant pressure.

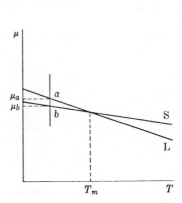

Fig. 12–2 μ versus T at constant pressure.

The thermodynamic conditions for equilibrium between phases at constant pressure are immediately apparent in Fig. 12–1. Solid and liquid coexist in equilibrium when $\mu_{\text{solid}} = \mu_{\text{liq}}$; that is, at the intersection point of curves S and L. The corresponding temperature is T_m, the melting point. Similarly, liquid and gas coexist in equilibrium at the temperature T_b, the intersection point of curves L and G at which $\mu_{\text{liq}} = \mu_{\text{gas}}$.

The temperature axis is divided into three intervals. Below T_m the solid has the lowest chemical potential. Between T_m and T_b the liquid has the lowest chemical potential. Above T_b the gas has the lowest chemical potential. *The phase with the lowest value of the chemical potential is the stable phase.* If liquid were present in a system at a temperature below T_m, Fig. 12–2, the chemical potential of the liquid would have the value μ_a while the solid has the value μ_b. Thus, liquid could freeze

---

[†]   The curves have been drawn as straight lines; they should be slightly concave downward. However, this refinement does not affect the argument.

spontaneously at this temperature, since freezing will decrease the free energy. At a temperature above T_m the situation is reversed; the μ of the solid is greater than that of the liquid and the solid melts spontaneously to decrease the free energy of the system. At T_m the chemical potentials of solid and liquid are equal, so neither phase is preferred; they coexist in equilibrium. The situation is much the same near T_b. Just below T_b liquid is stable, while just above, the gas is the stable phase.

The diagram illustrates the familiar sequence of phases observed if a solid is heated under constant pressure. At low temperatures the system is completely solid; at a definite temperature T_m the liquid forms; the liquid is stable until it vaporizes at a temperature T_b. This sequence of phases is a consequence of the sequence of entropy values, and so is an immediate consequence of the fact that heat is absorbed in the transformation from solid to liquid, and from liquid to gas.

12–3   Pressure Dependence of μ Versus *T* Curves

At this point it is natural to ask what happens to the curves if the pressure is changed. This question is answered using Eq. (12–2b) in the form $d\mu = \overline{V} dp$. If the pressure is decreased, dp is negative, $\overline{V}$ is positive; hence $d\mu$ is negative, and the chemical potential decreases in proportion to the volume of the phase. Since the molar volumes of the liquid and solid are very small, the value of μ is decreased only slightly; for the solid from a to a', for the liquid from b to b'; Fig. 12–3(a). The volume of the gas is roughly 1000 times larger than that of the solid or liquid, so the μ of the gas decreases greatly; from c to c'. The curves at the lower pressure are shown as dashed lines parallel to the original lines in Fig. 12–3(b). (The figure has been drawn for the case $\overline{V}_{liq} > \overline{V}_{solid}$.) Figure 12–3(b) shows that both equilibrium temperatures (both intersection points) have shifted; the shift in the melting point is small, while the shift in the boiling point is relatively large. The melting point shift has been exaggerated for emphasis; it is actually very small. The decrease in boiling point of a

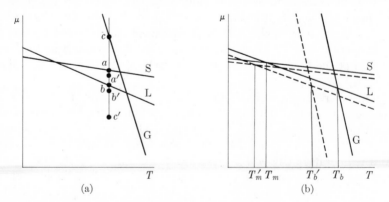

(a)   (b)

Fig. 12–3 Effect of pressure on melting and boiling points. Solid line indicates high pressure; dashed line low pressure.

liquid with decrease in pressure is neatly illustrated. At the lower pressure the range of stability of the liquid is noticeably decreased. If the pressure is reduced to a sufficiently low value, the boiling point of the liquid may even fall below the melting point of the solid; Fig. 12–4. Then there is no temperature at which the liquid is stable; the solid sublimes. At the temperature T_s, the solid and vapor coexist in equilibrium. The temperature T_s is the sublimation temperature of the solid. It is very dependent on the pressure.

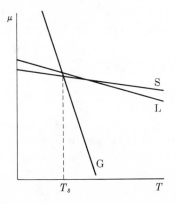

Fig. 12–4 μ versus T for a substance which sublimes.

Whether or not a particular material will sublime under reduced pressure rather than melt depends entirely on the individual properties of the substance. Water, for example, sublimes at pressures below 4.58 mm. The higher the melting point, and the smaller the difference between the melting point and boiling point at 1 atm pressure, the higher will be the pressure below which sublimation is observed. The pressure (in atm) below which sublimation is observed can be estimated for substances obeying Trouton's rule by the formula

$$\ln p = -10.5\left(\frac{T_b - T_m}{T_m}\right).\tag{12–4}$$

12–4 The Clapeyron Equation

The condition for equilibrium between two phases, α and β, of a pure substance is

$$\mu_\alpha(T, p) = \mu_\beta(T, p).\tag{12–5}$$

If the analytical forms of the functions μ_α and μ_β were known, it would be possible, in principle at least, to solve Eq. (12–5) for

$$T = f(p)\tag{12–6a}$$

or

$$p = g(T).\tag{12–6b}$$

Equation (12–6a) expresses the fact, illustrated in Fig. 12–3(b), that the equilibrium temperature depends on the pressure.

In the absence of this detailed knowledge of the functions μ_α and μ_β, it is possible nonetheless to obtain a value for the derivative of the temperature with respect to pressure. Consider the equilibrium between two phases α and β under a pressure p; the equilibrium temperature is T. Then, at T and p, we have

$$\mu_\alpha(T, p) = \mu_\beta(T, p). \tag{12–7}$$

If the pressure is changed to a value $p + dp$, the equilibrium temperature will change to $T + dT$, and the value of each μ will change to $\mu + d\mu$. Hence at $T + dT, p + dp$ the equilibrium condition is

$$\mu_\alpha(T, p) + d\mu_\alpha = \mu_\beta(T, p) + d\mu_\beta. \tag{12–8}$$

Subtracting Eq. (12–7) from Eq. (12–8), we obtain

$$d\mu_\alpha = d\mu_\beta. \tag{12–9}$$

We write $d\mu$ explicitly in terms of dp and dT using the fundamental equation, Eq. (12–1):

$$d\mu_\alpha = -\bar{S}_\alpha\, dT + \bar{V}_\alpha\, dp \qquad d\mu_\beta = -\bar{S}_\beta\, dT + \bar{V}_\beta\, dp. \tag{12–10}$$

Using Eqs. (12–10) in Eq. (12–9), we get

$$-\bar{S}_\alpha\, dT + \bar{V}_\alpha\, dp = -\bar{S}_\beta\, dT + \bar{V}_\beta\, dp.$$

Rearranging, we have

$$(\bar{S}_\beta - \bar{S}_\alpha)\, dT = (\bar{V}_\beta - \bar{V}_\alpha)\, dp. \tag{12–11}$$

If the transformation is written $\alpha \to \beta$, then $\Delta S = \bar{S}_\beta - \bar{S}_\alpha$, and $\Delta V = \bar{V}_\beta - \bar{V}_\alpha$, and Eq. (12–11) becomes

$$\frac{dT}{dp} = \frac{\Delta V}{\Delta S} \tag{12–12a}$$

or

$$\frac{dp}{dT} = \frac{\Delta S}{\Delta V}. \tag{12–12b}$$

Either of Eqs. (12–12) is called the Clapeyron equation.

The Clapeyron equation is fundamental to any discussion of the equilibrium between two phases of a pure substance. Note that the left-hand side is an ordinary derivative and not a partial derivative. The reason for this should be apparent from Eqs. (12–6).

Figure 12–3(b) shows that the equilibrium temperatures depend on the pressure, since the intersection points depend on pressure. The Clapeyron equation expresses the quantitative dependence of the equilibrium temperature on pressure, Eq. (12–12a), or the variation in the equilibrium pressure with temperature, Eq. (12–12b). Using this equation, we can plot the equilibrium pressure versus temperature schematically for any phase transformation.

12–5   Application of the Clapeyron Equation

The solid–liquid equilibrium

Applying the Clapeyron equation to the transformation solid → liquid, we have

$$\Delta S = \bar{S}_{\text{liq}} - \bar{S}_{\text{solid}} = \Delta S_{\text{fus}} \qquad \Delta V = \bar{V}_{\text{liq}} - \bar{V}_{\text{solid}} = \Delta V_{\text{fus}}.$$

At the equilibrium temperature, the transformation is reversible; hence $\Delta S_{\text{fus}} = \Delta H_{\text{fus}}/T$. The transformation from solid to liquid always entails an absorption of heat, (ΔH_{fus} is $+$); hence

$$\Delta S_{\text{fus}} \text{ is } + \qquad \text{(all substances).}$$

The quantity ΔV_{fus} may be positive or negative, depending on whether the density of the solid is greater or less than that of the liquid; therefore,

$$\Delta V_{\text{fus}} \text{ is } + \qquad \text{(most substances)}$$

$$\Delta V_{\text{fus}} \text{ is } - \qquad \text{(a few substances, e.g., } H_2O\text{).}$$

The ordinary magnitudes of these quantities are

$$\Delta S_{\text{fus}} = 2 \text{ to } 6 \text{ eu/mole} \qquad \Delta V_{\text{fus}} = \pm(1 \text{ to } 10) \text{ cm}^3/\text{mole.}$$

If, for illustration, we choose: $\Delta S_{\text{fus}} = 4 \text{ eu/mole}$ and $\Delta V_{\text{fus}} = \pm 4 \text{ cm}^3/\text{mole}$, then for the solid–liquid equilibrium line,

$$\frac{dp}{dT} = \frac{4 \text{ eu/mole}}{\pm 4 \text{ cm}^3/\text{mole}} = \pm 1 \text{ eu/cm}^3 = \pm 1 \text{ cal·cm}^{-3}\text{·deg}^{-1} = \pm 41 \text{ atm/deg.}$$

Inverting, we obtain $dT/dp = \pm 0.02 \text{ deg/atm}$. This value shows that a change in pressure of 1 atm alters the melting point by a few hundredths of a degree. In a plot of pressure as a function of temperature, the slope is given by Eq. (12–12b); (41 atm/deg in the example); this slope is large and the curve is nearly vertical. The case dp/dT is $+$ is shown in Fig. 12–5(a); over a moderate range of pressure the curve is linear.

The line in Fig. 12–5(a) is the locus of all points (T, p) at which the solid and liquid can coexist in equilibrium. Points which lie to the left of the line correspond

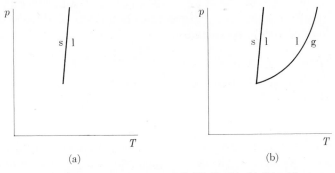

                       (a)   (b)

Fig. 12–5 Equilibrium lines. a) Solid–liquid. b) Liquid–vapor.

to temperatures below the melting point; these points are conditions (T, p) under which only the solid is stable. Points immediately to the right of the line correspond to temperatures above the melting point; hence these points are conditions (T, p) under which the liquid is stable.

The liquid–gas equilibrium

Application of the Clapeyron equation to the transformation liquid $\rightarrow$ gas yields

$$\Delta S = \bar{S}_{\text{gas}} - \bar{S}_{\text{liq}} = \frac{\Delta H_{\text{vap}}}{T} \text{ is } + \qquad \text{(all substances)}$$

$$\Delta V = \bar{V}_{\text{gas}} - \bar{V}_{\text{liq}} \text{ is } + \qquad \text{(all substances)}$$

and

$$\frac{dp}{dT} = \frac{\Delta S}{\Delta V} \text{ is } + \qquad \text{(all substances)}.$$

The liquid–gas equilibrium line always has a positive slope. At ordinary T and p the magnitudes are

$$\Delta S \approx +20 \text{ eu} \qquad \Delta V \approx +20{,}000 \text{ cm}^3.$$

However, ΔV depends strongly on T and p because $\bar{V}_{\text{gas}}$ depends strongly on T and p. The slope of the liquid–gas curve is small compared with that of the solid–liquid curve:

$$\left(\frac{dp}{dT}\right)_{\text{liq, gas}} \approx \frac{20}{20{,}000} = 10^{-3} \text{ cal·deg}^{-1}\text{·cm}^{-3} = 0.04 \text{ atm/deg}.$$

Figure 12–5(b) shows the l-g curve as well as the s-l curve. In Fig. 12–5(b), curve l-g is the locus of all points (T, p) at which liquid and gas coexist in equilibrium. Points just to the left of l-g are below the boiling point and so are conditions under which the liquid is stable. Points to the right of l-g are conditions under which the gas is stable.

The intersection of curves s-l and l-g corresponds to a temperature and pressure at which solid, liquid, and gas all coexist in equilibrium. The values of T and p at this point are determined by the conditions

$$\mu_{\text{solid}}(T, p) = \mu_{\text{liq}}(T, p) \qquad \text{and} \qquad \mu_{\text{liq}}(T, p) = \mu_{\text{gas}}(T, p). \qquad (12\text{–}13)$$

Equations (12–13) can, in principle at least, be solved for definite numerical values of T and p. That is,

$$T = T_t \qquad p = p_t, \qquad (12\text{–}14)$$

where T_t and p_t are the triple-point temperature and pressure. There is only one such triple point at which a specific set of three phases (e.g., solid–liquid–gas) can coexist in equilibrium.

The solid–gas equilibrium

For the transformation solid → gas, we have

$$\Delta S = \bar{S}_{\text{gas}} - \bar{S}_{\text{solid}} = \frac{\Delta H_{\text{sub}}}{T} \text{ is } + \quad \text{(all substances)}$$

$$\Delta V = \bar{V}_{\text{gas}} - \bar{V}_{\text{solid}} \text{ is } + \quad \text{(all substances)},$$

and the Clapeyron equation is

$$\left(\frac{dp}{dT}\right)_{\text{s-g}} = \frac{\Delta S}{\Delta V} = \frac{+\text{ve quantity}}{+\text{ve quantity}} \text{ is } + \quad \text{(all substances)}.$$

The slope of the s-g curve is steeper at the triple point than the slope of the l-g curve. At the triple point, $\Delta H_{\text{sub}} = \Delta H_{\text{fus}} + \Delta H_{\text{vap}}$. Then

$$\left(\frac{dp}{dT}\right)_{\text{l-g}} = \frac{\Delta H_{\text{vap}}}{T\,\Delta V} \quad \text{and} \quad \left(\frac{dp}{dT}\right)_{\text{s-g}} = \frac{\Delta H_{\text{sub}}}{T\,\Delta V}.$$

The ΔV's in the two equations are very nearly equal. Since ΔH_{sub} is greater than ΔH_{vap}, the slope of the s-g curve in Fig. 12–6 is steeper than that of the l-g curve.

Points on the curve s-g are those sets of temperatures and pressures at which solid coexists in equilibrium with vapor. Points to the left of the line lie below the sublimation temperature, and so correspond to conditions under which the solid is stable. Those points to the right of s-g are points above the sublimation temperature, and so are conditions under which the gas is the stable phase. The s-g curve must intersect the others at the triple point because of the conditions expressed by Eqs. (12–13).

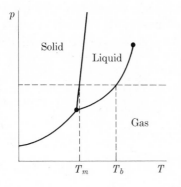

Fig. 12–6 Phase diagram for a simple substance.

12–6   The Phase Diagram

Examination of Fig. 12–6 at a constant pressure, indicated by the dashed horizontal line, shows the melting point and boiling point of the substance as the intersections of the horizontal line with the s-l and l-g curves. These intersection points correspond to the intersections of the μ-T curves in Fig. 12–1. At temperatures below T_m, the solid is stable; at the points between T_m and T_b the liquid is stable, while above T_b the gas is stable. Illustrations such as Fig. 12–6 convey more information than those such as 12–1 and 12–3(b). Figure 12–6 is called a *phase diagram*, or an *equilibrium diagram*.

The phase diagram shows at a glance the properties of the substance; melting point, boiling point, transition points, triple points. Every point on the phase diagram represents a state of the system, since it describes values of T and p.

The lines on the phase diagram divide it into regions, labeled *solid*, *liquid*, and *gas*. If the point which describes the system falls in the solid region, the substance exists as a solid. If the point falls in the liquid region, the substance exists as a liquid. If the point falls on a line such as l-g, the substance exists as liquid and vapor in equilibrium.

The l-g curve has a definite upper limit at the critical pressure and temperature, since it is not possible to distinguish between liquid and gas above this pressure and temperature.

12–7 The Phase Diagrams for CO_2 and H_2O

The phase diagram for carbon dioxide is shown schematically in Fig. 12–7. The solid–liquid line slopes slightly to the right, since $\bar{V}_{liq} > \bar{V}_{solid}$. Note that liquid CO_2 is not stable at pressures below 5 atm. For this reason "dry ice" is dry under ordinary atmospheric pressure. When carbon dioxide is confined to a cylinder under pressure at 25°C, the diagram shows that if the pressure reaches 67 atm, liquid CO_2 will form. Commercial cylinders of CO_2 commonly contain liquid and gas in equilibrium; the pressure in the cylinder is about 67 atm at 25°C.

Figure 12–8 is the phase diagram for water under moderate pressure. The solid–liquid line leans slightly to the left because $\bar{V}_{liq} < \bar{V}_{solid}$. The triple point is at 0.0098°C and 4.58 mm pressure. An increase in pressure decreases the melting point

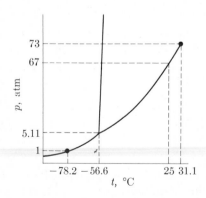

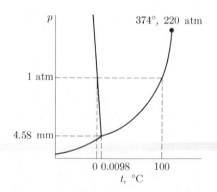

Fig. 12.7 Phase diagram for CO_2.   Fig. 12–8 Phase diagram for water.

of water. Ice skating is possible because of this lower melting point under the pressure exerted by the weight of the skater through the knife edge of the skate blade. This effect together with the heat developed by friction combine to produce a lubricating layer of liquid water between the ice and the blade. In this connection, it is interesting to note that if the temperature is too low, the skating is not good.

If water is studied under very high pressures, several crystalline modifications of ice are observed. The equilibrium diagram is shown in Fig. 12–9. Ice I is ordinary ice; ices II, III, V, VI, VII are modifications which are stable at higher pressures. The range of pressure is so large in Fig. 12–9 that the s-g and l-g curves lie but slightly above the horizontal axis; they are not shown in the figure. It is remarkable that under very high pressures, melting ice is quite hot! Ice VII melts at about 100°C under a pressure of 25,000 atm.

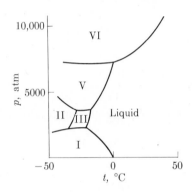

Fig. 12–9 Phase diagram for water at high pressures. (Redrawn by permission of the National Academy of Sciences from *International Critical Tables of Numerical Data.*)

12–8 The Phase Diagram for Sulfur

Figure 12–10 shows two phase diagrams for sulfur. The stable form of sulfur at ordinary temperatures and under 1 atm pressure is rhombic sulfur, which, if heated slowly, transforms to solid monoclinic sulfur at 95.4°C; see Fig. 12–10(a). Above 95.4°C monoclinic sulfur is stable, until 119°C is reached; monoclinic sulfur melts at 119°C. Liquid sulfur is stable up to the boiling point, 444.6°C. Since the transformation of one crystalline modification to another is often very slow, if rhombic sulfur is heated quickly to 114°C, it melts. This melting point of rhombic sulfur is shown as a function of pressure in Fig. 12–10(b). The equilibrium S(rhombic) $\rightleftharpoons$ S(l) is an example of a *metastable* equilibrium, since the line lies in the region of stability of monoclinic sulfur, shown by dashed lines in Fig. 12–10(b). In this region the reactions

$$S(rh) \rightarrow S(mono)$$

and

$$S(liq) \rightarrow S(mono)$$

both can occur with a decrease in free energy.

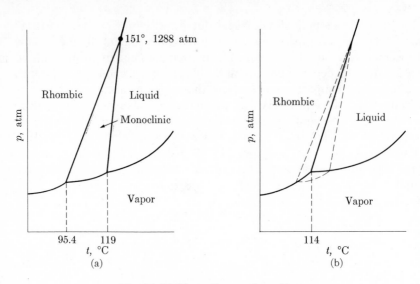

Fig. 12–10 Phase diagram for sulfur.

In Fig. 12–10(a) there are three triple points. The equilibrium conditions are

at 95.4°C: $\mu_{rh} = \mu_{mono} = \mu_{gas}$,

at 119°C: $\mu_{mono} = \mu_{liq} = \mu_{gas}$,

at 151°C: $\mu_{rh} = \mu_{mono} = \mu_{liq}$.

12–9   The Integration of the Clapeyron Equation

Solid–liquid equilibrium

The Clapeyron equation reads

$$\frac{dp}{dT} = \frac{\Delta S_{fus}}{\Delta V_{fus}}.$$

Then

$$\int_{p_1}^{p_2} dp = \int_{T_m}^{T'_m} \frac{\Delta H_{fus}}{\Delta V_{fus}} \frac{dT}{T}.$$

If ΔH_{fus} and ΔV_{fus} are nearly independent of T and p, the equation integrates to

$$p_2 - p_1 = \frac{\Delta H_{fus}}{\Delta V_{fus}} \ln \frac{T'_m}{T_m}, \tag{12–15}$$

where T'_m is the melting point under p_2; T_m is the melting point under p_1. Since

$T'_m - T_m$ is usually quite small, the logarithm can be expanded to

$$\ln (T'_m/T_m) = \ln \left(\frac{T_m + T'_m - T_m}{T_m} \right) = \ln \left(1 + \frac{T'_m - T_m}{T_m} \right) \approx \frac{T'_m - T_m}{T_m};$$

then Eq. (12–15) becomes

$$\Delta p = \frac{\Delta H_{\text{fus}}}{\Delta V_{\text{fus}}} \frac{\Delta T}{T_m}, \tag{12–16}$$

where ΔT is the increase in melting point corresponding to the increase in pressure Δp. [In using Eq. (12–16) the various quantities must be expressed in compatible units!]

Condensed-phase–gas equilibrium

For the equilibrium of a condensed phase, either solid or liquid, with vapor, we have

$$\frac{dp}{dT} = \frac{\Delta S}{\Delta V} = \frac{\Delta H}{T(\overline{V}_g - \overline{V}_c)},$$

where ΔH is either the molar heat of vaporization of the liquid or the molar heat of sublimation of the solid, and $\overline{V}_c$ is the molar volume of the solid or liquid. In most circumstances, $\overline{V}_g - \overline{V}_c \approx \overline{V}_g$, and this, assuming that the gas is ideal, is equal to RT/p. Then the equation becomes

$$\frac{d \ln p}{dT} = \frac{\Delta H}{RT^2}, \tag{12–17}$$

which is the Clausius-Clapeyron equation, relating the vapor pressure of the liquid (solid) to the heat of vaporization (sublimation) and the temperature. Integrating between limits, under the additional assumption that ΔH is independent of temperature yields

$$\int_{p_0}^{p} d \ln p = \int_{T_0}^{T} \frac{\Delta H}{RT^2} dT,$$

$$\ln \frac{p}{p_0} = \frac{\Delta H}{R} \left[-\frac{1}{T} + \frac{1}{T_0} \right] = \frac{\Delta H}{RT_0} - \frac{\Delta H}{RT}, \tag{12–18}$$

where p_0 is the vapor pressure at T_0, and p is the vapor pressure at T. (In Section 5–4, this equation was derived in a different way.) If $p_0 = 1$ atm, then T_0 is the normal boiling point of the liquid (normal sublimation point of the solid). Then

$$\ln p = \frac{\Delta H}{RT_0} - \frac{\Delta H}{RT}, \qquad \log_{10} p = \frac{\Delta H}{2.303RT_0} - \frac{\Delta H}{2.303RT}. \tag{12–19}$$

According to Eq. (12–19), if $\ln p$ or $\log_{10} p$ is plotted against $1/T$, a linear curve is obtained with a slope equal to $-\Delta H/R$ or $-\Delta H/2.303R$. The intercept at $1/T = 0$

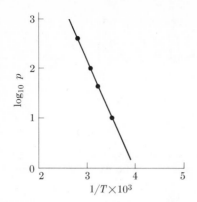

Fig. 12–11 $\log_{10} p$ (mm) versus $1/T$ for water.

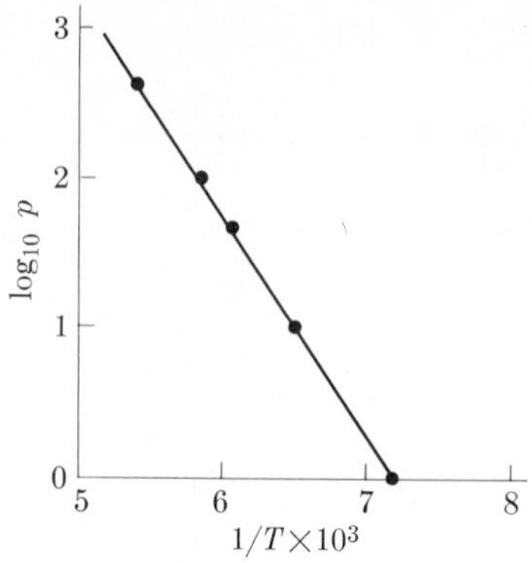

Fig. 12–12 $\log_{10} p$ (mm) versus $1/T$ for solid CO_2.

yields a value of $\Delta H/RT_0$. Thus, from the slope and intercept both ΔH and T_0 can be calculated. Heats of vaporization and sublimation are often determined through the measurement of the vapor pressure of the substance as a function of temperature. Figure 12–11 shows a plot of $\log_{10} p$ versus $1/T$ for water, while Fig. 12–12 is the same plot for solid CO_2 (dry ice).

Compilations of data on vapor pressure frequently use an equation of the form $\log_{10} p = A + B/T$, and tabulate values of A and B for various substances. This equation has the same functional form as Eq. (12–19).

For substances which obey Trouton's rule, Eq. (12–19) takes a particularly simple form which is useful for estimating the vapor pressure of a substance at any temperature T from a knowledge of the boiling point only (Problem 12–2).

12–10 Effect of Pressure on the Vapor Pressure

In the preceding discussion of the liquid–vapor equilibrium it was implicitly assumed that the two phases were under the same pressure p. If by some means it is possible to keep the liquid under a pressure P and the vapor under the vapor pressure p, then the vapor pressure depends upon P. Suppose that the liquid is confined in the container shown in Fig. 12–13. In the space above the liquid, the vapor is confined together with a foreign gas which is insoluble in the liquid. The vapor pressure p plus the pressure of the foreign gas is P, the total pressure exerted on the liquid. As usual, the equilibrium condition is

$$\mu_{\text{vap}}(T, p) = \mu_{\text{liq}}(T, P). \tag{12–20}$$

At constant temperature this equation implies that $p = f(P)$. To discover the functionality, Eq. (12–20) is differentiated with respect to P, keeping T constant:

$$\left(\frac{\partial \mu_{\text{vap}}}{\partial p}\right)_T \left(\frac{\partial p}{\partial P}\right)_T = \left(\frac{\partial \mu_{\text{liq}}}{\partial P}\right)_T.$$

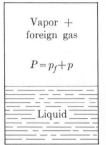

Figure 12–13

Using the fundamental equation, Eq. (12–2b), this becomes

$$\overline{V}_{\text{vap}}\left(\frac{\partial p}{\partial P}\right)_T = \overline{V}_{\text{liq}}$$

or

$$\left(\frac{\partial p}{\partial P}\right)_T = \frac{\overline{V}_{\text{liq}}}{\overline{V}_{\text{vap}}}. \tag{12–21}$$

The Gibbs equation, Eq. (12–21), shows that the vapor pressure increases with the total pressure on the liquid; the rate of increase is very small since $\overline{V}_{\text{liq}}$ is very much less than $\overline{V}_{\text{vap}}$. If the vapor behaves ideally, Eq. (12–21) can be written

$$\frac{RT}{p}\,dp = \overline{V}_{\text{liq}}\,dP, \qquad RT\int_{p_0}^{p}\frac{dp}{p} = \overline{V}_{\text{liq}}\int_{p_0}^{P} dP,$$

where p is the vapor pressure under a pressure P, p_0 is the vapor pressure when liquid and vapor are under the same pressure p_0, the orthobaric pressure. Thus,

$$RT\ln(p/p_0) = \overline{V}_{\text{liq}}(P - p_0). \tag{12–22}$$

We shall use Eqs. (12–21) and (12–22) in discussing the osmotic pressure of a solution.

12–11 The Phase Rule

The coexistence of two phases in equilibrium implies the condition

$$\mu_\alpha(T, p) = \mu_\beta(T, p), \tag{12–23}$$

which means that the two intensive variables ordinarily needed to describe the state of a system are no longer independent, but are related. Because of this relation, only one intensive variable, either temperature or pressure, is needed to describe the state of the system. The system has one *degree of freedom*, or is *univariant*, whereas if only one phase is present, two variables are needed to describe the state, and the system has two degrees of freedom, or is *bivariant*. If three phases are present, two relations exist between T and p:

$$\mu_\alpha(T, p) = \mu_\beta(T, p) \qquad \mu_\alpha(T, p) = \mu_\gamma(T, p). \tag{12–24}$$

These two relations determine T and p completely. No other information is necessary for the description of the state of the system. Such a system is *invariant*; it has no degrees of freedom. Table 12–1 shows the relation between the number of degrees of freedom and the number of phases present for a one-component system. The table suggests a rule relating the number of degrees of freedom, F, to the number of phases, P, present.

$$F = 3 - P, \tag{12–25}$$

which is the *phase rule* for a one-component† system.

Table 12–1

Number of phases present	Degrees of freedom
1	2
2	1
3	0

It would be helpful to have a simple rule by which we can decide how many independent variables are required for the description of the system. Particularly in the study of systems in which many components and many phases are present, any simplification of the problem is welcome.

We begin by finding the total conceivable number of intensive variables which would be needed to describe the state of the system containing C components and P phases. These are listed in Table 12–2. Each equation which connects these variables implies that one variable is dependent rather than independent. So we must find the total number of equations connecting the variables. These are listed in Table 12–3.

The number of independent variables, F, is obtained by subtracting the total number of equations from the total number of variables:

$$F = PC + 2 - P - C(P - 1),$$
$$F = C - P + 2. \tag{12–26}$$

Equation (12–26) is the phase rule of J. Willard Gibbs. The best way to remember the phase rule is by realizing that increasing the number of components increases

---

† The term "component" is defined in Section 12–12.

Table 12–2

Kind of variable	Total number of variables
Temperature and pressure:	2
Composition variables:	
(in each phase the mole fraction of each component must be specified; thus, C mole fractions are required to describe one phase; PC are needed to describe P phases):	
	PC
Total number of variables:	$PC + 2$

Table 12–3

Kind of equation	Total number of equations
In each phase there is a relation between the mole fractions:	
$$x_1 + x_2 + \cdots + x_C = 1$$	
For P phases, there are P equations:	P
The equilibrium conditions:	
For each component there exists a set of equations:	
$$\mu_i^{\alpha} = \mu_i^{\beta} = \mu_i^{\gamma} = \cdots = \mu_i^{P}$$	
There are $P - 1$ equations in the set.	
Since there are C components, there are $C(P - 1)$ equations:	$C(P - 1)$
Total number of equations:	$P + C(P - 1)$

the number of variables, therefore C enters with a positive sign. Increasing the number of phases increases the number of equilibrium conditions and the number of equations, thus eliminating some of the variables; therefore P enters with a negative sign.

In a one-component system, $C = 1$, so $F = 3 - P$. This result is, of course, the same as Eq. (12–25) obtained by inspection of Table 12–1. Equation (12–25) shows that the greatest number of phases which can coexist in equilibrium in a one-component system is three. In the sulfur system, for example, it is not possible for rhombic, monoclinic, liquid, and gaseous sulfur to coexist in equilibrium with one another. Such a quadruple equilibrium would imply three independent conditions on two variables, which is an impossibility.

For a system of only one component it is possible to derive, as was done in Table 12–1, the consequences of the phase rule quite easily. The equilibria are readily represented by lines and their intersections in a two-dimensional diagram of the type we have used in this chapter. It hardly seems necessary to have the phase rule for such a situation. However, if the system has two components, then three variables are required and the phase diagram consists of surfaces and their intersections in three dimensions. If three components are present, surfaces in a four-dimensional space are required. Visualization of the entire situation is difficult in three dimensions, impossible for four or more dimensions. Yet the phase rule, with exquisite simplicity, expresses the limitations which are placed on the intersections of the surfaces in these multidimensional spaces. For this reason, the Gibbs phase rule is counted among the truly great generalizations of physical science.

12–12 The Problem of Components

The number of components in a system is defined as the least number of *chemically independent* species which is required to describe the composition of every phase in the system. At face value, the definition seems simple enough, and in ordinary practice it is simple. A number of examples will show up the joker in the deck, that little phrase, "chemically independent."

Example 1 The system contains the *species* PCl_5, PCl_3, Cl_2. There are *three species* present but only *two components*, because the equilibrium

$$PCl_5 \rightleftharpoons PCl_3 + Cl_2$$

is established in this system. One can alter the number of moles of any two of these chemical individuals arbitrarily; the alteration in the number of moles of the third species is then fixed by the equilibrium condition, $K_x = x_{PCl_3} x_{Cl_2}/x_{PCl_5}$. Consequently, any two of these species are chemically independent; the third is not. There are only two components.

Example 2 Liquid water presumably contains an enormous number of chemical species: H_2O, $(H_2O)_2$, $(H_2O)_3, \ldots, (H_2O)_n$. Yet there is only one component, because, as far as is known, all of the equilibria

$$H_2O + H_2O \rightleftharpoons (H_2O)_2,$$

$$H_2O + (H_2O)_2 \rightleftharpoons (H_2O)_3,$$

$$\cdot$$
$$\cdot$$
$$\cdot$$

$$H_2O + (H_2O)_{n-1} \rightleftharpoons (H_2O)_n$$

are established in the system; thus, if there are n species, there are $n - 1$ equilibria connecting them, and so only one species is chemically independent. There is only one component, and we may choose the simplest species, H_2O, as that component.

Example 3 In the system water–ethyl alcohol, two species are present. No known equilibrium connects them at ordinary temperature; thus, there are two components also.

Example 4 In the system $CaCO_3$–CaO–CO_2, there are three species present; also, there are three distinct phases: solid $CaCO_3$, solid CaO, and gaseous CO_2. Because the equilibrium $CaCO_3 \rightleftharpoons CaO + CO_2$ is established, there are only two components. These are most simply chosen as CaO and CO_2; the composition of the phase $CaCO_3$ is then described as one mole of component CO_2 plus one mole of component CaO. If $CaCO_3$ and CO_2 were chosen as components, the composition of CaO would be described as one mole of $CaCO_3$ minus one mole of CO_2.

There is still another point to be made concerning the number of components. Our criterion is the fact of a chemical equilibrium being established in a system; the existence of such an equilibrium reduces the number of components. There are instances where this test is not very clear-cut. Take the example of water, ethylene, and ethyl alcohol; at high temperatures several equilibria are established in this system; we consider only one: $C_2H_5OH \rightleftharpoons C_2H_4 + H_2O$. The question arises as to the temperature at which the system shifts from a three-component system, which it surely is at room temperature, to the two-component system that it is at high temperature. The answer lies in how long it takes us to make successive measurements on the system! If we measure a certain property of the system at a series of pressures, then if the time we require to make the measurements is very short compared with the time required for the equilibrium to shift under the change in pressure, the system is effectively a three-component system; the equilibrium may as well not be there at all. On the other hand, if the equilibrium shifts very quickly under the change in pressure, shifts in a time which is very short in comparison with the time we need to make the measurement, then the fact of the equilibrium matters very much, and the system is indeed a two-component system.

Liquid water is a good example of both types of behavior. The equilibria between the various polymers of water shift very rapidly, within 10^{-11} second at most. Ordinary measurements require much longer times, so the system is effectively a one-component system. In contrast to this behavior, the system H_2, O_2, H_2O, is a three-component system. The equilibrium which could reduce the number of components is $H_2 + \frac{1}{2}O_2 \rightleftharpoons H_2O$. In the absence of a catalyst, eons are required for this equilibrium to shift from one position to another. For practical purposes the equilibrium is not established.

It is clear that an accurate assignment of the number of components in a system presupposes some experimental knowledge of the system. This is an unavoidable pitfall in the use of the phase rule. Failure to realize that an unsuspected equilibrium has been established in a system sometimes leads an investigator to rediscover, the hard way, the second law of thermodynamics.

Problems

12–1. At 25°C, we have

	ΔG_f^0, kcal/mole	S^0, eu
Rhombic sulfur	0	7.62
Monoclinic sulfur	0.023	7.78

Assuming that the entropies do not vary with temperature, sketch the value of μ versus T for the two forms of sulfur. From the data determine the equilibrium temperature for the transformation, rhombic sulfur to monoclinic sulfur.

12–2. a) From the boiling point T_b of a liquid and the assumption that the liquid follows Trouton's rule, calculate the value of the vapor pressure at any temperature T.

b) The boiling point of diethyl ether is 34.6°C. Calculate the vapor pressure at 25°C.

12–3. The vapor pressures of liquid sodium are

p, mm	1	10	100
t, °C	439	549	701

By plotting these data appropriately, determine the boiling point of liquid sodium, the heat of vaporization, and the entropy of vaporization at the boiling point.

12–4. Naphthalene, $C_{10}H_8$, melts at 80°C. If the vapor pressure of the liquid is 10 mm at 85.8°C and 40 mm at 119.3°C, and that of the solid is 1 mm at 52.6°C, calculate

a) the ΔH_{vap} of the liquid, the boiling point, and ΔS_{vap} at T_b,

b) the vapor pressure at the melting point.

c) Assuming the melting-point and triple-point temperatures are the same, calculate ΔH_{sub} of the solid, and ΔH_{fus}.

d) What must the temperature be if the vapor pressure of the solid is to be less than 10^{-5} mm Hg?

12–5. Iodine boils at 183.0°C, and the vapor pressure at 116.5°C is 100 mm. If the heat of fusion is 3.74 kcal/mole and the vapor pressure of the solid is 1 mm at 38.7°C calculate the triple-point temperature and pressure.

12–6. Given the data:

Substance	ΔS_{vap}^0, eu
S	3.5
P	5.37

The molecular formulas of these substances are S_8 and P_4. Show that if the correct molecular weights were used, these entropies of vaporization would have more normal values.

12-7. If the vapor is an ideal gas, there is a simple relation between the vapor pressure p and the concentration c (moles/liter) in the vapor. Consider a liquid in equilibrium with its vapor. Derive the expression for the temperature dependence of c in such a system.

12-8. Assuming that the vapor is ideal and that ΔH_{vap} is independent of temperature, calculate

a) The molar concentration of the vapor at the boiling point T_b of the liquid.

b) Using the result in Problem 12-7, find the expression for T_H in terms of ΔH_{vap} and T_b. The Hildebrand temperature, T_H, is that temperature at which the vapor concentration is 1/22.4 moles/liter.

c) The Hildebrand entropy, $\Delta S_H = \Delta H_{vap}/T_H$, is very nearly constant for many normal liquids. If $\Delta S_H = 22.1$ eu, use the result in (b) to compute values of T_b for various values of T_H. Plot T_H as a function of T_b. (Choose values of $T_H = 50, 100, 200, 300, 400°K$ to compute T_b.)

d) For the following liquids compute ΔS_H and the Trouton entropy, $\Delta S_T = \Delta H_{vap}/T_b$. Note that ΔS_H is more constant than ΔS_T (Hildebrand's rule).

Liquid	ΔH_{vap}, kcal/mole	T_b, °K
Argon	1.558	87.29
Krypton	2.158	119.93
Xenon	3.021	165.1
Oxygen	1.630	90.19
Methane	1.955	111.67
Carbon disulfide	6.40	319.41

12-9. The density of diamond is 3.52 gm/cm³, while that of graphite is 2.25 gm/cm³. At 25°C the free energy of formation of diamond from graphite is 0.6850 kcal/mole. At 25°C what pressure must be applied to bring diamond and graphite into equilibrium? (Be careful of units!)

12-10. The blade of an ice skate is ground to a knife edge on each side of the skate.

a) If the width of the knife edge is 0.001 in., and the length of the skate in contact with the ice is 3 in., calculate the pressure exerted on the ice by a 150-lb man.

b) What is the melting point of ice under this pressure? ($\Delta H_{fus} = 1.4363$ kcal/mole, $T_0 = 273.16°K$, density of ice is 0.92 gm/cm³, that of liquid water is 1.00 gm/cm³.)

12-11. Liquid water under an air pressure of 1 atm at 25°C has a larger vapor pressure than it would in the absence of air pressure. Calculate the increase in vapor pressure produced by the pressure of the atmosphere on the water. The density of water = 1 gm/cm³; the vapor pressure (in the absence of the air pressure) = 23.756 mm.

12-12. Derive Eq. (12-4).

Chapter Thirteen

Solutions

I. The Ideal Solution and Colligative Properties

13–1 Kinds of Solutions

A solution is a homogeneous mixture of chemical species dispersed on a molecular scale. By this definition, a solution is a single phase. A solution may be gaseous, liquid, or solid. *Binary* solutions are composed of two constituents, *ternary* solutions three, *quaternary* four. The constituent present in the greatest amount is ordinarily called the *solvent*, while those constituents, one or more, present in relatively small amounts are called the *solutes*. The distinction between solvent and solute is an arbitrary one. If it is convenient, the constituent present in relatively small amount may be chosen as the solvent. We shall employ the words *solvent* and *solute* in the ordinary way, realizing that nothing fundamental distinguishes them. Examples of kinds of solution are listed in Table 13–1.

Gas mixtures have been discussed in some detail in Chapter 11. The discussion in this chapter and in Chapter 14 is devoted to liquid solutions. Solid solutions are dealt with as they occur in connection with other topics.

Table 13–1

Gaseous solutions	Mixtures of gases or vapors
Liquid solutions	Solids, liquids, or gases, dissolved in liquids
Solid solutions	
Gases dissolved in solids	H_2 in palladium, N_2 in titanium
Liquids dissolved in solids	Mercury in gold
Solids dissolved in solids	Copper in gold, zinc in copper (brasses), alloys of many kinds

13–2 Definition of the Ideal Solution

The ideal gas law is an important example of a *limiting* law. As the pressure approaches zero, the behavior of any real gas approaches that of the ideal gas as a limit. Thus, all real gases behave ideally at zero pressure, and for practical purposes they are ideal at low finite pressures. From this generalization of experimental behavior, the ideal gas is defined as one which behaves ideally at any pressure whatsoever.

We arrive at a similar limiting law from observation of the behavior of solutions. For simplicity, we consider a solution composed of a volatile solvent and one or more involatile solutes, and examine the equilibrium between the solution and the vapor. If a pure liquid is placed in a container which is initially evacuated, the liquid evaporates until the space above the liquid is filled with vapor. The temperature of the system is kept constant. At equilibrium, the pressure established in the vapor is p^0, the vapor pressure of the pure liquid; Fig. 13–1(a). If an involatile material is dissolved in the liquid, the equilibrium vapor pressure p over the solution is observed to be less than over the pure liquid; Fig. 13–1(b).

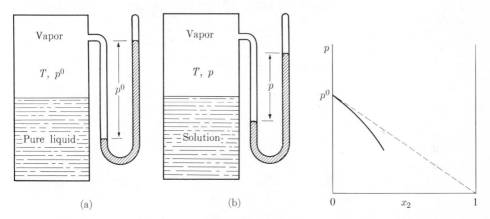

Fig. 13–1 Vapor pressure lowering by an involatile solute.   **Fig. 13–2** Vapor pressure as a function of x_2.

Since the solute is involatile, the vapor consists of pure solvent. As more involatile material is added, the pressure in the vapor phase decreases. A schematic plot of the vapor pressure of the solvent against the mole fraction of the involatile solute in the solution, x_2, is shown by the solid line in Fig. 13–2. At $x_2 = 0$, $p = p^0$; as x_2 increases, p decreases, The important feature of Fig. 13–2 is that the vapor pressure of the dilute solution (x_2 near zero) approaches the dashed line connecting p^0 and zero. Depending on the particular combination of solvent and solute, the experimental vapor-pressure curve at higher concentrations of solute may fall below the dashed line, as in Fig. 13–2, or above it, or even lie exactly on it. However, for all solutions the experimental curve is tangent to the dashed line at $x_2 = 0$, and approaches the dashed line very closely as the solution becomes more and more dilute. The equation of the ideal line (the dashed line) is

$$p = b + mx_2.$$

At $x_2 = 0$, $p = p^0$, so $b = p^0$, while at $x_2 = 1$, $p = 0$, so that $0 = b + m$, or $m = -b = -p^0$. Therefore, $p = (1 - x_2)p^0$. If x is the mole fraction of solvent in the solution, then $x + x_2 = 1$, so the equation becomes

$$p = xp^0, \tag{13-1}$$

which is *Raoult's law*. It states that the vapor pressure of the solvent over a solution is equal to the vapor pressure of the pure solvent multiplied by the mole fraction of the solvent in the solution.

Raoult's law is another example of a limiting law. Real solutions follow Raoult's law more closely as the solution becomes more dilute. The *ideal solution* is defined as one which follows Raoult's law over the entire range of concentrations. The vapor pressure of the solvent over an ideal solution of an involatile solute is shown in Fig. 13–3. All real solutions behave ideally as the concentration of the solutes approaches zero.

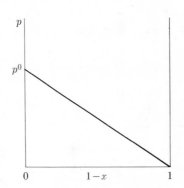

Fig. 13–3 Raoult's law for the solvent.

From Eq. (13–1) the *vapor pressure lowering*, $p^0 - p$, can be computed:

$$p^0 - p = p^0 - xp^0 = (1 - x)p^0,$$

$$p^0 - p = x_2 p^0. \tag{13-2}$$

The vapor pressure lowering is proportional to the mole fraction of the *solute*. If several solutes, $2, 3, \ldots$, are present, then it is still true that $p = xp^0$; but, in this case, $1 - x = x_2 + x_3 + \cdots$ and

$$p^0 - p = (x_2 + x_3 + \cdots)p^0. \tag{13-3}$$

In a solution containing several involatile solutes, the vapor pressure lowering depends on the sum of the mole fractions of the various solutes. Note particularly that it does not depend on what kinds of solutes are present, except that they be involatile. The vapor pressure depends only on the relative numbers of solute molecules.

13–3   Analytical Form of the Chemical Potential in Ideal Liquid Solutions

As a generalization of the behavior of real solutions the ideal solution follows Raoult's law over the entire range of concentration. Taking this definition of an ideal liquid solution and combining it with the general equilibrium condition leads to the analytical expression of the chemical potential of the solvent in an ideal solution. If the solution is in equilibrium with vapor, the requirement of the second law is that the chemical potential of the solvent have the same value in the solution as in the vapor;

$$\mu_{\text{liq}} = \mu_{\text{vap}}, \tag{13–4}$$

where μ_{liq} is the chemical potential of the solvent in the liquid phase, μ_{vap} the chemical potential of the solvent in the vapor. Since the vapor is pure solvent under a pressure p, the expression for μ_{vap} is given by Eq. (10–47); assuming the vapor is an ideal gas $\mu_{\text{vap}} = \mu_{\text{vap}}^0 + RT \ln p$. Then Eq. (13–4) becomes

$$\mu_{\text{liq}} = \mu_{\text{vap}}^0 + RT \ln p.$$

Using Raoult's law, $p = xp^0$, in this equation and expanding the logarithm, we obtain

$$\mu_{\text{liq}} = \mu_{\text{vap}}^0 + RT \ln p^0 + RT \ln x.$$

If pure solvent were in equilibrium with vapor, the pressure would be p^0; the equilibrium condition is

$$\mu_{\text{liq}}^0 = \mu_{\text{vap}}^0 + RT \ln p^0,$$

where μ_{liq}^0 signifies the chemical potential of the pure liquid solvent. Subtracting this equation from the preceding one, we obtain

$$\mu_{\text{liq}} - \mu_{\text{liq}}^0 = RT \ln x.$$

In this equation, nothing pertaining to the vapor phase appears, so omitting the subscript liq, the equation becomes

$$\mu = \mu^0 + RT \ln x. \tag{13–5}$$

The significance of the symbols in Eq. (13–5) must be clearly understood: μ is the chemical potential of the solvent in the solution, μ^0 is the chemical potential of the pure liquid solvent, a function of T and p, and x is the mole fraction of solvent in the solution. This equation is the result we suggested in Section 11–5, as a generalization from the form obtained for the μ of an ideal gas in a mixture.

13-4 Chemical Potential of the Solute in a Binary Ideal Solution; Application of the Gibbs-Duhem Equation

The Gibbs-Duhem equation can be used to calculate the chemical potential of the solute from that of the solvent in a binary ideal system. The Gibbs-Duhem equation, Eq. (11–93), for a binary system is (T, p constant)

$$n\,d\mu + n_2\,d\mu_2 = 0. \tag{13-6}$$

The symbols without subscripts in Eq. (13–6) refer to the solvent, those with the subscript 2 refer to the solute. From Eq. (13–6), $d\mu_2 = -(n/n_2)\,d\mu$; or, since $n/n_2 = x/x_2$, we have

$$d\mu_2 = -\frac{x}{x_2}\,d\mu.$$

Differentiating Eq. (13–5) keeping T and p constant, we obtain for the solvent $d\mu = (RT/x)\,dx$, so that $d\mu_2$ becomes

$$d\mu_2 = -RT\frac{dx}{x_2}.$$

However, $x + x_2 = 1$, so that $dx + dx_2 = 0$, or $dx = -dx_2$. Then $d\mu_2$ becomes

$$d\mu_2 = RT\frac{dx_2}{x_2}.$$

Integrating, we have

$$\mu_2 = RT\ln x_2 + C, \tag{13-7}$$

where C is the constant of integration; since T and p are kept constant throughout this manipulation, C can be a function of T and p and still be a constant for this integration. If the value of x_2 in the liquid is increased until it is unity, the liquid becomes pure *liquid* solute, and μ_2 must be μ_2^0, the chemical potential of pure *liquid* solute. So if $x_2 = 1$, $\mu_2 = \mu_2^0$. Using these values in Eq. (13–7), we find $\mu_2^0 = C$, and Eq. (13–7) becomes

$$\mu_2 = \mu_2^0 + RT\ln x_2. \tag{13-8}$$

Equation (13–8) relates the chemical potential of the solute to the mole fraction of the solute in the solution. This expression is analogous to Eq. (13–5), and the symbols have corresponding significances. Since the solute has the same form for μ as the solvent, the solute behaves ideally. This implies that in the vapor over the solution the partial pressure of the solute is given by Raoult's law:

$$p_2 = x_2 p_2^0. \tag{13-9}$$

If the solute is involatile, p_2^0 is immeasurably small so that Eq. (13–9) is not capable of experimental proof, and in such a case, has academic interest only.

13-5 Colligative Properties

Since the second term in Eq. (13–5) is negative, the chemical potential of the solvent in solution is less than the chemical potential of the pure solvent by an amount $-RT\ln x$. Several related properties of the solution have their origin in this lower value of the chemical potential. These properties are: (1) the vapor pressure lowering, discussed in Section 13–2; (2) the freezing-point depression; (3) the boiling-point elevation; and (4) the osmotic pressure. Since these properties are all bound together through their common origin, they are called *colligative*† properties. All of these properties have the common characteristic that they do not depend upon the nature of the solute present but only on the number of solute molecules relative to the total number of molecules present.

    The μ versus T diagram displays the freezing-point depression and the boiling-point elevation clearly. In Fig. 13–4(a) the solid lines refer to the pure solvent. Since the solute is involatile, it does not appear in the gas phase, so the curve for the gas is the same as for the pure gas. If we assume that the solid contains only the solvent, then the curve for the solid is unchanged. However, because the liquid contains a solute, the μ of the solvent is lowered at each temperature by an amount $-RT\ln x$. The dashed curve in Fig. 13–4(a) is the curve for the solvent in an ideal solution.

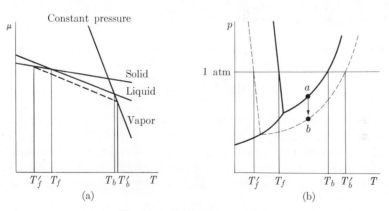

Fig. 13–4 Colligative properties.

The diagram shows directly that the intersection points with the curves for the solid and the gas have shifted. The new intersection points are the freezing point, T_f', and the boiling point, T_b', of the solution. It is apparent that the boiling point of the solution is higher than that of the pure solvent (boiling-point elevation), while the freezing point of the solution is lower (freezing-point depression). From the figure it is obvious that the change in the freezing point is greater than the change in the boiling point for a solution of the same concentration.

---

†   Colligative: from Latin: *co*-, together, *ligare*, to bind.

The freezing-point depression and boiling-point elevation can be illustrated on the ordinary phase diagram of the solvent, shown for water by the solid curves in Fig. 13–4(b). If an involatile material is added to the liquid solvent, then the vapor pressure is lowered at every temperature as, for example, from point a to point b. The vapor-pressure curve for the solution is shown by the dotted line. The dashed line shows the new freezing point as a function of pressure. At 1 atm pressure, the freezing points and boiling points are given by the intersections of the full and dashed lines with the horizontal line at 1 atm pressure. This diagram also shows that a given concentration of solute produces a greater effect on the freezing point than on the boiling point.

The freezing point and boiling point of a solution depend upon the equilibrium of the solvent in the solution with pure solid solvent or pure solvent vapor. The remaining possible equilibrium is that between solvent in solution and pure liquid solvent. This equilibrium can be established by increasing the pressure on the solution sufficiently to raise the μ of the solvent in solution to the value of the μ of the pure solvent. The additional pressure on the solution which is required to establish the equality of the μ of the solvent both in the solution and in the pure solvent is called the *osmotic pressure* of the solution.

13–6 The Freezing-point Depression

Consider a solution which is in equilibrium with pure solid solvent. The equilibrium condition requires that

$$\mu(T, p, x) = \mu_{\text{solid}}(T, p), \qquad (13\text{--}10)$$

where $\mu(T, p, x)$ is the chemical potential of the solvent in the solution, $\mu_{\text{solid}}(T, p)$ is the chemical potential of the pure solid. Since the solid is *pure*, μ_{solid} does not depend on any composition variable. In Eq. (13–10), T is the equilibrium temperature, the freezing point of the solution; from the form of Eq. (13–10), T is some function of pressure and x, the mole fraction of solvent in the solution. If the pressure is constant, then T is a function only of x.

If the solution is ideal, then $\mu(T, p, x)$ in the solution is given by Eq. (13–5), so that Eq. (13–10) becomes

$$\mu^0(T, p) + RT \ln x = \mu_{\text{solid}}(T, p).$$

Rearrangement yields

$$\ln x = -\frac{\mu^0(T, p) - \mu_{\text{solid}}(T, p)}{RT}. \qquad (13\text{--}11)$$

Since μ^0 is the chemical potential of the pure liquid, $\mu^0(T, p) - \mu_{\text{solid}}(T, p) = \Delta G_{\text{fus}}$, where ΔG_{fus} is the molar free energy of fusion of the pure solvent at the temperature

T. Equation (13–11) becomes

$$\ln x = -\frac{\Delta G_{\text{fus}}}{RT}.$$ (13–12)

To discover how T depends on x, we evaluate $(\partial T/\partial x)_p$. Differentiating Eq. (13–12) with respect to x, p being constant, we obtain

$$\frac{1}{x} = -\frac{1}{R}\left[\frac{\partial(\Delta G_{\text{fus}}/T)}{\partial T}\right]_p \left(\frac{\partial T}{\partial x}\right)_p.$$

Using the Gibbs–Helmholtz equation, Eq. (10–54), $[\partial(\Delta G/T)/\partial T]_p = -\Delta H/T^2$, we obtain

$$\frac{1}{x} = \frac{\Delta H_{\text{fus}}}{RT^2}\left(\frac{\partial T}{\partial x}\right)_p.$$ (13–13)

In Eq. (13–13), ΔH_{fus} is the heat of fusion of the *pure* solvent at the temperature T. The procedure is now reversed and we write Eq. (13–13) in differential form and integrate:

$$\int_1^x \frac{dx}{x} = \int_{T_0}^T \frac{\Delta H_{\text{fus}}}{RT^2}\,dT.$$ (13–14)

The lower limit $x = 1$ corresponds to pure solvent having a freezing point T_0. The upper limit x corresponds to a solution which has a freezing point T. The first integral can be evaluated immediately; the second integration is possible if ΔH_{fus} is known as a function of temperature. For simplicity we assume that ΔH_{fus} is a constant in the temperature range from T_0 to T; then Eq. (13–14) becomes

$$\ln x = \frac{\Delta H_{\text{fus}}}{R}\left(-\frac{1}{T} + \frac{1}{T_0}\right).$$ (13–15)

This equation can be solved for the freezing point T, or rather more conveniently for $1/T$,

$$\frac{1}{T} = \frac{1}{T_0} - \frac{R\ln x}{\Delta H_{\text{fus}}},$$ (13–16)

which relates the freezing point of an ideal solution to the freezing point of the pure solvent, T_0, the heat of fusion of the solvent, and the mole fraction of the solvent in the solution, x.

The relation between freezing point and composition of a solution can be simplified considerably if the solution is dilute. To begin it is desirable to express the freezing-point depression $-dT$ in terms of the total molality of the solutes present, m. In a solution containing 1000 gm of solvent, the mole numbers of the solutes are equal to their molalities; $n_2 = m_2$, $n_3 = m_3, \ldots, n_i = m_i$. The number of moles of solvent is $n = 1000/M$, where M is the molecular weight of the solvent. If m is the total molality of all the solutes present, then $m = m_2 + m_3 + \cdots + m_i$. The mole fraction of the solvent is given by

$$x = \frac{n}{n + n_2 + n_3 + \cdots + n_i} = \frac{1000/M}{(1000/M) + m_2 + m_2 + \cdots + m_i}$$

$$x = \frac{1}{1 + (Mm/1000)}. \tag{13-17}$$

Taking logarithms and differentiating, we obtain

$$\ln x = -\ln [1 + (Mm/1000)]$$

$$d \ln x = -\frac{(M/1000)}{[1 + (Mm/1000)]} dm. \tag{13-18}$$

Equation (13–13) can be written

$$dT = \frac{RT^2}{\Delta H_{\text{fus}}} d \ln x.$$

Replacing $d \ln x$ by the value in Eq. (13–18), we obtain

$$dT = -\frac{MRT^2}{1000 \, \Delta H_{\text{fus}}} \frac{dm}{[1 + (Mm/1000)]}. \tag{13-19}$$

If the solution is very dilute in all solutes, then m approaches zero and T approaches T_0, and Eq. (13–19) becomes

$$-\left(\frac{\partial T}{\partial m}\right)_{p,0} = \frac{MRT_0^2}{1000 \, \Delta H_{\text{fus}}} = K_f. \tag{13-20}$$

The subscript zero designates the limiting value of the derivative, and K_f is the freezing-point depression constant. The freezing-point depression $\theta_f = T_0 - T$, $d\theta_f = -dT$, so for dilute solutions we have

$$\left(\frac{\partial \theta_f}{\partial m}\right)_{p,0} = K_f, \tag{13-21}$$

which integrates immediately, if m is small, to

$$\theta_f = K_f m. \tag{13-22}$$

The constant K_f depends only on the properties of the pure solvent. For water: $M = 18.016$, $T_0 = 273.15°$, $\Delta H_{\text{fus}} = 1436.3$ cal/mole,

$$K_f = \frac{18.016(1.9872)(273.15)^2}{1000(1436.3)} = 1.860 \text{ deg·kgm·mole}^{-1}.$$

Equation (13–22) provides a simple relation between the freezing-point depression and the molal concentration of solute in a dilute ideal solution, which is often used to determine the molecular weight of a dissolved solute. If w_2 gm of a solute of unknown molecular weight, M_2, are dissolved in w gm of solvent, then the molality of solute is $m = 1000w_2/wM_2$. Using this value for m in Eq. (13–22) and solving for M_2 yields

$$M_2 = \frac{1000K_f}{\theta_f} \frac{w_2}{w}.$$

The measured values of θ_f, w_2, and w, together with a knowledge of K_f of the solvent, suffice to determine M_2. It is clear that for a given value of m, the larger the value of K_f, the greater will be θ_f. This increases the ease and accuracy of the measurement of θ_f; consequently, it is desirable to choose a solvent having a large value of K_f. By examining Eq. (13–20) we can decide what sorts of compounds will have large values of K_f. First of all, we replace ΔH_{fus} by $T_0 \Delta S_{fus}$; this reduces Eq. (13–20) to

$$K_f = \frac{R}{1000 \, \Delta S_{fus}} M T_0, \tag{13–23}$$

which shows that K_f increases as the product MT_0 increases. Since T_0 increases as M increases, K_f increases rapidly as the molecular weight of the substance increases. The increase is not very uniform, simply because ΔS_{fus} may vary a good deal, particularly when M is very large. Table 13–2 illustrates the behavior of K_f with increasing M. Because of variations in the value of ΔS_{fus}, marked exceptions occur; the general trend is apparent, however.

Table 13–2 Freezing-point depression constants

Compound	M	m.p., °C	K_f
Water	18.0	0	1.86
Acetic acid	60.0	16.6	3.57
Benzene	78.1	5.45	5.07
Dioxane	88.1	11.7	4.71
Naphthalene	128.3	80.1	6.98
p-dichlorobenzene	147.0	52.7	7.11
Camphor	152.2	178.4	37.7
p-dibromobenzene	235.9	86	12.5

13–7 Solubility

The equilibrium between solid solvent and solution has been considered in Section 13–6. The same equilibrium may be considered from a different point of view. The word "solvent" as we have seen is ambiguous. Suppose we consider the equilibrium between solute in solution and pure solid solute. In this condition the solution is *saturated* with respect to the solute. The equilibrium condition is that the μ of the solute must be the same everywhere, that is

$$\mu_2(T, p, x_2) = \mu_{2(\text{solid})}(T, p), \tag{13–24}$$

where x_2 is the mole fraction of solute in the saturated solution, and is, therefore, the *solubility* of the solute expressed as a mole fraction. If the solution is ideal then

$$\mu_2^0(T, p) + RT \ln x_2 = \mu_{2(\text{solid})}(T, p),$$

where $\mu_2^0(T, p)$ is the chemical potential of the pure *liquid* solute. The argument then proceeds in exactly the same way as for the freezing-point depression; the symbols refer to the solute, however. The equation corresponding to Eq. (13–15) is

$$\ln x_2 = -\frac{\Delta H_{\text{fus}}}{R}\left(\frac{1}{T} - \frac{1}{T_0}\right); \tag{13–25}$$

ΔH_{fus} is the heat of fusion of pure solute, T_0 the freezing point of pure solute. Using $\Delta H_{\text{fus}} = T_0 \Delta S_{\text{fus}}$ in Eq. (13–25), we obtain

$$\ln x_2 = \frac{\Delta S_{\text{fus}}}{R}\left(1 - \frac{T_0}{T}\right). \tag{13–26}$$

Either Eq. (13–25) or Eq. (13–26) is an expression of the *ideal law of solubility*. According to this law, the solubility of a substance is the same in all solvents with which it forms an ideal solution. The solubility of a substance in an ideal solution depends on the properties of that substance only. Low melting point T_0 and a low heat of fusion both favor enhanced solubility. Figure 13–5 shows the variation of the solubility, x, as a function of temperature for two substances with the same entropy of fusion but different melting points.

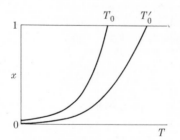

Fig. 13–5 Ideal solubility versus T.

The use of Eq. (13–25) can be illustrated by the solubility of naphthalene. The melting point is 80.0°C; the heat of fusion is 4560 cal·mole^{-1}. Using these data we find from Eq. (13–25) that the ideal solubility $x = 0.264$ at 20°C. The measured solubilities in various solvents are given in Table 13–3.

The ideal law of solubility is frequently in error if the temperature of interest is far below the melting point of the solid, since the assumption that ΔH_{fus} is independent of temperature is not very good in this circumstance. The law is never accurate for

Table 13-3†

Solvent	Solubility x_2	Solvent	Solubility x_2
Chlorobenzene	0.256	Aniline	0.130
Benzene	0.241	Nitrobenzene	0.243
Toluene	0.224	Acetone	0.183
CCl_4	0.205	Methyl alcohol	0.0180
Hexane	0.090	Acetic acid	0.0456

† By permission from J. H. Hildebrand and R. L. Scott, *The Solubility of Nonelectrolytes*, 3rd ed. Reinhold Publishing Corp., New York, 1950, p. 283.

solutions of ionic materials in water, since the saturated solutions of these materials are far from being ideal and are far below their melting points. As the table of solubilities of naphthalene shows, hydrogen-bonded solvents are poor solvents for a substance which cannot form hydrogen bonds.

13-8 Elevation of the Boiling Point

Consider a solution which is in equilibrium with the vapor of the pure solvent. The equilibrium condition is that

$$\mu(T, p, x) = \mu_{vap}(T, p). \tag{13-27}$$

If the solution is ideal,

$$\mu^0(T, p) + RT \ln x = \mu_{vap}(T, p),$$

and

$$\ln x = \frac{\mu_{vap} - \mu^0(T, p)}{RT}.$$

The molar free energy of vaporization is

$$\Delta G_{vap} = \mu_{vap}(T, p) - \mu^0(T, p),$$

so that

$$\ln x = \frac{\Delta G_{vap}}{RT}. \tag{13-28}$$

Note that Eq. (13-28) has the same functional form as Eq. (13-12) except that the sign is changed on the right-hand side. The algebra which follows is identical to that used for the derivation of the formulas for the freezing-point depression except that the sign is reversed in each term which contains either ΔG or ΔH. This difference in sign simply means that while the freezing point is depressed, the boiling point is elevated. Differentiating Eq. (13–28) with respect to x, keeping the pressure constant, we obtain

$$\frac{1}{x} = \frac{1}{R}\left[\frac{\partial(\Delta G_{vap}/T)}{\partial T}\right]_p \left(\frac{\partial T}{\partial x}\right)_p.$$

Using the Gibbs-Helmholtz equation, this reduces to

$$\frac{1}{x} = -\frac{\Delta H_{vap}}{RT^2}\left(\frac{\partial T}{\partial x}\right)_p. \tag{13–29}$$

The equation may be integrated to yield

$$\int_1^x \frac{dx}{x} = -\frac{1}{R}\int_{T_0}^T \frac{\Delta H_{vap}}{T^2}\, dT, \tag{13–30}$$

where the lower limit corresponds to pure solvent with a boiling point T_0 and the upper limit corresponds to a solution in which the mole fraction of solvent is x and which has a boiling point T. The integration is done assuming that ΔH_{vap} is independent of temperature:

$$\ln x = \frac{\Delta H_{vap}}{R}\left(\frac{1}{T} - \frac{1}{T_0}\right), \tag{13–31}$$

which becomes

$$\frac{1}{T} = \frac{1}{T_0} + \frac{R \ln x}{\Delta H_{vap}}. \tag{13–32}$$

The boiling point T of the solution is expressed in terms of the heat of vaporization and the boiling point of the pure solvent, ΔH_{vap} and T_0, and the mole fraction x of solvent in the solution.

To express the boiling-point elevation in terms of the total molality of the solutes, we write Eq. (13–29) in the form

$$dT = -\frac{RT^2}{\Delta H_{vap}}\, d \ln x.$$

Using Eq. (13–18) this becomes

$$dT = \frac{MRT^2}{1000\, \Delta H_{vap}}\frac{dm}{[1 + (Mm/1000)]}. \tag{13–33}$$

If the solution is dilute in all solutes, then m approaches zero and T approaches T_0. The boiling-point elevation constant is defined by

$$K_b = \left(\frac{\partial T}{\partial m}\right)_{p,0} = \frac{MRT_0^2}{1000\, \Delta H_{vap}}. \tag{13–34}$$

The boiling-point elevation, $\theta_b = T - T_0$, so that $d\theta_b = dT$. So long as m is small,

Eq. (13–34) integrates to

$$\theta_b = K_b m. \tag{13–35}$$

For water: $M = 18.016$, $T_0 = 373.15°$, $\Delta H_{\text{vap}} = 9717.1$ cal/mole, and we obtain $K_b = 0.513$ deg·kgm·mole^{-1}.

The relation, Eq. (13–35), between boiling-point elevation and the molality of a dilute ideal solution corresponds to that between freezing-point depression and molality; for any liquid, the constant K_b is smaller than K_f.

The elevation of the boiling point is used to determine the molecular weight of a solute in the same way as is the freezing-point depression. It is desirable to use a solvent which has a large value of K_b. In Eq. (13–34) if ΔH_{vap} is replaced by $T_0 \Delta S_{\text{vap}}$ then

$$K_b = \frac{R}{1000\,\Delta S_{\text{vap}}} M T_0.$$

But many liquids follow Trouton's rule: $\Delta S_{\text{vap}} = 21$ eu/mole. Since $R = 2$ cal· deg^{-1}·mole^{-1}, then, approximately, $K_b = 10^{-4} M T_0$. The higher the molecular weight of the solvent, the larger the value of K_b. The data in Table 13–4 illustrate the relationship.

Since the boiling point T_0 is a function of pressure, K_b is a function of pressure. The effect is rather small but must be taken into account in precise measurements. The Clausius-Clapeyron equation yields the connection between T_0 and p which is needed to calculate the magnitude of the effect.

Table 13–4 Boiling-point elevation constants

Compound	M	b.p., °C	K_b
Water	18.0	100	0.51
Methyl alcohol	32.0	64.7	0.86
Ethyl alcohol	46.1	78.5	1.23
Acetone	58.1	56.1	1.71
Acetic acid	60.0	118.3	3.07
Benzene	78.1	80.2	2.53
Cyclohexane	84.2	81.4	2.79
Ethyl bromide	109.0	38.3	2.93

13–9 Osmotic Pressure

The phenomenon of osmotic pressure is illustrated by the apparatus shown in Fig. 13–6. A collodion bag is tied to a rubber stopper through which a piece of glass capillary tubing is inserted. The bag is filled with a dilute solution of sugar in water, and immersed in a beaker of pure water. The level of the sugar solution in the tube

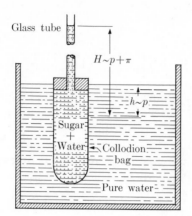

Glass tube

$H \sim p + \pi$

$h \sim p$

Sugar + Water

Collodion bag

Pure water

Fig. 13–6 Simple osmotic pressure experiment.

is observed to rise until it reaches a definite height, which depends on the concentration of the solution. The hydrostatic pressure resulting from the difference in levels of the sugar solution in the tube and the surface of the pure water is the *osmotic pressure* of the solution. Observation shows that no sugar has escaped through the membrane into the pure water in the beaker. The increase in volume of the solution which caused it to rise in the tube is a result of the passage of water through the membrane into the bag. The collodion functions as a *semipermeable* membrane which allows water to pass freely through it but does not allow sugar to pass. When the system reaches equilibrium, the sugar solution at any depth below the level of the pure water is under an excess hydrostatic pressure due to the extra height of the sugar solution in the tubing. The problem is to calculate the magnitude of this pressure difference and its relation to the concentration of the solution.

The equilibrium requirement is that the chemical potential of the water must have the same value on each side of the membrane at every depth in the beaker. This equality of the chemical potential is achieved by a pressure difference on the two sides of the membrane. Consider the situation at the depth h in Fig. 13–6. At this depth the solvent is under a pressure p, while the solution is under a pressure $p + \pi$. If $\mu(T, p + \pi, x)$ is the chemical potential of the solvent in the solution under the pressure $p + \pi$, and $\mu^0(T, p)$ that of the pure solvent under the pressure p, then the equilibrium condition is

$$\mu(T, p + \pi, x) = \mu^0(T, p), \tag{13–36}$$

$$\mu^0(T, p + \pi) + RT \ln x = \mu^0(T, p). \tag{13–37}$$

The problem is to express the μ of the solvent under a pressure $p + \pi$ in terms of the μ of the solvent under a pressure p. From the fundamental equation at constant T, we have $d\mu^0 = \bar{V}^0 \, dp$. Integrating, we have

$$\mu^0(T, p + \pi) - \mu^0(T, p) = \int_p^{p+\pi} \bar{V}^0 \, dp. \tag{13–38}$$

This reduces Eq. (13–37) to

$$\int_{p}^{p+\pi} \overline{V}^0 \, dp + RT \ln x = 0. \tag{13–39}$$

In Eq. (13–39), $\overline{V}^0$ is the molar volume of the pure solvent. If the solvent is incompressible, then $\overline{V}^0$ is independent of pressure and can be removed from the integral. Then

$$\overline{V}^0 \pi + RT \ln x = 0, \tag{13–40}$$

which is the relation between the osmotic pressure π and the mole fraction of solvent in the solution. Two assumptions are involved in Eq. (13–40); the solution is ideal and the solvent is incompressible.

In terms of the solute concentration, $\ln x = \ln(1 - x_2)$. If the solution is dilute, then $x_2 \ll 1$; the logarithm may be expanded in series. Keeping only the first term, we obtain

$$\ln(1 - x_2) = -x_2 = -\frac{n_2}{n + n_2} \approx -\frac{n_2}{n},$$

since $n_2 \ll n$ in the dilute solution. Then Eq. (13–40) becomes

$$\pi = \frac{n_2 RT}{n \overline{V}^0}. \tag{13–41}$$

By the addition rule the volume of the ideal solution is $V = n \overline{V}^0 + n_2 \overline{V}_2^0$. If the solution is dilute, n_2 is very small, so that $V \approx n \overline{V}^0$. This result reduces Eq. (13–41) to

$$\pi = n_2 RT/V \qquad \text{or} \qquad \pi = cRT. \tag{13–42}$$

In Eq. (13–42), $c = n_2/V$, the molar concentration of solute in the solution. Equation (13–42) is the van't Hoff equation for osmotic pressure.

The striking formal analogy between the van't Hoff equation and the ideal gas law should not go unnoticed. In the van't Hoff equation, n_2 is the number of moles of *solute*. The solute molecules dispersed in the solvent are analogous to the gas molecules dispersed in empty space. The solvent is analogous to the empty space between the gas molecules. In the experiment shown in Fig. 13–7, the membrane is attached to a movable piston. As the solvent diffuses through the membrane, the piston is pushed to the right; this continues until the piston is flush against the

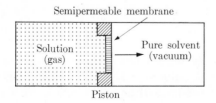

Fig. 13–7 Osmotic analog of the Joule experiment.

right-hand wall. The observed effect is the same *as if* the solution exerted a pressure against the membrane to push it to the right. The situation is comparable to the free expansion of a gas into vacuum. If the volume of the solution doubles in this experiment, the dilution will reduce the final osmotic pressure to half the original value, just as the pressure of a gas is reduced to half its value by doubling in volume.

In spite of the analogy, it is deceptive to consider the osmotic pressure as a sort of pressure which is somehow exerted by the solute. Osmosis, the passage of solvent through the membrane, is due to the inequality of the chemical potential on the two sides of the membrane. The kind of membrane does not matter; it is required only that it be permeable only to the solvent. Nor does the nature of the solute matter; all that is necessary is that the solvent contain dissolved foreign matter which is not passed by the membrane.

The mechanism by which the solvent permeates the membrane may be different for each different kind of membrane. A membrane could conceivably be like a sieve which allows small molecules such as water to pass through the pores while it blocks larger molecules. Another membrane might dissolve the solvent and so be permeated by it, while the solute is not soluble in the membrane. The elucidation of the mechanism by which the solvent passes through the membrane is a question which must be examined for every membrane–solvent pair using the methods of chemical kinetics. Thermodynamics cannot provide an answer, since the equilibrium result is the same for all membranes.

The measurement of osmotic pressure is useful for determining the molecular weights of materials which are only slightly soluble in the solvent, or which have very high molecular weights, e.g., proteins, polymers of various types, colloids. These are convenient measurements because of the large magnitude of the osmotic pressure. At 25°C, the product $RT = 24.4$ liter·atm·mole^{-1}. Thus for a 1 molar solution, $\pi = 24.4$ atm. This pressure corresponds to a height of a column of water of the order of 800 ft. Simply to keep the experiment in the laboratory, the solutions must be less than 0.01 molar, and are preferably of the order of 0.001 molar. This assumes that we are using an apparatus of the type shown in Fig. 13–6. Very precise measurements of osmotic pressures up to several hundred atmospheres have been made by H. N. Morse, and J. C. W. Frazer, and by Lord Berkeley and E. G. J. Hartley using special apparatus of different design.

In a molecular weight determination, if w_2 grams of solute is dissolved in 1 liter of solution, then $\pi = (w_2/M_2)RT$, or

$$M_2 = \frac{w_2 RT}{\pi}.$$

Even when w_2 is small and M_2 large, the value of π is measurable and can be translated into a value of M_2.

Osmosis plays a significant role in the function of organisms. A cell which is immersed in pure water undergoes plasmolysis. The cell wall permits water to flow into it; thereupon the cell becomes distended, the wall stretches until it ultimately ruptures, or becomes leaky enough to allow the solutes in the cellular material to escape from the interior. On the other hand, if the cell is immersed in a concentrated solution of salt, the water from the cell flows into the more concentrated salt solution and the cell shrinks. A salt solution which is just concentrated enough so that the cell neither shrinks nor is distended is called an *isotonic* solution.

Osmosis might be called the "principle of the prune." The skin of the prune acts as a membrane permeable to water. The sugars in the prune are the solutes. Water diffuses through the skin, the fruit swells until the skin ruptures or becomes leaky. Many plant and animal membranes function in this way although only rarely are they strictly semipermeable. Frequently, their function in the organism requires that they pass other materials in addition to water. Medicinally, the osmotic effect is utilized in, for example, the prescription of a salt-free diet in some cases of abnormal fluid retention by the body.

Problems

13–1. Suppose that a series of solutions is prepared using 180 gm of H_2O as a solvent and 10 gm of an involatile solute. What will be the relative vapor pressure lowering if the molecular weight of the solute is: 100 gm/mole, 200 gm/mole, 10,000 gm/mole?

13–2. a) For an ideal solution plot the value of p/p^0 as a function of x_2, the mole fraction of the solute.

b) Sketch the plot of p/p^0 as a function of the molality of the solute, if water is the solvent.

c) Suppose the solvent has a higher molecular weight; e.g., toluene. How does this affect the plot of p/p^0 versus m? How does it affect p/p^0 versus x_2?

d) Evaluate the derivative of $(p^0 - p)/p^0$ with respect to m, as $m \to 0$.

13–3. Twenty grams of a solute is added to 100 gm of water at 25°C. The vapor pressure of pure water is 23.76 mm; the vapor pressure of the solution is 22.41 mm.

a) Calculate the molecular weight of the solute.

b) What weight of this solute is required in 100 gm of water to reduce the vapor pressure to one-half the value for pure water?

13–4. The heat of fusion of water at the freezing point is 1.4363 kcal/mole; calculate the freezing point of water in solutions having a mole fraction of water equal to: 1.0, 0.8, 0.6, 0.4, 0.2. Plot the values of T versus x.

13–5. The heat of fusion of acetic acid is 44.7 cal/gm at the melting point 16.58°C. Calculate the freezing-point depression constant for acetic acid.

13–6. Two grams of benzoic acid dissolved in 25 gm of benzene, $K_f = 4.90$, produce a freezing-point depression of 1.62°. Calculate the molecular weight. Compare this with the molecular weight obtained from the formula for benzoic acid, C_6H_5COOH.

13–7. a) The heat of fusion per gram of p-dibromobenzene, $C_6H_4Br_2$, is 20.5 cal; the melting point is 86°C. Calculate the ideal solubility at 25°C.

b) For p-dichlorobenzene, $C_6H_4Cl_2$, the heat of fusion is 29.7 cal/gm and the melting point is 52.7°C. Calculate the ideal solubility at 25°C.

13-8. Calculate the boiling-point elevation constant for each of the following substances.

Substance	t_b, °C	Q_{vap}, cal/gm
Acetone, $(CH_3)_2CO$	56.1	124.5
Benzene, C_6H_6	80.2	94.3
Chloroform, $CHCl_3$	61.5	59.0
Methane, CH_4	−159	138
Ethyl acetate, $CH_3CO_2C_2H_5$	77.2	102.0

Plot the values of K_b versus the product MT_b.

13-9. The addition of 3 gm of a substance to 100 gm of CCl_4 raises the boiling point of CCl_4 by 0.60°; $K_b = 5.03$. Calculate the freezing-point depression $K_f = 31.8$, the relative vapor-pressure lowering, the osmotic pressure of the solution at 25°C, and the molecular weight of the substance. The density of CCl_4 is 1.59 gm/cm³.

13-10. Consider a vertical tube with a cross-sectional area of 1 cm². The bottom of the tube is closed with a semipermeable membrane and 1 gm of glucose, $C_6H_{12}O_6$, is placed in the tube. The end of the tube closed with the membrane is immersed in pure water. What will the height of the liquid level in the tube be at equilibrium? The density of the solution may be taken as 1 gm/cm³; the sugar concentration is assumed to be uniform in the solution. What is the osmotic pressure at equilibrium? (25°C; assume a negligible depth of immersion.)

13-11. If 6 gm of urea, $(NH_2)_2CO$, is dissolved in a liter of solution, calculate the osmotic pressure of the solution at 27°C.

13-12. Assume that ΔH_{fus} is independent of the temperature and that the thermometer available can measure a freezing-point depression to an accuracy of $\pm 0.01°$. The simple law for freezing-point depression, $\theta_f = K_f m$, is based on the limiting condition that $m = 0$. At what molality will this approximation no longer predict the result within the experimental error?

13-13. If the heat of fusion depends on temperature through the expression

$$\Delta H_{fus} = \Delta H_0 + \Delta C_p(T - T_0),$$

where ΔC_p is constant, then the value of θ_f can be expressed in the form $\theta_f = am + bm^2 + \cdots$, where a and b are constants. Calculate the values of a and b. [Hint: This is a Taylor series, so evaluate $(\partial^2\theta/\partial m^2)$ at $m = 0$.]

13-14. a) The complete expression for the osmotic pressure is given by Eq. (13-40). Since $c = n_2/V$, and $V = n\bar{V}^0 + n_2\bar{V}_2^0$, where $\bar{V}^0$ and $\bar{V}_2^0$ are constants, the mole numbers n and n_2 can be expressed in terms of V, $\bar{V}^0$, and $\bar{V}_2^0$, and c. Compute the value of $x = n/(n + n_2)$ in these terms. Then evaluate $(\partial\pi/\partial c)_T$ at $c = 0$ and show that it is equal to RT.

b) By evaluating $(\partial^2\pi/\partial c^2)_T$ at $c = 0$, show that $\pi = cRT(1 + V'c)$, where $V' = \bar{V}_2^0 - \frac{1}{2}\bar{V}^0$. Note that this is equivalent to writing a modified van der Waals equation, $\pi = n_2RT/(V - n_2V')$, and expanding it in a power series.

Chapter Fourteen

Solutions

II. More than One Volatile Component; the Ideal Dilute Solution

14–1 General Characteristics of the Ideal Solution

The discussion in Chapter 13 was restricted to those ideal solutions in which the solvent was the only volatile constituent present. The concept of an ideal solution extends to solutions containing several volatile constituents. As before, the concept is based on a generalization of the experimental behavior of real solutions and represents a limiting behavior which is approached by all real solutions.

Consider a solution composed of several volatile substances in a container which is initially evacuated. Since the components are all volatile, some of the solution evaporates to fill the space above the liquid with vapor. When the solution and the vapor come to equilibrium at the temperature T, the total pressure within the container is the sum of the partial pressures of the several components of the solution:

$$p = p_1 + p_2 + \cdots + p_i + \cdots. \tag{14–1}$$

These partial pressures are measurable, as are the equilibrium mole fractions, $x_1, \ldots x_i, \ldots$, in the liquid. Let one of the components, i, be present in relatively large amount compared with any of the others. Then it is found experimentally that

$$p_i = x_i p_i^0, \tag{14–2}$$

where p_i^0 is the vapor pressure of the pure liquid component i. Equation (14–2) is Raoult's law, and experimentally it is followed in any solution as x_i approaches unity regardless of which component is present in great excess. When any solution is dilute in all components but the solvent, the solvent always follows Raoult's law. Since all of the components are volatile, any one of them can be designated as the solvent. Therefore, the ideal solution is defined by the requirement that each component obey Raoult's law, Eq. (14–2), over the entire range of composition. The significance of the symbols is worth reiterating: p_i is the *partial* pressure of i in the vapor phase; p_i^0 is the vapor pressure of the pure liquid i; and x_i is the mole fraction of i in the *liquid* mixture.

The ideal solution has two other important properties. The heat of mixing the pure components to form the solution is zero, and the volume of mixing is zero. These properties are observed as the limiting behavior in all real solutions. If additional solvent is added to a solution which is dilute in all of the solutes, the heat of mixing approaches zero as the solution becomes more and more dilute. In the same circumstances the volume of mixing of all real solutions approaches zero.

14–2 The Chemical Potential in Ideal Solutions

Consider an ideal solution in equilibrium with its vapor at a fixed temperature T. For each component, the equilibrium condition is $\mu_i = \mu_{i(\text{vap})}$, where μ_i is the chemical potential of i in the solution, $\mu_{i(\text{vap})}$ is the chemical potential of i in the vapor phase. If the vapor is ideal, then by the same argument as in Section 13–3, the value of μ_i is

$$\mu_i = \mu_i^0(T, p) + RT \ln x_i, \tag{14–3}$$

where $\mu_i^0(T, p)$ is the chemical potential of the pure liquid i at temperature T and under pressure p. The chemical potential of each and every component of the solution is given by the expression in Eq. (14–3). Figure 14–1 shows the variation of $\mu_i - \mu_i^0$ as a function of x_i. As x_i becomes very small, the value of μ_i decreases very rapidly. At all values of x_i, the value of μ_i is less than that of μ_i^0.

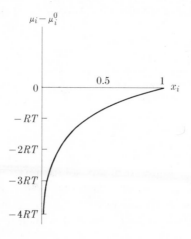

Fig. 14–1 $(\mu_i - \mu_i^0)$ versus x_i.

Since Eq. (14–3) is formally the same as Eq. (11–14) for the μ of an ideal gas in a gas mixture, by the same reasoning as in Section 11–6 it follows that in mixing

$$\Delta G_{\mathrm{mix}} = NRT \sum_i x_i \ln x_i, \qquad (14\text{–}4)$$

$$\Delta S_{\mathrm{mix}} = -NR \sum_i x_i \ln x_i, \qquad (14\text{–}5)$$

$$\Delta H_{\mathrm{mix}} = 0, \qquad \Delta V_{\mathrm{mix}} = 0, \qquad (14\text{–}6)$$

where N is the total number of moles in the mixture. The three properties of the ideal solution (Raoult's law, zero heat of mixing, and zero volume of mixing) are intimately related. If Raoult's law obtains for every component, then the heat and volume of mixing will be zero. (This statement cannot be reversed; if the volume of mixing and heat of mixing are both zero, it does not follow that Raoult's law will be obeyed.)

14–3 Binary Solutions

We turn our attention now to the consequences of Raoult's law in binary solutions in which both components are volatile. In a binary solution $x_1 + x_2 = 1$. We have

$$p_1 = x_1 p_1^0, \qquad (14\text{–}7)$$

$$p_2 = x_2 p_2^0 = (1 - x_1) p_2^0 \qquad (14\text{–}8)$$

If the total pressure over the solution is p, then

$$p = p_1 + p_2 = x_1 p_1^0 + (1 - x_1) p_2^0,$$

$$p = p_2^0 + (p_1^0 - p_2^0) x_1, \qquad (14\text{–}9)$$

which relates the total pressure over the mixture to the mole fraction of component 1 in the liquid. It shows that p is a linear function of x_1; Fig. 14–2(a). It is clear from Fig. 14–2(a) that the addition of a solute may raise or lower the vapor pressure of the solvent depending upon which is the more volatile.

The total pressure can also be expressed in terms of y_1, the mole fraction of component 1 in the vapor. From the definition of the partial pressure,

$$y_1 = p_1/p. \qquad (14\text{–}10)$$

Using the values of p_1 and p from Eqs. (14–7) and (14–9), we obtain

$$y_1 = \frac{x_1 p_1^0}{p_2^0 + (p_1^0 - p_2^0) x_1}.$$

Solving for x_1 yields,

$$x_1 = \frac{y_1 p_2^0}{p_1^0 + (p_2^0 - p_1^0) y_1}. \qquad (14\text{–}11)$$

Using the value of x_1 from Eq. (14–11) in Eq. (14–9), we obtain, after collecting terms,

$$p = \frac{p_1^0 p_2^0}{p_1^0 + (p_2^0 - p_1^0)y_1}. \tag{14–12}$$

Equation (14–12) expresses p as a function of y_1, the mole fraction of component 1 in the vapor. This function is plotted in Fig. 14–2(b). The relation in Eq. (14–12) can be rearranged to the more convenient, symmetrical form

$$1/p = y_1/p_1^0 + y_2/p_2^0. \tag{14–12a}$$

To describe a two-component system, the phase rule shows, since $C = 2$, that $F = 4 - P$. Since P is 1 or greater, three variables at most must be specified to describe the system. Since Fig. 14–2(a) and (b) are drawn at a specified temperature, only two additional variables are required to describe the state of the system completely. These two variables may be (p, x_1) or (p, y_1). As a consequence, the points in Fig. 14–2(a) or (b) describe states of the system.

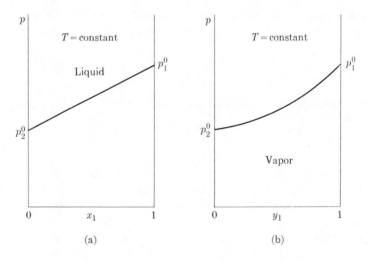

Fig. 14–2 Vapor pressure as a function of composition.

There is a difficulty here. The variable x_1, being a mole fraction in the liquid, is not capable of describing states of the system which are completely gaseous. Similarly, y_1 is incapable of describing a completely liquid state of the system. Hence, only liquid states and those states on the line in which liquid and vapor coexist are described by Fig. 14–2(a). Similarly, only gaseous states and those states, on the curve, in which vapor and liquid coexist are described by Fig. 14–2(b). The completely liquid states are those at high pressures, those above the line in Fig.

14–2(a). The completely gaseous states are stable at low pressures, below the curve
in Fig. 14–2(b). These regions of stability have been so labeled in the figures.

Life would be much simpler if we could represent all the states on one diagram.
If only liquid is present, x_1 describes the composition of the liquid and also the
composition of the entire system. If only vapor is present, y_1 describes the composition of the vapor and at the same time the composition of the entire system.
In view of that, it seems reasonable to plot the pressure against X_1, the mole fraction of component 1 in the entire system. In Fig. 14–3(a), p is plotted against X_1;
the two curves of Fig. 14–2(a) and (b) are drawn in. The upper curve is called the
liquid curve; the lower curve is the vapor curve. The system is neatly represented
by one diagram; the liquid is stable above the liquid curve, the vapor below the
vapor curve. What significance is attributed to the points which lie between the
curves? The points lying just above the liquid curve correspond to the lowest
pressures at which liquid can exist by itself, since vapor appears if the point lies on
the curve. Liquid cannot be present alone below the liquid curve. By the same
argument vapor cannot be present alone above the vapor curve. The only possible
meaning to the points between the curves is that they represent states of the system
in which liquid and vapor coexist in equilibrium. The enclosed region is the liquid–
vapor region.

Consider the point a in the liquid–vapor region; Fig. 14–3(b). The value X_1 corresponding to a is the mole fraction of component 1 in the entire system, liquid +
vapor. What composition of liquid can coexist with vapor at the pressure p in
question? Drawing a horizontal line, a *tie line*, at constant pressure, the intersection
of this line with the liquid curve at l yields the value of x_1 which describes the composition of the liquid, while the intersection with the vapor curve at v yields the value
of y_1 which describes the composition of the vapor.

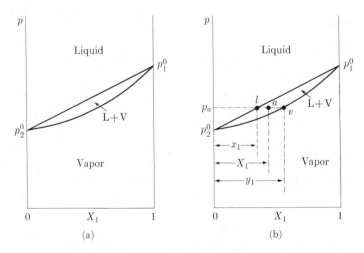

Fig. 14–3 Interpretation of the p-X diagram.

If two phases, liquid and vapor, are present in equilibrium, then the variance of the system is $F = 4 - 2 = 2$. Since the temperature is fixed, one other variable, any one of p, x_1, y_1, suffices to describe the system. So far we have used x_1 or y_1 to describe the system; since $x_1 + x_2 = 1$, and $y_1 + y_2 = 1$, we could equally well have chosen x_2 and y_2. If the pressure is chosen to describe the two-phase system, the intersections of the horizontal line at that pressure with the liquid and vapor curves yield the values of x_1 and y_1 directly. If x_1 is the describing variable, the intersection of the vertical line at x_1 with the liquid curve yields the value of p; from p the value of y_1 is obtained immediately.

14–4 The Lever Rule

In any two-phase region, such as L–V in Fig. 14–3(b), the composition of the entire system may vary between the limits x_1 and y_1, depending upon the relative amounts of liquid and vapor present. If the state point a is very near the liquid line, the system would consist of a large amount of liquid and a relatively small amount of vapor. If a is near the vapor line, the amount of liquid present is relatively small compared with the amount of vapor present.

The relative amounts of liquid and vapor present are calculated by the lever rule. Let the length of the line segment between a and l in Fig. 14–3(b) be $(\overline{al})$ and that between a and v be $(\overline{av})$; then let $n_{1(\text{liq})}$ and $n_{1(\text{vap})}$ be the number of moles of component 1 in the liquid and in the vapor, respectively; let $n_1 = n_{1(\text{liq})} + n_{1(\text{vap})}$. If n_{liq} and n_{vap} are the total number of moles of liquid and vapor present, respectively, and if $n = n_{\text{liq}} + n_{\text{vap}}$, then from Fig. 14–3(b), we have

$$(\overline{al}) = X_1 - x_1 = \frac{n_1}{n} - \frac{n_{1(\text{liq})}}{n_{\text{liq}}}, \qquad (\overline{av}) = y_1 - X_1 = \frac{n_{1(\text{vap})}}{n_{\text{vap}}} - \frac{n_1}{n}.$$

Multiply $(\overline{al})$ by n_{liq} and $(\overline{av})$ by n_{vap} and subtract:

$$n_{\text{liq}}(\overline{al}) - n_{\text{vap}}(\overline{av}) = \frac{n_1}{n}(n_{\text{liq}} + n_{\text{vap}}) - (n_{1(\text{liq})} + n_{1(\text{vap})}) = n_1 - n_1 = 0.$$

Therefore,

$$n_{\text{liq}}(\overline{al}) = n_{\text{vap}}(\overline{av}) \qquad \text{or} \qquad \frac{n_{\text{liq}}}{n_{\text{vap}}} = \frac{(\overline{av})}{(\overline{al})}; \tag{14–13}$$

this is called the lever rule, the point a being the "fulcrum" of the lever; the number of moles of liquid times the length $(\overline{al})$ from a to the liquid line is equal to the number of moles of vapor present times the length, $(\overline{av})$, from a to the vapor line. The ratio of the number of moles of liquid to the number of moles of vapor is given by the ratio of lengths of the line segments connecting a to v and l. Thus if a lies very close to v, $(\overline{av})$ is very small and $n_{\text{liq}} \ll n_{\text{vap}}$; the system consists mainly of vapor. Similarly when a lies close to l, $n_{\text{vap}} \ll n_{\text{liq}}$; the system consists mainly of liquid.

Since the derivation of the lever rule depends only on a mass balance, the rule is valid for calculating the relative amounts of the two phases present in any two-phase region of a two-component system. If the diagram is drawn in terms of mass fraction instead of mole fraction, the lever rule is valid and yields the relative masses of the two phases rather than the relative mole numbers.

14–5 Changes in State as the Pressure is Reduced Isothermally

The behavior of the system is now examined as the pressure is reduced from a very high value to a very low value, keeping the overall composition constant at a mole fraction of component 1 equal to X. At point a, Fig. 14–4, the system is entirely liquid, it remains so as the pressure is reduced until the point l is reached; at point l, the first trace of vapor appears having a composition y. Note that the first vapor to appear is considerably richer in 1 than the liquid; component 1 is the more volatile. As the pressure is reduced further, the point reaches a'; during this reduction of pressure, the composition of the liquid moves along the line ll', while the composition of the vapor moves along vv'. At a', liquid has the composition x' while vapor has the composition y'. The ratio of number of moles of liquid to vapor at the point a' is $\overline{(a'v')}/\overline{(a'l')}$, from the lever rule. Continued reduction of pressure brings the state point to v''; at this point, only a trace of liquid of composition x'' remains; the vapor has the composition X. Note that the liquid which remains is richer in the less volatile component 2. As the pressure is reduced, the state point moves into the vapor region, and the reduction of pressure from v'' to a'' simply corresponds to an expansion of the vapor. In the final state, a'', the vapor has, of course, the same composition as the original liquid.

The fact that the vapor which forms over a liquid as the pressure is reduced is richer in a particular component than the liquid is used as the basis of a method of separation, isothermal distillation. The method is useful for those mixtures which

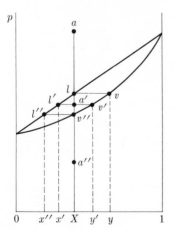

Fig. 14–4 Isothermal change in pressure.

would decompose if distilled by the ordinary method; it is sufficiently inconvenient so that it is used only if other methods are not suitable.

The system described above is an ideal solution. If the deviations from ideality are not very large, the figure will appear much the same except that the liquid composition curve is not a straight line. The interpretation is precisely the same as for the ideal solution.

14–6 Temperature–Composition Diagrams

In the diagrams shown in Section 14–5, the temperature was constant. The equilibrium pressure of the system was then a function of either x_1 or y_1, according to Eqs. (14–9) or (14–12). In those equations, the values of p_1^0 and p_2^0 are functions of temperature. If, in Eqs. (14–9) and (14–12), we consider the total pressure p to be constant, then the equations are relations between the equilibrium temperature, the boiling point, and either x_1 or y_1. The relations $T = f(x_1)$ and $T = g(y_1)$ are not such simple ones as between pressure and composition, but they may be determined theoretically through the Clapeyron equation or, ordinarily, experimentally through determination of the boiling points and vapor compositions corresponding to liquid mixtures of various compositions.

The plot at constant pressure of boiling points versus compositions for the ideal solution corresponding to that in Fig. 14–3 is shown in Fig. 14–5. Neither the liquid nor the vapor curve is a straight line; otherwise, the figure resembles Fig. 14–3. However, the lenticular liquid–vapor region is tilted down from left to right. This corresponds to the fact that component 1 had the higher vapor pressure, therefore it has the lower boiling point. Also in Fig. 14–5 the liquid region is at the bottom of the diagram, since under a constant pressure, the liquid is stable at low temperatures. The lower curve describes the liquid composition; the upper curve describes the vapor composition. The regions in the p-X diagram are sometimes

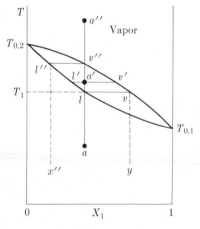

Fig. 14–5 Isobaric change in temperature.

thoughtlessly confused with those in the T-X diagram. A little common sense tells us that the liquid is stable at low temperatures, the lower part of the T-X diagram, and under high pressures, the upper part of the p-X diagram. Attempting to memorize the location of the liquid or vapor regions is foolish when it is so easy to figure it out.

The principles applied to the discussion of the p-X diagram can be applied in much the same way to the T-X diagram. The pressure on the system is constant; from the phase rule, at most two additional variables are needed to describe the system. Every point in the T-X diagram describes a state of the system. The points in the uppermost portion of the diagram are gaseous states of the system; those points in the lowest part are liquid states. The points in the middle region describe states in which liquid and vapor coexist in equilibrium. The tie line in the liquid–vapor region connects the composition of vapor and the composition of liquid which coexist at that temperature. The lever rule applies to the T-X diagram, of course.

14–7 Changes in State with Increase in Temperature

We examine now the sequence of events as a liquid mixture under a constant pressure is heated from a low temperature, corresponding to point a, Fig. 14–5, to a high temperature corresponding to point a''. At a, the system consists entirely of liquid; as the temperature rises, the system remains entirely liquid until the point l is reached; at this temperature T_1 the first trace of vapor appears having composition y. The vapor is much richer than the liquid in component 1, the lower boiling component. This fact is the basis for the separation of volatile mixtures by distillation. As the temperature is raised, the state point moves to a', and the liquid composition changes continuously along the line ll'; the vapor composition changes continuously along the line vv'. At a', the relative number of moles of liquid and vapor present is given by the ratio $(a'v')/(a'l')$. If the temperature is raised further, at v'' the last trace of liquid, of composition x'', disappears. At a'' the system exists entirely as a vapor.

14–8 Fractional Distillation

The sequence of events described in Section 14–7 is observed if no material is removed from the system as the temperature is increased. If some of the vapor formed in the early stages of the process is removed from the system and condensed, the condensate, or distillate, is enriched in the more volatile constituent, while the residue is impoverished in the more volatile constituent. Suppose that the temperature of a mixture M is increased until half the material is present as vapor and half remains as liquid; Fig. 14–6(a). The vapor has the composition v; the residue R has the composition l. The vapor is removed and condensed yielding a distillate D of composition v. Then the distillate is heated until half exists as vapor and half as liquid; Fig. 14–6(b). The vapor is removed and condensed yielding distillate D' with composition v', and residue R' with composition l'. The original residue R is treated in the same

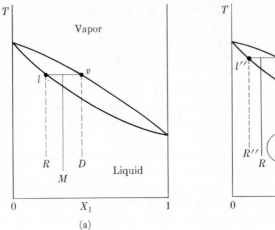
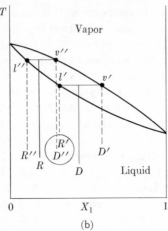

(a)   (b)

Fig. 14–6 Distillation.

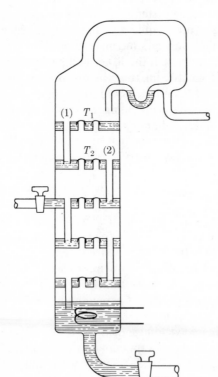

Fig. 14–7 Bubble-cap distilling column. (By permission from Findlay, Campbell, Smith, *The Phase Rule and Its Applications*, 9th ed. Dover Publications, Inc., New York.)

way to yield distillate D'', and residue R''. Since D'' and R' have about the same composition, they are combined; the process is now repeated on the three fractions, R'', $(D'' + R')$, and D'; continuation of this process ultimately yields a distillate which approaches the composition of the more volatile liquid and a residue close to the composition of the less volatile liquid, together with a series of fractions of intermediate composition.

The time and labor involved in this batch type of separation is prohibitive and is eliminated through the use of a continuous method using a *fractionating column*; Fig. 14–7. The type of column illustrated is a bubble-cap column. The column is heated at the bottom; there is a temperature gradient along the length of the column, the top being cooler than the bottom. Let us suppose that the temperature at the top of the column is T_1, and the vapor issuing at this point is in equilibrium with the liquid held up on the top plate, plate 1; the compositions of liquid and vapor are shown in Fig. 14–8 as l_1, and v_1. On the next plate, plate 2, the temperature is slightly higher, T_2, and the vapor issuing from it has the composition v_2. As this vapor passes upward to plate 1, it is cooled to temperature T_1, to point a. This means that some of the vapor v_2 condenses to form l_1; since l_1 is richer in the less volatile constituent, the remaining vapor is richer in the more volatile constituent and at equilibrium attains the composition v_1. This happens at every plate in the column. As the vapor moves up the column, it cools; this cooling condenses the less volatile component preferentially, so the vapor becomes increasingly enriched in the more volatile component as it moves upward. If at each position in the column the liquid is in equilibrium with vapor, then the composition of the vapor will be given by the vapor composition curve in Fig. 14–8. It is understood that the temperature is some function of the position in the column.

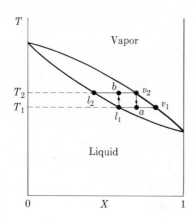

Fig. 14–8 Liquid and vapor exchange in a distilling column.

As the liquid l_1 on the top plate flows down to the next plate, the temperature rises to T_2, and the state point of the liquid reaches b; Fig. 14–8. Some of the more volatile component vaporizes to yield v_2; the liquid shifts to the composition l_2.

As it flows downward through the column, the liquid becomes richer in the less volatile component.

As vapor moves up the column and the liquid moves down, there is a continuous redistribution of the two components between the liquid and vapor phases to establish equilibrium at each position (i.e., each temperature) in the column. This redistribution must take place quickly if the equilibrium is to be established at every position. There must be efficient contact between the liquid and vapor. In the bubble-cap column, efficient contact is obtained by forcing the ascending vapor to bubble through the liquid on each plate. In the laboratory Hempel column, the liquid is spread out over glass beads and the vapor is forced upward through the spaces between the beads; intimate contact is achieved in this way. Industrial stills use a variety of packings, saddle-shaped pieces of ceramic being frequently used. Packing materials or arrangements which permit the liquid to channel, that is, to flow downward through the column along special paths, must be avoided. The aim is to spread liquid evenly in relatively thin layers so that redistribution of the components may occur quickly.

It should be noted that if a certain portion of the column is held at a particular temperature, then, at equilibrium, the composition of liquid and vapor have the values appropriate to that temperature. Under constant pressure, the variance of the system is $F = 3 - P$; since two phases are present, $F = 1$. Consequently, fixing the temperature at every position in the column fixes the liquid and vapor composition at every position in the column at equilibrium. Therefore, by imposing an arbitrary temperature distribution along the column, an equally arbitrary composition distribution of vapor and liquid along the column results "at equilibrium."

The phase "at equilibrium" or "in equilibrium" is commonly used to describe a distilling column which is not in equilibrium at all but rather in a *steady state*. Since there are inequalities of temperature along the column, the system cannot be truly in equilibrium in the thermodynamic sense. For this reason, the phase rule does not apply rigorously to this situation. It can be used as a guide, however. Other difficulties occur as well: the pressure is higher at the bottom of the column than at the top; the countercurrent flow of liquid and vapor is an additional nonequilibrium phenomenon.

In practice, equilibrium is not established at every position of the column, but rather the vapor at any position has a composition in equilibrium with the liquid at a slightly lower position. If the distance between these two positions is h, the column is said to have one *theoretical plate* in the length h. The number of theoretical plates in a column depends on its geometry, the kind and arrangement of the packing, and the manner in which the column is operated. It must be determined experimentally for a given set of operating conditions.

If the individual components have boiling points which are far apart, a distilling column with only a few theoretical plates suffices to separate the mixture. On the other hand, if the boiling points are close together, a column with a large number of theoretical plates is required.

14–9 Azeotropes

Mixtures which are ideal or depart only slightly from ideality can be separated into their constituents by fractional distillation. On the other hand, if the deviations from Raoult's law are so large as to produce a maximum or a minimum in the vapor pressure curve, then a corresponding minimum or maximum appears in the boiling point curve. Such mixtures cannot be completely separated into their constituents by fractional distillation. It can be shown that if the vapor pressure curve has a maximum or minimum, then at that point the liquid and vapor curves must be tangent to one another and the liquid and vapor must have the same composition; Gibbs-Konovalov theorem. The mixture having the maximum or minimum vapor pressure is called an *azeotrope.*†

Consider the system shown in Fig. 14–9 which exhibits a maximum boiling point. If a mixture described by point *a*, having the azeotropic composition, is heated, the vapor will first form at temperature *t*; that vapor has the same composition as the liquid; consequently, the distillate obtained has exactly the same composition as the original liquid; no separation is effected. If a mixture described by *b*, Fig. 14–9, is heated, the first vapor forms at *t'*, and has a composition *v'*. This vapor is richer in the *higher boiling component*. Fractionation would separate the mixture into pure component 1 in the distillate and leave the azeotropic mixture in the pot. A mixture described by *c* would boil first at *t''*; the vapor would have the composition *v''*. Fractionation of this mixture would yield pure component 2 in the distillate and azeotrope in the pot.

The behavior of minimum boiling azeotropes shown in Fig. 14–10 is analogous. The azeotrope itself distills unchanged. A mixture described by *b* boils first at

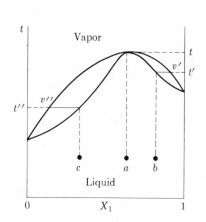

Fig. 14–9 *t-X* diagram with maximum boiling point.

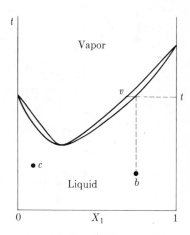

Fig. 14–10 *t-X* diagram with minimum boiling point.

---

† From Greek: to boil unchanged.

temperature t, the vapor having a composition v. Fractionation of this mixture produces azeotrope in the distillate; pure component 1 remains in the pot. Similarly, fractionation of a mixture described by c will produce azeotrope in the distillate and leave pure component 2 in the pot.

In Table 14–1, a number of azeotropic mixtures are listed, together with their properties. The azeotrope resembles a pure compound in the property of boiling at a constant temperature, while ordinary mixtures boil over a range of temperatures. However, changes in pressure produce changes in the *composition* of the

Table 14–1(a)† Minimum boiling azeotropes (1 atm)

Component A	b.p., °C	Component B	b.p., °C	Azeotrope	
				Wt % A	b.p., °C
H_2O	100	C_2H_5OH	78.3	4.0	78.174
H_2O	100	$CH_3COC_2H_5$	79.6	11.3	73.41
CCl_4	76.75	CH_3OH	64.7	79.44	55.7
CS_2	46.25	CH_3COCH_3	56.15	67	39.25
$CHCl_3$	61.2	CH_3OH	64.7	87.4	53.43

Table 14–1(b) Maximum boiling azeotropes (1 atm)

Component A	b.p., °C	Component B	b.p., °C	Azeotrope	
				Wt % A	b.p., °C
H_2O	100	HCl	− 80	20.222	108.584
H_2O	100	HNO_3	86	68	120.5
$CHCl_3$	61.2	CH_3COCH_3	56.10	78.5	64.43
C_6H_5OH	182.2	$C_6H_5NH_2$	184.35	42	186.2

† By permission from *Azeotropic Data*; Advances in Chemistry Series No. 6, American Chemical Society; Washington, D.C., 1952.

Table 14–2† Dependence of azeotropic temperature and composition on pressure

Pressure, mm	Wt % HCl	b.p., °C
500	20.916	97.578
700	20.360	106.424
760	20.222	108.584
800	20.155	110.007

† W. D. Bonner, R. E. Wallace, *J. Amer. Chem. Soc.*, **52**, 1747 (1930).

azeotrope, as well as changes in the boiling point, and so it cannot be a pure compound. The constant boiling hydrochloric acid is a case in point. The variation in composition with pressure is illustrated by the data in Table 14–2. These compositions have been determined accurately enough so that a standard HCl solution may be prepared by dilution of the constant boiling acid.

14–10 The Ideal Dilute Solution

The rigid requirement of the ideal solution that every component obey Raoult's law over the entire range of composition is relaxed in the definition of the *ideal dilute solution*. To arrive at the laws governing dilute solutions, we must examine the experimental behavior of these solutions. The vapor-pressure curves for three systems are described below.

Benzene–toluene

Figure 14–11 shows the vapor pressure versus mole fraction of benzene for the benzene–toluene system, which behaves ideally to a good degree of accuracy over the entire range of composition. The partial pressures of benzene and toluene, also shown in the figure, are linear functions of the mole fraction of benzene, since Raoult's law is obeyed.

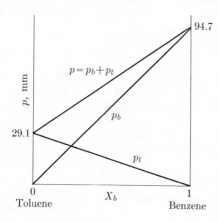

Fig. 14–11 Vapor pressures in the benzene–toluene system.

Acetone–carbon disulfide

Figure 14–12(a) shows the partial-pressure curves and the total vapor pressure of mixtures of carbon disulfide and acetone. In this system the individual partial-pressure curves fall well above the Raoult's law predictions indicated by the dashed lines. The system exhibits positive deviations from Raoult's law. The total vapor pressure exhibits a maximum which lies above the vapor pressure of either component.

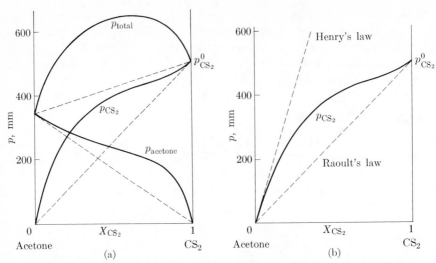

Fig. 14–12 Vapor pressure in the acetone–carbon disulfide system (35.17°C).
[J. v. Zawidski, *Z. physik Chem.*, **35**, 129 (1900).]

Figure 14–12(b) displays another interesting feature of this system. In this figure only the partial pressure of carbon disulfide is shown; in the region near $X_{CS_2} = 1$, when CS_2 is the solvent, the partial-pressure curve is tangent to the Raoult's law line. However, in the region near $X_{CS_2} = 0$, when CS_2 is the solute present in low concentration, the partial-pressure curve is linear:

$$p_{CS_2} = K_{CS_2}X_{CS_2}, \tag{14–14}$$

where K_{CS_2} is a constant. The slope of the line in this region is different from the Raoult's law slope. The solute obeys *Henry's law*; Eq. (14–14), where K_{CS_2} is the Henry's law constant. Inspection of the partial-pressure curve of the acetone discloses the same type of behavior:

$$p_{\text{acetone}} = X_{\text{acetone}}p^0_{\text{acetone}} \qquad \text{near } X_{\text{acetone}} = 1,$$

$$p_{\text{acetone}} = K_{\text{acetone}}X_{\text{acetone}} \qquad \text{near } X_{\text{acetone}} = 0.$$

Note that if the solution were ideal, then K would equal p^0 and both Henry's law and Raoult's law would convey the same information.

Acetone–chloroform

In the system, acetone–chloroform, shown in Fig. 14–13, the vapor pressure curves fall below the Raoult's law predictions. This system exhibits *negative* deviations from Raoult's law. The total vapor pressure has a minimum value which lies below the vapor pressure of either of the pure components. The Henry's law lines, fine dashed lines in the figure, also lie below the Raoult's law lines for this system.

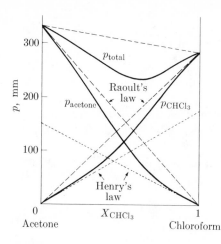

Fig. 14-13 Vapor pressure in the acetone–chloroform system (35.17°C). (J. v. Zawidski, *Z. physik Chem.*, **35**, 129 (1900).]

Algebraically, we can express the properties of the ideal dilute solution by the following equations:

$$\text{Solvent (Raoult's law):} \quad p_1 = x_1 p_1^0, \tag{14-15}$$

$$\text{Solutes (Henry's law):} \quad p_j = K_j x_j, \tag{14-16}$$

where the subscript j denotes any of the solutes, and the subscript 1 denotes the solvent. All real solutions approach the behavior described by Eqs. (14–15) and (14–16), provided that the solution is sufficiently dilute. The same is true if several solutes are present, but the solution must be dilute in all solutes; each solute has a different value of K_j.

14-11 The Chemical Potentials in the Ideal Dilute Solution

Since the solvent follows Raoult's law, the chemical potential of the solvent is given by Eq. (14–3), repeated here for easy comparison:

$$\mu_1 = \mu_1^0(T, p) + RT \ln x_1.$$

For the solutes, we require, as usual, equality of the chemical potential in the liquid, μ_j, with that in the vapor, $\mu_{j(\text{vap})}$:

$$\mu_j = \mu_{j(\text{vap})} = \mu_{j(\text{vap})}^0 + RT \ln p_j.$$

Using Henry's law value, Eq. (14–16), for p_j, this becomes

$$\mu_j = \mu_{j(\text{vap})}^0 + RT \ln K_j + RT \ln x_j.$$

We define a standard free energy μ_j^* by

$$\mu_j^* = \mu_{j(\text{vap})}^0 + RT \ln K_j. \tag{14-17}$$

In general, μ_j^* is a function of temperature and pressure, but not of composition. The expression for μ_j becomes

$$\mu_j = \mu_j^* + RT \ln x_j. \tag{14-18}$$

According to Eq. (14–18), μ_j^* is the chemical potential the solute j would have in the hypothetical state in which $x_j = 1$ if Henry's law were obeyed over the entire range from $x_j = 0$ to $x_j = 1$.

The concept of the ideal solution is extended to include nonvolatile solutes by requiring that the chemical potential of such solutes also have the form given by Eq. (14–18).

In Chapter 13 we obtained the relation between the mole fraction of the solvent x_1 and the total molality m of the solutes; Eq. (13–17). By the same reasoning, for x_j we obtain

$$x_j = \frac{m_j}{(1000/M) + m}, \tag{14-19}$$

where m_j is the molality of j. As the total molality of all the solutes approaches zero. we have

$$x_j = \frac{M m_j}{1000}, \qquad m \to 0. \tag{14-20}$$

Using this value of x_j in Eq. (14–18), we obtain

$$\mu_j = \mu_j^* + RT \ln \left(\frac{M}{1000} \right) + RT \ln m_j.$$

Defining $\mu_j^{**} = \mu_j^* + RT \ln (M/1000)$, this becomes

$$\mu_j = \mu_j^{**} + RT \ln m_j, \tag{14-21}$$

which expresses the μ_j in a dilute solution as a convenient function of m_j. The μ_j^{**} is the value μ_j would have in the hypothetical state of unit molality if the solution had the properties of the ideal dilute solution in the entire range from $m_j = 0$ to 1.

The chemical potential of the solutes can also be expressed in terms of the molar concentrations c_j. Choosing the quantity of solution which contains 1000 gm of solvent, then $n_1 = 1000/M$, and $n_j = m_j$; so we have, if V is the volume in liters,

$$c_j = \frac{n_j}{V} = \frac{m_j}{V}.$$

If ρ_s is the density of the solution, then $V = w/1000\rho_s$, where the weight of the solution $w = 1000 + \Sigma_j m_j M_j$. Thus,

$$V = \frac{1}{\rho_s} \left(1 + \frac{\Sigma_j m_j M_j}{1000} \right)$$

and

$$c_j = \frac{\rho_s m_j}{1 + (\Sigma_j m_j M_j / 1000)}. \tag{14-22}$$

As all the m_j approach zero, we have

$$c_j = \rho m_j, \tag{14–23}$$

where ρ is the density of the pure solvent. Using Eq. (14–23) in Eq. (14–21), we obtain

$$\mu_j = \mu_j^\square + RT \ln c_j. \tag{14–24}$$

In Eq. (14–24),

$$\mu_j^\square = \mu_j^{**} + RT \ln (1/\rho) = \mu_j^* + RT \ln \left(\frac{M}{1000\rho} \right). \tag{14–25}$$

Equation (14–24) relates the chemical potential of component j in a dilute solution to the molar concentration c_j. It is not used as often as Eq. (14–21); $\mu_j^\square$ is the chemical potential in the hypothetical state of unit molarity.

14–12   Henry's Law and the Solubility of Gases

Henry's law, Eq. (14–16), relates the partial pressure of the solute in the vapor phase to the mole fraction of the solute in the solution. Viewing the relation in another way, Henry's law relates the equilibrium mole fraction, the solubility of j in the solution, to the partial pressure of j in the vapor:

$$x_j = \frac{1}{K_j} p_j. \tag{14–26}$$

Equation (14–26) says that the solubility x_j of a volatile constituent is proportional to the partial pressure of that constituent in the gaseous phase in equilibrium with the liquid. Equation (14–26) is used to correlate the data on solubility of gases in liquids. If the solvent and gas do not react chemically, the solubility of gases in liquids is usually small so that the condition of diluteness is fulfilled. Here we have another example of the physical significance of the partial pressure.

The solubility of gases is often expressed in terms of the Bunsen absorption coefficient α, which is the number of cubic centimeters of gas measured at 0°C and 1 atm dissolved by 1 cm^3 of liquid if the partial pressure of the gas is 1 atm. The two ways of expressing solubility are related. Suppose that we apply Eq. (14–26) to a system consisting of 1 cm^3 of liquid having a gas j at 1 atm pressure over it; let x_j^0 be the solubility under these conditions. Then

$$x_j^0 = \frac{n_j^0}{n + n_j^0} = \frac{1}{K_j}.$$

But n is the number of moles of solvent in 1 cm^3, $n = \rho/M$; ρ and M are the density and molecular weight of the solvent. Furthermore, n is much larger than n_j^0 in a dilute solution, so that the equation becomes

$$\frac{Mn_j^0}{\rho} = \frac{1}{K_j}.$$

What volume α_j will n_j^0 moles of gas occupy at $T_0 = 273.15°K$ and $p_0 = 1$ atm?

$$\alpha_j = n_j^0 \left(\frac{RT_0}{p_0}\right) = n_j^0(22{,}414). \tag{14–27}$$

Using this result in the preceding equation, we obtain

$$\alpha_j K_j = \frac{22{,}414\rho}{M}, \tag{14–28}$$

which is the relation between the Henry's law constant K_j and the Bunsen absorption coefficient α_j; knowing one, we can calculate the other. The solubility of the gas in moles per cubic centimeter of solvent, n_j^0, is directly proportional to α_j, Eq. (14–27); this makes α_j more convenient than K_j for the discussion of solubility.

Some values of α for various gases in water are given in Table 14–3. Note the increase in α with increase in boiling point of the gas.

Table 14–3 Bunsen absorption coefficients in water at 25°C

Gas	b.p., °C	α
Helium	−268.9	0.0087
Hydrogen	−252.8	0.0175
Nitrogen	−195.8	0.0143
Oxygen	−182.96	0.0283
Methane	−161.5	0.0300
Ethane	−88.3	0.0410

14–13 Distribution of a Solute Between Two Solvents

If a dilute solution of iodine in water is shaken with carbon tetrachloride, the iodine is distributed between the two immiscible solvents. If μ and μ' are the chemical potentials of iodine in water and carbon tetrachloride, respectively, then at equilibrium $\mu = \mu'$. If both solutions are ideal dilute solutions, then, choosing Eq. (14–18) to express μ and μ', the equilibrium condition becomes $\mu^* + RT \ln x = \mu'^* + RT \ln x'$, which can be rearranged to

$$RT \ln \frac{x'}{x} = -(\mu'^* - \mu^*). \tag{14–29}$$

Since both μ'^* and μ^* are independent of composition, it follows that

$$x'/x = K, \tag{14–30}$$

where K, the distribution coefficient or partition coefficient, is independent of the concentration of iodine in the two layers. The quantity $\mu'^* - \mu^*$ is the standard free energy change ΔG^* for the transformation

$$I_2 \text{ (in } H_2O) \rightarrow I_2 \text{ (in } CCl_4).$$

Equation (14–29) becomes

$$RT \ln K = -\Delta G^*, \tag{14–31}$$

which is the usual relation between the standard free energy change and the equilibrium constant of a chemical reaction.

If the solutions are quite dilute, then the mole fractions are proportional to the molalities or the molarities; so we have

$$K' = m'/m \quad \text{and} \quad K'' = c'/c, \tag{14–32}$$

where K' and K'' are independent of the concentrations in the two layers. Equation (14–32) was originally proposed by W. Nernst and is called the Nernst distribution law.

14–14 Chemical Equilibrium in the Ideal Solution

In Section 11–7 it was shown that the condition of chemical equilibrium is

$$0 = \left[\sum_P \nu_P \mu_P - \sum_R \nu_R \mu_R \right]_{eq}, \tag{14–33}$$

the ν's being the stoichiometric coefficients. To apply this condition to chemical equilibrium in the ideal solution, we simply insert the proper form of the μ's from Eq. (14–3). This yields immediately

$$0 = \sum_P \nu_P \mu_P^0 - \sum_R \nu_R \mu_R^0 + RT \ln \frac{(x_{P_1})_e^{\nu_{P_1}}(x_{P_2})_e^{\nu_{P_2}} \cdots}{(x_{R_1})_e^{\nu_{R_1}}(x_{R_2})_e^{\nu_{R_2}} \cdots},$$

which can be written in the usual way

$$\Delta G^0 = -RT \ln K, \tag{14–34}$$

where ΔG^0 is the standard free energy change for the reaction, and K is the equilibrium quotient of mole fractions. Thus, in an ideal solution, the proper form of the equilibrium constant is a quotient of mole fractions.

If the solution is an ideal dilute solution, then for a reaction between solutes only each μ is given by Eq. (14–18),

$$\mu_i = \mu_i^* + RT \ln x_i,$$

so that the equilibrium condition is

$$\Delta G^* = -RT \ln K, \tag{14-35}$$

K being again a quotient of equilibrium mole fractions. Quite obviously, we could equally well have chosen either Eq. (14–21) or (14–24) to express μ_j. In that event we would obtain

$$\Delta G^{**} = -RT \ln K' \qquad \text{or} \qquad \Delta G^{\square} = -RT \ln K''; \tag{14-36}$$

K' is a quotient of the equilibrium molalities; K'' is a quotient of equilibrium molarities; ΔG^{**} and $\Delta G^{\square}$ are the appropriate standard free energy changes.

Values of standard free energy changes are obtained from the measurement of equilibrium constants in the same way as were those for reactions in the gas phase. Values of individual standard free energies of solutes in solution are obtained, as they are for gaseous reactions, by combining the free energy changes for several reactions.

Differentiation of Eq. (14–34) yields

$$\left[\frac{\partial (\Delta G^0 / T)}{\partial T} \right]_p = -R \left(\frac{\partial \ln K}{\partial T} \right)_p;$$

using the Gibbs-Helmholtz equation, this becomes

$$\left(\frac{\partial \ln K}{\partial T} \right)_p = \frac{\Delta H^0}{RT^2}, \tag{14-37}$$

which is the usual relation between the standard enthalpy change of the reaction and the derivative of $\ln K$ with respect to T.

If the chemical reaction involves the solvent, the equilibrium constant has a slightly modified form. For example, suppose the equilibrium

$$CH_3COOH + C_2H_5OH \rightleftharpoons CH_3COOC_2H_5 + H_2O$$

is studied in water solution; then if the solution is dilute enough to use molarities to describe the free energy of the solutes, the equilibrium constant has the form

$$K'' = \frac{c_{EtAc} x_{H_2O}}{c_{HAc} c_{EtOH}}, \tag{14-38}$$

since in dilute solution Raoult's law holds for the solvent. In dilute solution $x_{H_2O} \approx 1$, so K'' becomes

$$K'' = \frac{c_{EtAc}}{c_{HAc} c_{EtOH}}. \tag{14-39}$$

The standard free energy change for K'' is $\Delta G^{\square}$, by Eq. (14–36), and must include $\mu_{H_2O}^0$; therefore,

$$\Delta G^{\square} = \mu_{EtAc}^{\square} + \mu_{H_2O}^0 - \mu_{HAc}^{\square} - \mu_{EtOH}^{\square}. \tag{14-40}$$

The $\mu_{H_2O}^0$ is the molar free energy of pure water; the $\mu_i^{\square}$ are the chemical potentials of the solutes in the hypothetical ideal solution of unit molarity.

Problems

14-1. Show that while the vapor pressure in a binary ideal solution is a linear function of the mole fraction of either component in the liquid, the reciprocal of the pressure is a linear function of the mole fraction of either component in the vapor.

14-2. Suppose that the vapor over an ideal solution contains n_1 moles of 1 and n_2 moles of 2 and occupies a volume V under the pressure $p = p_1 + p_2$. If we define $\bar{V}_2^0 = RT/p_2^0$ and $\bar{V}_1^0 = RT/p_1^0$, then show that Raoult's law implies $V = n_1 \bar{V}_1^0 + n_2 \bar{V}_2^0$.

14-3. Benzene and toluene form nearly ideal solutions. If, at $300°K$, $p_{toluene}^0 = 32.06$ mm and $p_{benzene}^0 = 103.01$ mm,

a) compute the vapor pressure of a solution containing 0.60 mole fraction toluene,

b) calculate the mole fraction of toluene in the vapor for this composition of liquid.

14-4. The composition of the vapor over a binary ideal solution is determined by the composition of the liquid. If x_1 and y_1 are the mole fractions of 1 in the liquid and vapor, respectively find the value of x_1 for which $y_1 - x_1$ has a maximum. What is the value of the pressure at this composition?

14-5. a) A liquid mixture of benzene and toluene is composed of 1 mole of toluene and 1 mole of benzene. If the pressure over the mixture at $300°K$ is reduced, at what pressure does the first vapor form? (The vapor pressures are given in Problem 14-3.)

b) What is the composition of the first trace of vapor formed?

c) If the pressure is reduced further, at what pressures does the last trace of liquid disappear?

d) What is the composition of the last trace of liquid?

e) What will be the pressure, the composition of the liquid, the composition of the vapor, when 1 mole of the mixture has been vaporized? [*Hint*: Lever rule.]

14-6. Some nonideal systems can be represented by the equations $p_1 = x_1^a p_1^0$ and $p_2 = x_2^a p_2^0$. Show that if the constant a is greater than unity, the total pressure exhibits a minimum, while if a is less than unity, the total pressure exhibits a maximum.

14-7. a) The boiling points of pure benzene and pure toluene are $80.1°$ and $110.6°C$ under 1 atm. Assuming the entropies of vaporization at the boiling points are the same, 21 eu, and by applying the Clausius-Clapeyron equation to each, derive an implicit expression for the boiling point of a mixture of the two liquids as a function of the mole fraction of benzene, x_b.

b) What is the composition of the liquid which boils at $95°C$?

14-8. a) In an ideal dilute solution, if p_1^0 is the vapor pressure of the solvent and K_h is the Henry's law constant for the solute, write the expression for the total pressure over the solution as a function of x, the mole fraction of the solute.

b) Find the relation between y_1 and the total pressure of the vapor.

14-9. The Bunsen absorption coefficients of oxygen and nitrogen in water are 0.0283 and 0.0143 at $25°C$. Supposing that air is 20% oxygen and 80% nitrogen, how many cubic centimeters of gas, measured at STP, will be dissolved by $100\ cm^3$ of water in equilibrium with air at 1 atm pressure? How many if the pressure is 10 atm? What is the mole ratio, N_2/O_2, of the dissolved gas?

14-10. The Henry's law constant for argon in water is 2.17×10^4 atm at 0°C and 3.97×10^4 atm at 30°C. Calculate the standard heat of solution of argon in water in kcal/mole.

14-11. Suppose that a 250-cm^3 bottle of carbonated water at 25°C contains CO_2 under 2 atm pressure. If the Bunsen absorption coefficient of CO_2 is 0.76, what is the total volume of CO_2, measured at STP, which is dissolved in the water?

Chapter Fifteen

Equilibria Between Condensed Phases

15–1 Liquid–Liquid Equilibria

If small amounts of toluene are added to a beaker containing pure benzene, it is observed that regardless of the amount of toluene added, the mixture obtained remains as one liquid phase. The two liquids are *completely miscible*. In contrast to this behavior, if water is added to nitrobenzene, two separate liquid layers are formed; the water layer contains only a trace of dissolved nitrobenzene, while the nitrobenzene layer contains only a trace of dissolved water. Such liquids are *immiscible*. If small amounts of phenol are added to water, at first the phenol dissolves to yield one phase; however, at some point in the addition the water becomes saturated and further addition of phenol yields two liquid layers, one rich in water, the other rich in phenol. Such liquids are *partially miscible*. It is these systems which presently engage our attention.

Consider a system in equilibrium which contains two liquid layers, two liquid phases. Let one of these liquid layers consist of pure liquid A, the other layer is a saturated solution of A in liquid B. The thermodynamic requirement for equilibrium is that the chemical potential of A in the solution, μ_a, be equal to that in the pure liquid, μ_a^0. So $\mu_a = \mu_a^0$, or

$$\mu_a - \mu_a^0 = 0. \tag{15–1}$$

First, we ask if Eq. (15–1) can be satisfied for an ideal solution. In an ideal solution, by Eq. (14–3),

$$\mu_a - \mu_a^0 = RT \ln x_a. \tag{15–2}$$

It is clear from Eq. (15–2) that $RT \ln x_a$ is never zero unless the mixture of A and

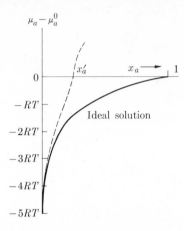

Fig. 15–1 Chemical potential in a nonideal solution.

B has $x_a = 1$, that is, unless the mixture contains no B. In Fig. 15–1, $\mu_a - \mu_a^0$ is plotted against x_a for the ideal solution (full line). The value of $\mu_a - \mu_a^0$ is negative for all compositions of the ideal solution. This implies that pure A can always be transferred into an ideal solution with a decrease in free energy. Consequently, substances which form ideal solutions are completely miscible in each other.

For partial miscibility the value for $\mu_a - \mu_a^0$ must be zero at some intermediate composition of the solution; thus, $\mu_a - \mu_a^0$ must follow some such curve as the dashed line in Fig. 15–1. At the point x_a', the value of $\mu_a - \mu_a^0$ is zero, and the system can exist as a solution having mole fraction of $A = x_a'$ and a separate layer of pure liquid A. The value x_a' is the solubility of A in B expressed as a mole fraction. If the mole fraction of A in B were to exceed this value, then Fig. 15–1 shows that $\mu_a - \mu_a^0$ would be positive, that is $\mu_a > \mu_a^0$. In this circumstance, A would flow spontaneously from the solution into the pure liquid A, thus reducing x_a until it reached the equilibrium value x_a'.

Liquids which are only partially miscible form solutions which are far from ideal, as the curves in Fig. 15–1 show. Rather than explore the mathematical side of this situation in great detail, we restrict ourselves to a description of the experimental results interpreted in the light of the phase rule.

Suppose that at a given temperature T_1, small amounts of liquid A are added successively to liquid B. The first amount of A dissolves completely, as do the second and the third; the state points can be represented on a T-X diagram such as Fig. 15–2(a), which is drawn at constant pressure. The points a, b, c represent the composition after the addition of three amounts of A to pure B. Since all the A dissolves, these points lie in a one-phase region. After a certain amount of A has been added, the solubility limit is reached, point l_1. If more A is added, a second layer forms, since no more A will dissolve. The region to the right of point l_1 is therefore a two-phase region.

The same could be done on the right side by adding B to A. At first B dissolves to yield a homogeneous (one-phase) system, points d, e, f. The solubility limit of B

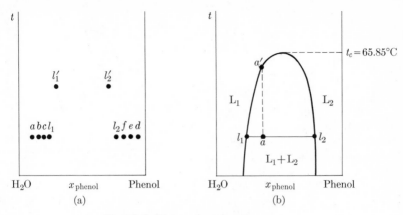

Fig. 15–2 Water–phenol system.

in A is reached at l_2. Points to the left of l_2 represent a two-phase system. In the region between l_1 and l_2 two liquid layers, called *conjugate solutions*, coexist. Layer l_1 is a saturated solution of A in B in equilibrium with layer l_2 which is a saturated solution of B in A. If the experiment were done at a higher temperature, different values of the solubility limits, l_1' and l_2', would be obtained.

The T versus X diagram for the system phenol–water is shown in Fig. 15–2(b). As the temperature increases, the solubility of each component in the other increases. The solubility curves join smoothly at the *upper consolute temperature*, also called the *critical solution temperature*, t_c. Above t_c, water and phenol are completely miscible. Any point a under the loop is the state point of a system consisting of two liquid layers: L_1 of compositon l_1 and L_2 of composition l_2. The relative mass of the two layers is given by the lever rule, by the ratio of the segments of the tie line $(\overline{l_1 l_2})$.

$$\frac{\text{moles of } l_1}{\text{moles of } l_2} = \frac{(\overline{al_2})}{(\overline{al_1})}.$$

If the temperature of this system is raised, the state point follows the dashed line aa'; L_1 becomes richer in phenol, while L_2 becomes richer in water. As the temperature increases, the ratio $(\overline{al_2})/(\overline{al_1})$ becomes larger; the amount of L_2 decreases. At point a' the last trace of L_2 disappears and the system becomes homogeneous.

Systems are known in which the solubility *decreases* with increase in temperature. In some of these systems, a *lower consolute temperature* is observed; Fig. 15–3(a) shows schematically the triethylamine–water system. The lower consolute temperature is at 18.5°C. The curve is so flat that it is difficult to determine the composition of the solution corresponding to the consolute temperature; it seems to be about 30% by weight of triethylamine. If a solution having a state point a is heated, it remains homogeneous until the temperature is slightly above 18.5°C; at this point, a', it splits into two layers. At a higher temperature a'', the solutions have the compositions given by l_1 and l_2. In view of the lever rule, l_1 will be present in somewhat greater

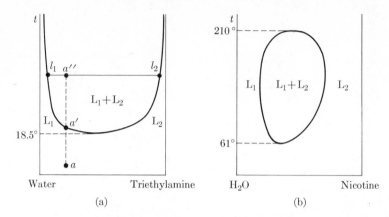

Fig. 15–3 (a) Lower consolute temperature. (b) Upper and lower consolute temperature.

amount than l_2. As a rule, the liquid pairs which have solubility diagrams of this type tend to form loosely bound compounds with each other; this enhances solubility at low temperatures. As the temperature is increased, the compound is dissociated and the mutual solubility is diminished.

Some substances exhibit both upper and lower consolute temperatures. The diagram for the system nicotine–water is shown schematically in Fig. 15–3(b). The lower consolute temperature is about 61°, the upper one about 210°. At all points in the closed loop two phases are present, while the points outside the loop represent homogeneous states of the system.

The phase rule for a system at constant pressure is $F' = C - P + 1$, in which F' is the number of variables in addition to the pressure needed to describe the system. For two-component systems, $F' = 3 - P$. If two phases are present, only one variable is required to describe the system. In the two-phase region, if the temperature is described, then the intersections of the tie line with the curve yield the compositions of *both* conjugate solutions. Similarly, the composition of one of the conjugate solutions is sufficient to determine the temperature and the composition of the other conjugate solution. If only one phase is present, $F' = 2$ and both the temperature and the composition of the solution must be specified.

15–2 Distillation of Partially Miscible and Immiscible Liquids

The discussion in Section 15–1 assumed that the pressure is high enough so that vapor does not form in the temperature range of interest. For this reason the liquid–vapor curves were omitted from the diagrams. A typical situation at lower pressures is shown in Fig. 15–4(a) in which the liquid–vapor curves are also shown, still with the assumption that the pressure is fairly high. Figure 15–4(a) presents no new problem in interpretation. The upper and lower portions of the diagram can be discussed

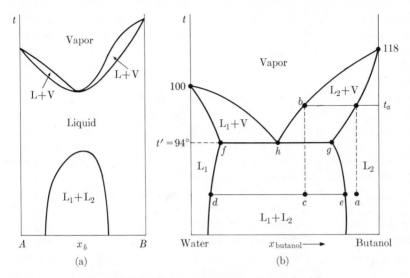

Fig. 15–4 Distillation of partially miscible liquids.

independently using the principles described before. Partial miscibility at low temperatures usually, though not always, implies a minimum boiling azeotrope, as is
shown in Fig. 15–4(a). The partial miscibility implies that when mixed the two
components have a greater escaping tendency than in an ideal solution. This greater
escaping tendency may lead to a maximum in the vapor pressure–composition curve,
and correspondingly to a minimum in the boiling point–composition curve.

If the pressure on the system shown in Fig. 15–4(a) is lowered, the boiling points
will all be shifted downward. At a low enough pressure, the boiling point curves will
intersect the liquid–liquid solubility curves. The result is shown in Fig. 15–4(b),
which represents schematically the system water-*n*-butanol under 1 atm pressure.

Figure 15–4(b) presents several new features. If the temperature of a homogeneous liquid, point a, is increased, vapor having the composition b forms at t_a.
This behavior is ordinary enough; however, if this vapor is chilled and brought to
point c, the condensate will consist of two liquid layers, since c lies in the two-liquid
region. So the first distillate produced by the distillation of the homogeneous liquid
a will separate into two liquid layers having compositions d and e. Similar behavior
is exhibited by mixtures having compositions in the region L_1.

As the temperature of the two-liquid system of overall composition c is increased,
the compositions of the conjugate solutions shift slightly. The system is univariant,
$F' = 3 - P = 1$ in this region. At the temperature t', the conjugate solutions have
the compositions f and g, and vapor, composition h, appears. Three phases are
present, liquids f and g, and vapor h. Then $F' = 0$; the system is invariant. As long
as these three phases remain, their compositions and the temperature are fixed. For
example, the flow of heat into the system does not change the temperature, but

simply produces more vapor at the expense of the two solutions. The vapor, h, which forms is richer in water than the original composition c; therefore, the water-rich layer evaporates preferentially. After the water-rich layer disappears, the temperature rises, the vapor composition changing along the curve hb. The last liquid, which has the composition a, disappears at t_a.

If a two-phase system in the composition range between f and h is heated, then at t' liquids f and g are present, and vapor h appears. The system at t' is invariant. Since the vapor is richer in butanol than the original overall composition, the butanol-rich layer evaporates preferentially, leaving liquid f and vapor h. As the temperature rises, the liquid is depleted in butanol; finally only vapor remains.

The point h has the azeotropic property; a system of this composition distills unchanged. It cannot be separated into its components by distillation.

The distillation of *immiscible* substances is most easily discussed from a different standpoint. Consider two immiscible liquids in equilibrium with vapor at a specified temperature; Fig. 15–5. The barrier only keeps the liquids apart; since they are immiscible, removing the barrier would not change anything. The total vapor pressure is the sum of the vapor pressures of the pure liquids; $p = p_a^0 + p_b^0$. The mole fractions y_a and y_b in the vapor are

$$y_a = p_a^0/p \qquad y_b = p_b^0/p.$$

If n_a and n_b are the number of moles of A and B in the vapor, then

$$\frac{n_a}{n_b} = \frac{y_a}{y_b} = \frac{p_a^0/p}{p_b^0/p} = \frac{p_a^0}{p_b^0}.$$

The masses of A and B are $w_a = n_a M_a$, and $w_b = n_b M_b$, so that

$$w_a/w_b = M_a p_a^0/M_b p_b^0, \tag{15–3}$$

which relates the relative weights of the two substances present in the vapor to their molecular weights and vapor pressures. If this vapor were condensed, Eq. (15–3) would express the relative weights of A and B in the condensate. Suppose we choose the system aniline (A)-water (B) at 98.4°C. The vapor pressure of aniline at this temperature is about 42 mm, while that of water is about 718 mm. The total vapor pressure is $718 + 42 = 760$, so this mixture boils at 98.4°C under 1 atm pressure. The weight of aniline which distills for each 100 gm of water which comes over is

$$w_a = 100\frac{(94)(42)}{(18)(718)} \approx 31 \text{ gm.}$$

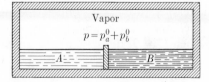

Fig. 15–5 Immiscible liquids in equilibrium with vapor.

Equation (15–3) can be applied to the steam distillation of liquids. Some liquids which decompose if distilled in the ordinary way can be steam distilled if they have fair volatility near the boiling point of water. In the laboratory, steam is passed through the liquid to be steam distilled. Since the vapor pressure is greater than that of either component, it follows that the boiling point is below the boiling points of both liquids. Furthermore, the boiling point is an invariant temperature so long as the two liquid phases and the vapor are present.

If the vapor pressure of the substance is known over a range of temperatures near 100°C, measurement of the temperature at which the steam distillation occurs and the weight ratio in the distillate yield, through Eq. (15–3), a value of the molecular weight of the substance.

15–3 Solid–Liquid Equilibria. The Simple Eutectic Diagram

If a liquid solution of two substances A and B is cooled, at a sufficiently low temperature a solid will appear. This temperature is the freezing point of the solution; it depends on the composition. In the discussion of freezing-point depression, Section 13–6, we obtained the equation

$$\ln x_a = -\frac{\Delta H_{\text{fus},a}}{R}\left(\frac{1}{T} - \frac{1}{T_{0a}}\right),\tag{15–4}$$

assuming that pure solid A is in equilibrium with an ideal liquid solution. Equation (15–4) relates the freezing point of the solution to x_a, the mole fraction of A in the solution. A plot of this function is shown in Fig. 15–6(a). The point above the curve represents liquid states of the system, those below the curve represent states in which

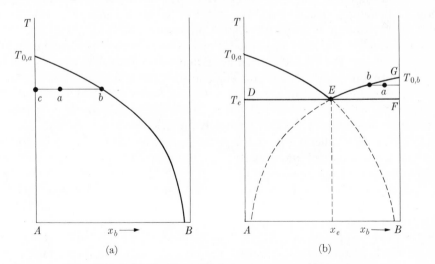

(a)   (b)

Fig. 15–6 Solid–liquid equilibria in a two-component system.

pure solid A coexists in equilibrium with solution. The curve is called the *liquidus* curve.

A point such as a represents solution of composition b in equilibrium with solid of composition c, i.e., pure A. By the lever rule, the ratio of the number of moles of solution to the number of moles of solid A present is equal to the ratio of segments of the tie line $\overline{ac}/\overline{ab}$. The lower the temperature, the greater the relative amount of solid for a specified overall composition.

This curve cannot represent the situation over the entire range of composition. As $x_b \to 1$, we would expect solid B to freeze out far above the temperatures indicated by the curve in this region. If the solution is ideal, the same law holds for substance B:

$$\ln x_b = -\frac{\Delta H_{\text{fus},b}}{R}\left(\frac{1}{T} - \frac{1}{T_{0b}}\right), \tag{15–5}$$

where T is the freezing point of B in the solution. This curve is drawn in Fig. 15–6(b) together with the curve for A from Fig. 15–6(a). The curves intersect at a temperature T_e, the *eutectic* temperature. The composition x_e is the eutectic composition. The line GE is the freezing point versus composition curve for B. Points such as a below this curve represent states in which pure solid B is in equilibrium with solution of composition b. A point on EF represents pure solid B in equilibrium with solution of composition x_e. However, a point on DE represents pure solid A in equilibrium with solution of composition x_e. *Therefore, the solution having the eutectic composition x_e is in equilibrium with both pure solid A and pure solid B.* If three phases are present, then $F' = 3 - P = 3 - 3 = 0$; the system is invariant at this temperature. If heat flows out of such a system, the temperature remains the same until one phase disappears; thus, the relative amounts of the three phases change as heat is withdrawn. The amount of liquid diminishes while the amounts of the two solids present increase. Below the line DEF are the states of the system in which only the two solids, two phases, pure A and pure B, are present.

The system lead–antimony has the simple eutectic type of phase diagram; Fig. 15–7. The regions are labeled; L signifies liquid, Sb or Pb signifies pure solid antimony or pure solid lead. The eutectic temperature is 246°C; the eutectic composition is 87 weight percent lead. In the lead–antimony system, the values of t_e and x_e calculated from Eqs. (15–4) and (15–5) agree satisfactorily with the experimental values. This implies that the liquid is nearly an ideal solution.

Consider the isothermal behavior of the system at 300°C, the horizontal line, *abcdfg*. The point a represents pure solid antimony at 300°. Suppose sufficient solid lead is added to bring the composition to point b. This point b lies in the region Sb + L, therefore solid antimony coexists with liquid of composition c. All the added lead melts and the molten lead dissolves enough of the solid antimony to bring the liquid to the composition c. The lever rule shows that the relative amount of liquid present at point b is quite small, so the liquid may not be visible; nonetheless it is present at equilibrium. On further addition of lead, the lead continues to melt and dissolve more of the solid antimony to form solution c; meanwhile the state

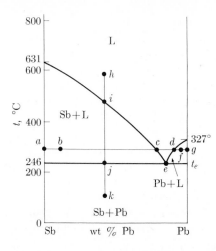

Fig. 15-7 The antimony–lead system.

point moves from *b* to *c*. When the state point reaches *c*, sufficient lead has been added to dissolve all of the original antimony present to form the saturated solution of antimony in lead. Further addition of lead simply dilutes this solution as the state point moves through the liquid region *c* to *d*. At *d* the solution is saturated with lead; further addition of lead produces no change. The state point meanwhile has moved to *f*. If we had reached *f* by starting with pure lead at *g* and adding antimony, all of the antimony would have melted, 330° below its melting point, and dissolved sufficient lead to form the solution *d*.

An *isopleth* is a line of constant composition such as *hijk* in Fig. 15–7. At *h*, the system is entirely liquid. As the system cools, solid antimony appears at *i*; as the antimony crystallizes out the saturated liquid becomes richer in lead, so the liquid composition moves along the curve *ice*. At *j* the solution has the eutectic composition *e* and is saturated with respect to lead also, so lead begins to precipitate. The temperature remains constant even though heat may be flowing out since, in this condition, the system is invariant. The amount of liquid diminishes and the amounts of solid lead and antimony increase. Finally the liquid solidifies, and the temperature of the mixed solids decreases along the line *jk*. If the process is done in reverse, heating a mixture of solid lead and solid antimony from *k*, the state point moves from *k* to *j*. At *j*, liquid forms having the composition *e*. Note that the liquid formed has a different composition than the solid mixture. The system is invariant, so the temperature remains at 246° until all of the lead melts; since the liquid was richer in lead than the original mixture, the lead melts completely leaving a residue of solid antimony. After the lead has melted the temperature rises, and the antimony which melts moves the liquid composition from *e* to *i*. At *i* the last bit of antimony melts and the system is homogeneous above *i*.

The eutectic (Greek: easily melted) point gets its name from the fact that the eutectic composition has the lowest melting point. The eutectic mixture melts sharply at t_e to form a liquid of the same composition, while other mixtures melt over a range of temperature. Because of the sharp melting point, the eutectic mixture was originally thought to be a compound. In aqueous systems, this "compound" was called a cryohydrate; the eutectic point was called the cryohydric point. Microscopic examination of the eutectic under high magnification discloses its heterogeneous character; it is a mixture, not a compound. In alloy systems, such as the lead–antimony system, the eutectic is often particularly fine-grained; however, under the microscope the separate crystals of lead and antimony can be discerned.

15–4 Thermal Analysis

The shape of the freezing point curves can be determined experimentally by *thermal analysis*. In this method, a mixture of known composition is heated to a high enough temperature so that it is homogeneous. Then it is allowed to cool at a regulated rate. The temperature is plotted as a function of time. The curves obtained for various compositions are shown schematically for a system A–B in Fig. 15–8. In the first curve the homogeneous liquid cools along the curve ab; at b the primary crystals of component A form. This releases the latent heat of fusion: the rate of cooling slows, and a kink in the curve appears at b. The temperature t_1 is a point on the liquidus curve for this composition. The cooling continues along bc; at c the liquid has the eutectic composition, and solid B appears. Since the system is invariant, the temperature remains constant at the eutectic temperature until all the liquid solidifies at d. The horizontal plateau cd is called the *eutectic halt*. After the liquid solidifies, the two solids cool quickly along the curve df. The second curve is for a liquid somewhat richer in B; the interpretation is the same; however, the eutectic halt is longer; t_2 is the point on the liquidus curve. The third curve illustrates the cooling of the

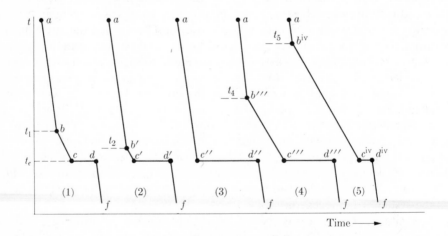

Fig. 15–8 Cooling curves.

eutectic mixture; the eutectic halt has its maximum length. The fourth and fifth curves are for compositions on the B-rich side of the eutectic point; t_4 and t_5 are the corresponding points on the liquidus curve. The length of the eutectic halt diminishes as the composition departs from the eutectic composition. The temperatures t_1, t_2, t_4, and t_5, and t_e are plotted against composition in Fig. 15–9(a). The eutectic composition can be determined as the intersection of the two solubility curves if sufficient points are taken; otherwise the length (in time) of the eutectic halt is plotted as a function of composition, Fig. 15–9(b). The intersection of the two curves yields the maximum value of the eutectic halt, and thus the eutectic composition.

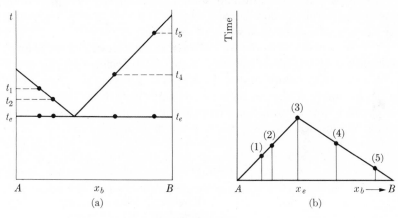

Figure 15–9

15–5   Other Simple Eutectic Systems

Many binary systems, both ideal and nonideal, have phase diagrams of the simple eutectic type. The phase diagram, water–salt, is the simple eutectic type if the salt does not form a stable hydrate. The diagram for H_2O–NaCl is shown in Fig. 15–10.

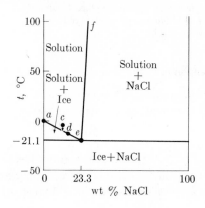

Fig. 15–10 Freezing points in the H_2O–NaCl system.

The curve *ae* is the freezing-point curve for water, while *ef* is the solubility curve, or the freezing-point curve, for sodium chloride.

The invariance of the system at the eutectic point allows eutectic mixtures to be used as constant temperature baths. Suppose solid sodium chloride is mixed with ice at 0°C in a vacuum flask. The composition point moves from 0% NaCl to some finite value. However, at this composition the freezing point of ice is below 0°C; hence, some ice melts. Since the system is in an insulated flask, the melting of the ice reduces the temperature of the mixture. If sufficient NaCl has been added, the temperature will drop to the eutectic temperature, −21.1°C. At the eutectic temperature, ice, solid salt, and saturated solution can coexist in equilibrium. The temperature remains at the eutectic temperature until the remainder of the ice is melted by the heat which leaks slowly into the flask.

The action of rock salt or calcium chloride in melting ice on sidewalks and streets can be interpreted by the phase diagram. Suppose sufficient solid salt is added to ice at −5°C to move the state point of the system to *c*, Fig. 15–10. At *c* the solution is stable; the ice will melt completely if the system is isothermal. If the system were adiabatic, the temperature would fall until the state point reached *d*. The eutectic temperatures of a few ice–salt systems are given in Table 15–1.

Table 15–1†

Salt	Eutectic temperature, °C	Weight percent anhydrous salt in eutectic
Sodium chloride	−21.1	23.3
Sodium bromide	−28.0	40.3
Sodium sulfate	−1.1	3.84
Potassium chloride	−10.7	19.7
Ammonium chloride	−15.4	19.7

† By permission from A. Findlay, A. N. Campbell, N. O. Smith, *The Phase Rule and Its Applications*, 9th ed. Dover Publications, Inc., New York, 1951, p. 141.

15–6 Freezing-point Diagram with Compound Formation

If two substances form one or more compounds, the freezing-point diagram often has the appearance of two or more simple eutectic diagrams in juxtaposition. Figure 15–11 is the freezing-point–composition diagram for the system in which a compound, AB_2, is formed. We can consider this diagram as two simple eutectic diagrams joined at the position of the arrows in Fig. 15–11. If the state point lies to the right of the arrows, the interpretation is based on the simple eutectic diagram for the system AB_2–B; if it lies to the left of the arrows, we discuss the system A–AB_2. In the

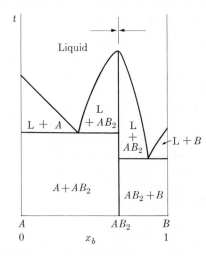

Fig. 15–11 Compound formation.

composite diagram there are two eutectics: one of the A–AB_2–liquid; the other of AB_2–B–liquid. The melting point of the compound is a maximum in the curve; a maximum in the melting-point–composition curve is almost always indicative of compound formation. Only a few systems are known in which the maximum occurs for other reasons. The first solid deposited on cooling a melt of any composition between the two eutectic compositions is the solid compound.

It is conceivable that more than one compound is formed between the two substances; this is often the case with salts and water. The salt forms several hydrates. An extreme example of this behavior is shown by the system ferric chloride–water; Fig. 15–12. This diagram could be split into five simple eutectic diagrams.

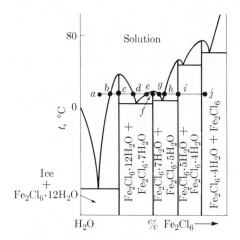

Fig. 15–12 Freezing points in the system H_2O–Fe_2Cl_6 (schematic).

15–7   Compounds Having Incongruent Melting Points

In the system in Fig. 15–11, the compound has a higher melting point than either component. In this situation the diagram has always the shape shown in Fig. 15–11; two eutectics appear on the diagram. However, if the melting point of the compound lies below the melting point of one of the constituents, two possibilities arise. The first of these is illustrated in Fig. 15–12; each part of the diagram is a simple eutectic diagram just as in the simpler case in Fig. 15–11. The second possibility is illustrated by the alloy system potassium–sodium shown schematically in Fig. 15–13. In this system, the solubility curve of sodium does not drop rapidly enough to intersect the other curve between the composition of Na_2K and pure Na. Instead it swings to the left of the composition Na_2K and intersects the other solubility curve at point c, the *peritectic* point. For the system Na–K it is at 7°C.

First we examine the behavior of the pure solid compound. If the temperature is raised, the state point moves along the line ab. At b liquid having the composition c forms. Since this liquid is richer in potassium than the original compound, some solid sodium d is left unmelted. Thus, on melting, the compound undergoes the reaction

$$Na_2K(s) \rightarrow Na(s) + c(l).$$

This is a *peritectic reaction* or a *phase reaction*. The compound is said to melt *incongruently*, since the melt differs from the compound in composition. (The compounds illustrated in Figs. 15–11 and 15–12 melt *congruently*, without change in composition.) Since three phases, solid Na_2K, solid sodium and liquid are present, the system is invariant; as heat flows into the system, the temperature remains the same until the solid compound melts completely. Then the temperature rises; the state point moves along the line bef, the system consisting of solid sodium plus liquid. At f the last trace of solid sodium melts, and above f the system consists of one liquid phase. Cooling the composition g reverses these changes. At f solid sodium appears; the liquid composition moves along fc. At b liquid of composition c coexists with solid sodium and solid Na_2K. The reverse of the phase reaction occurs until liquid and solid sodium are both consumed simultaneously; only Na_2K remains and the state point moves along ba.

If a system of composition i is cooled, primary crystals of sodium form at j; the liquid composition moves along jc as more sodium crystallizes. At k solid Na_2K forms because of the peritectic reaction.

$$c(l) + Na(s) \rightarrow Na_2K(s).$$

The amount of sodium in the composition i is insufficient to convert the liquid c completely into compound. Hence the primary crystals of sodium are consumed completely. After the sodium is consumed, the temperature drops, Na_2K crystallizing and the liquid composition moving along cm; at l, the tie line shows that Na_2K, n, coexists with liquid m. When the temperature reaches o, pure potassium begins to crystallize; the liquid has the eutectic composition p; the system is invariant until the liquid disappears, leaving a mixture of solid potassium and solid Na_2K.

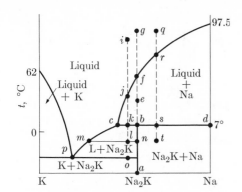

Fig. 15–13 Compound with incongruent melting point.

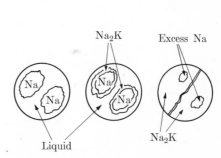

Fig. 15–14 Peritectic crystallization with excess Na.

If liquid of composition q is cooled, primary crystals of sodium form at r; continued cooling crystallizes more sodium, the liquid composition moves along rc. At s, solid Na_2K forms by the peritectic reaction. The liquid is consumed entirely, and the state point drops to t, the system consisting of a mixture of solids, Na_2K and sodium. Because the compound is formed by the reaction of liquid with the primary crystals of sodium, the structure of the solid mixture is unusual. The steps in the reaction are illustrated in Fig. 15–14. The final mixture has a kernel of the primary sodium crystals within a shell of the compound. Since the phase reaction occurs between the primary crystal which is shielded from the liquid by a layer of compound, it is difficult to establish equilibrium in a system such as this unless the experiments are prolonged to allow time for one reactant or the other to diffuse through the layer of compound. An interesting sidelight on this particular system is the wide range of composition in which the alloys of sodium and potassium are liquid at room temperature.

The system sodium sulfate–water forms an incongruently melting compound, $Na_2SO_4 \cdot 10H_2O$; Fig. 15–15(a). The line eb is the solubility curve for the decahydrate, while the line ba is the solubility curve for the anhydrous salt. The figure shows that the solubility of the decahydrate increases, while that of the anhydrous salt decreases with temperature. The peritectic point is at b. On the line bc, three phases coexist: Na_2SO_4, $Na_2SO_4 \cdot 10H_2O$, saturated solution; the system is invariant and the peritectic temperature, 32.383°C, is fixed. This temperature is frequently used as a calibration point for thermometers. If a small amount of water is added to anhydrous Na_2SO_4 in a vacuum bottle at room temperature, the salt and water react to form the decahydrate; this reaction is exothermic so that the temperature of the system rises to 32.383° and remains at that temperature as long as the three phases are present.

If an unsaturated solution of composition g is heated, anhydrous salt will crystallize at f; if it is cooled, the decahydrate will crystallize at h. It is possible to supercool the solution to a temperature below h; then the heptahydrate will crystallize at i; Fig.

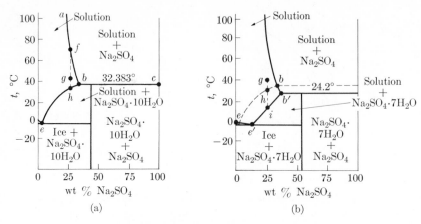

Fig. 15–15 The sodium sulfate–water system.

15–15(b). The curve $e'b'$ is the solubility curve for the heptahydrate, $Na_2SO_4 \cdot 7H_2O$. The peritectic temperature for anhydrous salt–heptahydrate-saturated solution is at 24.2°C. In Fig. 15–15(b), the dashed lines are the curves for the decahydrate. The solubility curve for the heptahydrate lies for the most part in the region of stability of solid decahydrate-saturated solution. Therefore, the equilibrium between solid heptahydrate and its saturated solution is a metastable one; the system in such a state can precipitate the less soluble decahydrate spontaneously.

15–8 Miscibility in the Solid State

In the system described so far only pure solids have been involved. Many solids are capable of dissolving other materials to form *solid solutions*. Copper and nickel, for example, are soluble in each other in all proportions in the solid state. The phase diagram for the system copper–nickel is shown in Fig. 15–16.

The upper curve in Fig. 15–16 is the *liquidus* curve; the lower curve, the *solidus* curve. If a system represented by point a is cooled to b, a solid solution of composition c appears. At point d the system consists of liquid of composition b' in equilibrium with solid solution of composition c'. The interpretation of the diagram is similar to the interpretation of the liquid–vapor diagrams in Section 14–6. An experimental difficulty arises in working with this type of system. Suppose the system were chilled quickly from a to e. If the system managed to stay in equilibrium, then the last vestige of liquid b'' would be in contact with a solid having a uniform composition e throughout. However, in a sudden chilling there is not time for the composition of the solid to become uniform throughout. The first crystal had the composition c and layers having compositions from c to e are built up on the outside of the first crystal. The average composition of the solid which has crystallized lies perhaps at the point f; the solid is richer in nickel than it should be; it lies to the right of e. Hence the liquid is richer in copper than it should be; its composition point lies perhaps

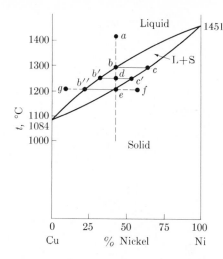

Fig. 15–16 The copper–nickel system.

at g. Thus, some liquid is left at this temperature and further cooling is required before the system solidifies completely. This difficulty poses a severe experimental problem. The system must be cooled extremely slowly to allow time for the solid to adjust its composition at each temperature to a uniform value. In the discussion of these diagrams we assume that equilibrium has been attained and disregard the experimental difficulty which this implies.

Binary systems are known which form solid solutions over the entire range of composition and which exhibit either a maximum or a minimum in the melting point. The liquidus–solidus curves have an appearance similar to that of the liquid–vapor curves in systems which form azeotropes. The mixture having the composition at the maximum or minimum of the curve melts sharply and simulates a pure substance in this respect just as an azeotrope boils at a definite temperature and distills over unchanged. Mixtures having a maximum in the melting-point curve are comparatively rare.

15–9 Freezing-point Elevation

In Section 13–6 we showed that the addition of a foreign substance always lowered the melting point of a pure solid. Figure 15–16 illustrates a system in which the melting point of one component, copper, is *increased* by the addition of a foreign substance. This increase in the melting point can only occur if the solid in equilibrium with the liquid is not pure but a solid solution.

Suppose that the solid solution is an *ideal solid solution*, defined, in analogy to ideal gaseous and ideal liquid solutions, by requiring that for every component, $\mu_i = \mu_i^0 + RT \ln x_i$, where μ_i^0 is the chemical potential of the pure solid, x_i its mole fraction in the solid solution. The equilibrium condition for solid solution in equilibrium with liquid solution for one of the components is $\mu_{\text{solid}} = \mu_{\text{liq}}$. Assuming both solutions are ideal, we obtain

$$\mu^0_{\text{solid}} + RT \ln x_{\text{solid}} = \mu^0_{\text{liq}} + RT \ln x_{\text{liq}}. \tag{15-6}$$

Let $\Delta G^0 = \mu^0_{\text{liq}} - \mu^0_{\text{solid}}$, the free energy of fusion of the pure component at temperature T. Then, Eq. (15-6) becomes

$$R \ln \left(\frac{x_{\text{solid}}}{x_{\text{liq}}} \right) = \frac{\Delta G^0}{T}. \tag{15-7}$$

Since $\Delta G^0 = \Delta H^0 - T\Delta S^0$; and at the melting point, T_0, of the pure substance, $\Delta S^0 = \Delta H^0/T_0$, this equation becomes

$$R \ln \left(\frac{x_{\text{solid}}}{x_{\text{liq}}} \right) = \Delta H^0 \left(\frac{1}{T} - \frac{1}{T_0} \right).$$

Solving this equation for T, we obtain

$$T = T_0 \left[\frac{\Delta H^0}{\Delta H^0 + RT_0 \ln (x_{\text{solid}}/x_{\text{liq}})} \right]. \tag{15-8}$$

If the *pure* solid were present, then $x_{\text{solid}} = 1$; in this case the second term of the denominator in Eq. (15-8) would be positive so that the fraction in the bracket would be less than unity. The freezing point T is therefore less than T_0. If a solid solution is present in equilibrium then if $x_{\text{solid}} < x_{\text{liq}}$, the second term in the denominator will be negative, the fraction in the bracket will be greater than unity and the melting point will be greater than T_0.

Figure 15-16 shows that the mole fraction of copper in the solid solution x_{solid} is always less than the mole fraction of copper in the liquid solution x_{liq}. Consequently, the melting point of copper is elevated. An analogous set of equations can be written for the second component; from which we would conclude that the melting point of nickel is depressed. In the argument we have assumed that the ΔH^0 and ΔS^0 do not vary with temperature; this is incorrect but does not affect the general conclusion.

15-10   Partial Miscibility in the Solid State

It is usual to find that two substances are neither completely miscible nor immiscible in the solid state, but rather each substance has a limited solubility in the other. For this case, the most common type of phase diagram is shown in Fig. 15-17. The points in region α describe solid solutions of B in A, while those in β describe solid solutions of A in B. The points in region $\alpha + \beta$ describe states in which the two saturated solid solutions, *two phases*, α and β, coexist in equilibrium. Cooling a system described by a, at b solid solution α having composition c appears. As the temperature drops, the compositions of solid and liquid shift; at d compositions f and g are in equilibrium. At h the liquid has the eutectic composition e; solid β

appears, α, β, and liquid coexist and the system is invariant. On cooling to i, two solid solutions coexist; α of composition j, β of composition k.

A different type of system in which solid solutions appear is shown in Fig. 15–18. This system has a transition point rather than a eutectic point. Any point on the line abc describes an invariant system in which α, β, and melt of composition c coexist. The temperature of abc is the transition temperature. If the point lies between a and b, cooling will cause melt to disappear, $\alpha + \beta$ remaining. If the point lies between b and c, cooling first causes α to disappear, $\beta + L$ remaining; further cooling causes liquid to disappear and only β remains. If the temperature increases, any point on abc goes into $\alpha + L$; β disappears.

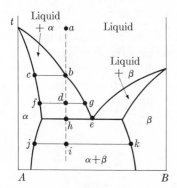

Fig. 15–17 Partial miscibility in the solid state.

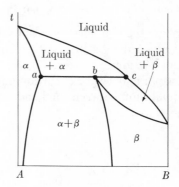

Fig. 15–18 System with transition point.

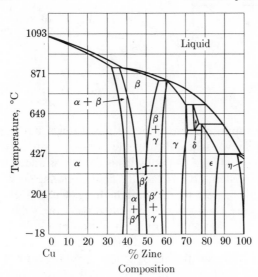

Fig. 15–19 The brass diagram. (From A. G. Guy, *Physical Metallurgy for Engineers*, Addison-Wesley Publishing Co., Inc., Reading, Mass., 1962.)

An interesting example of a system in which many solid solutions occur is the Cu–Zn diagram (brass diagram) in Fig. 15–19. The symbols α, β, γ, δ, ε, η refer to homogeneous solid solutions, while regions labeled $\alpha + \beta$, $\beta + \gamma$ indicate regions in which two solid solutions coexist. Note that there is a whole series of transition temperatures in this diagram and no eutectic temperatures.

It is useful for phase diagrams to contain several features: solid solutions, compound formation, eutectic points, transition points, etc. Once the interpretation of the individual features is understood, the interpretation of complex diagrams poses no difficulty.

15–11 Gas–Solid Equilibria. Vapor Pressure of Salt Hydrates

In describing the equilibria between solids and liquids, we assumed implicitly that the pressure on the system was high enough to prevent the appearance of vapor in the system. At lower pressures, if one or more of the components of the system is volatile, vapor may be present at equilibrium. A common and an important example of the equilibrium between solid and vapor is the equilibrium between salt hydrates and water vapor.

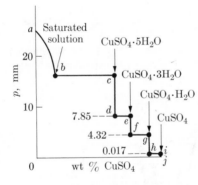

Fig. 15–20 Vapor pressure of $CuSO_4$–H_2O (25°C).

We examine the vapor pressure of the system, water–$CuSO_4$, at a fixed temperature. Figure 15–20 shows schematically the vapor pressure as a function of the concentration of copper sulfate. As anhydrous $CuSO_4$ is added to liquid water, the vapor pressure of the system drops (Raoult's law) along the curve ab. At b the solution is saturated with respect to the pentahydrate, $CuSO_4 \cdot 5H_2O$. The system is invariant along bc, since three phases, saturated solution, solid $CuSO_4 \cdot 5H_2O$, and vapor are present at constant temperature. Addition of anhydrous $CuSO_4$ does not change the pressure but converts some of the solution to pentahydrate. At c all of the water has been combined with $CuSO_4$ to form pentahydrate. Further addition of $CuSO_4$ drops the pressure to the value at de, with the formation of some trihydrate;

$$2CuSO_4 + 3CuSO_4 \cdot 5H_2O \rightarrow 5CuSO_4 \cdot 3H_2O.$$

The system is invariant along de; the three phases present are: vapor, $CuSO_4 \cdot 5H_2O$, $CuSO_4 \cdot 3H_2O$. At e the system consists entirely of $CuSO_4 \cdot 3H_2O$; addition of $CuSO_4$ converts some of the trihydrate to monohydrate; the pressure drops to the value at fg. Finally along hi the invariant system is vapor, $CuSO_4 \cdot H_2O$, $CuSO_4$.

The establishment of a constant pressure in a salt hydrate system requires the presence of three phases; a single hydrate does not have a definite vapor pressure. For example, the trihydrate can coexist in equilibrium with any water vapor pressure in the range from e to f. If the pentahydrate *and* the trihydrate are present, then the pressure is fixed at the value de.

As we have seen in Chapter 11, the equilibrium constant for the reaction

$$CuSO_4 \cdot 5H_2O(s) \rightarrow CuSO_4 \cdot 3H_2O(s) + 2H_2O(g)$$

can be written $K = p_e^2$, where p_e is the equilibrium vapor pressure of water over the mixture of tri- and pentahydrates. The dependence of the vapor pressure on temperature is readily obtained from this equation combined with the Gibbs–Helmholtz equation.

15–12 Systems of Three Components

In a system of three components the variance is $F = 3 - P + 2 = 5 - P$. If the system consists only of one phase, then four variables are required to describe the system; these may conveniently be taken as T, p, x_1, x_2. It is not possible to give a complete graphic representation of these systems in three dimensions, much less in two dimensions. Consequently, it is customary to represent the system at constant pressure and at constant temperature. The variance then becomes $F' = 3 - P$, so that the system has, at most, a variance of two, and can be represented in the plane. After fixing the temperature and pressure, the remaining variables are composition variables x_1, x_2, x_3, related by $x_1 + x_2 + x_3 = 1$. Specifying any two of them fixes the value of the third. The method of Gibbs and Roozeboom uses an equilateral triangle for graphical representation. Figure 15–21 illustrates the principle of the

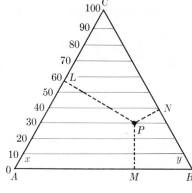

Fig. 15–21 The triangular diagram.

method. The points *A*, *B*, *C* at the apices of the triangle represent 100% of *A*, 100% *B*, 100% *C*. The lines parallel to *AB* represent the various percentages of *C*. Any point on the line *AB* represents a system containing 0% *C*; any point on *xy* represents a system containing 10% *C*, etc. Point *P* represents a system containing 30% *C*. The length perpendicular to a given side of the triangle represents the percent of the component at the vertex opposite to that side. Thus the length *PM* represents the percent of *C*, the length *PN* represents the percent of *A*, the length *PL* represents the percent of *B*. (The lines parallel to *AC* and *CB* have been omitted for clarity.) The sum of the lengths of these perpendiculars is always equal to the length of the height of the triangle which is taken as 100%. By this method any composition of a three-component system can be represented by a point within the triangle.

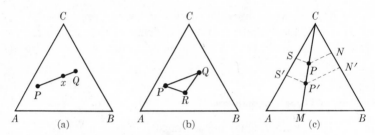

Fig. 15–22 Properties of the triangular diagram.

Two other properties of this diagram are important. The first is illustrated in Fig. 15–22(a). If two systems whose compositions are represented by *P* and *Q* are mixed together, the composition of the mixture obtained will be represented by a point *x* somewhere on the line connecting points *P* and *Q*. It follows immediately that if three systems, represented by points *P*, *Q*, *R*, are mixed, the composition of the mixture will lie within the triangle *PQR*. The second important property is that all systems represented by points on a line through a vertex contain the other two components in the same ratio. For example, all systems represented by points on *CM* contain *A* and *B* in the same ratio. In Fig. 15–22(c), by erecting the perpendiculars from two points *P* and *P'*, we obtain, using the properties of similar triangles:

$$\frac{PS}{P'S'} = \frac{CP}{CP'} \quad \text{and} \quad \frac{PN}{P'N'} = \frac{CP}{CP'}.$$

Therefore

$$\frac{PS}{P'S'} = \frac{PN}{P'N'} \quad \text{or} \quad \frac{PS}{PN} = \frac{P'S'}{P'N'},$$

which was to be proved. This property is important in discussing the addition or removal of a component to the system without change in the amount of the other two components present.

15–13 Liquid–Liquid Equilibria

Among the simplest examples of the behavior of three-component systems is the system chloroform–water–acetic acid. The pairs chloroform–acetic acid and water–acetic acid are completely miscible. The pair chloroform–water is not. Figure 15–23 shows schematically the liquid–liquid equilibrium for this system. Points a and b represent the conjugate liquid layers in the absence of acetic acid. Suppose that the overall composition of the system is c so that by the lever rule there is more of layer b than layer a. If a little acetic acid is added to the system, the composition moves along the line connecting c with the acetic acid apex to the point c'. The addition of acetic acid changes the composition of the two layers to a' and b'. Note that the acetic acid goes preferentially into the water-rich layer b', so that the tie line connecting the conjugate solutions a' and b' is not parallel to ab. The relative amounts of a' and b' are given by the lever rule; that is, by the ratio of the segments of the tie line $a'b'$. Continued addition of acetic acid moves the composition further along the dashed line cC; the water-rich layer grows in size while the chloroform-rich layer diminishes. At c'' only a trace of the chloroform-rich layer remains, while above c'' the system is homogeneous.

Since the tie lines are not parallel, the point at which the two conjugate solutions have the same composition does not lie at the top of the binodal curve but off to one side at the point k, the *plait* point. If the system has the composition d and acetic acid is added, the composition will move along dk; just below k the two layers will *both* be present in *comparable* amounts; at k the boundary between the two solutions vanishes as the system becomes homogeneous. Compare this behavior with that at c'' where only a trace of one of the conjugate layers remained.

If the temperature is increased, the shape and extent of the two-phase region alters. A typical example for a system in which increase in temperature increases

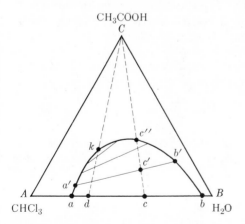

Fig. 15–23 Two partially miscible liquids.

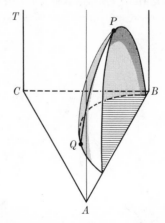

Fig. 15–24 Effect of temperature on a partially miscible pair.

the mutual solubility is shown in Fig. 15–24. If temperature were plotted as a third coordinate, the two-phase region would be a loaf-shaped region; Fig. 15–24. In the figure P is the consolute temperature for the two-component system A–B. The line PQ connects the plait points at the various temperatures.

If two of the pairs A–B and B–C are partially miscible, the situation becomes more complex. Two binodal curves can appear as in Fig. 15–25(a). At lower temperatures, the two binodal curves in Fig. 15–25(a) may overlap. If they do so in such a way that the plait points join one another, then the two-phase region becomes a band, as in Fig. 15–25(b). If the binodal curves do not join at the plait points, the resultant diagram has the form shown in Fig. 15–25(c). Points within the triangle abc represent states of the system in which *three* liquid layers having compositions a, b, and c coexist. Such a system is isothermally invariant.

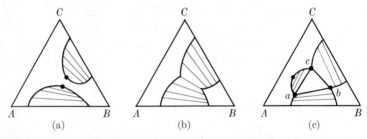

Fig. 15–25 Two partially miscible pairs.

15–14   Solubility of Salts. Common-ion Effect

Systems which contain two salts with a common ion and water have great interest from a practical standpoint. Each salt influences the solubility of the other. The schematic diagram for NH_4Cl, $(NH_4)_2SO_4$, H_2O at $30°$ is shown in Fig. 15–26. Point a represents the saturated solution of NH_4Cl in water in the absence of

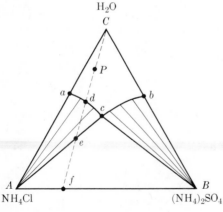

Fig. 15–26 Common-ion effect.

$(NH_4)_2SO_4$. Points between A and a represent various amounts of solid NH_4Cl in equilibrium with saturated solution a. Points between a and C represent the unsaturated solution of NH_4Cl. Similarly, b represents the solubility of $(NH_4)_2SO_4$ in the absence of NH_4Cl. Points on Cb represent the unsaturated solution, while those on bB represent solid $(NH_4)_2SO_4$ in equilibrium with saturated solution. The presence of $(NH_4)_2SO_4$ changes the solubility of NH_4Cl along the line ac, while the presence of NH_4Cl changes the solubility of $(NH_4)_2SO_4$ along the line bc. Point c represents a solution which is saturated with respect to both NH_4Cl and $(NH_4)_2SO_4$. The tie lines connect the composition of the saturated solution and the solid in equilibrium with it. The regions of stability are shown in Table 15–2.

Table 15–2

Region	System	Variance
$Cacb$	Unsaturated solution	2
Aac	NH_4Cl + saturated solution	1
Bbc	$(NH_4)_2SO_4$ + saturated solution	1
AcB	NH_4Cl + $(NH_4)_2SO_4$ + saturated solution c	0

Suppose an unsaturated solution represented by P is evaporated isothermally; the state point must move along the line $Pdef$, which has been drawn through the apex C and the point P. At d, NH_4Cl crystallizes; the composition of the solution moves along the line dc. At point e, the solution composition is c, and $(NH_4)_2SO_4$ begins to crystallize. Continued evaporation deposits both NH_4Cl and $(NH_4)_2SO_4$ until the point f is reached, where the solution disappears completely.

15–15 Double-salt Formation

If it happens that the two salts can form a compound, a double salt, then the solubility of the compound may also appear as an equilibrium line in the diagram. Figure 15–27 shows two typical cases of compound formation. In both figures, ab is the solubility of A; bc that of the compound AB, cd that of B. The regions and what they represent are tabulated in Table 15–3.

The difference in behavior of the two systems can be demonstrated in two ways. First begin with the dry solid compound and add water; the state point moves along the line DC. In Fig. 15–27(a), this moves the point into the region of the compound plus saturated solution of the compound. Hence, this compound is said to be *congruently saturating*. Addition of water to the compound AB in Fig. 15–27(b) moves the state point along DC into the region of stability of $A + AB +$ saturated solution b. The addition of water, therefore, decomposes the compound into solid A and a solution. This compound is said to be *incongruently saturating*. Similarly the compound in Fig. 15–27(b) cannot be prepared by evaporating a solution containing A

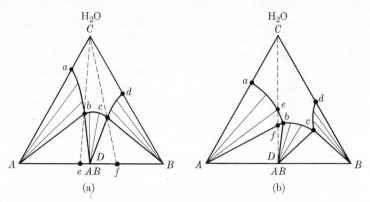

Fig. 15–27 (a) Congruently saturating compound. (b) Incongruently saturating compound.

Table 15–3

Region	System	Variance
$Cabcd$	Unsaturated solution	2
abA	A + saturated solution	1
AbD	$A + AB$ + saturated solution b	0
Dbc	AB + saturated solution	1
DcB	$AB + B$ + saturated solution c	0
cdB	B + saturated solution	1

and B in the equimolar ratio. Evaporation crystallizes solid A at point e; at point f the solid A reacts with the solution b to precipitate AB. When D is reached, all of A has disappeared and only the compound remains. If the solids are filtered off when the state point is between f and D, the crystals of compound will be mixed with

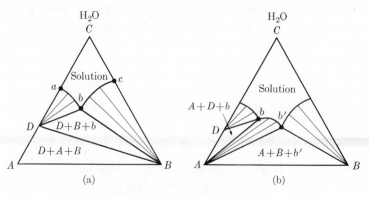

Fig. 15–28 Hydrate formation.

crystals of A. It is understandable how annoying this in the laboratory. A knowledge of the phase diagram for the double-salt system is very helpful in preparative problems.

If one of the salts forms a hydrate of composition D, then the diagram will have the appearance shown in Fig. 15–28(a). The interesting feature of this diagram is that if the state point lies in the triangle ADB, the system consists exclusively of the three solids, D, B, A. Under appropriate conditions, usually at a higher temperature, the anhydrous salt may make its appearance on the diagram as in Fig. 15–28(b).

15–16   The Method of "Wet Residues"

The determination of equilibrium curves in three-component systems in some respects is simpler than in two-component systems. Consider the diagram in Fig. 15–29.

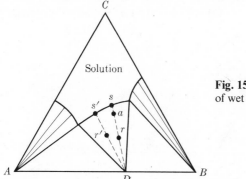

Fig. 15–29 The method of wet residues.

Suppose that the system consists of a solution in equilibrium with solid and that the state point is at a. We do not know the location of a, but we do know that it lies on a tie line connecting the solid composition with the liquid composition. We proceed as follows: some of the saturated liquid is removed and analysed for A and B. This fixes the point s on the equilibrium line. After the removal of some of the saturated solution, the state point of the remainder of the system must lie at point r. So the remainder, that is, the solids together with the supernatant liquid, called the "wet residue," is analysed for two of the components. This analysis determines the point r. A tie line is drawn through s and r. The procedure is repeated on a system which contains a slightly different ratio of two of the components. The solution analysis yields the point s', while the analysis of the wet residue yields a point r'. The tie line is drawn through s' and r'. These two tie lines must intersect at the composition of the solid which is present. In this system, they would intersect at point D. This intersection point yields the composition of the solid phase D, which is in equilibrium with the liquid.

The method of wet residues is superior to the procedure which is necessary in two-component systems, where the liquid and solid phase must be separated and analyzed individually. It is a practical impossibility to separate the solid phase from the liquid without some of the liquid adhering to the solid and thus contaminating it. For this reason, it is frequently easier to add a third substance to a two-component system, determine the equilibrium lines, as well as the composition of the solid phases by the method of wet residues, and infer the composition of the solid in the two-component system from the features of the triangular diagram.

15–17 "Salting Out"

In the practice of organic chemistry, it is common procedure to separate a mixture of an organic liquid in water by adding salt. For example, if the organic liquid and water are completely miscible, addition of salt to the system may produce a separation into two liquid layers—one rich in the organic liquid, the other rich in water. The phase relations may be illustrated as in Table 15–4 and by the diagram for K_2CO_3–H_2O–CH_3OH, Fig. 15–30, which is typical of the system salt–water–alcohol.

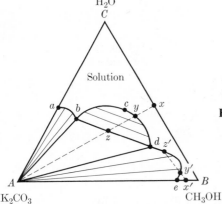

Fig. 15–30 Salt–alcohol–water diagram.

Table 15–4

Region	System
Aab	K_2CO_3 in equilibrium with water-rich saturated solution
Aed	K_2CO_3 in equilibrium with alcohol-rich saturated solution
bcd	two conjugate liquids joined by tie lines
Abd	K_2CO_3 in equilibrium with conjugate liquids b and d

The system is distinguished by the appearance of the two-liquid region bcd. Suppose that solid K_2CO_3 is added to a mixture of water and alcohol of composition x. The state point will move along the line $xyzA$. At y two layers form; at z K_2CO_3 ceases to dissolve so that solid K_2CO_3 and liquids b and d coexist. The liquid d is the alcohol-rich layer and may be separated from b, the water-rich layer.

Note that addition of salt after the solid ceases to dissolve does not produce any change in the composition of the layers b and d; this must be so since the system is isothermally invariant in the triangle Abd.

This diagram can also be used to show how additional salt may be precipitated by the addition of alcohol to a saturated solution; the state point moves from a, let us say, along a line connecting a and B. Since, in this particular case, only a little more salt is precipitated before two liquid layers form, the trick is not particularly useful. This system is curious in the effect of the addition of water to an unsaturated solution of K_2CO_3 in alcohol of composition x'. The line $x'y'z'$ connecting x' and C shows that K_2CO_3 will precipitate at y' if water is added to the alcoholic solution. Further addition of water will redissolve the K_2CO_3 at z'.

Problems

15–1. The melting point and heats of fusion of lead and antimony are

	Pb	Sb
t_0, °C	327.4	630.5
ΔH_{fus}, kcal/mole	1.22	4.8

Calculate the solid–liquid equilibrium lines; estimate the eutectic composition graphically; then calculate the eutectic temperature. Compare the result with the values given by Fig. 15–7.

15–2. From the melting points of the mixtures of Al and Cu, sketch the melting-point curve:

wt % Cu	0	20	40	60	80	100
t, °C	650	600	540	610	930	1084

15–3. The solubility of KBr in water is:

t, °C	0	20	40	60	80	100
gm KBr/gm H_2O	0.54	0.64	0.76	0.86	0.95	1.04

In a one molal solution, KBr depresses the freezing point of water by 3.29°C. Estimate the eutectic temperature for the system KBr–H_2O graphically.

15–4. KBr is recrystallized from water by saturating the solution at 100°C, then cooling to 20°C; the crystals obtained are redissolved in water and the solution evaporated until it is saturated at 100°C. Cooling to 20°C produces a second crop of crystals. What is the percent yield of pure KBr after these two crystallizations? Use data in Problem 15–3.

15-5. Two crops of KBr crystals are obtained as follows. A solution saturated at 100°C is cooled to 10°C; after filtering off the first crop, the mother liquid is evaporated until the solution is saturated at 100°C; cooling to 20°C produces the second crop. What fraction of the KBr is recovered in the two crops by this method? (Data in Problem 15–3.)

15-6. In Fig. 15–18, what is the variance in each region of the diagram? Keep in mind that the pressure is constant. What is the variance on the line abc?

15-7. Figure 15–16 shows the equilibrium between liquid and solid solutions in the copper–nickel system. If we suppose that both the solid and liquid solutions are ideal, then the equilibrium conditions lead to two equations of the form of Eq. (15–8); one of these applies to copper, the other to nickel. If we invert the equations they become

$$1/T = (1/T_{Cu})[1 + (R/\Delta S_{Cu}) \ln (x'_{Cu}/x_{Cu})]$$

$$1/T = (1/T_{Ni})[1 + (R/\Delta S_{Ni}) \ln (x'_{Ni}/x_{Ni})],$$

where x' is the mole fraction in the solid solution, x that in the liquid. In addition we have the relations, $x'_{Cu} + x'_{Ni} = 1$, and $x_{Cu} + x_{Ni} = 1$. There are four equations with five variables, T, x'_{Cu}, x'_{Ni}, x_{Cu}, x_{Ni}. Suppose that $x_{Cu} = 0.1$; calculate values for all the other variables. $T_{Cu} = 1356.2°K$, $T_{Ni} = 1728°K$; assume that $\Delta S_{Cu} = \Delta S_{Ni} = 2.35$ eu. [*Hint*: Use value of x_{Cu} in the first two equations, then eliminate T between them. By trial and error solve the resulting equation for either x'_{Cu} or x'_{Ni}. Then T is easily calculated. Repetition of this procedure for other values of x_{Cu} would yield the entire diagram.]

15-8. What is the variance in each region of Fig. 15–30?

15-9. a) Using Fig. 15–30, what changes will be observed if water is added to a system containing 50% K_2CO_3 and 50% CH_3OH?

b) What is observed if methanol is added to a system containing 90% water and 10% K_2CO_3? (or 30% water and 70% K_2CO_3?)

15-10. a) What is the variance in each of the regions of Fig. 15–15(a)?

b) Describe the changes which occur if an unsaturated solution of Na_2SO_4 is evaporated at 25°; at 35°.

15-11. Describe the changes which occur if water is evaporated isothermally along the line aj from the system in Fig. 15–12.

Chapter Sixteen

Equilibria in Nonideal Systems

16–1 The Concept of Activity

The mathematical discussions in the preceding chapters have been limited to systems which behave ideally; the systems were either pure ideal gases, or ideal mixtures, gaseous, liquid, solid. Many of the systems described in Chapter 15 are not ideal; the question that arises is how are we to deal mathematically with nonideal systems. These systems are handled conveniently using the concepts of fugacity and activity, first introduced by G. N. Lewis.

The chemical potential of a component in a mixture is in general a function of temperature, pressure, and the composition of the mixture. In gaseous mixtures we write the chemical potential of each component as a sum of two terms:

$$\mu_i = \mu_i^0(T) + RT \ln f_i. \tag{16–1}$$

The first term, μ_i^0, is a function of temperature only, while the fugacity f_i in the second term may depend on the temperature, pressure, and composition of the mixture. The fugacity is a measure of the chemical potential of the gas i in the mixture. In Section 10–9 a method of evaluating the fugacity for a pure gas was described.

Now we will confine our attention to liquid solutions, although most of what is said can be applied to solid solutions as well. For any component i in any liquid mixture, we write

$$\mu_i = g_i(T, p) + RT \ln a_i, \tag{16–2}$$

where $g_i(T, p)$ is a function only of temperature and pressure while a_i, the *activity of i*, may be a function of temperature, pressure, and composition. As it stands, Eq. (16–2) is not particularly informative; however, it does indicate that at a specified

357

temperature and pressure an increase in the activity of a substance means an increase in the chemical potential of that substance. *The equivalence of the activity to the chemical potential*, through an equation having the form of Eq. (16–2), is the fundamental property of the activity. The theory of equilibrium could be developed entirely in terms of the activities of substances instead of in terms of chemical potentials. In some problems it is convenient to do that.

To use Eq. (16–2), the significance of the function $g_i(T, p)$ must be accurately described; then a_i has a precise meaning. Two ways of describing $g_i(T, p)$ are in common use; each leads to a different system of activities. In either system it is still' true that the activity of a component is a measure of its chemical potential.

16–2 The Rational System

In the rational system, $g_i(T, p)$ is identified with the chemical potential of the pure liquid, $\mu_i^0(T, p)$:

$$g_i(T, p) = \mu_i^0(T, p). \tag{16–3}$$

Then Eq. (16–2) becomes

$$\mu_i = \mu_i^0 + RT \ln a_i. \tag{16–4}$$

As $x_i \to 1$, the system comes nearer to being pure i, and μ_i must approach μ_i^0, so that

$$\mu_i - \mu_i^0 = 0 \quad \text{as} \quad x_i \to 1.$$

Using this fact in Eq. (16–4), we have $\ln a_i = 0$, as $x_i \to 1$, or

$$a_i = 1 \quad \text{as} \quad x_i \to 1. \tag{16–5}$$

Therefore, the activity of the pure liquid is equal to unity.

If we compare Eq. (16–4) with the μ_i in an ideal liquid solution,

$$\mu_{i(\text{id})} = \mu_i^0 + RT \ln x_i, \tag{16–6}$$

by subtracting Eq. (16–6) from Eq. (16–4), we obtain

$$\mu_i - \mu_{i(\text{id})} = RT \ln (a_i/x_i). \tag{16–7}$$

The *rational activity coefficient of i*, γ_i, is defined by

$$\gamma_i = a_i/x_i. \tag{16–8}$$

With this definition, Eq. (16–7) becomes

$$\mu_i = \mu_{i(\text{id})} + RT \ln \gamma_i, \tag{16–9}$$

which shows that $\ln \gamma_i$ measures the extent of the deviation from ideality. From the relation in Eq. (16–5), and the definition of γ_i, we have

$$\gamma_i = 1 \quad \text{as} \quad x_i \to 1. \tag{16–10}$$

The rational activity coefficients are convenient for those systems in which the mole fraction of any component may vary from zero to unity; mixtures of liquids such as acetone and chloroform, for example.

The activity of volatile constituents in a liquid mixture can be readily measured by measuring the partial pressure of the constituent in the vapor phase in equilibrium with the liquid. Since at equilibrium the chemical potentials of each constituent must be equal in the liquid and the vapor phase, we have $\mu_{i(\text{liq})} = \mu_{i(\text{vap})}$. Using Eq. (16–4) for $\mu_{i(\text{liq})}$ and assuming the gas is ideal, component i having a partial pressure p_i, then

$$\mu_{i(\text{liq})}^0 + RT \ln a_i = \mu_{i(\text{gas})}^0 + RT \ln p_i.$$

For the pure liquid,

$$\mu_{i(\text{liq})}^0 = \mu_{i(\text{gas})}^0 + RT \ln p_i^0,$$

where p_i^0 is the vapor pressure of the pure liquid. Subtracting the last two equations and dividing by RT, we obtain $\ln a_i = \ln (p_i/p_i^0)$, or

$$a_i = p_i/p_i^0, \tag{16–11}$$

which is the analogue of Raoult's law for a nonideal solution. Thus, a measurement of p_i over the solution together with a knowledge of p_i^0 yields the value of a_i. From measurements at various values of x_i, the value of a_i can either be plotted or tabulated as a function of x_i. Similarly, the activity coefficient can be calculated and plotted as function of x_i. In Figs. 16–1 and 16–2, plots of a_i and γ_i versus x_i are shown for binary systems which exhibit positive and negative deviations from Raoult's law. If the solutions were ideal, then $a_i = x_i$, and $\gamma_i = 1$, for all values of x_i.

Depending on the system, the activity coefficient of a component may be greater or less than unity. In a system showing positive deviations from ideality, the activity

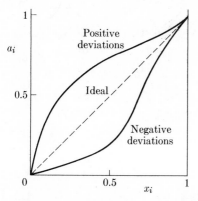

Fig. 16–1 Activity versus mole fraction.

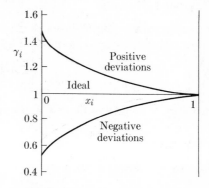

Fig. 16–2 Activity coefficient versus mole fraction.

coefficient, and therefore the escaping tendency, is greater than in an ideal solution of the same concentration. In a solution exhibiting negative deviations from Raoult's law, the substance has a lower escaping tendency than in an ideal solution of the same concentration; γ is less than unity.

16–3 Colligative Properties

The colligative properties of a solution of involatile solutes are simply expressed in terms of the rational activity of the solvent. If the vapor pressure over the solution is p, and the activity of the solvent is a, then $a = p/p^0$, from Eq. (16–11). If a is evaluated from measurements of vapor pressure at various concentrations, these values can be used to calculate the freezing-point depression, boiling-point elevation, and osmotic pressure for any concentration.

Freezing-point depression

If pure solid solvent is in equilibrium with solution, the equilibrium condition $\mu_{\text{liq}} = \mu_{\text{solid}}^0$ becomes, using Eq. (16–4), $\mu_{\text{liq}}^0 + RT \ln a = \mu_{\text{solid}}^0$; or,

$$\ln a = -\frac{\Delta G_{\text{fus}}^0}{RT}.$$

Repetition of the argument in Section 13–6 yields finally

$$\ln a = \frac{\Delta H_{\text{fus}}^0}{R}\left(\frac{1}{T_0} - \frac{1}{T}\right), \tag{16-12}$$

which is the analogue of Eq. (13–15) for the ideal solution. Knowing a from vapor pressure measurements, the freezing point can be calculated from Eq. (16–12); conversely, if the freezing point T is measured, a can be evaluated from Eq. (16–12).

Boiling-point elevation

The analogous argument shows that the boiling point is related to ΔH_{vap}^0 and T_0, the heat of vaporization and the boiling point of the pure solvent, by

$$\ln a = \frac{\Delta H_{\text{vap}}^0}{R}\left(\frac{1}{T} - \frac{1}{T_0}\right), \tag{16-13}$$

which is the analogue of Eq. (13–31) for the ideal solution.

Osmotic pressure

The osmotic pressure is given by

$$\bar{V}^0 \pi = -RT \ln a, \tag{16-14}$$

which is the analogue of Eq. (13–40).

In Eqs. (16–12), (16–13), and (16–14), a is the rational activity of the solvent. Measurements of any colligative property yield values of a through these equations.

16–4 The Practical System

The practical system of activities and activity coefficients is useful for solutions in which only the solvent has a mole fraction near unity; all of the solutes are present in relatively small amounts. For such a system we use the rational system for the *solvent* and the practical system for the *solutes*. As the concentration of solutes becomes very small, the behavior of any real solution approaches that of the *ideal dilute* solution. Using a subscript j to identify the solutes then in the ideal dilute solution (Section 14–11)

$$\mu_{j(\text{id})} = \mu_j^{**} + RT \ln m_j. \tag{16–15}$$

For a solute, Eq. (16–2) becomes

$$\mu_j = g_j(T, p) + RT \ln a_j. \tag{16–16}$$

If we subtract Eq. (16–15) from Eq. (16–16) and set $g_j(T, p) = \mu_j^{**}$, then

$$\mu_j - \mu_{j(\text{id})} = RT \ln \frac{a_j}{m_j}. \tag{16–17}$$

The identification of $g_j(T, p)$ with μ_j^{**} defines the practical system of activities; the practical activity coefficient γ_j is defined by

$$\gamma_j = a_j/m_j. \tag{16–18}$$

Equations (16–17) and (16–18) show that $\ln \gamma_j$ is a measure of the departure of a solute from its behavior in an ideal dilute solution. Finally, as $m_j \to 0$, the solute must behave in the ideal dilute way so that

$$\gamma_j = 1 \quad \text{as} \quad m_j \to 0. \tag{16–19}$$

It follows that $a_j = m_j$ as $m_j \to 0$. Thus, for the chemical potential of a solute in the practical system, we have

$$\mu_j = \mu_j^{**} + RT \ln a_j. \tag{16–20}$$

The μ_j^{**} is the chemical potential the solute would have in a 1 molal solution if that solution behaved according to the ideal dilute rule. This standard state is called the ideal solution of unit molality. It is a hypothetical state of a system. According to Eq. (16–20) the practical activity measures the chemical potential of the substance relative to the chemical potential in this hypothetical ideal solution of unit molality. Equation (16–20) is applicable to either volatile or involatile solutes.

16–5 The Solute is Volatile

The equilibrium condition for the distribution of a volatile solute j between solution and vapor is $\mu_{j(\text{vap})} = \mu_{j(\text{liq})}$. Using Eq. (16–20) and assuming the vapor is ideal, we have

$$\mu_j^0 + RT \ln p_j = \mu_j^{**} + RT \ln a_j.$$

Since μ_j^0 and μ_j^{**} depend only on T and p and not on composition, we can define a constant K_j', which is independent of composition, by

$$RT \ln K_j' = -(\mu_j^0 - \mu_j^{**}).$$

The relation between p_j and a_j becomes

$$p_j = K_j' a_j. \tag{16–21}$$

The constant K_j' is a modified Henry's law constant. If K_j' is known, values of a_j can be computed immediately from the measured values of p_j. Dividing Eq. (16–21) by m_j, we obtain

$$\frac{p_j}{m_j} = K_j'\left(\frac{a_j}{m_j}\right). \tag{16–22}$$

Measured values of the ratio p_j/m_j are plotted as a function of m_j. The curve is extrapolated to $m_j = 0$. The extrapolated value of p_j/m_j is equal to K_j', since $(a_j/m_j) = 1$ as $m_j \to 0$. Thus

$$\left(\frac{p_j}{m_j}\right)_{m_j = 0} = K_j'.$$

Having obtained the value of K_j', the values of a_j are computed from the measured p_j by Eq. (16–21).

16–6 The Solute is Involatile; Colligative Properties and the Activity of the Solute

In Section 16–3 we related the colligative properties to the rational activity of the *solvent*. These properties can also be related to the activity of the solute. Symbols without subscripts refer to the solvent; symbols with a subscript 2 refer to the solute, except that the molality m of the *solute* will not bear a subscript. For simplicity we assume that only one solute is present. The chemical potentials are

$$\text{Solvent:} \qquad \mu = \mu^0 + RT \ln a,$$

$$\text{Solute:} \qquad \mu_2 = \mu_2^{**} + RT \ln a_2.$$

These are related by the Gibbs-Duhem equation, Eq. (11–93),

$$d\mu = -\frac{n_2}{n} d\mu_2 \qquad (T, p \text{ constant}).$$

Differentiating μ and μ_2 keeping T and p constant, we obtain

$$d\mu = RT\, d\ln a \quad \text{and} \quad d\mu_2 = RT\, d\ln a_2.$$

Using these values in the Gibbs-Duhem equation, we have

$$d\ln a = -\frac{n_2}{n} d\ln a_2.$$

But $n_2/n = Mm/1000$, where M is the molecular weight of the solvent, and m is the molality of the solute. Therefore,

$$d\ln a = -\frac{Mm}{1000} d\ln a_2, \tag{16-23}$$

which is the required relation between the activities of solvent and solute.

Freezing-point depression

Differentiating Eq. (16–12) and using the value for $d\ln a$ given by Eq. (16–23), we obtain

$$d\ln a_2 = -\frac{1000\,\Delta H^0_{\text{fus}}}{MRT^2 m} dT = \frac{d\theta}{K_f m(1 - \theta/T_0)^2},$$

where $K_f = MRT_0^2/1000\,\Delta H^0_{\text{fus}}$, and the freezing-point depression, $\theta = T_0 - T$, $d\theta = -dT$, have been introduced. If $\theta/T_0 \ll 1$, then

$$d\ln a_2 = \frac{d\theta}{K_f m}. \tag{16-24}$$

A similar equation could be derived for the boiling-point elevation.

As it stands, Eq. (16–24) is not very sensitive to deviations from ideality. To arrange it in terms of more responsive functions, we introduce the *osmotic coefficient*, $1 - j$, defined by

$$\theta = K_f m(1 - j). \tag{16-25}$$

In an ideal dilute solution, $\theta = K_f m$, so that $j = 0$. In a nonideal solution, j is not zero. Differentiating Eq. (16–25), we have

$$d\theta = K_f[(1 - j)\, dm - m\, dj].$$

Using Eq. (16–18), we set $a_2 = \gamma_2 m$; and differentiate $\ln a_2$:

$$d\ln a_2 = d\ln \gamma_2 + d\ln m = d\ln \gamma_2 + \frac{dm}{m}.$$

Using these two relations in Eq. (16–24), it becomes

$$d\ln \gamma_2 = -dj - \left(\frac{j}{m}\right) dm.$$

This equation is integrated from $m = 0$ to m; at $m = 0$, $\gamma_2 = 1$, and $j = 0$; we obtain

$$\int_0^{\ln \gamma_2} d \ln \gamma_2 = -\int_0^j dj - \int_0^m \left(\frac{j}{m}\right) dm,$$

$$\ln \gamma_2 = -j - \int_0^m \left(\frac{j}{m}\right) dm. \tag{16-26}$$

The integral in Eq. (16–26) is evaluated graphically. From experimental values of θ and m, j is calculated from Eq. (16–25); j/m is plotted versus m; the area under the curve is the value of the integral. After obtaining the value of γ_2, the activity a_2 is obtained from the relation $a_2 = \gamma_2 m$.

We have assumed that ΔH_{fus}^0 is independent of temperature and that θ is much less than T_0. In precision work, more elaborate equations, not restricted by these assumptions, are used. Any of the colligative properties can be interpreted in terms of the activity of the solute.

16–7 Activities and Reaction Equilibrium

If a chemical reaction takes place in a nonideal solution, the chemical potentials in the form given by Eq. (16–4) or (16–20) must be used in the equation of reaction equilibrium. The practical system, Eq. (16–20), is more commonly used. The condition of equilibrium becomes

$$\Delta G^{**} = -RT \ln K_a, \tag{16-27}$$

where ΔG^{**} is the standard free energy change, and K_a is the proper quotient of equilibrium activities. Since ΔG^{**} is a function only of T and p, K_a is a function only of T and p, and is independent of the composition. Since each activity has the form $a_i = \gamma_i m_i$, we can write

$$K_a = K_\gamma K_m, \tag{16-28}$$

where K_γ and K_m are proper quotients of activity coefficients and of molalities, respectively. Since the γ's depend on composition, Eq. (16–28) shows that K_m depends on composition. In dilute real solutions all the γ's approach unity, K_γ approaches unity, and K_m approaches K_a. Except when we are particularly interested in the evaluation of activity coefficients, we shall treat K_m as if it were independent of composition, since doing so greatly simplifies the discussion of equilibria.

In most elementary treatments of equilibria in solution, the equilibrium constant is usually written as a quotient of equilibrium concentrations expressed as molarities, K_c. It is possible to develop an entire system of activities and activity coefficients using molar rather than molal concentrations. We could write $a = \gamma_c c$, where c is the molar concentration and γ_c the corresponding activity coefficient; as c approaches zero, γ_c must approach unity. We will not dwell on the details of this system except to show that in dilute aqueous solution the systems based on molarity and on molality are nearly the same. We have shown, Eq. (14–23), that the molarity c_j and the molality m_j of a solute in dilute solution are related by $c_j = \rho m_j$, where ρ is the density of the pure solvent. At 25°C the density of water is 0.997044 gm/cm^3. The error made by replacing molalities by molarities is therefore insignificant in ordinary circumstances. The concomitant error in the standard free energy is well below the experimental error in that value. In more concentrated solutions the relation between c_j and m_j is not so simple, Eq. (14–22), and the two systems of activities are different.

Ordinarily for purposes of illustration we shall use molar concentrations in the equilibrium constant, realizing that to be precise we should use the activities. One misunderstanding which arises because of this replacement of activity by concentration should be avoided. The activity is sometimes regarded as if it were an "effective concentration." This is a legitimate formal point of view; however, it is deceptive in that it conveys the incorrect notion that activity is designed to measure the concentration of a substance in a mixture. The activity is designed for one purpose only, namely to provide a convenient *measure of the chemical potential* of a substance in a mixture. The connection between activity and concentration in dilute solutions is not that one is a measure of the other, but that *either* one is a measure of the chemical potential of the substance. It would be better to think of the concentration in an ideal solution as being the effective activity.

16–8 Activities in Electrolytic Solutions

The problem of defining activities is somewhat more complicated in electrolytic solutions than in solutions of nonelectrolytes. Solutions of strong electrolytes exhibit marked deviations from ideal behavior even at concentrations well below those at which a solution of a nonelectrolyte would behave in the ideal dilute way. The determination of activities and activity coefficients has a correspondingly greater importance for solutions of strong electrolytes. To simplify the notation as much as possible a subscript s will be used for the properties of the solvent; symbols without subscript refer to the solute; subscripts $+$ and $-$ refer to the properties of the positive and negative ions.

Consider a solution of an electrolyte which is completely dissociated into ions. By the additivity rule the free energy of the solution should be the sum of the free energies of the solvent, the positive, and the negative ions:

$$G = n_s\mu_s + n_+\mu_+ + n_-\mu_-. \tag{16–29}$$

If each mole of the electrolyte dissociates into v_+ positive ions and v_- negative ions, then $n_+ = v_+n$, and $n_- = v_-n$, where n is the number of moles of electrolyte in the solution. Equation (16–29) becomes

$$G = n_s\mu_s + n(v_+\mu_+ + v_-\mu_-). \tag{16–30}$$

If μ is the chemical potential of the electrolyte in the solution, then we should also have

$$G = n_s\mu_s + n\mu. \tag{16-31}$$

Comparing Eqs. (16–30) and (16–31), we see that

$$\mu = v_+\mu_+ + v_-\mu_-. \tag{16-32}$$

Let the total number of moles of ions produced by one mole of electrolyte be $v = v_+ + v_-$. Then the mean ionic chemical potential $\mu_\pm$ is defined by

$$v\mu_\pm = v_+\mu_+ + v_-\mu_-. \tag{16-33}$$

Now we can proceed in a purely formal way to define the various activities. We write†

$$\mu = \mu^0 + RT\ln a, \tag{16-34}$$

$$\mu_\pm = \mu_\pm^0 + RT\ln a_\pm, \tag{16-35}$$

$$\mu_+ = \mu_+^0 + RT\ln a_+, \tag{16-36}$$

$$\mu_- = \mu_-^0 + RT\ln a_-. \tag{16-37}$$

In these relations, a is the activity of the electrolyte, $a_\pm$ is the mean ionic activity, and a_+ and a_- are the individual ion activities. To define the various activities completely we require the additional relations

$$\mu^0 = v_+\mu_+^0 + v_-\mu_-^0, \tag{16-38}$$

$$v\mu_\pm^0 = v_+\mu_+^0 + v_-\mu_-^0. \tag{16-39}$$

First we work out the relation between a and $a_\pm$. From Eqs. (16–32) and (16–33) we have $\mu = v\mu_\pm$. Using the values for μ and $\mu_\pm$ from Eqs. (16–34) and (16–35), we get

$$\mu^0 + RT\ln a = v\mu_\pm^0 + vRT\ln a_\pm.$$

Using Eqs. (16–38) and (16–39) this reduces to

$$a = a_\pm^v. \tag{16-40}$$

Next we want the relation between $a_\pm$, a_+, and a_-. Using the values of $\mu_\pm$, μ_+, and μ_- given by Eqs. (16–35), (16–36), and (16–37) in Eq. (16–33), we obtain

$$v\mu_\pm^0 + vRT\ln a_\pm = v_+\mu_+^0 + v_-\mu_-^0 + RT(v_+\ln a_+ + v_-\ln a_-).$$

From this equation we subtract Eq. (16–39); then it reduces to

$$a_\pm^v = a_+^{v_+} a_-^{v_-}. \tag{16-41}$$

The mean ionic activity is the geometric mean of the individual ion activities.

---

† Since we are using molalities, for consistency we should write μ^{**} for the standard value of μ. This would make the symbolism too forbidding.

The various activity coefficients are defined by the relations

$$a = \gamma m, \tag{16-42}$$

$$a_\pm = \gamma_\pm m_\pm, \tag{16-43}$$

$$a_+ = \gamma_+ m_+, \tag{16-44}$$

$$a_- = \gamma_- m_-, \tag{16-45}$$

where $\gamma_\pm$ is the mean ionic activity coefficient, $m_\pm$ is the mean ionic molality, etc. Using the values of $a_\pm$, a_+, and a_- from Eqs. (16–43), (16–44), and (16–45) in Eq. (16–41), we obtain

$$\gamma_\pm^\nu m_\pm^\nu = \gamma_+^{\nu_+} \gamma_-^{\nu_-} m_+^{\nu_+} m_-^{\nu_-}.$$

We then require that

$$\gamma_\pm^\nu = \gamma_+^{\nu_+} \gamma_-^{\nu_-}, \tag{16-46}$$

$$m_\pm^\nu = m_+^{\nu_+} m_-^{\nu_-}; \tag{16-47}$$

These equations show that $\gamma_\pm$ and $m_\pm$ are also geometric means of the individual ionic quantities. In terms of the molality of the electrolyte we have

$$m_+ = \nu_+ m \qquad m_- = \nu_- m,$$

so that the mean ionic molality is

$$m_\pm = (\nu_+^{\nu_+} \nu_-^{\nu_-})^{1/\nu} m. \tag{16-48}$$

Knowing the formula of the electrolyte, we obtain $m_\pm$ immediately in terms of m.

Example. In a 1:1 electrolyte such as NaCl, or in a 2:2 electrolyte such as $MgSO_4$

$$\nu_+ = \nu_- = 1, \qquad \nu = 2, \qquad m_\pm = m.$$

In a 1:2 electrolyte such as Na_2SO_4

$$\nu_+ = 2, \qquad \nu_- = 1, \qquad \nu = 3, \qquad m_\pm = (2^2 \cdot 1^1)^{1/3} m = \sqrt[3]{4} m = 1.587 m.$$

The expression for the chemical potential in terms of the mean ionic activity, from Eqs. (16–34) and (16–40), is

$$\mu = \mu^0 + RT \ln a_\pm^\nu. \tag{16-49}$$

Using Eqs. (16–43) and (16–48) this becomes

$$\mu = \mu^0 + RT \ln [\gamma_\pm^\nu (\nu_+^{\nu_+} \nu_-^{\nu_-}) m^\nu],$$

which can be written in the form

$$\mu = \mu^0 + RT \ln (\nu_+^{\nu_+} \nu_-^{\nu_-}) + \nu RT \ln m + \nu RT \ln \gamma_\pm. \tag{16-50}$$

In Eq. (16–50), the second term on the right is a constant, evaluated from the formula of the electrolyte; the third term depends on the molality; the fourth can be determined from measurements of the freezing point, or any other colligative property of the solution.

16–9 Freezing-point Depression and the Mean Ionic Activity Coefficient

The relation between the freezing-point depression θ and the mean ionic activity coefficient is obtained easily. Writing Eq. (16–24) using a for the activity of the *solute*, we have

$$d \ln a = d\theta / K_f m. \tag{16–51}$$

But from Section 16–8, we have

$$a = a_{\pm}^{\nu} = \gamma_{\pm}^{\nu} m_{\pm}^{\nu} = \gamma_{\pm}^{\nu} (\nu_{+}^{\nu_{+}} \nu_{-}^{\nu_{-}}) m^{\nu}.$$

Then,

$$d \ln a = \nu \, d \ln m + \nu \, d \ln \gamma_{\pm}. \tag{16–52}$$

So that Eq. (16–51) becomes

$$\nu \, dm/m + \nu \, d \ln \gamma_{\pm} = d\theta / K_f m. \tag{16–53}$$

If the solution were ideal, then $\gamma_{\pm} = 1$, and Eq. (16–53) becomes

$$d\theta = \nu K_f \, dm,$$

$$\theta = \nu K_f m, \tag{16–54}$$

which shows that the freezing-point depression in a very dilute solution of an electrolyte is the value for a nonelectrolyte multiplied by ν, the number of ions produced by the dissociation of one mole of the electrolyte.

The osmotic coefficient for an electrolytic solution is defined by

$$\theta = \nu K_f m (1 - j). \tag{16–55}$$

With this definition of j, Eq. (16–53) becomes, after repetition of the algebra in Section 16–6,

$$\ln \gamma_{\pm} = -j - \int_0^m \left(\frac{j}{m} \right) dm, \tag{16–56}$$

which has the same form as Eq. (16–26).

Values of the mean ionic activity coefficients for several electrolytes in water at $25°C$ are given in Table 16–1. Figure 16–3 shows a plot of $\gamma_{\pm}$ versus $m^{1/2}$ for different electrolytes in water at $25°C$.

The values of $\gamma_{\pm}$ are nearly independent of the kind of ions in the compound so long as the compounds are of the same valence type. For example, KCl and NaBr have nearly the same activity coefficients at the same concentration, as do K_2SO_4

Table 16–1† Activity coefficients of strong electrolytes

m	0.001	0.005	0.01	0.05	0.1	0.5	1.0
HCl	0.966	0.928	0.904	0.830	0.796	0.758	0.809
NaOH	—	—	—	0.82	—	0.69	0.68
KOH	—	0.92	0.90	0.82	0.80	0.73	0.76
KCl	0.965	0.927	0.901	0.815	0.769	0.651	0.606
NaBr	0.966	0.934	0.914	0.844	0.800	0.695	0.686
H_2SO_4	0.830	0.639	0.544	0.340	0.265	0.154	0.130
K_2SO_4	0.89	0.78	0.71	0.52	0.43	—	—
$Ca(NO_3)_2$	0.88	0.77	0.71	0.54	0.48	0.38	0.35
$CuSO_4$	0.74	0.53	0.41	0.21	0.16	0.068	0.047
$MgSO_4$	—	—	0.40	0.22	0.18	0.088	0.064
$La(NO_3)_3$	—	—	0.57	0.39	0.33	—	—
$In_2(SO_4)_3$	—	—	0.142	0.054	0.035	—	—

† By permission from Wendell M. Latimer, *The Oxidation States of the Elements and Their Potentials in Aqueous Solutions.* 2nd ed., Prentice-Hall, Inc., Englewood Cliffs, N.J., 1952, pp. 354–356.

and $Ca(NO_3)_2$. In Section 16–10 we shall see that the theory of Debye and Hückel predicts that in a sufficiently dilute solution the mean ionic activity coefficient should depend only on the charges on the ions and their concentration, but not upon any other individual characteristics of the ions.

Any of the colligative properties could be used to determine the activity coefficients of a dissolved substance whether it is an electrolyte or nonelectrolyte. The freezing-point depression is much used, since this experiment requires somewhat less elaborate equipment than any of the others. It has the disadvantage that the

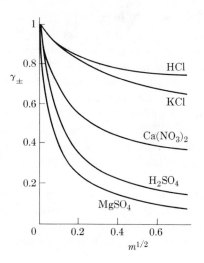

Fig. 16–3 Mean ionic activity coefficients as functions of $m^{1/2}$.

values of γ can be obtained only near the freezing point of the solvent. The measurement of vapor pressure does not have this drawback, but is more difficult to handle experimentally. In Chapter 17 the method of obtaining mean ionic activity coefficients from measurements of the emf of electrochemical cells is described. The emf method is easily handled experimentally, and it can be used at any convenient temperature.

16-10   The Debye-Hückel Theory of the Structure of Dilute Ionic Solutions

At this stage it is worthwhile to describe the constitution of ionic solutions in some detail. The solute in dilute solutions of nonelectrolytes is adequately described thermodynamically by the equation,

$$\mu = \mu^0 + RT \ln m. \tag{16-57}$$

The chemical potential is a sum of two terms: the first, μ^0, is independent of composition, and the second depends on the composition. Equation (16–57) is fairly good for most nonelectrolytes up to concentrations as high as $0.1\ m$, and for many others it does well at even higher concentrations. The simple expression in Eq. (16–57) is not adequate for electrolytic solutions; deviations are pronounced even at concentrations of $0.001\ m$. This is true even if Eq. (16–57) is modified to take account of the several ions produced.

To describe the behavior of an electrolyte in a dilute solution, the chemical potential must be written in the form, see Eq. (16–50),

$$\mu = \mu^0 + \nu RT \ln m + \nu RT \ln \gamma_\pm. \tag{16-58}$$

In Eq. (16–58) the second term on the right of Eq. (16–50) has been absorbed into the μ^0. The μ^0 is independent of the composition; the second and third terms depend on the composition.

The extra free energy represented by the term $\nu RT \ln \gamma_\pm$ in Eq. (16–58) is mainly the result of the energy of interaction of the electrical charges on the ions; since in one mole of the electrolyte there are νN_0 ions, this interaction energy is, on the average, $\mathbf{k}T \ln \gamma_\pm$ per ion, where the Boltzmann constant $\mathbf{k} = R/N_0$. The van der Waals forces acting between neutral particles of solvent and nonelectrolyte are weak and are effective only over very short distances, while the coulombic forces which act between ions and those between ions and neutral molecules of solvent are much stronger and act over greater distances. This difference in range of action accounts for the large deviations from ideality in ionic solutions even at high dilutions where the ions are far apart. Our object is to calculate this electrical contribution to the free energy.

For a model of the electrolyte solution we imagine that the ions are electrically charged, conducting spheres having a radius a, immersed in a solvent of dielectric constant ϵ. Let the charge on the ion be q. If the ion were not charged, $q = 0$, its μ could be represented by Eq. (16–57); since it is charged, its μ must be represented by Eq. (16–58). The extra term, which we are trying to calculate, must be the work expended in charging the ion, bringing q from zero to q. Let the electrical potential at the surface of the sphere be ψ_a, a function of q. By definition, the potential of the sphere is the work which must be expended to bring a unit positive charge from infinity to the surface of the sphere; if we bring a charge dq from infinity to the surface, the work will be $dW = \psi_a\, dq$. Integrating from zero to q, we obtain the work expended in charging the ion:

$$W = \int_0^q \psi_a\, dq, \qquad (16\text{–}59)$$

where W is the extra energy possessed by the ion in virtue of its charge; the free energy of an ion is greater than that of a neutral particle by W. This additional energy is made up of two contributions:

$$W = W_s + W_i. \qquad (16\text{–}60)$$

The energy required to charge an *isolated* sphere immersed in a dielectric medium is the *self-energy* of the charged sphere, W_s. Since W_s does not depend on the concentration of the ions, it will be absorbed in the value of μ^0. The additional energy beyond W_s needed to charge the ion in the presence of all the other ions is the interaction energy W_i, whose value depends very much on the concentration of the ions. It is W_i which we identify with the term, $\mathbf{k}T \ln \gamma_\pm$:

$$\mathbf{k}T \ln \gamma_\pm = W_i = W - W_s. \qquad (16\text{–}61)$$

The potential of an isolated conducting sphere immersed in a medium having a dielectric constant ϵ is given by the formula from classical electrostatics: $\psi_a = q/\epsilon a$. Using this value in the integral of Eq. (16–59), we obtain for W_s

$$W_s = \int_0^q \frac{q}{\epsilon a}\, dq = \frac{q^2}{2\epsilon a}. \qquad (16\text{–}62)$$

Having this value of W_s, we can obtain a value for W_i if we succeed in calculating W. To calculate W we must first calculate ψ_a; see Eq. (16–59). Before doing the calculation we can guess reasonably that W_i will be negative. Consider a positive ion; it attracts negative ions and repels other positive ions. As a result negative ions will be, on the average, a little closer to the positive ion than will be the other positive ions. This in turn gives the ion a lower energy than it would have if it were not charged; since we are interested in the energy relative to that of the uncharged species, W_i is negative. In 1923 P. Debye and E. Hückel succeeded in obtaining a value of ψ_a. The following is an abbreviated version of the method they used.

We locate the origin of a spherical coordinate system at the center of a positive ion; Fig. 16–4. Consider a point P at a distance r from the center of the ion. The potential ψ at the point P is related to the charge density ρ, the charge per unit volume,

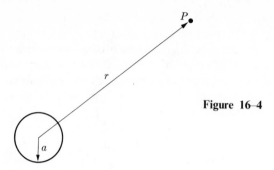

Figure 16–4

by the Poisson equation

$$\frac{1}{r^2}\frac{d}{dr}\left(r^2\frac{d\psi}{dr}\right) = -\frac{4\pi\rho}{\epsilon}.\tag{16–63}$$

If ρ can be expressed as a function of either ψ or r, then Eq. (16–63) can be integrated to yield ψ as a function of r, from which we can get ψ_a immediately.

To calculate ρ we proceed as follows. Let c_+ and c_- be the molar concentrations of positive and negative ions, respectively. If z_+ and z_- are the valences (complete with sign) of the ions and e is the magnitude of the charge on the electron, then the charge on one mole of positive ions is N_0ez_+, the positive charge in one liter is $N_0ez_+c_+$, and the positive charge in one cubic centimeter is $N_0ez_+c_+/1000$. The charge density ρ is the total charge, positive plus negative, in one cubic centimeter; therefore

$$\rho = \frac{N_0ec_+z_+}{1000} + \frac{N_0ec_-z_-}{1000} = \frac{N_0e}{1000}(c_+z_+ + c_-z_-).$$

If the electrical potential at P is ψ, then the potential energies of the positive and negative ions at P are $ez_+\psi$ and $ez_-\psi$, respectively. Debye and Hückel assumed that the distribution of the ions is a Boltzmann distribution; Section 4–11. Then

$$c_+ = c_+^0 e^{-ez_+\psi/kT} \quad \text{and} \quad c_- = c_-^0 e^{-ez_-\psi/kT},$$

where c_+^0 and c_-^0 are the concentrations in the region where $\psi = 0$; but where $\psi = 0$, the distribution is uniform and the solution must be electrically neutral; ρ must be zero. This requires that

$$c_+^0 z_+ + c_-^0 z_- = 0.$$

Putting the values of c_+ and c_- in the equation for ρ yields

$$\rho = \frac{N_0e}{1000}(c_+^0 z_+ e^{-ez_+\psi/kT} + c_-^0 z_- e^{-ez_-\psi/kT}).$$

Assuming that $ez\psi/kT \ll 1$, the exponentials are expanded in series: $e^{-x} \approx 1 - x$.

This reduces ρ to

$$\rho = \frac{N_0 e}{1000}\left[c_+^0 z_+ + c_-^0 z_- - \frac{e\psi}{kT}(c_+^0 z_+^2 + c_-^0 z_-^2) \right].$$

The condition of electrical neutrality drops out the first two terms; the ionic strength in a solution μ, is defined by

$$\mu = \tfrac{1}{2}\sum_i c_i z_i^2, \tag{16–64}$$

where the sum is over all the different kinds of ions in the solution. The value of ρ becomes

$$\rho = -\frac{2N_0 e^2}{1000kT}\mu\psi.$$

We define a quantity, κ^2:

$$\kappa^2 = \frac{8\pi N_0 e^2}{1000\epsilon kT}\mu; \tag{16–65}$$

then $\rho = -(\epsilon/4\pi)\kappa^2\psi$, and the Poisson equation reduces to

$$\frac{1}{r^2}\frac{d}{dr}\left(r^2\frac{d\psi}{dr}\right) - \kappa^2\psi = 0. \tag{16–66}$$

If we substitute $\psi = v/r$ in Eq. (16–66), it reduces to

$$\frac{d^2 v}{dr^2} - \kappa^2 v = 0, \tag{16–67}$$

which has the solution†

$$v = Ae^{-\kappa r} + Be^{\kappa r},$$

where A and B are arbitrary constants. The value of ψ is

$$\psi = A\frac{e^{-\kappa r}}{r} + B\frac{e^{\kappa r}}{r}. \tag{16–68}$$

As $r \to \infty$, the second term on the right approaches infinity.‡ The potential must remain finite as $r \to \infty$, so this second term cannot be part of the physical solution; therefore, we set $B = 0$ and obtain

$$\psi = A\frac{e^{-\kappa r}}{r}. \tag{16–69}$$

---

† The reader should verify this by substitution and work out the transformation of Eq. (16–66) into (16–67) in detail.

‡ Verify using L'Hospital's rule.

Expanding the exponential in series and retaining only the first two terms, we have

$$\psi = A\left(\frac{1 - \kappa r}{r}\right) = \frac{A}{r} - A\kappa. \tag{16-70}$$

If the ionic strength is zero, then $\kappa = 0$, and the potential at point P should be that due to the central positive ion only; $\psi = z_+ e/\epsilon r$. But when $\kappa = 0$, Eq. (16–70) reduces to $\psi = A/r$; hence, $A = z_+ e/\epsilon$; Eq. (16–70) becomes:

$$\psi = \frac{z_+ e}{\epsilon r} - \frac{z_+ e\kappa}{\epsilon}. \tag{16-71}$$

At $r = a$, we have

$$\psi_a = \frac{z_+ e}{\epsilon a} - \frac{z_+ e\kappa}{\epsilon}. \tag{16-72}$$

If, with the exception of our central positive ion, all the other ions in the solution are fully charged, then the work to charge this positive ion in the presence of all the others is, Eq. (16–59),

$$W_+ = \int_0^q \psi_a \, dq;$$

but $q = z_+ e$, so that $dq = e \, dz_+$. Using Eq. (16–72) for ψ_a, we obtain

$$W_+ = \int_0^{z_+} \left(\frac{z_+ e^2}{\epsilon a} - \frac{z_+ e^2 \kappa}{\epsilon}\right) dz_+ = \left(\frac{e^2}{\epsilon a} - \frac{e^2 \kappa}{\epsilon}\right) \int_0^{z_+} z_+ \, dz_+,$$

$$W_+ = \frac{(ez_+)^2}{2\epsilon a} - \frac{(ez_+)^2 \kappa}{2\epsilon}, \tag{16-73}$$

where the first term is the self-energy $W_{s,+}$, and the second is the interaction energy $W_{i,+}$, the extra free energy of a single positive ion which is due to the presence of the others. Using Eq. (16–61), we have

$$kT \ln \gamma_+ = -\frac{(ez_+)^2 \kappa}{2\epsilon}. \tag{16-74}$$

For a negative ion we would get

$$kT \ln \gamma_- = -\frac{(ez_-)^2 \kappa}{2\epsilon}. \tag{16-75}$$

The mean ionic activity coefficient can be calculated using Eq. (16–46):

$$\gamma_\pm^\nu = \gamma_+^{\nu_+} \gamma_-^{\nu_-}.$$

Taking logarithms, we obtain

$$\nu \ln \gamma_\pm = \nu_+ \ln \gamma_+ + \nu_- \ln \gamma_-.$$

Using Eqs. (16–74) and (16–75) this becomes

$$v \ln \gamma_\pm = -\frac{e^2 \kappa}{2\epsilon kT}(v_+ z_+^2 + v_- z_-^2).$$

Since the electrolyte itself is electrically neutral, we must have

$$v_+ z_+ + v_- z_- = 0:$$

Multiplying by z_+: $v_+ z_+^2 = -v_- z_+ z_-$
Multiplying by z_-: $v_- z_-^2 = -v_+ z_+ z_-$
Adding: $v_+ z_+^2 + v_- z_-^2 = -(v_+ + v_-)z_+ z_- = -vz_+ z_-.$

Using this result we obtain finally:

$$\ln \gamma_\pm = \frac{e^2 \kappa}{2\epsilon kT} z_+ z_-. \tag{16–76}$$

Converting to common logarithms and introducing the value of κ from Eq. (16–65), we obtain

$$\log_{10} \gamma_\pm = \left[\frac{(2\pi N_0)^{1/2}}{2.303}\left(\frac{e^2}{10\epsilon kT}\right)^{3/2} \right] z_+ z_- \mu^{1/2}. \tag{16–77}$$

The factor enclosed in the brackets is made up of universal constants, and the value of ϵ and T. At a specified temperature, ϵ and T are constant, and Eq. (16–77) can be written as

$$\log_{10} \gamma_\pm = A z_+ z_- \mu^{1/2}. \tag{16–78}$$

In water solution at 25°C, $A = 0.50$.

Either of Eqs. (16–77) or (16–78) is the *Debye-Hückel limiting law*. The limiting law predicts that the logarithm of the mean ionic activity coefficient should be a linear function of the ionic strength and that the slope of the line should be proportional to the product of the valences of the positive and negative ions. (The slope is negative, since z_- is negative.) These predictions are confirmed by experiment in dilute solutions of strong electrolytes. Figure 16–5 shows the variation of $\log_{10} \gamma_\pm$ with μ; the solid curves are the experimental data; the dashed lines are the values predicted by the limiting law, Eq. (16–78).

The approximations required in the theory restrict its validity to solutions which are very dilute. In practice deviations from the limiting law become appreciable in the concentration range from 0.005 to 0.01 molar. More accurate equations have been derived which extend the theory to slightly higher concentrations. However, as yet there is no satisfactory theoretical equation which can predict the behavior in solutions of concentration higher than 0.01 M.

The Debye-Hückel theory provides an accurate representation of the limiting behavior of the activity coefficient in dilute ionic solutions. In addition, it yields a picture of the structure of the ionic solution. We have alluded to the fact that the

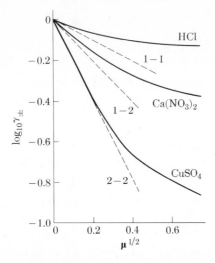

Fig. 16–5 $\log_{10} \gamma_{\pm}$ versus $\mu^{1/2}$.

negative ions cluster a little closer to a positive ion than do positive ions, which are pushed away. In this sense every ion is surrounded by an atmosphere of oppositely charged ions; the total charge on this atmosphere is equal, but opposite in sign, to that on the ion. The mean radius of the ionic atmosphere is given by $1/\kappa$, which is called *the Debye length*. Since κ is proportional to the ionic strength, at high ionic strengths the atmosphere is closer to the ion than at low ionic strengths. This concept of the ionic atmosphere and the mathematics associated with it have been extraordinarily fruitful in clarifying many aspects of the behavior of electrolytic solutions.

The concept of the ionic atmosphere can be made clearer by calculating the charge density as a function of the distance from the ion. By combining the final expression for the charge density in terms of ψ with Eq. (16–69) and the value of A, we obtain

$$\rho = -\frac{z_+ e\kappa^2}{4\pi} \frac{e^{-\kappa r}}{r}. \tag{16–79}$$

The total charge contained in a spherical shell bounded by spheres of radii r and $r + dr$ is the charge density multiplied by the volume of the shell, $4\pi r^2\, dr$:

$$-z_+ e\kappa^2 re^{-\kappa r}\, dr.$$

By integrating this quantity from zero to infinity we obtain the total charge on the atmosphere which is $-z_+ e$. The fraction of this total charge which is in the spherical shell, per unit width dr of the shell, we will call $f(r)$. Then

$$f(r) = \kappa^2 re^{-\kappa r}. \tag{16–80}$$

The function $f(r)$ is the distribution function for the charge in the atmosphere. A plot of $f(r)$ versus r is shown in Fig. 16–6. The maximum in the curve appears at

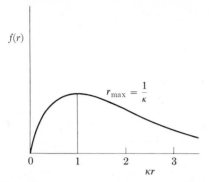

Fig. 16–6 Charge distribution in the ionic atmosphere.

$r_{max} = 1/\kappa$, the Debye length. In an electrolyte of a symmetrical valence type, $1:1$, $2:2$, etc., we may say that $f(r)$ represents the probability per unit width dr of finding the balancing ion in the spherical shell at the distance r from the central ion. In solutions of high ionic strength the mate to the central ion is very close, $1/\kappa$ is small; at lower ionic strengths $1/\kappa$ is large and the mate is far away.

16–11   Equilibria in Ionic Solutions

From the Debye-Hückel limiting law, Eq. (16–78), we find a negative value of $\ln \gamma_\pm$, which confirms the physical argument that the interaction with other ions lowers the free energy of an ion in an electrolytic solution. This lower free energy means that the ion is more stable in solution than it would be if it were not charged. The extra stability is measured by the term, $\mathbf{k}T \ln \gamma_\pm$, in the expression for the chemical potential. Now we examine the consequences of this extra stability in two simple cases; the ionization of a weak acid, and the solubility of a sparingly soluble salt.

Consider the dissociation equilibrium of a weak acid, HA:

$$HA \rightleftharpoons H^+ + A^-.$$

The equilibrium constant is the quotient of the activities,

$$K = \frac{a_{H^+} a_{A^-}}{a_{HA}}. \tag{16–81}$$

By definition, we have

$$a_{H^+} = \gamma_+ m_{H^+}, \qquad a_{A^-} = \gamma_- m_{A^-}, \qquad a_{HA} = \gamma_{HA} m_{HA},$$

so that

$$K = \left(\frac{\gamma_+ \gamma_-}{\gamma_{HA}} \right) \frac{m_{H^+} m_{A^-}}{m_{HA}} = \frac{\gamma_\pm^2}{\gamma_{HA}} \cdot \frac{m_{H^+} m_{A^-}}{m_{HA}},$$

where we have used the relation, $\gamma_+ \gamma_- = \gamma_\pm^2$. If the total molality of the acid is m, and the degree of dissociation is α, then

$$m_{H^+} = \alpha m, \qquad m_{A^-} = \alpha m, \qquad m_{HA} = (1 - \alpha)m.$$

Then,

$$K = \frac{\gamma_\pm^2 \alpha^2 m}{\gamma_{HA}(1 - \alpha)}. \tag{16-82}$$

If the solution is dilute, we may set $\gamma_{HA} = 1$, since HA is an uncharged species. Also if K is small, $1 - \alpha \approx 1$. Then, Eq. (16–82) yields

$$\alpha = \left(\frac{K}{m}\right)^{1/2} \frac{1}{\gamma_\pm}. \tag{16-83}$$

If we ignored the ionic interactions, we would set $\gamma_\pm = 1$, and calculate $\alpha_0 = (K/m)^{1/2}$. Then Eq. (16–83) becomes

$$\alpha = \alpha_0/\gamma_\pm. \tag{16-84}$$

From the limiting law, $\gamma_\pm < 1$; hence the correct value of α given by Eq. (16–84) is greater than the rough value α_0, which ignored the ionic interactions. The stabilization of the ions by the presence of the other ions shifts the equilibrium to produce more ions; hence the degree of dissociation is increased.

If the solution is dilute enough in ions, $\gamma_\pm$ can be obtained from the limiting law, Eq. (16–78), which for a 1:1 electrolyte can be written

$$\gamma_\pm = 10^{-0.50(\alpha_0 m)^{1/2}} = e^{-1.15(\alpha_0 m)^{1/2}},$$

where the ionic strength $\mu = \alpha_0 m$. The value of α_0 can be used to compute μ, since α and α_0 are not greatly different. Using this expression, Eq. (16–84) becomes

$$\alpha = \alpha_0 e^{1.15(\alpha_0 m)^{1/2}} = \alpha_0[1 + 1.15(\alpha_0 m)^{1/2}].$$

In the last equality, the exponential has been expanded in series. The computation for 0.1 m acetic acid, $K = 1.75 \times 10^{-5}$, shows that the degree of dissociation is increased by about 4%. The effect is small because the dissociation does not produce many ions.

If a large amount of an inert electrolyte, one which does not contain either H^+ or A^- ions, is added to the solution of the weak acid, then a comparatively large effect on the dissociation is produced. Consider a solution of a weak acid in 0.1 m KCl, for example. The ionic strength of this solution is too large to use the limiting law, but the value of $\gamma_\pm$ can be estimated from Table 16–1. The table shows that for 1:1 electrolytes the value of $\gamma_\pm$ in 0.1 m solution is about 0.8. We may assume that this is a reasonable value for H^+ and A^- ions in the 0.1 m KCl solution. Then by Eq. (16–84),

$$\alpha = \alpha_0/0.8 = 1.25\alpha_0.$$

Thus, the presence of a large amount of inert electrolyte exerts an appreciable influence, *salt effect*, on the degree of dissociation. The salt effect is larger the higher the concentration of the electrolyte.

Consider the equilibrium of a slightly soluble salt, such as silver chloride, with its ions:

$$AgCl(s) \rightleftharpoons Ag^+ + Cl^-.$$

The solubility product constant is

$$K_{sp} = a_{Ag^+} a_{Cl^-} = (\gamma_+ m_+)(\gamma_- m_-).$$

If s is the solubility of the salt in moles per kilogram of water, then $m_+ = m_- = s$, and

$$K_{sp} = \gamma_\pm^2 s^2.$$

If s_0 is the solubility calculated ignoring ionic interaction, then $s_0^2 = K_{sp}$, and we have

$$s = s_0/\gamma_\pm,$$

which shows that the solubility is increased by the ionic interaction. By the same reasoning as we used in discussing the dissociation of a weak acid, we can show that in 0.1 m solution of an inert electrolyte such as KNO_3 the solubility would be increased by 25%. This increase in solubility produced by an inert electrolyte is sometimes called "salting in." The effect of an inert electrolyte on the solubility of a salt such as $BaSO_4$ would be much larger because of the larger charges on the Ba^{2+} and SO_4^{2-} ions. The salt effect on solubility produced by an inert electrolyte should not be confused with the *decrease* in solubility effected by an electrolyte which contains an ion in common with the sparingly soluble salt. In addition to acting in the opposite sense, the "common ion" effect is enormous compared with the effect of an inert electrolyte.

Problems

16-1. The *apparent* value of K_f in sucrose $(C_{12}H_{22}O_{11})$ solutions of various concentrations is

m	0.10	0.20	0.50	1.00	1.50	2.00
K_f	1.88	1.90	1.96	2.06	2.17	2.30

a) Calculate the activity of water in each solution.

b) Calculate the activity coefficient of water in each solution.

c) Plot the values of a and γ against the mole fraction of water in the solution.

d) Calculate the activity and the activity coefficient of sucrose in a 1 molal solution.

16–2. The Henry's law constant for chloroform in acetone at 35.17°C is 6.64×10^{-3} if the vapor pressure is in millimeters, and concentration of chloroform is in mole fraction. The partial pressure of chloroform at several values of mole fraction is:

x_{CHCl_3}	0.059	0.123	0.185
p_{CHCl_3}, mm	9.2	20.4	31.9

If $a = \gamma x$, and $\gamma \to 1$ as $x \to 0$, calculate the values of a and γ for chloroform in the three solutions.

16–3. At the same concentrations as in Problem 16–2, the partial pressures of acetone are 323.2, 299.3, and 275.4 mm, respectively. The vapor pressure of pure acetone is 344.5 mm. Calculate the activities of acetone and the activity coefficients in these three solutions; $a = \gamma x; \gamma \to 1$ as $x \to 1$.

16–4. A regular binary liquid solution is defined by the equation

$$\mu_i = \mu_i^0 + RT \ln x_i + w(1 - x_i)^2,$$

where w is a constant.

a) What is the significance of the function μ_i^0?

b) Express $\ln \gamma_i$ in terms of w; γ_i is the *rational* activity coefficient.

c) At 25°C, $w = 77.5$ cal/mole for mixtures of benzene and carbon tetrachloride. Calculate γ for CCl_4 in solutions with $x_{CCl_4} = 0, 0.25, 0.50, 0.75$, and 1.0.

16–5. The freezing-point depression of aqueous solutions of NaCl is:

m	0.001	0.002	0.005	0.01	0.02	0.05	0.1
θ, °C	0.003676	0.007322	0.01817	0.03606	0.07144	0.1758	0.3470

a) Calculate the value of j for each of these solutions.

b) Plot j/m versus m, and evaluate $-\log_{10} \gamma_\pm$ for each solution. $K_f = 1.858$ deg·kgm·mole^{-1}. From the Debye-Hückel limiting law it can be shown that $\int_0^{0.001} (j/m) \, dm = 0.0226$. [G. Scatchard and S. S. Prentice, *J.A.C.S.*, **55**, 4355 (1933).]

16–6. From the data in Table 16–1, calculate the activity of the electrolyte and the mean activity of the ions in 0.1 molal solutions of

a) KCl, b) H_2SO_4, c) $CuSO_4$, d) $La(NO_3)_3$, e) $In_2(SO_4)_3$.

16–7. a) Calculate the mean ionic molality, $m_\pm$, in 0.05 molal solutions of $Ca(NO_3)_2$, NaOH, and $MgSO_4$.

b) What is the ionic strength of each of the solutions in (a)?

16–8. Calculate the values of $1/\kappa$ at 25°C, in 0.01 and 1 molal solutions of KBr. For water, $\epsilon = 81$.

16–9. a) What is the total probability of finding the balancing ion at a distance greater than $1/\kappa$ from the central ion?

b) What is the radius of the sphere around the central ion such that the probability of finding the balancing ion within the sphere is 0.5?

16–10. Using the limiting law, calculate the value of $\gamma_\pm$ in 10^{-4} and 10^{-3} molar solutions of HCl, $CaCl_2$, and $ZnSO_4$.

16–11. At 25°C the dissociation constant for acetic acid is 1.75×10^{-5}. Using the limiting law, calculate the degree of dissociation in 0.1 m solution. Compare this value with the rough value obtained by ignoring ionic interaction.

16–12. For silver chloride at 25°C, $K_{sp} = 1.56 \times 10^{-10}$. Using the data in Table 16–1, estimate the solubility of AgCl in 0.001 m, 0.01 m, 0.1 m, and 1.0 m KNO_3 solution. Plot $\log_{10} s$ against $m^{1/2}$.

16–13. At 25°C for MgF_2, $K_{sp} = 7 \times 10^{-9}$. Calculate the solubility in moles/kgm of water in

a) water,   b) 0.01 m NaF,   c) 0.01 m $Mg(NO_3)_2$.

Chapter Seventeen

Equilibria in
Electrochemical Cells

17–1 Definitions

An electrochemical cell is a device which can produce electrical work in the surroundings. For example, the commercial dry cell is a sealed cylinder with two brass connecting terminals protruding from it. One terminal is stamped with a plus sign and the other with a minus sign. If the two terminals are connected to a small motor, electrons flow through the motor from the negative to the positive terminal of the cell. Work is produced in the surroundings and a chemical reaction, the *cell reaction*, occurs within the cell. Before deriving the thermodynamic relations for such a system we examine some fundamental equations of electrostatics.

The electrical potential of a point in space is defined as the work expended in bringing a unit positive charge from infinity, where the electrical potential is zero, to the point in question. Thus if $\mathscr{V}$ is the electrical potential at the point and W_{el} the work required to bring a charge Q from infinity to that point, then

$$\mathscr{V} = W_{el}/Q. \tag{17–1}$$

Similarly, if $\mathscr{V}_1$ and $\mathscr{V}_2$ are the electrical potentials of two points in space, and W_1 and W_2 are the corresponding quantities of work required to bring the charge Q to these points, then

$$W_1 + W_{12} = W_2, \tag{17–2}$$

where W_{12} is the work to bring Q from point 1 to point 2. This relation exists since the same quantity of work must be expended to bring Q to point 2, if we bring it directly, W_2, or bring it first to point 1 and then to point 2, $W_1 + W_{12}$. Therefore,

$W_{12} = W_2 - W_1$, and, from Eq. (17–1),

$$\mathscr{V}_2 - \mathscr{V}_1 = W_{12}/Q. \qquad (17\text{–}3)$$

The difference in electrical potential between two points is the work expended in taking a unit positive charge from point 1 to point 2.

The definition of the electrical potential difference between two points is based on classical electrostatics, where an electrical charge is a thing in itself. In a real situation, electrical charge is associated with material bodies, electrons, or ions of some sort. It is beyond the scope of this treatment to discuss the justification for applying Eq. (17–3) in a real situation. It can be shown that the definition of electrical potential difference in Eq. (17–3) has physical meaning in a real situation if the two points 1 and 2 are located in regions having the same chemical composition. In the dry cell the two brass terminals have the same chemical composition, the measurable potential difference between them has the meaning given by Eq. (17–3).

Applying Eq. (17–3) to the transfer of an infinitesimal quantity of charge, we obtain the element of work expended:

$$dW_{el} = \mathscr{E}\,dQ, \qquad (17\text{–}4)$$

where $\mathscr{E}$ has been written for the potential difference $\mathscr{V}_2 - \mathscr{V}_1$.

17–2 The Chemical Potential of Charged Species

The escaping tendency of a charged particle, an ion or an electron, in a phase depends on the electrical potential of that phase. Clearly, if we impress a large negative electrical potential on a piece of metal, the escaping tendency of negative particles will be increased. To find the relation between the electrical potential and the escaping tendency, the chemical potential, we consider a system of two balls M and M′ of the same metal. Let their electrical potentials be $\mathscr{V}$ and $\mathscr{V}'$. If we transfer a number of electrons carrying a charge, dQ, from M to M′, the work expended on the system is given by Eq. (17–4): $dW_{el} = (\mathscr{V}' - \mathscr{V})\,dQ$. The work *produced* is $-dW_{el}$. If the transfer is done reversibly, then by Eq. (10–13), the work produced is equal to the decrease in free energy of the system; $-dW_{el} = -dG$, so that

$$dG = (\mathscr{V}' - \mathscr{V})\,dQ.$$

But, in terms of the chemical potential of the electrons, $\bar{\mu}_{e-}$, if dn moles of electrons were transferred, we have

$$dG = \bar{\mu}'_{e-}\,dn - \bar{\mu}_{e-}\,dn.$$

The dn moles of electrons carry a negative charge $dQ = -\mathscr{F}\,dn$, where the charge per equivalent $\mathscr{F} = 96{,}490$ coulombs/equivalent; for our purposes, 96,500 coulombs/equivalent is sufficiently accurate. The two equations yield, after division by dn

$$\bar{\mu}'_{e-} - \bar{\mu}_{e-} = -\mathscr{F}(\mathscr{V}' - \mathscr{V}),$$

which rearranges to

$$\bar{\mu}_{e-} = \bar{\mu}'_{e-} + \mathscr{F}\mathscr{V}' - \mathscr{F}\mathscr{V}.$$

Let $\bar{\mu}_{e-}$ be the chemical potential of the electrons in M when $\mathscr{V}$ is zero; then, $\mu_{e-} = \bar{\mu}'_{e-} + \mathscr{F}\mathscr{V}'$. Subtracting this equation from the preceding one, we obtain

$$\bar{\mu}_{e-} = \mu_{e-} - \mathscr{F}\mathscr{V}. \tag{17-5}$$

Equation (17–5) is the relation between the escaping tendency of the electrons, $\bar{\mu}_{e-}$, in a phase and the electrical potential of the phase, $\mathscr{V}$. The escaping tendency is a linear function of $\mathscr{V}$. Note that Eq. (17–5) shows that if $\mathscr{V}$ is negative, $\bar{\mu}_{e-}$ is larger than when $\mathscr{V}$ is positive.

By a similar argument, it may be shown that for any charged species in a phase

$$\bar{\mu}_i = \mu_i + z_i\mathscr{F}\mathscr{V}, \tag{17-6}$$

where z_i is the valence, including the sign, of the charged species. For electrons, $z_{e-} = -1$, so Eq. (17–6) would reduce to Eq. (17–5). Equation (17–6) divides the chemical potential $\bar{\mu}_i$ of a charged species into two terms; the first term, μ_i, is the "chemical" contribution to the escaping tendency. The chemical contribution is produced by the chemical environment in which the charged species exists, and is the same in two phases of the same chemical composition since it is a function only of T, p, and composition. The second term, $z_i\mathscr{F}\mathscr{V}$, is the "electrical" contribution to the escaping tendency; it depends on the electrical condition of the phase which is manifested by the value of $\mathscr{V}$.

Because it is convenient to divide the chemical potential into these two contributions, $\bar{\mu}_i$, sometimes called the *electrochemical potential*, has been introduced to preserve μ_i for the ordinary chemical potential. For ions in aqueous solution we assign $\mathscr{V} = 0$ in the solution; then $\bar{\mu}_i = \mu_i$, and we can use the ordinary μ_i for these ions. This assignment is justified by the fact that the value of $\mathscr{V}$ in the electrolytic solution will drop out of our calculations; we have no way of determining its value, and so it might as well be zero and save us a little algebraic labor.

In the metallic parts of our system we do not throw away the electrical potential, since we wish to compare the electrical potentials of two different wires of the same composition, the two terminals of the cell. However, within a *single* piece of a metal it is evident that the division of the chemical potential into a "chemical" part and an "electrical" part is a purely arbitrary one, justified only by convenience. Since the "chemical" part of the escaping tendency arises from the interactions of the electrically charged particles which compose any piece of matter, there is no way to determine in a single piece of matter where the "chemical" part leaves off and the "electrical" part begins.

To make the arbitrary division of $\bar{\mu}_i$ as convenient as possible, we assign the "chemical" part of the $\bar{\mu}_{e-}$ the most convenient value, zero, in every metal. Thus, in every metal,

$$\mu_{e-} = 0. \tag{17-7}$$

So, for electrons in every metal,† Eq. (17–5) becomes

$$\bar{\mu}_{e^-} = -\mathscr{F}\mathscr{V}. \tag{17–8}$$

The arbitrary definition in Eq. (17–8) simplifies the form of the chemical potential of the metal ion in a metal. Within any metal there exists an equilibrium between metal atoms M, metal ions M^{+z}, and electrons:

$$M \rightleftharpoons M^{+z} + ze^-.$$

The equilibrium condition is

$$\mu_M = \bar{\mu}_{M^{+z}} + z\bar{\mu}_{e^-}.$$

Using Eq. (17–6) for $\bar{\mu}_{M^{+z}}$ and Eq. (17–8) for $\bar{\mu}_{e^-}$, we obtain

$$\mu_M = \mu_{M^{+z}} + z\mathscr{F}\mathscr{V} - z\mathscr{F}\mathscr{V} \quad \text{or} \quad \mu_M = \mu_{M^{+z}}.$$

For a *pure* metal at 1 atm and 25°C, we have $\mu_M^0 = \mu_{M^{+z}}^0$; by our earlier convention that $\mu^0 = 0$ for elements under these conditions, we obtain

$$\mu_{M^{+z}}^0 = 0. \tag{17–9}$$

The "chemical" part of the escaping tendency of the metal ion is zero in a pure metal under standard conditions; then using Eq. (17–6),

$$\bar{\mu}_{M^{+z}} = z\mathscr{F}\mathscr{V}. \tag{17–10}$$

Equations (17–8) and (17–10) are the conventional values of the chemical potential of the electrons and the metal ions *within any pure metal*.

17–3 Electrode Potentials; the Standard Hydrogen Electrode

Any electrochemical cell has two electrodes. If the cell is reversible, then at each of these electrodes an electrochemical equilibrium is established. Knowing the chemical potentials of the species taking part in the equilibrium, the electrical potential of each electrode, the electrode potential, can be calculated. Having the electrode potentials of the two electrodes, we can obtain the difference between them, the cell emf.

The hydrogen electrode is illustrated in Fig. 17–1. Purified hydrogen gas is passed over a platinum electrode which is in contact with an acid solution. At the electrode surface the equilibrium

$$H^+(aq) + e^-(Pt) \rightleftharpoons \tfrac{1}{2}H_2(g)$$

is established. The equilibrium condition is the usual one;

$$\mu_{H^+(aq)} + \bar{\mu}_{e^-(Pt)} = \tfrac{1}{2}\mu_{H_2(g)}.$$

---

†   The convention in Eq. (17–7) is convenient for the discussion of electrochemical cells; it would not do for a discussion of some other phenomena; e.g., Volta potentials.

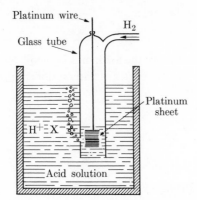

Platinum wire

Glass tube

H_2

Platinum sheet

$H^+ X$

Acid solution

Fig. 17–1 The hydrogen electrode.

Using Eq. (17–8) for $\bar{\mu}_{e-(\text{Pt})}$ and the usual forms of $\mu_{H^+(\text{aq})}$ and μ_{H_2}, we obtain

$$\mu^0_{H^+} + RT\ln a_{H^+} - \mathscr{F}\mathscr{V}_{H^+,H_2} = \tfrac{1}{2}\mu^0_{H_2} + \tfrac{1}{2}RT\ln f,$$

where f is the fugacity of H_2, a_{H^+} is the activity of the hydrogen ion in the aqueous solution. Then

$$\mathscr{V}_{H^+,H_2} = \frac{\mu^0_{H^+} - \tfrac{1}{2}\mu^0_{H_2}}{\mathscr{F}} - \frac{RT}{\mathscr{F}}\ln\frac{f^{1/2}}{a_{H^+}}. \qquad (17\text{–}11)$$

If the fugacity of the gas is unity, and the activity of H^+ in solution is unity, the electrode is in its standard state, and the potential is the standard potential, $\mathscr{V}^0_{H^+,H_2}$. Letting $f = 1$ and $a_{H^+} = 1$ in Eq. (17–11), we obtain

$$\mathscr{V}^0_{H^+,H_2} = \frac{\mu^0_{H^+} - \tfrac{1}{2}\mu^0_{H_2}}{\mathscr{F}} = \frac{\mu^0_{H^+}}{\mathscr{F}}, \qquad (17\text{–}12)$$

since $\mu^0_{H_2} = 0$. Subtracting Eq. (17–12) from Eq. (17–11) yields

$$\mathscr{V}_{H^+,H_2} = \mathscr{V}^0_{H^+,H_2} - \frac{RT}{\mathscr{F}}\ln\frac{f^{1/2}}{a_{H^+}}, \qquad (17\text{–}13)$$

which is the Nernst equation for the hydrogen electrode; it relates the electrode potential to a_{H^+} and f. Now the electrons in platinum of the standard hydrogen electrode are in a definite standard state. We choose the standard state of zero free energy for electrons as this state in the standard hydrogen electrode. Since, by Eq. (17–8), $\bar{\mu}_{e-} = -\mathscr{F}\mathscr{V}$, we have†

$$\bar{\mu}_{e-(\text{SHE})} = 0 \quad \text{and} \quad \mathscr{V}^0_{H^+,H_2} = 0. \qquad (17\text{–}14)$$

The free energy of the electrons in any metal is measured relative to the value in the standard hydrogen electrode. The assignment in Eq. (17–14) immediately yields the

---

† SHE = Standard Hydrogen Electrode.

conventional zero of free energy for ions in aqueous solution. Using Eq. (17–14) in Eq. (17–12), we obtain

$$\mu^0_{H^+} = 0. \tag{17–15}$$

Standard free energies of other ions in aqueous solution are measured relative to that of the H^+ ion, which has a standard free energy equal to zero.

The Nernst equation, Eq. (17–13), for the hydrogen electrode becomes

$$\mathscr{V}_{H^+,H_2} = -\frac{RT}{\mathscr{F}} \ln \frac{f^{1/2}}{a_{H^+}}. \tag{17–16}$$

Note that the argument of the logarithm is a proper quotient of fugacity and activity for the electrode reaction if the presence of the electrons is ignored in constructing the quotient. From Eq. (17–16) we can calculate the potential, relative to SHE, of a hydrogen electrode at which f_{H_2} and a_{H^+} have any values whatsoever.

17–4 Recapitulation of the Standard States

The choice of standard states of zero free energy and the general forms for the chemical potentials of various species are summarized here. In all the standard states, the temperature is 25°C, the pressure 1 atm.

1. Elements in their stable state of aggregation, 25°C, 1 atm pressure:

$$\mu^0_{elements} = 0.$$

2. Ions in aqueous solution:
   General form:

$$\mu = \mu^0 + RT \ln a. \tag{17–17}$$

   Standard state of zero free energy:

$$\mu^0_{H^+} = 0. \tag{17–15}$$

3. a) Electrons in any metal:
      General form:

$$\bar{\mu}_{e^-} = -\mathscr{F}\mathscr{V}. \tag{17–8}$$

   Standard state of zero free energy:

$$\bar{\mu}_{e^- (SHE)} = 0. \tag{17–14}$$

   b) Metal ions in a pure metal:
      General form:

$$\bar{\mu}_{M^{+z}} = z\mathscr{F}\mathscr{V}. \tag{17–10}$$

4. Conventional zero for electrode potential:

$$\mathscr{V}^0_{SHE} = \mathscr{V}^0_{H^+,H_2} = 0. \tag{17–14}$$

With these equations we can calculate the potential of any electrode from the ordinary condition of chemical equilibrium. Until we discuss the temperature dependence of the electrode potential we will speak only of the values at 25°C.

17–5 The Metal–Metal Ion Electrode

One of the most common kinds of electrode is the metal–metal ion electrode. Suppose a bar of zinc is in contact with a solution containing zinc ion at an activity $a_{Zn^{2+}}$. Then at the interface between the metal and the solution the equilibrium

$$Zn^{2+}(aq) \rightleftharpoons Zn^{2+}(zinc) \tag{17–18}$$

is established. The equilibrium condition is the equality of the chemical potentials of Zn^{2+} in the zinc and in the solution:

$$\mu_{Zn^{2+}(aq)} = \bar{\mu}_{Zn^{2+}(Zn)}.$$

But, by Eq. (17–10), $\bar{\mu}_{Zn^{2+}(Zn)} = 2\mathscr{F}\mathscr{V}_{Zn^{2+},Zn}$, so that

$$\mathscr{V}_{Zn^{2+},Zn} = \frac{\mu_{Zn^{2+}(aq)}}{2\mathscr{F}}, \tag{17–19}$$

which shows that a measurement of $\mathscr{V}_{Zn^{2+},Zn}$ relative to SHE yields the value of the free energy of the aqueous zinc ion. The quantity $\mu_{Zn^{2+}(aq)}/2$ is the free energy *per equivalent* of the aqueous zinc ion relative to hydrogen ion at unit activity ($\mu_{H^+}^0 = 0$).

Using Eq. (17–17) for the $\mu_{Zn^{2+}(aq)}$, we obtain

$$\mathscr{V}_{Zn^{2+},Zn} = \frac{\mu_{Zn^{2+}}^0}{2\mathscr{F}} + \frac{RT}{2\mathscr{F}} \ln a_{Zn^{2+}}.$$

The standard potential of the zinc–zinc ion electrode is defined by

$$\mathscr{V}_{Zn^{2+},Zn}^0 = \frac{\mu_{Zn^{2+}}^0}{2\mathscr{F}}, \tag{17–20}$$

so that

$$\mathscr{V}_{Zn^{2+},Zn} = \mathscr{V}_{Zn^{2+},Zn}^0 - \frac{RT}{2\mathscr{F}} \ln \frac{1}{a_{Zn^{2+}}}. \tag{17–21}$$

[The sign change in Eq. (17–21) is to compensate for changing $\ln a_{Zn^{2+}}$ to $\ln (1/a_{Zn^{2+}})$.] Equation (17–21) is the Nernst equation for the zinc–zinc ion electrode. The $\mathscr{V}_{Zn^{2+},Zn}^0 = -0.763$ volt. This value implies, through Eq. (17–20), that the standard free energy of zinc ion is highly negative relative to H^+ ion.

Example. Since $\mathscr{V}_{Zn^{2+},Zn}^0 = -0.763$ volt, by Eq. (17–20), we have
$$\mu_{Zn^{2+}}^0 = 2\mathscr{F}\mathscr{V}_{Zn^{2+},Zn}^0 = 2(96,500)(-0.763) \text{ volt·coulomb}$$
$$= -147.3 \text{ kJ/mole} = -35.20 \text{ kcal/mole}.$$

Chemically we can understand the negative potential of the zinc electrode in terms of the relative ease with which zinc ions enter the solution from the metal and leave electrons behind on the metal. Add two electrons (in zinc) to each side of the equilibrium (17–18); it becomes

$$Zn^{2+}(aq) + 2e^-(Zn) \rightleftharpoons Zn. \tag{17–22}$$

This does not change the free-energy relationship, since it adds the same free energy to both sides of the equation. The equilibrium (17–22) shows up the fact that the greater the stability of the Zn^{2+} ion in solution (that is, the lower the value of $\mu_{Zn^{2+}(aq)}$) the more electrons will be left on the zinc bar. The more electrons on the bar the more negative will be the potential of the bar.

The equilibrium at any metal–metal ion electrode can be written in analogy to (17–22);

$$M^{+n}(aq) + ne^-(M) \rightleftharpoons M. \tag{17–23}$$

The Nernst equation is, by analogy to Eq. (17–21),

$$\mathscr{V}_{M^{+n},M} = \mathscr{V}^0_{M^{+n},M} - \frac{RT}{n\mathscr{F}} \ln \frac{1}{a_{M^{+n}}}. \tag{17–24}$$

In Eq. (17–24) the proper quotient of activities in the equilibrium (17–23) appears as the argument of the logarithm. The activity of the pure metal is unity; the activity of the electrons does not appear in the proper quotient because it is included implicitly in the value of $\mathscr{V}_{M^{+n},M}$; see Eq. (17–8).

We compare briefly the values of $\mathscr{V}^0$ for several metal–metal ion electrodes:

$$Na^+(aq) + e^- \rightleftharpoons Na, \quad \mathscr{V}^0 = -2.714$$
$$Zn^{2+}(aq) + 2e^- \rightleftharpoons Zn, \quad \mathscr{V}^0 = -0.763,$$
$$Fe^{2+}(aq) + 2e^- \rightleftharpoons Fe, \quad \mathscr{V}^0 = -0.440,$$
$$Pb^{2+}(aq) + 2e^- \rightleftharpoons Pb, \quad \mathscr{V}^0 = -0.126,$$
$$Cu^{2+}(aq) + 2e^- \rightleftharpoons Cu, \quad \mathscr{V}^0 = +0.337,$$
$$Ag^+(aq) + e^- \rightleftharpoons Ag, \quad \mathscr{V}^0 = +0.7991.$$

From this list it is apparent that Na^+ ion has the lowest value of standard free energy, while Ag^+ ion has the highest value. This has the chemical implication that sodium is an active metal, silver a noble metal.

17–6 The Electrochemical Cell; Conventional Representation and Electromotive Force

Consider an electrochemical cell having two electrodes. We imagine that the cell is constructed so that if it is placed before us, one of the electrodes is on our right and one is on our left. Then the *electromotive force* (emf) of the cell is defined by

$$\mathscr{E} = \mathscr{V}_{\text{right}} - \mathscr{V}_{\text{left}}. \tag{17-25}$$

This is again a purely arbitrary definition; it has the consequence that $\mathscr{E}$ may be positive or negative, depending on which way we happened to turn the cell. This causes no difficulty as long as we do not turn the cell around (either physically or mentally) in the middle of our problem.

Suppose our cell consists of a zinc–zinc ion electrode and a hydrogen electrode. The two electrode compartments are separated by a partition which physically segregates the acid solution from the solution of the zinc salt but permits electrical contact between the two solutions; a tube of agar jelly saturated with KCl or NH_4NO_3, a *salt bridge*, is often used as a partition; Fig. 17–2. We choose the zinc–zinc ion electrode as the one on the right; then the cell emf is

$$\mathscr{E} = \mathscr{V}_{Zn^{2+},Zn} - \mathscr{V}_{H^+,H_2}. \tag{17-26}$$

The cell is symbolized by the notation

$$Pt|H_2|H^+ \parallel Zn^{2+}|Zn, \tag{17-27}$$

where the single vertical bar signifies a phase boundary, and the double vertical bar signifies the junction between the two electrode systems (*half-cells*) if the junction does not contribute to the emf of the cell. Later we will consider cells in which a junction potential contributes to the cell emf. Usually the activities of the species are also specified in the conventional symbolism; we would write instead of (17–27)

$$Pt|H_2(f)|H^+(a_{H^+}) \parallel Zn^{2+}(a_{Zn^{2+}})|Zn. \tag{17-28}$$

For the present we will be content with the simpler notation of (17–27).

If we had put the hydrogen electrode on the right (call this cell, "cell prime"), then the symbolism would be

$$Zn|Zn^{2+} \parallel H^+|H_2|Pt, \tag{17-29}$$

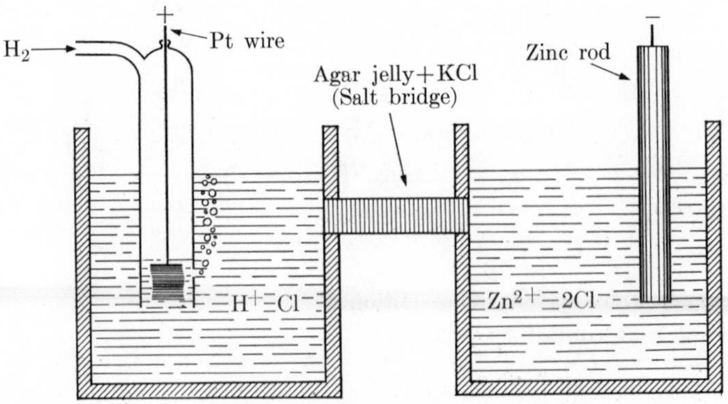

Fig. 17–2 The zinc–hydrogen cell.

and the emf of cell prime is

$$\mathscr{E}' = \mathscr{V}_{H^+,H_2} - \mathscr{V}_{Zn^{2+},Zn} = -\mathscr{E}. \tag{17-30}$$

It makes no difference which way we choose; but once we choose in a particular problem we must not change until we finish.

17–7 The Cell Reaction

From the equilibria involved we can easily deduce the cell reaction for the zinc–hydrogen cell. Choosing the zinc electrode on the right, we have

$$Pt|H_2|H^+ \parallel Zn^{2+}|Zn.$$

So that the cell emf is

$$\mathscr{E} = \mathscr{V}_{Zn^{2+},Zn} - \mathscr{V}_{H^+,H_2}.$$

Then *both* equilibria are written *with the electrons on the reactant side.*

Right electrode: $Zn^{2+}(aq) + 2e^-(Zn) \rightleftharpoons Zn,$

Left electrode: $2H^+(aq) + 2e^-(Pt) \rightleftharpoons H_2(g).$

Subtracting these equilibria in the same sense as the potentials, right − left, we obtain

$$Zn^{2+}(aq) + H_2(g) + 2e^-(Zn) \rightleftharpoons Zn + 2H^+(aq) + 2e^-(Pt).$$

This reaction summarizes the electrical and the chemical changes which take place in the cell. The two changes can be written separately:

$$2e^-(Zn) \rightarrow 2e^-(Pt), \qquad\qquad \Delta G_{el} = 2\bar{\mu}_{e^-}(Pt) - 2\bar{\mu}_{e^-}(Zn),$$

$$Zn^{2+}(aq) + H_2(g) \rightarrow Zn + 2H^+(aq), \qquad \Delta G_{chem} = \Delta G.$$

If two moles of electrons are consumed at the zinc electrode by combination with aqueous zinc ion and produced at the platinum by dissociation of H_2 then one mole of Zn^{2+} and one mole of H_2 are consumed to yield one mole of Zn and 2 moles of H^+. This chemical reaction which accompanies the electrical transformation is the *cell reaction.* Since the cell as a whole is in equilibrium, the condition $\Delta G_{el} + \Delta G_{chem} = 0$ must be satisfied. Therefore,

$$2\bar{\mu}_{e^-}(Pt) - 2\bar{\mu}_{e^-}(Zn) + \Delta G = 0.$$

Using Eq. (17–8) for $\bar{\mu}_{e^-}$ and $\mathscr{E} = \mathscr{V}_{Zn^{2+},Zn} - \mathscr{V}_{H^+,H_2}$, this becomes

$$2\mathscr{F}\mathscr{E} + \Delta G = 0, \qquad \Delta G = -2\mathscr{F}\mathscr{E}, \tag{17-31}$$

where ΔG is the reaction free energy for the cell reaction. This relation connects the emf of the cell to the reaction free energy. For any cell reaction involving n

equivalents,

$$\Delta G = -n\mathscr{F}\mathscr{E} \qquad (17\text{--}32)$$

is the fundamental relation between the cell emf and the reaction free energy. The cell emf is proportional to $\Delta G/n$; consequently it is an intensive property of the cell and does not depend on the size of the cell.

If the cell had been written with the hydrogen electrode on the right, then we would obtain as the cell reaction the reverse of the one obtained above; it would still be true that $\Delta G' = -n\mathscr{F}\mathscr{E}'$ but $\Delta G' = -\Delta G$ and $\mathscr{E}' = -\mathscr{E}$.

The spontaneity of a reaction can be judged by the corresponding cell emf. Through Eq. (17–32) it follows that if ΔG is negative, $\mathscr{E}$ is positive. Thus we have the criteria:

ΔG	$\mathscr{E}$	Cell reaction is
$-$	$+$	Spontaneous
$+$	$-$	Nonspontaneous
0	0	At equilibrium

17–8 The Nernst Equation

For any chemical reaction the reaction free energy is written

$$\Delta G = \Delta G^0 + RT \ln Q, \qquad (17\text{--}33)$$

where Q is the proper quotient of activities. Combining this with Eq. (17–32), we obtain

$$-n\mathscr{F}\mathscr{E} = \Delta G^0 + RT \ln Q.$$

The standard emf of the cell is defined by

$$-n\mathscr{F}\mathscr{E}^0 = \Delta G^0. \qquad (17\text{--}34)$$

Introducing this value of ΔG^0 and dividing by $-n\mathscr{F}$, we obtain

$$\mathscr{E} = \mathscr{E}^0 - \frac{RT}{n\mathscr{F}} \ln Q, \qquad (17\text{--}35a)$$

$$\mathscr{E} = \mathscr{E}^0 - \frac{2.303\, RT}{n\mathscr{F}} \log_{10} Q, \qquad (17\text{--}35b)$$

$$\mathscr{E} = \mathscr{E}^0 - \frac{0.05915}{n} \log_{10} Q \quad \text{(at 25°C)}. \qquad (17\text{--}35c)$$

Equations (17–35) are various forms of the Nernst equation for the cell. The Nernst equation relates the cell emf to a standard value, $\mathscr{E}^0$, and the activities of the species taking part in the cell reaction. Knowing the values of $\mathscr{E}^0$ and the activities, we can calculate the cell emf.

17–9 Standard Free Energies and Half-cell EMFs

We consider the zinc–hydrogen cell with the zinc electrode on the right, as before. Then the cell emf is

$$\mathscr{E} = \mathscr{V}_{Zn^{2+},Zn} - \mathscr{V}_{H^+,H_2}.$$

If the hydrogen electrode is the standard hydrogen electrode, then $\mathscr{V}_{H^+,H_2} = \mathscr{V}^0_{H^+,H_2} = 0$ and the cell emf is

$$\mathscr{E}_{Zn^{2+},Zn} = \mathscr{V}_{Zn^{2+},Zn}. \tag{17–36}$$

The corresponding cell reaction is

$$Zn^{2+}(aq) + H_2(f = 1) \rightleftharpoons Zn + 2H^+(a_{H^+} = 1). \tag{17–37}$$

In the SHE the equilibrium is

$$H_2(f = 1) \rightleftharpoons 2H^+(a_{H^+} = 1) + 2e^-_{SHE}.$$

All of the species in this reaction have zero free energy by our conventional choice. Subtracting this equation from Eq. (17–37), we obtain

$$Zn^{2+}(aq) + 2e^-_{SHE} \rightleftharpoons Zn, \tag{17–38}$$

which is simply a shorthand way of writing Eq. (17–37). Since the emf of this cell depends only upon the conventional free energies of the zinc and zinc ion, it is called a *half-cell* emf; the reaction (17–38) is called a half-cell reaction. Using Eq. (17–21) in Eq. (17–36) we obtain

$$\mathscr{E}_{Zn^{2+},Zn} = \mathscr{V}^0_{Zn^{2+},Zn} - \frac{RT}{2\mathscr{F}} \ln \frac{1}{a_{Zn^{2+}}}.$$

We may define the standard half-cell emf by

$$\mathscr{E}^0_{Zn^{2+},Zn} = \mathscr{V}^0_{Zn^{2+},Zn}, \tag{17–39}$$

so that the Nernst equation becomes

$$\mathscr{E}_{Zn^{2+},Zn} = \mathscr{E}^0_{Zn^{2+},Zn} - \frac{RT}{2\mathscr{F}} \ln \frac{1}{a_{Zn^{2+}}}. \tag{17–40}$$

This half-cell emf is related to the reaction free energy of (17–38) by

$$\Delta G = -2\mathscr{F}\mathscr{E}_{Zn^{2+},Zn}. \tag{17–41}$$

(Keep in mind that (17–38) is an abbreviation of (17–37) and that electrons in SHE have zero free energy.)

By measuring the emf of the cell at various concentrations of Zn^{2+} ion, we can determine $\mathscr{E}^0_{Zn^{2+},Zn}$. This standard emf is tabulated along with the standard emfs of other half-cells. Such a table of half-cell emfs is equivalent to a table of standard free energies from which we can calculate values of equilibrium constants for chemical reactions in solution.

The situation may be summarized as follows: if the half-cell reaction is written with the electrons in SHE *on the reactant side*,† then any electrode system can be represented by

$$\text{oxidized species} + ne^-_{\text{SHE}} \rightleftharpoons \text{reduced species}.$$

Then the following relations obtain:

$$\mathscr{E} = \mathscr{V}, \tag{17–42}$$

$$\Delta G = -n\mathscr{F}\mathscr{E}, \tag{17–43}$$

$$\mathscr{E} = \mathscr{E}^0 - \frac{RT}{n\mathscr{F}}\ln Q. \tag{17–44}$$

Values of $\mathscr{E}^0 = \mathscr{V}^0$ for a number of half-cell reactions are given in Table 17–1.

17–10   Kinds of Electrodes

At this point it is worthwhile to describe the various kinds of electrode systems. We write the half-cell reaction in its conventional form, the electrons being in SHE, the state of zero free energy. In the Nernst equation, since we will always write the electrons as reactants, the $\mathscr{E}$ and $\mathscr{E}^0$ are equal to the $\mathscr{V}$ and the $\mathscr{V}^0$, Eq. (17–42). We will symbolize these electrodes as right-hand ones.

Gas–ion electrode

In the gas–ion electrode an inert collector of electrons, platinum or graphite, is in contact with the gas and an ion (see Fig. 17–1).

Reaction: $2H^+ + 2e^- \rightleftharpoons H_2(g)$

Symbol: $H^+|H_2|Pt$

Nernst equation: $\mathscr{E}_{H^+,H_2} = \mathscr{E}^0_{H^+,H_2} - \dfrac{RT}{2\mathscr{F}}\ln\dfrac{f_{H_2}}{a^2_{H^+}}.$

---

†   If the electrons in SHE are written on the product side, this reverses the sign of ΔG and therefore the sign of $\mathscr{E}$. But $\mathscr{V}$ does not depend on how we write the reaction; $\mathscr{V}$ is a definite physical quantity. Thus, if electrons are written as products of the reaction the emf is $\mathscr{E}' = -\mathscr{V} = -\mathscr{E}$. The most extensive compilation of half-cell emfs (Latimer's) tabulates values of $\mathscr{E}'$. Equations (17–43) and (17–44) are still correct for $\mathscr{E}'$, if in Eq. (17–44) we use $Q' = 1/Q$; that is, if we simply construct the activity quotient properly.

Table 17–1† Standard electrode potentials at 25°C

Electrode reaction	$\mathcal{V}^0$, volts
$K^+ + e^- = K$	-2.925
$Na^+ + e^- = Na$	-2.714
$H_2 + 2e^- = 2H^-$	-2.25
$Al^{3+} + 3e^- = Al$	-1.66
$Zn(CN)_4^{2-} + 2e^- = Zn + 4CN^-$	-1.26
$ZnO_2^{2-} + 2H_2O + 2e^- = Zn + 4OH^-$	-1.216
$Zn(NH_3)_4^{2+} + 2e^- = Zn + 4NH_3$	-1.03
$Sn(OH)_6^{2-} + 2e^- = HSnO_2^- + H_2O + 3OH^-$	-0.90
$Fe(OH)_2 + 2e^- = Fe + 2OH^-$	-0.877
$2H_2O + 2e^- = H_2 + 2OH^-$	-0.828
$Fe(OH)_3 + 3e^- = Fe + 3OH^-$	-0.77
$Zn^{2+} + 2e^- = Zn$	-0.763
$Ag_2S + 2e^- = 2Ag + S^{2-}$	-0.69
$Fe^{2+} + 2e^- = Fe$	-0.440
$Bi_2O_3 + 3H_2O + 6e^- = 2Bi + 6OH^-$	-0.44
$PbSO_4 + 2e^- = Pb + SO_4^{2-}$	-0.356
$Ag(CN)_2^- + e^- = Ag + 2CN^-$	-0.31
$Ni^{2+} + 2e^- = Ni$	-0.250
$AgI + e^- = Ag + I^-$	-0.151
$Sn^{2+} + 2e^- = Sn$	-0.136
$Pb^{2+} + 2e^- = Pb$	-0.126
$Cu(NH_3)_4^{2+} + 2e^- = Cu + 4NH_3$	-0.12
$Fe^{3+} + 3e^- = Fe$	-0.036
$2H^+ + 2e^- = H_2$	0.000
$AgBr + e^- = Ag + Br^-$	$+0.095$
$HgO(r) + H_2O + 2e^- = Hg + 2OH^-$	$+0.098$
$Sn^{4+} + 2e^- = Sn^{2+}$	$+0.15$
$AgCl + e^- = Ag + Cl^-$	$+0.222$
$Hg_2Cl_2 + 2e^- = 2Hg + 2Cl^-$	$+0.2676$
$Cu^{2+} + 2e^- = Cu$	$+0.337$
$Ag(NH_3)_2^+ + e^- = Ag + 2NH_3$	$+0.373$
$Hg_2SO_4 + 2e^- = 2Hg + SO_4^{2-}$	$+0.6151$
$Fe^{3+} + e^- = Fe^{2+}$	$+0.771$
$Ag^+ + e^- = Ag$	$+0.7991$
$O_2 + 4H^+ + 4e^- = 2H_2O$	$+1.229$
$PbO_2 + SO_4^{2-} + 4H^+ + 2e^- = PbSO_4 + 2H_2O$	$+1.685$
$O_3 + 2H^+ + 2e^- = O_2 + H_2O$	$+2.07$

† Values in this table are printed by permission from W. M. Latimer, *The Oxidation States of the Elements and Their Potentials in Aqueous Solution.* 2nd ed., Prentice-Hall, Inc., Englewood Cliffs, N.J., 1952.

For the hydrogen electrode $\mathscr{E}^0_{H^+,H_2} = 0$; also since hydrogen is very nearly ideal, $f = p$, and the Nernst equation becomes

$$\mathscr{E}_{H^+,H_2} = -\frac{RT}{2\mathscr{F}}\ln\frac{p_{H_2}}{a^2_{H^+}}. \qquad (17\text{–}45)$$

The chlorine electrode:

Reaction: $Cl_2 + 2e^- \rightleftharpoons 2Cl^-$

Symbol: $Cl^-|Cl_2|graphite$

Nernst equation: $\mathscr{E}_{Cl_2,Cl^-} = \mathscr{E}^0_{Cl_2,Cl^-} - \dfrac{RT}{2\mathscr{F}}\ln\dfrac{a^2_{Cl^-}}{p_{Cl_2}}. \qquad (17\text{–}46)$

Metal–metal ion electrode

This electrode consists of a bar of metal immersed in a solution containing the metal ion:

Reaction: $M^{+n} + ne^- \rightleftharpoons M$

Symbol: $M^{+n}|M$

Nernst equation: $\mathscr{E}_{M^{+n},M} = \mathscr{E}^0_{M^{+n},M} - \dfrac{RT}{n\mathscr{F}}\ln\dfrac{1}{a_{M^{+n}}}. \qquad (17\text{–}47)$

Metal–insoluble salt-anion electrode

This electrode is sometimes called an "electrode of the second kind." It consists of a bar of metal immersed in a solution containing a solid insoluble salt of the metal and anions of the salt. There are a dozen common electrodes of this kind; we cite only two examples.

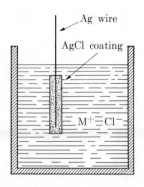

Fig. 17–3 Silver–silver chloride electrode.

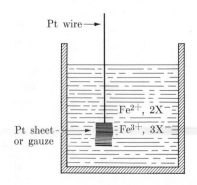

Fig. 17–4 The ferric–ferrous electrode.

The silver–silver chloride electrode (Fig. 17–3):

Reaction: $AgCl(s) + e^- \rightleftharpoons Ag(s) + Cl^-(aq)$

Symbol: $Cl^- | AgCl | Ag$

Nernst equation: $\mathscr{E}_{AgCl,Ag,Cl^-} = \mathscr{E}^0_{AgCl,Ag,Cl^-} - \dfrac{RT}{\mathscr{F}} \ln a_{Cl^-}.$ (17–48)

The activity of AgCl does not appear in the quotient, since AgCl is a pure solid. Note that the emf is sensitive to the concentration of chloride ion.

The mercury–mercuric oxide electrode: A pool of mercury is covered with a paste of solid HgO and a solution of a base:

Reaction: $HgO(s) + H_2O(l) + 2e^- \rightleftharpoons Hg(l) + 2OH^-(aq)$

Symbol: $OH^- | HgO | Hg$

Nernst equation: $\mathscr{E}_{HgO,Hg,OH^-} = \mathscr{E}^0_{HgO,Hg,OH^-} - \dfrac{RT}{2\mathscr{F}} \ln \dfrac{a^2_{OH^-}}{a_{H_2O}}.$ (17–49)

Ordinarily the solution is dilute enough so that $a_{H_2O} = 1$.

"Oxidation–reduction" electrodes

Any electrode involves oxidation and reduction in its operation, but these electrodes have had that superfluous phrase attached to them. An oxidation–reduction electrode has an inert metal collector, usually platinum, immersed in a solution which contains two ions of the same element in different states of oxidation.

The ferric–ferrous ion electrode (Fig. 17–4):

Reaction: $Fe^{3+} + e^- \rightleftharpoons Fe^{2+}$

Symbol: $Fe^{3+}, Fe^{2+} | Pt$

Nernst equation: $\mathscr{E}_{Fe^{3+},Fe^{2+}} = \mathscr{E}^0_{Fe^{3+},Fe^{2+}} - \dfrac{RT}{\mathscr{F}} \ln \dfrac{a_{Fe^{2+}}}{a_{Fe^{3+}}}.$ (17–50)

Note that a comma separates two species in the aqueous phase in the conventional symbol.

The stannic–stannous electrode:

Reaction: $Sn^{4+} + 2e^- \rightleftharpoons Sn^{2+}$

Symbol: $Sn^{4+}, Sn^{2+} | Pt$

Nernst equation: $\mathscr{E}_{Sn^{4+},Sn^{2+}} = \mathscr{E}^0_{Sn^{4+},Sn^{2+}} - \dfrac{RT}{2\mathscr{F}} \ln \dfrac{a_{Sn^{2+}}}{a_{Sn^{4+}}}.$ (17–51)

There are somewhat more complicated kinds of electrodes, but their treatment presents no difficulty. These four kinds encompass all the common electrodes met in practice.

17–11   Equilibrium Constants from Standard Half-cell EMFs

The equilibrium constant of a reaction is related to the standard reaction free energy by $\Delta G^0 = -RT \ln K$; so from Eq. (17–34), we obtain

$$\mathscr{E}^0 = \frac{RT}{n\mathscr{F}} \ln K, \tag{17–52}$$

and, at 25°C

$$\mathscr{E}^0 = \frac{0.05915}{n} \log_{10} K, \tag{17–53}$$

or

$$\log_{10} K = \frac{n\mathscr{E}^0}{0.05915}. \tag{17–54}$$

It follows from Eq. (17–54) that by combining the two appropriate standard half-cell emfs, it is possible to compute the equilibrium constant for any reaction whatsoever in an electrolytic solution. As we shall see, it is not necessary that the reaction be an oxidation–reduction reaction. First we examine examples of oxidation–reduction equilibria.

Example 1. $2Fe^{3+} + Sn^{2+} \rightleftharpoons 2Fe^{2+} + Sn^{4+}$.

This reaction is broken down into two half-cell reactions as follows: choose the oxidized species Fe^{3+} on the left-hand side of the equation and write its equilibrium with the appropriate reduced species; select the appropriate $\mathscr{E}^0$ from Table 17–1:

$$Fe^{3+} + e^- \rightleftharpoons Fe^{2+} \qquad \mathscr{E}^0_{Fe^{3+},Fe^{2+}} = 0.771.$$

Then write the equilibrium involving the remaining species:

$$Sn^{4+} + 2e^- \rightleftharpoons Sn^{2+} \qquad \mathscr{E}^0_{Sn^{4+},Sn^{2+}} = 0.15.$$

Both half-reactions are written with the electrons on the reactant side. The half-cell reactions are now balanced so that each involves the same number of electrons. In this example, the first reaction is multiplied by two. Subtracting the reactions yields the cell reaction; subtracting the potentials in the same sense yields the standard cell emf:

$$2Fe^{3+} + Sn^{2+} \rightleftharpoons 2Fe^{2+} + Sn^{4+} \qquad \mathscr{E}^0 = 0.771 - 0.15 = 0.62.$$

Note that the half-cell emf is *not* multiplied by two; the emf is an *intensive* property. The fact that the standard cell emf is positive means that the equilibrium position will be predominantly on the product side of the reaction as written. The equilibrium constant is now obtained from the relation $RT \ln K = n\mathscr{F}\mathscr{E}^0$, where n is the number of equivalents of reaction as written, the number of electrons in either of the half-cell reactions in the second writing; in this case $n = 2$. Since the $\mathscr{E}^0$'s are given at 25°C, we have

$$\log_{10} K = \frac{n\mathscr{E}^0}{0.05915} = \frac{2(+0.62)}{0.05915} = 20.96,$$

$$K = 9.1 \times 10^{20}.$$

This very large value for K means that stannous ion will reduce ferric ion quantitatively and leave only infinitesimal traces of Fe^{3+} present at equilibrium.

Before citing additional examples, a summary of the procedure is given here so that its application can be followed in the succeeding examples.

Step 1. Break the cell reaction into two half-cell reactions. Write both half-cell reactions with the electrons on the reactant side. Balance the half-cell reactions so that the same number of electrons, n, appears in each.

Step 2. Subtract the two half-cell reactions to obtain the cell reaction; subtract the $\mathscr{E}^0$'s in the same sense to obtain the standard emf of the cell.

Step 3. The equilibrium constant is given by

$$\log_{10} K = \frac{n\mathscr{F}\mathscr{E}^0}{2.303\,RT},$$

where R is in $J \cdot deg^{-1} \cdot mole^{-1}$, $\mathscr{F}$ is in coulombs/equivalent, $\mathscr{E}^0$ in volts. At 25°C, $\log_{10} K = n\mathscr{E}^0/0.05915$.

Example 2. $2MnO_4^- + 6H^+ + 5H_2C_2O_4 \rightleftharpoons 2Mn^{2+} + 8H_2O + 10CO_2$.

The half-reactions are (choose the oxidized species, MnO_4^-, on reactant side for the first half-reaction)

$$MnO_4^- + 8H^+ + 5e^- \rightleftharpoons Mn^{2+} + 4H_2O, \qquad \mathscr{E}^0 = 1.51,$$

$$2CO_2 + 2H^+ + 2e^- \rightleftharpoons H_2C_2O_4, \qquad \mathscr{E}^0 = -0.49.$$

Multiplying the coefficients of the first reaction by 2 and those in the second reaction by 5, we obtain

$$2MnO_4^- + 16H^+ + 10e^- \rightleftharpoons 2Mn^{2+} + 8H_2O, \qquad \mathscr{E}^0 = 1.51,$$

$$10CO_2 + 10H^+ + 10e^- \rightleftharpoons 5H_2C_2O_4, \qquad \mathscr{E}^0 = -0.49.$$

Subtracting, we have

$$2MnO_4^- + 6H^+ + 5H_2C_2O_4 \rightleftharpoons 2Mn^{2+} + 8H_2O + 10CO_2,$$

$$\mathscr{E}^0 = 1.51 - (-0.49) = 1.51 + 0.49 = 2.00 \text{ volts}.$$

Since $n = 10$,

$$\log_{10} K \doteq \frac{10(2.00)}{0.05915} = 338.$$

Example 3. $\qquad$ $Cd^{2+} + 4NH_3 \rightleftharpoons Cd(NH_3)_4^{2+}$.

This reaction is not an oxidation–reduction reaction; nonetheless, it may be decomposed into two half-cell reactions:

$$Cd^{2+} + 2e^- \rightleftharpoons Cd, \qquad\qquad \mathscr{E}^0 = -0.40,$$

$$Cd(NH_3)_4^{2+} + 2e^- \rightleftharpoons Cd + 4NH_3, \qquad \mathscr{E}^0 = -0.61.$$

Subtracting, we obtain

$$Cd^{2+} + 4NH_3 \rightleftharpoons Cd(NH_3)_4^{2+}, \qquad \mathscr{E}^0 = -0.40 - (-0.61) = +0.21,$$

$$\log_{10} K = \frac{2(0.21)}{0.05915} = 7.10, \qquad \text{or} \quad K = 1.26 \times 10^7.$$

This is the *stability* constant of the complex ion.

Example 4. $\qquad$ $Cu(OH)_2 \rightleftharpoons Cu^{2+} + 2OH^-,$

$$Cu(OH)_2 + 2e^- \rightleftharpoons Cu + 2OH^-, \qquad \mathscr{E}^0 = -0.224,$$

$$Cu^{2+} + 2e^- \rightleftharpoons Cu, \qquad\qquad \mathscr{E}^0 = +0.337.$$

Subtracting, we obtain

$$Cu(OH)_2 \rightleftharpoons Cu^{2+} + 2OH^-, \qquad \mathscr{E}^0 = -0.224 - (+0.337) = -0.561,$$

$$\log_{10} K = \frac{2(-0.561)}{0.05915} = -18.95, \qquad \text{or} \quad K = 1.1 \times 10^{-19}.$$

This is the solubility product constant of copper hydroxide.

17–12   Computation of the Cell EMF

If we wished to calculate the cell emf for an arbitrary combination of electrodes, we would proceed in the same way as in Section 17–11, but we would use the Nernst equation for half-cell emf. The examples are those in Section 17–11.

Example 1. The cell emf is given by $\mathscr{E} = \mathscr{E}_{Fe^{3+}, Fe^{2+}} - \mathscr{E}_{Sn^{4+}, Sn^{2+}}$, but

$$\mathscr{E}_{Fe^{3+}, Fe^{2+}} = \mathscr{E}^0_{Fe^{3+}, Fe^{2+}} - \frac{RT}{\mathscr{F}} \ln \frac{a_{Fe^{2+}}}{a_{Fe^{3+}}},$$

$$\mathscr{E}_{Sn^{4+}, Sn^{2+}} = \mathscr{E}^0_{Sn^{4+}, Sn^{2+}} - \frac{RT}{2\mathscr{F}} \ln \frac{a_{Sn^{2+}}}{a_{Sn^{4+}}}.$$

Subtracting, we obtain

$$\mathscr{E} = \mathscr{E}^0_{Fe^{3+}, Fe^{2+}} - \mathscr{E}^0_{Sn^{4+}, Sn^{2+}} - \frac{RT}{\mathscr{F}} \left[\ln \frac{a_{Fe^{2+}}}{a_{Fe^{3+}}} - \frac{1}{2} \ln \frac{a_{Sn^{2+}}}{a_{Sn^{4+}}} \right].$$

To combine the logarithmic terms, the coefficients must be the same, therefore we write

$$\ln \frac{a_{Fe^{2+}}}{a_{Fe^{3+}}} = \frac{1}{2} \ln \frac{a_{Fe^{2+}}^2}{a_{Fe^{3+}}^2},$$

so that we obtain for the cell potential

$$\mathscr{E} = \mathscr{E}_{Fe^{3+},Fe^{2+}}^0 - \mathscr{E}_{Sn^{4+},Sn^{2+}}^0 - \frac{RT}{2\mathscr{F}} \ln \frac{a_{Fe^{2+}}^2 \cdot a_{Sn^{4+}}}{a_{Fe^{3+}}^2 \cdot a_{Sn^{2+}}}.$$

The same result could be obtained from inspection of the cell reaction.

Example 2. After obtaining $\mathscr{E}^0 = \mathscr{E}_{MnO_4^-,Mn^{2+}}^0 - \mathscr{E}_{CO_2,H_2C_2O_4}^0$; then

$$\mathscr{E} = \mathscr{E}^0 - \frac{RT}{10\mathscr{F}} \ln Q.$$

The quotient is found by inspection of the cell reaction, so that

$$\mathscr{E} = \mathscr{E}^0 - \frac{RT}{10\mathscr{F}} \ln \frac{a_{Mn^{2+}}^2 \cdot p_{CO_2}^{10}}{a_{MnO_4^-}^2 \cdot a_{H^+}^6 \cdot a_{H_2C_2O_4}^5}.$$

17–13 The Measurement of Cell EMFs

The simplest method of measuring the emf of an electrochemical cell is to balance it against an equal and opposite potential difference in the slidewire of a potentiometer. Figure 17–5 shows a potentiometer circuit with the cell connected in it. The battery B sends a current i through the slidewire R. The contact S is adjusted until no deflection is observed on the galvanometer G. At the null point, the emf of the cell is balanced by the potential difference between the points S and P of the slidewire. The slidewire is calibrated so that the potential drop ir between the points S and P can be read directly. If the resistance of the cell is very large, the potentiometer setting may be moved over a wide range without producing a sensible deflection on the galvanometer. In this case a more elaborate instrument, the electron tube voltmeter, must be used.

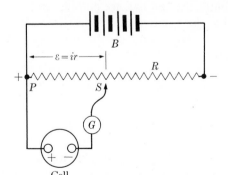

Fig. 17–5 Potentiometer circuit.

17–14   Reversibility

In the foregoing treatment of electrodes and cells we have assumed implicitly that the electrode or cell was in equilibrium with respect to certain chemical and electrical transformations. By definition such an electrode or cell is *reversible*. To correlate measured values of cell emfs with the ones calculated by the Nernst equation, the measured values must be equilibrium or reversible values; the potentiometric measurement in which no current is drawn from the cell is ideally suited for the measurement of reversible emfs. Since no current passes when the circuit is in balance, the measured emf is the equilibrium or reversible value of the emf.

Another aspect of the property of reversibility can be displayed by considering the transformations in the zinc–hydrogen cell as in Section 17–7:

$$2e^-(Zn) \to 2e^-(Pt) \quad \text{and} \quad Zn^{2+}(aq) + H_2(g) \to Zn + 2H^+(aq).$$

The reaction consumes electrons in zinc; hence, electrons must be supplied to the zinc from the external circuit to make the reaction proceed as written. The flow of electrons *in the external circuit* is from the platinum to the zinc. To supply electrons to the zinc the potentiometer must impose a slightly higher emf than that of the cell. If, on the other hand, the potentiometer imposes a slightly lower emf, then the electronic current will flow in the opposite sense, from the zinc to the platinum in the external circuit; the chemical reaction will be reversed, zinc will be consumed and gaseous hydrogen will be evolved. In this latter instance, the cell produces work while in the former work is destroyed.

The cell behaves reversibly if moving the potentiometer contact slightly to one side of the balance point and then to the other reverses the current and the direction of the chemical reaction. In practice it is not necessary to analyze for the amounts of the reactants and products after each of the adjustments to decide whether the reaction is behaving in the required way. If the cell is irreversible, then throwing the potentiometer slightly out of balance ordinarily results in the flow of a comparatively large current, while reversibility demands that only a small current flow when the imbalance between the emfs is slight. Furthermore, in the irreversible cell, after disturbing the balance in the circuit slightly, the new balance point is usually significantly different from the original one. For these reasons, the irreversible cell exhibits what is apparently an erratic behavior and often it is impossible to bring a potentiometer into balance with such a cell.

As an example of an irreversible cell consider zinc and silver as electrodes which dip into a solution of sodium chloride. When this cell produces work, it does so through the reaction

$$Zn + 2H^+ \to Zn^{2+} + H_2.$$

The zinc electrode dissolves to form zinc ion, and hydrogen is liberated on the silver electrode. However, if the current is reversed so that work is destroyed, then the reaction is

$$Ag + H^+ + Cl^- \rightarrow AgCl + \tfrac{1}{2}H_2.$$

Hydrogen is liberated at the zinc electrode and the silver corrodes to silver chloride.

17–15   The determination of the $\mathscr{E}^0$ of a Half-cell

Since the values of equilibrium constants are obtained from the standard half-cell emfs, the method of obtaining the $\mathscr{E}^0$ of a half-cell has great importance. Suppose we wish to determine the $\mathscr{E}^0$ of the silver–silver ion electrode. Then we set up a cell which includes this electrode and another electrode the potential of which is known; for simplicity we choose the SHE as the other electrode. Then the cell is

$$\text{SHE} \parallel \text{Ag}^+|\text{Ag}.$$

The cell reaction is $Ag^+ + e^-_{SHE} \rightleftharpoons Ag$, and the cell emf is

$$\mathscr{E} = \mathscr{E}_{Ag^+,Ag} = \mathscr{E}^0_{Ag^+,Ag} - \frac{RT}{\mathscr{F}}\ln\frac{1}{a_{Ag^+}}.$$

At 25°C,

$$\mathscr{E} = \mathscr{E}^0_{Ag^+,Ag} + 0.05915\log_{10} a_{Ag^+}. \tag{17–55}$$

If the solution were an ideal dilute solution, we could replace a_{Ag^+} by $m_+ = m$, the molality of the silver salt. Equation (17–55) would become

$$\mathscr{E} = \mathscr{E}^0_{Ag^+,Ag} + 0.05915\log_{10} m.$$

By measuring $\mathscr{E}$ at several values of m and plotting $\mathscr{E}$ versus $\log_{10} m$, a straight line of slope 0.05915 would be obtained, as in Fig. 17–6(a). The intercept on the vertical axis, $m = 1$, would be the value of $\mathscr{E}^0$. However, life is not so simple. We cannot replace a_{Ag^+} by m and hope for any real accuracy in our equation. In an ionic solution, the activity of an ion can be represented by the mean ionic activity $a_\pm = \gamma_\pm m_\pm$. If the solution contains only silver nitrate, then $m_\pm = m$; and Eq. (17–55) becomes

$$\mathscr{E} = \mathscr{E}^0_{Ag^+,Ag} + 0.05915\log_{10} m + 0.05915\log_{10}\gamma_\pm.$$

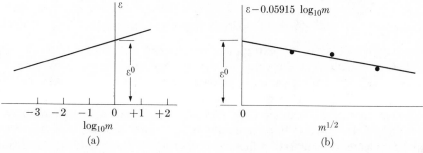

Fig. 17–6 (a) "Ideal" dependence of $\mathscr{E}$ on m. (b) Plot to obtain $\mathscr{E}^0$ by extrapolation.

If the measurements are carried to solutions dilute enough so that the Debye-Hückel limiting law, Eq. (16–78), is valid, then $\log_{10} \gamma_\pm = -0.50\, m^{1/2}$, and we can reduce the equation to

$$\mathscr{E} - 0.05915 \log_{10} m = \mathscr{E}^0_{Ag^+,Ag} - 0.02958\, m^{1/2}. \tag{17–56}$$

From the measured values of $\mathscr{E}$ and m, the left-hand side of this equation can be calculated. The left-hand side is plotted against $m^{1/2}$; extrapolation of the curve to $m^{1/2} = 0$ yields an intercept equal to $\mathscr{E}^0_{Ag^+,Ag}$. The plot is shown schematically in Fig. 17–6(b). It is by this method that accurate values of $\mathscr{E}^0$ are obtained from the measured values of the $\mathscr{E}$ of any half-cell.

17–16   The Determination of Activities and Activity Coefficients from Cell EMFs

Once an accurate value of $\mathscr{E}^0$ has been obtained for a cell, then the emf measurements yield values of the activity coefficients directly. Consider the cell

$$Pt|H_2(f = 1)|H^+, Cl^-|AgCl|Ag.$$

The cell reaction is

$$AgCl(s) + \tfrac{1}{2}H_2(f = 1) \rightleftharpoons Ag + H^+ + Cl^-.$$

The cell emf is

$$\mathscr{E} = \mathscr{E}^0 - \frac{RT}{\mathscr{F}} \ln (a_{H^+} a_{Cl^-}). \tag{17–57}$$

According to Eq. (17–57), the emf of the cell does not depend on the individual ion activities but on the product $a_{H^+} a_{Cl^-}$. As it turns out there is no measurable quantity which depends on an individual ion activity. Consequently we replace the product $a_{H^+} a_{Cl^-}$ by $a_\pm^2$. Since in HCl, $m_\pm = m$, we have $a_\pm^2 = (\gamma_\pm m)^2$; this reduces Eq. (17–57) to

$$\mathscr{E} = \mathscr{E}^0 - \frac{2RT}{\mathscr{F}} \ln m - \frac{2RT}{\mathscr{F}} \ln \gamma_\pm. \tag{17–58}$$

At 25°C

$$\mathscr{E} = \mathscr{E}^0 - 0.1183 \log_{10} m - 0.1183 \log_{10} \gamma_\pm. \tag{17–59}$$

Having determined $\mathscr{E}^0$ by the extrapolation described in Section 17–15, we see that the values of $\mathscr{E}$ determine the values of $\gamma_\pm$ at every value of m. Conversely, if the value of $\gamma_\pm$ is known at all values of m, the cell potential $\mathscr{E}$ can be calculated from Eq. (17–58) or (17–59) as a function of m.

The measurement of cell emfs is the most powerful method of obtaining values of activities of electrolytes. Experimentally it is, in many cases at least, much easier to handle than measurements of colligative properties. It has the additional advantage that it can be used over a wide range of temperatures. Although cell emfs can be measured in nonaqueous solvents, the electrode equilibria often are not as easily established so that the experimental difficulties are much greater.

17–17 Significance of the Half-cell Potential

The half-cell potential is a measure of the tendency of the reaction

$$Zn^{2+} + 2e_{\overline{SHE}} \to Zn$$

to occur. It is thus a measure of the tendency of Zn^{2+}, the oxidized form of the couple, to be reduced by H_2 under 1 atm pressure to form H^+ ions at unit activity. Since $\mathscr{V}^0_{Zn^{2+},Zn} = -0.763$ volt, the negative sign indicates that Zn^{2+} is not reduced by H_2 under these conditions. If we had chosen the silver ion–silver couple

$$Ag^+ + e_{\overline{SHE}} \to Ag, \qquad \mathscr{V}^0_{Ag^+,Ag} = +0.799 \text{ volt.}$$

The positive $\mathscr{V}^0_{Ag^+,Ag}$ indicates that silver ion can be reduced by H_2 under standard conditions.

Active metals such as Zn, Na, or Mg have highly negative standard potentials. Their compounds are not reduced by hydrogen, but rather the metal itself can be oxidized by H^+ to yield H_2. Noble metals such as Cu and Ag have positive $\mathscr{V}^0$'s. Compounds of these metals are readily reduced by H_2; the metals themselves are not oxidized by hydrogen ion.

Since the potential of a metal depends on the activity of the metal ion in solution, factors which influence the activity of the ion will ipso facto influence the electrode potential. For silver,

$$\mathscr{V}_{Ag^+,Ag} = 0.799 - 0.05915 \log_{10} \frac{1}{a_{Ag^+}}.$$

The pH is defined by $pH = -\log_{10} a_{H^+}$; suppose we define pAg in the same way, $pAg = -\log_{10} a_{Ag^+}$, then

$$\mathscr{V}_{Ag^+,Ag} = +0.799 - 0.05915 \, pAg.$$

The lower the value of a_{Ag^+}, the greater pAg and the lower will be the value of $\mathscr{V}_{Ag^+,Ag}$. This implies that Ag^+ is less readily reduced by H_2 when the concentration of Ag^+ is small. Conversely, silver metal is more easily oxidized if the concentration of Ag^+ is small. By diluting the solution of $AgNO_3$ in contact with the electrode, the electrode potential can be lowered. This method of reducing the electrode potential is effective until the concentration of Ag^+ ion reaches 10^{-6} or 10^{-7} molar; $pAg = 6$ or 7. At this point if $pAg = 6$, the electrode potential drops to $\mathscr{V} = 0.799 - 0.355 = 0.444$ volt. However, impurities in the solution are present in comparable concentrations, 10^{-7} molar, and may produce spurious effects at the electrode surface.

A more effective method of changing the electrode potential is to add to a reasonably concentrated solution of the $AgNO_3$ some reagent which will combine with the silver ion and reduce its activity in this way. Suppose we add an excess of NaCl to the solution, more than enough to precipitate the silver as silver chloride; then the equilibrium is established

$$AgCl(s) \rightleftharpoons Ag^+ + Cl^-, \qquad a_{Ag^+} a_{Cl^-} = K_{sp},$$

where K_{sp} is the solubility product constant of AgCl. Taking logarithms, we obtain

$$\log_{10} a_{Ag^+} = \log_{10} K_{sp} - \log_{10} a_{Cl^-}, \qquad -pAg = -pK_{sp} + pCl.$$

The electrode potential becomes (AgCl present)

$$\mathscr{V}_{Ag^+,Ag} = \mathscr{V}^0_{Ag^+,Ag} - 0.05915(pK_{sp} - pCl). \tag{17-60}$$

If K_{sp} is a very small number, the pK_{sp} is fairly large. In this case $pK_{sp} = 9.75$, so that (AgCl present)

$$\mathscr{V}_{Ag^+,Ag} = 0.799 - 0.577 + 0.05915\,pCl$$

$$= 0.222 + 0.05915\,pCl.$$

The electrode is now sensitive to the Cl^- concentration through the second term. The overall equilibrium is the sum of the two equilibria:

$$Ag^+ + e^- \rightleftharpoons Ag$$

$$\underline{\qquad AgCl(s) \rightleftharpoons Ag^+ + Cl^- \qquad}$$

Summing: $$AgCl(s) + e^- \rightleftharpoons Ag + Cl^-$$

This electrode is the silver chloride–silver electrode. It has the Nernst equation

$$\mathscr{V}_{AgCl,Ag,Cl^-} = \mathscr{V}^0_{AgCl,Ag,Cl^-} - 0.05915 \log_{10} a_{Cl^-}.$$

Comparing this equation with Eq. (17–60), we see that

$$\mathscr{V}^0_{AgCl,Ag,Cl^-} = \mathscr{V}^0_{Ag^+,Ag} - 0.05915\,pK_{sp}$$

$$\mathscr{V}^0_{AgCl,Ag,Cl^-} = \mathscr{V}^0_{Ag^+,Ag} + 0.05915 \log_{10} K_{sp}. \tag{17-61}$$

From this equation, the solubility product constant can be calculated from measured values of the $\mathscr{V}^0$'s of the silver–silver ion electrode and the silver–silver chloride–chloride ion electrode. This method of determining solubility product constants is much used for substances which have only a tiny solubility.

From the argument above it can be seen that the more stable the species in which the silver ion is bound, the lower will be the electrode potential of the silver. A group of $\mathscr{V}^0$'s for various silver couples is given in Table 17–2. From the values in Table 17–2, it is clear that iodide ion ties up Ag^+ more effectively than bromide or chloride; AgI is less soluble than AgCl or AgBr. The fact that the silver iodide–silver couple has a negative potential means that silver should dissolve in HI with the liberation of hydrogen. This occurs in fact, but the action ceases promptly due to the layer of insoluble AgI which forms and protects the Ag surface from further attack.

Table 17–2

Couples	$\mathscr{V}^0$, volts
$Ag^+ + e^- \rightleftharpoons Ag$	0.7991
$AgCl(s) + e^- \rightleftharpoons Ag + Cl^-$	0.2222
$AgBr(s) + e^- \rightleftharpoons Ag + Br^-$	0.03
$AgI(s) + e^- \rightleftharpoons Ag + I^-$	−0.151
$Ag_2S(s) + 2e^- \rightleftharpoons 2Ag + S^=$	−0.69

Substances which form soluble complexes with the metal ion also lower the electrode potential. Two examples are

$$Ag(NH_3)_2^+ + e^- \rightleftharpoons Ag + 2NH_3, \qquad \mathscr{V}^0 = +0.373,$$

$$Ag(CN)_2^- + e^- \rightleftharpoons Ag + 2CN^-, \qquad \mathscr{V}^0 = -0.31.$$

The discussion above is not restricted to silver but applies to any metal. Reagents which complex or precipitate the oxidized member of the couple lower the electrode potential. Whether a metal is a noble metal or an active metal depends upon its environment. Ordinarily silver is a noble metal; in the presence of iodide, sulfide, or cyanide ion, it is an active metal, if we consider the zero of potential as the dividing line between active and noble metals.

In an analogous way, if the activity of the reduced member of the couple is lowered by some means, the potential of the electrode is increased. If the reduced member is a metal, the most usual method of lowering its activity is to alloy it (dilute it) with another metal. In amalgam electrodes mercury is used for this purpose.

17–18 Temperature Dependence of the Cell EMF

By differentiating the equation, $n\mathscr{F}\mathscr{E} = -\Delta G$ with respect to temperature, we obtain

$$n\mathscr{F}\left(\frac{\partial \mathscr{E}}{\partial T}\right)_p = -\left(\frac{\partial \Delta G}{\partial T}\right)_p = \Delta S,$$

$$\left(\frac{\partial \mathscr{E}}{\partial T}\right)_p = \frac{\Delta S}{n\mathscr{F}}. \tag{17–62}$$

To use Eq. (17–62), ΔS must be expressed in J/deg rather than in eu. If the cell does not contain a gas electrode, then since the entropy changes of reactions in solution are frequently rather small, less than 10 eu, the temperature coefficient of the cell emf is usually of the order of 10^{-4} or 10^{-5} volt/deg. As a consequence, if only routine equipment is being used to measure the cell emf and the temperature coefficient is sought, the measurements should cover as wide a range of temperature as is feasible.

The value of ΔS is independent of temperature to a good approximation; by integrating Eq. (17–62) between a reference temperature T_0 and any temperature T, we obtain

$$\mathscr{E} = \mathscr{E}_{T_0} + \frac{\Delta S}{n\mathscr{F}}(T - T_0), \tag{17-63a}$$

or

$$\mathscr{E} = \mathscr{E}_{25°C} + \frac{\Delta S}{n\mathscr{F}}(t - 25), \tag{17-63b}$$

where t is in °C. The cell emf is a linear function of temperature.

The temperature coefficient of the cell emf yields, through Eq. (17–62), the value of ΔS. From this and the value of $\mathscr{E}$ at any temperature we can calculate ΔH for the cell reaction. Since $\Delta H = \Delta G + T \Delta S$, then

$$\Delta H = -n\mathscr{F}\left[\mathscr{E} - T\left(\frac{\partial \mathscr{E}}{\partial T}\right)_p\right]. \tag{17-64}$$

Thus, by measuring $\mathscr{E}$ and $(\partial \mathscr{E}/\partial T)_p$ we can obtain the thermodynamic properties of the cell reaction, ΔG, ΔH, ΔS.

Example. The cell reaction

$$Hg_2Cl_2(s) + H_2(1 \text{ atm}) \rightarrow 2Hg(l) + 2H^+(a = 1) + 2Cl^-(a = 1)$$

has the emf, $\mathscr{E}^0_{298.15} = +0.2676$ volt, and $(\partial \mathscr{E}^0/\partial T)_p = -3.19 \times 10^{-4}$ volt/deg. Since $n = 2$,

$$\Delta H^0 = -2(96,490)[0.2676 + (298.15)(3.19)(10^{-4})]$$

$$= -192,980[0.2676 + 0.0951] = -192,980(0.3627)$$

$$= -70,000 \text{ J} = -16.730 \text{ cal}$$

$$\Delta S^0 = 2(96,490)(-3.19)(10^{-4}) \text{ J/deg}$$

$$= -61.6 \text{ J/deg} = -14.7 \text{ eu}.$$

17–19 Heat Effects in the Operation of a Reversible Cell

In Section 17–18, we computed the ΔH^0 for the cell reaction from the cell potential and its temperature coefficient. If the reaction were carried out irreversibly by simply mixing the reactants together, ΔH^0 is the heat which flows into the system in the transformation by the usual relation, $\Delta H = Q_p$. However, if the reaction is brought about reversibly in the cell, electrical work in the amount U_{rev} is produced. Then, by Eq. (9–4),

$$Q_{p(rev)} = T \Delta S \tag{17-65}$$

which is the definition of ΔS. Using the example in Section 17–18, we have $Q_{p(rev)} = 298.15(-14.7) = -4380$ cal. Consequently, in the operation of the cell only 4380 cal of heat flow to the surroundings, while if the reactants are mixed directly, 16,730 cal of heat pass to the surroundings. The ΔH^0 for the transformation is $-16,730$ cal and is independent of the way the reaction is carried out.

17–20   Electrodes for the Measurement of pH

If the equilibrium at the electrode involves either H^+ ion or OH^- ion, then the electrode potential depends on the pH of the solution. Among the electrodes used to measure pH are the hydrogen electrode, the quinhydrone electrode, and the glass electrode.

The hydrogen electrode

The hydrogen electrode is illustrated in Fig. 17–1. Using the definition,† pH = $-\log_{10} a_{H^+}$, the Nernst equation, Eq. (17–45), at $p = 1$ atm and 25°C becomes

$$\mathscr{V}_{H^+,H_2} = -0.05915\,\text{pH}. \tag{17–66}$$

The principal difficulty in the use of the hydrogen electrode is that the platinum is easily poisoned by the adsorption of impurities from the solution and the gas. The adsorption of impurities hinders the establishment of the equilibrium at the surface and the electrode no longer behaves reversibly. An elaborate purification train is required to purify the hydrogen before it is passed through the cell; this requirement makes the use of the hydrogen electrode cumbersome. The presence of oxidants in the solution displaces the equilibrium and alters the potential. A more trivial annoyance is that changes in barometric pressure or in the depth of immersion of the electrode in the solution (since both vary the pressure of H_2 at the active surface) produce a small variation in the potential of the electrode.

The quinhydrone electrode

Quinone and hydroquinone establish the equilibrium‡

$$Q + 2H^+ + 2e^- \rightleftharpoons QH_2.$$

The Nernst equation is

$$\mathscr{V}_{Q,QH_2} = \mathscr{V}^0_{Q,QH_2} - \frac{RT}{2\mathscr{F}} \ln \frac{a_{QH_2}}{a_Q a_{H^+}^2}.$$

The compound quinhydrone has the formula $Q \cdot QH_2$ and is sparingly soluble in water. In a saturated solution of quinhydrone the concentrations of quinone and

---

†   Originally, pH was defined by pH = $-\log_{10} c_{H^+}$. This definition is often extended to pH = $-\log_{10} a_{H^+}$. We shall use this extension in spite of the fact that it introduces a number of difficulties into the concept of pH.

‡   Q stands for O=⟨≡⟩=O; QH_2 for HO⟨ ⟩OH.

hydroquinone are equal and presumably the activities are very equal; hence $a_Q = a_{QH_2}$, so that

$$\mathscr{V}_{Q,QH_2} = \mathscr{V}^0_{Q,QH_2} - \frac{2.303\,RT}{\mathscr{F}}\text{pH}.$$

In use, the quinhydrone electrode consists of a spiral of gold wire immersed in the solution of unknown pH, which has been saturated with quinhydrone. The combination of the quinhydrone electrode with a suitable reference electrode is calibrated using a standard buffer solution. In solutions of pH greater than about 8.5 the potential of this electrode is no longer a linear function of pH.

The glass electrode

A diagram of the glass electrode is shown in Fig. 17–7. The glass envelope is filled with a dilute solution of hydrochloric acid, the electrode is silver covered with a coating of silver chloride. The lowest part of the bulb is blown very thin. The potential of a cell consisting of this electrode against a reference electrode is a linear function of the pH of the solution in which the glass electrode is immersed. Because of the very high resistance of the glass membrane it is not possible to use an ordinary potentiometer circuit to measure the emf of a cell which includes the glass electrode. A vacuum tube voltmeter which is capable of measuring the potential difference in spite of the high resistance of the cell is required.

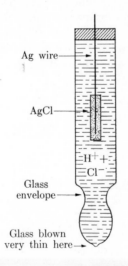

Ag wire

AgCl

$H^+ +$
Cl^-

Glass
envelope

Glass blown
very thin here

Fig. 17–7 The glass electrode.

The exact mode of operation of the glass electrode is not completely understood. It apparently functions as a concentration cell; Section 17–22. If the solutions inside and outside the glass membrane differ in H^+ ion concentration, then a potential difference develops across the membrane; this potential difference depends on the pH of the exterior solution through the equation

$$\mathscr{E} = \mathscr{E}' - \frac{2.303\,RT}{\mathscr{F}}\text{pH},$$

where $\mathscr{E}'$ is a constant depending on the reference electrode used and on the characteristics of each individual glass electrode. Each glass electrode is calibrated against several standard buffer solutions.

Of the common electrodes used for the measurement of pH, the glass electrode is the most convenient. It is unaffected by oxidizing or reducing agents, and almost any other material which may be present in aqueous solution. In highly alkaline solutions, electrodes made with special glasses must be used. The glass electrode functions only in aqueous solutions.

17–21 Reference Electrodes

The choice of a reference electrode in a particular situation depends on the substances in the solution; it is often desirable to choose an electrode which will not require a salt bridge. If the electrode system of interest involves a metal–metal ion couple, then the reference electrode can conveniently be a metal–insoluble salt–anion electrode. There are several of these which are adaptable to various situations, the mercury–mercurous salt electrodes being very satisfactory because of their reproducibility and ease of construction.

The calomel electrodes

The calomel electrodes are suitable for use as reference electrodes in solutions containing chloride ion if a salt bridge is not used. The half-cell reaction is

$$Hg_2Cl_2(s) + 2e^- \rightleftharpoons 2Hg(l) + 2Cl^-.$$

At 25°C, the potential of this electrode is

$$\mathscr{V} = 0.2676 - 0.05915 \log_{10} a_{Cl^-}.$$

If the calomel electrode is used with a bridge, its potential is fixed by the concentration of the chloride solution immediately around it. The potentials of calomel electrodes in KCl solutions of various concentrations are given together with their temperature coefficients in Table 17–3. The saturated calomel electrode is more easily prepared than the others, since it does not require the use of standard solutions.

Table 17–3

Solution	$\mathscr{V}$
Saturated KCl	$+0.242 - 7.6 \times 10^{-4}(t - 25)$
$1M$ KCl	$+0.280 - 2.4 \times 10^{-4}(t - 25)$
$0.1M$ KCl	$+0.334 - 0.7 \times 10^{-4}(t - 25)$

The silver–silver chloride electrode

The silver–silver chloride electrode is another convenient reference electrode which functions reversibly. It consists of silver wire coated with silver chloride immersed in a solution containing chloride ion; Fig. 17–3. The electrode reaction is

$$AgCl(s) + e^- \rightleftharpoons Ag(s) + Cl^-.$$

At 25°C

$$\mathscr{V} = 0.2222 - 0.05915 \log_{10} a_{Cl^-}.$$

In a saturated KCl solution the potential of the electrode at 25°C is +0.197 volt.

The mercury–mercurous sulfate electrode

The mercury–mercurous sulfate electrode is convenient for use in solutions containing sulfate ion. The electrode consists of mercury in a solution containing sulfate ion which has been saturated with mercurous sulfate. The electrode reaction is

$$Hg_2SO_4(s) + 2e^- \rightleftharpoons 2\,Hg(l) + SO_4^{2-}.$$

At 25°C

$$\mathscr{V} = 0.6151 - \frac{0.05915}{2} \log_{10} a_{SO_4^{2-}}.$$

The mercury–mercurous sulfate electrode is the positive electrode in the Weston standard cell. This cell is used throughout the world as a laboratory standard of emf. The negative electrode consists of a saturated cadmium amalgam in a cadmium sulfate solution. In the saturated cell, the electrolyte is saturated in $CdSO_4 \cdot \frac{8}{3}H_2O$ and in Hg_2SO_4. The emf of the saturated Weston cell in absolute volts is

$$\mathscr{E} = 1.018646 - 4.06(10^{-5})(t - 20) - 0.95 \times 10^{-6}(t - 20)^2.$$

The usual standard cell is not saturated with cadmium sulfate. The unsaturated cell has a lower temperature coefficient than the saturated cell, but the emf is not as stable and drifts over a narrow range of values.

The mercury–mercuric oxide electrode

In alkaline solutions the mercury–mercuric oxide electrode, which is reversible to OH^- ion, is a convenient reference electrode. It consists of a pool of mercury in contact with a solution of NaOH or KOH which is saturated with HgO. The reaction is

$$HgO(s) + H_2O + 2e^- \rightleftharpoons Hg(l) + 2OH^-.$$

At 25°C

$$\mathscr{V} = 0.098 - 0.05915 \log_{10} a_{OH^-}.$$

17–22 Concentration Cells

If the two electrode systems which compose a cell involve electrolytic solutions of different composition, then there will be a potential difference across the boundary between the two solutions. This potential difference is called the liquid junction potential, or the diffusion potential. To illustrate how such a potential difference arises, consider two silver–silver chloride electrodes, one in contact with a concentrated HCl solution, activity $= a_1$, the other in contact with a dilute HCl solution, activity $= a_2$; Fig. 17–8(a). If the boundary between the two solutions is open, the H^+ and Cl^- ions in the more concentrated solution diffuse into the more dilute solution. The H^+ ion diffuses much more rapidly than does the Cl^- ion; Fig. 17–8(b). As the H^+ ion begins to outdistance the Cl^- ion, an electrical double layer develops at the interface between the two solutions; Fig. 17–8(c). The potential difference across the double layer produces an electrical field which slows the faster moving ion and speeds the slower moving ion. A steady state is established in which the two ions migrate at the same speed; the ion which moved faster initially leads the march.

Ag, AgCl electrode

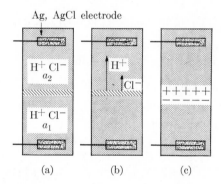

(a)   (b)   (c)

Fig. 17–8 Development of the junction potential.

The diffusion from the concentrated to the dilute solution is an irreversible change; however, if it is very slow, slow enough that the interface does not move appreciably in the time we require to make the measurements, then we may consider the system at "equilibrium" and ignore the motion of the boundary. However, the additional potential difference in the liquid junction will show up in the measurements of the emf of the cell.

We can calculate the emf of the cell if we assume that on the passage of one equivalent of electricity through the cell all of the changes take place reversibly. Then the emf of the cell is given by

$$-\mathscr{F}\mathscr{E} = \sum \Delta G_i, \qquad (17\text{–}67)$$

where $\sum \Delta G_i$ is the sum of all the free energy changes in the cell which accompany the passage of one faraday of positive current upward through the cell. These free energy changes are:

Lower electrode: $\qquad$ $Ag(s) + Cl^-(a_1) \rightarrow AgCl(s) + e^-$

Upper electrode: $\qquad$ $AgCl(s) + e^- \rightarrow Cl^-(a_2) + Ag(s)$

Net change at two electrodes: $\qquad$ $Cl^-(a_1) \rightarrow Cl^-(a_2)$

In addition, at the boundary of the two solutions a fraction t_+ of the current is carried by H^+ and a fraction t_- is carried by Cl^-. The fractions t_+ and t_- are the transference numbers, or transport numbers, of the ions. One equivalent of positive current passing through the boundary requires that t_+ equivalents of H^+ ion are moved upward from the solution a_1 to the solution a_2, and t_- equivalents of Cl^- are moved downward from a_2 to a_1. Thus, at the boundary:

$$t_+H^+(a_1) \rightarrow t_+H^+(a_2), \qquad t_-Cl^-(a_2) \rightarrow t_-Cl^-(a_1).$$

The total change within the cell is the sum of the changes at the electrodes and at the boundary:

$$t_+H^+(a_1) + Cl^-(a_1) + t_-Cl^-(a_2) \rightarrow t_+H^+(a_2) + Cl^-(a_2) + t_-Cl^-(a_1).$$

The sum of the fractions must be unity, so that $t_- = 1 - t_+$. Using this value of t_- in the equation, after some rearrangement, it reduces to

$$t_+H^+(a_1) + t_+Cl^-(a_1) \rightarrow t_+H^+(a_2) + t_+Cl^-(a_2). \tag{17–68}$$

The cell reaction (17–68) is the transfer of t_+ equivalents of HCl from the solution a_1 to the solution a_2. The total free energy change is

$$\Delta G = t_+[\mu_{H^+}^0 + RT\ln(a_{H^+})_2 + \mu_{Cl^-}^0 + RT\ln(a_{Cl^-})_2$$

$$- \mu_{H^+}^0 - RT\ln(a_{H^+})_1 - \mu_{Cl^-}^0 - RT\ln(a_{Cl^-})_1],$$

$$\Delta G = t_+RT\ln\frac{(a_{H^+}a_{Cl^-})_2}{(a_{H^+}a_{Cl^-})_1} = 2t_+RT\ln\frac{(a_\pm)_2}{(a_\pm)_1},$$

since $a_{H^+}a_{Cl^-} = a_\pm^2$. Using Eq. (17–67), we have for the emf of the cell *with transference*,

$$\mathscr{E}_{wt} = -\frac{2t_+RT}{\mathscr{F}}\ln\frac{(a_\pm)_2}{(a_\pm)_1}. \tag{17–69}$$

If the boundary between the two solutions did not contribute to the cell potential, then the only change would be that contributed by the electrodes, which is

$$Cl^-(a_1) \rightarrow Cl^-(a_2).$$

The corresponding value of ΔG is

$$\Delta G = \mu_{Cl^-}^0 + RT\ln(a_{Cl^-})_2 - \mu_{Cl^-}^0 - RT\ln(a_{Cl^-})_1$$

$$= RT\ln\frac{(a_\pm)_2}{(a_\pm)_1},$$

where a_{Cl^-} has been replaced by the mean ionic activity $a_\pm$. This cell is without transference and has the emf,

$$\mathscr{E}_{wot} = -\frac{\Delta G}{\mathscr{F}} = -\frac{RT}{\mathscr{F}} \ln \frac{(a_\pm)_2}{(a_\pm)_1}. \tag{17–70}$$

The total emf of the cell with transference is that of the cell without transference plus the emf of the junction, $\mathscr{E}_j$. Thus, $\mathscr{E}_{wt} = \mathscr{E}_{wot} + \mathscr{E}_j$, so that

$$\mathscr{E}_j = \mathscr{E}_{wt} - \mathscr{E}_{wot}. \tag{17–71}$$

Using Eqs. (17–69) and (17–70), this becomes

$$\mathscr{E}_j = (1 - 2t_+)\frac{RT}{\mathscr{F}} \ln \frac{(a_\pm)_2}{(a_\pm)_1}. \tag{17–72}$$

From Eq. (17–72) it is apparent that if t_+ is near 0.5, the liquid junction emf will be small; this relation is correct only if the two electrolytes in the cell produce two ions in solution. By measuring the emf of the cells with and without transference it is possible to evaluate $\mathscr{E}_j$ and t_+. Note, by comparing Eqs. (17–69) and (17–70), that

$$\mathscr{E}_{wt} = 2t_+\mathscr{E}_{wot}. \tag{17–73}$$

The trick in all of this is how to establish a sharp boundary so as to obtain reproducible measurements of $\mathscr{E}_{wt}$ and how to construct a cell which eliminates $\mathscr{E}_j$ so that $\mathscr{E}_{wot}$ can be measured. There are several clever ways of establishing a sharp boundary between the two solutions; however, they will not be described here. The second problem of constructing a cell without a liquid boundary is more pertinent to our discussion.

A concentration cell without transference, that is, without a liquid junction, is shown in Fig. 17–9. The cell consists of two cells in series, which can be symbolized by

$$Pt|H_2(p)|H^+, Cl^-(a_\pm)_1|AgCl|Ag \quad \overline{\quad\quad} \quad Ag|AgCl|Cl^-, H^+(a_\pm)_2|H_2(p)|Pt.$$

The emf is the sum of the emfs of the two cells separately;

$$\mathscr{E} = (\mathscr{V}_{AgCl,Ag,Cl^-} - \mathscr{V}_{H^+,H_2})_1 + (\mathscr{V}_{H^+,H_2} - \mathscr{V}_{AgCl,Ag,Cl^-})_2.$$

Writing the Nernst equation for each potential, we obtain

$$\mathscr{E} = \left[\mathscr{V}^0_{AgCl,Ag,Cl^-} - \frac{RT}{\mathscr{F}} \ln (a_{Cl^-})_1 + \frac{RT}{\mathscr{F}} \ln \frac{p^{1/2}}{(a_{H^+})_1}\right]$$

$$+ \left[-\frac{RT}{\mathscr{F}} \ln \frac{p^{1/2}}{(a_{H^+})_2} - \mathscr{V}^0_{AgCl,Ag,Cl^-} + \frac{RT}{\mathscr{F}} \ln (a_{Cl^-})_2\right],$$

$$\mathscr{E} = \frac{RT}{\mathscr{F}} \ln \frac{(a_{H^+}a_{Cl^-})_2}{(a_{H^+}a_{Cl^-})_1} = \frac{2RT}{\mathscr{F}} \ln \frac{(a_\pm)_2}{(a_\pm)_1}.$$

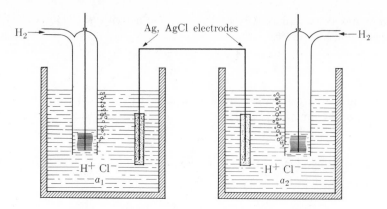

Fig. 17–9 Concentration cell without transference.

By comparison with Eq. (17–70), we see that

$$\mathscr{E} = -2\mathscr{E}_{\text{wot}}. \tag{17–74}$$

Measurement of the emf of this double cell yields the value of $\mathscr{E}_{\text{wot}}$ through Eq. (17–74).

Every measurement of the emf of a cell whose two electrodes require different electrolytes raises the problem of the liquid junction potential between the electrolytes. The problem can be solved in two ways. Either measure the junction potential or eliminate it. The junction potential can be eliminated by designing the experiment, as above, so that no liquid junction appears. Or, rather than using two cells, choose a reference electrode which uses the same electrolyte as the electrode being investigated. This is often the best way to eliminate the liquid junction; however, it is not always feasible.

The salt bridge, an agar jelly saturated with either KCl or NH_4NO_3, is often used to connect the two electrode compartments. This device introduces two liquid junctions, whose emfs are often opposed to one another, and the net junction emf is very small. The physical reason for the cancellation of the two emfs is complex. The use of a jelly has some advantages in itself; it prevents siphoning if the electrolyte levels differ in the two electrode compartments, and it slows the ionic diffusion very much so that the junction emfs, whatever they may be, settle down to reproducible values very quickly.

17–23 Oxidation–Reduction Indicators

If a platinum wire is immersed in a solution containing both ferrous and ferric ions, the potential relative to SHE at 25°C is given by

$$\mathscr{V} = 0.771 - 0.05915 \log_{10} \frac{c_{Fe^{2+}}}{c_{Fe^{3+}}}.$$

To simplify the argument we have replaced activities by concentrations. This equation may be taken in the sense that the potential is determined by the ratio $c_{Fe^{2+}}/c_{Fe^{3+}}$. On the other hand, if we write the equation in the form

$$\log_{10}\frac{c_{Fe^{2+}}}{c_{Fe^{3+}}} = \frac{0.771 - \mathscr{V}}{0.05915},$$

we may say that the concentration ratio is determined by the value of $\mathscr{V}$.

Suppose, for example, we have a solution which is 0.1 molar in both ferrous and ferric ion. Then the potential of the electrode will be 0.771 volt.† If we add to this solution a drop of dilute solution of ceric ion, Ce^{4+}, then some of the ferrous ion will be oxidized according to the reaction

$$Fe^{2+} + Ce^{4+} \rightarrow Fe^{3+} + Ce^{3+}.$$

This does not change the concentrations of ferrous and ferric ions even if all the ceric ion is reduced, since the amount of ceric ion added was so small in comparison with the amount of ferrous ion present. The potential remains at 0.771 volt. Meanwhile, the potential of the ceric–cerous couple is

$$\mathscr{V} = 1.61 - 0.05915 \log_{10}\frac{c_{Ce^{3+}}}{c_{Ce^{4+}}},$$

or

$$\log_{10}\frac{c_{Ce^{3+}}}{c_{Ce^{4+}}} = \frac{1.61 - \mathscr{V}}{0.05915}.$$

The ratio of ceric to cerous ion concentrations is determined by the potential $\mathscr{V}$, which in turn is determined by the ferric–ferrous ratio. Since, in this case, $\mathscr{V} = 0.771$, we have

$$\log_{10}\frac{c_{Ce^{3+}}}{c_{Ce^{4+}}} = \frac{1.61 - 0.77}{0.0592} = \frac{0.84}{0.059} = 14,$$

$$\frac{c_{Ce^{3+}}}{c_{Ce^{4+}}} = 10^{14}.$$

In adjusting itself to the potential of the ferric–ferrous couple, the concentration of the ceric ion has been reduced to a fantastically small fraction of the total amount of cerium present. It is always true that the substance present in insignificant amount, cerium in this instance, adjusts the ratio of reduced to oxidized forms to conform to the potential established by a substance present in relatively large amounts, iron in this example.

The substance present in insignificant amount can function as an indicator. In this case, the ceric ion is brilliant orange, while the cerous ion is nearly colorless.

† Not exactly 0.771 volt because of the approximation of using concentrations rather than activities.

The addition of a drop of ceric solution containing a large amount of ferric and ferrous ions in equal concentration would decolorize the ceric solution, thus indicating that the potential of the ferrous–ferric system was far below the standard potential of the cerous–ceric system.

A large number of substances are available which are suitable to indicate the potential of a couple in solution. If the oxidized and reduced forms of a substance differ in color, then the substance can be an oxidation–reduction indicator. Suppose the indicator has oxidized and reduced forms O and R; then

$$O + e^- \rightleftharpoons R, \qquad \log_{10}\frac{c_R}{c_O} = \frac{\mathscr{V}^0 - \mathscr{V}}{0.0592}.$$

At $\mathscr{V} = \mathscr{V}^0$, $c_R = c_O$; at $\mathscr{V} = \mathscr{V}^0 + 0.0592$, $c_R = (\frac{1}{10})c_O$; and at $\mathscr{V} = \mathscr{V}^0 - 0.0592$, $c_R = 10\,c_O$. From this it is clear that at 59 millivolts above $\mathscr{V}^0$ the indicator is about 10% reduced form and 90% oxidized form, while at 59 millivolts below $\mathscr{V}^0$, the indicator is about 90% reduced form and 10% oxidized form. Figure 17–10 shows the fraction present in the oxidized form as a function of $\mathscr{V}$. As the potential shifts through this range of about $2(59) = 118$ millivolts, the indicator changes color.

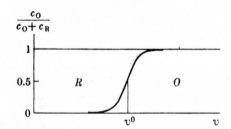

Fig. 17–10 Fraction of oxidized form as a function of potential.

17–24 Potentiometric Titrations

The potential of an electrode can be used to detect the equivalence point in a titration. As an example we choose the titration of ferrous ion with ceric ion. A platinum electrode is placed in a solution of ferrous sulfate along with a calomel reference electrode. A solution of ceric sulfate is added from a buret. Using $\mathscr{V}_1^0$ for the standard potential of the ferric–ferrous couple, and $\mathscr{V}_2^0$ for that of the ceric–cerous couple, we have two expressions for the potential of the platinum:

$$\mathscr{V} = \mathscr{V}_1^0 - \frac{RT}{\mathscr{F}}\ln\frac{c_{Fe^{2+}}}{c_{Fe^{3+}}}$$

$$\mathscr{V} = \mathscr{V}_2^0 - \frac{RT}{\mathscr{F}}\ln\frac{c_{Ce^{3+}}}{c_{Ce^{4+}}}.$$

The first expression is more convenient to use before the equivalent point, while the second is more convenient after the equivalence point. The reaction is

$$Fe^{2+} + Ce^{4+} \rightarrow Fe^{3+} + Ce^{3+},$$

the equilibrium constant being given by

$$K = \frac{c_{Fe^3} + c_{Ce^3+}}{c_{Fe^2} + c_{Ce^4+}}$$

and by

$$\frac{RT}{\mathscr{F}} \ln K = \mathscr{V}_2^0 - \mathscr{V}_1^0.$$

Let C be the original concentration of ferrous ion in the solution; then at any stage of the titration

$$C = c_{Fe^{2+}} + c_{Fe^{3+}}.$$

Let C' be the total concentration of cerium in the solution at any stage in the titration; then

$$C' = c_{Ce^{3+}} + c_{Ce^{4+}}.$$

A ferric ion is produced only if a cerous ion is produced, thus at each stage, let $c = c_{Fe^{3+}} = c_{Ce^{3+}}$. Then we have

$$c_{Fe^{2+}} = C - c \quad \text{and} \quad c_{Ce^{4+}} = C' - c.$$

The equilibrium constant becomes

$$K = \frac{c^2}{(C - c)(C' - c)}.$$

At the equivalence point (ep) the cerium and iron are present in equal concentrations, so that $C = C'_{ep}$ and

$$K = \left(\frac{c_{ep}}{C - c_{ep}}\right)^2.$$

Thus, at the ep

$$\left(\frac{c_{Fe^{2+}}}{c_{Fe^{+3}}}\right)_{ep} = \left(\frac{C - c_{ep}}{c_{ep}}\right) = \frac{1}{\sqrt{K}},$$

$$\ln\left(\frac{c_{Fe^{2+}}}{c_{Fe^{3+}}}\right)_{ep} = -\tfrac{1}{2}\ln K = -\frac{\mathscr{F}(\mathscr{V}_2^0 - \mathscr{V}_1^0)}{2\,RT}.$$

Using this value of the ratio in the expression for $\mathscr{V}$, we obtain

$$\mathscr{V}_{ep} = \mathscr{V}_1^0 + \tfrac{1}{2}(\mathscr{V}_2^0 - \mathscr{V}_1^0) = \tfrac{1}{2}(\mathscr{V}_1^0 + \mathscr{V}_2^0).$$

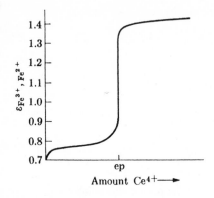

Fig. 17–11 Potentiometric titration curve.

The potential at the equivalence point is, in this example, the mean of the standard potentials of the two couples. A plot of the potential of the platinum electrode against the amount of ceric ion added is shown in Fig. 17–11. The potential increases sharply at the equivalence point, and this sharp increase is easily picked up by the deflection of a needle on an electronic voltmeter. In some instruments the sharp increase is signalled by the "wink" of an electronic tuning eye.

In the discussion we have assumed that the volume of the solution does not change as the titrant is added. If this assumption is set aside, the algebra becomes considerably more complicated, but the curve which is obtained differs from that in Fig. 17–11 only in the noncritical regions. The essential physical fact remains that there is a sharp increase in potential at the equivalence point.

Problems

Units and Conversion Factors
$$R = 8.3147 \text{ J·deg}^{-1}\text{·mole}^{-1}$$
$$\mathscr{F} = 96{,}490\,(\approx 96{,}500)\text{ coulomb/equivalent}$$
$$1\text{ J} = 1\text{ V·coulomb}$$
$$1\text{ cal} = 4.1840\text{ J}$$

Unless otherwise noted, the temperature is to be taken as 25°C in the following problems.

17–1. From the values of the standard potentials in Table 17–1, calculate the standard molar free energy μ^0 of the ions Na^+, Pb^{2+}, Ag^+.

17–2. Calculate $\mu^0_{Fe^{3+}}$ from the data: $\mathscr{V}^0_{Fe^{3+},Fe^{2+}} = +0.771$ V, $\mathscr{V}^0_{Fe^{2+},Fe} = -0.440$ V.

17–3. Compute the cell potential and find the cell reaction for each of the cells (data in Table 17–1):

a) $Ag|Ag^+(a = 0.01) \| Zn^{2+}(a = 0.1)|Zn$,

b) $Pt|Fe^{2+}(a = 1.0), Fe^{3+}(a = 0.1) \| Cl^-(a = 0.001)|AgCl|Ag$,

c) $Zn|ZnO_2^{2-}(a = 0.1), OH^-(a = 1)|HgO|Hg$.

In each case is the cell reaction as written spontaneous or not?

17-4. Consider the lead storage cell

$$Pb|PbSO_4|H_2SO_4(a)| PbSO_4|PbO_2|Pb,$$

in which $\mathscr{V}^0_{SO_4^{2-},PbSO_4,Pb} = -0.356$ V, and $\mathscr{V}^0_{PbSO_4,PbO_2,Pb} = +1.685$ V.

a) If the cell emf is 2.016 volts, compute the activity of the sulfuric acid.

b) Write the cell reaction; is this reaction spontaneous for the cell as written above?

c) If the cell produces work (discharge), the reaction goes in one direction, while if work is destroyed, (charge) the reaction goes in the opposite direction. How much work must be destroyed per equivalent of electrode materials produced if the average potential during charge is 2.15 volts?

d) Sketch the dependence of the cell potential on the activity of the sulfuric acid.

17-5. The Edison storage cell is symbolized

$$Fe|FeO|KOH(a)|Ni_2O_3|NiO|Ni.$$

The half-cell reactions are

$$Ni_2O_3(s) + H_2O(l) + 2e^- = 2NiO(s) + 2OH^-, \qquad \mathscr{V}^0 = 0.4\ V,$$
$$FeO(s) + H_2O(l) + 2e^- = Fe + 2OH^-, \qquad \mathscr{V}^0 = -0.87\ V.$$

a) What is the cell reaction?

b) How does the cell potential depend on the activity of the KOH?

17-6. a) Calculate the potential of the Ag^+, Ag electrode; $\mathscr{V}^0 = 0.80$ V, for activities of $Ag^+ = 1, 0.1, 0.01, 0.001.$

b) For AgI, $K_{sp} = 1 \times 10^{-16}$; what will be the potential of the Ag^+, Ag electrode in a saturated solution of AgI?

c) Calculate the standard potential of the AgI, Ag, I^- electrode.

17-7. Consider the couple $O + e^- = R$, with all of the oxidized and reduced species at unit activity. What must be the value of $\mathscr{V}^0$ of the couple if the reductant R is to liberate hydrogen from

a) an acid solution, $a_{H^+} = 1,$

b) water at pH $= 7$?

c) Is hydrogen a better reducing agent in acid or in basic solution?

17-8. Consider the same couple under the same conditions as in Problem 17-7. What must be the value of $\mathscr{V}^0$ of the couple if the oxidant is to liberate oxygen by the half-cell reaction

$$O_2(g) + 2H_2O(l) + 4e^- = 4OH^-, \qquad \mathscr{V}^0 = 0.401\ V,$$

a) from a basic solution, $a_{OH^-} = 1,$

b) from an acid solution, $a_{H^+} = 1,$

c) from water at pH $= 7$?

d) Is oxygen a better oxidizing agent in acid or in basic solution?

17-9. From the data in Table 17-1, calculate the equilibrium constant for each of the reactions.

a) $Cu^{2+} + Zn = Cu + Zn^{2+}$

b) $Zn^{2+} + 4CN^- = Zn(CN)_4^{2-}$

c) $3H_2O + Fe = Fe(OH)_3(s) + \frac{3}{2}H_2$

d) $Fe + 2Fe^{3+} = 3Fe^{2+}$

e) $3HSnO_2^- + Bi_2O_3 + 6H_2O + 3OH^- = 2Bi + 3Sn(OH)_6^{2-}$

f) $PbSO_4(s) = Pb^{2+} + SO_4^{2-}$.

17–10. The standard potential of the quinhydrone electrode is $\mathscr{V}^0 = 0.6994$ V. The half-cell reaction is

$$Q(s) + 2H^+ + 2e^- \rightleftharpoons QH_2(s).$$

Using a calomel electrode as a reference electrode, $\mathscr{V}^0 = 0.2676$ V, we have the cell

$$Hg|Hg_2Cl_2(s)|HCl(a)|Q\cdot QH_2|Au.$$

Using the values of the activity coefficients for HCl given in Table 16–1, compute the potential of this cell at $m_{HCl} = 0.001, 0.005, 0.01$.

17–11. H. S. Harned and W. J. Hamer [*J. Amer. Chem. Soc.* **57**, 33 (1935)] present values for the emf of the cell

$$Pb|PbSO_4|H_2SO_4(a)|PbSO_4|PbO_2|Pt$$

over a wide range of temperature and concentration of H_2SO_4. In 1 m H_2SO_4 they found between 0 and 60°C:

$$\mathscr{E} = 1.91737 + 56.1(10^{-6})t + 108(10^{-8})t^2,$$

where t is the celsius temperature.

a) Calculate ΔG, ΔH, and ΔS for the cell reaction at 0° and 25°C.

b) For the half-cells at 25°C:

$$PbO_2(s) + SO_4^{2-} + 4H^+ + 2e^- \rightleftharpoons PbSO_4(s) + 2H_2O, \qquad \mathscr{V}^0 = 1.6849 \text{ V},$$

$$PbSO_4(s) + 2e^- \rightleftharpoons Pb(s) + SO_4^{2-}, \qquad \mathscr{V}^0 = -0.3553 \text{ V}.$$

Calculate the mean ionic activity coefficient in 1 m H_2SO_4 at 25°C. Assume that the activity of water is unity.

17–12. At 25°C, the emf of the cell

$$H_2|H_2SO_4(a)|Hg_2SO_4|Hg$$

is 0.61201 V in 4 m H_2SO_4; $\mathscr{E}^0 = 0.61515$ V. Calculate the mean ionic activity coefficient in 4 m H_2SO_4. [H. S. Harned and W. J. Hamer, *J. Amer. Chem. Soc.* **57**, 27 (1933).]

17–13. In 4 m H_2SO_4, the emf of the cell in Problem 17–11 is 2.0529 V at 25°C. Calculate the value of the activity of water in 4 m H_2SO_4 using the result in Problem 17–12.

17–14. A 0.1 M solution of NaCl is titrated with $AgNO_3$. The titration is followed potentiometrically, using a silver wire as the indicating electrode and a suitable reference electrode. Calculate the potential of the silver wire when the amount of $AgNO_3$ added is 50%, 90%, 99%, 99.9%, 100%, 100.1%, 101%, 110%, 150% of the equivalent amount; ignore the change in volume of the solution.

$$\mathscr{V}^0_{AgCl,Ag,Cl^-} = 0.222 \text{ V}, \qquad \mathscr{V}^0_{Ag^+,Ag} = 0.799 \text{ V}.$$

$K_{sp} = 1.7 \times 10^{-10}$ for silver chloride.

17–15. Write the cell reaction and compute the emf of the following cells without transference.

a) $Pt|H_2(p = 1\ atm)|HCl(a)|H_2(p = 0.5\ atm)|Pt$

b) $Zn|Zn^{2+}(a = 0.01) \parallel Zn^{2+}(a = 0.1)|Zn$

17–16. The emf of the cell with transference

$$H_2|HCl(a_\pm = 0.009048) \vdots HCl(a_\pm = 0.01751)|H_2$$

at 25°C is 0.02802 V. The corresponding cell without transference has an emf of 0.01696 V. Calculate the transference number of H^+ ion and the value of the junction potential.

Chapter Eighteen

Surface Phenomena

18–1 Surface Energy and Surface Tension

Consider a solid composed of spherical molecules in a close-packed arrangement. The molecules are bound by a cohesive energy E per mole and $\epsilon = E/N$ per molecule. Each molecule is bonded to twelve others; the bond strength is $\epsilon/12$. If the surface layer is close packed, a molecule on the surface is bonded to a total of only nine neighbors. Then the total binding energy of the surface molecule is $9\epsilon/12 = \frac{3}{4}\epsilon$. From this rather crude picture we conclude that the surface molecule is bound with only 75% of the binding energy of a molecule in the bulk. The energy of a surface molecule is therefore higher than that of a molecule in the interior of the solid and energy must be expended to move a molecule from the interior to the surface of a solid; this is also true of liquids.

Suppose that a film of liquid is stretched on a wire frame; Fig. 18–1. If the area of the film is increased by dA, the energy of the film increases by $\gamma\, dA$, where γ is the surface energy per square centimeter. The energy increase implies that the motion of the wire is opposed by a force f; if the wire moves a distance dx, the energy expended is $f\, dx$. These two energy increments are equal, so that

$$f\, dx = \gamma\, dA.$$

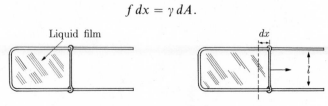

Liquid film

dx

Fig. 18–1 Stretched film.

But the increase in area is $2(l\,dx)$; the factor 2 appears because the film has two sides. Thus, $f\,dx = \gamma 2l\,dx$, or, $f = \gamma 2l$. The length of film in contact with the wire is l cm on each side, a total of $2l$ cm; the force acting per unit length of the wire in contact with the film is $f/2l = \gamma$. The force acting per cm length of the film is the *surface tension* of the liquid; the surface tension acts as a force which opposes the increase in area of the liquid. The surface tension (dyne/cm) is numerically equal to the rate of increase of the surface energy with area (erg/cm^2). Table 18–1 lists the values of the surface tension of a few liquids.

Table 18–1 Surface tension of liquids at 20°C

Liquid	γ, dynes/cm	Liquid	γ, dynes/cm
Acetone	23.70	Ethyl ether	17.01
Benzene	28.85	n-Hexane	18.43
Carbon tetrachloride	26.95	Methyl alcohol	22.61
Ethyl acetate	23.9	Toluene	28.5
Ethyl alcohol	22.75	Water	72.75

18–2 Magnitude of the Surface Tension

The rough estimate in Section 18–1 showed that the surface atoms have an energy roughly 25 % above that of those in the bulk. This excess energy does not show up in systems of ordinary size since the number of molecules on the surface is an insignificant fraction of the total number of molecules present. Consider a cube 1 cm on a side. If the molecules are 10^{-8} cm in diameter, then 10^8 molecules can be placed on an edge; the number of molecules in the cube is $(10^8)^3 = 10^{24}$. On each face there will be $(10^8)^2 = 10^{16}$ molecules; there are 6 faces making a total of 6×10^{16} molecules on the surface. The fraction of molecules on the surface is $6 \times 10^{16}/10^{24} = 6 \times 10^{-8}$. Thus only 6 molecules in 100 million are on the surface. As a consequence, unless we are particularly concerned with the surface energy, we may ignore its presence as we have in all the earlier thermodynamic discussions.

If the ratio of surface to volume of the system is very large, the surface energy shows up willy nilly. We can calculate the size of particle for which the surface energy will contribute, let us say 1 % of the total energy. We write the energy in the form,

$$E = E_v V + E_s A,$$

where V and A are the volume and area, E_v and E_s are the energy per unit volume and the energy per unit area. But, $E_v = \epsilon_v N_v$, and $E_s = \epsilon_s N_s$, where ϵ_v and ϵ_s are energies per molecule. Then,

$$E = E_v V\left(1 + \frac{E_s A}{E_v V}\right) = E_v V\left(1 + \frac{N_s \epsilon_s A}{N_v \epsilon_v V}\right).$$

As we have seen, for a cube $(N_s/N_v) \approx 6 \times 10^{-8}$; the ratio $(\epsilon_s/\epsilon_v) \approx 1$. So we have

$$E = E_v V \left[1 + (6 \times 10^{-8}) \frac{A}{V} \right].$$

If the second term is to have 1 % of the value of the first, then $0.01 = 6 \times 10^{-8}(A/V)$. This requires $(A/V) \approx 2 \times 10^5$. If a cube has a side a, the area is $6a^2$, and the volume is a^3, so that $(A/V) = 6/a$. Therefore, $6/a \approx 2 \times 10^5$, and $a \approx 3 \times 10^{-5}$ cm. This gives us a very rough but quite reasonable estimate of the maximum size of particle for which the effect of the surface energy becomes noticeable. In practice, surface effects are significant for particles having diameters less than 10^{-4} cm.

18–3 Thermodynamic Formulation

Consider two phases and the interface between them. We choose as the system the portions of the two phases M_1 and M_2, and the portion of the interface S enclosed by a bounding surface B; Fig. 18–2(a). Suppose that the interface moves to a new position S'. The changes in energy are:

$$\text{For } M_1: \qquad dE_1 = T\,dS_1 - p_1\,dV_1. \qquad (18\text{–}1)$$

$$\text{For } M_2: \qquad dE_2 = T\,dS_2 - p_2\,dV_2. \qquad (18\text{–}2)$$

$$\text{For the surface:} \quad dE_s = T\,dS_s + \gamma\,dA. \qquad (18\text{–}3)$$

The last equation is written in analogy to the others, since $dW = -\gamma\,dA$. There is no $p\,dV$ term for the surface, since the surface obviously has no volume. The total change in energy is

$$dE = dE_1 + dE_2 + dE_s = T\,d(S_1 + S_2 + S_s) - p_1\,dV_1 - p_2\,dV_2 + \gamma\,dA$$

$$= T\,dS - p_1\,dV_1 - p_2\,dV_2 + \gamma\,dA.$$

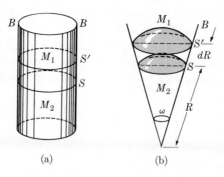

(a)   (b)

Fig. 18–2 Displacement of the interface.
(a) Plane interface. (b) Spherical interface.

Since the total volume $V = V_1 + V_2$, then $dV_1 = dV - dV_2$, and

$$dE = T\,dS - p_1\,dV + (p_1 - p_2)\,dV_2 + \gamma\,dA. \tag{18–4}$$

If the entropy and volume are constant, $dS = dV = 0$, then at equilibrium the energy is a minimum, $dE = 0$. This reduces the equation to

$$(p_1 - p_2)\,dV_2 + \gamma\,dA = 0. \tag{18–5}$$

If, as is shown in Fig. 18–2(a), the interface is plane and the bounding surface B is a cylinder having sides perpendicular to the interface, then the area of the interface does not change, $dA = 0$. Since $dV_2 \neq 0$, Eq. (18–5) requires that $p_1 = p_2$. Consequently, the pressure has the same value in two phases which are separated by a *plane* dividing surface.

If the interface is not a plane, then a displacement of the interface will involve a change in area. This in turn will lead to an inequality of the pressures in the two phases. Suppose that the bounding surface is a conical one and that the interface is a spherical cap having a radius of curvature R; Fig. 18–2(b). Then the area of the cap is $A = \omega R^2$, and the volume of M_2 enclosed by the cone and the cap is $V_2 = \omega R^3/3$, where ω is the solid angle subtended by the cap. But $dV_2 = \omega R^2\,dR$ and $dA = 2\omega R\,dR$; therefore, Eq. (18–5) becomes

$$(p_2 - p_1)\omega R^2\,dR = \gamma 2\omega R\,dR,$$

which reduces immediately to

$$p_2 = p_1 + \frac{2\gamma}{R}. \tag{18–6}$$

Equation (18–6) expresses the fundamental result that the pressure inside a phase which has a convex surface† is greater than that outside. The difference in pressure in passing across a curved surface is the physical reason for capillary rise and capillary depression, which we consider in the next section.

18–4 Capillary Rise and Capillary Depression

If an open tube having a very small inner diameter (a capillary tube) is partially immersed in a liquid, the liquid stands at different levels inside and outside the tube; this behavior is a consequence of the fact that the interface between liquid and vapor is curved inside the tube and flat outside. By considering Eq. (18–6) and the effect of gravity on the system, we can determine the relation between the difference in liquid levels, the surface tension, and the relative densities of the two phases.

Figure 18–3 shows two phases, 1 and 2, separated by an interface which is plane for the most part but has a portion in which phase 2 is convex; the levels of the

---

†   If the surface is not spherical but has principal radii of curvature R, and R', then Eq. (18–6) would have the form $p_2 = p_1 + \gamma(1/R + 1/R')$.

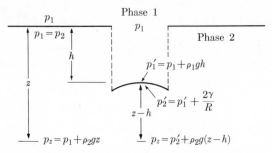

Fig. 18–3 Pressures under plane and curved portions of a surface.

interface are different under the plane and curved portions. The densities of the two phases are ρ_1 and ρ_2. Let p_1 be the pressure in phase 1 at the plane surface separating the two phases; this position is taken as the origin ($z = 0$) of the z-axis which is directed downward. The pressures at the other positions are as indicated in the figure; p'_1 and p'_2 are the pressures just inside phases 1 and 2 at the curved interface; p'_1 and p'_2 are related by Eq. (18–6). The condition of equilibrium is that the pressure at the depth z, which lies below both the plane and curved parts of the interface, must have the same value everywhere. Otherwise a flow of material would occur from one region at the depth z to another. Equality of the pressures at the depth z requires that

$$p_1 + \rho_2 gz = p'_2 + \rho_2 g(z - h). \tag{18–7}$$

Since $p'_2 = p'_1 + 2\gamma/R$, and $p'_1 = p_1 + \rho_1 gh$, Eq. (18–7) reduces to

$$(\rho_2 - \rho_1)gh = \frac{2\gamma}{R}, \tag{18–8}$$

which relates the capillary depression h to the surface tension, the densities of the two phases, and the radius of curvature of the curved surface. We have assumed that the surface of phase 2, the liquid phase, is convex. In this case there is a capillary depression. If the surface of the liquid is concave, this is equivalent to R being negative, which makes the capillary depression h negative. Therefore, a liquid which has a concave surface will exhibit a capillary elevation. Water rises in a glass capillary, while mercury in a glass tube is depressed.

To use Eq. (18–8) to calculate the surface tension from the capillary depression, the radius of curvature must be related to the radius of the tube. Figure 18–4 shows the relation between the radius of curvature R, the radius of the tube r, and the contact angle θ, which is the angle within the liquid between the wall of the tube and the tangent to the liquid surface at the wall of the tube. From Fig. 18–4, we have $r/R = \sin \varphi = \sin (\theta - 90°) = -\cos \theta$, so that $R = -r/\cos \theta$. In terms of the radius of the tube, Eq. (18–8) becomes

$$-\gamma \cos \theta = \tfrac{1}{2}(\rho_2 - \rho_1)grh.$$

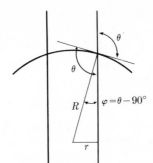

Fig. 18–4 Contact angle.

Now since h is the capillary depression, it is convenient to replace it by the capillary rise $-h$. This removes the negative sign and we have

$$\gamma \cos \theta = \tfrac{1}{2}(\rho_2 - \rho_1)grH. \tag{18–9}$$

In Eq. (18–9), H is the capillary rise. If $\theta < 90°$, the liquid meniscus is concave and H is positive. When $\theta > 90°$, the meniscus is convex and $\cos \theta$ and H are negative.

Liquids which wet the tube have values of θ less than 90°, while those which do not wet the tube have values greater than 90°. In making measurements the tube is chosen narrow enough so that $\theta = 0°$ (or 180°). This is necessary since it is difficult to establish other values of θ reproducibly.

18–5   Surface Tension and Adsorption

Consider the system of the type shown in Fig. 18–2(a): two phases with a plane interface between them. Since the interface is plane, we have $p_1 = p_2 = p$ and the free energy becomes a convenient function. If we have a multicomponent system the chemical potential of each component must have the same value in each phase and at the interface. The variation in total free energy of the system is given by

$$dG = -S\,dT + V\,dp + \gamma\,dA + \sum_i \mu_i\,dn_i, \tag{18–10}$$

in which $\gamma\,dA$ is the increase in free energy of the system associated with a variation in area. The free-energy increments for the two phases are given by

$$dG_1 = -S_1\,dT + V_1\,dp + \sum_i \mu_i\,dn_i^{(1)}$$

$$dG_2 = -S_2\,dT + V_2\,dp + \sum_i \mu_i\,dn_i^{(2)}.$$

Subtracting these two equations from the equation for the change in total free energy yields

$$d(G - G_1 - G_2) = -(S - S_1 - S_2)\,dT + (V - V_1 - V_2)\,dp + \gamma\,dA$$
$$+ \sum_i \mu_i\,d(n_i - n_i^{(1)} - n_i^{(2)}).$$

If the presence of the interface produced no physical effect, then the difference between the total free energy, G, and the sum of the free energies of the bulk phases, $G_1 + G_2$, would be zero. Since the presence of the interface does produce physical effects, we ascribe the difference $G - (G_1 + G_2)$ to the presence of the surface and define it as the surface free energy, G_s. Then,

$$G_s = G - G_1 - G_2, \qquad S_s = S - S_1 - S_2, \qquad n_i^{(s)} = n_i - n_i^{(1)} - n_i^{(2)}.$$

Note that the presence of the interface cannot affect the geometric requirement that $V = V_1 + V_2$. The differential equation becomes

$$dG_s = -S_s\, dT + \gamma\, dA + \sum_i \mu_i\, dn_i^{(s)}. \qquad (18\text{--}11)$$

At constant temperature, pressure, and composition, let the bounding surface, the cylinder B in Fig. 18–2a, increase in radius from zero to some finite value. Then the interfacial area increases from zero to A and the $n_i^{(s)}$ increase from zero to $n_i^{(s)}$, while γ and all the μ_i are constants. Then Eq. (18–11) integrates

$$\int_0^{G_s} dG_s = \gamma \int_0^A dA + \sum_i \mu_i \int_0^{n_i^{(s)}} dn_i^{(s)}$$

$$G_s = \gamma A + \sum_i \mu_i n_i^{(s)}. \qquad (18\text{--}12)$$

This equation is similar to the usual additivity rule for free energy, but contains the additional term, γA. Differentiation yields

$$dG_s = \gamma\, dA + A\, d\gamma + \sum_i \mu_i\, dn_i^{(s)} + \sum_i n_i^{(s)}\, d\mu_i. \qquad (18\text{--}13)$$

By subtracting Eq. (18–11) from Eq. (18–13), we obtain

$$0 = S_s\, dT + A\, d\gamma + \sum_i n_i^{(s)}\, d\mu_i.$$

Division by A, and introduction of the entropy per unit area $\sigma = S_s/A$, and the surface excess, $\Gamma_i = n_i^{(s)}/A$, reduces this relation to

$$d\gamma = -\sigma\, dT - \sum_i \Gamma_i\, d\mu_i. \qquad (18\text{--}14)$$

At constant temperature and pressure, this becomes

$$d\gamma = -(\Gamma_i\, d\mu_i + \Gamma_2\, d\mu_2 + \cdots). \qquad (18\text{--}15)$$

This equation relates the change in surface tension, γ, to change in the μ_i which, at constant T and p, are determined by the variation in composition.

To obtain a clearer meaning for the surface excesses, consider a column having a constant cross-sectional area, A. Phase 1 fills the space between height $z = 0$ and z_0, and has a volume, $V_1 = Az_0$. Phase 2 fills from z_0 to Z, and has a volume $V_2 = A(Z - z_0)$. The molar concentration, c_i, of species i is shown as a function of height, z, in Fig. 18–5. The interface between the two phases is located approximately

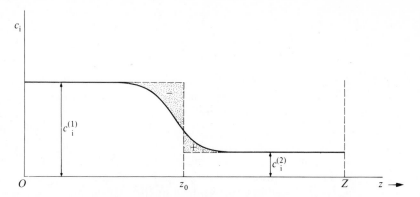

Fig. 18–5 Concentration as a function of position.

at z_0. In the region near z_0 the concentration changes smoothly from $c_i^{(1)}$, the value in the bulk of phase 1, to $c_i^{(2)}$, the value in the bulk of phase 2; the width of this region has been enormously exaggerated in Fig. 18–5. To calculate the actual number of moles of species i in the system, we multiply c_i by the volume element, $dV = A\,dz$ and integrate over the entire length of the system from zero to Z:

$$n_i = \int_0^Z c_i A\,dz = A\int_0^Z c_i\,dz. \qquad (18\text{–}16)$$

The concentration c_i is the function of z shown in Fig. 18–5. It is clear that the value of n_i calculated in this way is the correct value and does not depend in the least upon the position chosen for the reference surface, z_0.

Now if we define the total number of moles of i in phase 1, $n_i^{(1)}$ and the total number in phase 2, $n_i^{(2)}$, in terms of the *bulk* concentrations, $c_i^{(1)}$ and $c_i^{(2)}$, we obtain

$$n_i^{(1)} = c_i^{(1)}V_1 = c_i^{(1)}Az_0 = A\int_0^{z_0} c_i^{(1)}\,dz$$

$$n_i^{(2)} = c_i^{(2)}V_2 = c_i^{(2)}A(Z - z_0) = A\int_{z_0}^Z c_i^{(2)}\,dz.$$

Using the definition of $n_i^{(s)}$,

$$n_i^{(s)} = n_i - n_i^{(1)} - n_i^{(2)}$$

$$= A\left[\int_0^Z c_i\,dz - \int_0^{z_0} c_i^{(1)}\,dz - \int_{z_0}^Z c_i^{(2)}\,dz\right].$$

Since $\Gamma_i = n_i^{(s)}/A$, and since

$$\int_0^Z c_i\,dz = \int_0^{z_0} c_i\,dz + \int_{z_0}^Z c_i\,dz,$$

we have

$$\Gamma_i = \int_0^{z_0} (c_i - c_i^{(1)}) \, dz + \int_{z_0}^Z (c_i - c_i^{(2)}) \, dz. \tag{18-17}$$

The first of these integrals is the negative of the shaded area to the left of the line z_0 in Fig. 18–5, while the second integral is the shaded area to the right of z_0. It is clear from the manner in which this figure is drawn that Γ_i, sum of the two integrals, is negative. However, it is also clear that this value of Γ_i depends critically on the position chosen for the reference plane, z_0. By moving z_0 slightly to the left, Γ_i would have a positive value; moving z_0 to the right would decrease the value to zero; moving z_0 further to the right would make Γ_i negative. We may vary the numerical values of the surface excesses arbitrarily by adjusting the position of the reference surface z_0. Suppose we adjust the position of the reference surface in such a way that the surface excess of one of the components is made equal to zero. This component is usually chosen as the solvent and labeled component 1. Then, by this adjustment

$$\Gamma_1 = 0.$$

However, in general this location for the reference surface will not yield zero values for the surface excesses for the other components. Hence, Eq. (18–15) for a two-component system takes the form

$$-d\gamma = \Gamma_2 \, d\mu_2. \tag{18-18}$$

In an ideal dilute solution, $\mu_2 = \mu_2^0 + RT \ln c_2$, and $d\mu_2 = RT(dc_2/c_2)$, so that

$$-\left(\frac{\partial \gamma}{\partial c_2}\right)_{T,p} = \Gamma_2 \frac{RT}{c_2}$$

or

$$\Gamma_2 = -\frac{1}{RT}\left(\frac{\partial \gamma}{\partial \ln c_2}\right)_{T,p}. \tag{18-19}$$

This is the Gibbs adsorption isotherm. If the surface tension of the solution decreases with increase in concentration of solute, then $(\partial \gamma / \partial c_2)$ is negative and Γ_2 is positive; there is an excess of solute at the interface. This is the usual situation with surface active materials; if they accumulate at the interface, they lower the surface tension. The Langmuir surface films described in the following section are a classical example of this.

18–6 Surface Films

Certain insoluble substances will spread upon the surface of a liquid such as water until they form a monomolecular layer. Long-chain fatty acids, stearic and oleic acid, are classical examples. The —COOH group at one end of the molecule is strongly attracted to the water, while the long hydrocarbon chain is hydrophobic.

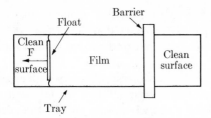

Fig. 18–6 Langmuir film experiment.

A shallow tray, the Langmuir tray, is filled to the brim with water; (Fig. 18–6). The film is spread in the area between the float and the barrier by adding a drop of a dilute solution of stearic acid in benzene. The benzene evaporates leaving the stearic acid on the surface. The float is attached rigidly to a superstructure which allows any lateral force, indicated by the arrow, to be measured by means of a torsion wire.

By moving the barrier, we can vary the area confining the film. If the area is reduced, the force on the barrier is practically zero until a critical area is reached, whereupon the force rises rapidly; Fig. 18–7(a). The extrapolated value of the critical area is 20.5 Å^2 per molecule. This is the area at which the film becomes close packed. In this state the molecules in the film have the polar heads attached to the surface and the hydrocarbon tails extended upward. The cross-sectional area of the molecule is therefore 20.5 Å^2.

The force F is a consequence of the lower surface tension on the film-covered surface as compared with that of the clean surface. If the length of the barrier is l, and it moves a distance dx, then the area of the film decreases by $l\,dx$ and that of the clean surface behind the barrier increases by $l\,dx$. The energy increase is $\gamma_0 l\,dx - \gamma l\,dx$, where γ_0 and γ are the surface tensions of the water and the film-covered surface. This energy is supplied by the barrier moving a distance dx against a force Fl, so that $Fl\,dx = (\gamma_0 - \gamma)l\,dx$, or

$$F = \gamma_0 - \gamma. \qquad (18\text{–}20)$$

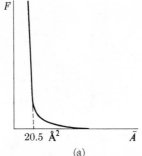
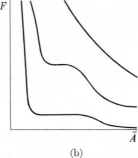

(a)   (b)

Fig. 18–7 Force–area curves. (a) High surface pressure. (b) Low surface pressure.

Note that F is a force per unit length of the barrier, which is equal to that on the float. From curve 1 in Fig. 18–7(a) and Eq. (18–20), we see that the surface tension of the film-covered surface is not very different from that of the clean surface until the film becomes close packed.

Figure 18–7(b) shows the behavior of the surface pressure at very high areas and very low surface pressures F. The curves look very much like the isotherms of a real gas. In fact, the uppermost curve follows a law which is much like the ideal gas law,

$$F\bar{A} = \mathbf{k}T, \tag{18–21}$$

where $\bar{A}$ is the area per molecule, and $\mathbf{k}$ is the Boltzmann constant. Equation (18–21) is easily derived from kinetic theory by supposing that the "gas" is two dimensional. The plateaus in Fig. 18–7(b) represent a phenomenon which is analogous to liquefaction.

We can obtain Eq. (18–21) by writing the Gibbs adsorption isotherm in the form

$$d\gamma = -RT\Gamma_2 \frac{dc_2}{c_2}$$

and considering the difference in surface tension in comparing the film-covered surface, γ, with the clean surface, γ_0. Then $d\gamma = \gamma - \gamma_0$ and $dc_2 = c_2^* - 0$, since c_2^* is the concentration of stearic acid in the bulk liquid phase and this is extremely small. We have $dc_2/c_2 = (c_2^* - 0)/c_2^* = 1$, and consequently

$$\gamma - \gamma_0 = -RT\Gamma_2.$$

Since $F = \gamma_0 - \gamma$ and $R = N_0\mathbf{k}$, we have

$$F = \mathbf{k}T(N_0\Gamma_2).$$

But $N_0\Gamma_2$ is the number of molecules per cm^2, then the area/molecule $= \bar{A} = 1/N_0\Gamma_2$; thus $F\bar{A} = \mathbf{k}T$, which is the result in Eq. (18–21).

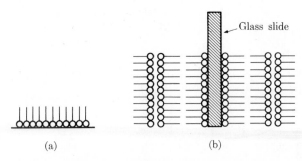

Fig. 18–8 Surface films. (a) Monolayer of stearic acid on a surface. (b) Multilayer obtained by dipping a glass slide through a monolayer.

If a glass slide is dipped through the close-packed film, as it is withdrawn the polar heads of the stearic acid molecules attach themselves to the glass. Pushing the slide back in allows the hydrocarbon tails on the water surface to join with the tails on the glass slide. Figure 18–8 shows the arrangement of molecules on the surface and on the slide. By repeated dipping, a layer of stearic acid containing a known number of molecular layers can be built up on the slide. After about twenty dippings the layer is thick enough to show interference colors, from which the thickness of the layer is calculated. Knowing the number of molecular layers on the slide from the number of dippings, we can calculate the length of the molecule. This method of Langmuir and Blodgett is an incredibly simple method for the direct measurement of the size of molecules. The results agree well with those obtained from x-ray diffraction.

18–7 Adsorption on Solids

If a finely divided solid is stirred into a dilute solution of a dye, it is observed that the depth of color in the solution is much decreased. If a finely divided solid is exposed to a gas at low pressure, the pressure decreases noticeably. In these situations the dye or the gas is *adsorbed* on the surface. The magnitude of the effect depends on the temperature, the nature of the adsorbed substance (the adsorbate), the nature and state of subdivision of the adsorbent (the finely divided solid), and the concentration of the dye or pressure of the gas.

The Freundlich isotherm is one of the first equations proposed to relate the amount of material adsorbed to the concentration of the material in the solution

$$m = kc^{1/n}, \tag{18–22}$$

where m is the number of grams adsorbed per gram of adsorbent, c is the concentration, and k and n are constants. By measuring m as a function of c and plotting $\log_{10} m$ versus $\log_{10} c$, the values of n and k can be determined from the slope and intercept of the line. The Freundlich isotherm fails if the concentration (or pressure) of the adsorbate is too high.

We can represent the process of adsorption by a chemical equation. If the adsorbate is a gas, then we write the equilibrium

$$A(g) + S \rightleftharpoons AS,$$

where A is the gaseous adsorbate, S is a vacant site on the surface, and AS represents an adsorbed molecule of A or an occupied site on the surface. The equilibrium constant can be written

$$K = \frac{x_{AS}}{x_S p},$$

where x_{AS} is the mole fraction of occupied sites on the surface, x_S is the mole fraction of vacant sites on the surface, and p is the pressure of the gas. It is more common to use θ for x_{AS}. Then $x_S = (1 - \theta)$ and the equation can be written

$$\frac{\theta}{1 - \theta} = Kp, \tag{18–23}$$

which is the Langmuir isotherm; K is the equilibrium constant for the adsorption. Solving for θ, we obtain

$$\theta = \frac{Kp}{1 + Kp}. \tag{18–24}$$

If we are speaking of adsorption of a substance from solution, Eq. (18–24) is correct if p is replaced by the molar concentration c.

The amount of the substance adsorbed, m, will be proportional to θ for a specified adsorbent, so $m = b\theta$, where b is a constant. Then

$$m = \frac{bKp}{1 + Kp}, \tag{18–25}$$

which, if inverted, yields

$$\frac{1}{m} = \frac{1}{b} + \frac{1}{bKp}. \tag{18–26}$$

By plotting $1/m$ against $1/p$, the constants K and b can be determined from the slope and intercept of the line. Knowing K, we can calculate the fraction of the surface covered from Eq. (18–24).

The Langmuir isotherm, in the form of Eq. (18–24), is generally more successful in interpreting the data than is the Freundlich isotherm if only a monolayer is formed. A plot of θ versus p is shown in Fig. 18–9. At low pressures, $Kp \ll 1$ and $\theta = Kp$, so that θ increases linearly with pressure. At high pressures, $Kp \gg 1$, so that $\theta \approx 1$. The surface is nearly fully covered with a monomolecular layer at high pressures, so that change in pressure produces little change in the amount adsorbed.

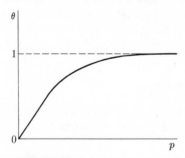

Fig. 18–9 Langmuir isotherm.

18–8 Physical and Chemisorption

If the adsorbate and the surface of the adsorbent interact only by van der Waals forces, then we speak of physical adsorption, or van der Waals adsorption. The adsorbed molecules are weakly bound to the surface and heats of adsorption are low, a few kilocalories at most, and are comparable to the heat of vaporization of the adsorbate. Increase in temperature markedly decreases the amount of adsorption.

Since the van der Waals forces are the same as those which produce liquefaction, adsorption does not occur at temperatures which are much above the critical temperature of the gaseous adsorbate. Also, if the pressure of the gas has values near the equilibrium vapor pressure of the liquid adsorbate, then a more extensive adsorption occurs, multilayer adsorption. A plot of the amount of material adsorbed versus p/p_0, where p_0 is the vapor pressure of the liquid, is shown in Fig. 18–10. Near $p/p_0 = 1$ more and more of the gas is adsorbed; this large increase in adsorption is a preliminary to outright liquefaction of the gas, which would occur at p_0 if the solid were not present.

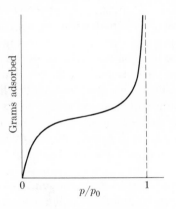

Fig. 18–10 Multilayer adsorption.

If the adsorbed molecules react chemically with the surface, the phenomenon is called *chemisorption*. Since chemical bonds are broken and formed in the process of chemisorption, the heat of adsorption has the same range of values as for chemical reactions, from a few kilocalories to as high as 100 kcal. Chemisorption does not go beyond the formation of a monolayer on the surface. For this reason an isotherm of the Langmuir type which predicts a monolayer and nothing more is well suited for interpreting the data. The Langmuir adsorption isotherm predicts a heat of adsorption which is independent of θ, the fraction of the surface covered at equilibrium. It is observed that for many systems the heat of adsorption decreases with increasing coverage of the surface. If the heat adsorption depends on the coverage, then we must use an isotherm more elaborate than the Langmuir isotherm.

The difference between physical and chemisorption is typified by the behavior of nitrogren on iron. At the temperature of liquid nitrogen, $-190°C$, nitrogen is adsorbed physically on iron as nitrogen molecules, N_2. The amount of N_2 adsorbed decreases rapidly as the temperature rises. At room temperature iron does not adsorb nitrogen at all. At high temperatures, $\sim 500°C$, nitrogen is chemisorbed on the iron surface as nitrogen atoms.

18–9   Electrical Phenomena at Interfaces; the Double Layer

If two phases of different chemical constitution are in contact, an electrical potential difference develops between the two phases. This potential difference is accompanied by a charge separation, one side of the interface being positively charged, the other negatively.

   If one of the phases is a solid and the other is an electrolytic solution, then several structures for the double layer are possible. Suppose that the solid surface is positively charged and the electrolytic solution has a matching negative charge. The first possibility is that the negative charge is located entirely in a plane a short distance δ away from the solid surface. The variation of the electrical potential ψ within the solution is shown in Fig. 18–11(a) as a function of distance x from the solid surface. This fixed double layer is called the *Helmholtz double layer*. A second possibility is a fixed layer of negative charge at the distance δ, not enough to balance the positive charge on the solid, and a diffuse distribution of the remainder of the negative charge; the potential variation corresponding to this situation is shown in Fig. 18–11(b). The diffuse part of the double layer is called the *Gouy layer*. Finally, the fixed layer may contain more negative charge than is required to balance the positive charge on the solid, so that the diffuse Gouy layer is positively charged; the potential distribution is as shown in Fig. 18–11(c). Either combination of fixed and diffuse layers is called a *Stern double layer*. Finally, the distribution of charge in the solution may be entirely diffuse, a "pure" Gouy layer; Fig. 18–11(d) shows this potential distribution. If the solid were negatively charged, four additional possibilities analogous to these could be realized.

   The interaction between the solid surface and the ions in the solution determines the structure of the double layer in a particular situation. If the negative ions are

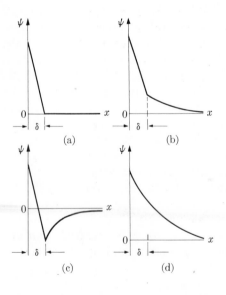

Fig. 18–11 Potential variation in different types of double layers.

not specifically adsorbed on the positive solid, then the double layer is entirely diffuse, as in Fig. 18–11(d). If there is a slight specific adsorption of negative ions, then some negative ions will be located at the distance δ (approximately equal to the molecular diameter), and the situation is as shown in Fig. 18–11(b). These specifically adsorbed ions are called *gegenions*, or *counter ions*. If negative ions are strongly adsorbed, then many negative ions will be located in the plane at δ, and Fig. 18–11(c) represents the situation. Since the amount of adsorption on a surface depends on the concentration of electrolyte, it is possible in some cases to adjust the concentration so that just enough negative ions are adsorbed to balance the positive charge on the surface; then the situation in Fig. 18–11(a) can be realized.

18–10 Electrokinetic Effects

The existence of the double layer has as consequences four electrokinetic effects; these are electro-osmosis, streaming potential, electrophoresis and sedimentation potential (or Dorn effect). All of these effects depend on the fact that part of the double layer is only loosely attached to the solid surface and therefore is mobile. Consider the device in Fig. 18–12, which has a porous quartz disc fixed in position and is filled with water. If an electrical potential is applied between the electrodes, a flow of water to the cathode compartment occurs. In the case of quartz and water, the diffuse (mobile) part of the double layer in the liquid is positively charged. This positive charge moves to the negative electrode and the water flows with it (*electro-osmosis*). Conversely, if water is forced through fine pores of a plug, it carries the charge from one side of the plug to the other, and a potential difference, the *streaming potential*, develops between the electrodes.

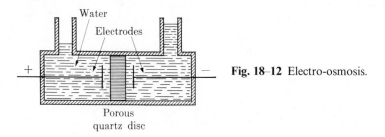

Fig. 18–12 Electro-osmosis.

Very finely divided particles suspended in a liquid carry an electrical charge which is equivalent to the charge on the particle itself plus the charge on the fixed portion of the double layer. If an electrical field is applied to such a suspension, the particles move in the field in the direction determined by the charge on the particle (*electrophoresis*). The diffuse part of the double layer, since it is mobile, has the opposite sign and is attracted to the other electrode. Conversely, if a suspension of particles is allowed to settle, they carry their charge toward the bottom of the vessel and leave the charge on the diffuse layer in the upper portion of the vessel. A potential difference, the *sedimentation potential*, develops between the top and bottom of the container.

The magnitude of all of the electrokinetic effects depends on how much of the electrical charge resides in the mobile part of the double layer. The potential at the dividing line between the fixed and mobile portions of the double layer is the *zeta-potential* (ζ-potential). The charge in the mobile portion of the double layer depends on the ζ-potential and therefore the magnitude of all of the electrokinetic effects depends on ζ. It is commonly assumed that the entire diffuse portion of the double layer is mobile; if this is so, then the ζ-potential is the value of ψ at the position $x = \delta$ in Fig. 18–11. It is more likely that part of the diffuse layer is fixed so that the value of ζ corresponds to the value of ψ at a distance of perhaps two or three times δ. In any case, ζ has the same sign and same general magnitude as the value of ψ at $x = \delta$.

18–11 Colloids

A colloidal dispersion is a suspension of finely divided particles in a continuous medium. The particles themselves are called the disperse phase, or the colloid, and the medium is the dispersing medium. The colloidal dispersion differs from an ordinary solution in that the size of the particle lies in the range between 10^{-4} to 10^{-7} cm. Since the particles are very small, they expose a very large area per unit volume. Thus, surface effects are magnified and the behavior of colloidal dispersions is necessarily interpreted mainly on the basis of such properties as surface tension and the ζ-potential.

The dispersing medium may be a solid, liquid, or a gas, and the disperse phase may be solid, liquid, or gaseous. (Gases cannot be colloidally dispersed in gases, obviously.) The types of dispersion and their names are shown in Table 18–2.

Table 18–2 Types of colloidal dispersion

Dispersing medium	Disperse phase		
	Gas	Liquid	Solid
Gas	—	Fogs	Smokes
Liquid	Foam	Emulsion	Suspension
Solid	Solid foam	?	Suspension

18–12 Lyophilic and Lyophobic Colloids

There are two general classes of colloids, *lyophilic*, or *solvent-loving* and *lyophobic*, or *solvent-hating*, colloids. The lyophilic colloids are sometimes called *gels*, while the lyophobic are called *sols*. The lyophobic colloids are substances which ordinarily are insoluble but which are dispersed in the solvent by special methods. The Bredig arc, in which an arc is passed between two metal electrodes immersed in the solvent, is a common device used to prepare a metal sol. The metal vaporized in the arc is condensed in finely divided form by the solvent. Among other substances which form sols are the silver halides, the metal sulfides, and the metal hydroxides. Lyophilic colloids are substances which ordinarily are soluble and are generally formed by allowing the solvent to stand in contact with the solid material. Gelatin, for example, is a protein with many polar groups such as $-C=O$, and $-N-H$ which can form hydrogen bonds with water as a solvent. Starch, a carbohydrate, is also thoroughly solvated in water.

The colloidal particles carry a net electrical charge which is the sum of the charge on the particle and the charge on the tightly bound gegenions. Because of this electrical charge, the particle migrates in an electrical field (electrophoresis). At the electrode the charge is neutralized. The stability of a lyophobic suspension is intimately connected with the fact that the particle is charged; since they have charges of the same sign, the particles repel each other and thus tend to remain in suspension. If the charge is neutralized, as at an electrode, the particles come together and the suspension is destroyed. A colloidal suspension of rubber latex is used to form rubber articles having complex shapes, e.g., rubber gloves. The particles migrate to the electrode and are precipitated in the shape of the electrode when the charge is neutralized.

The charge on the colloidal particles can be neutralized by adding an electrolyte to the suspension. The ions having the opposite charge to that on the particle are attracted to the particle; on close approach these ions lower the charge on the particle (lower than ζ-potential) and reduce the repulsion between the particles. A positive sol such as ferric oxide is precipitated by negative ions as Cl^-, SO_4^{2-}, etc. A negative sol such as arsenious sulfide is precipitated by positive ions such as K^+, Mg^{2+}, Al^{3+}. The highly charged ions are much more effective in coagulating the colloid than those of lower charge (Schulz–Hardy rule), the minimum concentration of electrolyte needed to produce rapid coagulation being very roughly in the ratio $1:10:500$ for triply, doubly, and singly charged ions. The ion having the same charge as the colloid does not exert very much effect on the coagulation.

The stability of lyophilic colloids is greater than that of lyophobic colloids. Upon addition of water to solid gelatin, the solid swells slowly as the gelatin imbibes the water (imbibition). The protein molecules, which are long and twisted, entangle in a network with much open space between the molecules. The water molecules nearest the protein molecules are relatively tightly bound, while those water molecules in the interstices are in much the same condition as in ordinary liquid water. Only small concentrations of the colloid, often as low as 1 % or less, are needed to produce the gel. Addition of an electrolyte will ultimately precipitate the gel because the ions of the electrolyte remove the solvent which is bound to the lyophilic particle. It follows that large amounts of electrolyte are required compared with that required to precipitate a lyophobic suspension, where a mere trace of electrolyte will often produce coagulation. Those ions which are heavily hydrated, Li^+, for example, are more effective than others. Lyophilic colloids are easily precipitated by the addition of a liquid (alcohol if water is the solvent) in which the solvent is soluble. The solvent is thereby removed from the colloidal particle and aggregation occurs.

If a lyophilic colloid is adsorbed on the particles of a lyophobic sol, the stability of the sol is increased. The lyophilic substance in this situation is called a protective colloid.

The particles in a lyophilic suspension are ordinarily much longer in one direction than in another, as we would expect of polymeric molecules. In addition to being able to form a tangled network characteristic of a gel, these elongated molecules in the suspension produce unusual optical and viscous effects. If a lyophilic suspension flows through a capillary, the molecules tend to line up with their axes parallel to the axis of the tube. The suspension has different values of the index of refraction in the two directions parallel and perpendicular to the axis of the tube. This results in a double refraction, and the suspension exhibits *streaming birefringence*. If the suspension is not flowing, the molecules orient randomly, and the suspension is optically isotropic having a uniform value of the refractive index.

The resistance to flow, the viscosity, of a lyophilic suspension is very high if the rate of shear is small. At high rates of shear, the molecules line up with their long axes parallel to the direction of shear, there is less tangling of the molecules and the suspension flows more easily. When the resistance to flow depends on the rate of shear, the flow is said to be non-Newtonian, and the system is thixotropic. Some paints are thixotropic; they flow easily when spread by a brush, which produces a high rate of shear, but flow very slowly (down a wall, for example) under their own weight, since the rate of shear is very small.

18–13 Colloidal Electrolytes; Soaps and Detergents

The metal salt of a long-chain fatty acid is a soap, the most common example being sodium stearate, $C_{17}H_{35}COO^-\ Na^+$. At low concentrations the solution of sodium stearate consists of individual sodium and stearate ions dispersed throughout the solution in the same way as in any ordinary salt solution. At a rather definite concentration the stearate ions aggregate into clumps, called *micelles*. The micelle contains perhaps 50 to 100 individual stearate ions. The micelle is roughly spherical and the hydrocarbon chains are in the interior leaving the polar $-COO^-$ groups on the outer surface. It is the outer surface which is in contact with the water, and the polar groups on the outer surface stabilize the micelle in the water solution. The micelle is the size of a colloidal particle; since it is charged, it is a colloidal ion. The micelle binds a fairly large number of positive ions to its surface as gegenions, which reduces its charge considerably.

The formation of micelles results in a sharp drop in the electrical conductivity per equivalent of the electrolyte. Suppose 100 sodium and 100 stearate ions were present individually. If the stearate ions aggregate into a micelle and the micelle binds 70 Na^+ as gegenions, then there will be 30 Na^+ ions and 1 micellar ion having a charge of -30 units; a total of 31 ions. The same quantity of sodium stearate would produce 200 ions as individuals but only 31 ions if the micelle is formed. This sharp reduction in the number of ions sharply reduces the conductivity. The formation of micelles also reduces the osmotic pressure of the solution. The average molecular weight, and thus an estimate of the average number of stearate ions in the micelle, can be obtained from the osmotic pressure.

By incorporating molecules of hydrocarbon in the hydrocarbon interior of the micelle, the soap solution can act as a solvent for hydrocarbons. The action of soap as a cleanser depends in part on this ability to hold grease in suspension.

The detergents are similar in structure to the soaps. The typical anionic detergent is an alkyl sulfonate, $ROSO_3^- Na^+$. For good detergent action, R should have at least 16 carbon atoms. Cationic detergents are often quaternary ammonium salts, in which one alkyl group is a long chain; $(CH_3)_3 RN^+ Cl^-$ is a typical example if R has between 12 and 18 carbon atoms.

18–14 Emulsions and Foams

Water and oil can be whipped or beaten mechanically to produce a suspension of finely divided oil droplets in water, an *emulsion*. Mayonnaise is a common household example. It is also possible to produce an emulsion consisting of water droplets in a continuous oil phase, e.g., butter. In either type of emulsion, the large interfacial tension between water and oil coupled with the very large interfacial area implies that the emulsion has a high free energy compared with the separated phases. To supply this free energy an equal amount of mechanical work must be expended in the whipping or beating.

The addition of a surface active agent, such as a soap or detergent, or any molecule with a polar end and a large hydrocarbon end, to the separated system of oil and water lowers the interfacial tension markedly. In this way the free energy requirement for formation of the emulsion can be lowered. Such additives are called emulsifying agents. The interfacial tension is lowered because of the adsorption of the surface active agent at the interface with the polar end in the water and the hydrocarbon end in the oil. The interfacial tension decreases just as it does when a monomolecular film of stearic acid is spread on a water surface in the Langmuir experiment.

Foams consist of a large number of tiny gas bubbles in a continuous liquid phase. A thin film of liquid separates any two gas bubbles. As in the case of emulsions, the surface energy is high and foaming agents are added to lower the interfacial tension between liquid and gas. The foaming agents are the same type of surface active agents as the emulsifying agents. Since the bubbles in the foam are fragile, other additives are needed to give the foam an elasticity to stabilize the foam against mechanical shock. Long chain alcohols (or if a soap is the foaming agent, the undissociated acid) can serve as foam stabilizers.

Problems

18–1. One cm^3 of water is broken into droplets having a radius of 10^{-5} cm. If the surface tension of water is 72.8 dynes/cm, calculate the free energy of the fine droplets relative to that of the water.

18–2. In the duNouy tensiometer, the force required to pull up a ring of fine wire lying in the surface of the liquid is measured. If the diameter of the ring is 1.0 cm and the force to pull the ring up, with the surface of the liquid attached to the inner *and* outer periphery of the ring, is 677 dynes, what is the surface tension of the liquid?

18–3. As a vapor condenses to liquid and a droplet grows in size, the free energy of the droplet varies with its size. For a bulk liquid, $G_{vap} - G_{liq} = \Delta H_{vap} - T \Delta S_{vap}$; if ΔH_{vap} and ΔS_{vap} are independent of temperature, then $\Delta S_{vap} = \Delta H_{vap}/T_0$, where T_0 is the boiling point. If we take $G_{vap} = 0$, then $G_{liq} = -\Delta H_{vap} (1 - T/T_0)$. If G_{liq} and ΔH_{vap} refer to the values per cm^3 of liquid, then the total free energy of the volume V of bulk liquid is $G' = V G_{liq} = -V \Delta H_{vap} (1 - T/T_0)$. If we speak of a fine droplet rather than the bulk liquid then a term γA, where A is the area of the droplet, must be added to this expression : $G' = -V \Delta H_{vap} (1 - T/T_0) + \gamma A$.

a) Show that for a spherical droplet, the free energy of the droplet is positive when the drop is small, then passes through a maximum, then decreases rapidly as the radius increases. If $T < T_0$, at what value of the radius r does $G' = 0$? Show that at larger values of r, G' is negative. Keeping in mind that we chose $G_{vap} = 0$, what radius must the droplet have before it can grow spontaneously by condensation from the vapor?

b) For water $\gamma = 73$ ergs/cm^2, $\Delta H_{vap} = 540$ cal/gm, and the density is unity. What radius must a water droplet at 25°C have before it grows spontaneously?

18–4. The density of mercury is 13.5 gm/cm^3 and $\gamma = 480$ dynes/cm. What would be the capillary depression of mercury in a glass tube of 1 mm inner diameter if we assume that $\theta = 180°$? Neglect the density of air.

18–5. In a glass tube, water exhibits a capillary rise of 2 cm. If $\rho = 1$ gm/cm^3 and $\gamma = 73$ dynes/cm, calculate the diameter of the tube. ($\theta = 0°$.)

18–6. Stearic acid, $C_{17}H_{35}COOH$, has a density of 0.85 gm/cm^3. The molecule occupies an area of 20.5 $Å^2$ in a close-packed surface film. Calculate the length of the molecule.

18–7. The number of cubic centimeters of methane, measured at STP, adsorbed on 1 gm of charcoal at 0°C and several different pressures is

p, cm Hg	10	20	30	40
cm^3 adsorbed	9.75	14.5	18.2	21.4

Plot the data using the Freundlich isotherm and determine the constants k and $1/n$.

18–8. a) The adsorption of ethyl chloride on a sample of charcoal at 0°C and at several different pressures is

p, cm Hg	2	5	10	20	30
grams adsorbed	3.0	3.8	4.3	4.7	4.8

Using the Langmuir isotherm, determine the fraction of the surface covered at each pressure.

b) If the area of the ethyl chloride molecule is $10 \, \text{Å}^2$, what is the area of the charcoal?

18–9. By considering the derivation of the Langmuir isotherm on the basis of a chemical reaction between the gas and the surface, show that if a diatomic gas is adsorbed as atoms on the surface, then $\theta = K^{1/2}p^{1/2}/(1 + K^{1/2}p^{1/2})$.

18–10. An emulsion of toluene in water can be prepared by pouring a solution of toluene in alcohol into water. The alcohol diffuses into the water and leaves the toluene behind in small droplets. If 10 gm of a solution which is 15% ethanol by weight and 85% toluene is poured into 10 gm of water, an emulsion forms spontaneously. The interfacial tension between the suspended toluene droplets and the water–alcohol mixture is 36 dynes/cm, the average diameter of the droplets is 10^{-4} cm, and the density of toluene is 0.87 gm/cm^3. Calculate the increase in free energy associated with the formation of the droplets. Compare this increase with the free energy of mixing of the alcohol and water (25°C).

Chapter Nineteen

The Structure of Matter

19–1 Introduction

The notion that matter consists of discrete, indivisible particles (atoms) is quite ancient. Pre-Christian writers, Lucretius and Democritus, constructed elaborate speculative natural philosophies based on the supposition of the atomicity of matter. The absence of experimental evidence permitted one to believe as he would, so that these early atomic theories bore no fruit.

Modern atomic theory is based on the quantitative observation of nature; its first proposal by Dalton came after a period in which quantitative measurement had risen to importance in scientific investigation. In contrast to the ancient theories, modern atomic theory has been exceedingly fruitful.

To put modern theory in some perspective, it is worthwhile to trace some of its development, at least in bare outline. We shall not attempt anything which could be dignified by the name of history, but only call attention to some major mileposts and courses of thought.

19–2 Nineteenth Century

In the period 1775–1780, Lavoisier established chemistry as a quantitative science by proving that in the course of a chemical reaction the total mass is unaltered. The conservation of mass in chemical reactions proved ultimately to be a death blow to the phlogiston theory. Shortly after Lavoisier, Proust and Dalton proposed the laws of definite and of multiple proportions. In 1803 Dalton proposed his atomic theory. Matter was made up of very small particles called atoms. Every kind of atom has a definite weight. The atoms of different elements have different weights. Compounds are formed by atoms which combine in definite ratios of (usually small) whole numbers. This theory could give a satisfying interpretation of the quantitative data available at the time.

Gay-Lussac's experiments on gas volumes in 1808 led to the law of combining volumes. The volumes of the reactant gases are related to those of the product gases by simple ratios of whole numbers. Gay-Lussac suggested that equal volumes of different gases contained the same number of atoms. This suggestion was rejected. At that time attempts to construct a table of atomic weights were mired in contradictions, since it was supposed that the "atom" of the simplest compound of two elements was formed by combination of two single atoms of the elements; the formation of water and of ammonia would be written

$$H + O \rightarrow OH \quad \text{and} \quad H + N \rightarrow NH.$$

This would require a ratio of atomic weights $N/O = 7/12$. No compound of nitrogen and oxygen exhibiting such a ratio of combining weights was known.

By distinguishing between an atom, the smallest particle which can take part in a chemical change, and a molecule, the smallest particle which can exist permanently, Avogadro (1811) removed the contradictions in the weight ratios by supposing that the molecules of certain elementary gases were diatomic; e.g., H_2, N_2, O_2, Cl_2. He also proposed what is now Avogadro's law: under the same conditions of temperature and pressure equal volumes of all gases contain the same number of molecules. These ideas were ignored and forgotten until 1858 when Cannizaro used them with the law of Dulong and Petit (ca. 1816) to establish the first consistent table of atomic weights. Chaos reigned in the realm of chemical formulas in the fifty-five year interval between the announcement of the atomic theory and the construction of a table of atomic weights substantially the same as the modern one.

A parallel development began in 1832 when Faraday announced the laws of electrolysis. First law: the weight of material formed at an electrode is proportional to the quantity of electricity passed through the electrolyte. Second law: the weights of different materials formed at an electrode by the same quantity of electricity are in the same ratio as their chemical equivalent weights. It was not until 1881 when Helmholtz wrote that acceptance of the atomic hypothesis and Faraday's laws compelled the conclusion that both positive and negative electricity were divided into definite elementary portions, "atoms" of electricity; a conclusion which today seems obvious waited fifty years to be drawn. From 1880 onward, intensive study of electrical conduction in gases led to the discovery of the free electron (J. J. Thomson, 1897), positive rays, and x-rays (Roentgen, 1895). The direct measurement of the charge on the electron was made by Millikan, 1913.

Another parallel development began with Count Rumford's experiment (ca. 1798) of rubbing a blunt boring tool against a solid plate. (He was supposed to be boring cannon at the time; no doubt his assistants thought him a bit odd.) The tool and plate were immersed in water and the water finally boiled. This suggested to Rumford that "heat" was not a fluid, "caloric," but a form of motion. Later experiments, particularly the careful work of Joule in the 1840's, culminated in the recognition of the first law of thermodynamics in 1847. Independently, Helmholtz in 1847 proposed the law of conservation of energy. The second law of thermodynamics, founded on the work of Carnot in 1824, was formulated by Kelvin and Clausius in the 1850's.

In the late 1850's the kinetic theory of gases was intensively developed and met with phenomenal success. Kinetic theory is based on the atomic hypothesis and depends importantly on Rumford's idea of the relation between "heat" and motion.

The chemical achievements, particularly in synthetic and analytical chemistry, in the 19th century are staggering in number; we mention only a few. The growth of organic chemistry after Woehler's synthesis of urea, 1824. The stereochemical studies of van't Hoff, LeBel, and Pasteur. The chemical proof of the tetrahedral arrangement of the bonds about the carbon atom; Kekule's structure for benzene. Werner's work on the stereochemistry of inorganic complexes. The work of Stas on exact atomic weights. The Arrhenius theory of electrolytic solutions. Gibbs' treatise on heterogeneous equilibria and the phase rule. And in fitting conclusion, the observation of chemical periodicity: Döbereiner's triads, Newland's octaves, climaxed by the periodic law of Mendeleev and Meyer, 1869–1870.

In the chronicles above, those developments which supported the atomic idea have been stressed. On the other hand, Maxwell's development of electromagnetic theory, an undulatory theory, is an important link in the chain. Another fact of great consequence is that in the latter part of the 19th century a great amount of experimental work was devoted to the study of spectra of all kinds.

Today it is difficult to imagine the complacency of the physicist of 1890. Classical physics was a house in order: mechanics, thermodynamics, kinetic theory, optics, electromagnetic theory were the main foundations; an imposing display. By choosing tools from the appropriate discipline any problem could be solved. Of course, there were one or two problems which were giving some trouble, but everyone was confident that these would soon yield under the usual attack. There were two parts in this house of physics, the corpuscular and the undulatory, or the particle and the wave. Matter was corpuscular, light was undulatory, and that was that. The joint between matter and light did not seem very smooth.

19–3 The Earthquake

It is difficult to describe what happened next because everything happened so quickly. Within thirty-five years classical physics was shaken to the very foot of the cellar stairs. When the dust settled, the main foundations remained, not too much the worse for wear. But entirely new areas of physics were opened. Again only the barest mention of these events must suffice for the moment: the discovery of the photoelectric effect by Hertz in 1887. The discovery of x-rays by Roentgen in 1895. The discovery of radioactivity by Becquerel in 1896. The discovery of the electron by J. J. Thomson in 1897. The quantum hypothesis in blackbody radiation by Planck in 1900. The quantum hypothesis in the photoelectric effect by Einstein in 1905. Thomson's model of the atom in 1907. The scattering experiment with α-particles by Geiger, Marsden, and Rutherford in 1909. The nuclear model of the atom of Rutherford in 1911. Quantitative confirmation of Rutherford's calculations on scattering by Geiger and Marsden in 1913. The quantum hypothesis applied to the atom, the Bohr model of the atom, in 1913. Another development in the first decade which does not concern us directly here is the Einstein theory of relativity.

The year 1913 marks a major climax in the history of science. The application of Planck's quantum hypothesis to blackbody radiation, and later by Einstein to the photoelectric effect, had met with disbelief and in some quarters even with scorn. Bohr's application to the theory of the hydrogen atom compelled belief and worked a revolution in thought. In the following ten years this new knowledge was quickly assimilated and applied with spectacular success to the interpretation of spectra and chemical periodicity.

A new series of discoveries was made in the third decade of the 20th century. The theoretical prediction of the wave nature of matter by de Broglie in 1924. Experimental verification; measurement of the wavelength of electrons by Davisson and Germer in 1927. The quantum mechanics of Heisenberg and Schrödinger in 1927. Since 1927 the quantum mechanics has been successful in all of its applications to atomic problems. In principle any chemical problem can be solved on paper using the Schrödinger equation. In practice, the computations are so laborious for most chemical problems that experimental chemistry is, and will be for many years, a very active field. This attitude must be distinguished from that of the complacent physicist of 1890. Although the theoretical basis for attacking chemical problems is well understood today and it is unlikely that this foundation will be overturned, we recognize our limitations. Our understanding of nuclear structure is much less satisfactory. Nuclear physics has not yet had its Niels Bohr in the way that atomic physics had Niels Bohr in 1913.

We break off the chronology in 1927. Those discoveries since 1927 which concern us will be dealt with as they are needed.

Looking back on the developments before 1927 we see two main consequences. Radiation, which was a wave phenomenon in classical physics, was endowed with a particle aspect by the work of Planck, Einstein, and Bohr. Electrons and atoms, which were particles in the classical view, were given a wave aspect by the work of de Broglie, Schrödinger, and Heisenberg. The two parts of classical physics which did not join smoothly are brought together in a unified way in the quantum mechanics. The dual nature of matter and of light, the wave-particle nature, permits this unification.

19–4   Discovery of the Electron

From the time of Dalton, atoms were indivisible. The discovery of the electron by J. J. Thomson in 1897 was the first hint of the existence of particles smaller than atoms.

Thomson's discovery allowed speculation about the interior structure of the atom and extended the hope that such speculation could be verified experimentally.

The studies of electrical conduction in gases had led to the discovery of cathode rays. If a glass tube fitted with two electrodes connected to a source of high potential is evacuated, a spark will jump between the electrodes. At lower pressures the spark broadens to a glow which fills the tube; at still lower pressure various dark spaces appear in the glowing gas. At very low pressures the interior of the tube is dark, but its walls emit a fluorescence whose color depends only on the kind of glass. It was soon decided that the cathode was emitting some kind of ray, a cathode ray, which on impinging on the glass wall produced the fluorescence. Objects placed in the path of these cathode rays cast a shadow on the walls of the tube; the rays are deflected by electric and magnetic fields. Figure 19–1 shows the device used by J. J. Thomson in his famous experiment which showed that the cathode ray was a stream of particles, later called electrons.

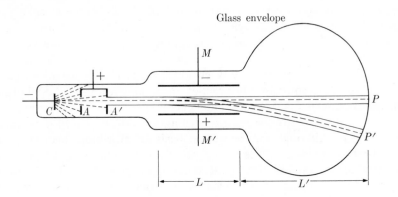

Fig. 19–1 Device to measure e/m for cathode rays.

In the highly evacuated tube, cathode rays are emitted from the cathode C. Two slotted metal plates A and A' serve as anodes. Passage through the two slots collimates the beam, which then moves in a straight line to hit the spot P at which the fluorescence appears. An electric field can be applied between the plates M and M'; a magnetic field can be applied in the region of M and M' but perpendicular to the plane of the drawing. The forces produced on the ray by the fields act in the vertical direction only; the horizontal component of velocity is unaffected by the fields. Two experiments are done.

The electrical field E is applied which pulls the beam downward and deflects the spot to P'; the magnetic field H is applied and adjusted so the spot returns to the original position P. If the beam consists of particles of charge e and mass m, then the force on the beam due to the electrical field is eE, and that due to the magnetic field is Hev/c, where v is the horizontal component of velocity of the particle and c

is the velocity of light (H is in electrostatic units). Since these forces are in balance, $eE = Hev/c$, and we obtain the horizontal velocity component in terms of E and H:

$$v = Ec/H. \tag{19–1}$$

In the second experiment, the magnetic field is turned off, and the deflection PP' under the electrical field only is measured. Since the force is eE, the vertical acceleration is eE/m. The time to pass through the field is $t = L/v$. After this time, the vertical component of velocity $w = (eE/m)t$; in this same time the vertical displacement is $s = (eE/m)t^2/2$. The value of s can be computed from the displacement PP' and the length L'. Using the value for t, we have

$$e/m = 2sv^2/EL^2,$$

and using the value for v from Eq. (19–1)

$$e/m = 2sEc^2/H^2L^2. \tag{19–2}$$

The experiment yields the value of e/m for the particles. The modern value of this ratio is

$$e/m = 5.2742 \times 10^{17}\text{esu/gm}.$$

From the direction of the deflection it is apparent that e is negative.

Earlier experiments on electrolysis had measured the ratio of charge to mass of hydrogen, the lightest atom. The modern value is

$$\left(\frac{e}{m}\right)_{\text{H}} = 2.8718 \times 10^{14} \text{ esu/gm}.$$

The e/m for the cathode particles was about 1840 times larger than that of hydrogen. At the time it was not known whether this was because of a difference in charge or mass or both. In 1913, R. A. Millikan measured the charge on the electron directly, the "oil-drop" experiment. The modern value is

$$e = 4.80286 \times 10^{-10} \text{ esu}.$$

Combining this with the e/m value, we obtain for the electron mass

$$m = 9.1083 \times 10^{-28} \text{ gm}.$$

From the atomic weight of hydrogen, and the value of the Avogadro number from kinetic theory, the mass of the hydrogen atom could be estimated. The modern value is

$$m_{\text{H}} = 1.6724 \times 10^{-24} \text{ gm}.$$

It was finally apparent that the charge on the hydrogen ion was equal and opposite to that on the electron, while the mass of the electron was very much less, 1836.57 times less, than that of the hydrogen atom. Being less massive, the electron was a more elementary particle than the atom. Presumably atoms were composites of negative electrons and positively charged matter, which was much more massive. After Thomson's work it was possible to think of how atoms could be built with such materials.

19–5 Positive Rays and Isotopes

The discovery of positive rays, canal rays, by Goldstein in 1886 is another important result of the studies of electrical conduction in gases. The device is shown in Fig. 19–2.

The cathode C has a hole, a canal, drilled through it. In addition to the usual discharge between A and C, a luminous stream emerges from the canal to the left of the cathode. This ray is positively charged and, reasonably enough, is called a positive ray. The systematic study of positive rays was long delayed, but it was determined at an early date that their characteristics depended on the kind of residual gas in the tube. In contrast, the cathode ray did not depend on the residual gas.

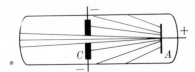

Fig. 19–2 Simple positive-ray tube.

Thomson was engaged in the measurement of the e/m of positive rays by the same general method as he used for the electron when, in 1913, he discovered that neon consisted of two different kinds of atoms; one having a mass of 20, the other having a mass of 22. These different atoms of the same element are called *isotopes*, meaning "same place," that is, in the periodic table. Since this discovery that an element may contain atoms of different mass, the isotopic constitution of all the elements has been determined. Moreover, as is well known, in recent years many artificial isotopes have been synthesized by the high-energy techniques of physics.

Isotopes of an element are almost indistinguishable chemically, since the external electron configurations are the same. Their physical properties differ slightly because of the difference in mass. The differences are most pronounced with the lightest elements, since the relative difference in mass is greatest. The differences in properties of isotopes are most marked in the positive-ray tube itself, where the strengths of the applied electrical and magnetic fields can be adjusted to spread the rays having different values of e/m into a pattern resembling a spectrum, called a mass spectrum. The modern mass spectrometer is a descendant of Thomson's e/m apparatus.

19–6 Radioactivity

In 1896, shortly after the discovery of x-rays, H. Becquerel tried to discover if fluorescent substances emitted x-rays. He found that a fluorescent salt of uranium emitted a penetrating radiation which was not connected with the fluorescence of the salt. The radiation could pass through several thicknesses of the black paper used to protect photographic plates and through thin metal foils. The radiation differed from x-radiation in that it could be resolved into three components, α-, β-, and γ-rays, by the imposition of a strong magnetic field. The β-ray has the same e/m as the electron; the γ-ray is undeflected in the field; the α-ray is positively charged, with an e/m value of one-half that of hydrogen. The β-ray is a stream of electrons; the α-ray is a stream of helium nuclei; the γ-ray is a light ray of extremely short wavelength. A great deal of effort was devoted to the study of radioactivity in the years that followed. The discovery of two new elements, polonium and radium, by Pierre and Marie Curie, being one of the notable accomplishments.

The striking fact about radioactivity is that the rate of emission of the rays is completely unaffected by even the most drastic changes in external conditions such as chemical environment, temperature, pressure, and electrical and magnetic fields. The rays are emitted from the nucleus; the lack of influence of external variables on this process shows that the situation in the nucleus is independent of these variables. Secondly, the energies of the emitted rays are of the order of one-million electron volts, (1 ev = 23 kcal). This energy is enormously greater than that associated with any chemical transformation.

The law governing the rate of radioactive decay of a nucleus is described in Section 31–5.

19–7 Alpha-ray Scattering

In 1908 Thomson proposed a model of the atom: the positive charge was uniformly spread throughout a sphere of definite radius; to confer electrical neutrality, electrons were imbedded in the sphere. For stability according to classical theory, the electrons had to be at rest. This requirement could be met for the hydrogen atom by having the electron at the center of the sphere. This model failed the crucial test provided by the scattering of α-rays by thin metal foils.

In 1909 Geiger and Marsden discovered that if a beam of α-particles was directed at a thin gold foil, some of the α-particles were scattered back toward the source. Figure 19–3 shows the experiment in scheme. The majority of the α-particles pass

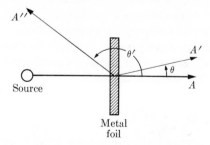

Fig. 19–3 α-ray scattering experiment.

through the foil and can be detected at A. Some are scattered through small angles θ and are detected at A'; remarkably, quite a few are scattered through large angles such as θ' and can be detected at A''.

The scattering occurs because of the repulsion between the positive charge on the α-particle and the positive charges on the atoms of the foil. If the positive charge on the atoms were spread uniformly, as in the Thomson model, the scattering would be the result of a gradual deflection of the particle as it progressed through the foil. The scattering angle would be very small. Rutherford reasoned that the scattering at large angles was due to a very close approach of the α-particle to a positively charged center with subsequent rebound; a single scattering event. By calculation he could show that to be scattered through a large angle in a single event, the α-particle would have to approach the positive part of the scattering atom very closely, to within 10^{-12} cm. The sizes of atoms were known to be about 10^{-8} cm. Since the mass of the atom is associated with the positively charged part of the atom, Rutherford's calculation implied that the positive charge and the mass of the atom are concentrated in a space which is very much smaller than that occupied by the atom as a whole.

The nuclear model of the atom proposed by Rutherford supposed that the atom was a sphere of negative charge, not having much mass, having a tiny kernel or nucleus at the center in which the mass and positive charge are concentrated. Using the nuclear model, Rutherford calculated the angular distribution of scattered α-particles. Later experiments of Geiger and Marsden confirmed the predicted distribution in all its particulars.

The Rutherford model had its difficulties. The sphere, uniformly filled with negative charge, was incompatible with the concept of the electron as a particle which should be localized in space. But it is not possible to take a positive discrete particle and a negative discrete particle, place them a certain distance apart and ask them to stay put. Being oppositely charged, they will attract one another; the electron will fall into the nucleus. Thomson's model did not have this type of instability. Matters are not helped by whirling the electron around in an orbit to achieve the stability of a satellite in orbit around a planet. The electrical attractive force could be balanced by the centrifugal force, but a fundamental objection arises. An electron in orbit is subject to a continual acceleration toward the center, else the orbit would not be stable. Classical electromagnetic theory, confirmed by Hertz's discovery of radio waves, required an accelerated electrical charge to emit radiation. The consequent loss of energy should bring the electron down in a spiral to the nucleus. This difficulty seemed insuperable. But less than two years later Niels Bohr found a way out. To appreciate Bohr's contribution we must return to 1900 and follow the course of another series of discoveries.

19–8 Radiation and Matter

By 1900 the success of Maxwell's electromagnetic theory had firmly established the wave nature of light. One puzzle which remained was the distribution of wavelengths in a cavity, or blackbody; the observed distribution had eluded explanation on accepted principles. In 1900 Max Planck calculated the distribution, within the experimental error, in a completely mysterious way. Planck's work proved ultimately to be the key to the entire problem of atomic structure; yet at first glance it seems to have little bearing on that problem.

A perfect blackbody is one which adsorbs all the radiation, light, which falls upon it. Experimentally the most nearly perfect blackbody is a pin hole in a hollow object. Radiation falling on the pin hole enters the cavity and is trapped (absorbed) within the cavity. Let the radiation in the cavity be brought to thermal equilibrium with the walls at a temperature T. Since there is energy in the radiation, there is a certain energy density in the cavity, $u = E/V$, where E is the energy, V the volume, and u the energy density. From electromagnetic theory, the pressure exerted by the radiation is $p = u/3$, and experiment shows that the energy density is independent of the volume, that is, $u = u(T)$. The relation between u and T is obtained from the thermodynamic equation of state, Eq. (10–31):

$$\left(\frac{\partial E}{\partial V}\right)_T = T\left(\frac{\partial p}{\partial T}\right)_V - p.$$

Since $E = u(T)V$, $(\partial E/\partial V)_T = u$. Also $p = u(T)/3$, so that $(\partial p/\partial T)_V = \frac{1}{3}(du/dT)$. The equation of state becomes $du/dT = 4u/T$. Integration yields

$$u = \alpha T^4, \tag{19–3}$$

where the constant $\alpha = 7.569 \times 10^{-15}$ erg·cm^{-3}·deg^{-4}. The rate of emission of energy from a cavity per cm^2 of opening is proportional to the energy density within; this rate is the total emissive power, e_t; thus,

$$e_t = \sigma T^4, \tag{19–4}$$

where the Stefan-Boltzmann constant $\sigma = 5.672 \times 10^{-5}$ erg·cm^{-2}·sec^{-1}·deg^{-4}. Equation (19–4) is the Stefan-Boltzmann law; among other things it is used to establish the absolute temperature scale at very high temperatures.

So far everything is fine; we may keep our confidence in the second law of thermodynamics. The difficulty is this: the energy in the cavity is the sum of the energies of light waves of many different wavelengths. Let $u_\lambda \, d\lambda$ be the energy density contributed by light waves having wavelengths in the range λ to $\lambda + d\lambda$. Then the total energy density u is

$$u = \int_0^\infty u_\lambda \, d\lambda, \tag{19–5}$$

where we sum the contribution of all wavelengths from zero to infinity. It is rather easy to measure the distribution function u_λ, shown in Fig. 19–4. Experimentally it has been shown that the wavelength at the maximum of this spectral distribution is inversely proportional to the temperature:

$$\lambda_m T = \text{constant}.$$

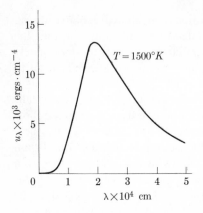

Fig. 19-4 Spectral distribution in blackbody radiation.

This is Wien's displacement law. Classical principles had failed to explain the shape of the curve in Fig. 19–4 and failed to predict the displacement law of Wien. The application of the classical law of equipartition of energy between the various degrees of freedom by Rayleigh and Jeans was satisfactory at long wavelengths but failed at short wavelengths, in the ultraviolet; "ultraviolet catastrophe."

The Rayleigh-Jeans treatment assigned the classical value kT to the average energy of each mode of oscillation in the cavity; $\frac{1}{2}kT$ for kinetic and $\frac{1}{2}kT$ for potential energy. The number of modes of oscillation dn in the wavelength range from λ to $\lambda + d\lambda$ per unit volume of the cavity is† $dn = 8\pi \, d\lambda/\lambda^4$. The energy density in the same wavelength range is $u_\lambda \, d\lambda$ and is equal to the number of modes of oscillation multiplied by kT. Therefore, $u_\lambda \, d\lambda = 8\pi kT \, d\lambda/\lambda^4$, so that

$$u_\lambda = 8\pi kT/\lambda^4, \qquad (19\text{--}6)$$

which is the Rayleigh-Jeans formula. It predicts an infinite energy density as $\lambda \to 0$; hence an infinite value of the total energy density in the cavity, an absurdity.

If a mode of oscillation can possess any arbitrary amount of energy from zero to infinity, then there is no reason for the Rayleigh-Jeans formula to be incorrect. Let us suppose, for the sake of argument, that an oscillator cannot have any arbitrary energy but may have energy only in integral multiples of a certain unit of energy ϵ. Then the distribution of a collection of these oscillators is discrete and we can represent it by

Energy	0	ϵ	2ϵ	3ϵ	4ϵ	...
Number	n_0	n_1	n_2	n_3	n_4	...

We further suppose that the distribution is governed by the Boltzmann law;

---

† The derivation of this formula is beyond the scope of the treatment here.

$n_i = n_0 e^{-\epsilon_i/kT}$. Using these ideas we compute the total number of particles N and the total energy:

$$N = \sum_i n_i = n_0 + n_0 e^{-\epsilon/kT} + n_0 e^{-2\epsilon/kT} + n_0 e^{-3\epsilon/kT} + \cdots.$$

If we set $x = e^{-\epsilon/kT}$, this expression becomes

$$N = n_0(1 + x + x^2 + x^3 + \cdots).$$

The series is the expansion of $1/(1 - x)$, so we obtain

$$N = n_0/(1 - x). \tag{19-7}$$

The average energy $\bar{E}$ is given by

$$N\bar{E} = n_0(0) + n_1\epsilon + n_2(2\epsilon) + n_3(3\epsilon) + \cdots$$

$$= n_0\epsilon(x + 2x^2 + 3x^3 + \cdots) = n_0\epsilon x(1 + 2x + 3x^2 + \cdots).$$

But $(1 + 2x + 3x^2 + \cdots) = d(1 + x + x^2 + x^3 + \cdots)/dx = d[1/(1 - x)]/dx = 1/(1 - x)^2$, so that

$$N\bar{E} = \frac{n_0\epsilon x}{(1 - x)^2}.$$

Putting in the values of N and x, this becomes

$$\bar{E} = \frac{\epsilon e^{-\epsilon/kT}}{1 - e^{-\epsilon/kT}} = \frac{\epsilon}{e^{\epsilon/kT} - 1}. \tag{19-8}$$

If we use the value given by Eq. (19–8) for the average energy in a mode of vibration, then multiplying by the number of modes in the wavelength range to compute the spectral distribution, we obtain for u_λ

$$u_\lambda = \frac{8\pi}{\lambda^4}\left(\frac{\epsilon}{e^{\epsilon/kT} - 1}\right). \tag{19-9}$$

Now if ϵ is a constant in Eq. (19–9), then we are no better off than were Rayleigh and Jeans.

Planck took the extraordinary step of setting ϵ inversely proportional to the wavelength, recognizing that the Wien displacement law would come out of the resulting equation. Since the frequency times the wavelength is equal to the velocity, we have $1/\lambda = v/c$, where v is the frequency and c the velocity. Setting ϵ proportional to $1/\lambda$ is equivalent to setting it proportional to the frequency:

$$\epsilon = hv = hc/\lambda, \tag{19-10}$$

where Planck's constant $h = 6.6252 \times 10^{-27}$ erg·sec. Putting the value of ϵ from

Eq. (19–10) into Eq. (19–9) yields the distribution function

$$u_\lambda = \frac{8\pi hc}{\lambda^5} \frac{1}{e^{hc/\lambda kT} - 1}.$$ (19–11)

By properly choosing the value of the constant h, Eq. (19–11) agreed with the measured distribution within the experimental error! To find the maximum, we set $du_\lambda/d\lambda = 0$; the Wien displacement law is obtained in the form

$$\lambda_m T = hc/4.965\mathbf{k}.$$ (19–12)

The value of ϵ is the energy gap separating the energies of the various groups of oscillators; classically this gap should be zero to yield a continuous energy distribution. Planck's assumption that $\epsilon = h\nu$ required the gap to be finite, approaching zero, the classical value, only at infinitely long wavelengths.

The worst part of this is that it lacks the logic of classical physics and it has far-reaching implications. If the radiation in a cavity can possess energy only in multiples of a certain unit $h\nu$, then it can exchange energy with the oscillators in the cavity walls only in multiples of this unit. Therefore the interchange of energy had to be discontinuous also; energy had to be exchanged in small lumps or bundles called *quanta*. The quantum of energy for an oscillator is $h\nu$.

The nature of light seemed no longer to be simple. Light was a wave motion, but with Planck's work it acquired a corpuscular aspect. The light wave contains energy in elementary discrete units, quanta.

19–9 The Photoelectric Effect

As may be imagined, Planck's discovery excited very little interest and no controversy. The prevailing attitude seemed to be "if we ignore it, it will go away." Perhaps it might have gone away but for Einstein's interpretation of the photoelectric effect, another longstanding thorn in the side of classical physics.

If a beam of light falls on a clean metal plate in vacuum, the plate emits electrons. This effect, discovered by Hertz in 1887, had been thoroughly investigated. Two aspects of the phenomenon were the rocks on which classical physics foundered.

1. Whether or not electrons are emitted from the plate depends only on the frequency of the light and not at all on the intensity of the beam. The number of electrons emitted is proportional to the intensity.
2. There is no time lag between the light beam striking the plate and the emission of the electrons.

An electron in a metal is bound by a potential energy ω, which must be supplied to bring the electron outside the metal. If, in addition, the electron outside the metal has kinetic energy, then the total energy of the electron is

$$E_t = \tfrac{1}{2}mv^2 + \omega.$$ (19–13)

Presumably the electron acquires this energy from the beam of light. Classically, the energy of the light beam depends on its intensity, and that energy should be absorbed continuously by the metal plate. It can be shown that for weak intensities and reasonable values of ω that after the onset of illumination, a long time period, days or even years, should intervene before any electron would soak up enough energy to be kicked out of the metal. After this time interval many electrons should be energetic enough to escape and a steady current should flow from then on. Increasing the intensity should lessen the time interval. No time interval has ever been observed. The proportionality of the current to intensity is reasonable on classical grounds, but the absence of a time interval could not be explained.

In 1905 Albert Einstein took a different view of the problem. Classically the energy of the light beam is absorbed continuously by the metal and that energy is divided among all the electrons in the plate, each electron receiving only a tiny share of the total. Suppose that the energy of the light beam is concentrated in Planck's quanta of energy $h\nu$. Suppose further that the entire quantum of energy must be accepted by a single electron, and cannot be divided among all the electrons present. Then the energy of the electron after accepting the quantum must be $h\nu$, and this must be the total energy after emission, Eq. (19–13). Therefore,

$$h\nu = \tfrac{1}{2}mv^2 + \omega. \tag{19–14}$$

Equation (19–14) is the Einstein photoelectric equation. It is apparent from the equation that below a critical frequency, ν_0, given by $h\nu_0 = \omega$, the electron does not gain sufficient energy from the light quantum to escape the metal. This explains the "cut-off" frequency ν_0 which is observed. A greater light intensity means only that more quanta are absorbed per unit time and more electrons are emitted; the energies of the emitted electrons are completely independent of the intensity.

Using the same value of h as had been obtained by Planck in the treatment of blackbody radiation, the Einstein equation provided a completely satisfactory explanation of the photoelectric effect. Satisfactory? Yes, but very unsettling! Einstein spoke of photons, corpuscles of light, each carrying energy $h\nu$. Planck's idea seemed to be gaining ground; a most distressing turn of events.

19–10 Bohr's Model of the Atom

Throughout the 19th century, spectroscopy was a very popular field of study. A great number of precise measurements of wavelengths of lines had been made and catalogued. Regularities in the spacing between lines had been observed and correlated by empirical formulas. One of the most famous of these formulas is that given by Balmer in 1885. Balmer found that the wavelengths, in angstrom units, of nine lines in the visible and near-ultraviolet spectrum of hydrogen could be expressed by the formula

$$\lambda = 3645.6\left(\frac{n^2}{n^2 - 2^2}\right), \tag{19–15}$$

where n had the integral values $3, 4, 5, \ldots, 11$. Each integral value of n corresponds through Eq. (19–15) to a wavelength. The computed wavelengths agreed excellently with the measured values. Somewhat later, Ritz proposed a more general formula which, for hydrogen, takes the form

$$\frac{1}{\lambda} = \tilde{\nu} = R_{\mathrm{H}}\left(\frac{1}{k^2} - \frac{1}{n^2}\right), \tag{19–16}$$

where both k and n are integers, R_{H} is the Rydberg constant for hydrogen. The wave number $\tilde{\nu}$ is the reciprocal of the wavelength. If $k = 2$, the Rydberg formula reduces to the Balmer formula. The Rydberg formula is remarkably accurate, and with slight modification it represents the wavelengths in the spectra of many different atoms. Because of the accuracy of the formula and the precision with which wavelengths can be measured, the Rydberg constant was known with great accuracy. Today the value is known to within less than one part in ten million. The modern value is $R_{\mathrm{H}} = 109{,}677.581 \pm 0.007 \mathrm{~cm}^{-1}$.

The spectrum emitted by an atom is presumably related to the structure of the atom. Until 1913, attempts to relate the spectrum to a definite atomic model were unsuccessful. By 1913 it was known that the atom had a positively charged nucleus, but the nuclear model of Rutherford was unstable according to classical electromagnetic theory. This Gordian knot was cut by Niels Bohr in 1913.

In the Bohr model the hydrogen atom consists of a central nucleus with a charge $+e$, and an electron of charge $-e$ whirling about the nucleus with velocity v in an orbit of radius r; Fig. 19–5. For mechanical stability, the electrical force of attraction $-e^2/r^2$ must balance the centrifugal force mv^2/r:

$$-\frac{e^2}{r^2} + \frac{mv^2}{r} = 0,$$

or

$$mv^2 = \frac{e^2}{r}. \tag{19–17}$$

The total energy E is the sum of the kinetic energy $mv^2/2$, and the potential energy $-e^2/r$:

$$E = \frac{mv^2}{2} - \frac{e^2}{r}.$$

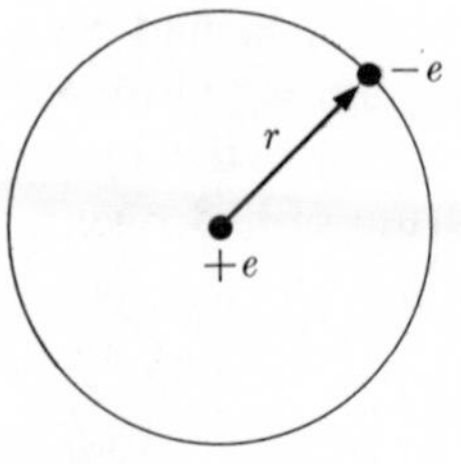

Figure 19–5

Using Eq. (19–17), this becomes

$$E = -\frac{e^2}{2r}. \tag{19–18}$$

Classically, since the electron is accelerated, this system should radiate. To avoid this difficulty, Bohr broke completely with tradition. Bohr assumed: (1) that the electron can move around the nucleus only in certain orbits, and not in others (classically, no particular orbit is preferable to any other); (2) that these allowed orbits correspond to definite stationary states of the atom, and in such a stationary state the atom is stable and does not radiate (Bohr avoided the classical difficulty by simply *assuming* that it was not a difficulty in these special circumstances!†); (3) that in the transition of the electron from one stable orbit to another, radiation is emitted or absorbed, the frequency of the radiation being given by

$$h\nu = \Delta E,$$

where ΔE is the energy difference between the two stationary states and h is Planck's constant. (There was nothing quite so nonclassical as a formula with Planck's constant in it.)

The problem of choosing these special orbits out of all the possible ones remained. Bohr's condition is that the angular momentum mvr be an integral multiple of $\hbar = h/2\pi$:

$$mvr = n\hbar, \qquad n = 1, 2, \ldots . \tag{19–19}$$

This condition is equivalent in a certain sense to Planck's condition on an oscillator.

Solving Eqs. (19–17) and (19–19) for v and r, we obtain

$$v = e^2/n\hbar \qquad \text{and} \qquad r = n^2\hbar^2/me^2 .$$

If $n = 1$, then $r = a_0$, the radius of the first Bohr orbit;

$$a_0 = \hbar^2/me^2 = 0.52917 \times 10^{-8} \text{ cm} \tag{19–20}$$

then

$$r = n^2 a_0. \tag{19–21}$$

Using this value of r in Eq. (19–18) for the total energy yields

$$E_n = -\frac{e^2}{2a_0}\left(\frac{1}{n^2}\right), \tag{19–22}$$

where the subscript on E indicates that the energy depends on the integer n. Equation (19–22) expresses the energy entirely in terms of fundamental constants, e, h, m, and the integer n. Consider two stationary states, one described by the integer n and the other by the integer k. The difference in energy of these two states is

$$\Delta E_{nk} = E_n - E_k = \frac{e^2}{2a_0}\left(\frac{1}{k^2} - \frac{1}{n^2}\right) .$$

---

† Bohr's approach cannot be recommended for solving *standard* problems in physical chemistry!

By Bohr's third assumption, this difference should equal hv:

$$hv = \frac{e^2}{2a_0}\left(\frac{1}{k^2} - \frac{1}{n^2}\right).$$

If we replace v by $v = c/\lambda = c\tilde{v}$, the equation becomes

$$\tilde{v} = \frac{e^2}{2a_0hc}\left(\frac{1}{k^2} - \frac{1}{n^2}\right), \tag{19-23}$$

which is the Rydberg formula. Bohr's argument yields a value of the Rydberg constant:

$$R_H = \frac{e^2}{2a_0hc}. \tag{19-24}$$

Calculating the value of R_H from Eq. (19–24), Bohr obtained a value of R_H which agreed with the empirical value within the uncertainty of the knowledge of the constants.

Bohr had calculated the most accurately known experimental constant in physics by a method which was, to use a mild description, simply an outrage! The corpuscular nature of light had come to stay; it could no longer be ignored. No evangelist ever made so many converts in so short a time as did Bohr.

The connection between matter and radiation soon became firmly established. In the decade following Bohr's discovery, what is now called the quantum theory or the "old quantum theory" burst into full flower. The systematic interpretation of the data in the catalogues of spectra went forward by leaps and bounds. The Bohr-Sommerfeld atom model, which used elliptical as well as circular orbits, was introduced and found useful. From studies of spectra Bohr constructed a theoretical periodic chart which agreed with that of the chemists. A detail was different; according to Bohr, element 72, which chemists had sought among the rare earths, was not a rare earth, but a member of the fourth family, with titanium and zirconium. Shortly thereafter, von Hevesy looked at the spectrum of zirconium and found that many of the lines should be ascribed to element 72. Therefore, zirconium was a mixture of zirconium and element 72. The new element was named hafnium, after the ancient name of Copenhagen, in honor of Bohr who is Danish. The discovery of hafnium ended a long controversy over the atomic weight of zirconium; samples used by different investigators contained different amounts of hafnium, so the discrepancies were rather large.

The Bohr theory of the atom destroyed the last pockets of resistance to the quantum concept. Yet the wave attributes of light were there too. The nature of light took on a dual aspect. This duality in the nature of light is accepted now, though to some in the beginning it was a bitter pill.

19–11   Particles and Louis de Broglie

In 1924, Louis de Broglie argued on theoretical grounds that particles should have a wavelength associated with them. The de Broglie formula for the wavelength is

$$\lambda = h/p = h/mv, \qquad (19\text{–}25)$$

where $p = mv$ is the momentum of the particle. We cannot reproduce de Broglie's argument here, since it requires some knowledge of electromagnetic theory as well as relativity theory. However, if a particle does have a wavelength, then that fact must be capable of experimental demonstration.

The experimental test was made by Davisson and Germer in 1927. A beam of light reflected from a ruled grating produces a diffraction pattern; diffraction is a property exclusively of wave motion. Davisson and Germer directed an electron beam at a nickel crystal. The rows of nickel atoms serve as the ruling. The intensity of the diffracted beam was measured as a function of the diffraction angle. They found maxima in the intensity at the special values of the diffraction angle. From these values of the diffraction angle and the usual diffraction formula they computed the wavelength of the electrons. This value of the wavelength agreed with that predicted by the de Broglie formula for electrons having the experimental velocity.

This confirmation of de Broglie's prediction brought duality into the nature of fundamental particles. A particle was not simply a particle but had a wave aspect to its nature. This idea led very quickly to the development of the wave mechanics, or quantum mechanics, by Heisenberg and Schrödinger. All of our modern ideas on atomic and molecular structure are based on wave mechanics.

The distinct concepts, wave or particle, of 19th century physics are now inseparably mingled. The wave mechanics, so essential to our ideas now, would have been a contradiction in terms in the 19th century. The question of whether an electron, or a photon, is a wave or particle has lost all meaning. We can say with precision in what circumstances it is useful to treat the electron as a classical particle or the photon as a classical wave. We know when we must consider the wave aspects of the classical particles and the particle aspects of the classical wave. Any final classification into particle or wave would be artificial. Both particles and waves have a more general nature than their names indicate. We use the old names fully realizing the more general character of the entity in question.

In 1927 Werner Heisenberg and Erwin Schrödinger independently formulated the law which governs the motion of a particle. The discussion here will be more closely related to Schrödinger's treatment.

19–12   The Classical Wave Equation

The classical law governing wave motion is the wave equation

$$\frac{\partial^2 D}{\partial x^2} + \frac{\partial^2 D}{\partial y^2} + \frac{\partial^2 D}{\partial z^2} = \frac{1}{v^2}\frac{\partial^2 D}{\partial t^2}, \qquad (19\text{–}26)$$

where x, y, z are the coordinates, t the time, v the velocity of propagation, and D the displacement of the wave. If v does not depend on the time, then the displacement is the product of a function of the coordinates only, $\psi(x, y, z)$, and a periodic function of time, $e^{i2\pi vt}$, where v is the frequency of the wave, and $i = \sqrt{-1}$. Then

$$D = \psi(x, y, z)e^{i2\pi vt}, \tag{19-27}$$

which means that if we sit at a fixed position x, y, z, and observe the value of D, then at an arbitrary time $t = 0$, $e^{i2\pi v0} = 1$ and $D = \psi$. At a later time,

$$t = 0 + 1/v, \qquad e^{i2\pi v/v} = e^{i2\pi} = 1 \dagger$$

and the displacement $D = \psi$, the same value as at $t = 0$. Thus the value of D at any point varies with a period, $t_0 = 1/v$. By the mean value theorem, we can calculate the average value of the displacement D in the time interval t_0:

$$\bar{D} = \frac{1}{t_0} \int_0^{t_0} D \, dt = \frac{1}{t_0} \psi(x, y, z) \int_0^{t_0} e^{i2\pi vt} \, dt \, .$$

Evaluation of the integral yields $\bar{D} = 0$. This result is physically obvious, since in one complete period D is positive for half the time and negative for half the time; the values sum to zero. We avoid this difficulty by computing the average value of the square of the absolute value of the function:

$$\overline{|D|^2} = \frac{1}{t_0} \int_0^{t_0} |D|^2 \, dt \, .$$

Since D may be a complex function, to compute $|D|^2$ we use the formula $|D|^2 = D^*D$, where D^* is the complex conjugate of D obtained by replacing i in the function by $-i$. Then

$$|D|^2 = \psi^*\psi e^{i2\pi vt} e^{-i2\pi vt} = \psi^*\psi \, .$$

Putting this value in the integral, we obtain

$$\overline{|D|^2} = \psi^*\psi \frac{1}{t_0} \int_0^{t_0} dt = \psi^*\psi = |\psi|^2 \, . \tag{19-28}$$

By Eq. (19–28) the *time average* of the square of the absolute value of the displacement is equal to the square of the absolute value of the space-dependent part ψ. The function ψ is called the *amplitude* of the wave.

Using the value of D given by Eq. (19–27), we can form the second derivatives and put them into the wave equation, Eq. (19–26). The result, after division by $e^{i2\pi vt}$, is

$$\frac{\partial^2 \psi}{\partial x^2} + \frac{\partial^2 \psi}{\partial y^2} + \frac{\partial^2 \psi}{\partial z^2} + \frac{4\pi^2 v^2 \psi}{v^2} = 0. \tag{19-29}$$

---

† Remember that $e^{ix} = \cos x + i \sin x$. Therefore, e^{ix} has the period 2π.

In Eq. (19–29) only the space coordinates appear; this equation governs the spatial dependence of the amplitude ψ. After solving this differential equation for ψ, we can write down the value of ψ^* immediately. Multiplying ψ and ψ^* yields, through Eq. (19–28), a value of the time average of the square of the absolute value of the displacement. It is the time average which is of interest in the discussion of stationary states.

19–13 The Schrödinger Equation

Now we can make an argument for the Schrödinger equation.† If, as de Broglie says, the particle has some of the properties of a wave, then it seems likely that it will have some property which is analogous to the displacement of a classical wave. Since in an atomic system we cannot follow the detailed motion of a particle in time, perhaps we should concentrate on the time average value of the displacement analogue, which can be calculated if we know the amplitude. The classical amplitude ψ is governed by Eq. (19–29); this equation can be translated into a nonclassical one by using de Broglie's equation, $\lambda = h/mv$. But for any wave, the frequency times the wavelength is the velocity, $\lambda v = v$. Combining this with de Broglie's relation, we have $v/v = mv/h$. Using this in Eq. (19–29), it becomes

$$\frac{\partial^2 \psi}{\partial x^2} + \frac{\partial^2 \psi}{\partial y^2} + \frac{\partial^2 \psi}{\partial z^2} + \frac{4\pi^2 m^2 v^2}{h^2}\psi = 0. \tag{19–30}$$

For the moment we will think of ψ as some analogue of the classical amplitude. As yet we have no meaning for ψ, but using Eq. (19–30), we can express a familiar mechanical variable, the velocity v, in terms of ψ and its derivatives. Suppose we solve Eq. (19–30) algebraically for v and then calculate the kinetic energy, $\frac{1}{2}mv^2$. This yields

$$E_{\text{kin}}\psi = \frac{1}{2m}\left[-\hbar^2\frac{\partial^2 \psi}{\partial x^2} - \hbar^2\frac{\partial^2 \psi}{\partial y^2} - \hbar^2\frac{\partial^2 \psi}{\partial x^2}\right], \tag{19–31}$$

where we have used $\hbar = h/2\pi$. If it were not for the ψ function in Eq. (19–31), we would be tempted to see a similarity between this equation and the classical one,

$$E_{\text{kin}} = \frac{1}{2m}(p_x^2 + p_y^2 + p_z^2). \tag{19–32}$$

Since ψ bothers us, suppose we just leave a blank where it appears in Eq. (19–31), then

$$\mathbf{K} = \frac{1}{2m}\left[-\hbar^2\frac{\partial^2}{\partial x^2} - \hbar^2\frac{\partial^2}{\partial y^2} - \hbar^2\frac{\partial^2}{\partial z^2}\right], \tag{19–33}$$

where $\mathbf{K}$ has replaced E_{kin}, since the right-hand side of Eq. (19–33) is an *operator*;

---

† This must be regarded as "argument for," not "proof of" or "derivation of."

this operator tells us to perform the operation of taking the second partial derivatives of some function, multiply each by $-\hbar^2/2m$, and add them together. We can see an analogy between the classical equation, Eq. (19–32), and Eq. (19–33). Corresponding to the classical E_{kin}, in the quantum mechanics there is an operator $\mathbf{K}$. Corresponding to the classical momentum p_x, there is an operator $\mathbf{p}_x$. We find the momentum operator easily using p_x as an example. Comparing Eqs. (19–32) and (19–33), we get

$$\mathbf{p}_x^2 = -\hbar^2 \frac{\partial^2}{\partial x^2} = i^2 \hbar^2 \frac{\partial^2}{\partial x^2} = \left(-i\hbar \frac{\partial}{\partial x}\right)\left(-i\hbar \frac{\partial}{\partial x}\right),$$

where $i = \sqrt{-1}$. Therefore, the operators corresponding to the momenta in the three directions are

$$\mathbf{p}_x = -i\hbar \frac{\partial}{\partial x}, \qquad \mathbf{p}_y = -i\hbar \frac{\partial}{\partial y}, \qquad \mathbf{p}_z = -i\hbar \frac{\partial}{\partial z}. \qquad (19\text{–}34)$$

All of this is very puzzling, but we go a step further. Using Eq. (19–31), we compute the total energy $E = E_{kin} + V$, where V is the potential energy and is usually a function of the coordinates; then we have

$$E\psi = \frac{1}{2m}\left[-\hbar^2 \frac{\partial^2 \psi}{\partial x^2} - \hbar^2 \frac{\partial^2 \psi}{\partial y^2} - \hbar^2 \frac{\partial^2 \psi}{\partial z^2}\right] + V(x, y, z)\psi, \qquad (19\text{–}35)$$

which could be written in operator form as

$$E\psi = \mathbf{K}\psi + \mathbf{V}\psi. \qquad (19\text{–}36)$$

Either of Eqs. (19–35) or (19–36) is the Schrödinger equation, which bears some similarity to the classical equation for the conservation of energy:

$$E = E_{kin} + E_{pot}. \qquad (19\text{–}37)$$

Indeed, the Schrödinger equation is the quantum-mechanical analogue of this classical equation.

19–14 The Interpretation of ψ

A fly in the ointment of Eq. (19–36) is that this sort of partial differential equation often has complex solutions, solutions with real and imaginary parts. The physical quantities we measure are real quantities. We can convert the equation into one containing only real quantities if we multiply both sides by the complex conjugate of ψ. Then we have (E is a constant!)

$$\psi^* E\psi = \psi^* \mathbf{K}\psi + \psi^* \mathbf{V}\psi.$$

Since $\psi^* \psi = |\psi|^2$, and $\mathbf{V}\psi = V(x, y, z)\psi$, this equation can be written as

$$E|\psi|^2 = \psi^* \mathbf{K}\psi + V(x, y, z)|\psi|^2. \qquad (19\text{–}38)$$

Having derived this equation which contains only real quantities, let us divert our attention for a few moments.

Suppose we wish to compute the average potential energy of a classical particle (moving in one dimension) over a time interval T in which it traverses a distance X. We multiply the potential energy V by the probability of finding the particle between x and dx. This probability is dt/T, the fraction of the total time spent in the interval dx. The average potential energy is

$$\overline{V} = \int_0^T V \frac{dt}{T}.$$

If the particle travels a distance dx in the interval dt, and a distance X in the interval T, then the average value is

$$\overline{V} = \int_0^X V \left(\frac{1}{X}\right) dx$$

if V is a function only of x. If V is a function of the three coordinates, we would have an expression such as

$$\overline{V} = \int V \left(\frac{1}{\varphi}\right) d\tau,$$

where $d\tau$ is the small element of volume, and $1/\varphi$ is the probability per unit volume of finding the particle at the position x, y, z.

Now it seems that we could do something similar with Eq. (19–38). First multiply by the small volume element $d\tau$,

$$E|\psi|^2 \, d\tau = \psi^* \mathbf{K} \psi \, d\tau + V(x, y, z)|\psi|^2 \, d\tau.$$

If we interpret $|\psi|^2$ as a probability per unit volume, then the second term on the right is the potential energy of the particle at the position x, y, z, multiplied by the probability of finding it in the volume element at that position. If we integrate over the entire coordinate space, that second term should be the average potential energy. Thus,

$$E \int |\psi|^2 \, d\tau = \int \psi^* \mathbf{K} \psi \, d\tau + \int V(x, y, z)|\psi|^2 \, d\tau. \qquad (19\text{–}39)$$

The integration extends over all space, and E is removed from the integral, since it is a constant. The sum of the probability density $|\psi|^2$ times $d\tau$ must be the total probability of finding the particle and this must be unity; the particle must be somewhere!

$$\int |\psi|^2 \, d\tau = 1. \qquad (19\text{–}40)$$

Equation (19–40) is called the normalization condition; if ψ fulfills this condition, ψ is called a normalized wave function. Then Eq. (19–39) becomes

$$E = \int \psi^* \mathbf{K} \psi \, d\tau + \int V|\psi|^2 \, d\tau,$$

or, more symmetrically,

$$E = \int \psi^* \mathbf{K} \psi \, d\tau + \int \psi^* \mathbf{V} \psi \, d\tau. \tag{19–41}$$

This equation can be written

$$E = \bar{E}_{kin} + \bar{V}, \tag{19–42}$$

where

$$\bar{E}_{kin} = \int \psi^* \mathbf{K} \psi \, d\tau \quad \text{and} \quad \bar{V} = \int \psi^* \mathbf{V} \psi \, d\tau. \tag{19–43}$$

The $\bar{E}_{kin}$ and $\bar{V}$ are average values, and these are more sensible analogues of the classical mechanical properties. Now we have a name for the wave function. Since $|\psi|^2$ is a probability density, ψ is called a *probability amplitude*.

The Schrödinger equation, Eq. (19–36), can be written in the form

$$\mathbf{H}\psi = E\psi, \tag{19–44}$$

in which

$$\mathbf{H} = \mathbf{K} + \mathbf{V}. \tag{19–45}$$

The Hamiltonian operator, $\mathbf{H}$, is the sum of the operators for the kinetic energy and the potential energy. The Hamiltonian operator is the operator for the total energy of the system.

19–15   Retrospection

The Schrödinger equation can be written in the form

$$\mathbf{H}\psi = E\psi.$$

This differential equation can be solved for a property of the particle, ψ, the wave function of the particle. The wave function itself does not have any immediate physical significance. However, the quantity $|\psi|^2 \, d\tau$ can be interpreted as the probability of finding the particle in the volume element $d\tau$ at the position x, y, z. Since the wave function is continuous, this interpretation yields the extraordinary result that everywhere in space there is a finite probability of finding the particle. This is in marked contrast to the classical picture of a strictly localized particle. For example, the Bohr model and the Schrödinger model of the hydrogen atom are quite different pictorially. If we could shrink ourselves and sit on the nucleus of the Bohr atom, we would see the electron moving around the nucleus in a circular orbit having a radius of precisely a_0. Sitting on the nucleus of the Schrödinger atom, we would see a fog of negative charge.† Near the nucleus the density of the fog is high, but if we walk out along a radius, the fog thins out. When we are several atomic diameters away

†   This supposes that our eyes take a "time exposure" of the electron motion.

from the nucleus, we look back and see a spherical cloud of negative charge; with keen eyesight the nucleus can be discerned in the center of the cloud. (Rutherford's model of a uniform sphere of negative charge with the nucleus imbedded in it was not so far wrong after all!) The electron in the Bohr atom moves in an orbit much like a satellite about a planet; in the Schrödinger atom the electron is smeared out into an electron cloud, much like a puff of cotton candy.

Another important property of the wave function is its relation to average properties of physically observable quantities, illustrated by Eq. (19–43). Knowing the wave function of a system, we can calculate the average value of any measurable quantity. Thus, in a somewhat mysterious way, the wave function has hidden in it all of the physically important properties of the system.

The Schrödinger equation opened the way to the systematic mathematical treatment of all atomic and molecular phenomena. The predictions of this equation for atoms and molecules have been confirmed without exception. It is therefore the basis for any modern discussion of atomic and molecular structure.

Problems

19–1.   Compute the energy density of the radiation in a cavity at 100°K, 300°K, and 1000°K.

19–2.   a) In a cavity at 1000°K estimate the fraction of the energy density that is provided by light in the region between 7800 and 8000 Å.

   b) Repeat the calculation for 2500°K.

19–3.   At what wavelength does the maximum in the energy density distribution function for a blackbody occur if a) $T = 300°K$? b) $T = 500°K$?

19–4.   If the energy density distribution function is to have a maximum at 6000 Å what must be the temperature?

19–5.   About 5 eV are required to remove an electron from the interior of platinum. What is the minimum frequency of light required to observe the photoelectric effect using platinum?

19–6.   If an electron falls through an electrical potential difference of one volt, it acquires an energy of one electron volt. If the electron is to have a wavelength of one angstrom, what potential difference must it pass through?

19–7.   Through what electrical potential difference must a proton pass if it is to have a wavelength of 1 Å?

19–8.   What is the wavelength of a ball bearing, $m = 10$ gm, and velocity = 10 cm/sec?

19–9.   What is the kinetic energy of an electron which has a wavelength of 100 Å?

19–10. Use the value of D given in Eq. (19–27) in Eq. (19–26) and prove Eq. (19–29).

19–11. Derive the Wien displacement law from Eq. (19–11).

Chapter Twenty

Introduction to Quantum
Mechanical Principles

20–1 Introduction

Although in Chapter 19 we patched the De Broglie relation onto the equation for classical wave motion to lend plausibility to the time-independent Schrödinger equations, this procedure has dubious merit. In a certain sense it is as if we were to attempt to justify Newtonian mechanics by appealing to the Pythagorean music of the spheres. Experience has shown that the Schrödinger equation is the correct one for atomic and molecular problems and whether or not it is "derivable" from other equations is not a matter which need concern us greatly here. For a systematic treatment of atomic and molecular problems we can most easily proceed by stating a series of postulates and using them to discuss the behavior of a system.

20–2 Postulates of the Quantum Mechanics

Postulate I

There exists a function, $\Psi(x, y, z, t)$, of the coordinates and time which we call a wave function and describe as a probability amplitude. This wave function is in general a complex function; that is,

$$\Psi(x, y, z, t) = u(x, y, z, t) + iv(x, y, z, t), \tag{20–1}$$

where $i = \sqrt{-1}$, and u and v are real functions of coordinates and time. The complex conjugate of Ψ is designated by Ψ^* and is obtained from Ψ by replacing i by $-i$;

$$\Psi^*(x, y, z, t) = u(x, y, z, t) - iv(x, y, z, t). \tag{20–2}$$

The product, $\Psi^*\Psi$, is a purely real function of x, y, z and t,

$$\Psi^*\Psi = |\Psi|^2 = u^2(x, y, z, t) + v^2(x, y, z, t), \tag{20-3}$$

and is equal to the square of the absolute value of Ψ. The product, $\Psi^*\Psi\, dx\, dy\, dz = \Psi^*\Psi\, d\tau$, is the probability at time t that the system will be in the volume element $d\tau$ at the position x, y, z. Thus, $\Psi^*\Psi$ is a *probability density*. In view of this, if the probability of finding the system in the volume element $d\tau$ is summed over all possible positions of the volume element, the result is unity. We must have unit probability of finding the system somewhere. Thus, we have the property,

$$\int \Psi^*\Psi\, d\tau = \int_{-\infty}^{\infty} dx \int_{-\infty}^{\infty} dy \int_{-\infty}^{\infty} \Psi^*(x, y, z, t)\Psi(x, y, z, t)\, dz = 1, \tag{20-4}$$

where the limits in the first integral are understood to be such as to cover the entire coordinate space. The value of the integral in Eq. (20-4) must be independent of the time, t. This implies that the time dependence of the wave function must have the form

$$\Psi(x, y, z, t) = \psi(x, y, z)e^{if(q, t)}, \tag{20-5}$$

in which $f(q, t)$ is some real function of the coordinates, symbolized by q, and time. Using Eq. (20-5) in Eq. (20-4) yields

$$\int \psi^*(x, y, z)\psi(x, y, z)\, d\tau = 1. \tag{20-6}$$

The requirement expressed by Eqs. (20-4) and (20-6), namely that the wave function be quadratically integrable, imposes severe restrictions on ψ. The wave function must be single-valued, continuous, and may not have singularities anywhere of a character which result in the nonconvergence of the integral in Eq. (20-6). In particular, at the extremes of the Cartesian coordinates, $x = \pm\infty$, $y = \pm\infty$, and $z = \pm\infty$, the wave function, as well as its first derivative, must vanish.

Postulate II

The expectation value, $\langle A \rangle$, of any observable is related to the wave function of the system by

$$\langle A \rangle = \frac{\int \psi^*\mathbf{A}\psi\, d\tau}{\int \psi^*\psi\, d\tau}, \tag{20-7}$$

in which $\mathbf{A}$ is an operator corresponding to the observable A. If the wave functions have been *normalized*, then Eq. (20-6) is fulfilled; Eq. (20-7) becomes simply

$$\langle A \rangle = \int \psi^*\mathbf{A}\psi\, d\tau. \tag{20-8}$$

Since we are dealing with wave functions which are functions only of coordinates, then, as was pointed out in Section 19-14, to obtain the expectation value of any function of the coordinates we multiply that function by $\psi^*\psi$, the probability density, and integrate over the entire space. Thus, for a function, $f(x, y, z)$, we have

$$\langle f \rangle = \int f(x, y, z)\psi^*\psi \, d\tau = \int \psi^* f(x, y, z)\psi \, d\tau = \int \psi^* \mathbf{f}\psi \, d\tau. \qquad (20\text{--}9)$$

So we may conclude that the operation corresponding to any function of the coordinates only is multiplication by that function, e.g.,

$$\mathbf{f}\psi = f(x, y, z)\psi. \qquad (20\text{--}10)$$

The situation is not quite so simple if the observable is a momentum component. In this case, it turns out that we must have, for example,

$$\mathbf{p}_x\psi = -i\hbar\frac{\partial\psi}{\partial x}. \qquad (20\text{--}11)$$

The proof of this statement is beyond our scope here; an argument for plausibility appeared in Section (19–13). For the component of momentum along any Cartesian coordinate q, we associate the differential operator,

$$\mathbf{p}_q = \frac{h}{2\pi i}\frac{\partial}{\partial q} = -i\hbar\frac{\partial}{\partial q}. \qquad (20\text{--}12)$$

The quantity $\langle A \rangle$ in Eq. (20–8) is sometimes called the "average" value of A; for example, in Section 19–4 we used this terminology. The meaning of "average" in this context can be misinterpreted; see the discussion of this point in Section 20–4 in connection with Eqs. (20–29) through (20–31).

20–3 Operator Algebra

An operator changes a function into another function according to a rule. Suppose we have a function $w(x, y, z)$ and an operator $\mathbf{x}$, defined by

$$\mathbf{x}w(x, y, z) = xw(x, y, z). \qquad (20\text{--}13)$$

The function $w(x, y, z)$ is the *operand*; the operator acting on w changes w into xw. This is one example of the type of operator described in Eq. (20–10). Similarly we may have a differential operator such as the one in Eq. (20–12)

$$\mathbf{p}_x w = -i\hbar\frac{\partial w}{\partial x}. \qquad (20\text{--}14)$$

This operator replaces the function w by its partial derivative with respect to x multiplied by the constant, $-i\hbar$.

Operators may be combined by addition; if $\boldsymbol{\alpha}$ and $\boldsymbol{\beta}$ are two operators on the same function, then

$$(\boldsymbol{\alpha} + \boldsymbol{\beta})w = \boldsymbol{\alpha}w + \boldsymbol{\beta}w, \qquad (20\text{--}15)$$

since the new functions αw and βw are simply functions, it is clear that operator addition is commutative, i.e.,

$$(\alpha + \beta)w = (\beta + \alpha)w. \tag{20-16}$$

Operators may also be combined by multiplication which may be defined by

$$\alpha\beta w = \alpha(\beta w). \tag{20-17}$$

This equation states that to form the function corresponding to $\alpha\beta w$, we first form the function βw, then perform the operation α on the new function βw. For example, suppose that for $w(x, y, z)$

$$\alpha w = xw$$

and

$$\beta w = \frac{\partial w}{\partial x},$$

then

$$\alpha\beta w = \alpha(\beta w) = \alpha\left(\frac{\partial w}{\partial x}\right) = x\frac{\partial w}{\partial x}; \tag{20-18}$$

but, note that

$$\beta\alpha w = \beta(\alpha w) = \beta(xw) = \frac{\partial(xw)}{\partial x} = w + x\frac{\partial w}{\partial x}. \tag{20-19}$$

It is apparent that in general α and β do not commute:

$$\alpha\beta w \neq \beta\alpha w. \quad .$$

The *commutator* γ is defined by $\alpha\beta - \beta\alpha = \gamma$. In this example, $(\alpha\beta - \beta\alpha)w = -w$ so that the commutator, γ, is multiplication by -1; $\gamma w = -w$. If, for all w,

$$(\alpha\beta - \beta\alpha)w = 0 \tag{20-20}$$

then the operators commute. As will be seen in Section 20–4, the commutation properties of quantum-mechanical operators have great significance for the properties of a system.

Repeated applications of an operator are handled in the manner of Eq. (20–17):

$$\alpha^2 w = \alpha(\alpha w).$$

If, for example, $\alpha w = xw$, then

$$\alpha^2 w = \alpha(\alpha w) = \alpha(xw) = x(xw) = x^2 w.$$

If $\alpha w = (\partial w/\partial x)$, then

$$\alpha^2 w = \alpha(\alpha w) = \alpha\left(\frac{\partial w}{\partial x}\right) = \frac{\partial}{\partial x}\left(\frac{\partial w}{\partial x}\right) = \frac{\partial^2 w}{\partial x^2}.$$

If an operator α is such that in operating on two different operands, v and w, the relation

$$\alpha[c_1 v + c_2 w] = c_1 \alpha v + c_2 \alpha w \tag{20-21}$$

is fulfilled, in which c_1 and c_2 are constants, the operator is said to be *linear*. For example, let $\alpha = \partial/\partial x$, then

$$\frac{\partial}{\partial x}(c_1 v + c_2 w) = c_1 \frac{\partial v}{\partial x} + c_2 \frac{\partial w}{\partial x},$$

so that the differential operator is linear. All the quantum mechanical operators are linear.†

20-4   The Schrödinger Equation

Postulate III

The probability amplitude, $\Psi(x, y, z, t)$, must satisfy the differential equation

$$\mathbf{H}\Psi - i\hbar \frac{\partial \Psi}{\partial t} = 0, \tag{20-22}$$

in which $\mathbf{H}$ is the Hamiltonian operator. To construct the Hamiltonian operator for one particle we write down the classical expression for the total energy,

$$\frac{1}{2m}(p_x^2 + p_y^2 + p_z^2) + V(x, y, z).$$

The first three terms are the kinetic energy, the fourth term is the potential energy. Then we replace the classical momentum components by their quantum-mechanical operators:

$$p_x \rightarrow \mathbf{p}_x = -i\hbar \frac{\partial}{\partial x}$$

$$p_x^2 \rightarrow \mathbf{p}_x^2 = -i\hbar \frac{\partial}{\partial x}\left(-i\hbar \frac{\partial}{\partial x}\right) = -\hbar^2 \frac{\partial^2}{\partial x^2},$$

so that

$$\mathbf{H} = -\frac{\hbar^2}{2m}\left[\frac{\partial^2}{\partial x^2} + \frac{\partial^2}{\partial y^2} + \frac{\partial^2}{\partial z^2}\right] + V(x, y, z). \tag{20-23}$$

We can conveniently introduce the abbreviation

$$\nabla^2 \equiv \frac{\partial^2}{\partial x^2} + \frac{\partial^2}{\partial y^2} + \frac{\partial^2}{\partial z^2}, \tag{20-24}$$

---

† As an example of a nonlinear operator, suppose that $\alpha f = e^f$ and $\alpha g = e^g$. Then $\alpha(c_1 f + c_2 g) = e^{c_1 f + c_2 g}$, which is obviously not equal to $c_1 e^f + c_2 e^g$. This operator is therefore not linear.

where ∇^2 is the Laplacian operator, to obtain

$$\mathbf{H} = -\frac{\hbar^2}{2m}\nabla^2 + V(x, y, z). \tag{20–25}$$

If there are two particles, masses m_1 and m_2, with coordinates x_1, y_1, z_1 and x_2, y_2, z_2, then the classical energy would have the form

$$E = \frac{1}{2m_1}(p_{x_1}^2 + p_{y_1}^2 + p_{z_1}^2) + \frac{1}{2m_2}(p_{x_2}^2 + p_{y_2}^2 + p_{z_2}^2) + V(x_1, y_1, z_1, x_2, y_2, z_2).$$

The corresponding Hamiltonian operator would be

$$\mathbf{H} = -\frac{\hbar^2}{2}\left[\frac{1}{m_2}\nabla_1^2 + \frac{1}{m_2}\nabla_2^2\right] + V(x_1, y_1, z_1, x_2, y_2, z_2),$$

in which ∇_1^2 is the Laplacian operator containing the coordinates of the first particle, ∇_2^2 is the operator containing the coordinates of the second particle. The extension to a system of many particles is obvious.

In the simple case where the potential energy is time independent, then the solution to Eq. (20–22) has the form

$$\Psi(x, y, z, t) = \psi(x, y, z)f(t). \tag{20–26}$$

Then, $\mathbf{H}\Psi = f(t)\mathbf{H}\psi$, and the Schrödinger equation, Eq. (20–22), becomes

$$f(t)\mathbf{H}\psi - i\hbar\psi\frac{df}{dt} = 0.$$

Dividing by ψf and transposing we obtain

$$\frac{1}{\psi}\mathbf{H}\psi = \frac{i\hbar}{f(t)}\frac{df}{dt}. \tag{20–27}$$

The left-hand side of this equation is a function only of the coordinates, while the right-hand side is a function only of time. If we vary the coordinates keeping time constant, the left-hand side would appear to vary but in fact it does not since the right-hand side remains constant. It follows that both the members of Eq. (20–27) are equal to a constant, which we designate by E. Then

$$\frac{1}{f}\frac{df}{dt} = \frac{E}{i\hbar} = -\frac{iE}{\hbar}$$

and therefore

$$f = Ae^{-iEt/\hbar}, \tag{20–28}$$

where A is a constant; also,

$$\mathbf{H}\psi = E\psi. \tag{20–29}$$

Equation (20–29) is a differential equation, the time-independent Schrödinger equation. By solving the Schrödinger equation, we obtain the function $\psi(x, y, z)$ from which we can compute expectation values of the observables associated with the system by use of the appropriate forms of Eq. (20–8).

The Schrödinger equation has a special form. Whenever we have an operator, α, such that

$$\alpha f = af \tag{20–30}$$

where a is a constant, then a is called an *eigenvalue* of the operator and f is called an eigenfunction of the operator α. (Rather less frequently these are called characteristic values and characteristic functions.) Since the Hamiltonian operator corresponds to the total energy of the system, we compute the expectation value of the total energy, $\langle E \rangle$, by applying Eq. (20–8)

$$\langle E \rangle = \int \psi^* \mathbf{H} \psi \, d\tau.$$

Substituting for $\mathbf{H}$ from Eq. (20–29) we obtain

$$\langle E \rangle = \int \psi^* E \psi \, d\tau = E \int \psi^* \psi \, d\tau,$$

$$\langle E \rangle = E, \tag{20–31}$$

where the second form follows since E is constant and the final form since $\int \psi^* \psi \, d\tau = 1$. Eq. (20–31) says that the expectation value for the total energy of the system is precisely equal to the constant E introduced in solving Eq. (20–22), or Eq. (20–27).

The apparently obvious result in Eq. (20–31) is important enough to make us digress for a moment. Suppose we compute the expectation value of the square of the energy of the system:

$$\langle E^2 \rangle = \int \psi^* \mathbf{H}^2 \psi \, d\tau = \int \psi^* \mathbf{H}(\mathbf{H}\psi) \, d\tau$$

$$= \int \psi^* \mathbf{H}(E\psi) \, d\tau = E \int \psi^* (\mathbf{H}\psi) \, d\tau = E \int \psi^* E \psi \, d\tau$$

$$\langle E^2 \rangle = E^2 = \langle E \rangle^2. \tag{20–32}$$

In Eq. (20–32) we have the result that the expectation value of the square of the energy is equal to the square of the expectation value of the energy. This could not be correct if the energy were in some way distributed. The reader will recall that in dealing with the Maxwell distribution of molecular speeds we found that

$$\overline{c^2} = 3RT/M, \qquad \bar{c} = \sqrt{8RT/\pi M}.$$

Of these, the first, $\overline{c^2}$, is not equal to the square, $\bar{c}^2$, of the second. In averaging the square of a value over a distribution, the higher values are always accentuated. The fact that Eq. (20–32) is correct means that we are not averaging bits of energy here and there in space using $\psi^*\psi$ as some kind of distribution function; if we were, Eq. (20–32) could not be correct. The energy of the system in the state described by ψ has a precise value, determined by the Schrödinger equation as well as additional conditions which we shall discuss below. Such a state of the system is called an eigenstate (or characteristic state) of the system.

This result is general in the following sense. Whenever the wave function of a system is an eigenfunction of an operator corresponding to an observable then that observable has a precise value.

Suppose that in a given state, described by ψ, two different observables, A and B, have precise values. This implies that ψ is an eigenfunction of *both* of the operators **A** and **B**; that is

$$\mathbf{A}\psi = a\psi, \qquad \mathbf{B}\psi = b\psi,$$

where a and b are the eigenvalues. Then if ψ is normalized,

$$\langle A \rangle = \int \psi^*(\mathbf{A}\psi)\,d\tau = \int \psi^* a\psi\,d\tau = a,$$

$$\langle B \rangle = \int \psi^*(\mathbf{B}\psi)\,d\tau = \int \psi^* b\psi\,d\tau = b.$$

If we construct the commutator, **AB** − **BA**, and operate on the wave function, we have

$$(\mathbf{AB} - \mathbf{BA})\psi = \mathbf{A}(\mathbf{B}\psi) - \mathbf{B}(\mathbf{A}\psi) = \mathbf{A}(b\psi) - \mathbf{B}(a\psi)$$

$$= b(\mathbf{A}\psi) - a(\mathbf{B}\psi) = ba\psi - ab\psi = (ba - ab)\psi = 0.$$

Thus,

$$\mathbf{AB}\psi = \mathbf{BA}\psi;$$

the operators **A** and **B** commute.

Conversely, it can be shown that if the operators for two observables commute then the two observables can have precise values simultaneously. If the two operators do not commute, then it is not possible for the corresponding observables to have precise values simultaneously. Consider the operators for the x-coordinate of position and the x-component of momentum; then $\mathbf{x}\psi = x\psi$ and $\mathbf{p}_x\psi = -i\hbar\,\partial\psi/\partial x$, so that

$$\mathbf{xp}_x\psi - \mathbf{p}_x\mathbf{x}\psi = -i\hbar\left[x\frac{\partial\psi}{\partial x} - \frac{\partial}{\partial x}(x\psi)\right]$$

$$= -i\hbar[-\psi] = i\hbar\psi$$

These operators do not commute, consequently x and p_x cannot simultaneously have precise values. This is the basis for the Heisenberg uncertainty principle which we will discuss in more detail later.

One final remark on the Schrödinger equation. For physical sense we require that the constant E, the energy, be a real quantity. This requires that $E^* = E$; consequently, since

$$E = \int \psi^*(\mathbf{H}\psi)\, d\tau, \quad \text{and} \quad E^* = \int \psi(\mathbf{H}\psi)^*\, d\tau,$$

then we must have

$$\int \psi^*(\mathbf{H}\psi)\, d\tau = \int \psi(\mathbf{H}\psi)^*\, d\tau. \tag{20–33}$$

An operator which satisfies the condition in Eq. (20–33) is said to be *Hermitian*. Conversely, *the eigenvalues of Hermitian operators are all real quantities.* In the quantum mechanics we deal only with Hermitian operators. The definition of an Hermitian operator is somewhat more general than Eq. (20–33) would imply. The operator $\mathbf{H}$ is Hermitian if

$$\int \psi_1^*(\mathbf{H}\psi_2)\, d\tau = \int \psi_2(\mathbf{H}\psi_1)^*\, d\tau \tag{20–34}$$

in which ψ_1 and ψ_2 are the same or different operands.

20–5 The Eigenvalue Spectrum

The Schrödinger equation

$$\mathbf{H}\psi = \mathbf{K}\psi + \mathbf{V}\psi = E\psi$$

of itself admits a variety of solution, depending on the nature of the potential function V. However, we must impose further restrictions. Even with a particular potential function, among the possible solutions we frequently find ones which are inadmissible; for example, the requirement that the wave function be single valued everywhere eliminates some functions; the requirement of quadratic integrability is a very restrictive condition; the imposition of particular boundary conditions in some problems reduces the number of acceptable solutions.

The net effect of these restrictions is that in some cases, E may have any value; then we speak of a continuous spectrum of eigenvalues of E. In other cases, E may be restricted to certain particular values; then we have a discrete spectrum of eigenvalues. In these latter cases we say that E is *quantized*. Ordinarily for each restriction which we impose we introduce a quantization of some observable.

If the energy is quantized and restricted to certain values, $E_1, E_2, E_3, \ldots, E_n, \ldots$, then to each of these values there corresponds at least one eigenfunction ψ_n, such that

$$\mathbf{H}\psi_n = E_n\psi_n$$

so that in general we deal with a set of eigenfunctions, $\psi_1, \psi_2, \ldots, \psi_n$. If for each energy level (eigenvalue), there is only one eigenfunction, then the set of eigenfunctions and the set of eigenstates is *nondegenerate*. If for the nth eigenstate there are g_n eigenfunctions, $\psi_{n1}, \psi_{n2}, \ldots$, such that

$$\mathbf{H}\psi_{nk} = E_n\psi_{nk} \qquad k = 1, 2, \ldots, g_n, \tag{20–35}$$

then the nth eigenstate is g_n-fold degenerate. To speak of a three-fold degenerate level simply means that corresponding to one particular energy value there are three distinct eigenfunctions.

The existence of a degeneracy poses a problem in describing the state. Suppose the eigenstate with energy E is three-fold degenerate, with eigenfunctions ψ_1, ψ_2, and ψ_3. This means that

$$\mathbf{H}\psi_1 = E\psi_1, \qquad \mathbf{H}\psi_2 = E\psi_2, \qquad \mathbf{H}\psi_3 = E\psi_3.$$

Since the Hamiltonian operator is linear, this implies that if we construct a linear combination,

$$\varphi_1 = c_{11}\psi_1 + c_{12}\psi_2 + c_{13}\psi_3,$$

where c_{11}, c_{12}, and c_{13} are constants, then in view of the linear character of $\mathbf{H}$,

$$\begin{aligned}
\mathbf{H}\varphi_1 &= c_{11}\mathbf{H}\psi_1 + c_{12}\mathbf{H}\psi_2 + c_{13}\mathbf{H}\psi_3 \\
&= c_{11}E\psi_1 + c_{12}E\psi_2 + c_{13}E\psi_3 \\
&= E(c_{11}\psi_1 + c_{12}\psi_2 + c_{13}\psi_3) \\
\mathbf{H}\varphi_1 &= E\varphi_1.
\end{aligned}$$

The linear combination φ_1 is also an eigenfunction of the Hamiltonian operator with E as an eigenvalue; therefore, φ_1 is an appropriate description of the eigenstate. We may construct two additional independent linear combinations of the same type:

$$\varphi_2 = c_{21}\psi_1 + c_{22}\psi_2 + c_{23}\psi_3,$$

$$\varphi_3 = c_{31}\psi_1 + c_{32}\psi_2 + c_{33}\psi_3.$$

In general, we cannot prefer the description ψ_1, ψ_2, ψ_3 to the description $\varphi_1, \varphi_2, \varphi_3$. Out of the entire collection of possibilities, we are required for completeness to choose any three linearly independent eigenfunctions.† Beyond that, it is a matter only of convenience. As we shall see later, in systems with certain symmetry properties, certain combinations are preferred over others.

The foregoing paragraph provides an example of the *principle of superposition*. Because of the linear character of the Schrödinger equation (linear character of the Hamiltonian), if a state is equally well described by either of two functions, for

---

† Linearly independent functions are such that no relation $a_1\varphi_1 + a_2\varphi_2 + a_3\varphi_3 = 0$ exists (the a_i are constants) other than the trivial one, $a_1 = a_2 = a_3 = 0$.

example, then it is equally well described by any two independent linear combinations of those functions.

20-6 Expansion Theorem

The Hermitian property of quantum mechanical operators leads to an important result. Consider two eigenfunctions, ψ_n and ψ_k, of the Hamiltonian operator; we have

$$\mathbf{H}\psi_n = E_n\psi_n \quad\text{and}\quad \mathbf{H}\psi_k = E_k\psi_k. \tag{20-36}$$

We take the complex conjugate of the second equation, $(\mathbf{H}\psi_k)^* = E_k\psi_k^*$, then we multiply the first equation by ψ_k^* and the second by ψ_n, and integrate over all space; this yields

$$\int \psi_k^* \mathbf{H}\psi_n \, d\tau = E_n \int \psi_k^* \psi_n \, d\tau, \qquad \int \psi_n (\mathbf{H}\psi_k)^* \, d\tau = E_k \int \psi_k^* \psi_n \, d\tau.$$

Subtracting these two equations and transposing,

$$(E_n - E_k) \int \psi_k^* \psi_n \, d\tau = \int \psi_k^* (\mathbf{H}\psi_n) \, d\tau - \int \psi_n (\mathbf{H}\psi_k)^* \, d\tau.$$

By Eq. (20–34), the Hermitian property, the two integrals on the right are equal, hence

$$(E_n - E_k) \int \psi_k^* \psi_n \, d\tau = 0.$$

If $E_n \neq E_k$, then

$$\int \psi_k^* \psi_n \, d\tau = 0 \qquad k \neq n. \tag{20-37}$$

Equation (20–37) is the orthogonality relation. Two eigenfunctions of a linear Hermitian operator corresponding to distinct eigenvalues are *orthogonal*.†

---

†   This concept of orthogonality can be obtained by extension from the concept of orthogonality of two ordinary vectors in three-dimensional space. If the x-, y-, and z-components are a_x, a_y, a_z for the first vector and b_x, b_y, b_z for the second vector, then the condition for orthogonality is

$$a_x b_x + a_y b_y + a_z b_z = 0 \qquad\text{or}\qquad \sum_{i=1}^{3} a_i b_i = 0.$$

As a simple illustration, take the first vector along the x-axis and the second along the y-axis, then $a_y = a_z = 0$ and $b_x = b_z = 0$; the left-hand side becomes $a_x \cdot 0 + 0 \cdot b_y + 0 \cdot 0$ which is clearly equal to zero. The sum of products on the left-hand side is called the scalar product of the two vectors, the summation is taken over the components. The integral in Eq. (20–37) is called the Hermitian scalar product of the two functions ψ_k and ψ_n, summation over the components in the ordinary vector is replaced by integration over the variables of the functions.

Note that if $k = n$, our normalization requirement is

$$\int \psi_n^* \psi_n \, d\tau = 1. \tag{20–38}$$

These two conditions are usually written

$$\int \psi_k^* \psi_n \, d\tau = \delta_{nk}, \tag{20–39}$$

where the function δ_{nk} (of n and k) is called the Kronecker delta and is defined by

$$\begin{aligned} \delta_{nk} &= 1, \qquad n = k, \\ \delta_{nk} &= 0, \qquad n \neq k. \end{aligned} \tag{20–40}$$

(Note that $\delta_{nk} = \delta_{kn}$.) The set of functions ψ_n which satisfies Eq. (20–39) is called an orthonormal set. This property of the eigenfunctions of Hermitian operators allows us to expand an arbitrary function in the domain of definition of the orthonormal set in terms of the members of that set.

Suppose φ is an arbitrary† function in the domain of the orthonormal set; then assume that we can write φ as a series with c_n as constant coefficients

$$\varphi = \sum_{n=1}^{\infty} c_n \psi_n. \tag{20–41}$$

To determine the coefficients of this series we multiply by ψ_k^* and integrate over the entire space, so that

$$\int \psi_k^* \varphi \, d\tau = \sum_{n=1}^{\infty} c_n \int \psi_k^* \psi_n \, d\tau.$$

By Eq. (20–39) this becomes

$$\int \psi_k^* \varphi \, d\tau = \sum_{n=1}^{\infty} c_n \delta_{nk}.$$

In view of the properties of δ_{nk}, the sum on the right-hand side reduces to one term, c_k, so we have for the coefficients:

$$c_k = \int \psi_k^* \varphi \, d\tau. \tag{20–42}$$

This very simple and very elegant means of expanding a function in terms of an orthonormal set is extremely useful not only in the quantum mechanics but in many other areas of theoretical physics.

---

†   Not entirely arbitrary. However, we will not deal with questions of convergence, uniqueness, or completeness here. In most practical cases no difficulty arises.

20–7 Small Perturbations

A typical example of the use of the expansion theorem is in the treatment of systems in which the Hamiltonian differs only slightly from the Hamiltonian of a system for which we know exact solutions.

Suppose we know the exact solutions to the equation

$$\mathbf{H}^0 \psi_n^0 = E_n^0 \psi_n^0, \tag{20–43}$$

and we wish to know the solutions to the equation

$$\mathbf{H}\varphi_n = E_n \varphi_n, \tag{20–44}$$

in which $\mathbf{H}$ is not very different from $\mathbf{H}^0$; we write

$$\mathbf{H} = \mathbf{H}^0 + \mathbf{H}^{(1)} + \mathbf{H}^{(2)} \tag{20–45}$$

where $\mathbf{H}^{(1)}$ and $\mathbf{H}^{(2)}$ are small. We can frequently expand the Hamiltonian in terms of increasing powers of some small parameter λ; then $\mathbf{H}^{(1)}$ represents a term proportional to the first power of λ, $\mathbf{H}^{(2)}$ is proportional to λ^2, etc. We can expand E_n and φ_n in the same way to obtain

$$E_n = E_n^0 + E_n^{(1)} + E_n^{(2)}, \tag{20–46}$$

$$\varphi_n = \psi_n^0 + \varphi_n^{(1)} + \varphi_n^{(2)}, \tag{20–47}$$

Inserting Eqs. (20–45), (20–46), (20–47) in Eq. (20–44) we obtain

$$[\mathbf{H}^0 + \mathbf{H}^{(1)} + \mathbf{H}^{(2)}][\psi_n^0 + \varphi_n^{(1)} + \varphi_n^{(2)}] = [E_n^0 + E_n^{(1)} + E_n^{(2)}][\psi_n^0 + \overset{\bullet}{\varphi}_n^{(1)} + \varphi_n^{(2)}].$$

We expand the expression on the left as follows:

$$\mathbf{H}^0 \psi_n^0 + \mathbf{H}^0 \varphi_n^{(1)} + \mathbf{H}^0 \varphi_n^{(2)} + \mathbf{H}^{(1)} \psi_n^0$$
$$+ \mathbf{H}^{(1)} \varphi_n^{(1)} + \mathbf{H}^{(1)} \varphi_n^{(2)} + \mathbf{H}^{(2)} \psi_n^0 + \mathbf{H}^{(2)} \varphi_n^{(1)} + \mathbf{H}^{(2)} \varphi_n^{(2)}.$$

The terms $\mathbf{H}^0 \varphi_n^{(1)}$, $\mathbf{H}^{(1)} \psi_n^0$ are proportional to λ; the terms $\mathbf{H}^0 \varphi_n^{(2)}$, $\mathbf{H}^{(1)} \varphi_n^{(1)}$, $\mathbf{H}^{(2)} \psi_n^0$ are proportional to λ^2, etc. We drop the third and higher order terms such as $\mathbf{H}^{(1)} \varphi_n^{(2)}$ (third order in λ). Then

$$\mathbf{H}^0 \psi_n^0 + \mathbf{H}^0 \varphi_n^{(1)} + \mathbf{H}^{(1)} \psi_n^0 + \mathbf{H}^0 \varphi_n^{(2)} + \mathbf{H}^{(1)} \varphi_n^{(1)} + \mathbf{H}^{(2)} \psi_n^0$$
$$= E_n^0 \psi_n^0 + E_n^0 \varphi_n^{(1)} + E_n^{(1)} \psi_n^0 + E_n^0 \varphi_n^{(2)} + E_n^{(1)} \varphi_n^{(1)} \varphi_n^{(1)} + E_n^{(2)} \psi_n^0. \tag{20–48}$$

The terms corresponding to the various powers of λ must balance separately so that this equation corresponds to three equations:

$$\mathbf{H}^0 \psi_n^0 = E_n^0 \psi_n^0$$

$$\mathbf{H}^0 \varphi_n^{(1)} + \mathbf{H}^{(1)} \psi_n^0 = E_n^0 \varphi_n^{(1)} + E_n^{(1)} \psi_n^0 \tag{20–49}$$

$$\mathbf{H}^0 \varphi_n^{(2)} + \mathbf{H}^{(1)} \varphi_n^{(1)} + \mathbf{H}^{(2)} \psi_n^0 = E_n^0 \varphi_n^{(2)} + E_n^{(1)} \varphi_n^{(1)} + E_n^{(2)} \psi_n^0. \tag{20–50}$$

The first of these equations is the zero-order equation whose solution we already know. To solve the second equation we expand the function $\varphi_n^{(1)}$ in terms of the orthonormal set ψ_n^0;

$$\varphi_n^{(1)} = \sum_k a_{nk}\psi_k^0 .$$

Then

$$\mathbf{H}^0\varphi_n^{(1)} = \sum_k a_{nk}\mathbf{H}^0\psi_k^0 = \sum_k a_{nk}E_k^0\psi_k^0 ,$$

where the last form results from Eq. (20–43). Placing these results in Eq. (20–49), we obtain

$$\sum_k a_{nk}E_k^0\psi_k^0 + \mathbf{H}^{(1)}\psi_n^0 = E_n^0 \sum_k a_{nk}\psi_k^0 + E_n^{(1)}\psi_n^0 .$$

We multiply this equation by $\psi_m^0{}^*$ and integrate over the entire coordinate space:

$$\sum_k a_{nk}E_k^0 \int \psi_m^0{}^*\psi_k^0\, d\tau + \int \psi_m^0{}^*\mathbf{H}^{(1)}\psi_n^0 d\tau = E_n^0 \sum_k a_{nk} \int \psi_m^0{}^*\psi_k^0\, d\tau + E_n^{(1)} \int \psi_m^0{}^*\psi_n^0\, d\tau .$$

Employing Eq. (20–39) this reduces to

$$\sum_k a_{nk}E_k^0\delta_{mk} + H_{mn}^{(1)} = E_n^0 \sum_k a_{nk}\delta_{mk} + E_n^{(1)}\delta_{mn},$$

where we have introduced

$$H_{mn}^{(1)} = \int \psi_m^0{}^*\mathbf{H}^{(1)}\psi_n^0\, d\tau . \tag{20–51}$$

In view of the property of the Kronecker delta, every term in the two sums is zero except for the term in which $k = m$; then

$$a_{nm}E_m^0 + H_{mn}^{(1)} = E_n^0 a_{nm} + E_n^{(1)}\delta_{mn},$$

or

$$E_n^{(1)}\delta_{mn} + (E_n^0 - E_m^0)a_{nm} = H_{mn}^{(1)} . \tag{20–52}$$

This is in reality two equations; if $n = m$, it becomes

$$E_n^{(1)} = H_{nn}^{(1)} = \int \psi_n^0{}^*\mathbf{H}^{(1)}\psi_n^0\, d\tau . \tag{20–53}$$

This equation gives us a first-order correction to the energy of the perturbed system; we can evaluate the integral knowing the unperturbed eigenfunctions and the perturbation operator $\mathbf{H}^{(1)}$. This integral is often easy to evaluate in practical situations. The energy of the system is given by Eq. (20–46),

$$E_n = E_n^0 + H_{nn}^{(1)} . \tag{20–54}$$

If $n \neq m$, then Eq. (20–52) becomes

$$a_{nm} = \frac{H_{mn}^{(1)}}{E_n^0 - E_m^0} . \tag{20–55}$$

To evaluate the coefficient, a_{nn}, we require that $\int \varphi_n^* \varphi_n \, d\tau = 1$. Then

$$1 = \int \varphi_n^* \varphi_n \, d\tau = \int \left[\psi_n^{0*} + \sum_k a_{nk} \psi_k^{0*} \right] \left[\psi_n^0 + \sum_m a_{nm} \psi_m^0 \right] d\tau,$$

$$1 = \int \psi_n^{0*} \psi_n^0 \, d\tau + \sum_m a_{nm} \int \psi_n^{0*} \psi_m^0 \, d\tau + \sum_k a_{nk} \int \psi_k^{0*} \psi_n^0 \, d\tau + \sum_k \sum_m a_{nk} a_{nm} \int \psi_k^{0*} \psi_m^0 \, d\tau,$$

$$1 = 1 + \sum_m a_{nm} \delta_{nm} + \sum_k a_{nk} \delta_{nk} + \sum_k \sum_m a_{nk} a_{nm} \delta_{mk},$$

or

$$0 = a_{nn} + a_{nn} + \sum_k a_{nk}^2.$$

The term $\sum_k a_{nk}^2$ is a second-order term in λ so it is ignored; then we have $a_{nn} = 0$; and the expression for the corrected wave function, $\varphi_n = \psi_n^0 + \varphi_n^{(1)}$ is

$$\varphi_n = \psi_n^0 + \sum_{m \neq n} \frac{H_{mn}^{(1)}}{E_n^0 - E_m^0} \psi_m^0. \tag{20–56}$$

Note that for a small perturbation, only the states with energies near that of state n contribute much to the wave function in Eq. (20–56); for states m in which $E_n^0 - E_m^0$ is large, the coefficients in the series become very small and little contribution is made to the overall state.

The treatment of the second-order equation, Eq. (20–50), goes in the same way. We expand $\varphi_n^{(2)} = \sum_k b_{nk} \psi_k^0$; then construct $\mathbf{H}^0 \varphi_n^{(2)} = \sum_k b_{nk} E_k^0 \psi_k^0$. These we insert in Eq. (20–50) along with the results from Eqs. (20–53) and (20–55), then multiply through by ψ_m^{0*} and integrate. The result is

$$E_n^{(2)} \delta_{mn} + b_{nm} [E_n^0 - E_m^0] = \sum_{k \neq n} a_{nk} H_{mk}^{(1)} - H_{nn}^{(1)} a_{nm} + H_{mn}^{(2)},$$

in which $H_{mn}^{(2)} = \int \psi_m^{0*} \mathbf{H}^{(2)} \psi_n^0 \, d\tau$. When $n = m$, we obtain

$$E_n^{(2)} = \sum_{k \neq n} \frac{H_{nk}^{(1)} H_{kn}^{(1)}}{E_n^0 - E_k^0} + H_{nn}^{(2)}, \tag{20–57}$$

while if $n \neq m$ we obtain

$$b_{nm} = \frac{1}{E_n^0 - E_m^0} \left[\sum_{k \neq n} \frac{H_{mk}^{(1)} H_{kn}^{(1)}}{E_n^0 - E_k^0} - \frac{H_{mn}^{(1)} H_{nn}^{(1)}}{E_n^0 - E_m^0} + H_{mn}^{(2)} \right]. \tag{20–58}$$

The normalization requirement yields

$$b_{nn} = -\frac{1}{2} \sum_{k \neq n} \left(\frac{H_{kn}^{(1)}}{E_n^0 - E_k^0} \right)^2. \tag{20–59}$$

20–8   The Variation Method

Another very neat example of the use of the expansion theorem lies in the proof of the variation theorem. Suppose that we have the Schrödinger equation in the form

$$\mathbf{H}\psi_n = E_n\psi_n, \tag{20–60}$$

but we do not know the ψ_n explicitly; perhaps we cannot solve the differential equation. Having a practical turn of mind, we assume that the solutions ψ_n exist, and if they exist they must form an orthornomal set in view of the argument in Section 20–6. In desperation we choose an arbitrary function, φ, which is subject only to the general requirements for acceptability of a wave function, namely φ must be single-valued, continuous, and quadratically integrable.

In view of the expansion theorem, we can write

$$\varphi = \sum_n c_n\psi_n, \qquad \varphi^* = \sum_k c_k^*\psi_k^*. \tag{20–61}$$

Quadratic integrability allows us to normalize the function:

$$1 = \int \varphi^*\varphi \, d\tau. \tag{20–62}$$

Replacing φ^* and φ with the expressions from Eqs. (20–61) yields

$$1 = \int \left(\sum_k c_k^*\psi_k^* \right)\left(\sum_n c_n\psi_n \right) d\tau = \sum_k \sum_n c_k^* c_n \int \psi_k^*\psi_n \, d\tau.$$

But the ψ_n are an orthonormal set so we have, by Eq. (20–39),

$$1 = \sum_k \sum_n c_k^* c_n \delta_{kn}.$$

Summing over n, we obtain

$$1 = \sum_k c_k^* c_k = \sum_k |c_k|^2. \tag{20–63}$$

Next we consider the integral, $\int \varphi^*\mathbf{H}\varphi \, d\tau$. If we operate on φ with $\mathbf{H}$ using Eq. (20–61), we see that

$$\mathbf{H}\varphi = \sum_n c_n(\mathbf{H}\psi_n) = \sum_n c_n E_n\psi_n.$$

In the second writing, we have used the value of $\mathbf{H}\psi_n$ from Eq. (20–60). Multiplying this equation by φ^* and integrating, we have

$$\int \varphi^*\mathbf{H}\varphi \, d\tau = \sum_n c_n E_n \int \varphi^*\psi_n \, d\tau = \sum_n c_n E_n \int \left(\sum_k c_k^*\psi_k^* \right) \psi_n \, d\tau.$$

To obtain the final form the value of φ^* from Eq. (20–61) has been used. Then

$$\int \varphi^*\mathbf{H}\varphi \, d\tau = \sum_n \sum_k c_n E_n c_k^* \delta_{kn}.$$

After the summation over n is carried out, this yields

$$\int \varphi^* \mathbf{H} \varphi \, d\tau = \sum_k c_k^* c_k E_k = \sum_k |c_k|^2 E_k. \tag{20–64}$$

Now we number the energies in order, beginning with the lowest energy, E_1. Then $E_1 < E_2 < E_3 < \cdots$, etc.; Eq. (20–63) is multiplied by E_1 and the result is subtracted from Eq. (20–64):

$$\int \varphi^* \mathbf{H} \varphi \, d\tau - E_1 = \sum_k |c_k|^2 E_k - \sum_k |c_k|^2 E_1,$$

$$\int \varphi^* \mathbf{H} \varphi \, d\tau - E_1 = \sum_k |c_k|^2 (E_k - E_1).$$

Since $(E_k - E_1) > 0$, every term in the sum on the right-hand side is positive (the first term, $k = 1$, is zero); then,

$$\int \varphi^* \mathbf{H} \varphi \, d\tau - E_1 \geq 0$$

$$\int \varphi^* \mathbf{H} \varphi \, d\tau \geq E_1. \tag{20–65}$$

In (20–65), we must include the equality since we might fortuitously choose $\varphi = \psi_1$. The relation (20–65) is the variation theorem. It states that the integral on the left, which involves an arbitrary function and the Hamiltonian of the system in question, must exceed the value of the lowest energy of the system. Note that to obtain this very general result we needed no information whatsoever about the correct wave functions ψ_n, except that they form an orthonormal set.

The value of the variation theorem lies in our ability to guess at an approximate function and include adjustable parameters, $a_1, a_2, \ldots$, in the function. Then the integral will have the form:

$$\int \varphi^* \mathbf{H} \varphi \, d\tau = f(a_1, a_2, \ldots).$$

The values of $a_1, a_2, \ldots$ are adjusted to yield a minimum value of the integral. This minimum value will still exceed E_1. Thus, for a given functional form we can achieve the best approximation to the ground-state energy. The great majority of the treatments of chemical bonding depend on the variation theorem and/or on perturbation calculations.

20–9   Concluding Remarks on the General Equations

Thus far we have developed equations which are not restricted to particular systems, although we have kept in the main to systems which are in stationary states. These are systems whose energy is precise and unchanging in time.

A great deal more could be said on a strictly general level. As yet nothing has been said of the uncertainty principle, for example. The postulates of the quantum mechanics have not been exhausted by the list given here. However, at this point we take up particular examples, in the belief that the skeleton so far presented will have a less repulsive aspect if fleshed out a bit.

Finally, a remark or two about the treatment at the end of Chapter 19 which attempts to relate the classical wave equation and the Schrödinger equation. It should be clear that whether or not the Schrödinger equation is correct depends only on its predictions of behavior and not in the least on whether or not there is some means of transforming the classical wave equation into the Schrödinger equation. On the other hand, the Schrödinger treatment of a system is required to reduce to Newtonian mechanics in the limit as Planck's constant approaches zero, or in the limit of large masses and distances. Suffice it to say that the Schrödinger equation does not reduce properly in these circumstances.

In passing, it may be mentioned that since $|\psi|^2$ is a probability per unit volume, it follows that ψ has dimension $(length)^{-3/2}$ in three-dimensional space. For a one-dimensional problem, the volume element is simply a length, so the dimension of ψ is $(length)^{-1/2}$.

Problems

20–1. Show that the function Ae^{ax} is an eigenfunction of the differential operator, (d/dx).

20–2. If $f = x^n$, show that $x(df/dx) = nf$, and thus that f is an eigenfunction of the operator $x(d/dx)$.

20–3. Find the commutator for the operators, x^2 and d^2/dx^2.

20–4. Consider the differential equation, $d^2u/dx^2 + k^2u = 0$. Show that two possible solutions are: $u_1 = \sin kx$ and $u_2 = \cos kx$. Then show that if a_1 and a_2 are constants, then $a_1u_1 + a_2u_2$ is also a solution.

20–5. Show that in the interval $-1 \leq x \leq +1$ the polynomials, $P_0(x) = a_0$, $P_1(x) = a_1 + b_1x$, $P_2(x) = a_2 + b_2x + c_2x^2$ are the first members of an orthogonal set of functions. Evaluate the constants $a_0, a_1, b_1, \ldots$, etc.

20–6. Show that in the interval $0 \leq \varphi \leq 2\pi$ the functions, $e^{in\varphi}$, where $n = 0, \pm 1, \pm 2, \ldots$, form an orthogonal set.

20–7. The operators for the components of angular momentum are:

$$\mathbf{M}_x = -i\hbar\left(y\frac{\partial}{\partial z} - z\frac{\partial}{\partial y}\right), \qquad \mathbf{M}_y = -i\hbar\left(z\frac{\partial}{\partial x} - x\frac{\partial}{\partial z}\right), \qquad \mathbf{M}_z = -i\hbar\left(x\frac{\partial}{\partial y} - y\frac{\partial}{\partial x}\right).$$

Show that: $\mathbf{M}_x\mathbf{M}_y - \mathbf{M}_y\mathbf{M}_x = i\hbar\mathbf{M}_z$, and that $\mathbf{M}^2\mathbf{M}_z = \mathbf{M}_z\mathbf{M}^2$, in which $\mathbf{M}^2 \equiv \mathbf{M}_x^2 + \mathbf{M}_y^2 + \mathbf{M}_z^2$. Derive the corresponding commutation rules for $\mathbf{M}_y$ and $\mathbf{M}_z$, for $\mathbf{M}_z$ and $\mathbf{M}_x$, and for $\mathbf{M}^2$ with $\mathbf{M}_x$ and with $\mathbf{M}_y$.

20–8. From the description in the text, derive Eqs. (20–57), (20–58), and (20–59).

Chapter Twenty-one

The Quantum Mechanics of
Some Simple Systems

21-1 Introduction

In the quantum mechanical discussion of a system the following general scheme should be kept in mind. Firstly, in principle we obtain the wave function for the system by solving the Schrödinger equation for that system. In practice we may have to guess at the form of the wave function. Secondly, after obtaining the wave function we calculate expectation values for any observable by application of the equation

$$\langle a \rangle = \int \psi^* \alpha \psi \, d\tau, \tag{21-1}$$

in which α is the operator corresponding to the observable a. It follows that, in spite of its lack of direct physical significance, *the wave function is implicitly a complete description of the system*. In fact, we will often refer to ψ as "the description of the system" rather than the "wave function of the system."

Thirdly, the wave function is such that the product $\psi^* \psi = |\psi|^2$ is the probability density. Therefore, $|\psi|^2 \, d\tau$ is the probability of finding the particle in the volume element $d\tau$. Since the particle has unit probability of being somewhere,

$$\int |\psi|^2 \, d\tau = 1, \tag{21-2}$$

where the integration is carried over the entire coordinate space.

Finally, the Schrödinger equation is a linear differential equation. The solutions of the equation are an entire set of functions: $\psi_1, \psi_2, \psi_3, \psi_4, \ldots$, each of which describes a state of the system. The property of linearity in the differential equation means that linear combinations of these descriptions, or of a subset of them, are also

488

descriptions of the system. The descriptions of a system can thus be superposed to obtain new descriptions of the system; principle of superposition. For example, if ψ_1 and ψ_2 are two descriptions of the system, then the linear combinations,

$$\varphi_1 = a_1\psi_1 + a_2\psi_2, \qquad \varphi_2 = b_1\psi_1 + b_2\psi_2, \qquad (21\text{–}3)$$

where the a's and the b's are arbitrary constants, are also descriptions of the system.

With these four fundamental properties in mind, we can understand a great deal of the consequences of the quantum mechanics. To begin, we discuss a few simple systems in detail.

21–2 The Free Particle

Consider a particle of mass m which moves in the absence of external forces along the x-axis only; the absence of forces implies that the potential energy is constant, so for convenience we may choose $V = 0$. The components of momentum along the y- and z-axes are zero; that along the x-axis is p_x. The total energy is a constant and is equal to the kinetic energy; the classical description is

$$E = p_x^2/2m. \qquad (21\text{–}4)$$

Replacing p_x by $\mathbf{p}_x$ as in Section 20–4, we obtain the Schrödinger equation for this system,

$$E\psi = -\frac{\hbar^2}{2m}\frac{d^2\psi}{dx^2},$$

or

$$\frac{d^2\psi}{dx^2} + \left(\frac{2mE}{\hbar^2}\right)\psi = 0. \qquad (21\text{–}5)$$

Since E is a constant, this differential equation has two solutions:†

$$\psi_1 = Ae^{i\sqrt{2mE}\,x/\hbar} \qquad \text{and} \qquad \psi_2 = Be^{-i\sqrt{2mE}\,x/\hbar},$$

where A and B are arbitrary constants.

If we operate on ψ_1 with the momentum operator, we obtain

$$\mathbf{p}_x\psi_1 = -i\hbar\frac{d\psi_1}{dx} = -i\hbar(i\sqrt{2mE}/\hbar)\psi_1,$$

$$\mathbf{p}_x\psi_1 = \sqrt{2mE}\,\psi_1.$$

Similarly,

$$\mathbf{p}_x\psi_2 = -\sqrt{2mE}\,\psi_2.$$

---

† If the reader is not familiar with Eq. (21–5), the solutions are readily verified by substituting them into the equation.

These equations are typical eigenvalue relations; the constant $\sqrt{2mE}$ appearing on the right is an eigenvalue of the momentum operator.

The interpretation is that in a state described by ψ_1 the momentum of the particle is a fixed precise value, $\sqrt{2mE}$. The classical values of momentum according to Eq. (21–4) are $p_x = \pm\sqrt{2mE}$. Thus, ψ_1 describes a particle moving in the $+x$-direction (p_x is positive) with the classical momentum. On the other hand, ψ_2 describes the particle moving in the $-x$-direction (p_x is negative) with the classical momentum. Since no other conditions are specified, the energy may have any value whatsoever, and so may the momentum. The spectra of eigenvalues of energy and momentum are continuous.

Using ψ_1, suppose we calculate the probability density $\psi_1^*\psi_1$. Since

$$\psi_1 = Ae^{i\sqrt{2mE}x/\hbar} \qquad \text{then} \qquad \psi_1^* = A^*e^{-i\sqrt{2mE}x/\hbar}.$$

Therefore

$$|\psi_1|^2 = \psi_1^*\psi_1 = A^*e^{-i\sqrt{2mE}x/\hbar}Ae^{i\sqrt{2mE}x/\hbar} = A^*A = |A|^2. \tag{21–6}$$

But A is a constant, therefore $|A|^2$ is a constant and is *independent of the value of x*. This means that the probability of finding the particle is the same everywhere along its path. We can therefore make no statement as to its position. The momentum has a definite value p_x, but we find that we can make no statement about the position of the particle. If we use ψ_2, the momentum is definite, $-p_x$, but, as with ψ_1, the position is completely indefinite.

21–3 Particle in a "Box"

In view of the indefiniteness in the position of the free particle, suppose that we enclose the particle in a "box" so that we know that its position lies within the boundaries of the "box." The "box" is made in the following way: let the potential energy of the particle be zero inside the box and infinitely large at the walls and everywhere outside the box. Since the particle cannot have an infinite potential energy, it will stay in the box where its potential energy is zero. Again we restrict the particle to move only along the x-axis. A plot of the potential energy as a function of x is shown in Fig. 21–1; L is the width of the box.

Since $V = 0$ inside the box, the Schrödinger equation has the same form as for the free particle, so the solutions are

$$\psi_1 = Ae^{i\sqrt{2mE}x/\hbar} \qquad \psi_2 = Be^{-i\sqrt{2mE}x/\hbar}.$$

However, we must place the following boundary conditions on the wave function (see Fig. 21–1):

$$\psi = 0, \qquad \text{when} \quad x \le 0,$$

$$\psi = 0, \qquad \text{when} \quad x \ge L.$$

If these conditions were not fulfilled, the probability density $\psi^*\psi$ would be finite at

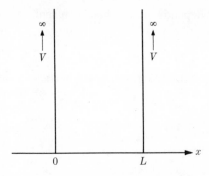

Fig. 21–1 The potential-energy "box."

the edges and outside of the box, where the potential energy is infinite. This is not possible.

These conditions cannot be satisfied by ψ_1 or ψ_2 individually. For example, at $x = 0$, $\psi_1(0) = A$; to satisfy the condition $\psi_1 = 0$, A would have to be zero. But if A is zero, then ψ_1 is zero everywhere. This would mean that $|\psi_1|^2 = 0$ everywhere; the particle isn't anywhere! The same difficulty appears with ψ_2. To avoid this we use the principle of superposition and construct a more general description,

$$\psi = Ae^{i\sqrt{2mE}\,x/\hbar} + Be^{-i\sqrt{2mE}\,x/\hbar}, \tag{21–7}$$

and apply our conditions to ψ.

At $x = 0$, $\psi = 0$, so Eq. (21–7) becomes $0 = A + B$, or $B = -A$. Using this result, we obtain

$$\psi = A(e^{i\sqrt{2mE}\,x/\hbar} - e^{-i\sqrt{2mE}\,x/\hbar}).$$

By the Euler equation, $e^{iy} - e^{-iy} = 2i \sin y$, this becomes

$$\psi = 2iA \sin\left(\frac{\sqrt{2mE}\,x}{\hbar}\right) = C \sin\left(\frac{\sqrt{2mE}\,x}{\hbar}\right). \tag{21–8}$$

The first condition being satisfied, we look to the second; that $\psi = 0$ when $x = L$. At $x = L$, ψ becomes

$$\psi(L) = C \sin\left(\sqrt{2mE}\,L/\hbar\right) = 0.$$

This condition cannot be met by setting $C = 0$, because again the particle would be nowhere. The condition must be met by requiring that $\sin\left(\sqrt{2mE}\,L/\hbar\right) = 0$. If the argument of the sine is an integral multiple of π, then the sine is zero: that is,

$$\sin(\pm n\pi) = 0, \qquad n = 1, 2, 3, 4, \ldots.$$

The boundary condition is therefore fulfilled if, and only if,

$$\sqrt{2mE_n}\,L/\hbar = \pm n\pi. \tag{21–9}$$

Since Eq. (21–9) indicates that the energy depends on n, E has been replaced by E_n. Using this result in Eq. (21–8), we obtain

$$\psi_n = \pm C \sin (n\pi x/L). \tag{21–10}$$

The final condition is that the total probability of finding the particle in the box is unity:

$$\int_0^L \psi_n^* \psi_n \, dx = 1.$$

Since $\psi_n^* = \pm C^* \sin (n\pi x/L)$, we have

$$C^* C \int_0^L \sin^2 \left(\frac{n\pi x}{L} \right) dx = 1.$$

The integral is equal to $L/2$ so $C^*CL/2 = 1$ or $|C|^2 = 2/L$. Choosing C as a real number, we have $C = \sqrt{2/L}$. The final description is

$$\psi_n = \sqrt{2/L} \sin (n\pi x/L). \tag{21–11}$$

There are several curious things about this problem.

First of all, we write Eq. (21–9) in the form

$$E_n = \frac{n^2 \pi^2 \hbar^2}{2mL^2}, \qquad n = 1, 2, 3, \ldots . \tag{21–12}$$

Since n may have only integral values, E may have only the special value given by Eq. (21–12) and may not have any other value. The energy in this system is *quantized*, and the integer n is called a quantum number. The spectrum of energy eigenvalues is discrete. In contrast, the energy of the free particle could have any value. In retrospect, we see that the quantization entered when we restricted the particle to the interior of the box.

The classical momentum corresponding to the energy value E_n is given formally by

$$p_x = \pm \frac{n\pi \hbar}{L} = \pm \frac{nh}{2L}, \qquad n = 1, 2, 3, \ldots . \tag{21–13}$$

Another aspect of the situation is displayed if the de Broglie wavelength is introduced in Eq. (21–13). The de Broglie relation is $\lambda = h/|p_x|$, where we have used the absolute value of p_x, since λ is not a vector quantity; then Eq. (21–13) becomes

$$L = n(\lambda/2), \tag{21–14}$$

which requires that an integral number of half-wavelengths fit exactly in the length L. This situation is analogous to the possible vibrations of a string which is clamped at positions 0 and L. The permissible modes of vibration of the string are given by the same formula as Eq. (21–14). The value of the wave function for several values of n is shown in Fig. 21–2(a), while the probability density $\psi^*\psi$ is shown in Fig. 21–2(b). Note that the values of ψ in Fig. 21–2(a) look exactly like the displacements of a vibrating string clamped at 0 and L in its various modes of vibration.

The probability density in Fig. 21–2(b) is curious. When $n = 1$ the most probable position is at $L/2$, but all positions in the box have fairly large probability density. When $n = 2$, $\psi^*\psi$ vanishes at $L/2$! The particle has zero probability of being at $L/2$. Yet it has high probability of being at either side of the midpoint. This situation can be legitimately viewed in two ways.

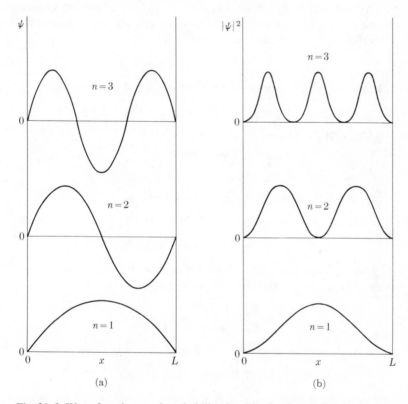

Fig. 21–2 Wave functions and probability densities for the particle in the box.

1. The particle can get from one side of the midpoint to the other without ever passing through the midpoint. Surprising as this may seem, it is correct.

2. The particle is "smeared" over the entire space, the density of the smear being large on both sides and exactly zero at the center. This is the point of view we shall usually adopt. After getting used to the notion of a "smeared" particle, this idea is quite comfortable. It is also a bit easier to become accustomed to than the idea of the particle being now here, then there, but never in between. This latter notion

also tends to keep us tangled with classical ideas of mechanical motion. The smearing of the particle may be regarded as what we would see if we attached a light to the particle and took a time exposure of the motion. The probability densities in Fig. 21–2(b) represent this time average.

21–4 Energy Levels

The allowed values of energy given by Eq. (21–12) are called the energy levels of the system. The least energy the particle may possess is called the zero-point energy, or the ground-state energy E_0. Note that if $n = 0$, the wave function vanishes everywhere; we lose our particle. For the particle in the box the least energy value is that for $n = 1$, so that $E_0 = \pi^2\hbar^2/2mL^2 = h^2/8mL^2$. Then

$$E_n = n^2 E_0, \tag{21–15}$$

and the spacing, or energy gap, between the levels $n + 1$ and n is

$$E_{n+1} - E_n = [(n + 1)^2 - n^2]E_0 = (2n + 1)E_0. \tag{21–16}$$

Since $h = 6.625 \times 10^{-27}$ erg·sec,

$$E_0 = \frac{5.48 \times 10^{-54}}{mL^2} \text{ erg.} \tag{21–17}$$

We apply Eqs. (21–15) through (21–17) to three cases.

Case I. Consider a ball bearing, $m = 1$ gm, in a box 10 cm in length; then

$$E_0 = 5.48 \times 10^{-56} \text{ erg.}$$

If the ball bearing has a velocity of 1 cm/sec, then its kinetic energy is $\frac{1}{2}(1$ gm) $(1$ cm/sec$)^2 = 0.5$ erg. The quantum number n would be

$$n^2 = \frac{E_n}{E_0} = \frac{0.5}{5.48(10^{-56})} \approx 10^{55},$$

so that $n = 3 \times 10^{27}$. The spacing between levels for this value of n is

$$[2(3 \times 10^{27}) + 1]5.48(10^{-56}) \approx 3 \times 10^{-28} \text{ erg.}$$

Thus, to observe the quantization in this system would require us to distinguish between an energy of 0.5 erg and $0.5 + 3 \times 10^{-28}$ erg. This type of precision is impossible, of course, so we do not observe the quantization in the kinetic energy of the ball bearing; it behaves as though it could have any value of kinetic energy at all and the most convenient way to treat this system is to use the classical mechanical laws of Newton. This example illustrates the Bohr "correspondence principle," a rough statement of which is: in the limit of high quantum numbers, the results of the quantum mechanics approach those of the classical mechanics. Whenever we deal with massive particles such as ball bearings and golf balls, the quantum mechanics reduces to classical mechanics.

Case II. Consider an electron, $m \approx 10^{-27}$ gm, in a box the size of an atom, 10^{-8} cm; then

$$E_0 = 5.5 \times 10^{-11} \text{ erg} \approx 34 \text{ eV}.$$

This 34 eV is in the energy region of very short x-rays. The spacing between levels with $n = 1$ and with $n = 2$ is $E_2 - E_1 = 102$ eV. If the electron dropped from level 2 to level 1, a quantum of x-radiation of energy $= 102$ eV would be emitted. The energy quantization is readily observed here.

Case III. Consider an electron, $m \approx 10^{-27}$ gm, in a box the size of the nucleus, 10^{-12} cm; then

$$E_0 = 5.5 \times 10^{-3} \text{ erg}$$

$$= 3.5 \times 10^9 \text{ eV} = 3500 \text{ MeV}$$

(1 MeV $= 10^6$ eV). This is a fantastic energy. The coulombic energy which might be expected to hold an electron in the nucleus would be

$$V = -e^2/r = -(4.8)^2(10^{-20})/10^{-12} = -2.3 \times 10^{-7} \text{ erg} = -0.14 \text{ MeV}.$$

It is clear that if the electron were confined in the nucleus, it would have an inordinately large kinetic energy which would not be compensated by the electrostatic attraction of the positive charge. It could not be held there. For this reason, among others, only heavy particles such as protons, neutrons, etc., are supposed to be present within the nucleus. This argument also explains why the electron in the hydrogen atom does not fall into the nucleus; to exist in the nucleus this fantastically high kinetic energy is required.

Finally, we observe that for the allowed energy levels in the box, the half-wavelength of the particle must fit in the box exactly an integral number of times. If we are to fit the electron in the nucleus, its half-wavelength must be the size of the nucleus or smaller. This very small wavelength implies, by the de Broglie equation $p = h/\lambda$, a very high momentum and consequently the inordinately high kinetic energy.

21–5 Position and Momentum

We return now to the composite description of the system given by Eq. (21–7). The first term in Eq. (21–7) represents motion of the particle along the $+x$-axis with momentum $+p_x$, while the second term represents motion of the particle in the $-x$-direction with momentum $-p_x$. Confining the particle in the box forces us to use a composite description embodying motion in both directions. In a certain sense the particle is moving in both directions! Or rather, we cannot decide from the description whether the particle is coming or going!

To calculate the expectation value of the momentum of the particle in the box, we rewrite Eq. (21–10) in the form

$$\psi_n = A(e^{in\pi x/L} - e^{-in\pi x/L}).$$

Normalization requires $A = (1/\sqrt{2L})$. Hence

$$\psi_n = \frac{1}{\sqrt{2L}}(e^{in\pi x/L} - e^{-in\pi x/L}). \tag{21–18}$$

At this point we observe that the functions

$$\varphi_n = \frac{1}{\sqrt{L}}e^{in\pi x/L}, \qquad \varphi_{-n} = \frac{1}{\sqrt{L}}e^{-in\pi x/L} \tag{21–19}$$

are eigenfunctions of the momentum operator corresponding to different discrete eigenvalues; i.e.,

$$\mathbf{p}_x\varphi_n = -i\hbar\frac{\partial\varphi_n}{\partial x} = \frac{n\pi\hbar}{L}\varphi_n = p_n\varphi_n,$$

$$\mathbf{p}_x\varphi_{-n} = -i\hbar\frac{\partial\varphi_{-n}}{\partial x} = -\frac{n\pi\hbar}{L}\varphi_{-n} = -p_n\varphi_n.$$

Consequently, φ_n and φ_{-n} are members of an orthonormal set in the interval $0 \leq x \leq L$. (This was not true in the case of the free particle where the spectrum of eivenvalues of $\mathbf{p}_x$ was continuous.) Thus, we can write Eq. (21–18) in the form

$$\psi_n = \frac{1}{\sqrt{2}}(\varphi_n - \varphi_{-n}),$$

$$\psi_n = c_n\varphi_n + c_{-n}\varphi_{-n},$$

which is the series expansion of ψ_n in terms of the appropriate orthonormal set of functions; the coefficients, $c_n = -c_{-n} = 1/\sqrt{2}$, are the appropriate series coefficients required to normalize ψ_n.

The expectation value of p_x is given by

$$\langle p_x\rangle = \int_0^L \psi_n^*\mathbf{p}_x\psi_n\,dx = \int_0^L (c_n^*\varphi_n^* + c_{-n}^*\varphi_{-n}^*)(c_n\mathbf{p}_x\varphi_n + c_{-n}\mathbf{p}_x\varphi_{-n})\,dx$$

$$= \int_0^L (c_n^*\varphi_n^* + c_{-n}^*\varphi_{-n})(c_np_n\varphi_n + c_{-n}(-p_n)\varphi_{-n})\,dx$$

$$= c_n^*c_np_n\int_0^L \varphi_n^*\varphi_n\,dx + c_n^*c_{-n}(-p_n)\int_0^L \varphi_n^*\varphi_{-n}\,dx$$

$$+ c_{-n}^*c_np_n\int_0^L \varphi_{-n}^*\varphi_n\,dx + c_{-n}^*c_{-n}(-p_n)\int_0^L \varphi_n^*\varphi_{-n}\,dx.$$

In view of the orthonormality condition, we have

$$\int_0^L \varphi_n^* \varphi_n \, dx = \int_0^L \varphi_{-n}^* \varphi_{-n} \, dx = 1$$

$$\int_0^L \varphi_{-n}^* \varphi_n \, dx = \int_0^L \varphi_n^* \varphi_{-n} \, dx = 0.$$

This reduces the expression for $\langle p_x \rangle$ to

$$\langle p_x \rangle = c_n^* c_n p_n + c_{-n}^* c_{-n}(-p_n)$$

$$= |c_n|^2 p_n + |c_{-n}|^2 (-p_n).$$

Putting in the values of $|c_n|^2 = |c_{-n}|^2 = \frac{1}{2}$, we have

$$\langle p_x \rangle = \tfrac{1}{2} p_n + \tfrac{1}{2}(-p_n) = 0. \tag{21–20}$$

The expectation value of p_x is zero; we could have shown this most easily by using the value of ψ_n from Eq. (21–11):

$$\langle p_x \rangle = \int_0^L \sqrt{\frac{2}{L}} \sin\left(\frac{n\pi x}{L}\right)\left(-i\hbar \frac{d[\sqrt{2/L}\,\sin\,(n\pi x/L)]}{dx}\right) dx$$

$$= -\frac{i\hbar}{L} \sin^2\left(\frac{n\pi x}{L}\right)\Bigg|_0^L = 0.$$

However, the lengthier procedure given above is more instructive. The short computation might erroneously be taken to mean that the momentum of the particle is precisely zero, while the longer computation shows quite clearly that the zero value is composed equally of a 50% probability of the particle moving with $p_x = +n\pi\hbar/L$ and a 50% probability of the particle moving with $p_x = -n\pi\hbar/L$. (The square of the absolute value of the series coefficient $|c_n|^2$, is the probability of finding the situation described by the function φ_n.)

Compare the classical description in which at some time, t_0, we would specify precisely a value of momentum, p_x. At any subsequent time the momentum can be calculated from the classical laws. In the quantum mechanics this precision is replaced by a probability. At any randomly chosen time, if by some magic we could ask the particle in the box for its momentum, it could only reply "Plus $n\pi\hbar/L$, with 50% probability", or "Minus $n\pi\hbar/L$, with 50% probability". Thus, a statistical element is present in the quantum mechanics. Certainty in classical mechanics is replaced by probability, or uncertainty, in the quantum mechanics.

We can give precision to the meaning of uncertainty in an observable by defining it as the root-mean-square deviation from the expectation value. Thus, if Δp_x is the uncertainty in p_x, then

$$(\Delta p_x)^2 = \langle (p_x - \langle p_x \rangle)^2 \rangle \tag{21–21}$$

$$= \langle (p_x^2 - 2p_x \langle p_x \rangle + \langle p_x \rangle^2) \rangle = \langle p_x^2 \rangle - \langle p_x \rangle^2$$

$$\Delta p_x = \sqrt{\langle p_x^2 \rangle - \langle p_x \rangle^2}.$$

In view of the fact that $\langle p_x \rangle = 0$, and that $\mathbf{p}_x^2 = 2m\mathbf{H}$, it follows that

$$\langle p_x^2 \rangle = \int_0^L \psi_n^* \mathbf{p}_x^2 \psi_n \, dx = 2m \int_0^L \psi_n^* \mathbf{H} \psi_n \, dx = 2mE_n \int_0^L \psi_n^* \psi_n \, dx$$

$$= 2mE_n = \frac{n^2 \pi^2 \hbar^2}{L^2},$$

and we have

$$\Delta p_x = \sqrt{\frac{n^2 \pi^2 \hbar^2}{L^2}} = \frac{n\pi\hbar}{L} = \frac{nh}{2L}. \tag{21-22}$$

In a similar way we compute the average value of the particle position,

$$\langle x \rangle = \int_0^L \psi_n^* (\mathbf{x}\psi_n) \, dx = \int_0^L \psi_n^* x \psi_n \, dx.$$

Proceeding in a straightforward way, we find

$$\langle x \rangle = \frac{2}{L} \int_0^L x \sin^2 \frac{n\pi x}{L} \, dx = \frac{1}{L} \int_0^L x \left[1 - \cos\left(\frac{2\pi n x}{L} \right) \right] dx$$

$$= \frac{1}{L} \left[\frac{x^2}{2} \Big|_0^L - \int_0^L x \cos\left(\frac{2\pi n x}{L} \right) dx \right] = \tfrac{1}{2}L - \frac{1}{L} \left(\frac{L}{2\pi n} \right)^2 \int_0^{2\pi n} y \cos y \, dy$$

$$= \tfrac{1}{2}L - L \left(\frac{1}{2\pi n} \right)^2 \left[y \sin y \Big|_0^{2\pi n} - \int_0^{2\pi n} \sin y \, dy \right],$$

which reduces to

$$\langle x \rangle = \tfrac{1}{2}L. \tag{21-23}$$

Not suprisingly, the expectation value of the position of the particle is at the middle of the box. If we calculate $\langle x^2 \rangle$, we obtain, using Eq. (21-11) for ψ_n,

$$\langle x^2 \rangle = \int_0^L \psi_n^* x^2 \psi_n \, dx = \frac{2}{L} \int_0^L x^2 \sin^2 \left(\frac{n\pi x}{L} \right) dx.$$

Direct evaluation of the integral yields the expression

$$\langle x^2 \rangle = \left(\frac{L}{2n\pi} \right)^2 \left[\frac{(2n\pi)^2}{3} - 2 \right].$$

Then, defining the uncertainty in the position in the analogous way to the definition of Δp_x,

$$\Delta x = \sqrt{\langle x^2 \rangle - \langle x \rangle^2}$$

$$\Delta x = \sqrt{\left(\frac{L}{2n\pi} \right)^2 \left[\frac{(2n\pi)^2}{3} - 2 \right] - \frac{L^2(n\pi)^2}{4(n\pi)^2}} = \frac{L}{2n\pi} \sqrt{\frac{(n\pi)^2}{3} - 2} = \frac{L}{2n\pi} \sqrt{1 + \frac{n^2\pi^2 - 3^2}{3}}$$

Multiplying Δp_x by Δx, we obtain

$$\Delta p_x \cdot \Delta x = \frac{h}{4\pi}\sqrt{1 + \frac{n^2\pi^2 - 3^2}{3}}.$$

Since the value of the radical is greater than unity for all values of n, we have the result,

$$\Delta p_x \cdot \Delta x > \frac{h}{4\pi}. \tag{21–24}$$

The inequality (21–24) is the statement of the Heisenberg uncertainty principle for the particle in the box.

21–6   The Uncertainty Principle

The situation for the free particle compared with the particle in the box may be summarized as follows.

1. The free particle has an exactly defined momentum, but the position is completely indefinite.

2. When we try to gain information about the position of the particle by confining it within the length L, an indefiniteness, or uncertainty, is introduced in the momentum. The product of these uncertainties is given by the inequality (21–24) $\Delta p_x \Delta x > h/4\pi$.

3. If we attempt to give the particle a precise position by letting $L \to 0$, then to satisfy (21–24), $\Delta p_x \to \infty$; the momentum becomes completely indefinite.

These facts are given general expression by the Heisenberg uncertainty principle, which we may state in the form: the product of the uncertainty in a coordinate and the uncertainty in the conjugate momentum is at least as large as $h/4\pi$. (By the conjugate momentum of a coordinate we mean the component of momentum along that coordinate.) In Cartesian coordinates we can state the uncertainty principle by the relations

$$\Delta p_x \cdot \Delta x \geq h/4\pi, \qquad \Delta p_y \cdot \Delta y \geq h/4\pi, \qquad \Delta p_z \cdot \Delta z \geq h/4\pi \tag{21–25}$$

In passing, we reiterate that the operators for p_x and x do not commute. Variables whose operators do not commute are subject to uncertainties which are related as in (21–25). It follows from this principle that it is not possible to measure exactly and simultaneously both the x-position and the x-component of the momentum of a particle. *Either* the position *or* the conjugate momentum may be measured as precisely as we please, but increase in precision in the knowledge of one results in a loss of precision in the knowledge of the other.

Suppose that we attempt a precise measurement of the position of a particle using a microscope. The resolving power of a microscope is limited by the wavelength of the light used to illuminate the object; the shorter the wavelength (the higher the frequency) of the light used, the more accurately the position of the particle can be defined. If we wish to measure the position very accurately, then light of very high frequency would be required; a γ-ray, for example. To be seen, the γ-ray must be scattered from the particle into the objective of the microscope. However, a γ-ray of such high frequency has a large momentum; it if hits the particle, some of this momentum will be imparted to the particle, which will be kicked off in an arbitrary direction. The very process of measurement of position introduces an uncertainty in the momentum of the particle.

The scattering of a γ-ray by a small particle and the accompanying recoil of the particle is called the Compton effect; Fig. 21–3. Let p be the momentum of the particle after the collision; the momentum of the γ-ray is obtained from the energy $h\nu$, which according to the Einstein equation, must be equal to $m_\gamma c^2$. Therefore, the momentum of the γ-ray is $m_\gamma c = h\nu/c$ before the collision and $h\nu'/c$ after the collision. Energy conservation requires that

$$h\nu = h\nu' + p^2/2m,$$

while momentum conservation requires that for the x-component

$$\frac{h\nu}{c} = \frac{h\nu'}{c} \cos\varphi + p \cos\theta,$$

and for the y-component

$$0 = \frac{h\nu'}{c} \sin\varphi - p \sin\theta.$$

In addition to the original frequency ν, these three equations involve four variables: ν', p, φ, and θ. Using two of these equations, we can eliminate θ and ν', and reduce the third to a relation between p and φ. If the particle is to be seen, the γ-ray must be scattered into the objective of the microscope, that is, within a range $\Delta\varphi$. Since $\Delta\varphi$ is finite, there is a corresponding finite range of values, an uncertainty, in the particle momentum p. The process of measurement itself perturbs the system so that the momentum becomes indefinite even if it were not indefinite before the measurement. Since no method of measurement has been devised which is free from this difficulty, the uncertainty principle is an accepted physical principle. Examination of the equations for the Compton effect shows that this difficulty is a practical one only for particles with a mass of the order of that of the electron or of that of individual atoms. It does not give trouble with golf balls.

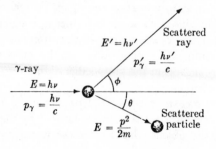

Fig. 21–3 The Compton effect.

Another important uncertainty relation occurs in time-dependent systems. We can take the result for the particle in a box, which classically is written $E = p_x^2/2m$; if the momentum has an uncertainty Δp_x, then there is a corresponding uncertainty in the energy, $\Delta E = (\partial E/\partial p_x)\Delta p_x$, thus

$$\Delta E = (1/m)p_x \,\Delta p_x = (1/m)(mv_x) \,\Delta p_x = v_x \,\Delta p_x.$$

However, the velocity, v_x, can be written $v_x = \Delta x/\Delta t$; using this result in the equation for ΔE yields $\Delta E \cdot \Delta t = \Delta p_x \cdot \Delta x$; extending this argument to the general case we have

$$\Delta E \cdot \Delta t \geq h/4\pi. \tag{21–26}$$

This relation says that there is an uncertainty in the energy of a particle and an uncertainty in the time at which the particle passes a given point in space; the product of these uncertainties must equal or exceed $h/4\pi$.

Heisenberg's development of the quantum mechanics began with the uncertainty relations and led to the quantum mechanical equation. Schrödinger's treatment began with the wave equation and, as we have seen, one can argue from that to the uncertainty relations.

21–7 The Harmonic Oscillator

The particle in the "box" was strictly confined to a particular region of space by the infinitely high potential energy "walls" erected at the boundaries. We now ask how the system behaves if the walls are not infinitely high at any particular point in space but rise gradually to infinity. The simplest potential energy function which has this property is $V(x) = \frac{1}{2}kx^2$, in which k is a constant. The potential energy is parabolic; Fig. 21–4. Choice of this potential-energy function has a double advantage. It displays the behavior if the walls are not infinitely high and, since this is the potential function for the harmonic oscillator, the results will be applicable to real physical oscillators insofar as they are harmonic oscillators. For example, the vibration of a diatomic molecule such as N_2, or O_2, in the lower energy states is nearly harmonic. We begin with a brief outline of the classical mechanical problem and then discuss the quantum mechanical behavior.

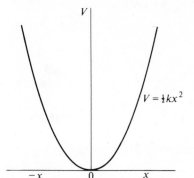

$V = \frac{1}{2}kx^2$ **Fig. 21–4** Potential energy for the harmonic oscillator.

Classical mechanics

Consider a particle moving in one dimension, along the x-axis, and bound to the origin ($x = 0$) by a Hooke's law restoring force, $-kx$. Newton's law, $ma = F$, then reads $m(d^2x/dt^2) = -kx$, or

$$\frac{d^2x}{dt^2} + \frac{k}{m}x = 0. \tag{21–27}$$

Define a circular frequency, ω, such that: $\omega^2 = k/m$; this is related to the frequency ν, by $\omega = 2\pi\nu$ or $\nu = (1/2\pi)\sqrt{k/m}$. Equation (21–27) becomes:

$$\frac{d^2x}{dt^2} + \omega^2 x = 0.$$

This equation has the solution,

$$x = Ae^{i\omega t} + Be^{-i\omega t}. \tag{21–28}$$

The velocity, $v = dx/dt$, is obtained by differentiating Eq. (21–28):

$$v = i\omega[Ae^{i\omega t} - Be^{-i\omega t}]. \tag{21–29}$$

The constants A and B are determined by specifying the position, x, and the velocity, v, at some time t_0. For a simple solution suppose that at $t = t_0$, $x = x_0$, and $v = 0$; then Eqs. (21–28) and (21–29) become

$$x_0 = Ae^{i\omega t_0} + Be^{-i\omega t_0},$$

$$0 = Ae^{i\omega t_0} - Be^{-i\omega t_0}.$$

Solving for A and B, we obtain

$$A = \tfrac{1}{2}x_0 e^{-i\omega t_0} \quad \text{and} \quad B = \tfrac{1}{2}x_0 e^{i\omega t_0}.$$

Then, using these values in Eq. (21–28),

$$x = \tfrac{1}{2}x_0(e^{i\omega(t-t_0)} + e^{-i\omega(t-t_0)}).$$

Since $\cos y = (e^{iy} + e^{-iy})/2$ and $\sin y = (e^{iy} - e^{-iy})/2i$, we obtain

$$x = x_0 \cos \omega(t - t_0) \tag{21–30}$$

and

$$v = \frac{dx}{dt} = -\omega x_0 \sin \omega(t - t_0), \tag{21–31}$$

Equation (21–30) shows that the particle moves between $-x_0$ and $+x_0$ with a sinusoidal motion. Equation (21–31) shows that the velocity varies between $-\omega x_0$ and $+\omega x_0$ and is 90° out of phase with respect to the position. The total energy is

$$E = \tfrac{1}{2}mv^2 + \tfrac{1}{2}kx^2, \tag{21–32}$$

or, using Eqs. (21–30), (21–31),

$$E = \tfrac{1}{2}m[\omega^2 x_0^2 \sin^2\omega(t - t_0)] + \tfrac{1}{2}kx_0^2 \cos^2\omega(t - t_0),$$

which reduces to

$$E = \tfrac{1}{2}kx_0^2. \qquad (21\text{--}33)$$

Note that the energy of the classical oscillator depends only on the force constant, k, and on the maximum displacement, x_0, which is an arbitrary quantity. The oscillator may have any total energy, depending on how large we make x_0.

In terms of the momentum, $p = mv$, we can write the total energy at any time in the form

$$E = \frac{p^2}{2m} + \frac{k}{2}x^2. \qquad (21\text{--}34)$$

The total energy, E, is a constant throughout the motion. When the particle reaches the extreme value, $\pm x_0$, then the momentum, $p = 0$. When $x = 0$, the momentum has its largest value. Thus, the particle moves very slowly at the extremes of the displacement and moves very quickly near $x = 0$.

Quantum mechanics

The classical energy of the harmonic oscillator is given by Eq. (21–34). We obtain the Hamiltonian operator by replacing p by $\mathbf{p} = -i\hbar(d/dx)$. Then $\mathbf{H} = (-\hbar^2/2m)(d^2/dx^2) + (k/2)x^2$. The Schrödinger equation becomes

$$-\frac{\hbar^2}{2m}\frac{d^2\psi}{dx^2} + \tfrac{1}{2}kx^2\psi = E\psi. \qquad (21\text{--}35)$$

Before we can deal with a differential equation such as Eq. (21–35) we must remove the garbage. To do this we introduce dimensionless variables. Let $x = \beta\xi$, where the constant, β, is some unit of length and ξ is dimensionless. Then Eq. (21–35) becomes

$$-\left(\frac{\hbar^2}{2m\beta^2}\right)\frac{d^2\psi}{d\xi^2} + (\tfrac{1}{2}k\beta^2)\xi^2\psi = E\psi.$$

Now we observe that each term in the equation contains ψ so the dimensions of ψ, whatever they are, do not matter. However, since ψ is multiplied by E on the right side, it follows that every term in the equation is multiplied by a quantity of energy. Thus, both $\hbar^2/2m\beta^2$ and $\tfrac{1}{2}k\beta^2$ must have the dimensions of energy. We observe that by adjusting the value of β properly we can make these two quantities of energy equal; we determine β by the condition

$$\frac{\hbar^2}{2m\beta^2} = \tfrac{1}{2}k\beta^2, \quad \text{or} \quad \beta^4 = \frac{\hbar^2}{mk}, \quad \text{or} \quad \beta^2 = \frac{\hbar}{\sqrt{mk}}. \qquad (21\text{--}36)$$

Then we find that our unit of energy is

$$\tfrac{1}{2}k\beta^2 = \tfrac{1}{2}k\left(\frac{h}{2\pi}\right)\frac{1}{\sqrt{mk}} = \left(\frac{h}{2}\right)\frac{1}{2\pi}\sqrt{\frac{k}{m}} = \tfrac{1}{2}h\nu. \tag{21-37}$$

Next we write E as a multiple of this unit of energy;

$$E = \tfrac{1}{2}h\nu(2n + 1) \tag{21-38}$$

in which n is a dimensionless parameter. (We could have written $E = \tfrac{1}{2}h\nu\alpha$, with α dimensionless, of course. The choice of $2n + 1$ rather than α is convenient because it reduces the equation to a well-known standard form.) Introducing Eqs. (21–36) through (21–38) into the Schrödinger equation, dividing by $\tfrac{1}{2}h\nu$ and rearranging yields

$$\frac{d^2\psi}{d\xi^2} + (2n + 1 - \xi^2)\psi = 0, \tag{21-39}$$

which is considerably less cumbersome than Eq. (21–35).

To solve this equation we observe that at very large values of ξ, such that $\xi^2 \gg 2n + 1$, it becomes approximately

$$\frac{d^2\psi}{d\xi^2} = \xi^2\psi.$$

This equation has approximate solutions

$$\psi_1 = e^{\xi^2/2} \quad \text{and} \quad \psi_2 = e^{-\xi^2/2}.$$

we note that $d^2\psi_1/d\xi^2 = (1 + \xi^2)\psi_1 \approx \xi^2\psi_1$, and $d^2\psi_2/d\xi^2 = (-1 + \xi^2)\psi_1 \approx \xi^2\psi_1$ when $\xi^2 \gg 1$. The first solution ψ_1 must be rejected since it would yield a probability density, $\psi_1^2 = e^{\xi^2}$, which becomes infinitely large as ξ (or x) approaches $\pm\infty$. In brief, ψ_1^2 is not quadratically integrable and is unacceptable as a wave function. On the other hand, $\psi_2^2 = e^{-\xi^2}$ becomes zero at $\xi = \pm\infty$ and is quadratically integrable.

Having an approximate solution, $e^{-\xi^2/2}$, we attempt an exact solution of Eq. (21–39) by choosing a solution in the form

$$\psi = u(\xi)e^{-\xi^2/2}. \tag{21-40}$$

Then, using prime and double prime for first and second derivatives,

$$\psi'' = u''e^{-\xi^2/2} - 2\xi u'e^{-\xi^2/2} + u(-1 + \xi^2)e^{-\xi^2/2}.$$

Using this value for ψ'' and the value of ψ from Eq. (21–40), Eq. (21–39) becomes, after dividing out $e^{-\xi^2/2}$,

$$u'' - 2\xi u' + 2nu = 0. \tag{21-41}$$

This is Hermite's differential equation. Two cases occur.

Case I. The parameter n is either nonintegral or is a negative integer.

In the first of these situations, the solution of Eq. (21–41) behaves as e^{ξ^2} for large values of ξ. Then we would have $\psi = e^{\xi^2}e^{-\xi^2/2} = e^{\xi^2/2}$ for large values of ξ. This function is not quadratically integrable and is therefore unacceptable. The case for negative integral values of n also yields functions which are not quadratically integrable.

Case II. The parameter n is either zero or a positive integer.

In this case, the solutions of Eq. (21–41) are polynomials of nth degree, the Hermite polynomials, usually written $H_n(\xi)$. Then the wave function has the form

$$\psi_n = \beta^{-1/2}A_nH_n(\xi)e^{-\xi^2/2}, \tag{21–42}$$

in which A_n is a constant. The integral

$$\int_{-\infty}^{\infty} \psi_n^2 \, dx = A_n^2 \int_{-\infty}^{\infty} H_n^2(\xi)e^{-\xi^2} \, d\xi = 1$$

converges since $H_n^2(\xi)$ is a polynomial of $2n$th degree and any polynomial multiplied by $e^{-\xi^2}$ will yield a convergent integral. The constant A_n is determined by requiring that the integral equal unity. Evaluation of the integral yields

$$A_n = \left(\frac{1}{\sqrt{\pi}2^n n!}\right)^{1/2}. \tag{21–43}$$

The grand conclusion from all of this is that the condition of quadratic integrability requires n to be a positive integer or zero. Looking back, we realize that n governs the energy through Eq. (21–38), which can be written

$$E_n = (n + \tfrac{1}{2})h\nu \qquad n = 0, 1, 2, \ldots. \tag{21–44}$$

As in the case of the particle in a box, the energy is quantized. Only the special values given by Eq. (21–44) are permitted; these are the energy levels of the harmonic oscillator. The lowest value of energy permitted is $(n = 0)$ $E_0 = \tfrac{1}{2}h\nu$, the zero-point energy. This is required by the uncertainty principle. An energy eigenvalue of zero would require $p_x = 0$ and $x = 0$ precisely. This is not permitted. Note that the energy levels are evenly spaced, $E_{n+1} - E_n = h\nu$, while in the case of the particle in a box the spacing increased with the value of n.

The wave functions have some interest, of course. The general expression for the Hermite polynomials is

$$H_n(\xi) = (-1)^n e^{\xi^2} \frac{d^n(e^{-\xi^2})}{d\xi^n}. \tag{21–45}$$

Explicitly, we may write

$$H_n(\xi) = \sum_{k=0} \frac{(-1)^k n!(2\xi)^{n-2k}}{(n-2k)!k!}. \tag{21–46}$$

Since the factorial of a negative integer is infinite, all terms for which $2k > n$ have vanishing coefficients so the series represents a polynomial. When n is even, the upper limit of k is $n/2$; when n is odd, the upper limit is $(n - 1)/2$. The first few Hermite polynomials are given in Table 21–1.

Table 21–1

n even	n odd
$H_0(\xi) = 1$	$H_1(\xi) = 2\xi$
$H_2(\xi) = 4\xi^2 - 2$	$H_3(\xi) = 8\xi^3 - 12\xi$
$H_4(\xi) = 16\xi^4 - 48\xi^2 + 12$	$H_5(\xi) = 32\xi^5 - 160\xi^3 + 120\xi$
$H_6(\xi) = 64\xi^6 - 480\xi^4 + 720\xi^2 - 120$	$H_7(\xi) = 128\xi^7 - 1344\xi^5 + 3360\xi^3 - 1680\xi$

Since the functions $\psi_n = \beta^{-1/2} A_n H_n(\xi) e^{-\xi^2/2}$ are eigenfunctions of an Hermitian operator, they form an orthonormal set in the interval $-\infty < \xi < +\infty$. Thus

$$\int_{-\infty}^{\infty} \psi_m \psi_n \, dx = \delta_{mn}$$

or

$$\int_{-\infty}^{\infty} A_m A_n H_m(\xi) H_n(\xi) e^{-\xi^2} \, d\xi = \delta_{mn}. \tag{21–47}$$

Note that the product of two Hermite polynomials by themselves would not yield a convergent integral, but when each is weighted by the function $e^{-\xi^2/2}$, the integral does converge. The polynomials are said to be orthonormal with respect to the weighting function $e^{-\xi^2/2}$.

It is easily shown using either Eq. (21–45) or (21–46) that $H_n(\xi) = (-1)^n H_n(\xi)$; the polynomial is even or odd according as n is even or odd.

For the evaluation of integrals involving Hermite polynomials two formulas are very useful. The differential relation

$$\frac{dH_n(\xi)}{d\xi} = 2n H_{n-1}(\xi), \tag{21–48}$$

and the recurrence formula

$$\xi H_n(\xi) = n H_{n-1}(\xi) + \tfrac{1}{2} H_{n+1}(\xi). \tag{21–49}$$

Equation (21–48) is easily derived from either Eq. (21–45) or Eq. (21–46); Eq. (21–49) is then obtained by evaluating derivatives, using Eq. (21–48) and combining with the Hermite equation, Eq. (21–41).

For example, to evaluate $\langle \xi \rangle$ we need an integral of the form

$$\langle \xi \rangle = \int_{-\infty}^{\infty} A_n^2 H_n(\xi) \xi H_n(\xi) e^{-\xi^2} \, d\xi.$$

Using the value for $\xi H_n(\xi)$ from Eq. (21–49), this becomes

$$\langle \xi \rangle = A_n^2 \int_{-\infty}^{\infty} H_n(\xi)[nH_{n-1}(\xi) + \tfrac{1}{2}H_{n+1}(\xi)]e^{-\xi^2}\,d\xi,$$

$$\langle \xi \rangle = nA_n^2 \int_{-\infty}^{\infty} H_n(\xi)H_{n-1}(\xi)e^{-\xi^2}\,d\xi + \tfrac{1}{2}A_n^2 \int_{-\infty}^{\infty} H_n(\xi)H_{n+1}(\xi)e^{-\xi^2}\,d\xi.$$

By the orthogonality relation, Eq. (21–47), both of these integrals vanish, so $\langle \xi \rangle = 0$. We could have predicted this from the first form of the integral since $H_n^2(\xi)e^{-\xi^2}$ is an even function of ξ, so that when multiplied by ξ it becomes an odd function which vanishes on integration over the symmetrical interval.

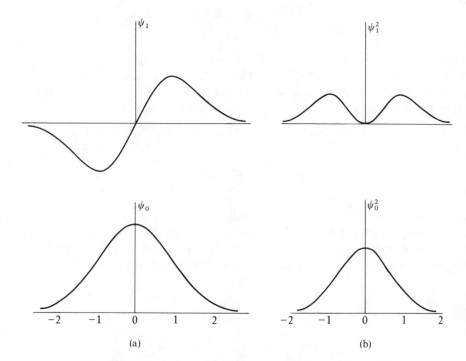

Fig. 21–5 Harmonic oscillator: (a) wave functions; (b) densities.

The wave functions shown in Fig. (21–5) indicate that there is a finite probability of finding the particle at very large distances from $x = 0$. Classically the particle may not pass beyond the point at which the kinetic energy is zero; that is, where the potential energy is equal to the total energy. If x_0 is the maximum displacement allowed classically, then $\tfrac{1}{2}kx_0^2 = E$, or $\tfrac{1}{2}k\beta^2\xi_0^2 = (n + \tfrac{1}{2})h\nu$, which yields $\xi_0^2 = 2n + 1$. Thus, for $n = 0$, the maximum displacement allowed classically would correspond to $\xi_0 = \pm 1$. The density function shown in Fig. 21–5(b) is substantial at $\xi_0 = \pm 1$.

The total probability of finding the particle in the classically forbidden region, $P(n)$, is obtained by integrating the probability density over the forbidden region.

$$P(n) = \int_{-\infty}^{-\sqrt{2n+1}} A_n^2 H_n^2(\xi) e^{-\xi^2} d\xi + \int_{+\sqrt{2n+1}}^{\infty} A_n^2 H_n^2(\xi) e^{-\xi^2} d\xi.$$

From symmetry these two integrals are equal, so we have

$$P(n) = 2A_n^2 \int_{\sqrt{2n+1}}^{\infty} H_n^2(\xi) e^{-\xi^2} d\xi.$$

If $n = 0$, we have $A_0^2 = 1/\sqrt{\pi}$, $H_0^2(\xi) = 1$,

$$P(0) = \frac{2}{\sqrt{\pi}} \int_1^{\infty} e^{-\xi^2} d\xi = 1 - \operatorname{erf}(1),$$

in which $\operatorname{erf}(x)$ is the error function of x. From Table 4–2 of $\operatorname{erf}(x)$ we find $\operatorname{erf}(1) = 0.8427$, so that $P(0) = 0.1573$. This indicates that in the ground state the particle spends about 15.73% of its time in the classically forbidden region. This is by no means a negligible figure. The values for some other values of n are:

n	0	1	2	3	4
$P(n)$	0.1573	0.1116	0.0951	0.0855	0.0785

As n increases, $P(n)$ decreases and approaches the classical value, zero, as $n \to \infty$.

The ability of a particle to penetrate into a classically forbidden region is the basis of the quantum mechanical "tunnel effect." Consider a particle for which the potential function looks as in Fig. 21–6. Two regions of low potential energy are

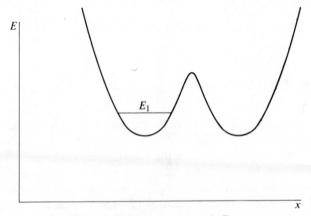

Fig. 21–6 Barrier for tunnel effect.

separated by a barrier. Assume that the particle has a total energy corresponding to E_1 and is in the left-hand region. The energy E_1 is less than the height of the potential barrier so that classically the particle would be confined to the left side. However, quantum mechanically there is a finite probability of finding the particle in the forbidden region and therefore there is a probability of the particle leaking or "tunneling" through the barrier. The probability of this event decreases as the mass of the particle increases and as the barrier gets higher and wider.

21–8 Multidimensional Problems

The majority of interesting problems involve more than one coordinate and momentum. Immediately the Schrödinger equation becomes a partial differential equation and the solutions become more complicated. One of the simplest cases which illustrates a general method of solving the partial differential equation is the example of the particle in a three-dimensional box.

We assume that the potential energy is defined by

$$V = 0 \quad 0 < x < L_1; \qquad V \to +\infty \quad x \le 0 \text{ and } x \ge L_1$$

$$V = 0 \quad 0 < y < L_2; \qquad V \to +\infty \quad y \le 0 \text{ and } y \ge L_2$$

$$V = 0 \quad 0 < z = L_3; \qquad V \to +\infty \quad z \le 0 \text{ and } z \ge L_3.$$

Since the particles cannot exist in a region of infinite potential energy, we know that $\psi = 0$ outside of and at the walls of the box. Since $V = 0$ in the interior of the box, we have the Schrödinger equation in the form:

$$-\frac{\hbar^2}{2m}\left(\frac{\partial^2 \psi}{\partial x^2} + \frac{\partial^2 \psi}{\partial y^2} + \frac{\partial^2 \psi}{\partial z^2}\right) = E\psi. \tag{21–50}$$

We now assume that ψ is a product of functions of the individual coordinates; that is,

$$\psi(x, y, z) = X(x) \cdot Y(y) \cdot Z(z), \tag{21–51}$$

then

$$\frac{\partial^2 \psi}{\partial x^2} = Y(y) \cdot Z(z) \cdot \frac{d^2 X}{dx^2}, \qquad \frac{\partial^2 \psi}{\partial y^2} = X(x) \cdot Z(z) \cdot \frac{d^2 Y}{dy^2}, \qquad \text{etc.}$$

We insert these expressions for the partial derivatives in the Schrödinger equation and divide through by ψ; this reduces the equation to the form

$$-\frac{\hbar^2}{2m}\left(\frac{1}{X(x)}\frac{d^2 X}{dx^2}\right) - \frac{\hbar^2}{2m}\left(\frac{1}{Y(y)}\frac{d^2 Y}{dy^2}\right) - \frac{\hbar^2}{2m}\left(\frac{1}{Z(z)}\frac{d^2 Z}{dz^2}\right) = E. \tag{21–52}$$

Now suppose we keep x and y constant, then the first terms in the equation, since they depend on x and on y respectively, remain constant. If we vary z, the third term would appear to vary since it depends on z, but in fact it cannot vary, since the addition of a varying third term to the two constant ones would make E vary, and

E is a constant. Thus, we may write

$$-\frac{\hbar^2}{2m}\left(\frac{1}{Z(z)}\frac{d^2Z}{dz^2}\right) = E_z,$$

where E_z is a constant. The analogous argument may be applied to show that the first and second terms in Eq. (21–52) must also be constants, E_x and E_y. The partial differential Schrödinger equation has thus been reduced to three ordinary differential equations which can be written

$$\frac{d^2X}{dx^2} + \frac{2mE_x}{\hbar^2}X = 0 \qquad \frac{d^2Y}{dy^2} + \frac{2mE_y}{\hbar^2}Y = 0 \qquad \frac{d^2Z}{dz^2} + \frac{2mE_z}{\hbar^2}Z = 0. \qquad (21\text{–}53)$$

Comparison of these equations with that for the one-dimensional particle in the box shows that both the equation and the boundary conditions are the same. The solutions are, therefore,

$$X(x) = \sqrt{\frac{2}{L_1}}\sin\left(\frac{n_x\pi x}{L_1}\right) \qquad n_x = 1, 2, 3, \ldots,$$

$$Y(y) = \sqrt{\frac{2}{L_2}}\sin\left(\frac{n_y\pi y}{L_2}\right) \qquad n_y = 1, 2, 3, \ldots,$$

$$Z(z) = \sqrt{\frac{2}{L_3}}\sin\left(\frac{n_z\pi z}{L_3}\right) \qquad n_z = 1, 2, 3, \ldots.$$

Thus, by Eq. (21–51),

$$\psi(x, y, z) = \sqrt{\frac{8}{L_1L_2L_3}}\sin\left(\frac{n_x\pi x}{L_1}\right)\sin\left(\frac{n_y\pi y}{L_2}\right)\sin\left(\frac{n_z\pi z}{L_3}\right). \qquad (21\text{–}54)$$

The energy is given by $E = E_x + E_y + E_z$, and since each term has the form for a particle in a box, we have

$$E_{n_x,n_y,n_z} = \frac{n_x^2 h^2}{8mL_1^2} + \frac{n_y^2 h^2}{8mL_2^2} + \frac{n_z^2 h^2}{8mL_3^2} \qquad \left.\begin{array}{l} n_x \\ n_y \\ n_z \end{array}\right\} = 1, 2, 3, \ldots. \qquad (21\text{–}55)$$

If the dimensions of the box are all equal, i.e. if $L_1 = L_2 = L_3 = L$, then

$$E_{n_x,n_y,n_z} = \frac{(n_x^2 + n_y^2 + n_z^2)h^2}{8mL^2}.$$

In this case, we have an interesting case of degeneracy; for example, the quantum number combinations $(n_x, n_y, n_z) = (112), (121), (211)$ represent different states of the system having the same energy. This energy state is three-fold degenerate.

We may generalize the result for the three-dimensional particle in the box in the following way. If the Hamiltonian operator can be written as a *sum* of groups of terms, each of which depends only on one coordinate or one set of coordinates, then the wave function can be written as a *product* of functions each of which depends only on the one coordinate or the one set of coordinates; correspondingly, the total energy is the sum of the energies associated with each coordinate or each set of coordinates.

For example, suppose there are two sets of coordinates, q_1 and q_2; further, suppose that the Hamiltonian operator can be arranged in the form

$$\mathbf{H} = \mathbf{H}_1 + \mathbf{H}_2,$$

where $\mathbf{H}_1$ depends only on the first set of coordinates q_1 and $\mathbf{H}_2$ depends only on a second set q_2. Then the wave function will have the form $\psi = f_1(q_1) \cdot f_2(q_2)$ and the energy will have the form $E = E_1 + E_2$. The proof is simple:

$$\mathbf{H}\psi = E\psi$$

$$(\mathbf{H}_1 + \mathbf{H}_2)f_1 f_2 = E f_1 f_2$$

$$f_2(\mathbf{H}_1 f_1) + f_1(\mathbf{H}_2 f_2) = E f_1 f_2.$$

Dividing by $f_1 f_2$ yields:

$$\frac{1}{f_1}(\mathbf{H}_1 f_1) + \frac{1}{f_2}(\mathbf{H}_2 f_2) = E.$$

The first term depends only on the set q_1, the second term depends only on the set q_2. Keeping the members of the set q_1 constant and varying the members of q_2 shows that the second term is a constant, E_2. Similarly, the first term must be a constant E_1. Thus we can write

$$\mathbf{H}_1 f_1 = E_1 f_1, \; + \; \mathbf{H}_2 f_2 = E_2 f_2, \tag{21–56}$$

$$E = E_1 + E_2. \tag{21–57}$$

We see that if the energy is made up of contributions from independent modes of motion, the wave function will be a product of functions each of which is a wave function describing an independent mode of motion. If, for example, as a first approximation, the internal energy of a diatomic molecule is made up of a sum of contributions

$$E = E_{\text{electronic}} + E_{\text{rotational}} + E_{\text{vibrational}}$$

then, in the first approximation, the wave function will be a product

$$\psi = \psi_{\text{electronic}} \cdot \psi_{\text{rotational}} \cdot \psi_{\text{vibrational}}$$

where $\psi_{\text{electronic}}$ depends only on the electronic coordinates; $\psi_{\text{rotational}}$ depends only on the angular coordinates, and $\psi_{\text{vibrational}}$ depends only on the vibrational coordinates, in this case the internuclear distance. We shall have numerous examples of the use of this theorem.

21–9 The Two–body Problem

The classical energy of a system of two point masses, m_1 and m_2, has the form

$$E = \frac{1}{2m_1}(p_{x_1}^2 + p_{y_1}^2 + p_{z_1}^2) + \frac{1}{2m_2}(p_{x_2}^2 + p_{y_2}^2 + p_{z_2}^2) + V(x_1, y_1, z_1, x_2, y_2, z_2),$$

in which x_1, y_1, z_1 and x_2, y_2, z_2 are the coordinates of the masses m_1 and m_2 respectively. Replacing the momenta by the corresponding quantum mechanical operators yields the Hamiltonian operator

$$\mathbf{H} = -\frac{\hbar^2}{2m_1}\left(\frac{\partial^2}{\partial x_1^2} + \frac{\partial^2}{\partial y_1^2} + \frac{\partial^2}{\partial z_1^2}\right) - \frac{\hbar^2}{2m_2}\left(\frac{\partial^2}{\partial x_2^2} + \frac{\partial^2}{\partial y_2^2} + \frac{\partial^2}{\partial z_2^2}\right) + V(x_1, y_1, z_1, x_2, y_2, z_2).$$

In general, the potential energy is not separable into terms involving only certain sets of the six Cartesian coordinates, so to simplify the problem we transform these six coordinates into three coordinates for the center of mass X, Y, and Z and three internal coordinates, x, y, z.

The center-of-mass coordinates (X, Y, Z) are determined by the condition that the sum of the first moments of mass about the center of mass vanish for each axis; that is,

$$m_1(x_1 - X) + m_2(x_2 - X) = 0 \qquad\qquad X = \frac{m_1 x_1 + m_2 x_2}{m_1 + m_2}$$

$$m_1(y_1 - Y) + m_2(y_2 - Y) = 0 \quad \text{or} \quad Y = \frac{m_1 y_1 + m_2 y_2}{m_1 + m_2}$$

$$m_1(z_1 - Z) + m_2(z_2 - Z) = 0 \qquad\qquad Z = \frac{m_1 z_1 + m_2 z_2}{m_1 + m_2}.$$

In addition, we define the three internal coordinates, x, y, z, by

$$x = x_2 - x_1, \qquad y = y_2 - y_1, \qquad z = z_2 - z_1.$$

Since x_1 and x_2 depend only on X and x, we have

$$\frac{\partial}{\partial x_1} = \left(\frac{\partial X}{\partial x_1}\right)\frac{\partial}{\partial X} + \left(\frac{\partial x}{\partial x_1}\right)\frac{\partial}{\partial x} = \left(\frac{m_1}{m_1 + m_2}\right)\frac{\partial}{\partial X} - \frac{\partial}{\partial x}.$$

In the second expression, the derivatives have been evaluated from the definitions of X and x. Then,

$$\frac{\partial^2}{\partial x_1^2} = \left[\left(\frac{m_1}{m_1 + m_2}\right)\frac{\partial}{\partial X} - \frac{\partial}{\partial x}\right]\left[\left(\frac{m_1}{m_1 + m_2}\right)\frac{\partial}{\partial X} - \frac{\partial}{\partial x}\right]$$

$$= \left(\frac{m_1}{m_1 + m_2}\right)^2 \frac{\partial^2}{\partial X^2} - \frac{2m_1}{m_1 + m_2}\frac{\partial^2}{\partial X \partial x} + \frac{\partial^2}{\partial x^2}.$$

Similarly, we find

$$\frac{\partial}{\partial x_2} = \left(\frac{m_2}{m_1 + m_2}\right)\frac{\partial}{\partial X} + \frac{\partial}{\partial x},$$

and

$$\frac{\partial^2}{\partial x_2^2} = \left(\frac{m_2}{m_1 + m_2}\right)^2 \frac{\partial^2}{\partial X^2} + \left(\frac{2m_2}{m_1 + m_2}\right)\frac{\partial^2}{\partial X \partial x} + \frac{\partial^2}{\partial x^2}.$$

These two terms are combined as they appear in the Hamiltonian,

$$\frac{1}{m_1}\frac{\partial^2}{\partial x_1^2} + \frac{1}{m_2}\frac{\partial^2}{\partial x_2^2} = \frac{m_1}{(m_1 + m_2)^2}\frac{\partial^2}{\partial X^2} - \frac{2}{m_1 + m_2}\frac{\partial^2}{\partial X \partial x} + \frac{1}{m_1}\frac{\partial^2}{\partial x^2}$$

$$+ \frac{m_2}{(m_1 + m_2)^2}\frac{\partial^2}{\partial X^2} + \frac{2}{m_1 + m_2}\frac{\partial^2}{\partial X \partial x} + \frac{1}{m_2}\frac{\partial^2}{\partial x^2}$$

$$= \frac{1}{m_1 + m_2}\frac{\partial^2}{\partial X^2} + \left(\frac{1}{m_1} + \frac{1}{m_2}\right)\frac{\partial^2}{\partial x^2}.$$

The algebra for y_1, y_2 and z_1, z_2 proceeds in exactly the same fashion, so the Hamiltonian becomes

$$\mathbf{H} = -\frac{\hbar^2}{2(m_1 + m_2)}\left(\frac{\partial^2}{\partial X^2} + \frac{\partial^2}{\partial Y^2} + \frac{\partial^2}{\partial Z^2}\right)$$

$$-\frac{\hbar^2}{2\mu}\left(\frac{\partial^2}{\partial x^2} + \frac{\partial^2}{\partial y^2} + \frac{\partial^2}{\partial z^2}\right) + V(x, y, z),$$

in which μ is the "reduced mass" of the system, defined by $1/\mu = 1/m_1 + 1/m_2$. In the problems of interest here, the potential energy will be independent of the position of the center of mass, hence we have written $V(x, y, z)$.

The transformation has separated the Hamiltonian into two groups of terms, the first group depending only on X, Y, Z, the second group depending on x, y, z. Thus we may write by the theorem, Eq. (21–56),

$$\psi_{total} = \psi_{trans}(X, Y, Z) \cdot \psi(x, y, z),$$

$$E_{total} = E_{trans} + E,$$

$$-\frac{\hbar^2}{2(m_1 + m_2)}\left(\frac{\partial^2}{\partial X^2} + \frac{\partial^2}{\partial Y^2} + \frac{\partial^2}{\partial Z^2}\right)\psi_{trans}(X, Y, Z) = E_{trans}\psi_{trans}(X, Y, Z),$$

and

$$-\frac{\hbar^2}{2\mu}\left(\frac{\partial^2\psi}{\partial x^2} + \frac{\partial^2\psi}{\partial y^2} + \frac{\partial^2\psi}{\partial z^2}\right) + V(x, y, z)\psi = E\psi. \tag{21–58}$$

The wave function, ψ_{trans}, is the wave function of a free particle of mass $m_1 + m_2$, moving with the center of mass of the system. Correspondingly, E_{trans} is the translational energy of the total system. This motion has no particular interest to us, so we may ignore it. We are interested in $\psi = \psi(x, y, z)$ which will provide a description of the internal motions of the system; E is the energy of these internal motions. For any system we can always separate out the center-of-mass coordinates in this way, discard the energy associated with it and consider only the internal coordinates. In the future we will assume that this has been done.

It is convenient now to transform Eq. (21–58) into spherical coordinates with m_1 at the center and m_2 at the position r, θ, φ. (See Fig. 21–7.) The transformation equations are

$$x = r \sin \theta \cos \varphi \qquad\qquad r = \sqrt{x^2 + y^2 + z^2}$$

$$y = r \sin \theta \sin \varphi \quad \text{or} \quad \cos \theta = z/\sqrt{x^2 + y^2 + z^2} \qquad (21\text{–}59)$$

$$z = r \cos \theta \qquad\qquad \tan \varphi = y/x.$$

The calculation of the differential operators goes as above, e.g.,

$$\frac{\partial}{\partial x} = \frac{\partial r}{\partial x}\frac{\partial}{\partial r} + \frac{\partial \theta}{\partial x}\frac{\partial}{\partial \theta} + \frac{\partial \varphi}{\partial x}\frac{\partial}{\partial \varphi}, \qquad \text{etc.,}$$

but is tedious. The final result takes the form

$$-\frac{\hbar^2}{2\mu}\left[\frac{1}{r^2}\frac{\partial}{\partial r}\left(r^2 \frac{\partial \psi}{\partial r}\right) + \frac{1}{r^2 \sin \theta}\frac{\partial}{\partial \theta}\left(\sin \theta \frac{\partial \psi}{\partial \theta}\right) + \frac{1}{r^2 \sin^2 \theta}\frac{\partial^2 \psi}{\partial \varphi^2}\right]$$

$$+ V(r, \theta, \varphi)\psi = E\psi. \qquad (21\text{–}60)$$

Depending on the form of the potential, this equation is applicable to a number of problems. We consider them in turn.

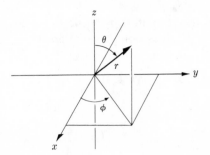

Fig. 21–7 Spherical coordinates.

21–10   The Rigid Rotator

Suppose the two masses are held rigidly apart at some fixed distance r_0. Since there is no momentum in the r-direction, the derivative with respect to r cannot appear in the equation. The potential energy is equal to zero since the system rotates freely.

Then Eq. (21–60) becomes

$$-\frac{\hbar^2}{2\mu r_0^2}\left[\frac{1}{\sin\theta}\frac{\partial}{\partial\theta}\left(\sin\theta\frac{\partial Y(\theta,\varphi)}{\partial\theta}\right)+\frac{1}{\sin^2\theta}\frac{\partial^2 Y(\theta,\varphi)}{\partial\varphi^2}\right]=EY(\theta,\varphi).\qquad(21\text{–}61)$$

In which we have written $Y(\theta,\varphi)$ for the wave function.

Suppose the center of mass is at the position R; then the sum of the second moments of mass about the center of mass yields the moment of inertia, I, about any axis perpendicular to the axis of the rotator; (Fig. 21–8). From the figure, we have $I = m_1(0 - R)^2 + m_2(r_0 - R)^2$. As usual, R is determined by the vanishing of the sum of the first moments.

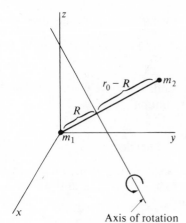

Fig. 21–8 Coordinates for calculation of moment of inertia.

Axis of rotation

$$0 = m_1(-R) + m_2(r_0 - R)\qquad\text{or}\qquad R = \frac{m_2}{m_1 + m_2}r_0$$

$$I = m_1\left(\frac{m_2}{m_1 + m_2}\right)^2 r_0^2 + m_2\left(1 - \frac{m_2}{m_1 + m_2}\right)^2 r_0^2 = \frac{m_1 m_2}{m_1 + m_2}r_0^2 = \mu r_0^2.$$

Introducing I for μr_0^2, Eq. (21–61) becomes

$$-\frac{\hbar^2}{2I}\left[\frac{1}{\sin\theta}\frac{\partial}{\partial\theta}\left(\sin\theta\frac{\partial Y}{\partial\theta}\right)+\frac{1}{\sin^2\theta}\frac{\partial^2 Y}{\partial\varphi^2}\right]=EY.\qquad(21\text{–}62)$$

The classical energy of a rigid rotator is given by

$$\frac{M^2}{2I}=E,$$

where M^2 is the square of the total angular momentum of the system. Comparing these equations, we conclude that the operator for the square of the total angular momentum is

$$\mathbf{M}^2 = -\hbar^2 \left[\frac{1}{\sin\theta} \frac{\theta}{\partial\theta} \left(\sin\theta \frac{\partial}{\partial\theta} \right) + \frac{1}{\sin^2\theta} \frac{\partial^2}{\partial\varphi^2} \right]. \tag{21-63}$$

We further observe from dimensionality that the energy E must be some multiple of $\hbar^2/2I$, so we write

$$E_J = J(J + 1)\frac{\hbar^2}{2I}, \tag{21-64}$$

in which J is, for the moment at least, an arbitrary parameter. After rearranging,

$$\frac{1}{\sin\theta} \frac{\partial}{\partial\theta} \left(\sin\theta \frac{\partial Y}{\partial\theta} \right) + \frac{1}{\sin^2\theta} \frac{\partial^2 Y}{\partial\varphi^2} + J(J + 1)Y = 0.$$

We observe that multiplying the equation through by $\sin^2\theta$ brings it to the form

$$\sin\theta \frac{\partial}{\partial\theta} \left(\sin\theta \frac{\partial Y}{\partial\theta} \right) + J(J + 1)\sin^2\theta \, Y + \frac{\partial^2 Y}{\partial\varphi^2} = 0.$$

By our earlier arguments, since only the last term depends on φ, we can substitute $Y(\theta, \varphi) = \Theta(\theta)\Phi(\varphi)$, and divide through by $\Theta(\theta)\Phi(\varphi)$ to obtain

$$\frac{1}{\Theta(\theta)} \sin\theta \frac{d}{d\theta} \left(\sin\theta \frac{d\Theta}{d\theta} \right) + J(J + 1)\sin^2\theta + \frac{1}{\Phi(\varphi)} \frac{d^2\Phi}{d\varphi^2} = 0.$$

The terms in this partial differential equation have been separated into a group depending on θ and a single term depending on φ. It follows that each of these sets of terms must be a constant so we write

$$\frac{1}{\Phi} \frac{d^2\Phi}{d\varphi^2} = -m^2 \quad \text{or} \quad \frac{d^2\Phi}{d\varphi^2} = -m^2\Phi.$$

This equation has the solution $\Phi = Ae^{im\varphi}$. We note that when $\varphi \to \varphi + 2\pi$ we return to the same set of points in space so that our boundary condition must be

$$\Phi(\varphi + 2\pi) = \Phi(\varphi),$$

$$Ae^{im(\varphi + 2\pi)} = Ae^{im\varphi}$$

$$e^{im2\pi} = 1.$$

This relation is satisfied only if m is zero or a positive or negative integer. We may write

$$\Phi_m(\varphi) = Ae^{im\varphi} \qquad m = 0, \quad \pm 1, \quad \pm 2, \ldots .$$

For normalization we require, since φ varies from 0 to 2π,

$$\int_0^{2\pi} \Phi_m^* \Phi_m \, d\varphi = 1 \qquad A^*A \int_0^{2\pi} e^{-im\varphi} e^{im\varphi} \, d\varphi = 1$$

$$|A|^2 \int_0^{2\pi} d\varphi = 1 \qquad\qquad A = \frac{1}{\sqrt{2\pi}}$$

where we have chosen A as a real number; then

$$\Phi_m(\varphi) = \frac{1}{\sqrt{2\pi}} e^{im\varphi}, \qquad m = 0, \quad \pm 1, \quad \pm 2, \quad \pm 3, \quad \ldots. \tag{21-65}$$

The remaining part of the equation can be written

$$\sin\theta \frac{d}{d\theta}\left(\sin\theta \frac{d\Theta}{d\theta}\right) + [J(J+1)\sin^2\theta - m^2]\Theta = 0. \tag{21-66}$$

In this equation we change variable to $\xi = \cos\theta$; then $d/d\theta = (d\xi/d\theta)(d/d\xi)$, but $d\xi/d\theta = -\sin\theta$, and $\sin^2\theta = 1 - \cos^2\theta = 1 - \xi^2$; then Eq. (21-66) becomes, with $\Theta(\theta) = P(\xi)$

$$(1 - \xi^2)\frac{d}{d\xi}\left[(1 - \xi^2)\frac{dP(\xi)}{d\xi}\right] + [J(J+1)(1 - \xi^2) - m^2]P(\xi) = 0.$$

Dividing by $(1 - \xi^2)$ we get

$$(1 - \xi^2)\frac{d^2P}{d\xi^2} - 2\xi\frac{dP}{d\xi} + \left[J(J+1) - \frac{m^2}{(1 - \xi^2)}\right]P = 0 \tag{21-67}$$

Since $\xi = \cos\theta$, and the limits of θ are 0 and π, the corresponding limits of ξ are $+1$ and -1. Eq. (21-67) is the associated Legendre equation and it may be shown that the only case in which this equation possesses continuous, single-valued, and quadratically integrable solutions in the interval $-1 \leq \xi \leq +1$ is the case in which J is a positive integer or zero. The solutions depend on the integers J and m and are written $P_J^{|m|}(\xi)$. It is clear from Eq. (21-67) which contains only m^2 that the function $P_J^{|m|}(\xi)$ can therefore depend only on the absolute value of m. The function of $P_J^{|m|}(\xi)$ is called the associated Legendre function of degree J and order $|m|$.

Comparing this result with Eq. (21-64) we find that the energy of the rigid rotator is quantized:

$$E_J = J(J+1)\frac{\hbar^2}{2I}, \qquad J = 0, 1, 2, 3, \ldots. \tag{21-68}$$

Similarly since $E = M^2/2I$, it follows that the square of the total angular momentum is quantized

$$M^2 = J(J+1)\hbar^2, \qquad J = 0, 1, 2, 3, \ldots. \tag{21-69}$$

The interesting result here is that, contrary to the case of the particle in the box and the harmonic oscillator, when $J = 0$ the lowest energy is zero, corresponding to a precise value, zero, for the angular momentum. This is possible since it allows the angles of orientation, θ and φ, to be completely unspecified, in conformity with the uncertainty principle.

The meaning of the integer m remains to be investigated. To do this we consider the classical z-component of angular momentum in Cartesian coordinates. This has the form

$$M_z = xp_y - yp_x.$$

The quantum mechanical operator for the z-component is then

$$\mathbf{M}_z = -i\hbar\left(x\frac{\partial}{\partial y} - y\frac{\partial}{\partial x}\right). \tag{21-70}$$

If we transform $\mathbf{M}_z$ into spherical coordinates using the same method as above, we obtain the very simple result

$$\mathbf{M}_z = -i\hbar\frac{\partial}{\partial\varphi}. \tag{21-71}$$

The wave function for the rigid rotator has the form

$$Y_{J,m}(\xi, \varphi) = A_{J,m}P_J^{|m|}(\xi)\frac{1}{\sqrt{2\pi}}e^{im\varphi}, \tag{21-72}$$

in which the normalization constant, $A_{J,m}$, is

$$A_{J,m} = \left[\left(\frac{2J+1}{2}\right)\frac{(J-|m|)!}{(J+|m|)!}\right]^{1/2}. \tag{21-73}$$

Then

$$\mathbf{M}_z Y_{J,m} = A_{J,m}P_J^{|m|}(\xi)\frac{1}{\sqrt{2\pi}}\mathbf{M}_z e^{im\varphi}$$

$$= -i\hbar(im)Y_{J,m};$$

$$\mathbf{M}_z Y_{J,m} = m\hbar Y_{J,m}. \tag{21-74}$$

Therefore, we find that $Y_{J,m}$ is an eigenfunction of $\mathbf{M}_z$ with the eigenvalue $m\hbar$. The z-component of the angular momentum is therefore quantized; the quantum number is $m = 0, \pm 1, \pm 2$, etc. Again, precise values of the z-component of angular momentum are permitted since the angle φ is totally unspecifiable. Repeating the application of $\mathbf{M}_z$ on Eq. (21-74), we obtain

$$\mathbf{M}_z^2 Y_{J,m} = m\hbar(\mathbf{M}_z Y_{J,m}) = m^2\hbar^2 Y_{J,m}, \tag{21-75}$$

so that the square of the z-component of angular momentum is $m^2\hbar^2$. The total angular momentum is the sum of squares of the components:

$$M^2 = M_x^2 + M_y^2 + M_z^2. \tag{21-76}$$

Replacing M^2 and M_z^2 by their values from Eqs. (21-69) and (21-75), we obtain after rearranging

$$[J(J+1) - m^2]\hbar^2 = M_x^2 + M_y^2.$$

Since the right-hand side is a sum of squares, it cannot be negative, hence we have the condition

$$J(J + 1) - m^2 \geq 0.$$

It is apparent that this condition is fulfilled so long as $|m| \leq J$. The values of m are therefore restricted to $m = 0, \pm 1, \pm 2, \ldots, \pm J$. For a given value of J, there are $2J + 1$ values of m. The energy is determined by J only, hence each energy level has a degeneracy of $2J + 1$.

The possible orientations of the angular-momentum vector for $J = 1$ are shown in Fig. 21–9. When $J = 1$,

$$\sqrt{M^2} = \sqrt{J(J + 1)}\hbar$$
$$= \sqrt{2}\hbar$$

Any position of the vector in the conical surface above the xy-plane in Fig. 21–9(a) will yield a projection of the vector equal to $+\hbar$ on the z-axis, while any vector in the conical surface below the plane will have a projection $-\hbar$ on the z-axis. The permissible projections for $J = 2$ are 2, 1 and 0.

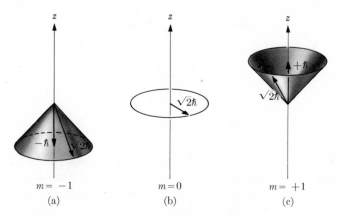

$$m = -1 \qquad m = 0 \qquad m = +1$$
$$\text{(a)} \qquad \text{(b)} \qquad \text{(c)}$$

Fig. 21–9 The z-components of angular momentum for $J = 1$: (a) $m = -1$; (b) $m = 0$; (c) $m = +1$.

Properties of the functions $P_n^{|m|}(\xi)$

We consider Eq. (21–67) under the following conditions: $m = 0$ and J is replaced by $n = 0, 1, 2, \ldots$. The equation becomes

$$(1 - \xi^2)\frac{d^2 P_n(\xi)}{d\xi^2} - 2\xi\frac{dP_n(\xi)}{d\xi} + n(n + 1)P_n(\xi) = 0. \qquad (21–77)$$

This is Legendre's equation and the solutions for positive integral values of n are defined by

$$P_n(\xi) = \frac{1}{2^n n!} \frac{d^n}{d\xi^n} (\xi^2 - 1)^n. \tag{21–78}$$

It is apparent from the definition that the function $(\xi^2 - 1)$ is a polynomial of $2n$th degree; differentiation n times will reduce it to a polynomial of nth degree. The functions $P_n(\xi)$ are called Legendre polynomials of degree n. More explicitly we may write

$$P_n(\xi) = \sum_{k=0} \frac{(-1)^k (2n - 2k)! \xi^{n-2k}}{2^n (n-k)! k! (n-2k)!}, \tag{21–79}$$

where the upper limit of k in the sum is $n/2$ if n is even and $(n-1)/2$ if n is odd.

We may summarize the properties of the Legendre polynomials as follows:

1. The functions $\sqrt{(n + \frac{1}{2})} P_n(\xi)$ compose an orthonormal set in the interval $-1 \le \xi \le +1$; therefore, we have

$$\sqrt{(n + \tfrac{1}{2})(j + \tfrac{1}{2})} \int_{-1}^{1} P_n(\xi) P_j(\xi) \, d\xi = \delta_{nj}. \tag{21–80}$$

2. The functions are symmetric or antisymmetric as n is even or odd

$$P_n(-\xi) = (-1)^n P_n(\xi). \tag{21–81}$$

3. The functions do not exceed 1 in absolute value

$$|P_n(\xi)| \le 1$$

$$P_n(1) = 1, \qquad P_n(-1) = (-1)^n. \tag{21–82}$$

Since the $P_n(\xi)$ are polynomials, there exist n roots, or n values of ξ for which $P_n(\xi)$ changes sign. The sign of $P_n(\xi)$ is often indicated by a circular diagram. At the north pole $\xi = \cos 0 = +1$; at the equator $\xi = \cos \pi/2 = 0$; at the south pole $\xi = \cos \pi = -1$. The lines on the circle indicate the values of θ at which the polynomial is zero (Fig. 21–10).

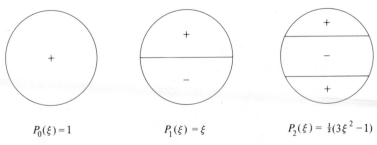

$$P_0(\xi) = 1 \qquad\qquad P_1(\xi) = \xi \qquad\qquad P_2(\xi) = \tfrac{1}{2}(3\xi^2 - 1)$$

Fig. 21–10 Signs and locations of zeros of Legendre polynomials.

4. Differential relations.

$$(1 - \xi^2)\frac{dP_n(\xi)}{d\xi} = -n[\xi P_n(\xi) - P_{n-1}(\xi)] = (n + 1)[\xi P_n(\xi) - P_{n+1}(\xi)], \qquad (21\text{–}83)$$

$$(1 - \xi^2)\frac{d^2 P_n(\xi)}{d\xi^2} - 2\xi\frac{dP_n(\xi)}{d\xi} + n(n + 1)P_n(\xi) = 0 \qquad \text{(Legendre equation)}. \qquad (21\text{–}84)$$

5. Recurrence formula:

$$(n + 1)P_{n+1}(\xi) + nP_{n-1}(\xi) - (2n + 1)\xi P_n(\xi) = 0. \qquad (21\text{–}85)$$

A typical application of the recurrence formula is in the evaluation of integrals of the type

$$\int_{-1}^{1} P_n(\xi)\xi P_j(\xi)\, d\xi;$$

replacing $\xi P_n(\xi)$ by its value from the recurrence formula we obtain

$$\int_{-1}^{1} P_n(\xi)\xi P_j(\xi)\, d\xi = \int_{-1}^{1}\left[\frac{(n + 1)P_{n+1}}{2n + 1} + \frac{nP_{n-1}}{2n + 1}\right]P_j\, d\xi$$

$$= \frac{1}{2n + 1}\left[(n + 1)\int_{-1}^{1} P_{n+1}P_j\, d\xi + n\int_{-1}^{1} P_{n-1}P_j\, d\xi\right].$$

Using the orthogonality property:

$$\int_{-1}^{1} P_n(\xi)\xi P_j(\xi)\, d\xi = \frac{(n + 1)\delta_{n+1,j}}{(2n + 1)(n + 3/2)} + \frac{n\delta_{n-1,j}}{(2n + 1)(n - 1/2)}.$$

The integral is zero unless $j = n + 1$ or $n - 1$. This type of integral furnishes the basis for spectral selection rules as we shall see later.

The associated Legendre functions, $P_n^{|m|}(\xi)$, are obtained from the Legendre polynomials by the definition

$$P_n^{|m|}(\xi) = (1 - \xi^2)^{|m|/2}\frac{d^m P_n(\xi)}{d\xi^m}. \qquad (21\text{–}86)$$

Since $P_n(\xi)$ is an nth degree polynomial, no more than n nonzero derivatives exist; consequently, $|m| \le n$. Note that the associated functions are *not* polynomials unless $|m|$ is even. These functions are solutions of the associated Legendre equation, Eq. (21–67). In terms of a finite sum we can write

$$P_n^{|m|}(\xi) = (1 - \xi^2)^{|m|/2}\sum_{k=0}\frac{(-1)^k(2n - 2k)!\,\xi^{n-|m|-2k}}{2^n(n - k)!k!(n - |m| - 2k)!}. \qquad (21\text{–}87)$$

The upper limit of the sum is $k = (n - |m|)/2$ if $n - |m|$ is even and $(n - |m| - 1)/2$ if $(n - |m|)$ is odd. The properties of the $P_n^{|m|}(\xi)$ are as follows:

1. The functions of the same order form an orthogonal set in the interval $-1 \leq \xi \leq +1$

$$\int_{-1}^{+1} P_n^{|m|}(\xi) P_j^{|m|}(\xi) \, d\xi = \left(\frac{1}{n + \frac{1}{2}}\right) \frac{(n + |m|)!}{(n - |m|)!} \delta_{nj}. \qquad (21\text{--}88)$$

2. Symmetry:

$$P_n^{|m|}(-\xi) = (-1)^{n-m} P_n^{|m|}(\xi) \qquad (21\text{--}89)$$

3. Differential relations:

$$(1 - \xi^2) \frac{dP_n^{|m|}(\xi)}{d\xi} = (n + 1)\xi P_n^{|m|}(\xi) - (n - |m| + 1)P_{n+1}^{|m|}(\xi), \qquad (21\text{--}90)$$

$$(1 - \xi^2) \frac{d^2 P_n^{|m|}(\xi)}{d\xi^2} - 2\xi \frac{dP_n^{|m|}(\xi)}{d\xi} + \left[n(n+1) - \frac{m^2}{1 - \xi^2} \right] P_n^{|m|}(\xi) = 0. \qquad (21\text{--}91)$$

4. Recurrence relations:

$$(2n + 1)\xi P_n^{|m|}(\xi) = (n - |m| + 1)P_{n+1}^{|m|}(\xi) + (n + |m|)P_{n-1}^{|m|}(\xi), \qquad (21\text{--}92)$$

$$(2n + 1)(1 - \xi^2)^{1/2} P_n^{|m|-1}(\xi) = P_{n+1}^{|m|}(\xi) - P_{n-1}^{|m|}(\xi), \qquad (21\text{--}93)$$

$$(2n + 1)(1 - \xi^2)^{1/2} P_n^{|m|+1}(\xi)$$
$$= (n + |m|)(n + |m| + 1)P_{n-1}^{|m|}(\xi) - (n - |m|)(n - |m| + 1)P_{n+1}^{|m|}(\xi). \qquad (21\text{--}94)$$

Table 21–2 Values of the Associated Legendre Polynomials

m \ n	0	1	2	3	4
0	1	ξ	$\frac{1}{2}(3\xi^2 - 1)$	$\frac{1}{2}(5\xi^3 - 3\xi)$	$\frac{1}{8}(35\xi^4 - 30\xi^2 + 3)$
1	—	$(1 - \xi^2)^{1/2}$	$(1 - \xi^2)^{1/2}3\xi$	$(1 - \xi^2)^{1/2}\frac{3}{2}(5\xi^2 - 1)$	$(1 - \xi^2)^{1/2}\frac{5}{2}(7\xi^3 - 3\xi)$
2	—	—	$(1 - \xi^2)3$	$(1 - \xi^2)15\xi$	$(1 - \xi^2)\frac{15}{2}(7\xi^2 - 1)$
3	—	—	—	$(1 - \xi^2)^{3/2}15$	$(1 - \xi^2)^{3/2}105\xi$
4	—	—	—	—	$(1 - \xi^2)^2 105$

Problems

21–1. Calculate the probability of finding the particle in the "box" in the region between $\frac{1}{4}L$ and $\frac{3}{4}L$.

21–2. The electrons in a vacuum tube are confined in a "box" between filament and plate which is perhaps 0.1 cm in width. Compute the spacing between the energy levels in this situation. Do the electrons behave more like waves or like golf balls? In a simple tube the energy of the electron is about 100 eV. What is the quantum number of the electrons?

21–3. If the energy of the electron is 5 eV, what size box must confine it so that the wave property will be exhibited? Assume we can observe 0.1 % of the total energy.

21–4. The muzzle velocity of a rifle bullet is about 3000 ft/sec. If the rifle bullet weighs 30 gm, with what accuracy can the position be measured without perturbing the momentum by more than one part in a million?

21–5. Derive Eq. (21–49) using the hint suggested in the text.

21–6. Evaluate $\langle x^2 \rangle$ for the harmonic oscillator and from this value obtain

$$\Delta x = [\langle x^2 \rangle - \langle x \rangle^2]^{1/2}.$$

21–7. Evaluate $\langle p_x^2 \rangle$ and $\langle p_x \rangle^2$ for the harmonic oscillator and calculate the uncertainty in the momentum, Δp_x.

21–8. By combining the results of Problems 21–6 and 21–7 find the uncertainty relation for the harmonic oscillator.

21–9. Show that for $n = 1$, the probability of finding the harmonic oscillator in the classically forbidden region is 0.1116.

21–10. For a particle in a cubical box, $L_1 = L_2 = L_3 = L$, tabulate the energy values in the lowest eight energy levels, (as multiples of $h^2/8mL^2$), and the degeneracy of each level.

21–11. Calculate the moment of inertia and the energy in the first rotational state above the ground level, $J = 1$, for

a) H_2 in which $r_0 = 0.746 \times 10^{-8}$ cm, and
b) O_2 in which $r_0 = 1.208 \times 10^{-8}$ cm.

21–12. For a particle in a one-dimensional box suppose the potential is not zero but is perturbed by a small amount. Calculate the effect on the energy if the potential has the following forms:

a) $V = bx$ (b is constant).
b) $V = 0$ $(0 \leq x < L/2)$
   $V = b$ $(L/2 \leq x < L)$
c) $V = 0$ $(0 \leq x < L/2 - a)$; $V = b$, $(L/2 - a \leq x \leq L/2 + a)$
   $V = 0$ $(L/2 + a < x \leq L)$.
d) $V = \frac{1}{2}k(x - L/2)^2$

21–13. Consider the harmonic oscillator and the following forms of the perturbing potential. Calculate the energy in each case.

a) $V = \alpha x^k$ (k is any odd integer),
b) $V = \alpha x^4$.

21–14. From the definitions in Eqs. (21–70) and (21–59), prove Eq. (21–71).

Chapter Twenty-two

The Hydrogen Atom

22–1 The Central-field Problem

Returning to Eq. (21–60), we consider the case in which $V(r, \theta, \varphi)$ is, in fact, a function only of r, the distance between the two bodies. Then any forces act only along the line of centers of the two bodies; this defines a "central-field" problem. To discuss the central-field problem, we multiply Eq. (21–60) by $2\mu r^2$, and rearrange to

$$-\hbar^2 \frac{\partial}{\partial r}\left(r^2 \frac{\partial \psi}{\partial r}\right) - \hbar^2 \left[\frac{1}{\sin\theta}\frac{\partial}{\partial\theta}\left(\sin\theta\frac{\partial\psi}{\partial\theta}\right) + \frac{1}{\sin^2\theta}\frac{\partial^2\psi}{\partial\varphi^2}\right] + 2\mu r^2[V(r) - E]\psi = 0. \quad (22\text{–}1)$$

We recognize immediately from Eq. (21–63) the operator for the square of the total angular momentum; then Eq. (22–1) becomes

$$-\hbar^2 \frac{\partial}{\partial r}\left(r^2 \frac{\partial \psi}{\partial r}\right) + \mathbf{M}^2\psi + 2\mu r^2[V(r) - E]\psi = 0.$$

Since the set of terms, $\mathbf{M}^2\psi$, depends only on θ and φ and not on r, we may write

$$\psi(r, \theta, \varphi) = R(r)Y_{J,m}(\theta, \varphi).$$

Then, since $\mathbf{M}^2 Y_{J,m} = J(J + 1)\hbar^2 Y_{J,m}$, we find that $\mathbf{M}^2\psi = R(r)\mathbf{M}^2 Y_{J,m}(\theta, \varphi) = J(J + 1)\hbar^2 R(r)Y_{J,m}(\theta, \varphi)$. Using this result in the equation and dividing through by $Y_{J,m}(\theta, \varphi)$ and $2\mu r^2$ yields

$$-\frac{\hbar^2}{2\mu r^2}\frac{d}{dr}\left(r^2 \frac{dR}{dr}\right) + \left[\frac{J(J + 1)\hbar^2}{2\mu r^2} + V(r) - E\right]R(r) = 0. \quad (22\text{–}2)$$

This result simply tells us that the angular momentum in the presence of a central field is quantized in exactly the same way as for the rigid rotator. The energy, on the other hand, will be affected by the fact that r is not constant; consequently, the moment of inertia is not constant.

Equation (22–2) also shows that the total energy is made up of three contributions: the first term in the equation is the contribution of the kinetic energy of the motion along the line of centers; the second term is the kinetic energy associated with the rotation; the third term is the potential energy, $V(r)$.

22–2 The Hydrogen Atom

The hydrogen atom is a typical case of the central-field problem. As was shown in Fig. 19–5, the proton is at the center with a charge $+e$ while the electron is at a distance r with a charge $-e$. The coulombic force acts along the line of centers and corresponds to a potential energy, $V(r) = -e^2/r$.

We can rewrite Eq. (22–2) in the form

$$-\frac{\hbar^2}{2\mu r^2}\frac{d}{dr}\left(r^2\frac{dR}{dr}\right) + \left[\frac{l(l+1)\hbar^2}{2\mu r^2} - \frac{e^2}{r}\right]R = ER. \tag{22–3}$$

We have replaced the rotational quantum number J by l, since this is the usual notation in atomic systems. The quantum number l is called the *azimuthal* quantum number and characterizes the total angular momentum of the atom,

$$M^2 = l(l+1)\hbar^2, \qquad l = 0, 1, 2, 3, \ldots.$$

The quantum number m has the same interpretation as before; it characterizes the z-component of the angular momentum.

$$M_z = mh, \qquad m = -l, -(l-1), \ldots, -1, 0, 1, \ldots, l-1, l.$$

In this situation, m is called the *magnetic* quantum number, for reasons which will be apparent later.

Again we introduce dimensionless variables; $r = \beta\rho$ and $E = -(1/n^2)(\hbar^2/2\mu\beta^2)$, where n is a parameter characterizing the energy. Note that the energy has been chosen as a negative quantity; this implies that the discussion will deal only with the bound states of the hydrogen atom. The zero of potential energy is at $r \to \infty$, in which state the two particles move independently. The equation becomes

$$-\frac{\hbar^2}{2\mu\beta^2}\left[\frac{1}{\rho^2}\frac{d}{d\rho}\left(\rho^2\frac{dR}{d\rho}\right)\right] + \left\{\frac{l(l+1)}{\rho^2}\left(\frac{\hbar^2}{2\mu\beta^2}\right) - \frac{2}{\rho}\left(\frac{e^2}{2\beta}\right)\right\}R = -\frac{1}{n^2}\left(\frac{\hbar^2}{2\mu\beta^2}\right)R.$$

It is convenient to determine β by requiring that

$$\frac{\hbar^2}{2\mu\beta^2} = \frac{e^2}{2\beta} \qquad \text{or} \qquad \beta = \frac{\hbar^2}{\mu e^2} = a_0,$$

in which a_0 is the first Bohr radius; compare Eq. (19–20). Then we have

$$\frac{1}{\rho^2}\frac{d}{d\rho}\left(\rho^2\frac{dR}{d\rho}\right) + \left(\frac{2}{\rho} - \frac{l(l+1)}{\rho^2} - \frac{1}{n^2}\right)R = 0,$$

or

$$\frac{d^2R}{d\rho^2} + \frac{2}{\rho}\frac{dR}{d\rho} + \left(\frac{2}{\rho} - \frac{l(l+1)}{\rho^2} - \frac{1}{n^2}\right)R = 0. \tag{22-4}$$

As $\rho \to \infty$, this equation becomes $d^2R/d\rho^2 = (1/n^2)R$ which yields values

$$R_\infty = e^{\rho/n} \qquad \text{and} \qquad R_\infty = e^{-\rho/n}$$

only the second value is finite at $\rho = \infty$, so we choose our solution in the form

$$R = u(\rho)e^{-\rho/n}. \tag{22-5}$$

Calculation of $d^2R/d\rho^2$ and $dR/d\rho$ and substitution in Eq. (22-4) yields, after dividing out $e^{-\rho/n}$,

$$\frac{d^2u}{d\rho^2} + \frac{1}{\rho}\left(2 - \frac{2\rho}{n}\right)\frac{du}{d\rho} + \frac{1}{\rho^2}\left[\frac{2\rho}{n}(n-1) - l(l+1)\right]u = 0.$$

If we set $(2\rho/n) = x$ in this equation it simplifies immediately to

$$\frac{d^2u}{dx^2} + \left(\frac{2-x}{x}\right)\frac{du}{dx} + \frac{1}{x^2}[(n-1)x - l(l+1)]u = 0. \tag{22-6}$$

Solutions to this equation have the form

$$u(x) = x^l L(x); \tag{22-7}$$

then

$$\frac{du}{dx} = x^l\frac{dL}{dx} + lx^{l-1}L,$$

$$\frac{d^2u}{dx^2} = x^l\frac{d^2L}{dx^2} + 2lx^{l-1}\frac{dL}{dx} + l(l-1)x^{l-2}L.$$

These values reduce the equation, after division by x^{l-1}, to

$$x\frac{d^2L}{dx^2} + [2(l+1) - x]\frac{dL}{dx} + [n - (l+1)]L = 0. \tag{22-8}$$

The only solutions, $L(x)$ of this equation which are quadratically integrable are those for which the coefficient of L is zero or a positive integer; this condition requires that the parameter n be an integer such that $n - (l+1) \geq 0$ or $n \geq l+1$. Since the least value of l is zero we have the quantization conditions

$$n = 1, 2, 3, \ldots, \qquad 0 \leq l \leq n - 1. \tag{22-9}$$

We recognize these conditions as the familiar requirements on the value of n, *the principal quantum number*, of the hydrogen atom.

The functions L are the associated Laguerre polynomials which depend on n and l; if we write $s = n + l$ and $t = 2l + 1$, then the equation becomes

$$\frac{xd^2L_s^t(x)}{dx^2} + (t + 1 - x)\frac{dL_s^t(x)}{dx} + (s - t)L_s^t(x) = 0. \tag{22–10}$$

The general form for the polynomial $L_s^t(x)$ is

$$L_s^t(x) = -\sum_{k=0}^{s-t} (-1)^k \frac{(s!)^2 x^k}{(s - t - k)!(t + k)!k!}. \tag{22–11}$$

It should be observed that in contrast to the Hermite and Legendre polynomials, the Laguerre polynomials contain both odd and even powers of x. The first few are given in Table 22–1.

Table 22–1 Associated Laguerre Polynomials

$n = 1$;	$l = 0$	$L_1^1(x) = -1$	$x = 2\rho$
$n = 2$;	$l = 0$	$L_2^1(x) = -2!(2 - x)$	$x = \rho$
	$l = 1$	$L_3^3(x) = -3!$	
$n = 3$;	$l = 0$	$L_3^1(x) = -3!(3 - 3x + \tfrac{1}{2}x^2)$	$x = \tfrac{2}{3}\rho$
	$l = 1$	$L_4^3(x) = -4!(4 - x)$	
	$l = 2$	$L_5^5(x) = -5!$	
$n = 4$;	$l = 0$	$L_4^1(x) = -4!(4 - 6x + 2x^2 - \tfrac{1}{6}x^3)$	$x = \tfrac{1}{2}\rho$
	$l = 1$	$L_5^3(x) = -5!(10 - 5x + \tfrac{1}{2}x^2)$	
	$l = 2$	$L_6^5(x) = -6!(6 - x)$	
	$l = 3$	$L_7^7(x) = -7!$	
$n = 5$;	$l = 0$	$L_5^1(x) = -5!(5 - 10x + 5x^2 - \tfrac{5}{6}x^3 + \tfrac{1}{24}x^4)$	$x = \tfrac{2}{5}\rho$
	$l = 1$	$L_6^3(x) = -6!(20 - 15x + 3x^2 - \tfrac{1}{6}x^3)$	
	$l = 2$	$L_7^5(x) = -7!(21 - 7x + \tfrac{1}{2}x^2)$	
	$l = 3$	$L_8^7(x) = -8!(8 - x)$	
	$l = 4$	$L_9^9(x) = -9!$	

The normalized radial wave functions have the form

$$R_{nl}(r) = -\frac{2}{n^2 a_0^{3/2}}\left[\frac{(n - l - 1)!}{[(n + l)!]^3}\right]^{1/2} e^{-x/2} x^l L_{n+l}^{2l+1}(x) \tag{22–12}$$

in which $x = 2r/na_0$.

The complete wave function for the hydrogen atom has the form

$$\psi_{n,l,m}(r, \theta, \varphi)$$

$$= -\frac{2}{n^2 a_0^{3/2}} \left[\frac{(n - l - 1)!(2l + 1)(l - |m|)!}{[(n + l)!]^3 4\pi(l + |m|)!} \right]^{1/2} e^{-x/2} x^l L_{n+l}^{2l+1}(x) P_l^{|m|}(\cos \theta) e^{im\varphi}. \quad (22\text{--}13)$$

A list of the complete hydrogen atom wave functions is given in Table 22–2.

22–3   Recapitulation on the Hydrogen Atom

The hydrogen atom consists of two particles, a proton and an electron. Six coordinates and six momenta, three for each particle, are needed to describe the mechanical state of this system. The six coordinates of such a system can always be transformed to three coordinates of the center of mass of the system and three internal coordinates. After this is done the Schrödinger equation separates into two independent equations; the first involves only the coordinates of the center of mass and the translational energy of the atom as a whole. Since the translational part has no interest to us, we discard it. The remaining equation involves the internal coordinates of the atom and the internal energy. It is this energy and this part of the description which is of interest. We will refer to this internal energy simply as the energy of the atom.

The internal coordinates are the usual spherical coordinates r, θ, and φ, displayed in their relation to the Cartesian coordinates in Fig. 21–7. The nucleus is at the origin, r is the distance between the nucleus and the electron, θ is the angle between the z-axis and the radius vector connecting nucleus and electron, and φ is the angle between the $+x$-axis and the projection of the radius vector on the xy-plane. The potential energy is $-e^2/r$, resulting from the electrical attraction of the charges $+e$ on the nucleus and $-e$ on the electron. The Schrödinger equation can be solved for ψ as a function of the coordinates r, θ, and φ. The wave functions obtained by solving this equation are descriptions of the states of the hydrogen atom.

If the solutions of the Schrödinger equation are to make physical sense, certain integers, quantum numbers, must be introduced. Just as in the case of the particle in the box, these integers enter because of the constraints which are placed on the system. For example, if the probability density $|\psi|^2$ is to have a unique value at every point in space, then the description ψ must have the same value at $\varphi = 2\pi$ and at $\varphi = 0$, since these values of φ correspond to the same set of points in space. This restriction together with the form of the equation requires ψ to depend on φ through either $e^{im\varphi}$ or $e^{-im\varphi}$, where m is an integer. Two other integers, n and l, are introduced by the requirement that the probability density be finite everywhere. It is clear that we may not have an infinite probability of finding the electron at any point in space.

The final description, $\psi_{nlm}(r, \theta, \varphi)$ or, more concisely, ψ_{nlm}, is a function of the coordinates r, θ, and φ, and of the quantum numbers n, l, and m. Since ψ_{nlm} depends

Table 22–2 Complete hydrogen-atom wave functions, $\psi_{n,l,m}(r, \theta, \varphi)$ $(\rho = r/a_0)$

$n = 1, \quad l = 0, \quad m = 0.$ $\psi_{100} = \left(\dfrac{1}{\pi a_0^3}\right)^{1/2} e^{-\rho}$

$n = 2, \quad l = 0, \quad m = 0.$ $\psi_{200} = \dfrac{1}{8}\left(\dfrac{2}{\pi a_0^3}\right)^{1/2} (2 - \rho) e^{-\rho/2}$

$n = 2, \quad l = 1, \quad m = 0.$ $\psi_{210} = \dfrac{1}{8}\left(\dfrac{2}{\pi a_0^3}\right)^{1/2} \rho\, e^{-\rho/2} \cos\theta$

$n = 2, \quad l = 1, \quad m = \pm 1.$ $\psi_{211} = \dfrac{1}{8}\left(\dfrac{1}{\pi a_0^3}\right)^{1/2} \rho\, e^{-\rho/2} \sin\theta\, e^{i\varphi}$

                                  $\psi_{21-1} = \dfrac{1}{8}\left(\dfrac{1}{\pi a_0^3}\right)^{1/2} \rho\, e^{-\rho/2} \sin\theta\, e^{-i\varphi}$

$n = 3, \quad l = 1, \quad m = 0.$ $\psi_{300} = \dfrac{1}{243}\left(\dfrac{3}{\pi a_0^3}\right)^{1/2} (27 - 18\rho + 2\rho^2)\, e^{-\rho/3}$

$n = 3, \quad l = 1, \quad m = 0.$ $\psi_{310} = \dfrac{1}{81}\left(\dfrac{2}{\pi a_0^3}\right)^{1/2} \rho(6 - \rho)\, e^{-\rho/3} \cos\theta$

$n = 3, \quad l = 1, \quad m = \pm 1.$ $\psi_{311} = \dfrac{1}{81}\left(\dfrac{1}{\pi a_0^3}\right)^{1/2} \rho(6 - p)\, e^{-\rho/3} \sin\theta\, e^{i\varphi}$

                                  $\psi_{31-1} = \dfrac{1}{81}\left(\dfrac{1}{\pi a_0^3}\right)^{1/2} \rho(6 - \rho)\, e^{-\rho/3} \sin\theta\, e^{-i\varphi}$

$n = 3, \quad l = 2, \quad m = 0.$ $\psi_{320} = \dfrac{1}{486}\left(\dfrac{6}{\pi a_0^3}\right)^{1/2} \rho^2 e^{-\rho/3}(3\cos^2\theta - 1)$

$n = 3, \quad l = 2, \quad m = \pm 1.$ $\psi_{321} = \dfrac{1}{81}\left(\dfrac{1}{\pi a_0^3}\right)^{1/2} \rho^2\, e^{-\rho/3} \sin\theta \cos\theta\, e^{i\varphi}$

                                  $\psi_{32-1} = \dfrac{1}{81}\left(\dfrac{1}{\pi a_0^3}\right)^{1/2} \rho^2\, e^{-\rho/3} \sin\theta \cos\theta\, e^{-i\varphi}$

$n = 3, \quad l = 2, \quad m = \pm 2.$ $\psi_{322} = \dfrac{1}{162}\left(\dfrac{1}{\pi a_0^3}\right)^{1/2} \rho^2 e^{-\rho/3} \sin^2\theta\, e^{i2\varphi}$

                                  $\psi_{32-2} = \dfrac{1}{162}\left(\dfrac{1}{\pi a_0^3}\right)^{1/2} \rho^2 e^{-\rho/3} \sin^2\theta\, e^{-i2\varphi}$

---

on the integers in a unique way, the integers by themselves constitute a convenient abbreviated description of the system. Knowing the integers, we can look up the corresponding ψ_{nlm} in a table such as Table 22–2 if we need it. For the most part we shall use only the quantum numbers to describe the system.

22–4 Significance of the Quantum Numbers

The integer n, the principal quantum number, describes the energy of the hydrogen atom through the equation

$$E_n = -\frac{1}{n^2}\left(\frac{e^2}{2a_0}\right), \tag{22–14}$$

with allowed values $n = 1, 2, 3, \ldots$. In Eq. (22–14), a_0 is the first Bohr radius. These energy values are the same as those given by Bohr's initial calculation, Eq. (19–22). However, in the Schrödinger model, n has nothing directly to do with angular momentum, while in the Bohr model, n was a measure first of the angular momentum of the system. This difference in interpretation should be kept in mind.

The energy of the hydrogen atom is quantized, so it has a system of energy levels. The lowest permissible energy is that corresponding to $n = 1$ in Eq. (22–14).

$$E_1 = -\frac{e^2}{2a_0} = -2.180 \times 10^{-11} \text{ erg} = -13.61 \text{ eV}.$$

Then the permitted energies are $E_1, \frac{1}{4}E_1, \frac{1}{9}E_1, \frac{1}{16}E_1, \ldots$. The energy levels and the possible transitions between them are shown in Fig. 22–1.

In making a transition between a high energy state and one of lower energy, the atom emits a quantum of light having a frequency determined by $h\nu = \Delta E$, where ΔE is the difference in energy of the two states. The spectrum of the atom therefore

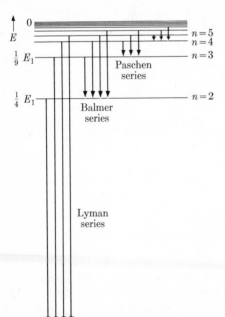

Fig. 22–1 Energy levels in the hydrogen atom.

consists of series of lines having frequencies corresponding to the possible values of the energy differences, represented by the lengths of the arrows in Fig. 22–1. There are several series of lines in the spectrum. The general form for the energy difference between two states, from Eq. (22–14), is

$$\Delta E = h v_{nk} = E_1 \left(\frac{1}{n^2} - \frac{1}{k^2} \right). \tag{22–15}$$

This is equivalent to the Rydberg formula, Eq. (19–16).

Transitions from the upper states to the ground state, $n = 1$, involve large differences in energy; the Lyman series of lines is in the ultraviolet region of the spectrum. Transitions from higher states to the level $n = 2$ involve smaller differences in energy; the Balmer series of lines lies in the visible and near ultraviolet. Transitions to the level $n = 3$ yield the Paschen series of lines in the infrared. Transitions to $n = 4$ and to $n = 5$ yield the Brackett and Pfund series in the far infrared. Note that as the transitions to any particular level occur from higher and higher levels, the energy difference, and therefore the frequency, does not change much. The lines in the spectrum come closer together and approach a *series limit*.

If a hydrogen atom absorbs light, only those frequencies will be absorbed which match the allowed energy differences. The absorption spectrum and the emission spectrum therefore have the same lines. The absorbed quantum lifts the hydrogen atom from one permitted energy level to another.

The integer l, the azimuthal quantum number, describes the total angular momentum of the hydrogen atom through the equation

$$M^2 = l(l + 1)\hbar^2, \tag{22–16}$$

with allowed values $l = 0, 1, 2, \ldots, n - 1$; M^2 is the square of the total angular momentum. The principal quantum number n could have any positive, nonzero, integral value; in contrast, l may be zero, and may not exceed $n - 1$. It is customary to designate the values of l by letter; the correspondence is

Value of l:	0	1	2	3	4	5
Letter designation :†	s	p	d	f	g	h

After $l = 3$, the letter designation proceeds alphabetically. The possible combinations of values of n and l for $n = 1$ to $n = 4$ are shown in the scheme:

Value of n:	1	2		3			4			
Value of l:	0	0	1	0	1	2	0	1	2	3
Notation:	$1s$	$2s$	$2p$	$3s$	$3p$	$3d$	$4s$	$4p$	$4d$	$4f$

The notation on the third line is that usually employed for the particular combination

---

† The letters s, p, d, f originated in the initial letters of *s*harp, *p*rincipal, *d*iffuse, and *f*undamental; words originally used to describe lines and series in spectra.

of values of n and l; the number is the value of n, the letter is the letter designation of the value of l.

In making a transition from one state to another in the absorption or emission of radiation, there is a restriction on l, called a selection rule. The value of l must change by ± 1. Thus, if a hydrogen atom in the ground state, 1s state, absorbs radiation and goes to level $n = 2$, then it must be finally in the 2p state. Any other transition between levels 1 and 2 is forbidden by the selection rule. The existence of selection rules helps enormously in the interpretation of spectra.

Although we know the magnitude of the total angular momentum from Eq. (22–16), we do not know the sign. Therefore, the orientation of the angular momentum vector is indefinite, and so the orientation of the orbit is indefinite.

Suppose we compare the values of angular momentum in the Bohr and the Schrödinger models:

$$\text{Bohr:} \qquad M^2 = n^2\hbar^2 \qquad n = 1, 2, \ldots$$
$$\text{Schrödinger:} \quad M^2 = l(l + 1)\hbar^2 \qquad l = 0, 1, 2, \ldots.$$

(Incidentally, the Bohr-Sommerfeld model required the Schrödinger value of the angular momentum.) In the Bohr atom, the angular momentum always had a nonzero value, while in the modern theory the angular momentum is zero in the s states for which $l = 0$. The absence of angular momentum in the s states makes it impossible to imagine the motion of the electron in these states in terms of a classical orbital motion. It is better not to try. Where it may help to visualize it, the electronic motion will sometimes be described *as if* it were moving in a classical orbit; this description must not be accepted literally, but analogically.

The integer m, the magnetic quantum number, describes the z-component of the angular momentum M_z through the equation

$$M_z = m\hbar; \tag{22–17}$$

with allowed value $m = -l, -l + 1, -l + 2, \ldots, -1, 0, +1, +2, \ldots, +l$. Any integral value from $-l$ to $+l$ including zero is a permitted value for m. There are $2l + 1$ values of m for a given value of l. If $l = 0$, then $m = 0$. But if $l = 1$, then m may be $-1, 0, +1$. If $l = 2$, then m may be $-2, -1, 0, +1, +2$. In the absorption or emission of a light quantum, the selection rules require either $\Delta m = 0$, or $\Delta m = \pm 1$.

22–5 Probability Distribution of the Electron Cloud in the Hydrogen Atom

The requirements of the uncertainty principle make it necessary to visualize the hydrogen atom as a nucleus imbedded in a "fog" of negative charge. This electron cloud has a different shape in the different states of the atom. To discover the shape of the cloud, we construct the probability density $|\psi_{nlm}|^2$, for the state in question.

For the ground state, $1s$ state, of the hydrogen atom, the wave function is

$$\psi_{1s} = \frac{1}{(\pi a_0^3)^{1/2}} e^{-r/a_0}. \tag{22–18}$$

This equation shows that in the $1s$ state, the wave function and the probability density are independent of the angles θ and φ. Consequently, the electron cloud is spherically symmetric. Let $P(r)$ be the probability density, then

$$P_{1s}(r) = \psi_{1s}^2 = \frac{e^{-2r/a_0}}{\pi a_0^3}. \tag{22–19}$$

This function is shown in Fig. 22–2(a). The probability density is high near the nucleus and decreases rapidly as r increases. Since the volume near the nucleus is very small, the total amount of the cloud near the nucleus is very small. So we ask a different question. How much of the cloud is contained in the spherical shell bounded by the spheres of radius r and $r + dr$?

The volume of this spherical shell is $dV_{\text{shell}} = 4\pi r^2 \, dr$, so that the amount of the cloud in the shell is $P_{1s}(r)4\pi r^2 \, dr$. The function $4\pi r^2 P(r) = f(r)$ is the radial distribution function:

$$f_{1s}(r) = 4\pi r^2 P_{1s}(r) = \frac{4r^2}{a_0^3} e^{-2r/a_0}. \tag{22–20}$$

The radial distribution function is the total probability of finding the electron in the spherical shell; Fig. 22–2(b). In the $1s$ state the probability of finding the electron is a maximum in the spherical shell which has a radius a_0, the radius of the first Bohr orbit. The probability of finding the electron in a spherical shell near the nucleus is very small, as is that of finding it very far away from the nucleus.

The distance of the electron from the nucleus in the $1s$ state can be computed using the theorem on expectation values, Eq. (20–7). For an s state the volume

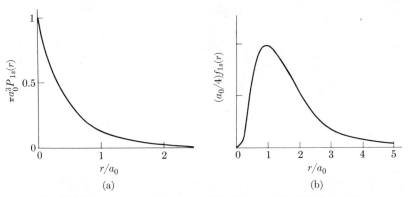

Fig. 22–2 The $1s$ state of the hydrogen atom. (a) Probability density. (b) Radial distribution function.

element may be taken as the volume of the spherical shell; the operator for r is simply multiplication of the wave function by r; so we get

$$\langle r_{1s} \rangle = \int \psi_{1s} r \psi_{1s} \, d\tau = \int \psi_{1s} r \psi_{1s} \, d\tau.$$

Putting in the values of ψ_{1s} and the volume element this becomes

$$\langle r_{1s} \rangle = \frac{a_0}{4} \int_0^\infty \left(\frac{2r}{a_0}\right)^3 e^{-2r/a_0} \, d\left(\frac{2r}{a_0}\right) = \frac{a_0}{4} 3! = \tfrac{3}{2}a_0 \tag{22-21}$$

By the same method, using the appropriate wave function, we can show that for any s state of principal quantum number n,

$$\langle r_{ns} \rangle = n^2 \tfrac{3}{2} a_0 \tag{22-22}$$

In states with larger values of n, higher energies, the average distance of the electron from the nucleus is larger. This is apparent in Fig. 22–3, which shows the radial distribution function for hydrogen in the 1s, 2s, and 3s states. Also, note that as n increases, the distribution function becomes "lumpier"; this "lumpiness' is characteristic of the larger values of the kinetic energy in these states.

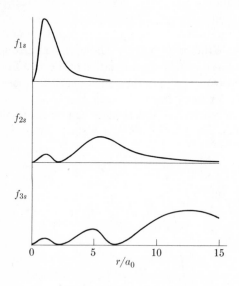

Fig. 22–3 Radial distribution function for 1s, 2s, and 3s states.

22–6 States with Angular Momentum

In states having angular momentum, the z-component has a precise value. This fact has the consequence, through the uncertainty principle, that the angle of orientation of the electron around the z-axis is completely indefinite. The electron has equal probability of having any orientation about the z-axis; therefore, the charge cloud is symmetric about the z-axis. In contrast to s states, which have spherical symmetry, states with angular momentum have axial symmetry, conventionally associated with the z-axis.

To be concrete, consider the p states. In these states $l = 1$, so that by Eq. (22–16) the total angular momentum is $M = \sqrt{l(l+1)}\hbar = \sqrt{2}\hbar$. Since m may be -1, 0, or $+1$, the possible values of the z-component are, by Eq. (22–17),

$$M_z = -\hbar, \qquad 0, \qquad +\hbar.$$

Figure 22–4(a) shows the possible orientations of the angular momentum vector, of magnitude $\sqrt{2}\hbar$, which have $M_z = -\hbar$. Any vector lying in the conical surface fulfills this requirement. In Fig. 22–4(b) it is apparent that any vector lying in the xy-plane has $M_z = 0$. Any vector lying in the conical surface of Fig. 22–4(c) has $M_z = +\hbar$.

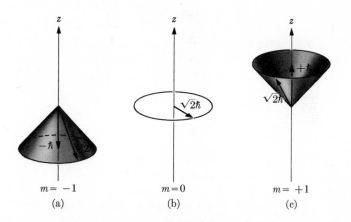

Fig. 22–4 The z-components of angular momentum for $l = 1$.
(a) $m = -1$. (b) $m = 0$. (c) $m = +1$.

The corresponding charge density distributions are shown in Fig. 22–5. Note that a large z-component, either positive or negative, squeezes the charge cloud nearer the xy-plane. If $M_z = 0$, the charge cloud density vanishes in the xy-plane. The only distinction we can make between the charge clouds for $M_z = +\hbar$ and $-\hbar$ is by supposing that for $M_z = -\hbar$ the rotation of the electron is in a clockwise sense, and for $M_z = +\hbar$ that the rotation of the electron is counterclockwise.

Since for $m = \pm 1$ we cannot distinguish the shapes of the clouds, we use the principle of superposition to construct new descriptions. Let the old p functions be designated by the proper values of m; we write p_{+1} and p_{-1}. By taking linear combinations of these descriptions, we obtain the new descriptions which we designate by p_x and p_y. The procedure is displayed in the scheme of Table 22–3.

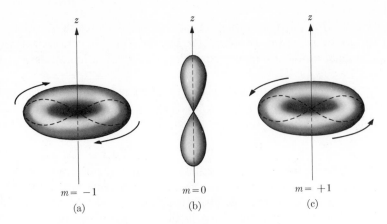

$m = -1$ $m = 0$ $m = +1$

(a)   (b)   (c)

Fig. 22–5 The charge clouds for p states.

Table 22–3

Old description	New description	Relation between old and new
$p_{+1}(m = +1)$	p_x	$p_x = \frac{1}{2}\sqrt{2}(p_{+1} + p_{-1})$
$p_{-1}(m = -1)$	p_y	$p_y = -i\frac{1}{2}\sqrt{2}(p_{+1} - p_{-1})$
$p_0(m = 0)$	p_z	$p_z = p_0$

The advantage of the new description is that the three charge clouds shown in Fig. 22–6 for p_x, p_y, p_z look equivalent; each consists of two lobes which lie along the x-, y-, and z-axis respectively. The function p_x corresponds to $M_x = 0$, p_y to $M_y = 0$, and p_z to $M_z = 0$. For each of the p functions in the new description, the maximum in the probability density is along the particular axis. The probability density is zero in the coordinate plane perpendicular to that axis; this is evident in Fig. 22–6.

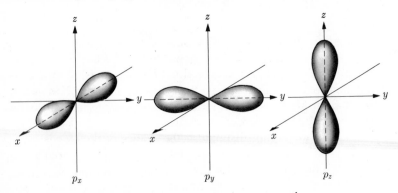

p_x p_y p_z

Fig. 22–6 Charge clouds for p_x, p_y, and p_z.

There are five d states corresponding to values of $m = -2, -1, 0, +1, +2$. The charge clouds for these states are shown in Fig. 22–7. The charge cloud has the same appearance for d_{+2} as for d_{-2}. The direction of motion is counterclockwise for d_{+2} and clockwise for d_{-2}; the same is true for d_{+1} and d_{-1}. We will not have need for alternative descriptions of the d functions. Note the axial symmetry for all values of m in Fig. 22–7, and the fact that the higher the value of m, the closer the charge cloud is to the xy-plane.

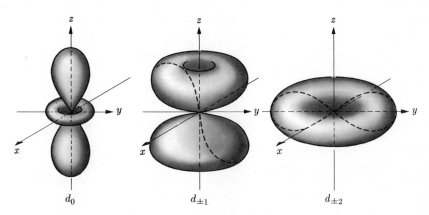

Fig. 22–7 Charge clouds in the d states.

We can compare the extension of these charge clouds in space by calculating the average distance of the electron from the nucleus for the state in question. Using the appropriate wave function, we do the calculation by the same method as was used to obtain $\langle r_{1s} \rangle$ in Eq. (22–21). We omit the tedious details and write only the result, which is quite simple. The average distance of the electron from the nucleus depends only on n and l:

$$\langle r_{nl} \rangle = \tfrac{1}{2}a_0[3n^2 - l(l + 1)]. \tag{22–23}$$

Equation (22–23) shows that for a specified value of n, in states having high angular momentum, high values of l, the average distance of the electron from the nucleus is less than in states of low angular momentum. As we shall see later, this is the underlying reason for the great similarity in the chemistries of the rare earth elements.

22–7 Electron Spin

Before the development of the Schrödinger equation it was shown by Uhlenbeck and Goudsmit that certain troublesome features of atomic spectra could be explained if the electron itself possessed an intrinsic angular momentum. If we do not take the picture too seriously, we may imagine the electron as a tiny ball of negative charge which is spinning on its axis. If the square of the total spin angular momentum is

$M^2_{\text{spin}} = s(s + 1)\hbar^2$, and if $s = \frac{1}{2}$, then the experimental data are explained. The z-component of the spin angular momentum has the value

$$M_{z(\text{spin})} = m_s\hbar. \tag{22-24}$$

The quantity m_s is the *spin quantum number*; it may have only the values $+\frac{1}{2}$ or $-\frac{1}{2}$. The Schrödinger equation in its usual form gives no indication of the existence of the electron spin. However, Dirac has shown that if the Schrödinger equation is cast into a form which satisfies certain requirements of relativity theory, then four quantum numbers, the fourth being the electron spin quantum number, appear in the solution for the hydrogen atom. Thus the spin is a coherent part of the fundamental theory and is not tacked on just to patch things up.

Hereafter we will describe an electron in an atom by four quantum numbers, n, l, m, and m_s.

22–8 Magnetic Properties of the Electron and the Hydrogen Atom

If the electron spins on its axis, the fact that it is electrically charged implies that there is a current flow around the axis. This flow of current gives the electron a magnetic moment, just as the flow of current in a coil of wire gives the coil a magnetic moment. The magnetic moment is perpendicular to the plane of the current flow and so is parallel to the angular momentum vector, but directed oppositely because of the negative charge on the electron. It is the magnetic moment of the electron which made the first observations of the spin property possible.

The magnetic moment μ of the electron is given by

$$\mu = -g\left(\frac{e}{2mc}\right)M_{\text{spin}}, \tag{22-25}$$

where g is a constant, and M_{spin} is the total spin angular momentum. Using the value of M_{spin} from Section 22–7, this becomes

$$\mu = -g\left(\frac{e\hbar}{2mc}\right)\sqrt{s(s + 1)}.$$

A natural unit of magnetic moment, the Bohr magneton μ_1, is defined by

$$\mu_1 = \frac{e\hbar}{2mc} = 0.9273 \times 10^{-20}\ \text{erg/gauss}, \tag{22-26}$$

so that $\mu = -g\sqrt{s(s + 1)}\mu_1$. For spin, $g = 2$ and $s = \frac{1}{2}$, so the total magnetic moment of the electron due to spin is $\mu = -\sqrt{3}\mu_1$.

The z-component of the magnetic moment is given by

$$\mu_z = -g\frac{\mu_1}{\hbar}M_{z(\text{spin})},$$

but $M_{z(\text{spin})} = m_s\hbar$, so that

$$\mu_z = -gm_s\mu_1 = -2m_s\mu_1. \qquad (22\text{–}27)$$

Since $m_s = \pm\frac{1}{2}$, $\mu_z = \pm\mu_1$. Therefore, the component of the spin magnetic moment along any specified axis is ± 1 Bohr magneton.

If in an atom, the orbital angular momentum is not zero, $l \neq 0$, then the atom has an orbital magnetic moment as well as a magnetic moment due to the spin of the electron. In states having angular momentum, the electronic motion constitutes a current flowing around the atom which produces the magnetic moment. The relation between the orbital magnetic moment and the orbital angular momentum is

$$\mu_{\text{orb}} = -\frac{\mu_1}{\hbar}M, \qquad \mu_{z(\text{orb})} = -\frac{\mu_1}{\hbar}M_z. \qquad (22\text{–}28)$$

Note that the factor $g = 2$ does not appear in these equations. Using the values for M and M_z, we obtain

$$\mu_{\text{orb}} = -\sqrt{l(l + 1)}\mu_1, \qquad \mu_{z(\text{orb})} = -m\mu_1. \qquad (22\text{–}29)$$

From Eq. (22–29) it is clear that the greater the z-component of the angular momentum, the greater will be the z-component of the magnetic moment. This is the reason for calling m the magnetic quantum number.

22–9   The Zeeman Effect

Suppose we apply a uniform magnetic field of strength H along the z-axis to a hydrogen atom. If the original energy of the atom is E_n, the application of the magnetic field lowers the energy by an amount $-\mu_z H$, where μ_z is the component of the magnetic moment of the atom in the direction of the field. In the presence of the field the energy is E_n'

$$E_n' = E_n - \mu_z H.$$

The magnetic moment is the sum of the contributions of the orbital moment and the spin moment: $\mu_z = \mu_{z(\text{orb})} + \mu_{z(\text{spin})}$. Using Eqs. (22–27) and (22–29) this becomes $\mu_z = -(m + 2m_s)\mu_1$. The energy is

$$E_n' = E_n + (m + 2m_s)\mu_1 H. \qquad (22\text{–}30)$$

In the presence of a uniform magnetic field the energy of the atom depends on m and the spin quantum number m_s.

Applying Eq. (22–30) to the 1s state of hydrogen, $E_n = E_1$, $m = 0$, $m_s = \pm\frac{1}{2}$; hence

$$E_1' = E_1 \pm \mu_1 H.$$

The imposition of the magnetic field splits the ground energy level of hydrogen into two levels, one above and one below the original level; Fig. 22–8. The same is true of any s level.

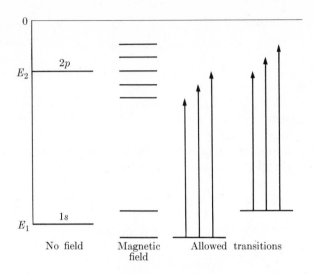

Fig. 22–8 The Zeeman splitting.

A p state is split into five levels, since there are five possible ways to construct the sum, $m + 2m_s$:

m	$+1$	0	-1	$+1$	0	-1
m_s	$\frac{1}{2}$	$\frac{1}{2}$	$\frac{1}{2}$	$-\frac{1}{2}$	$-\frac{1}{2}$	$-\frac{1}{2}$
$m + 2m_s$	2	1	0	0	-1	-2

The two levels having $m + 2m_s = 0$ have the same energy, so there are five distinct levels, not six.

If a hydrogen atom makes the transition from the $2p$ to the $1s$ state in the absence of a magnetic field, the spectral line emitted has the frequency $v = (E_2 - E_1)/h$. The presence of the magnetic field splits this line into three lines. The allowed transitions are shown by the vertical lines in Fig. 22–8. The selection rules are $\Delta m_s = 0$, and $\Delta m = 0, +1, -1$. The second rule does not impose any restriction in this example. However, the requirement that $\Delta m_s = 0$ means that in the transition the spin quantum number is unchanged. Therefore, from the lower level of the $1s$ state, transitions are permitted only to the three lowest levels of the $2p$; similarly from the upper level of the $1s$ only transitions to the three highest $2p$ levels are permitted. The selection rule together with the equality of the spacing between the levels in the $1s$ and $2p$ states have the result that there are only three distinct energy differences, and so only three frequencies are emitted.

The splitting of a spectral line into several lines by the application of a magnetic field was first observed by Zeeman in 1896 and is called the Zeeman effect. The Zeeman effect is a great help in the analysis of complex spectra. The example described above is a very simple one; a more complex splitting, the anomalous Zeeman effect, is more commonly observed.

22–10   The Structure of Complex Atoms

Using the hydrogen atom as a guide, we will assume that every electron in a complex atom (a complex atom is an atom with more than one electron) can be described by a set of four quantum numbers n, l, m, m_s. And we assume that the system of energy levels in a complex atom is generally similar to that of the hydrogen atom. Usually our attention will be confined to the state of lowest energy, the ground state, of the atom. It is worth noting that we are talking about the structure of the isolated atom; for example, the structure of an isolated sodium atom, not that of a sodium atom in metallic sodium, where many atoms are very close together.

If the electron is in the 1s state, the hydrogen atom is in its lowest state of energy. In a polyelectronic atom such as carbon, six electrons, or sodium, eleven electrons, it would not seem unreasonable if all the electrons were in the 1s level, thereby giving the atom the lowest possible energy. We might denote such a structure for carbon by the symbol $(1s)^6$ and for sodium, $(1s)^{11}$. This result is wrong, but from what has been said so far there is no apparent reason why it should be wrong. The reason lies in an independent and fundamental postulate of the quantum mechanics, the Pauli exclusion principle: *no two electrons may have the same set of four quantum numbers.* Only two sets of four quantum numbers exist for the 1s level; in the order, n, l, m, m_s; these sets are $(100|\frac{1}{2})$ and $(100| -\frac{1}{2})$. If more than two electrons are placed in the 1s level, at least one of these sets would be duplicated, a situation forbidden by the Pauli principle. In this light it is clear why the structures, $(1s)^6$ for carbon and $(1s)^{11}$ for sodium, are incorrect.

The construction principle (*Aufbau Prinzip*) for the electronic structure of complex atoms is as follows.

1. Each electron in a complex atom is described by a set of four quantum numbers, the quantum numbers being the same as those used to describe the states of the hydrogen atom.
2. The relative arrangement of energy levels in the complex atom is roughly the same as that in the hydrogen atom. To make up the structure of the complex atom, the electrons are arranged in the lowest possible energy levels consistent with the restriction imposed by the Pauli principle.

We divide the levels into *shells*, those levels with the same value of the principal quantum number, and *subshells*, those within a shell which have the same value of the azimuthal quantum number. For a specified value of l, there are $2l + 1$ values of m; for a specified value of m, the electron may have two values of m_s. Hence there are $2(2l + 1)$ distinct combinations of m and m_s. This is the maximum number of electrons permitted in any subshell. For an s subshell, $l = 0$, so only two electrons may occupy the subshell. For a p subshell, $l = 1$, and six electrons are required to fill the p subshell. Ten electrons fill a d subshell, $l = 2$, and so on. The shell with $n = 1$, is the K shell; that with $n = 2$, the L shell; $n = 3$, the M shell; and so on. The number of electrons required to fill the shells is shown in Table 22–4. The numbers $2, 8, 18, 32, \ldots$ in the last column are given by $2n^2$, where n is the principal quantum number. The numbers in this famous sequence are the numbers of elements in the periods of the periodic table.

Table 22–4

Value of n	Subshells present	Number of electrons in the filled shell
1 (K shell)	s	2
2 (L shell)	s, p	$2 + 6 = 8$
3 (M shell)	s, p, d	$2 + 6 + 10 = 18$
4 (N shell)	s, p, d, f	$2 + 6 + 10 + 14 = 32$

The number of electrons in a subshell is indicated by the superscript on the symbol of the subshell. Using the principles outlined above, we write the electronic configurations for hydrogen and helium as

$$\text{H:} \quad 1s, \quad \text{He:} \quad (1s)^2.$$

The K shell is complete with helium. The next electrons added must go into the shell with $n = 2$. The question is which subshell, the $2s$ or the $2p$, fills in first? In the hydrogen atom, the energy of these subshells is the same; but in complex atoms, the energy depends on l as well as on n. For a specified value of n, the order of the sublevels is $s, p, d, \ldots$, where s has the lowest energy. So in lithium the $2s$ level lies lower than the $2p$, and the structure is Li: $(1s)^2 2s$. Following Li we have Be: $(1s)^2(2s)^2$. Then in the six succeeding elements the $2p$ shell fills; B: $(1s)^2(2s)^2 2p$; C: $(1s)^2(2s)^2(2p)^2$, and so on until neon is reached; Ne: $(1s)^2(2s)^2(2p)^6$. With neon the L shell is filled. Atoms such as helium and neon which have filled electronic shells are chemically inert. The completed group of eight electrons $(ns)^2(np)^6$, the octet, is the configuration of the inert gases (except He) and is always a very stable configuration chemically. The stability of this configuration is one of the bases for the Lewis rules of chemical valency.

The electronic configurations of all the elements are shown in Table 22–5, in order of the number of electrons in the atom. Examination of the table shows immediately that chemically similar atoms have similar configurations of the outer electrons. For example, all the alkali metals have the configuration ns over a shell

Table 22-5 Electronic configurations of the atoms

Z	Element	1s	2s	2p	3s	3p	3d	4s	4p
1	H	1							
2	He	2							
3	Li	2	1						
4	Be	2	2						
5	B	2	2	1					
6	C	2	2	2					
7	N	2	2	3					
8	O	2	2	4					
9	F	2	2	5					
10	Ne	2	2	6					
11	Na		Neon shell		1				
12	Mg				2				
13	Al				2	1			
14	Si				2	2			
15	P				2	3			
16	S				2	4			
17	Cl				2	5			
18	Ar				2	6			
19	K		Argon shell					1	
20	Ca							2	
21	Sc						1	2	
22	Ti						2	2	
23	V						3	2	
24	Cr						5	1	
25	Mn						5	2	
26	Fe						6	2	
27	Co						7	2	
28	Ni						8	2	
29	Cu						10	1	
30	Zn						10	2	
31	Ga						10	2	1
32	Ge						10	2	2
33	As						10	2	3
34	Se						10	2	4
35	Br						10	2	5
36	Kr						10	2	6

(Continued)

Table 22–5 (*Continued*)

Z	Element		4d	4f	5s	5p	5d	5f	6s	6p
37	Rb	Krypton shell			1					
38	Sr				2					
39	Y		1		2					
40	Zr		2		2					
41	Nb		4		1					
42	Mo		5		1					
43	Tc		6		1					
44	Ru		7		1					
45	Rh		8		1					
46	Pd		10							
47	Ag		10		1					
48	Cd		10		2					
49	In		10		2	1				
50	Sn		10		2	2				
51	Sb		10		2	3				
52	Te		10		2	4				
53	I		10		2	5				
54	Xe		10		2	6				
55	Cs	Xenon shell							1	
56	Ba								2	
57	La						1		2	
58	Ce			2					2	
59	Pr			3					2	
60	Nd			4					2	
61	Pm			5					2	
62	Sm			6					2	
63	Eu			7					2	
64	Gd			7			1		2	
65	Tb			9					2	
66	Dy			10					2	
67	Ho			11					2	
68	Er			12					2	
69	Tm			13					2	
70	Yb			14					2	
71	Lu			14			1		2	
72	Hf			14			2		2	
73	Ta			14			3		2	
74	W			14			4		2	
75	Re			14			5		2	
76	Os			14			6		2	
77	Ir			14			9			
78	Pt			14			9		1	

(*Continued*)

Table 22–5 (*Continued*)

Z	Element		4f	5s	5p	5d	5f	6s	6p	6d	7s
79	Au	Xenon shell	14			10		1			
80	Hg		14			10		2			
81	Tl		14			10		2	1		
82	Pb		14			10		2	2		
83	Bi		14			10		2	3		
84	Po		14			10		2	4		
85	At		14			10		2	5		
86	Rn		14			10		2	6		
87	Fr	Radon shell									1
88	Ra										2
89	Ac									1	2
90	Th									2	2
91	Pa						2			1	2
92	U						3			1	2
93	Np						4			1	2
94	Pu						5			1	2
95	Am						6			1	2
96	Cm						7			1	2
97	Bk						8			1	2
98	Cf						9			1	2

of eight. The coinage metals have the configuration $[(n-1)d]^{10}ns$; the lone s electron lies over a completed d subshell rather than a shell of s^2p^6. This gives the coinage metals characteristically different properties from those of the alkalies. Other similarities can be picked out readily; e.g., the halogens, the inert gases, etc.

A number of points may be made about Table 22–5. Argon has the configuration $(1s)^2(2s)^2(2p)^6(3s)^2(3p)^6$. Logically one might expect that the $3d$ subshell would commence to fill with the element following argon. However, the interactions of the electrons give the $4s$ level a lower energy than the $3d$, so the $4s$ level fills first with the elements potassium and calcium; then the $3d$ levels start to fill with the *transition elements* from scandium through nickel. Transition elements have a partially completed d subshell. A similar thing happens after krypton; the $5s$ level fills first, then the $4d$. The $4f$ shell is not filled until the $5s$, $5p$, and $6s$ levels are complete. Following barium, the $5d$ level acquires one electron in lanthanum. Then the $4f$ level fills with 14 electrons in the *inner transition elements* from cerium to lutecium, the rare earths. The inner transition elements have *two* partially filled inner subshells, the $5d$ and the $4f$.

In a level of high angular momentum such as the f level, $l = 3$, the electron is much closer to the nucleus than the other electrons of the same principal quantum number; Eq. (22–23). These electrons are buried down in the interior of the atom. As a result, the exterior electrons which give the atom its chemical properties are not much affected by the number of f electrons present. The chemistries of lanthanum and the succeeding fourteen elements, the rare earths, are remarkably similar, since they differ only by the number of $4f$ electrons present.

22–11   The Magnetic Properties of Atoms

The $1s$ level of an atom is filled by two electrons having the quantum numbers:

	n	l	m	m_s
Electron 1:	1	0	0	$+\frac{1}{2}$
Electron 2:	1	0	0	$-\frac{1}{2}$

Two electrons in an atom which differ only in the value of the spin quantum number are called *paired* electrons. The z-component of spin angular momentum of the second electron is equal and opposite in sign to that of the first electron. The net z-component of spin angular momentum of the electron pair is the sum, $\frac{1}{2} + (-\frac{1}{2}) = 0$; thus, the pair of electrons has no net spin angular momentum and no magnetic moment along any axis. The pairing of the spins is frequently indicated by using a box to represent the energy level and arrows to represent the spin quantum number of the electron, as in the following diagram:

$1s$

A head-up arrow represents an electron with $m_s = +\frac{1}{2}$; a head-down arrow represents an electron with $m_s = -\frac{1}{2}$.

Similarly, the net orbital angular momentum of any filled subshell is zero. First of all, the orbital angular momentum in an s level is zero by definition. In a group of p levels, the possible values of m are $-1, 0, +1$. If we place a pair of electrons in each of these p levels, then two electrons have $m = -1$, two have $m = +1$, two have $m = 0$. The net z-component for all of these is zero since $2(-1) + 2(0) +$

$2(+1) = 0$. A filled subshell has no net component of orbital angular momentum around any specified axis, and so it contributes no magnetic moment due to orbital motion.

The magnetic moment of an atom is due entirely to *partially* filled subshells and unpaired spins. The completed subshells do not contribute to the permanent magnetic moment of an atom. In this connection we might remark that molecules ordinarily have an even number of electrons whose spins are all paired, and usually the subshells are complete so that there is no contribution to the magnetic moment from the orbital motion. It is unusual for a molecule to have a permanent magnetic moment; the possession of a permanent magnetic moment is an important key to the electronic structure of the molecule.

22–12   Some General Trends in the Periodic System

The diameter of the hydrogen atom in its ground state is about one angstrom unit. Using Eq. (22–23), we can compute that in the $3s$ state the diameter of the atom would be of the order of 25 to 30 Å. This is much larger than the diameter of the largest atoms. As we progress from hydrogen to the atoms of the heavier elements, the increased nuclear charge pulls the electrons much closer to the center than is possible with the simple hydrogen atom. Equation (22–23) therefore does not give an accurate indication of the radius of atoms other than hydrogen.

If we compare the sizes of the atoms in a vertical family of the periodic table such as the alkali metals, the size of the atom increases going downward in the table. The valence electron is in the $2s$, $3s$, $4s$, $5s$, $6s$, $7s$ level as we pass from lithium to francium. Thus we may retain the general statement that the radius of an atom increases with the principal quantum number of the electrons in the valence shell. The increase is not very rapid.

In passing through a horizontal row of the periodic table, the atom size is a maximum with the alkali metal, drops quickly to a minimum in about the third group, then rises irregularly to reach another maximum in the next alkali metal. Two tendencies are operative in this behavior. In passing, let us say from lithium to beryllium, the nuclear charge increases; this tends to pull the electrons in and make the atom smaller. But there is an additional electron whose presence increases the mutual repulsions in the electron cloud and tends to make the atom larger. In the early part of the period, the electrons easily keep out of each other's way, and the effect of the nuclear charge is dominant; the size decreases. Toward the middle and end of a period the mutual repulsion of the electrons in the shell increases and overbalances the effect of the increased nuclear charge; the size increases.

In speaking of sizes of atoms, it should be kept in mind that the electron cloud does not cut off at any definite distance from the nucleus. As an experimental quantity the radius of the atom is found to depend a good deal upon the environment of the atom during the measurement. The following values for the molecular diameter of argon are obtained by the method indicated:

Gas viscosity 2.97×10^{-8} cm,

Liquid density 4.04×10^{-8} cm,

Solid density 3.84×10^{-8} cm.

This sort of variation is to be expected because of the lack of definition of the radius of the atom. In making comparisons of the sizes of atoms, situations should be chosen in which the atoms have about the same environment. It is possible to construct consistent tables of ionic radii, for example, or of covalent radii, or metallic radii. Where it is possible, it is best to compare the radii for situations in which the atoms have the same number of neighbors.

22–13 Ionization Potentials

The ionization potential of an atom is a measure of how strongly an electron is bound to the atom. The first ionization potential (I.P.) of an atom is the energy required to remove an electron from the atom to an infinite distance.

$$A \rightarrow A^+ + e^-.$$

The second I.P. of an atom is the energy required to remove an electron from the singly charged ion:

$$A^+ \rightarrow A^{++} + e^-.$$

An atom has as many ionization potentials as it has electrons.

For comparison, the first (I) and second (II) ionization potentials of a number of atoms are recorded in Table 22–6. All of these elements have comparable electronic configurations: a single s electron over an inert gas shell of eight. It is clear that as the quantum number goes up, the electron is more easily removed. This is mainly because of the increase in the distance between nucleus and the outer electron as the quantum number goes up. The greatest difference is between hydrogen and lithium. Note that the second I.P.'s of these atoms are 6 to 15 times larger than the first I.P. There are two reasons for this. First, there is the unbalanced

Table 22–6 Ionization potentials

Element	H	Li	Na	K	Rb	Cs
Outer electron configuration	$1s$	$2s$	$3s$	$4s$	$5s$	$6s$
I, ev	13.595	5.390	5.138	4.339	4.176	3.893
II, ev		75.62	47.29	31.81	27.36	23.4

positive charge, which always increases a second I.P. compared with the first I.P. Secondly, to remove the second electron in these atoms requires that a very stable closed shell, the shell of eight, be opened. This increases the required energy enormously.

The I.P.'s for the inert gases are very high:

	He	Ne	Ar	Kr	Xe
I.P., ev	24.580	21.559	15.755	13.996	12.127

These high values account for the inability of these elements to form compounds involving ions such as He^+, Ne^+, etc.

If in removing the second electron it is not necessary to break into a closed shell, then the second ionization potential is not enormously greater than the first; generally it is two to three times the first ionization potential. The energy required to remove the second electron is always greater than that required to remove the first; the removal of the third requires more energy than the removal of the second. Roughly, provided that a closed shell of eight is not broken into, the second electron requires two to three times as much energy as the first; the third requires 1.5 to 2 times as much as the second. If a closed shell of eight must be opened, the energy required is *very much* larger. These facts are illustrated by the data in Table 22–7.

Table 22–7

Element	H	He	Li	Be	Na
Configuration	$1s$	$(1s)^2$	$(1s)^2 2s$	$(1s)^2 (2s)^2$	$(1s)^2 (2s)^2 (2p)^6 3s$
I, ev	15.595	24.580	5.390	9.320	5.138
II, ev		54.40	75.62	18.206	47.29
III, ev			122.4	153.8	
IV, ev				217.7	

22–14 Electron Affinity

The energy released when an atom and an electron are brought together from an infinite separation is called the electron affinity of the atom. Values for the halogens are given in Table 22–8. The process is symbolized

$$A + e^- \rightarrow A^-.$$

The ionization potential of an atom is readily measured spectroscopically as the convergence limit of a line series in the several types of spectra of the atom,

Table 22–8

Atom	F	Cl	Br	I
Electron affinity, ev	3.62	3.79	3.56	3.28

and I.P.'s are known with great accuracy. Such a direct measurement of electron affinity has not been devised; the experiments to determine electron affinity are complicated and the interpretation of the results is subject to some controversy.

Problems

22–1. Compute the wavelengths of the first three lines of the Lyman, Balmer, and Paschen series. Compute the series limit, shortest wavelength line, for each series.

22–2. The radial distribution function for the 1s state of hydrogen is given by Eq. (22–20). Show that the maximum of this function occurs at $r = a_0$.

22–3. Calculate the radius of the sphere which will contain 90 % of the hydrogen atom's electron cloud, if the atom is in the 1s state; the 2s state; the 3s state.

22–4. The p_z wave function for hydrogen has the form $f(r) \cos \theta$, where θ is the angle between the radius vector and the z-axis (Fig. 21–7) and may vary from 0 to π. For a fixed value of r, sketch the probability density as a function of θ.

22–5. By evaluating the appropriate integrals, compute $\langle r \rangle$ in the 2s and the 2p and the 3s states of the hydrogen atom; compare with the results from Eq. (22–22).

22–6. Compute the expectation value of the potential energy, $V(r) = -e^2/r$, for the hydrogen atom in the 1s state, the 2s state, the 2p state, and the 3s state. Also compute the expectation value of the kinetic energy in each of these states.

22–7. a) How many levels appear in the Zeeman splitting of the d level in hydrogen?
b) In the presence of a magnetic field, how many lines appear in the transition 3d to 2p?

22–8. Calculate the strength of magnetic field necessary to produce a Zeeman splitting of $\tilde{v} = 1/\lambda = 10 \, \text{cm}^{-1}$.

Chapter Twenty-three

The Covalent Bond

23–1 General Remarks

Until the advent of the quantum mechanics the reasons for the stability of molecules were unknown. The cohesive energy of ionic crystals could be adequately interpreted on the purely classical basis of the electrical attraction of the oppositely charged ions. Some attempts were made to interpret the interaction of all atoms on the basis of the electrical interaction of positive and negative charges, electrical dipoles, induced dipoles, and so on. These classical calculations indicated that the bonding between two like atoms, such as two hydrogen atoms, should be very much weaker than it is. This is another problem which classical physics failed to solve.

The quantum mechanical problem is to calculate the energies of the individual atoms which make up the molecule, then calculate the energy of the molecule itself. The molecule is stable if the energy of the molecule is less than the sum of the energies of the individual atoms. The difference in these energies is a measure of the strength of binding in the molecule. To state the problem is easy; to do the calculation in complete detail is apparently impossible. Fortunately, there are several simplifying circumstances.

First of all, consider the hydrogen molecule H_2. It consists of four bodies, two protons and two electrons. The classical problem of the motion of three bodies, and the quantum mechanical one as well, has escaped exact solution; a larger number of bodies only aggravates this difficulty. However, since the nuclei are so much heavier than the electrons, their motion is sluggish in comparison; the electronic motion is fast enough to adjust to any change in the position of the nuclei. Thus, to a good approximation the nuclear motions, vibrations and rotations of the molecule can be treated as a completely separate problem; the Born-Oppenheimer

551

approximation. The internuclear distances and the relative orientation of the nuclei thus enter the problem of the electronic motion as parameters; if we wish, we can explore how changing those parameters affects the energy of the molecule.

The energy of the molecule is given by the expression

$$E = \int \psi^* \mathbf{H} \psi \, d\tau, \tag{23-1}$$

where ψ is the wave function or description of all the electrons in the molecule, and $\mathbf{H}$ is the operator for the total energy, kinetic and potential, of the electrons in the nuclear skeleton of the molecule. And now a second fortunate thing happens. If instead of using the exact wave function in Eq. (23–1), we use an approximate one (which we might even obtain by guessing!) the Schrödinger equation has the property that the value of the integral is always *greater* than the energy of the ground state of the molecule; the *variation theorem*. This theorem allows us to take a guess at the description ψ, put adjustable constants in the mathematical form of the guess, and evaluate the integral. Then we adjust the constants to minimize the value of the integral; this minimum value is still greater (by the theorem) than the ground state energy. With experience, our guesses become more refined and we come closer to the correct value of the energy. The theorem is helpful, since it tells us that a "guessed" description will never give us an energy below the correct value.

Having agreed to be content with approximate descriptions of the system, we can gain some insight into the nature of the chemical bond. Two main approaches to this problem can be distinguished. The *valence bond* method, developed principally by Heitler, London, Slater, and Pauling, recognizes that two electrons are usually needed to form a chemical bond and then looks at the behavior of an electron pair. Each bonding pair in the molecule is described in a simple way, and a description of the molecule is built up by a description of its parts. The *molecular orbital*† method, developed by Hund and Mulliken, looks at the nuclear framework of the individual molecule and says that this system must have a system of energy levels just as the hydrogen atom has such a system of levels. Fit the electrons of the molecule into this system of levels, observing the Pauli principle, and obtain the description of the molecular electronic structure. This approach is the method, modified appropriately for molecules, that we used to describe the electronic structure of complex atoms.

The molecular orbital theory is more satisfying esthetically, perhaps, but its lack of emphasis on a localized chemical bond has led many chemists to prefer the valence bond method, which gives them a more pictorial grasp of the situation. The above distinction between the two methods is a primitive one. If all the refinements in the present day valence bond and molecular orbital theories are included, any distinction between them is probably more imagined than real.

---

†   "Orbital" is *not* a fancy word for "orbit." Orbital and wave function are used synonymously.

23–2   The Electron Pair

To describe the electron pair in a molecule, we investigate the behavior of two identical particles in the potential field supplied by the nuclei of the molecule. We begin by oversimplifying the problem. If we ignore the electrical repulsion between the two electrons, then each moves independently. The state of the first electron is described by a wave function $\psi_n(1)$, where the (1) is an abbreviation for the coordinates of electron 1. Similarly, the second electron is described by a wave function $\psi_k(2)$. The subscripts n and k indicate that the states of the two electrons may be different. Since the particles move independently, the energy of the pair is the sum of the energies of the individuals: $E = E_n + E_k$. If the energy of the system is given by this sum, then the Schrödinger equation requires that the wave function for the pair be the *product* of the individual descriptions; the electron pair is described by the function

$$\psi_I = \psi_n(1)\psi_k(2). \tag{23–2}$$

Since the electrons are indistinguishable, we have no way of discovering which is in state k and which in state n. An equally correct description is therefore

$$\psi_{II} = \psi_k(1)\psi_n(2), \tag{23–3}$$

where the coordinates of the particles have been exchanged. The description in Eq. (23–3) has the same energy as that in Eq. (23–2). (States with the same energy are *degenerate* states; these two states exhibit exchange degeneracy, since they differ only in the exchange of the coordinates.) If the particles do not interact, either description, or a superposition of them, is perfectly correct.

The curious feature of the problem is that if the repulsion between the electrons is introduced, we are forced to use a superposition of these descriptions. The permissible combinations are

$$\psi_S = \frac{1}{\sqrt{2}}[\psi_n(1)\psi_k(2) + \psi_n(2)\psi_k(1)], \tag{23–4}$$

$$\psi_A = \frac{1}{\sqrt{2}}[\psi_n(1)\psi_k(2) - \psi_n(2)\psi_k(1)]. \tag{23–5}$$

The two functions ψ_S and ψ_A have an important symmetry property. If we exchange the coordinates of electrons 1 and 2 (interchange the 1's and 2's in the parentheses), the function ψ_S is unaffected; ψ_S is symmetric under this operation. The function ψ_A changes sign under this operation and so is antisymmetric.

Now we ask which of these descriptions is likely to describe a bond between two atoms. Consider the hydrogen molecule H_2, with two protons rather close together and two electrons. By themselves the two protons would repel one another. To form a stable molecule this repulsion must be reduced. To reduce it, the electrons must be for the most part in the small space between the two nuclei which implies that the electrons must be rather close to one another. As the coordinates of 1 and

2 approach in value, then

$$\psi_n(1) \approx \psi_n(2) \quad \text{and} \quad \psi_k(1) \approx \psi_k(2).$$

Using these relations in Eqs. (23–4) and (23–5), we find

$$\psi_S \approx \frac{2}{\sqrt{2}}\psi_n(1)\psi_k(2) \quad \text{and} \quad \psi_A \approx 0.$$

Therefore, if the two electrons are described by ψ_A, the probability ψ_A^2 of finding the two electrons close together is very small, while if they are described by ψ_S, there is a sizable probability ψ_S^2 of finding them close together. We conclude that it is ψ_S which describes the state of the electrons in the electron-pair bond between two nuclei; this conclusion is confirmed by detailed calculation of the energy of the molecule.

All of this is fine; however, the Pauli exclusion principle requires that the wave function of a system be *antisymmetric* under this operation of interchanging the coordinates of the particles. We save the situation by realizing that the total wave function of an electron pair is the product of a space part, ψ_S or ψ_A, and a spin part. The spin part may also be symmetric, Σ_S, or antisymmetric, Σ_A, under interchange of the particles. The possible combinations of space and spin functions to yield an antisymmetric total wave function are

$$\psi_1 = \psi_S\Sigma_A \quad \text{and} \quad \psi_2 = \psi_A\Sigma_S.$$

The first, ψ_1, incorporates the function we need for the chemical bond. The antisymmetric spin function implies that the spins of the two electrons in the bonding pair have opposite orientations; hence, their magnetic moments cancel one another. For this reason, the majority of molecules have no net magnetic moment. The possession of a magnetic moment by a molecule indicates that one or more of the electrons in the molecule are unpaired.

The conclusions about the bonding electron pair can be summed up briefly. The requirement of the Pauli principle, antisymmetry of the wave function under exchange of identical particles, with the requirement that the electrons concentrate in a small region of space between the nuclei, forces us to describe the electron pair in a chemical bond by the function

$$\psi_1 = \psi_S\Sigma_A. \tag{23–6}$$

The symmetric space function ψ_S has a large electron cloud density between the nuclei and thus prevents electrical repulsion from driving the nuclei apart. The antisymmetric spin function requires the magnetic moments of the two electrons to be oppositely oriented (paired). Thus the proposal of G. N. Lewis in 1916 that atoms are held together by electron pairs is confirmed and given additional meaning by the quantum mechanics. Detailed calculation shows that the energy of the state described by ψ_S is very much lower than that of the state described by ψ_A. These conclusions are general and can be applied to the electron pair holding any two atoms together. First we examine the hydrogen molecule in more detail.

23–3 The Hydrogen Molecule; Valence Bond Method

First we label the protons a and b, and the electrons 1 and 2. If the two hydrogen atoms are infinitely far apart, then there is no interaction between the electrons or between the two protons. If electron 1 is with proton a, it is described by $\psi_a(1)$, which represents any wave function of hydrogen atom a. Similarly, $\psi_b(2)$ describes electron 2 with proton b; $\psi_b(2)$ is any wave function of hydrogen atom b. Since we are concerned only with the state of lowest energy, we choose ψ_a and ψ_b as $1s$ functions on the respective atoms. As we have seen, Section 23–2, the description of the two-electron system is given by either of the products, $\psi_a(1)\psi_b(2)$ or $\psi_a(2)\psi_b(1)$. Regardless of which description is used, the energy of the system at infinite separation is $E = 2E_{1s}$; the sum of the energies of the individual atoms in the $1s$ state.

It is customary to write a sort of "chemical" structure to correspond to each of these quantum mechanical descriptions.

Designation	"Chemical" structure		Description	Energy
I	$H_a^{\cdot 1}$	$^2{\cdot}H_b$	$\psi_I = \psi_a(1)\psi_b(2)$	$E_I = 2E_{1s}$
II	$H_a^{\cdot 2}$	$^{1\cdot}H_b$	$\psi_{II} = \psi_a(2)\psi_b(1)$	$E_{II} = 2E_{1s}$

As the two atoms approach one another, the electrons no longer move independently; they influence each other and are influenced by both nuclei. The descriptions ψ_I and ψ_{II} are no longer exact; furthermore, neither by itself is satisfactory as an approximate wave function. We are forced to choose between the linear combinations,

$$\psi_S = \frac{1}{\sqrt{2}}(\psi_I + \psi_{II}), \tag{23–7}$$

$$\psi_A = \frac{1}{\sqrt{2}}(\psi_I - \psi_{II}). \tag{23–8}$$

From what has been said, ψ_S is the description of the molecule with a stable bond between the two atoms. So far no one has devised a simple chemical representation of the description ψ_S. We write the structures I and II, which are called *resonance* structures, and then describe the correct structure as a *resonance hybrid* of the two.

From ψ_S, we can calculate the energy as a function of R, the internuclear distance; this energy, relative to that of the two atoms at infinite separation, is shown by the curve labeled ψ_S in Fig. 23–1. The wave function ψ_S predicts a minimum in the energy of the system at R_0, the equilibrium value of the internuclear distance. The existence of this minimum indicates that a stable molecule is formed; the depth of the minimum, E_D, is the binding energy or dissociation energy of the molecule.

In Fig. 23–1 the energy curve for ψ_A shows that at all values of R the energy of the system is greater than that of the separated atoms. The lowest energy is obtained if the atoms remain apart. This state is an antibonding or a repulsive state of the system.

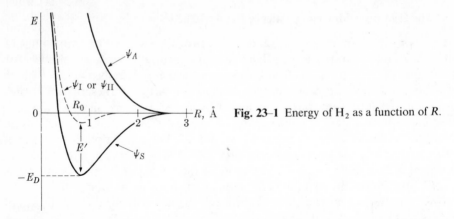

Fig. 23–1 Energy of H_2 as a function of R.

If we ignore our principles and calculate the energy using either ψ_I or ψ_{II} by itself, we obtain the dotted curve in Fig. 23–1. The difference in energy between this curve and that for ψ_S is the *resonance stabilization energy* or the *resonance energy*. It is apparent that the resonance energy accounts for the greater part of the stability of the molecule. Physically we can understand why the simple descriptions, ψ_I and ψ_{II}, are not adequate in a molecule. Both positive nuclei attract an electron on an atom which has been brought close to another atom. Therefore, the electron spreads itself over both nuclei. The remarkable thing is that spreading the electrons over both nuclei lowers the energy of the system so greatly.

Fig. 23–2 Electron densities in the two states of H_2.

The probability density of the electron cloud is obtained by squaring the wave functions. The density along the internuclear axis is shown for the two states ψ_S and ψ_A in Fig. 23–2. It is apparent that in the bound state described by ψ_S, the electron cloud is very dense in the region between the nuclei, while in the state described by ψ_A, the cloud is comparatively thin between the nuclei. The electron density which builds up between the nuclei in the bound state of the molecule can be thought of as the result of the overlapping and interpenetrating of the electron clouds on the individual atoms. Qualitatively, the greater the overlapping of the two electron clouds, the stronger the bond between the two atoms; this is Pauling's principle of maximum overlap.

23–4   Detailed Calculation of the Energy of the Hydrogen Molecule; Valence Bond Method

The coordinates which appear in the Hamiltonian operator for the hydrogen molecule are shown in Fig. 23–3. We write the Hamiltonian in atomic units; energy in multiples of (e^2/a_0) and distance in multiples of a_0. The terms in the Hamiltonian can be grouped in two different ways:

$$\mathbf{H} = \left(-\tfrac{1}{2}\nabla_1^2 - \frac{1}{r_{a1}} - \tfrac{1}{2}\nabla_2^2 - \frac{1}{r_{b2}}\right) + \left(-\frac{1}{r_{a2}} - \frac{1}{r_{b1}} + \frac{1}{r_{12}} + \frac{1}{R}\right) = \mathbf{H}_\mathrm{I}^0 + \mathbf{V}_\mathrm{I}, \qquad (23\text{–}9)$$

$$\mathbf{H} = \left(-\tfrac{1}{2}\nabla_2^2 - \frac{1}{r_{a2}} - \tfrac{1}{2}\nabla_1^2 - \frac{1}{r_{b1}}\right) + \left(-\frac{1}{r_{a1}} - \frac{1}{r_{b2}} + \frac{1}{r_{12}} + \frac{1}{R}\right) = \mathbf{H}_\mathrm{II}^0 + \mathbf{V}_\mathrm{II}. \qquad (23\text{–}10)$$

The second form is obtained from the first by interchanging the coordinates of the two electrons. It is apparent that no physical property is affected by interchanging the positions of two elementary particles.

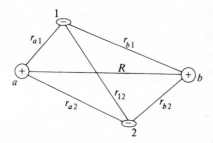

Fig. 23–3 Coordinates for the hydrogen molecule.

If we suppose that the two nuclei are infinitely far apart and that electron 1 is with proton a and electron 2 is with proton b, then $\mathbf{H} = \mathbf{H}_\mathrm{I}^0$.

Since $\mathbf{H}_\mathrm{I}^0$ is the Hamiltonian for two infinitely separated hydrogen atoms (proton a with electron 1) and (proton b with electron 2), the wave function is the product of two hydrogen-atom wave functions;

$$\psi_\mathrm{I} = \psi_a(1)\psi_b(2). \qquad (23\text{–}11)$$

The energy is the sum of the energies of the two hydrogen atoms:

$$E_\mathrm{I}^0 = E_a^0 + E_b^0 = E^0. \qquad (23\text{–}12)$$

The function $\psi_a(1)$ is a hydrogen-atom wave function for electron 1 on proton a; the function $\psi_b(2)$ is a hydrogen atom wave function for electron 2 on proton b. We obtain the energy as an eigenvalue of the Hamiltonian operator:

$$\mathbf{H}_\mathrm{I}^0\psi_\mathrm{I} = \left(-\tfrac{1}{2}\nabla_1^2 - \frac{1}{r_{a1}}\right)\psi_a(1)\psi_b(2) + \left(-\tfrac{1}{2}\nabla_2^2 - \frac{1}{r_{b2}}\right)\psi_a^{(1)}\psi_b(2)$$

$$= \psi_b(2)E_a^0\psi_a(1) + \psi_a(1)E_b^0\psi_b(2);$$

$$\mathbf{H}_\mathrm{I}^0\psi_\mathrm{I} = (E_a^0 + E_b^0)\psi_\mathrm{I} = E^0\psi_\mathrm{I}. \qquad (23\text{–}13)$$

Thus, ψ_I is an eigenfunction of $\mathbf{H}_I^0$ and $E_a^0 + E_b^0 = E^0$ is the corresponding energy eigenvalue.

In a similar way, if we associate electron 2 with proton a and electron 1 with proton b, the Hamiltonian reduces to $\mathbf{H} = \mathbf{H}_{II}^0$ as R becomes infinite. Then, the wave function is

$$\psi_{II} = \psi_a(2)\psi_b(1) \tag{23–14}$$

and the energy, E_{II}^0, is

$$E_{II}^0 = (E_a^0 + E_b^0) = E^0. \tag{23–15}$$

Consequently, ψ_I and ψ_{II} are two different eigenfunctions of the system when R is infinite. Since the eigenvalues are the same, the two eigenfunctions are degenerate. Furthermore, the wave functions are normalized; that is

$$\int \psi_I^2 \, d\tau = \int \psi_{II}^2 \, d\tau = 1. \tag{23–16}$$

To show this in more detail, we use the value of ψ_I in the integral:

$$\int \psi_I^2 \, d\tau = \int [\psi_a(1)\psi_b(2)]^2 \, d\tau_1 \, d\tau_2,$$

in which the general volume element $d\tau$ has been replaced by volume elements $d\tau_1$ and $d\tau_2$ for the two electrons. It is necessary to integrate over the entire coordinate space for each electron. Then the integral separates into two integrals:

$$\int \psi_I^2 \, d\tau = \int [\psi_a(1)]^2 \, d\tau_1 \int [\psi_b(2)]^2 \, d\tau_2,$$

but

$$\int [\psi_a(1)]^2 \, d\tau_1 = \int [\psi_b(2)]^2 \, d\tau_2 = 1,$$

since the $\psi_a(1)$ and $\psi_b(2)$ are the usual orthonormal hydrogen-atom wave functions.

Next we examine the question of orthogonality by evaluating the integral:

$$\int \psi_I \psi_{II} \, d\tau = \int \psi_a(1)\psi_b(2)\psi_a(2)\psi_b(1) \, d\tau_1 \, d\tau_2$$

$$= \int \psi_a(1)\psi_b(1) \, d\tau_1 \int \psi_a(2)\psi_b(2) \, d\tau_2.$$

Since the two integrals on the right differ only in the labeling of the coordinates or the variables of integration, they are equal. We define S, the *overlap integral*, by

$$S = \int \psi_a(1)\psi_b(1) \, d\tau_1 = \int \psi_a(2)\psi_b(2) \, d\tau_2; \tag{23–17}$$

then

$$\int \psi_I \psi_{II}\, d\tau = S^2. \tag{23–18}$$

If we choose any point, the wave functions extending from nucleus a and nucleus b each have a particular value at that point. The product of these values summed over the entire coordinate space is the overlap integral. If the two nuclei are far apart, then near a, where $\psi_a(1)$ is large, $\psi_b(1)$ is extremely small so that the product is extremely small; similarly, near b, $\psi_b(1)$ is large but $\psi_a(1)$ is extremely small so that the product is extremely small. Thus, when the nuclei are far apart, S is very small and, indeed, is zero when R is infinite. As the nuclei approach, S gets larger. We may think of S as a measure of the interpenetration or overlapping of the electron clouds on the two nuclei; thus the name, *overlap integral*. We shall discuss the overlap integral in more detail later.

Equation (23–18) shows that when S is not zero (R is not infinite), ψ_I and ψ_{II} are not orthogonal; consequently, when the atoms are not infinitely separated ψ_I and ψ_{II} are no longer proper wave functions for the system.

We develop the proper wave functions by requiring that they be symmetric or antisymmetric under interchange of the two electrons. If the operator, $\mathbf{I}$, interchanges the coordinates of the two electrons, then

$$\mathbf{I}\psi_I = \mathbf{I}[\psi_a(1)\psi_b(2)] = \psi_a(2)\psi_b(1) = \psi_{II}, \tag{23–19}$$

and similarly,

$$\mathbf{I}\psi_{II} = \psi_I. \tag{23–20}$$

If we construct the linear combination,

$$\psi_S = N\psi_I + \lambda\psi_{II},$$

in which N and λ are constants, and require that

$$\mathbf{I}\psi_S = \psi_S, \tag{23–21}$$

then

$$\mathbf{I}\psi_S = N(\mathbf{I}\psi_I) + \lambda(\mathbf{I}\psi_{II}) = N\psi_{II} + \lambda\psi_I.$$

If Eq. (23–21) is to be satisfied, it must be that

$$N\psi_{II} + \lambda\psi_I = N\psi_I + \lambda\psi_{II}$$

or

$$(N - \lambda)(\psi_{II} - \psi_I) = 0.$$

This condition can be satisfied for nonzero values of $\psi_{II} - \psi_I$ only if $\lambda = N$. Thus, we obtain

$$\psi_S = N(\psi_I + \psi_{II}). \tag{23–22}$$

The constant N is determined by the condition that ψ_S be normalized; $\int \psi_S^2 \, d\tau = 1$. This means

$$1 = \int N^2(\psi_\mathrm{I} + \psi_\mathrm{II})^2 \, d\tau$$

$$= N^2 \left[\int \psi_\mathrm{I}^2 \, d\tau + 2 \int \psi_\mathrm{I} \psi_\mathrm{II} \, d\tau + \int \psi_\mathrm{II}^2 \, d\tau \right].$$

Using Eqs. (23–16) and (23–18) yields $1 = N^2(1 + 2S^2 + 1)$ or

$$N = 1/\sqrt{2(1 + S^2)}.$$

If we require antisymmetry, that is $\mathbf{I}\psi_A = -\psi_A$, then by a similar argument we obtain

$$\psi_A = N'(\psi_\mathrm{I} - \psi_\mathrm{II}),$$

where

$$N' = 1/\sqrt{2(1 - S^2)}.$$

The energy corresponding to these wave functions is obtained by evaluating the integral

$$E = \int \psi \mathbf{H} \psi \, d\tau \tag{23–23}$$

for each of the wave functions. Since none of these wave functions is exact when R is finite, the variation theorem assures us that the energy obtained in this way is greater than the actual energy of the ground state of the system.

Beginning with ψ_I, we obtain

$$E_\mathrm{I} = \int \psi_\mathrm{I} \mathbf{H} \psi_\mathrm{I} \, d\tau = \int \psi_\mathrm{I}(\mathbf{H}_\mathrm{I}^0 \psi_\mathrm{I} + \mathbf{V}_\mathrm{I} \psi_\mathrm{I}) \, d\tau.$$

In view of Eq. (23–13) this becomes

$$E_\mathrm{I} = E^0 + \int \psi_\mathrm{I} \mathbf{V}_\mathrm{I} \psi_\mathrm{I} \, d\tau. \tag{23–24}$$

Using ψ_II, we obtain

$$E_\mathrm{II} = \int \psi_\mathrm{II} \mathbf{H} \psi_\mathrm{II} \, d\tau = \int \psi_\mathrm{II}(\mathbf{H}_\mathrm{II}^0 \psi_\mathrm{II} + \mathbf{V}_\mathrm{II} \psi_\mathrm{II}) \, d\tau.$$

Again, Eq. (23–15) reduces this to

$$E_\mathrm{II} = E^0 + \int \psi_\mathrm{II} \mathbf{V}_\mathrm{II} \psi_\mathrm{II} \, d\tau. \tag{23–25}$$

The two integrals in Eqs. (23–24) and (23–25) are equal, since they are converted into one another by the interchange operator; we define

$$J = \int \psi_\mathrm{I} V_\mathrm{I} \psi_\mathrm{I} \, d\tau = \int \psi_\mathrm{II} V_\mathrm{II} \psi_\mathrm{II} \, d\tau. \tag{23-26}$$

We see that ψ_I and ψ_II yield the same energy since

$$E_\mathrm{I} = E_\mathrm{II} = E^0 + J. \tag{23-27}$$

The energy, J, is a coulombic energy and is a sum of several potential-energy terms:

$$J = \int \psi_a(1)\psi_b(2)\left[-\frac{1}{r_{a2}} - \frac{1}{r_{b1}} + \frac{1}{r_{12}} + \frac{1}{R} \right]\psi_a(1)\psi_b(2) \, d\tau_1 \, d\tau_2,$$

$$J = \int \psi_a^2(1) \, d\tau_1 \int \psi_b(2)\left(-\frac{1}{r_{a2}} \right)\psi_b(2) \, d\tau_2 + \int \psi_b^2(2) \, d\tau_2 \int \psi_a(1)\left(-\frac{1}{r_{b1}} \right)\psi_a(1) \, d\tau_1$$

$$+ \int\int \frac{\psi_a^2(1) \, d\tau_1 \psi_b^2(2) \, d\tau_2}{r_{12}} + \frac{1}{R} \int \psi_a^2(1) \, d\tau_1 \int \psi_b^2(2) \, d\tau_2.$$

Note that since the internuclear distance R is independent of the electronic co-ordinates, the quantity $1/R$ can be taken out of the integral. Since $\psi_a(1)$ and $\psi_b(2)$ are normalized, J becomes

$$J = -\int \frac{\psi_b^2(2) \, d\tau_2}{r_{a2}} - \int \frac{\psi_a^2(1) \, d\tau_1}{r_{b1}} + \int\int \frac{\psi_a^2(1) \, d\tau_1 \, \psi_b^2(2) \, d\tau_2}{r_{12}} + \frac{1}{R}.$$

In the first integral, $\psi_b^2(2) \, d\tau_2$ is the electronic charge at the position $d\tau_2$ due to electron 2; when multiplied by $-1/r_{a2}$ this yields the potential energy of the coulombic attraction between electron 2 and nucleus a. Similarly, the second integral yields the potential energy of the coulombic attraction between electron 1 and nucleus b. These two integrals are equal. We define

$$\epsilon_{aa} = \int \frac{\psi_a^2(1) \, d\tau_1}{r_{b1}} = \int \frac{\psi_b^2(2) \, d\tau_2}{r_{a2}}; \tag{23-28}$$

then

$$J = -2\epsilon_{aa} + \frac{1}{R} + \int\int \frac{\psi_a^2(1) \, d\tau_1 \, \psi_b^2(2) \, d\tau_2}{r_{12}}. \tag{23-29}$$

The quantity $1/R$ is the internuclear repulsion energy. The integral is the product of the charge of the two electrons divided by r_{12}, the interelectronic distance. This is the coulombic repulsion between the two electrons. The quantity J is shown as a function of the internuclear distance by the dotted curve in Fig. 23–1.

When we introduce the symmetry requirement on the wave function, an entirely new set of terms appears in the expression for the energy:

$$E_S = \int \psi_S \mathbf{H} \psi_S \, d\tau = \int \psi_S N[(\mathbf{H}_I^0 + \mathbf{V}_I)\psi_I + (\mathbf{H}_{II}^0 + \mathbf{V}_{II})\psi_{II}] \, d\tau$$

$$= \int \psi_S N[E^0 \psi_I + \mathbf{V}_I \psi_I + E^0 \psi_{II} + \mathbf{V}_{II} \psi_{II}] \, d\tau$$

$$= \int \psi_S [E^0 \psi_S + N\mathbf{V}_I \psi_I + N\mathbf{V}_{II} \psi_{II}] \, d\tau$$

$$= E^0 + N^2 \int (\psi_I + \psi_{II})(\mathbf{V}_I \psi_I + \mathbf{V}_{II} \psi_{II}) \, d\tau$$

$$= E^0 + N^2 \left[\int \psi_I \mathbf{V}_I \psi_I \, d\tau + \int \psi_I \mathbf{V}_{II} \psi_{II} \, d\tau + \int \psi_{II} \mathbf{V}_I \psi_I \, d\tau + \int \psi_{II} \mathbf{V}_{II} \psi_{II} \, d\tau \right],$$

$$E_S = E^0 + N^2[2J + 2K], \tag{23–30}$$

in which we have used K for the two equal integrals

$$K = \int \psi_I \mathbf{V}_{II} \psi_{II} \, d\tau = \int \psi_{II} \mathbf{V}_I \psi_I \, d\tau. \tag{23–31}$$

The energy K is called the exchange energy. The total energy associated with ψ_S is given by

$$E_S = E^0 + \frac{1}{2(1 + S^2)}(2J + 2K) = E^0 + \frac{J}{1 + S^2} + \frac{K}{1 + S^2}. \tag{23–32}$$

By a similar computation, we find that ψ_A has the energy

$$E_A = E^0 + \frac{J}{1 - S^2} - \frac{K}{1 - S^2}. \tag{23–33}$$

The quantity $(J + K)/(1 + S^2)$ is shown as the lowest curve in Fig. 23–1, while $(J - K)/(1 - S^2)$ is the uppermost curve. It is clear that a substantial part of the binding energy (85 to 90%) is due to the exchange energy or resonance energy.

If we put the expressions for ψ_I and ψ_{II} in the exchange integral we find

$$K = \int \psi_{II} \mathbf{V}_I \psi_I \, d\tau$$

$$= \int \psi_a(2)\psi_b(1) \left[-\frac{1}{r_{a2}} - \frac{1}{r_{b1}} + \frac{1}{r_{12}} + \frac{1}{R} \right] \psi_a(1)\psi_b(2) \, d\tau_1 \, d\tau_2$$

$$= -\int \psi_a(1)\psi_b(1) \, d\tau_1 \int \psi_a(2)\frac{1}{r_{a2}}\psi_b(2) \, d\tau_2 - \int \psi_a(2)\psi_b(2) \, d\tau_2 \int \psi_a(1)\frac{1}{r_{b1}}\psi_b(1) \, d\tau_1$$

$$+ \iint \frac{\psi_a(1)\psi_b(1)\psi_a(2)\psi_b(2) \, d\tau_1 \, d\tau_2}{r_{12}} + \frac{1}{R} \int \psi_a(1)\psi_b(1) \, d\tau_1 \int \psi_a(2)\psi_b(2) \, d\tau_2.$$

Recalling the definition of S and introducing

$$\epsilon_{ab} = \int \frac{\psi_a(2)\psi_b(2)}{r_{a2}} \, d\tau_2 = \int \frac{\psi_a(1)\psi_b(1)}{r_{b1}} \, d\tau_1, \tag{23–34}$$

we obtain

$$K = -2S\epsilon_{ab} + \frac{S^2}{R} + \iint \frac{\psi_a(1)\psi_b(1)\psi_a(2)\psi_b(2) \, d\tau_1 \, d\tau_2}{r_{12}}. \tag{23–35}$$

The exchange energy is a pure quantum effect arising as a consequence of our inability to distinguish between identical particles, which requires that we choose a particular mathematical form for the wave function. This mathematical form produces the so-called exchange integrals; these integrals cannot be interpreted by any classical analogy. The last integral in Eq. (23–35) is roughly proportional to S^2 and ϵ_{ab} is roughly proportional to S; thus, the exchange energy depends on the square of the overlap integral. Therefore, the exchange effect is very important at short distances and negligible at large distances. Pauling has suggested a "principle of maximum overlap" which states that the strongest bond between two atoms will involve those electron clouds (wave functions) which have the greatest amount of interpenetration (overlap). Roughly, by this principle, we can gauge the strength of a bond by the value of the overlap integral.

23–5 The Hydrogen Molecule; Molecular Orbital Method

In the molecular orbital method we consider the motion of one electron in the potential field due to all the nuclei of the molecule. We first arrange the nuclei in specified fixed positions; e.g., in Fig. 23–4, we have four nuclei, a, b, c, d, with positive charges Z_a, Z_b, Z_c, Z_d. The Hamiltonian for one electron is written (in atomic units)

$$\mathbf{H} = -\tfrac{1}{2}\nabla_1^2 - \frac{Z_a}{r_{a1}} - \frac{Z_b}{r_{b1}} - \frac{Z_c}{r_{c1}} - \frac{Z_d}{r_{d1}} + C. \tag{23–36}$$

The constant C is the sum of the internuclear repulsions which is independent of the electronic coordinates; hence, if $\psi(1)$ is the appropriate wave function, the energy contribution from internuclear repulsion is $\int \psi C \psi \, d\tau = C \int \psi\psi \, d\tau = C$. As a result, at the beginning we can ignore the internuclear repulsion term in the Hamiltonian and simply add it in at the end of the calculation. In principle, we could solve the

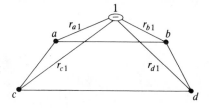

Fig. 23–4 Coordinates for one electron in the field of four nuclei.

Schrödinger equation and obtain a set of wave functions and energy levels appropriate to the motion of one electron in the molecular framework. These wave functions are called *molecular orbitals*. Having such a set of wave functions for a given molecular geometry, the electronic structure of the molecule could be built up in much the same way that one builds the structure of an atom on the basis of the hydrogenic wave functions. For example, we represent the structure of the carbon atom by putting two electrons in each atomic level until we have placed all six in the lowest energy orbitals.

For illustration, we choose the hydrogen molecular framework of two protons. Then, in Fig. 23–4, nuclei c and d are absent and $Z_a = Z_b = 1$; the Hamiltonian is (momentarily ignoring $1/R$, the internuclear repulsion)

$$\mathbf{H} = -\tfrac{1}{2}\nabla_1^2 - \frac{1}{r_{a1}} - \frac{1}{r_{b1}}. \tag{23–37}$$

If R is very large and the electron is on proton a, we have $\psi_a(1)$ as a solution, in which $\psi_a(1)$ is a hydrogen atom wave function centered on proton a. We could write the Hamiltonian in two ways:

$$\mathbf{H} = \mathbf{H}_a^0 - \frac{1}{r_{b1}} \quad \text{or} \quad \mathbf{H} = \mathbf{H}_b^0 - \frac{1}{r_{a1}},$$

with

$$\mathbf{H}_a^0\psi_a(1) = E_a^0\psi_a(1), \qquad \mathbf{H}_b^0\psi_b(1) = E_b^0\psi_b(1), \tag{23–38}$$

where $\psi_b(1)$ is a hydrogen atom wave function centered on atom b.

The requirement of symmetry in this case is that the molecular wave function be symmetric or antisymmetric under the interchange of nuclei. This is equivalent in the case of a homonuclear diatomic molecule to inversion of coordinates through the center of symmetry. A wave function which is symmetric under inversion is labeled g, while the antisymmetric function is labeled u. (From initial letters of the German words *gerade*, even, and *ungerade*, odd.) We obtain

$$\psi_g(1) = \frac{1}{\sqrt{2(1 + S)}}[\psi_a(1) + \psi_b(1)], \tag{23–39}$$

$$\psi_u(1) = \frac{1}{\sqrt{2(1 - S)}}[\psi_a(1) - \psi_b(1)]. \tag{23–40}$$

To calculate the energy we form the function

$$\mathbf{H}\psi_g = \frac{1}{\sqrt{2(1 + S)}}[\mathbf{H}\psi_a(1) + \mathbf{H}\psi_b(1)]$$

$$= \frac{1}{\sqrt{2(1 + S)}}\left[E_a^0\psi_a(1) - \frac{1}{r_{b1}}\psi_a(1) + E_b^0\psi_b(1) - \frac{1}{r_{a1}}\psi_b(1)\right].$$

$$E_g = \int \psi_g \mathbf{H} \psi_g \, d\tau$$

$$= \frac{1}{2(1 + S)} \int [\psi_a(1) + \psi_b(1)] \left[E_a^0 \psi_a(1) + E_b^0 \psi_b(1) - \frac{1}{r_{b1}} \psi_a(1) - \frac{1}{r_{a1}} \psi_b(1) \right] d\tau$$

$$= \frac{1}{2(1 + S)} \left[E_a^0 \int \psi_a^2(1) \, d\tau_1 + E_b^0 \int \psi_a(1)\psi_b(1) \, d\tau_1 - \int \frac{\psi_a^2(1)}{r_{b1}} \, d\tau_1 \right.$$

$$- \int \frac{\psi_a(1)\psi_b(1)}{r_{a1}} \, d\tau_1 + E_a^0 \int \psi_a(1)\psi_b(1) \, d\tau_1 + E_b^0 \int \psi_b^2(1) \, d\tau_1$$

$$\left. - \int \frac{\psi_b(1)\psi_a(1)}{r_{b1}} \, d\tau_1 - \int \frac{\psi_b^2(1)}{r_{a1}} \, d\tau_1 \right],$$

$$E_g = \frac{1}{2(1 + S)} [E_a^0 + SE_b^0 - \epsilon_{aa} - \epsilon_{ab} + SE_a^0 + E_b^0 - \epsilon_{ab} - \epsilon_{aa}],$$

$$E_g = \tfrac{1}{2}(E_a^0 + E_b^0) - \left(\frac{\epsilon_{aa} + \epsilon_{ab}}{1 + S} \right),$$

$$E_g = \tfrac{1}{2}E^0 - \left(\frac{\epsilon_{aa} + \epsilon_{ab}}{1 + S} \right). \tag{23–41}$$

This is the energy of an electron in this molecular orbital. If we add a second electron, the Hamiltonian becomes

$$\mathbf{H} = -\tfrac{1}{2}\nabla_1^2 - \frac{1}{r_{a1}} - \frac{1}{r_{b1}} - \tfrac{1}{2}\nabla_2^2 - \frac{1}{r_{a2}} - \frac{1}{r_{b2}} + \frac{1}{r_{12}}$$

$$= \mathbf{H}_1 + \mathbf{H}_2 + \frac{1}{r_{12}}. \tag{23–42}$$

If we ignore the electronic repulsion ($1/r_{12}$), then $\mathbf{H}$ is a sum of two equivalent sets of terms, and the wave function is a product:

$$\Psi = \frac{1}{2(1 + S)} [\psi_a(1) + \psi_b(1)][\psi_a(2) + \psi_b(2)]. \tag{23–43}$$

Here we have decided to put two electrons (they must have opposite spin) in the same molecular orbital. The energy must then be twice what is given by Eq. (23–41):

$$E = E^0 - \frac{2(\epsilon_{aa} + \epsilon_{ab})}{1 + S}. \tag{23–44}$$

If we wish to account for nuclear repulsion, we add a term $1/R$; similarly, to account for electron repulsion we add $\int \Psi(1/r_{12})\Psi \, d\tau$. This yields, for the total energy,

$$E = E^0 - \frac{2(\epsilon_{aa} + \epsilon_{ab})}{1 + S} + \frac{1}{R} + \int \Psi \left(\frac{1}{r_{12}} \right) \Psi \, d\tau. \tag{23–45}$$

If we expand the wave function in Eq. (23–43), we obtain

$$\Psi = \frac{1}{2(1 + S)}[\psi_a(1)\psi_a(2) + \psi_b(1)\psi_b(2) + \psi_a(1)\psi_b(2) + \psi_b(1)\psi_a(2)]. \qquad (23\text{–}46)$$

The last two terms are the function which we used in our earlier discussion of the electron-pair bond. The description contains two additional terms: the first, $\psi_a(1)\psi_a(2)$, corresponds to both electrons on proton a. The second, $\psi_b(1)\psi_b(2)$, has both electrons on proton b. The chemical structures would be written

III $H^{\cdot 1\ -}_{a\cdot 2}$ H^+_b,

IV H^+_a $^1_2\!\cdot\!H^-_b$.

Structures III and IV are ionic structures. These structures do contribute slightly, $\sim 3\%$, to the overall structure of the hydrogen molecule. In this simple molecular orbital approach the ionic structures III and IV are weighted equally with the covalent structures I and II. In more refined versions their contribution is cut down to a more realistic level.

23–6   The Covalent Bond

The covalent bond between any two atoms A and B can be described by a wave function similar to the ψ_S used for the hydrogen molecule. Consider the structures I and II:

I $A^{\cdot 1\ 2\cdot}B$ $\psi_I = \psi_a(1)\psi_b(2),$

II $A^{\cdot 2\ 1\cdot}B$ $\psi_{II} = \psi_a(2)\psi_b(1),$

where ψ_a and ψ_b are wave functions appropriate to atoms A and B, respectively. The structure is described by the symmetric combination of ψ_I and ψ_{II}:

$$\psi_S = \frac{1}{\sqrt{2(1 + S^2)}}(\psi_I + \psi_{II}). \qquad (23\text{–}47)$$

This description predicts a minimum energy corresponding to formation of a bond. The resonance energy is obtained by taking the difference between the energy computed for ψ_S and that computed for either ψ_I or ψ_{II}.

Since the stability of the bond depends principally on the resonance energy, it is important to know what factors influence the magnitude of this energy. The resonance energy has its largest possible value if the energies of the contributing structures are the same, or nearly so. The greater the energy difference between the contributing structures the less the stabilization due to resonance between them. In the case of any molecule AB, the structures I and II differ only by the exchange of the electron coordinates, so they have exactly the same energy. Consequently, the stabilization conferred by resonance between I and II is large.

The two ionic structures of the molecule AB,

III A^+ $\overset{..}{.}B^-$,

IV $A\overset{..}{.}{}^{-}$ B^+,

also contribute to the overall structure of the molecule; however, one of these structures is usually much lower in energy than the other; and, of course, the energies are different from the energy of I or II. In many molecules one or several ionic structures contribute to the overall structure of the molecule. In the molecule of hydrogen chloride three structures are important:

    I $H^{.1}\ {}^{2.}Cl$ II $H^{.2}\ {}^{1.}Cl$ III $H^+\ \overset{..}{.}Cl^-$.

The overall structure of the molecule is a resonance hybrid of the structures I, II, and III. The quantum-mechanical description is a linear combination,

$$\psi = c[\psi_I + \psi_{II} + \lambda\psi_{III}].$$

The coefficients of ψ_I and ψ_{II} in the composite description are equal, indicating that these two contribute equally to the structure. The coefficient of ψ_{III} differs from the other two, indicating that ψ_{III} contributes differently. The contributions of the three structures in HCl are estimated to have the values: I, 26%; II, 26%; III, 48%. The structures I and II are covalent structures, so we may say that the bond in HCl is 52% covalent and 48% ionic. A bond in which the ionic contribution is significant is called a covalent bond with partial ionic character.

Every covalent bond has more or less ionic character. Even if the two atoms are the same, there is a small contribution of ionic structures, $\sim 3\%$ in H_2. The bond between two like atoms is usually called a pure covalent bond, nonetheless.

There are restrictions on the structures which can contribute to the composite structure of a molecule. The structures which can "resonate" to produce a composite structure must: (1) have the same number of unpaired electrons; and (2) have the same arrangement of nuclei. For resonance to be effective, the structures should not differ greatly in energy.

23–7   Overlap and Directional Character of the Covalent Bond

To form a covalent bond two things are needed: a pair of electrons with spins opposed, and a stable orbital, an orbital in the valence shell, on each atom. The strength of a bond is qualitatively proportional to the extent of overlap of the charge clouds on the two atoms. The *overlap integral S* is a measure of the overlap of two charge clouds:

$$S = \int \psi_a(1)\psi_b(1)\,d\tau_1, \tag{23–48}$$

whose value depends on the direction of approach of the two clouds. To see how this comes about, we represent the atomic wave functions, the *s*, *p*, and *d* functions, by *boundary* surfaces, surfaces on which the value of the wave function is a constant and which has been chosen large enough so that most of the cloud, 90 % for example, is contained within it. Figure 23–5 shows the two-dimensional representation of the boundary surfaces for the s, p_y, and p_z orbitals. For an s orbital, the boundary surface is a sphere, so the circle is the two-dimensional representation. For the p orbital, the two-lobed surface shown in Fig. 22–6 is represented in two dimensions by a figure not unlike a figure eight. The signs + and − in Fig. 23–5 are the algebraic signs of the wave function in the respective regions.

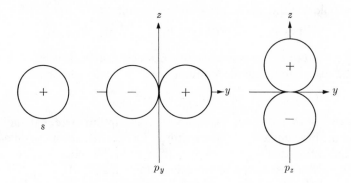

Fig. 23–5 Boundary surfaces for s and p functions.

The overlap of s orbitals

If the electrons on the two atoms both occupy s orbitals, then the extent of overlapping of the two clouds is independent of the direction of approach. Figure 23–6 shows the overlapping in H_2 for two different directions of approach. Since both functions are positive, the overlap integral, Eq. (23–48), is positive.

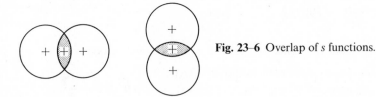

Fig. 23–6 Overlap of s functions.

Overlap between s and p orbitals

If the electrons which will form the bond are in an s orbital on one atom and in a p orbital on the other, then the overlap depends on the relative direction of approach of the two atoms. Figure 23–7(a) shows the approach of an s electron to a p electron. The p function changes sign on passing through the plane of the nucleus. The s function is positive everywhere in space. The integral $\int \psi_s \psi_p \, d\tau$ is the sum of the

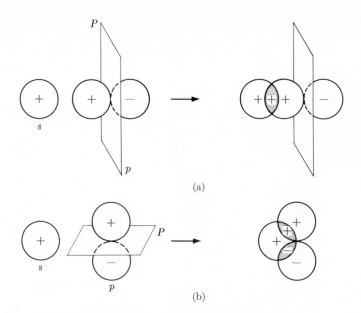

Fig. 23–7 Overlap of s and p functions.

values of the product $\psi_s\psi_p$ everywhere in space. In the region to the left of the plane P, the product $\psi_s\psi_p$ is always positive, since both ψ_s and ψ_p are positive in this region. The numerical value of $\psi_s\psi_p$ is moderately large, since part of this region is close to both nuclei where both wave functions have sizable values. The integration over this region yields a sum of positive contributions. To the right of plane P, ψ_p is negative and ψ_s is positive; their product is therefore negative and the integral is a sum of these negative contributions. The total integral has a positive contribution to which a small negative contribution is added. The positive contribution pre-dominates, because the value of ψ_s is smaller the greater the distance from the nucleus on which it is centered, and so is very small to the right of P. Thus, for this direction of approach of an s electron cloud to a p electron cloud, the overlap integral has a positive value and bond formation is possible.

Consider the approach of an s cloud to a p cloud along the direction in Fig. 23–7(b). The p function is positive above the plane and negative below it. Therefore the product $\psi_s\psi_p$ is positive above the plane and negative below it. Because of the symmetry of the s and p clouds, the positive contributions are exactly balanced by the negative contributions. The overlap integral is equal to zero. There is no overlap, therefore no possibility of forming a bond if the clouds approach in this orientation. It is readily shown that the maximum overlapping of the two charge clouds occurs if the approach is along the axis of the p cloud; Fig. 23–7(a). Therefore, the strongest bond is formed in this manner.

Overlap of a *p* orbital with a *p* orbital

If we consider the approach of two atoms each having a *p* electron, there are several possibilities; Fig. 23–8. By the same argument as above it may be shown that in the approach illustrated in Fig. 23–8(b), the overlap is zero, while maximum overlap is achieved in the configuration of Fig. 23–8(a), and a moderate value in the configuration of Fig. 23–8(c). It should be noted that in Fig. 23–8(c) the wave functions are both zero in the horizontal plane, and that the internuclear axis lies in this plane. This implies that the charge density is zero along the internuclear axis in this type of bond (π bond).

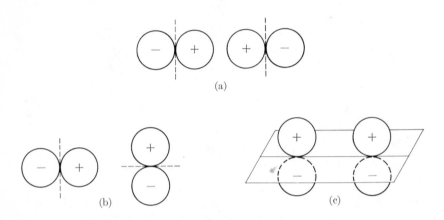

Fig. 23–8 Overlap of *p* functions. (a) Maximum overlap. (b) Zero overlap. (c) Moderate overlap.

The way in which electron clouds overlap gives the first indication of the reason an atom forms covalent bonds in a particular relative orientation. Directional character is a distinguishing attribute of the covalent bond; other types of bonds do not prefer special directions. The ability to explain and predict the number of bonds formed and their geometrical arrangement around the atom is one of the great triumphs of the quantum mechanics. If what follows note that very approximate methods suffice to provide the qualitative picture.

23–8　Elements in the First Row of the Periodic Table

The atoms in the first row of the periodic table have only four valence orbitals: $2s, 2p_x, 2p_y, 2p_z$. Since for every bond the atom must have an orbital in the valence shell, the number of bonds formed by these elements is limited to four.

The simplest example of the overlap of *s* and *p* orbitals is provided by the compounds HX, where X is a halogen. Fluorine, for example, has the electronic configuration

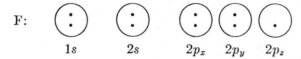

F: 1s   2s $2p_x$ $2p_y$ $2p_z$

where the circles indicate the orbitals, and the dots symbolize the electrons occupying them. The single unpaired electron available for bond formation with an electron from another atom is in a p orbital, which overlaps with the electron in the s orbital of hydrogen to form the molecule HF; Fig. 23–9(a). The paired electrons in the other p orbitals of fluorine are not indicated in the figure since they do not contribute to the bond. The molecule of fluorine, F_2, is formed by the overlap of the p orbitals on each fluorine atom; Fig. 23–9(b).

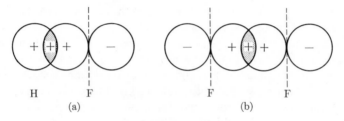

H   F   F   F
(a)   (b)

Fig. 23–9 Overlap in HF and F_2.

A more interesting case arises with oxygen, which has the electronic configuration

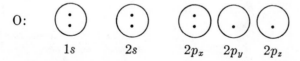

O: 1s   2s $2p_x$ $2p_y$ $2p_z$

Oxygen has two unpaired electrons and thus a valence of two; these electrons occupy two different p orbitals which lie along perpendicular axes; Fig. 23–10(a). If two hydrogen atoms are brought up to form bonds with the oxygen atoms, a maximum overlapping occurs if the configuration is as shown in Fig. 23–10(b). From this simple picture we expect the bond angle in water to be 90°. The measured value is 104.5°. Even considering the simplicity of this approach, the agreement is not good. We will return to the subject of the bond angle in water in Section 23–9. If we examine the bond angles in H_2S, H_2Se, H_2Te, for which, on the same basis, we would predict a bond angle of 90°, the agreement is better. The experimental values are as follows:

Compound:	H_2O	H_2S	H_2Se	H_2Te
Bond angle:	104.5°	92.2°	91.0°	89.5°

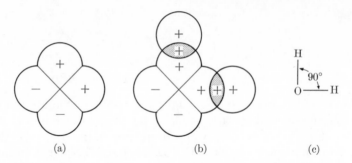

(a)   (b)   (c)

Fig. 23–10 (a) Oxygen atom. (b) Water molecule. (c) Predicted bond angle in water.

The bond between two oxygen atoms is illustrated in Fig. 23–11. The bond formed by the overlap along the line of centers of the atoms is called a σ bond, while the bond formed by the overlapping above and below the plane is called a π bond. Thus, the double bond in the oxygen molecule consists of a σ bond having its charge cloud along the internuclear axis, and a π bond having the charge cloud above and below the plane. Unfortunately, this simple picture does not account for the paramagnetism of O_2, a classic failure of the valence bond theory. The O_2 molecule has a magnetic moment which shows that there are two unpaired electrons in the molecule. In the picture presented above, all the electrons are paired, so the magnetic moment should be zero. The molecular orbital method is able to predict this paramagnetic property as well as the strength of the bond.

Fig. 23–11 Oxygen showing overlap to form σ and π bonds.

The nitrogen atom has the electronic configuration

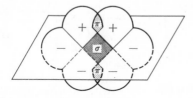

Three unpaired electrons are available so that three bonds are possible.† These

---

† Hund's rule of maximum multiplicity requires that the electrons be spread over as many of the orbitals having the same value of l, p orbitals here as is possible. Therefore, we do not pair up electrons in the p, or d or f, orbitals if it can be avoided by spreading them out.

three electron clouds are directed along the three axes; Fig. 23–12(a). Three H atoms
will form bonds of maximum strength if they approach at right angles; Fig. 23–12(b).
For ammonia we also predict bond angles of 90°. The actual value is 107.3°. The
molecule is a trigonal pyramid with the nitrogen atom at the apex and the three
hydrogen atoms at the base. The widening of the angle means only that the pyramid
is flattened out somewhat. The general shape is predicted correctly. In PH_3 the
angles are 93.3°, which is nearer the predicted 90° value.

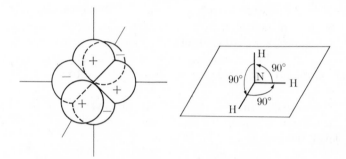

Fig. 23–12 (a) Nitrogen atom. (b) Predicted bond angles in
ammonia.

The formation of the nitrogen molecule is illustrated in Fig. 23–13. Two π bonds
are formed in addition to the σ bond. The formation of two π bonds results in the
charge cloud having the shape of a cylindrical sheath around the axis of the molecule
(compare Fig. 22–5 for $m = \pm 1$).

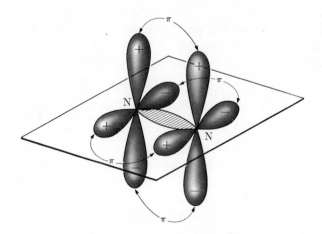

Fig. 23–13 The two π bonds in nitrogen.

23–9   Hybridization and the Valence of Carbon

If we attempt to apply the ideas developed so far to the valency states of carbon, a difficulty appears immediately. The configuration of the normal carbon atom is

C:

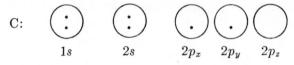

$1s$ $\qquad$ $2s$ $\qquad\quad$ $2p_x$ $\quad$ $2p_y$ $\quad$ $2p_z$

The presence of two unpaired electrons implies a valence of two, and one could expect the formula of the simplest hydride of carbon to be CH_2 with a bond angle of 90°. Such a conclusion would undoubtedly meet with some objection from the organic chemists who are firmly convinced that in the majority of cases carbon forms four bonds, not two, and that these four bonds are directed to the corners of a tetrahedron.

Knowing the answer which is required, we observe that in an excited state, denoted C*, the carbon atom will be tetravalent. If an electron is raised from the $2s$ to the $2p_z$ level, the excited configuration is

C*:

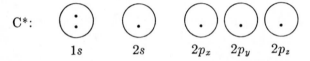

$1s$ $\qquad$ $2s$ $\qquad\quad$ $2p_x$ $\quad$ $2p_y$ $\quad$ $2p_z$

It is true that energy must be expended to excite the atom; however, the extra stability gained by forming four bonds instead of two more than compensates for this. Our problem is not yet solved; C* has four unpaired electrons and therefore is four-valent. But it appears that the three bonds using the p orbitals should be formed at 90° angles and a fourth, using the s orbital, should be formed in an arbitrary direction probably in such a way as to be out of the way of the other three bonds. This scheme is not satisfactory. However, the principle of maximum overlap comes to the rescue. Pauling suggested taking the original four descriptions, the $2s$, $2p_x$, $2p_y$, and $2p_z$, in linear combinations to obtain four new descriptions, which we label t_1, t_2, t_3, t_4:

$$t_1 = a_{11}\psi_{2s} + a_{12}\psi_{2p_x} + a_{13}\psi_{2p_y} + a_{14}\psi_{2p_z},$$
$$t_2 = a_{21}\psi_{2s} + a_{22}\psi_{2p_x} + a_{23}\psi_{2p_y} + a_{24}\psi_{2p_z},$$

and similarly for t_3 and t_4. The constants $a_{11}, a_{12}, \ldots$ are determined by the conditions: (1) the four new descriptions are to be equivalent in their extension in space; (2) the extension of the orbitals shall be as large as possible so that the overlap will be a maximum. It is possible to determine the coefficients $a_{11}, a_{12}, \ldots$ so that four equivalent orbitals with maximum extension are formed. These four new orbitals are directed to the apices of a tetrahedron! The shape of one of these orbitals is shown in Fig. 23–14(a), and the set of four is shown in Fig. 23–14(b). The orbitals

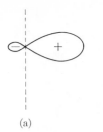 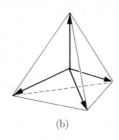

Fig. 23–14 (a) Tetrahedral hybrid.
(b) Set of four tetrahedral orbitals.

(a)   (b)

t_1, t_2, t_3, and t_4 are called hybrid (or mixed) orbitals. The process of making linear combinations is called hybridization (or mixing). These particular ones are called tetrahedral hybrids or sp^3 hybrids. The extension of the hybrid orbitals is much greater than that of either an s or p orbital by itself. The overlap and consequently the bond strength are correspondingly greater.

Every carbon atom which forms single bonds to four neighboring atoms has its valence orbitals hybridized tetrahedrally; e.g., the carbon atoms in saturated aliphatic compounds.

Hybridization is also important in the structures of water and ammonia. In water we consider that the orbitals on the oxygen atom are hybridized, sp^3, and that these hybrid orbitals are occupied by the four electron pairs of the octet. Then the bond angle in water should be the tetrahedral angle, 109.47°. However, the electron pairs in the bonds repel each other less strongly than do the unshared electron pairs. This effect tends to close the bond angle somewhat. The observed value is 104.5°. The same effect is operative in ammonia in which the observed angle, 107.3°, is only slightly less than the ideal tetrahedral value.

23–10   Carbon Bonded to Three Neighbors; the Double Bond

In many compounds some of the carbon atoms form bonds with only three neighboring atoms; e.g., ethylene, or acetone:

$$\begin{array}{cc} H & H \\ \diagdown & \diagup \\ & C{=}C \\ \diagup & \diagdown \\ H & H \end{array} \qquad \begin{array}{c} H_3C \\ \diagdown \\ C{=}O. \\ \diagup \\ H_3C \end{array}$$

In ethylene, both carbon atoms form bonds to only three neighbors, while in acetone the carbon in the carbonyl group is attached to only three atoms. If the s orbital is mixed with only two of the p orbitals, and the third p orbital left as it is, we obtain three new orbitals which lie in a plane and are directed to the corners of an equilateral triangle; these sp^2 hybrids also have greater extension in space than an s or a p orbital. The set is illustrated in Fig. 23–15. The remaining p_z orbital is perpendicular to the plane in which the hybrids lie.

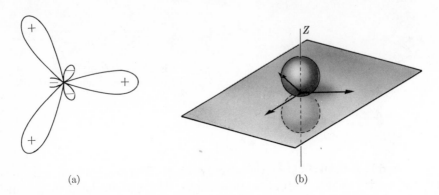

(a)   (b)

Fig. 23–15 (a) sp^2 hybrids. (b) sp^2 hybrids with p_z orbital.

In ethylene, both carbon atoms are hybridized in this way; a strong σ bond is formed by the overlap of a hybrid orbital from each carbon atom. The remaining two hybrids on each carbon form σ bonds with the s orbitals of the four hydrogen atoms. All of the atoms lie in one plane. The overlap of the p_z orbitals on each carbon atom forms the second bond between the two carbon atoms, the π bond. The charge cloud of the electrons in the π bond lies above and below the plane of the atoms. Figure 23–16(a) and (b) shows the relative locations of the bonds.

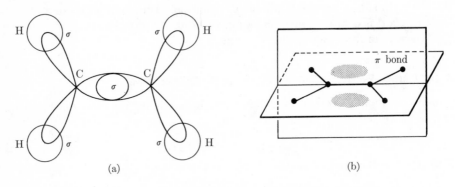

(a)   (b)

Fig. 23–16 (a) σ bonds in ethylene. (b) π bond in ethylene.

The stability gained by the molecule through the overlapping of the p_z orbitals in the π bond locks the molecule in a planar configuration. If the plane of one CH_2 group were at 90° to the plane of the other, the p_z orbitals would not overlap; the molecule would be much less stable in such a configuration. This accounts for the absence of rotation about the double bond and makes possible the existence of geometric isomers, the *cis* and *trans* forms of disubstituted ethylene.

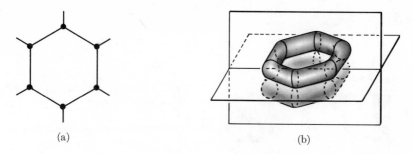

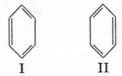

Fig. 23–17 (a) Carbon skeleton in benzene. (b) π bonds in benzene.

Any carbon atom bound to three atoms in a stable compound is hybridized in this fashion. The unsaturated aliphatic hydrocarbons are only one class of compound which includes this kind of bonding. Benzene is an important example of a compound in which each carbon atom is attached to only three other atoms. Each carbon atom in benzene is hybridized sp^2 so that the bonds are at 120° in a plane. The carbon skeleton is shown in Fig. 23–17(a). The p_z orbitals of the six carbon atoms project above and below the plane of the ring. The overlap of the p_z orbitals produces a doughnut-shaped cloud above and below the plane of the ring; Fig. 23–17(b). There are six electrons spread out in these "doughnuts," enough for only three bonds in the classical sense. These three bonds are spread over six positions so that each carbon–carbon bond in benzene has one-half double-bond character. The use of the two resonance formulas I and II

$$\text{I} \qquad\qquad \text{II}$$

to describe this situation is a lame attempt to reduce the structure as we know it to classical chemical terms.

23–11   Carbon Bonded to Two Neighbors; the Triple Bond

A carbon atom can be connected to only two other atoms if it forms a single and a triple bond or two double-bonds; $-C\equiv$ or $=C=$. The bond angle is 180°, the carbon and the two other atoms lying on a straight line. The hybridization involves only the s orbital and one p orbital to yield two sp hybrids which are oppositely directed along a straight line; Fig. 23–18(a). In the acetylene molecule, both carbon atoms are hybridized in this way. The σ bond and π bonds are shown in Fig. 23–18(b), (c), (d). One π bond is formed above and below the axis of the molecule, the other in front and in back of the axis; Fig. 23–18(c). The net result is the cylindrical sheath shown in Fig. 23–18(d). The description is similar to that for the nitrogen molecule.

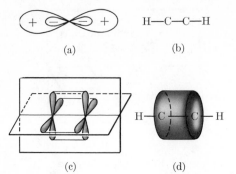

H—C—C—H

(b)

(a)

Fig. 23–18 (a) *sp* hybrids. (b) bonds in acetylene. (c) π bonds in acetylene. (d) Sheath.

H—C—C—H

(c)   (d)

In carbon dioxide, $O{=}C{=}O$, carbon is joined to two neighbors. The molecule is linear. The π bonds yield a cylindrical sheath of electron cloud around the nuclear axis.

23–12 Bond Order and Bond Length

It is a general rule that a double bond between two atoms is stronger than a single bond, and a triple bond is stronger than a double bond. The higher the bond order, the stronger the bond. It is also a general rule that increasing the strength of the bond shortens the bond length, the distance between the two atoms. This is illustrated by the carbon–carbon bond lengths and bond orders in some simple carbon compounds; see Table 23–1. A correlation such as shown in the table is useful if we do not know what contributions the various resonance structures make to the overall structure of the molecule. If we determine the bond length, the bond order can be estimated from a plot of bond order versus bond length. The estimate of the bond order may provide a clue to the contributions of the various resonance structures. The correlation must, of course, be worked out for the particular kind of bond in question.

Table 23–1

Compound	Diamond	Graphite	Benzene	Ethylene	Acetylene
Bond order	1.0	1.33	1.50	2.0	3.0
C—C bond length, Å	1.54	1.42	1.39	1.35	1.20

23–13 The Covalent Bond in Elements of the Second and Higher Periods

The elements in the second and higher periods have *d* orbitals in the valence shell in addition to the *s* and *p* orbitals. There are a total of nine orbitals; one *s*, three *p*, and five *d* orbitals, which could be used for bond formation. It is conceivable that an atom could be bonded to as many as nine other atoms. This coordination number is unknown, and ordinarily the number of atoms attached to a central atom does not exceed six, although there are a few compounds in which seven and eight atoms or groups are attached.

The most common higher coordination number in these elements is six; some of the fluorides of the first and second periods provide examples.

First period : CF_4 NF_3 OF_2 F_2
Second period : SiF_4 PF_3 S_2F_2 ClF_3
                   SiF_6^{2-} PF_5 SF_4
                              PF_6^- SF_6

The fluorides are chosen, since fluorine tends to bring out high coordination numbers.

In phosphorus, the electronic configuration is

P:
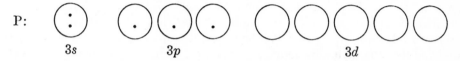

3s   3p   3d

Since there are three unpaired electrons, the valence is three. In this state phosphorus forms the same type of trivalent compounds as nitrogen: NH_3, PH_3, NF_3, PF_3. Due to the presence of vacant d orbitals *in the valence shell* a relatively small expenditure of energy is required to form pentavalent P* with five unpaired electrons:

P*:
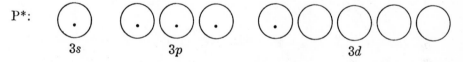

3s   3p   3d

In this state phosphorus can form bonds to five neighboring atoms as in PF_5, PCl_5. The orbitals used are hybridized, sp^3d hybrids, and are directed to the apices of a trigonal bipyramid; Fig. 23–19. The phosphorus and three of the fluorine atoms lie in a plane; the remaining two fluorine atoms are placed symmetrically above and

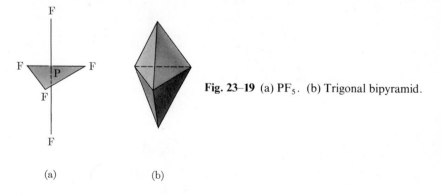

Fig. 23–19 (a) PF_5. (b) Trigonal bipyramid.

(a)   (b)

below this plane. To promote an electron in nitrogen, the electron would have to be moved out of the valence shell to a shell of higher principal quantum number. The energy required would be too large to be compensated by the formation of two additional bonds.

If we add an electron to pentavalent phosphorus, the hexavalent species P^- is obtained:

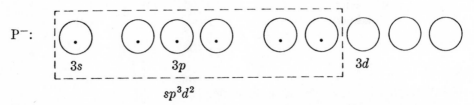

$$sp^3d^2$$

The ion PF_6^- may be regarded as a compound of P^- with six neutral fluorine atoms. The hybridization, sp^3d^2, yields six equivalent bonds directed to the apices of a regular octahedron;† Fig. 23–20(a). Similarly, the sulfur atom forms bonds to six atoms in SF_6; the hybridization is sp^3d^2 and the geometric configuration is octahedral; Fig. 23–20 (b).

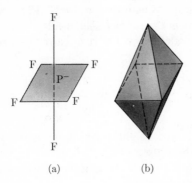

(a)   (b)

Fig. 23–20 (a) PF_6^-. (b) Regular octahedron

Table 23–2

Hybridization	Geometry
sp	Linear
sp^2	Trigonal planar
sp^3	Tetrahedral
sp^2d	Square planar
sp^3d	Trigonal bipyramidal
sp^3d^2	Octahedral

---

†   An *octa*hedron has *eight faces*, but only *six* apices.

The species PF_5, PF_6^-, SF_6 are exceptions to the octet rule. In PF_5 there are ten electrons in the valency group around the phosphorus atom; in PF_6^- and SF_6 there are twelve electrons in the valency groups. The elements in the first row, on the other hand, are bound rigidly to the "rule of eight."

Because of the availability of vacant d orbitals in the valence shell, the transition elements can form a variety of complex compounds. The electron pair for the bond is provided by a *donor* molecule or group such as NH_3 or CN^-. The most common hybridizations and their geometry are summarized in Table 23–2.

Problems

23–1. a) Sketch the double-bond system in 1,3-butadiene.
   b) Compare the double bonds in 1,3-butadiene with those in 1,4-pentadiene.
   c) Show that any hydrocarbon containing a conjugated system of double bonds is planar over the region of conjugation.

23–2. Nitrogen forms two distinct types of compounds in which it is attached to three neighbors. In ammonia and the amines, the configuration is pyramidal, while the NO_3^- ion is planar. Sketch the hybridization possibilities for the two situations. [*Hint*: N^+ is isoelectronic with carbon.]

23–3. The "one-electron" bond is stabilized in the species H_2^+ by resonance between the structures

$$H^+ \quad \cdot H \quad \text{and} \quad H\cdot \quad H^+.$$

Suggest a reason for the fact that the "one-electron" bond is not observed between two *unlike* atoms A and B to yield $(A\cdot B)^+$.

23–4. Nickel ion, Ni^{2+}, forms two types of four-coordinate complex compounds. One type is tetrahedral; the other is square. Which of these types will have a magnetic moment due to unpaired electron spins?

23–5. Iron, Fe^{3+}, forms two types of six-coordinate complexes. Both are octahedral, having a hybridization sp^3d^2. In one type (inner orbital complex) two $3d$ orbitals are used, and in the other type (outer orbital complex) two $4d$ orbitals are used in the hybridization. Which has the greater magnetic moment, the inner orbital or the outer orbital complex?

23–6. Using the definitions of V_I and V_{II} in Eqs. (23–9) and (23–10), show that the two integrals in Eq. (23–26) are converted into each other by the interchange operator. Show that the same is true for the two integrals in Eq. (23–31).

23–7. Let α and β be the two spin wave functions corresponding to the two possible values of the electron spin quantum number; then $\alpha(1)$ indicates that electron 1 has spin α. The possible spin functions for two electrons are: $\sigma_1 = \alpha(1)\alpha(2)$; $\sigma_2 = \alpha(1)\beta(2)$; $\sigma_3 = \beta(1)\alpha(2)$; $\sigma_4 = \beta(1)\beta(2)$. By making linear combinations where necessary show that three functions are symmetric (triplet state) and one is antisymmetric (singlet state) under the interchange of the two electrons.

Chapter Twenty-four

Fundamentals of
Spectroscopy

24–1 Introduction

From the spectrum of a molecule we can obtain experimental information about the geometry of the molecule (bond lengths), and the energy states from which bond strengths are ultimately obtained. The molecule spectrum depends on the characteristics of the nuclear motions as well as on the electronic motions. In Section 23–1, by invoking the Born-Oppenheimer approximation, we discussed the electronic motion which produces the bonding between the atoms as a problem separate from that of the nuclear motions.

We begin the discussion of spectroscopy with a brief recapitulation of the description of the nuclear motions and a consideration of molecular spectroscopy from an elementary standpoint. Then we discuss the fundamental quantum mechanical behavior of systems which undergo transitions from one state to another.

24–2 Nuclear Motions; Vibration and Rotation

The motions of the nuclei are of three kinds: the translational motion of the molecule as a whole, which we discard as uninteresting; the rotation of the molecule; and the vibrations of the nuclei within the molecule. To a good approximation these motions are independent and can be discussed separately.

A molecule containing N atoms has $3N$ nuclear coordinates and $3N$ nuclear momenta; therefore, there are $3N$ independent modes of motion or $3N$ *degrees of freedom*. Discarding three coordinates and three momenta which pertain to the translation of the whole molecule, there remain $3N - 3$ degrees of freedom. If the molecule is linear and the axis of the molecule is the z-axis, then two independent modes of rotation, about the x- and y-axis, are possible. For linear molecules the number of coordinates and momenta remaining to describe the vibrations is

$$3N - 3 - 2 = 3N - 5.$$

Nonlinear molecules have three independent modes of rotation about three mutually perpendicular axes, so the number of coordinates and momenta remaining to describe the vibrations is $3N - 3 - 3 = 3N - 6$. The number of modes of each type of motion is shown in Table 24–1.

Table 24–1

Molecule	Linear	Nonlinear
Total number of degrees of freedom	$3N$	$3N$
Number of translational degrees of freedom	3	3
Number of rotational degrees of freedom	2	3
Number of vibrational degrees of freedom	$3N - 5$	$3N - 6$

Vibrations

For simplicity we assume that each of the molecular vibrations is a simple harmonic vibration characterized by an appropriate mass m and Hooke's law constant k. The wave functions are determined by a single quantum number n, the vibrational quantum number. The energy of the oscillator is

$$E_n = (n + \tfrac{1}{2})h\nu_0, \qquad n = 0, 1, 2, \ldots, \tag{24–1}$$

where $\nu_0 = (1/2\pi)\sqrt{k/m}$ is the classical vibration frequency. Each vibrational degree of freedom has a characteristic value of the fundamental frequency ν_0.

The lowest permissible value of the energy is the zero-point energy, obtained by setting $n = 0$ in Eq. (24–1):

$$E_0 = \tfrac{1}{2}h\nu_0. \tag{24–2}$$

In this lowest energy state, there is still some motion of the oscillator. If the motion ceased altogether, the relative position of the two atoms would be precisely defined. That is not possible without giving the particles an infinitely large uncertainty in their relative momentum. A compromise is reached which leaves a small residual motion and uncertainties in both position and momentum in conformity with the uncertainty principle.

Diatomic molecules provide the simplest example of molecular vibration. There is only one mode of vibration, the oscillation of the two atoms along the line of centers.

The four modes of vibration of carbon dioxide, a symmetrical, linear, triatomic molecule, are shown in Fig. 24–1. The first vibration ν_1 is the totally symmetric

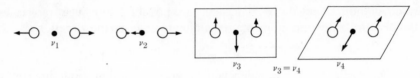

Fig. 24–1 Vibrations of CO_2.

stretching vibration. The second, v_2, is an asymmetric stretching vibration. Since they differ only in that the motion takes place in mutually perpendicular planes, the bending vibrations are *degenerate*, that is, they have the same frequency, $v_3 = v_4$. They are the two perpendicular components of a doubly degenerate bending vibration.

Rotations

For simplicity, we restrict the discussion to linear molecules. The origin of coordinates is fixed at the center of mass of the molecule; if the nuclei lie on the z-axis, then the two independent modes of motion are rotation about the x-axis and about the y-axis. For either mode the moment of inertia is

$$I = \sum_i m_i r_i^2, \tag{24-3}$$

where m_i is the mass of the ith atom, and r_i is its perpendicular distance from the axis of rotation.† Solution of the Schrödinger equation for this motion shows that the angular momentum M is quantized through the relation

$$M^2 = J(J + 1)\hbar^2, \qquad J = 0, 1, 2, \ldots, \tag{24-4}$$

where J, the rotational quantum number, may have any positive integral value or may be zero. The rotational energy is

$$E_J = \frac{M^2}{2I} = J(J + 1)\frac{\hbar^2}{2I}. \tag{24-5}$$

In Eq. (24–5) it is interesting to note that when $J = 0$, $E_0 = 0$. This was not possible for the harmonic oscillator, where if E_0 were zero, the atoms would have been fixed in definite positions in space, an impossibility according to the uncertainty principle. For the rotator, the momentum in the lowest state is definite, equal to zero, but the orientation of the rotator in space is completely undetermined, quite in accord with the uncertainty principle.

24–3 The Rotational Spectrum

If a molecule changes from one rotational state to another, the energy difference between the two states is made up by the emission or absorption of a quantum of

---

†   Since the masses are located on the z-axis, the moment of inertia about this axis is zero; all the $r_i = 0$. The mass of the electrons is negligible in this problem.

radiation. For transitions between rotational states of linear molecules, the selection rule requires that $\Delta J = \pm 1$. The energy difference between these neighboring states is

$$E_{J+1} - E_J = [(J+1)(J+2) - J(J+1)]\frac{\hbar^2}{2I} = (J+1)\frac{\hbar^2}{I}.$$

The frequency v_J associated with this transition is determined by $hv_J = E_{J+1} - E_J$; since $\hbar = h/2\pi$, we obtain

$$v_J = (J+1)\frac{h}{4\pi^2 I}. \tag{24-6}$$

For each value of J there is a line of frequency v_J in the rotational spectrum. These lines are in the far infrared and microwave region of the spectrum, an experimentally difficult region.

The spacing between the lines is $v_{J+1} - v_J = h/4\pi^2 I$. Therefore, from the measured spacing between the rotational lines the moment of inertia of the molecule can be determined. For a diatomic molecule the interatomic distance can be computed immediately from the value of the moment of inertia. For any molecule the rotational spectrum yields important information about the interatomic distances in the molecule.

The principles underlying the rotational spectra of polyatomic, nonlinear molecules are much the same as those for diatomic molecules. The selection rules for nonlinear molecules are somewhat less restrictive. The information obtained is the moment of inertia from which the interatomic distances can be calculated if the molecule is not too complicated.

Only molecules with a permanent dipole moment have a pure rotational spectrum; we deal with this restriction in somewhat greater detail later.

24-4 The Vibration–Rotation Spectrum

Molecules do not have a pure vibrational spectrum since the selection rules require a change in the vibrational state of the molecule to be accompanied by a change in the rotational state as well. As a result, in the infrared region of the spectrum there are vibration–rotation *bands*; each band consists of a number of closely spaced lines. The appearance of a band can be simply interpreted by supposing that the vibrational and rotational energies of the molecule are additive. For simplicity we consider a linear molecule; the energy is

$$E_{\text{vib-rot}} = (n + \tfrac{1}{2})hv_0 + J(J+1)hB'_e, \tag{24-7}$$

where $B = h/8\pi^2 I$. In the transition from the state with energy E' to that with energy E,

$$\Delta E = (E' - E)_{\text{vib-rot}} = (n' - n)hv_0 + [J'(J'+1) - J(J+1)]hB'_e.$$

The selection rule for vibration is $\Delta n = \pm 1$; since the frequency emitted is $v = \Delta E/h$, we have

$$v = v_0 + [J'(J' + 1) - J(J + 1)B'_e.$$

The selection rule for the rotational quantum number requires that either $J' = J + 1$ or $J' = J - 1$. Thus we obtain two sets of values for the frequency, designated by v_R and v_P.

$$\text{If } J' = J + 1: \qquad v_R = v_0 + 2(J + 1)B'_e. \qquad J = 0, 1, 2, \ldots.$$

$$\text{If } J' = J - 1: \qquad v_P = v_0 - 2JB'_e, \qquad J = 1, 2, 3, \ldots.$$

These formulas can be simplified by writing both in the form

$$\left. \begin{array}{l} v_R = v_0 + 2JB'_e \\ v_P = v_0 - 2JB'_e \end{array} \right\}, \qquad J = 1, 2, 3, \ldots,$$

and excluding the value $J = 0$. The vibration–rotation band is made up of two sets of lines, the P and the R branch. Since J may not be zero, the fundamental vibration frequency v_0 does not appear in the spectrum. The lines in the band appear on each side of v_0.

In addition to the selection rules imposed on the changes in the quantum numbers n and J, the presence or absence of a dipole moment in the molecule imposes a restriction on the appearance of lines and bands in the spectrum. If the transition between one vibrational or rotational state to another is to produce emission or absorption of radiation the vibration or rotation must be accompanied by an oscillation in the magnitude of the *dipole moment* of the molecule.

An electrical dipole consists of a positive and a negative charge, $+\epsilon$ and $-\epsilon$, separated by a distance r:

$$\overset{\longleftarrow \; r}{+\epsilon \quad -\epsilon}.$$

The dipole moment μ is defined by

$$\mu = \epsilon r, \tag{24–8}$$

and is a vector quantity; the direction is indicated by an arrow drawn from the negative to the positive charge. If the centers of positive and negative charge in a molecule do not coincide, the molecule has a permanent dipole moment.

Symmetrical (homonuclear) diatomic molecules such as H_2, O_2, N_2 do not have a permanent dipole moment, since an asymmetry in the electrical charge distribution is not possible. The symmetrical vibration does not alter the dipole moment, so these molecules do not emit or absorb in the infrared; the vibration is said to be *forbidden* in the infrared.

In a heteronuclear molecule such as HCl, the centers of positive and negative charge do not coincide, and the molecule has a permanent dipole moment. As this molecule vibrates, the displacement of the centers of charge varies and the magnitude of the dipole moment changes. The corresponding vibration–rotation band appears in the infrared. Rotation of the HCl molecule will produce an oscillation of the component of the dipole moment along a specified axis; hence, HCl has a pure rotational spectrum in the far infrared.

The vibration–rotation band for a molecule such as HCl is shown schematically in Fig. 24–2(a). There is no absorption at the fundamental frequency v_0. The spacing between the lines $\Delta v = v_{J+1} - v_J = B'_e$. Since B'_e contains the moment of inertia, measurement of the spacing yields a value of I immediately. For a diatomic molecule, $I = m_1 m_2 r^2/(m_1 + m_2)$, where m_1 and m_2 are the masses of the atoms and r is the internuclear distance. From the line spacing the internuclear distance is easily calculated.

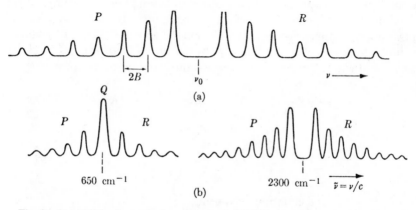

Fig. 24–2 (a) Vibration–rotation spectrum of HCl (schematic). (b) Vibration–rotation spectrum of CO_2 (schematic).

The CO_2 molecule has two principal absorption bands, shown schematically in Fig. 24–2(b). The totally symmetric vibration does not produce a band, since it does not produce an oscillation in the dipole moment. The antisymmetric stretching vibration v_2, at 2300 cm^{-1}, produces an oscillating dipole moment and therefore has a band with P and R branches much like the band for HCl. The band at 650 cm^{-1} corresponds to the degenerate bending vibration $v_3 = v_4$. This band has, in addition to the P and R branches, a Q branch at the fundamental frequency v_3. The fact that in the bending vibration the oscillating dipole moment is perpendicular to the molecular axis allows the fundamental frequency to appear. The presence of the Q branch in the band identifies this band as that belonging to the bending vibration. Many more lines appear in the bands than are indicated in these figures.

The analysis of the spectra of more complicated molecules is more difficult but can be done with profit.

24-5 Detailed Derivation of the Energy Levels of the Diatomic Molecule

In Section 22–1 we developed the Schrödinger equation for a system composed of two bodies having masses m_1 and m_2 with a force acting along the line of centers. If we assume that a Hooke's law force acts along the line of centers between the two masses, then the potential energy is $V(r) = \frac{1}{2}k(r - r_e)^2$ in which r is the distance between the two masses and r_e is the distance between them where the potential energy is a minimum. Then Eq. (22–2) takes the form

$$-\frac{\hbar^2}{2\mu r^2}\frac{d}{dr}\left(r^2\frac{dR}{dr}\right) + \left[\tfrac{1}{2}k(r - r_e)^2 + \frac{J(J + 1)\hbar^2}{2\mu r^2}\right]R = ER, \qquad (24\text{–}9)$$

and the system is a reasonable model of a rotating, vibrating diatomic molecule. Note that in Eq. (24–9) the energy is made up of the energy of vibration, described by the first two terms on the left side of the equation, and the energy of rotation described by the last term on the left of the equation. To remove the garbage we again transform to dimensionless variables $r = \beta\xi$ and $E = \epsilon(\hbar^2/2\mu\beta^2)$, where ξ and ϵ are dimensionless. We determine β by requiring that $\hbar^2/2\mu\beta^2 = \frac{1}{2}k\beta^2$; then $E = \epsilon(\frac{1}{2}h\nu)$, and $\nu = (1/2\pi)\sqrt{k/\mu}$. This brings the equation to the form

$$-\frac{1}{\xi^2}\frac{d}{d\xi}\left(\xi^2\frac{dR}{d\xi}\right) + \left[(\xi - \xi_e)^2 + \frac{J(J + 1)}{\xi^2}\right]R = \epsilon R.$$

Placing $R(\xi) = S(\xi)/\xi$, yields

$$-\frac{d^2S}{d\xi^2} + \left[(\xi - \xi_e)^2 + \frac{J(J + 1)}{\xi^2}\right]S = \epsilon S.$$

If we let $x = \xi - \xi_e$, the equation becomes

$$-\frac{d^2S}{dx^2} + x^2S + \frac{J(J + 1)}{(\xi_e + x)^2}S = \epsilon S. \qquad (24\text{–}10)$$

At this point we observe that if there is no rotation, $J = 0$ and the differential equation has the same form as Eq. (21–39) for the harmonic oscillator.† The term $J(J + 1)/(\xi_e + x)^2$ represents the influence of the rotation on the vibration; the centrifugal force tends to stretch the oscillator, effectively increasing its moment of inertia. It is useful to expand this rotational term in powers of x/ξ_e;

$$\frac{J(J + 1)}{(\xi_e + x)^2} = \frac{J(J + 1)}{\xi_e^2[1 + (x/\xi_e)]^2}$$

$$= \frac{J(J + 1)}{\xi_e^2[1 + (2x/\xi_e) + (x^2/\xi_e^2)]} = \frac{J(J + 1)}{\xi_e^2}\left[1 - \frac{2x}{\xi_e} + \frac{3x^2}{\xi_e^2} + \cdots\right].$$

---

† For the harmonic oscillator, the domain of x was $-\infty < x < +\infty$. In this case the least value of r (or ξ) is zero, hence the least value of x is $-\xi_e$. For a typical diatomic molecule, $-\xi_e \approx -10$; for all practical purposes the value of -10 is indistinguishable from $-\infty$.

At least in the lower vibrational states, $(x/\xi_e) \ll 1$, so that we may arrange Eq. (24–10) into the form

$$-\frac{d^2S}{dx^2} + x^2S + \frac{J(J + 1)}{\xi_e^2}\left(-\frac{2x}{\xi_e} + \frac{3x^2}{\xi_e^2} + \cdots\right)S = \left[\epsilon - \frac{J(J + 1)}{\xi_e^2}\right]S. \qquad (24–11)$$

Since the terms x/ξ_e and x^2/ξ_e^2 are small, we can regard them as a perturbation and write the equation in the form

$$H^0S + H^{(1)}S + H^{(2)}S = \epsilon'S, \qquad (24–12)$$

in which

$$H^0 = -\frac{d^2}{dx^2} + x^2; \qquad H^{(1)} = -\frac{2x}{\xi_e}\frac{J(J + 1)}{\xi_e^2};$$

$$H^{(2)} = \frac{3x^2}{\xi_e^2}\frac{J(J + 1)}{\xi_e^2}; \qquad \epsilon' = \epsilon - \frac{J(J + 1)}{\xi_e^2}.$$

Now we need only apply the perturbation formulas.

The unperturbed wave function is given by

$$H^0S^0 = \epsilon'^{(0)}S^0.$$

This is the equation for the simple harmonic oscillator; according to Section (21–7), the wave functions are

$$S^0 = A_nH_n(x)e^{-x^2/2}, \qquad (24–13)$$

the parameter $\epsilon'^{(0)}$ is

$$\epsilon'^{(0)} = 2n + 1, \qquad n = 0, 1, 2, \ldots.$$

Since the energy is $\epsilon = \epsilon' + J(J + 1)/\xi_e^2$, in this approximation the energy ϵ^0 is

$$\epsilon^0 = 2n + 1 + \frac{J(J + 1)}{\xi_e^2}. \qquad (24–14)$$

This equation says that as a first approximation, the energy of a diatomic molecule is simply the sum of the energy due to vibration represented by the terms $2n + 1$, plus that due to rotation, $J(J + 1)/\xi_e^2$, with a moment of inertia $I_e = \mu r_e^2$. If we return all of the constants to the energy equation we obtain the more familiar form:

$$E^0 = (n + \tfrac{1}{2})h\nu + \frac{J(J + 1)\hbar^2}{2I_e} = (n + \tfrac{1}{2})h\nu + J(J + 1)B'_e \qquad (24–15)$$

in which $B'_e = \hbar^2/2I_e$; B'_e is called the rotational constant. This is the relation we used for the energy value in Section 24–4.

To evaluate the effect of the perturbation, $H^{(1)}$ and $H^{(2)}$, we use Eqs. (20–46), (20–53), (20–57) to obtain

$$\epsilon' = \epsilon'^{(0)} + \epsilon'^{(1)} + \epsilon'^{(2)},$$

where

$$\epsilon'^{(1)} = H_{nn}^{(1)}$$

and

$$\epsilon'^{(2)} = \sum_{k \neq n} \frac{H_{nk}^{(1)} H_{kn}^{(1)}}{\epsilon_n'^{(0)} - \epsilon_k'^{(0)}} + H_{nn}^{(2)}.$$

Since

$$H_{nn}^{(1)} = \int_{-\infty}^{\infty} S_n^0 \left(-\frac{2x}{\zeta_e^3}\right) J(J+1) S_n^0 \, dx = -\frac{2}{\zeta_e^3} J(J+1) \int_{-\infty}^{\infty} x(S_n^0)^2 \, dx,$$

we observe that the integrand is an odd function of x; hence the integral must vanish; $H_{nn}^{(1)} = \epsilon'^{(1)} = 0$. There is no first-order effect on the energy.

To obtain the second-order effects we evaluate

$$H_{kn}^{(1)} = H_{nk}^{(1)} = \int_{-\infty}^{\infty} S_n^0 \left(-\frac{2x}{\zeta_e^3}\right) J(J+1) S_k^0 \, dx$$

$$= -\frac{2}{\zeta_e^3} J(J+1) A_n A_k \int_{-\infty}^{\infty} H_n(x) x H_k(x) e^{-x^2} \, dx.$$

According to Eq. (21–49), $xH_n(x) = nH_{n-1}(x) + \frac{1}{2}H_{n+1}$, so that

$$H_{nk}^{(1)} = -\frac{2}{\zeta_e^3} J(J+1) A_n A_k \int_{-\infty}^{\infty} [nH_{n-1}(x)H_k(x) + \frac{1}{2}H_{n+1}(x)H_k(x)] e^{-x^2} \, dx.$$

Using Eq. (21–47),

$$H_{nk}^{(1)} = -\frac{2}{\zeta_e^3} J(J+1) A_n A_k \left[n \left(\frac{\delta_{n-1,k}}{A_{n-1} A_k} \right) + \frac{1}{2} \left(\frac{\delta_{n+1,k}}{A_{n+1} A_k} \right) \right]$$

$$= -\frac{2}{\zeta_e^3} J(J+1) \left[\frac{nA_n}{A_{n-1}} \delta_{n-1,k} + \frac{1}{2} \frac{A_n}{A_{n+1}} \delta_{n+1,k} \right]$$

$$= -\frac{2}{\zeta_e^3} J(J+1) \left(\sqrt{\frac{n}{2}} \delta_{n-1,k} + \sqrt{\frac{n+1}{2}} \delta_{n+1,k} \right).$$

In the last expression the values of A_n, etc., have been inserted from Eq. (21–43). Then since $\epsilon_n'^{(0)} - \epsilon_k'^{(0)} = 2n + 1 - (2k + 1) = 2(n - k)$, we have

$$\sum_{k \neq n} \frac{H_{nk}^{(1)} H_{kn}^{(1)}}{\epsilon_n'^{(0)} - \epsilon_k'^{(0)}} = \sum_{k \neq n} \frac{(H_{nk}^{(1)})^2}{2(n - k)}$$

$$= \frac{2J^2(J+1)^2}{\zeta_e^6} \sum_{k \neq n} \frac{\left(\sqrt{\frac{n}{2}} \delta_{n-1,k} + \sqrt{\frac{n+1}{2}} \delta_{n+1,k} \right)^2}{n - k}.$$

In view of the Kronecker delta functions, the terms in the sum over k all vanish except two: the term for $k = n - 1$, when the term becomes

$$\frac{\left(\sqrt{\frac{n}{2}}\right)^2}{n - (n-1)} = \frac{n}{2};$$

and the term for $k = n + 1$, when the term becomes $(\sqrt{(n+1)/2})^2/[n - (n+1)] = -\frac{1}{2}(n+1)$; so we have finally

$$\sum_{k \neq n} \frac{H_{nk}^{(1)^2}}{n - k} = \frac{2J^2(J+1)^2}{\zeta_e^6}\left(\frac{n}{2} - \frac{n+1}{2}\right) = -\frac{1}{\zeta_e^2}\left(\frac{J(J+1)}{\zeta_e^2}\right)^2. \tag{24-16}$$

Next we need the integral,

$$H_{nn}^{(2)} = \int_{-\infty}^{\infty} S_n^0\left(\frac{3x^2}{\zeta_e^4}\right)J(J+1)S_n^0\, dx = \frac{3}{\zeta_e^4}J(J+1)A_n^2\int_{-\infty}^{\infty} H_n(x)x^2 H_n(x)e^{-x^2}\, dx$$

$$= \frac{3A_n^2}{\zeta_e^4}J(J+1)\int_{-\infty}^{\infty} [xH_n(x)]^2 e^{-x^2}\, dx$$

$$= \frac{3A_n^2}{\zeta_e^4}J(J+1)\int_{-\infty}^{\infty} [nH_{n-1}(x) + \tfrac{1}{2}H_{n+1}(x)]^2 e^{-x^2}\, dx$$

$$= \frac{3A_n^2}{\zeta_e^4}J(J+1)\int_{-\infty}^{\infty} \{n^2[H_{n-1}(x)]^2 + nH_{n-1}(x)H_{n+1}(x) + \tfrac{1}{4}[H_{n+1}(x)]^2\}e^{-x^2}\, dx.$$

Since $H_{n-1}(x)$ and $H_{n+1}(x)$ are orthogonal, the integral of the middle term vanishes. So we have

$$H_{nn}^{(2)} = \frac{3A_n^2}{\zeta_e^4}J(J+1)\left(\frac{n^2}{A_{n-1}^2} + \frac{1}{4A_{n+1}^2}\right)$$

$$= \frac{3}{\zeta_e^2}J(J+1)\left(\frac{n}{2} + \frac{n+1}{2}\right) = \frac{3}{\zeta_e^2}(n + \tfrac{1}{2})\left(\frac{J(J+1)}{\zeta_e^2}\right). \tag{24-17}$$

This yields for $\epsilon'^{(2)}$

$$\epsilon'^{(2)} = \left(\frac{J(J+1)}{\zeta_e^2}\right)\left(\frac{3(n+\frac{1}{2})}{\zeta_e^2}\right) - \frac{1}{\zeta_e^2}\left(\frac{J(J+1)}{\zeta_e^2}\right)^2.$$

The energy is then

$$\epsilon = (2n+1) + \frac{J(J+1)}{\zeta_e^2} + \left(\frac{J(J+1)}{\zeta_e^2}\right)\left[\frac{3(n+\frac{1}{2})}{\zeta_e^2} - \frac{1}{\zeta_e^2}\left(\frac{J(J+1)}{\zeta_e^2}\right)\right]. \tag{24-18}$$

Returning to the usual notation,

$$E_{n,J} = (n+\tfrac{1}{2})h\nu + J(J+1)\left[B_e' + \left(\frac{6B_e'^2}{h\nu}\right)(n+\tfrac{1}{2}) - \left(\frac{4B_e'^3}{h^2\nu^2}\right)J(J+1)\right]. \tag{24-19}$$

Comparing this with the simple form

$$E_{n,J} = (n + \tfrac{1}{2})h\nu + J(J + 1)B'_e, \tag{24-20}$$

we see that the effective rotational constant is now

$$B'_{e(\text{effective})} = B'_e + \left(\frac{6B'^{2}_e}{h\nu}\right)(n + \tfrac{1}{2}) - \left(\frac{4B'^{3}_e}{h^2\nu^2}\right)J(J + 1). \tag{24-21}$$

The second term represents the effect of an harmonic oscillation on the moment of inertia while the third term shows the effect of the rotation on the moment of inertia. Since $B'_e = \hbar^2/2I_e$, the harmonic vibration decreases I_e while the rotation increases I_e. At this point it should be noted that in actual molecules the effects of anharmonicity are sufficiently large to change the sign of the second term in Eq. (24–21).

24–6 The Electronic Spectrum

The light absorbed or emitted in the transition from one electronic state to another constitutes the electronic spectrum. The existence of vibrations and rotations in the molecule introduces a multiplicity of lines. In atoms the excitation of an electron from one state to another produces a single line or a set of two or three closely spaced lines. In molecules, a rather large number of closely spaced lines is produced, since in the electronic transition the vibrational quantum state and the rotational quantum state may change at the same time. Furthermore, the selection rule for vibration is swept away; as a result, there is no restriction on the change in the vibrational quantum number in an electronic transition. This fact increases the number of lines in the spectrum.

Consider two electronic states of a diatomic molecule between which a transition is permitted. The curves 1 and 2, Fig. 24–3, show the energy variation with the internuclear molecule in these states. In addition to the electronic energy, the molecule may have vibrational energy. The horizontal lines indicate the total energy, electronic plus vibrational. In electronic state 1, a molecule having a vibrational quantum number $n = 0$ has an energy given by the line aa', if $n = 1$, the energy is given by bb', and so on.

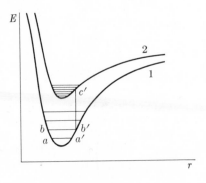

Fig. 24–3 Electronic transitions.

If $n = 1$, the vibration alters the internuclear distance from the value at b, to the value at point b'. Consider a jump from the level $n = 1$ in state 1 to state 2. This jump will most likely occur when r has either of the values b or b'. These are the points at which the relative velocity of the two atoms is zero; hence the molecule spends more time near these points. An electronic transition occurs very quickly in comparison to the time required for the nuclei to move appreciably, so that the transition takes place "vertically" from b' to the upper curve at point c'; this is the Franck-Condon principle. The new vibrational state of the molecule must be the one which is nearest the point c'. Counting levels, it appears that the excited molecule will be in the fourth or fifth vibrational level of state 2. The frequency of the quantum absorbed corresponds to the energy difference represented by the vertical connecting line. The rotational levels which are very closely spaced around each vibrational level are not shown in the diagram.

For a specified electronic and vibrational transition there are several rotational levels corresponding to different rotational states between which transitions are allowed. As a result several frequencies are absorbed. This group of closely spaced lines is called an *electronic band*. In any collection of molecules, there are some in various vibrational states; hence, transitions may occur between any vibrational state in the lower electronic state to whatever level matches properly in the upper state. Each one of these possible transitions produces a band; the collection of the bands is called a *band system*. Add to this fact that there are several electronic states, not just two, between which transitions may occur. The result is that the electronic spectrum of a molecule consists of *several* band systems. By detailed analysis of the wavelengths of the lines in the spectrum, one can calculate the energy curves for the various electronic states as functions of r, the internuclear distances, the force constants of the vibrations, and bond energies.

The important point about the electronic spectrum is that one gains information about any vibration in the molecule, even though the fundamental frequency of that vibration might not appear in the infrared spectrum. For example, the vibrational frequency of homonuclear diatomic molecules does not appear in the infrared, since the oscillating dipole moment is absent; the bands resulting from that vibration do show up in the electronic spectrum.

The energies involved in electronic transitions are comparatively large, so the bands and band systems appear in the visible and ultraviolet region of the spectrum; e.g., the violet color of iodine is due to an electronic transition.

24–7   The Raman Effect

If a substance is irradiated with monochromatic light, the substance absorbs the light if the quantum of energy $h\nu$ fits the gap between two energy levels of the molecule. If the quantum does not fit, the molecule cannot absorb it and the light passes through; the substance is transparent to light of that wavelength. The usual rule is that the quantum must fit. In spite of this rule, there is a small but finite probability of the molecule absorbing part of the energy of the quantum and releasing the remaining energy as light of a different frequency; Raman effect. Ordinarily the energy absorbed fits the energy gap between two vibrational states or two rotational states of the molecule. The difference between the frequency of the incident light and that of the scattered light therefore corresponds to a rotational or vibrational frequency.

Experimentally, a beam of light of frequency v passes through the substance. The scattered light is observed at a 90° angle from the direction of the incident beam. In addition to the incident line, very faint lines of slightly different frequency are observed; these lines are displaced on either side of the frequency of the incident light. Since the probability of this type of event is very small, the lines are very faint.

Observation of the Raman effect is advantageous in that vibrational and rotational energies can be measured in a convenient region of the spectrum, e.g., the visible region.

24–8 Quantum Mechanical Description of Time-dependent Systems

Heretofore in our quantum mechanical discussions we have described the stationary or time-independent states of a system. Furthermore, our language has been such as to imply that we had sufficient information about the system to know that it was in a particular state described by a particular set of quantum numbers. For example, in the case of the harmonic oscillator we spoke as though we knew that the oscillator was in the nth state with wave function ψ_n, and energy $E_n = (n + \frac{1}{2})hv$; or, in the case of the hydrogen atom, that it was in a state described by the set of numbers n, l, m. This approach is very useful in a first discussion of quantum mechanical properties of various kinds of systems. However, we do not have reason to presuppose that a system is in a particular quantum state.

Having obtained the set of particular solutions of the Schrödinger equation, the set of ψ_n, the general solution is a linear combination of the ψ_n; namely,

$$\psi = \sum_n a_n \psi_n. \tag{24–22}$$

The normalization requirement is

$$\int \psi^* \psi \, d\tau = 1. \tag{24–23}$$

This is the total probability of finding the system in some state. If we use Eq. (24–22) in Eq. (24–23), we have

$$1 = \int \psi^* \psi \, d\tau = \int \left(\sum_m a_m^* \psi_m^* \right) \left(\sum_n a_n \psi_n \right) d\tau$$

$$= \sum_m \sum_n a_m^* a_n \int \psi_m^* \psi_n \, d\tau.$$

Since the particular solutions are orthonormal we have

$$\int \psi_m^* \psi_n \, d\tau = \delta_{mn}, \tag{24-24}$$

so that

$$1 = \sum_m \sum_n a_m^* a_n \delta_{mn}.$$

Performing the summation over m yields

$$1 = \sum_n a_n^* a_n = \sum_n |a_n|^2. \tag{24-25}$$

Equation (24-25) says that the sum of the squares of the absolute value of the coefficients a_n in the series in Eq. (24-22) is unity. The manner in which we obtained Eq. (24-25) requires that we interpret the right-hand side as a sum of probabilities. Therefore, we interpret $|a_n|^2$ as the probability of finding the system in the state described by ψ_n. According to Eq. (24-25), the total probability of finding the system in one or another of the eigenstates is unity.

For example, suppose $|a_k|^2 = 1$, and $|a_n|^2 = 0$ for all $n \neq k$. Then the system is in state k and is described as completely as possible by ψ_k. This is the type of terminology, referred to above, which we have used so far. On the other hand, if we envision transitions occurring which take the system from one state to another, this implies that the a_n are functions of time, $a_n(t)$. Suppose that at $t = 0$ the system is in state k, while at some later time, t, there is a probability that the system is in state m, as a result of transitions occurring in the time interval t. This situation could be described by writing

$$|a_k(0)|^2 = 1 \qquad\qquad |a_k(t)|^2 < 1$$
$$|a_m(0)|^2 = 0 \qquad (m \neq k) \qquad |a_m(t)|^2 > 0.$$

Our problem is to discover how these coefficients, $a_n(t)$, depend on the time. Since one way for transitions to occur is by absorption or emission of radiation, this problem is central to any discussion of spectroscopy.

24-9 Interaction of Charged Particles with an Electrical Field

Since the largest interaction between a light wave and a molecule is a result of the effect of the oscillating electric field of the light wave on the electric charges in the molecule, we begin by discussing the energy of this interaction.

By definition, the electrical field, $\mathbf{E}$, is the force acting on a unit charge; the force acting on the jth particle having a charge ϵ_j is $\epsilon_j \mathbf{E}$; the work done by the field on the charged particle as it moves through a distance $d\mathbf{s}_j$ is $\epsilon_j \mathbf{E} \cdot d\mathbf{s}_j$. We recognize that both the field, $\mathbf{E}$, and the displacement, $d\mathbf{s}_j$, are vectors. If the field acts in the x-direction with a value E_x, and the displacement of the jth particle is dx_j, then we have for the interaction energy

$$\epsilon_j E_x \, dx_j.$$

In a uniform field, E_x is independent of x, so that the energy of a particle relative to its energy at the origin is $\epsilon_j E_x x_j$. Similarly, if the field has y- and z-components, E_y and E_z, these will contribute energy terms $\epsilon_j E_y y_j$ and $\epsilon_j E_z z_j$, to the total energy which will then have the form

$$\epsilon_j x_j E_x + \epsilon_j y_j E_y + \epsilon_j z_j E_z.$$

If there are several charged particles in our system, then we sum this expression over the various particles to obtain the total interaction energy, H'.

$$H' = \left(\sum_j \epsilon_j x_j \right) E_x + \left(\sum_j \epsilon_j y_j \right) E_y + \left(\sum_j \epsilon_j z_j \right) E_z. \tag{24–26}$$

The quantity $\sum_j \epsilon_j x_j$ defines the x-component of the dipole moment of the array of charges; so we have

$$\mu_x \equiv \sum_j \epsilon_j x_j, \qquad \mu_y \equiv \sum_j \epsilon_j y_j, \qquad \mu_z \equiv \sum_j \epsilon_j z_j, \tag{24–27}$$

where μ_x, μ_y, and μ_z are the x-, y-, and z-components of the dipole moment of the system. If we apply this definition of the dipole moment to the classic dipole array of charges on the x-axis, $+\epsilon$ at $x = +a/2$ and $-\epsilon$ at $x = -a/2$, then $\mu_y = \mu_z = 0$ and $\mu_x = (+\epsilon)(+a/2) + (-\epsilon)(-a/2) = \epsilon a$, which is the familiar formula, charge times separation, for the moment of a simple dipole. The expression for the total energy of interaction of the system with the electrical field can be written

$$H' = \mu_x E_x + \mu_y E_y + \mu_z E_z. \tag{24–28}$$

It is this expression, Eq. (24–28), which we shall use for the discussion of the influence of a light wave on a molecular system.

In the case of a light wave, the field is time-dependent so we would write, for example, $E_x = E_x^0 \cos \omega t$, where ω is the circular frequency, $\omega = 2\pi\nu$ (ν being the ordinary frequency of the light wave), and E_x^0 is the amplitude of the electric field vector of the wave. Strictly speaking, E_x^0 depends on the coordinates in a progressive wave; however, the wavelengths of interest are so large (> 1000 Å) compared with the size of the molecule (an Ångstrom or two), that we may take E_x^0 as a constant over the length of the molecule. Since E_y and E_z will also depend on $\cos \omega t$, we can write the energy of interaction in Eq. (24–28) in the form

$$H' = (\mu_x E_x^0 + \mu_y E_y^0 + \mu_z E_z^0) \cos \omega t, \tag{24–29}$$

to show explicitly how the energy of a system of charges in the presence of a light wave depends on time.

It is useful at this point to introduce some additional relations appropriate to an alternating electrical field. From the relation

$$E_x = E_x^0 \cos \omega t,$$

it is clear that the average value of E_x over the time of one oscillation, $(2\pi/\omega = 1/v)$, is zero $(\overline{E_x} = 0)$, since the cosine is positive over half the period and negative over the other half. The average value of $\overline{E_x^2}$ over a period is

$$\overline{E_x^2} = \frac{1}{(2\pi/\omega)} \int_0^{2\pi/\omega} E_x^2 \, dt = \frac{E_x^{0\,2}}{(2\pi/\omega)} \int_0^{2\pi/\omega} \cos^2 \omega t \, dt.$$

The value of the integral is (π/ω); so we have

$$\overline{E_x^2} = \tfrac{1}{2}E_x^{0\,2}, \qquad \overline{E_y^2} = \tfrac{1}{2}E_y^{0\,2}, \qquad \overline{E_z^2} = \tfrac{1}{2}E_z^{0\,2}, \tag{24–30}$$

where we have included the corresponding results for the y- and z-components.

Furthermore, as is correct for any vector quantity, the square of the magnitude of the field vector is equal to the sum of squares of its components;

$$E^2 = E_x^2 + E_y^2 = E_z^2 ; \tag{24–31}$$

averaging yields

$$\overline{E^2} = \overline{E_x^2} + \overline{E_y^2} = \overline{E_z^2} = \tfrac{1}{2}(E_x^{0\,2} + E_y^{0\,2} + E_z^{0\,2}). \tag{24–32}$$

If the radiation is isotropic, that is, if it is not polarized in a particular direction, we have equal amplitudes in each direction. Then, $E_x^0 = E_y^0 = E_z^0$. Combining with Eq. (24–32),

$$\overline{E_x^2} = \overline{E_y^2} = \overline{E_z^2} = \tfrac{1}{3}\overline{E^2} \tag{24–33}$$

and

$$E_x^{0\,2} = E_y^{0\,2} = E_z^{0\,2} = \tfrac{2}{3}\overline{E^2}. \tag{24–34}$$

These relations apply to the individual frequencies in a light beam which may contain a distribution of frequencies. To deal with this complication we need only recognize that the amplitudes are functions of frequency and write, for example, $E_x^0(v)$ and $\overline{E^2}(v)$. The radiation density, $\rho(v)$, in a light beam is defined by the relation

$$\overline{E^2}(v) = 4\pi\rho(v)\,dv. \tag{24–35}$$

Thus, $\rho(v)$ is the energy per unit volume, per unit frequency interval, due to the light beam; or $\rho(v)\,dv$ is the energy per unit volume due to the radiation with frequencies between v and $v + dv$.

24–10   Variation in the State of a System with Time

To discuss the variation in the state of a system with time, we use the time-dependent Schrödinger equation,

$$\mathbf{H}\Psi(q, t) = i\hbar \frac{\partial \Psi(q, t)}{\partial t} \tag{24–36}$$

in which q symbolizes the set of coordinates appropriate to the system and $\mathbf{H}$ is the complete Hamiltonian operator. In the presence of a beam of light, the Hamiltonian takes the form

$$\mathbf{H} = \mathbf{H}^0 + \mathbf{H}'$$

in which $\mathbf{H}'$ represents the energy of interaction of the system with the light beam; $\mathbf{H}^0$ is the Hamiltonian of the system in the absence of the light beam. Only $\mathbf{H}'$ depends on the time, through an equation such as Eq. (24–29). Since $\mathbf{H}'$ is small it seems reasonable to solve the problem in the absence of the light beam first. In all that follows we will restrict ourselves to systems with a discrete set of nondegenerate energy levels. This simplifies the mathematical treatment substantially.

In the absence of the light beam, $\mathbf{H}' = 0$, and we have

$$\mathbf{H}^0\Psi_k^0(q, t) = i\hbar\frac{\partial\Psi_k^0(q, t)}{\partial t}. \tag{24–37}$$

As we have seen, Eqs. (20–26) and (20–28), the solution to this equation has the form

$$\Psi_k^0(q, t) = \psi_k^0(q)e^{-iE_k t/\hbar}, \tag{24–38}$$

where $\psi_k^0(q)$ is a function only of the coordinates q, and we recognize by using the subscript k that the system may exist in any of a number of quantum states. The energy of the kth quantum state is E_k; the function $\psi_k^0(q)$ satisfies the time-independent Schrödinger equation,

$$\mathbf{H}^0\psi_k^0(q) = E_k\psi_k^0(q). \tag{24–39}$$

Under the influence of a light beam, the state of a system will change with time. We assume in this situation that the wave function describing the system can be represented by a superposition of the unperturbed wave functions $\Psi_k^0(q, t)$:

$$\Psi(q, t) = \sum_k a_k(t)\Psi_k^0(q, t), \tag{24–40}$$

in which the coefficients, $a_k(t)$, depend on time. The wave function, $\Psi(q, t)$, is a solution of the complete equation, Eq. (24–36), so we have

$$(\mathbf{H}^0 + \mathbf{H}')\Psi(q, t) = i\hbar\frac{\partial\Psi(q, t)}{\partial t}.$$

Using the value of $\Psi(q, t)$ from Eq. (24–40), this becomes

$$\sum_k a_k\mathbf{H}^0\Psi_k^0 + \sum_k a_k\mathbf{H}'\Psi_k^0 = i\hbar\sum_k\left[a_k\frac{\partial\Psi_k^0}{\partial t} + \frac{da_k}{dt}\Psi_k^0\right].$$

Replacing $\mathbf{H}^0\Psi_k^0$ in the first series on the left by its value from Eq. (24–37), we obtain

$$i\hbar\sum_k a_k\frac{\partial\Psi_k^0}{\partial t} + \sum_k a_k\mathbf{H}'\Psi_k^0 = i\hbar\sum_k a_k\frac{\partial\Psi_k^0}{\partial t} + i\hbar\sum_k\frac{da_k}{dt}\Psi_k^0,$$

which reduces immediately to

$$\sum_k a_k \mathbf{H}' \Psi_k^0 = i\hbar \sum_k \frac{da_k}{dt} \Psi_k^0.$$

We multiply this equation by $\Psi_m^{0}*$ and by $d\tau$, then integrate over the entire coordinate space;

$$\sum_k a_k \int \Psi_m^{0}* \mathbf{H}' \Psi_k^0 \, d\tau = i\hbar \sum_k \frac{da_k}{dt} \int \Psi_m^{0}* \Psi_k^0 \, d\tau.$$

However, the wave functions Ψ_k^0 are orthonormal, so that

$$\int \Psi_m^{0}* \Psi_k^0 \, d\tau = \delta_{mk}.$$

This brings the equation to the form

$$\sum_k a_k \int \Psi_m^{0}* \mathbf{H}' \Psi_k^0 \, d\tau = i\hbar \sum_k \frac{da_k}{dt} \delta_{mk}.$$

However, because of the factor δ_{mk}, all of the terms in the sum on the right are zero except for the one term in which $k = m$; the equation reduces to

$$\sum_k a_k \int \Psi_m^{0}* \mathbf{H}' \Psi_k^0 \, d\tau = i\hbar \frac{da_m}{dt}.$$

Rearranging, keeping in mind that $1/i = -i$, we obtain

$$\frac{da_m}{dt} = -\frac{i}{\hbar} \sum_k a_k \int \Psi_m^{0}* \mathbf{H}' \Psi_k^0 \, d\tau. \tag{24-41}$$

This equation tells us how the coefficient a_m depends on the time by giving an expression for the derivative (da_m/dt). From this equation we can obtain an expression for $a_m(t)$. This expression is the basis for the so-called *selection rules* in spectra. Since most of the integrals in the series on the right are zero, only certain special states can feed into the mth state to make a_m change with time.

According to Eq. (24-29), the first term in H' is proportional to μ_x. Since the integration in Eq. (24-41) is over coordinates only, the time-dependent factors in H' and in the wave functions can be removed from under the integral. Thus we see that the integrals in Eq. (24-41) are proportional to integrals of the type

$$(\mu_x)_{mn} = \int \psi_m^{0}* \mu_x \psi_n^0 \, d\tau,$$

$$(\mu_y)_{mn} = \int \psi_m^{0}* \mu_y \psi_n^0 \, d\tau, \tag{24-42}$$

$$(\mu_z)_{mn} = \int \psi_m^{0}* \mu_z \psi_n^0 \, d\tau.$$

These are called the transition-moment integrals. We define the quantity μ_{mn}^2 by

$$\mu_{mn}^2 = |(\mu_x)_{mn}|^2 + |(\mu_y)_{mn}|^2 + |(\mu_z)_{mn}|^2. \tag{24-43}$$

If we solve Eq. (24–41) under the condition that at $t = 0$ the system was in state n, then we finally obtain the result (see Appendix 1 for details) that

$$|a_m(t)|^2 = \frac{2\pi\rho(v_{mn})\mu_{mn}^2 t}{3\hbar^2} \tag{24-44}$$

in which $\rho(v_{mn})$ is the radiation density at the frequency v_{mn} defined by $hv_{mn} = |E_m - E_n|$.

All of the preceding mathematical labor is dedicated to the achievement of the fundamental result in Eq. (24–44). This equation is crucial to the understanding of spectroscopy.

For a system at which time $t = 0$ was in state n, in the presence of a light beam a short time t later, the probability of finding the system in state m is given by the value of $|a_m(t)|^2$ in Eq. (24–44). This probability is proportional to t. It is also proportional to the energy density, $\rho(v_{mn})$, of the light beam at the frequency which "fits" the transition from n to m; $v_{mn} = |(E_m - E_n)|/h$. If the light beam contains only a minor component at v_{mn}, then $\rho(v_{mn})$ will be small and the probability of transition from n to m will be small. Most importantly, the probability of finding the system in state m is proportional to the square of the absolute value of the *transition moment*, μ_{mn}. This last dependence is the basis for *selection rules* which govern the appearance or nonappearance of lines in a spectrum.

24–11 Selection Rules for the Harmonic Oscillator

Consider a diatomic molecule which has a dipole moment due to effective charges $+\epsilon$ and $-\epsilon$ separated by a distance r; then $\mu = \epsilon r$; we rewrite this as

$$\mu = \epsilon(r_e + r - r_e) = \epsilon r_e + \epsilon(r - r_e)$$

$$\mu = \mu_e + \epsilon(r - r_e) \tag{24-45}$$

where μ_e is the dipole moment which the molecule would have if the charges were at rest at the equilibrium separation r_e, and the term $\epsilon(r - r_e)$ is the variation in the dipole moment due to the variation in r. We may regard this expression as the first two terms of a Taylor series expansion of μ in terms of $r - r_e$; that is,

$$\mu = \mu_e + \left(\frac{d\mu}{dr}\right)_{r_e} (r - r_e).$$

Then the dipole moment integrals have the form (we take r in the direction of the field)

$$\mu_{mn} = \int_{-\infty}^{\infty} \psi_m^{0*} \left[\mu_e + \left(\frac{d\mu}{dr}\right)_{r_e} (r - r_e) \right] \psi_n^0 \, dr. \tag{24-46}$$

The normalized harmonic oscillator functions are given by

$$\psi_n^0 = \left(\frac{1}{\beta\sqrt{\pi}\,2^n n!}\right)^{1/2} H_n(\xi)e^{-\xi^2/2} = \frac{A_n}{\beta^{1/2}} H_n(\xi)e^{-\xi^2/2}, \tag{24-47}$$

in which $\xi = (r - r_e)/\beta$. Then

$$\mu_{mn} = \int_{-\infty}^{\infty} \frac{A_m}{\beta^{1/2}} H_m(\xi)e^{-\xi^2/2}\left[\mu_e + \left(\frac{d\mu}{dr}\right)_{r_e}\beta\xi\right]\frac{A_n}{\beta^{1/2}} H_n(\xi)e^{-\xi^2/2}\beta\,d\xi$$

$$= A_m A_n\left\{\mu_e \int_{-\infty}^{\infty} H_m(\xi)H_n(\xi)e^{-\xi^2}\,d\xi + \left(\frac{d\mu}{dr}\right)_{r_e}\beta\int_{-\infty}^{\infty} H_m(\xi)\xi H_n(\xi)e^{-\xi^2}\,d\xi\right\}.$$

In view of the orthogonality the first integral vanishes unles $m = n$, a case which is specifically excluded in our derivation. So we are left with only the second integral. To evaluate the second integral we use the recurrence relation Eq. (21-49), that is

$$\xi H_n(\xi) = nH_{n-1}(\xi) + \tfrac{1}{2}H_{n+1}(\xi).$$

This brings μ_{mn} to the form

$$\mu_{mn} = A_m A_n\left(\frac{d\mu}{dr}\right)_{r_e}\beta\left[n\int_{-\infty}^{\infty} H_m(\xi)H_{n-1}(\xi)e^{-\xi^2}\,d\xi + \tfrac{1}{2}\int_{-\infty}^{\infty} H_m(\xi)H_{n+1}(\xi)e^{-\xi^2}\,d\xi\right].$$

But

$$\int_{-\infty}^{\infty} H_m(\xi)H_{n-1}(\xi)e^{-\xi^2}\,d\xi = (1/A_{n-1}^2)\delta_{m,n-1},$$

and

$$\int_{-\infty}^{\infty} H_m(\xi)H_{n+1}(\xi)e^{-\xi^2}\,d\xi = (1/A_{n+1}^2)\delta_{m,n+1}.$$

This yields for μ_{mn}

$$\mu_{mn} = \left(\frac{d\mu}{dr}\right)_{r_e}\beta\left[n\frac{A_n}{A_{n-1}}\delta_{m,n-1} + \frac{1}{2}\frac{A_n}{A_{n+1}}\delta_{m,n+1}\right].$$

Using the value of A_n from Eq. (24-47), we find

$$\mu_{mn} = \beta\left(\frac{d\mu}{dr}\right)_{r_e}\left[\sqrt{\frac{n}{2}}\delta_{m,n-1} + \sqrt{\frac{n+1}{2}}\delta_{m,n+1}\right]. \tag{24-48}$$

When $m = n + 1$, the system is making the transition $n \to n + 1$, so the radiation is absorbed;

$$\mu_{n+1,n}^2 = \left(\frac{n+1}{2}\right)\beta^2\left(\frac{d\mu}{dr}\right)_{r_e}^2, \qquad n = 0, 1, 2, \ldots. \tag{24-49}$$

When $m = n - 1$, the final state m is lower in energy than the initial state n, so radiation is emitted (*stimulated emission*). The impinging light wave stimulates an excited molecule to emit radiation. We have

$$\mu_{n-1,n}^2 = \frac{n}{2}\beta^2\left(\frac{d\mu}{dr}\right)_{r_e}^2, \qquad n = 1, 2, \ldots. \tag{24-50}$$

Note that the lowest possible value of n in this last formula is $n = 1$.

The selection rule for the harmonic oscillator, Eq. (24-48), requires that $\Delta n = \pm 1$. Under the influence of a light beam the harmonic oscillator makes transitions only to states immediately above and below its original state. The existence of selection rules simplifies the interpretation of spectra enormously.

The other requirement on the transition is that the derivative $(d\mu/dr)_{r_e}$ be non-vanishing. Whether or not a molecule has a dipole moment is not significant; the dipole moment *must change as the vibration occurs*. For example, the molecule HCl has a permanent dipole moment and as the molecule vibrates this dipole moment varies. The molecule therefore has $(d\mu/dr) \neq 0$ and exhibits a vibrational spectrum. In contrast consider the CO_2 molecule, which has no permanent dipole moment because of the symmetrical distribution of positive and negative charges; the moment remains zero throughout this vibration. Hence, $d\mu/dr = 0$ and this symmetric vibration, Fig. 24-1, the charge symmetry is undisturbed and the dipole moment remains zero throughout this vibration. Hence, $d\mu/dr = 0$ and this vibration does not appear in the spectrum. In the second normal vibration, the symmetry is destroyed. The dipole moment varies during this vibration and therefore this vibration appears in the spectrum. In the remaining two bending vibrations, degenerate since they differ only in the plane in which the vibration occurs, the symmetry is destroyed and the dipole moment varies during the oscillation. These vibrations appear in the spectrum.

This case of the simple harmonic oscillator is not realizable in any natural system. The simplest example of a vibrating dipole is provided by gaseous diatomic molecules. However, the motion of a diatomic molecule always involves rotation, and, depending on the temperature, may or may not involve vibration. The simplest natural example is therefore the rotating dipolar diatomic molecule in its lowest vibrational state; e.g., HCl at a low enough temperature so that only a negligible number of molecules are in excited vibrational levels. Our mathematical model for this case can be the rigid rotator.

24-12 Selection Rules and Symmetry

For a system in state n, Eq. (24-44) shows that the probability of finding it in state m at a later time is proportional to μ_{mn}^2. The transition moment, μ_{mn}, has components which are given by integrals such as

$$(\mu_x)_{mn} = \epsilon \int \psi_m^{0*} x \psi_n^0 \, d\tau, \tag{24-51}$$

where x could represent any one of the coordinates x, y, or z. We can frequently decide from a consideration of the symmetry of the system for which combinations of m and n the integral will be nonvanishing. These combinations of m and n are the "allowed" transitions for the system. If the integral is zero for a particular combination of n and m, the transition has zero probability of occurring; it is a "forbidden" transition. As we have seen in the preceding sections via detailed computations, Eq. (24–51) provides the basis for establishing *selection rules*. The selection rules govern the type of states between which transitions may or may not occur.

From symmetry we can establish a general selection rule with a minimum of computation. To begin, we consider a general wave function, $\psi(x, y, z)$, for any system together with the integral for the total probability of finding the particle,

$$\int \psi^*(x, y, z)\psi(x, y, z)\, d\tau = 1. \tag{24–52}$$

Since the integral in Eq. (24–52) is a real physical quantity, its value cannot depend on the orientation of the coordinate system. Consequently, the integrand, $\psi^*\psi$, must be invariant under transformations of the coordinate system. This invariance can obtain only if ψ is invariant or merely changes sign under transformations of the coordinate system.

Suppose that we subject the system to the operation of *inversion*, symbolized by the operator **i**, which reverses the direction of all three axes. This operation simply changes the sign of all the coordinates; thus,

$$\mathbf{i}\psi(x, y, z) = \psi(-x, -y, -z).$$

If we operate on the integrand in Eq. (24–52), we obtain

$$\mathbf{i}[\psi^*(x, y, z)\psi(x, y, z)] = [\mathbf{i}\psi^*(x, y, z)][\mathbf{i}\psi(x, y, z)$$

$$= \psi^*(-x, -y, -z)\psi(-x, -y, -z).$$

If the integrand is to be unchanged by this operation, it is clear that the worst that may happen to the wave function is that it changes sign. Thus, we have the two possibilities alluded to above:

$$\mathbf{i}\psi_g(x, y, z) = \psi_g(-x, -y, -z) = \psi_g(x, y, z), \qquad \text{(symmetric)}$$

$$\mathbf{i}\psi_u(x, y, z) = \psi_u(-x, -y, -z) = -\psi_u(x, y, z), \qquad \text{(antisymmetric)}$$

and similarly for the complex conjugate. In the first case, the wave function is said to be symmetric under inversion or is an "even" function; in the second case, the wave function is antisymmetric under inversion or is an "odd" function. The subscripts g and u (from the initial letters of the German: *gerade* = even; *ungerade* = odd) are used to describe these two kinds of wave function.

If the inversion operation is applied to the integrand of the transition-moment integral in Eq. (24–51), we note that the coordinate x (or y or z) is an odd function under inversion; $\mathbf{i}x = -x$. Thus, if $\psi_m^* x \psi_n$ is to be invariant under inversion, the product, $\psi_m^* \psi_n$ must be an odd function. This can only be so if ψ_m is even and ψ_n is odd or vice versa. Thus, we have the important result that transitions are allowed only between odd and even states. Transitions between two odd states or between two even states are forbidden. This is a fundamental selection rule for dipole radiation. If the system contains several particles, the argument is unchanged and leads to the same result.

If we seek to apply this rule in the case of the simple harmonic oscillator, we write the transition moment integral

$$\int_{-\infty}^{\infty} H_m(\xi)\xi H_n(\xi)e^{-\xi^2}\, d\xi.$$

Applying the inversion operator to the Hermite polynomial yields $\mathbf{i}H_n(\xi) = H_n(-\xi) = (-1)^n H_n(\xi)$. The last equality is one of the properties of the Hermite polynomials which is easily obtained from the definition in Eq. (21–46). Thus, $H_n(\xi)$ is even or odd depending on whether n is even or odd. Therefore, the quantum number must change from even to odd or from odd to even in an allowed transition. Detailed evaluation of the integral (Section 24–11) using the recurrence formula for Hermite polynomials shows that the integral vanishes unless $m = n \pm 1$. The selection rule is generally written $\Delta n = \pm 1$.

To obtain the selection rules for the rigid rotator we must look at the symmetry of the problem in slightly greater detail. The rotator is described by the two angles θ and φ and by a wave function having the form (omitting normalization constants)

$$\psi_{J,m} = P_J^m(\cos\theta)e^{im\varphi}. \tag{24–53}$$

Suppose we fix the direction of the z-axis and then choose the position of the x- and y-axes so that the angle φ is established. Then it is clear that if we rotate the x- and y-axes about the z-axis to some new position which changes φ to $\varphi + \alpha$ in the new coordinate system, nothing is changed physically. We symbolize this rotation operation by $\mathbf{C}_\alpha$. This operation does not affect θ in the slightest. Examining the effect on the function $e^{im\varphi}$ we find

$$\mathbf{C}_\alpha e^{im\varphi} = e^{im(\varphi+\alpha)} = e^{im\alpha}e^{im\varphi}. \tag{24–54}$$

Similarly, for the complex conjugate, $e^{-im\varphi}$:

$$\mathbf{C}_\alpha e^{-im\varphi} = e^{-im\alpha}e^{-im\varphi}. \tag{24–55}$$

To find the effect of $\mathbf{C}_\alpha$ on $x = r\sin\theta\cos\varphi$ and $y = r\sin\theta\sin\varphi$, we construct the sum and difference of x and iy; since $\cos\varphi + i\sin\varphi = e^{i\varphi}$, we have

$$x + iy = r\sin\theta e^{i\varphi}$$

$$x - iy = r\sin\theta e^{-i\varphi}.$$

Then,

$$\mathbf{C}_\alpha(x + iy) = r\sin\theta\, \mathbf{C}_\alpha e^{i\varphi} = r\sin\theta e^{i(\varphi+\alpha)} = (x + iy)e^{i\alpha},$$

and similarly,

$$C_\alpha(x - iy) = (x - iy)e^{-i\alpha}.$$

Next we consider the combination of transition-moment integrals, defined by

$$\int \psi^*_{J',m'}(x + iy)\psi_{J,m}\, d\tau = (\mu_x) + i(\mu_y).$$

Then

$$(\mu_x) + i(\mu_y) = r \int_0^\pi P_{J'}^{|m'|}(\cos\theta)\sin\theta\, P_J^{|m|}(\cos\theta)\sin\theta\, d\theta \int_0^{2\pi} e^{-im'\varphi}e^{i\varphi}e^{im\varphi}\, d\varphi. \qquad (24\text{–}56)$$

Consider the integral over φ in Eq. (24–56); let

$$I_{+1}(m', m) = \int_0^{2\pi} e^{i(m - m' + 1)\varphi}\, d\varphi.$$

Then,

$$C_\alpha I_{+1}(m', m) = \int_0^{2\pi} C_\alpha e^{i(m - m' + 1)\varphi}\, d\varphi = e^{i(m - m' + 1)\alpha}I_{+1}(m', m).$$

Since the physical situation requires that this integral be independent of α, we must have

$$C_\alpha I_{+1}(m', m) = I_{+1}(m', m).$$

It follows that the integral will be independent of α only if $e^{i(m - m' + 1)\alpha} = 1$, or if $m - m' + 1 = 0$. Let $\Delta m = m' - m$; then $\Delta m = +1$ for the integral to be invariant under the rotation. Similarly, if we consider the integral

$$I_{-1}(m', m) = \int_0^{2\pi} e^{i(m - m' - 1)\varphi}\, d\varphi,$$

which will appear in the combination $(\mu_x) - i(\mu_y)$, then we find that the integral is invariant under rotation if $\Delta m = -1$. In the case of the z-component of the transition moment integral, $z = r\cos\theta$; consequently, z is independent of φ and the integral

$$I_0(m', m) = \int_0^{2\pi} e^{i(m - m')\varphi}\, d\varphi$$

must be invariant under rotation. This invariance requires $\Delta m = 0$.

The selection rules on m can be restated briefly; for the x- or y-component of the transition moment, $\Delta m = \pm 1$; while for the z-component, $\Delta m = 0$.

The second consideration of symmetry which arises is that the orientation of the z-axis cannot matter. The z-axis may point up or down; this is equivalent to saying that the integrals must be independent of a reflection in the horizontal plane, an operation we symbolize by σ_h. This operation changes θ into $\pi - \theta$ or $\xi = \cos\theta$ into $\cos(\pi - \theta) = -\cos\theta = -\xi$. The integrals for the x- and y-components of the transition moment both have the form

$$I_{x,y} = \int_0^\pi P_{J'}^{|m'|}(\cos\theta)\sin\theta \, P_J^{|m|}(\cos\theta)\sin\theta \, d\theta = \int_{-1}^1 P_{J'}^{|m'|}(\xi)(1-\xi^2)^{1/2} P_J^{|m|}(\xi) \, d\xi.$$

If we apply the horizontal reflection operator to the integral we obtain

$$\sigma_h I_{x,y} = \int_{-1}^1 P_{J'}^{|m'|}(-\xi)[1-(-\xi)^2]^{1/2} P_J^{|m|}(-\xi) \, d\xi.$$

But, by Eq. (21–89), $P_J^{|m|}(-\xi) = (-1)^{J-|m|} P_J^{|m|}(\xi)$. Using this relation we obtain

$$\sigma_h I_{x,y} = (-1)^{J'-|m'|+J-|m|} I_{x,y}.$$

If we replace $J' = J + \Delta J$ and $|m'| = |m| + \Delta|m|$, we obtain

$$\sigma_h I_{x,y} = (-1)^{2J+\Delta J-2|m|-\Delta|m|} I_{x,y}$$

$$= (-1)^{\Delta J-\Delta|m|} I_{x,y}.$$

If $I_{x,y}$ is to be invariant, $\Delta J - \Delta|m|$ must be an even number; since the requirement on $\Delta|m|$ for the x- or y-component is $\Delta|m| = \pm 1$, it follows that ΔJ must be odd. Detailed calculation using the recurrence formulas shows that $\Delta J = \pm 1$ only.

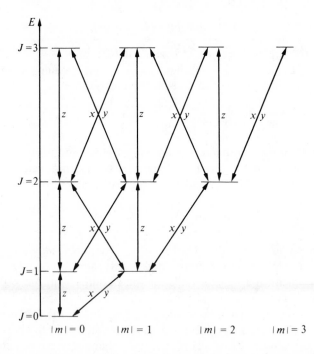

Fig. 24–4 Allowed transitions for the rigid rotator for various polarizations.

For the z-component, $z = r \cos \theta$, and since we must have $|m'| = |m|$, the integral has the form

$$I_z = \int_0^\pi P_{J'}^{|m|}(\cos \theta) \cos \theta \, P_J^{|m|}(\cos \theta) \sin \theta \, d\theta = \int_{-1}^1 P_{J'}^{|m|}(\xi)\xi P_J^{|m|}(\xi) \, d\xi;$$

then

$$\sigma_h I_z = \int_0^\pi P_{J'}^{|m|}(-\xi)(-\xi)P_J^{|m|}(-\xi) \, d\xi = (-1)^{J'-|m|+1+J-|m|}I_z$$

$$= (-1)^{\Delta J + 1}I_z.$$

Again this relation requires ΔJ to be odd if the integral is to be invariant under reflection. Detailed calculation using the recurrence formula shows that $\Delta J = \pm 1$ only.

The types of transitions and the polarizations which produce them are shown in Fig. 24–4; the vertical axis represents the energy; the horizontal axis is simply used to space out the values of $|m|$ corresponding to a particular value of J.

24–13 Selection Rules for the Hydrogen Atom

The wave functions for the hydrogen atom have the form of a product of a radial function (a function of r only) and the rigid rotator functions. The selection rules for the rotator functions must be the same as those obtained above; namely,

$$\Delta l = \pm 1$$

$$\Delta |m| = 0 \qquad (z\text{-component})$$

$$\Delta |m| = \pm 1 \qquad (x\text{- and }y\text{-component}).$$

Thus, the only question which remains is that concerning the radial functions. The transition-moment integral has the form

$$\int_0^\infty R_{nl}(r)r R_{n'l'}(r)r^2 \, dr.$$

However, the functions in the integral depend only on r, which is unaffected by any rotation of the axes or by any reflection in a plane. We conclude that any of the symmetry operations will leave r and functions of r unchanged.

Since symmetry imposes no restrictions, there is no selection rule for n, the principal quantum number. This result which is obtained so simply by symmetry considerations would be extremely cumbersome to prove by direct evaluation of the transition-moment integrals.

Problems

24–1. Spectral lines are described in several different ways; by frequency, ν; by wavelength, λ; and by wave number, $1/\lambda = \tilde{\nu}$. Note that $\nu\lambda = c$, the velocity of light, $c = 3.0 \times 10^{10}$ cm/sec. The difference in frequency between adjacent lines in the rotational spectrum of HCl is 6.33×10^{11} sec^{-1}. Express this number in terms of a difference in wavelength and in terms of a difference in wave number.

24-2. The spacing between two adjacent rotational lines in the spectrum of the HCl molecule is 6.33×10^{11} sec^{-1}. Calculate the moment of inertia of the HCl molecule and the internuclear spacing if the atomic masses are H = 1.008 and Cl^{35} = 34.97.

24-3. The iodine atom has an atomic mass of 126.90 while hydrogen has a mass of 1.008. If the internuclear separation in HI is 1.604 Å, calculate the moment of inertia and the separation of the rotational frequencies.

24-4. The fundamental vibration frequency of the chlorine molecule is 1.688×10^{13} sec^{-1}. Calculate the energies of the first three vibrational levels.

24-5. Calculate the position of the center of mass and the moment of inertia of each of the following linear molecules:

a) the assymetrical $\overset{-}{N}=\overset{+}{N}=O$, which has bond lengths $\overset{-}{N}=\overset{+}{N}$ = 1.126 Å; and N=O = 1.191 Å.

b) the symmetrical O=C=O, which has C=O = 1.162 Å.

c) H—C≡C—C≡N, in which C—H = 1.057 Å; C≡C = 1.203 Å; C—C = 1.382 Å; and C≡N = 1.157 Å.

24-6. The force constant in Br$_2$ is 2.4 dynes/cm. Calculate the fundamental vibrational frequency and the zero-point energy in Br$_2$.

24-7. The atomic masses are given: H = 1.0078; D = 2.0141; Cl^{35} = 34.9689; Cl^{37} = 36.9659. The fundamental vibrational frequency in HCl^{35} is 2891 cm^{-1}.

a) Assuming the force constant does not change, calculate the reduced masses and the fundamental vibrational frequencies in HCl^{37}, in DCl^{35}; and in DCl^{37}.

b) Given that the internuclear distance is 1.2746 Å and is the same for all, calculate the moments of inertia, and the separation of the rotational lines in HCl^{35}, HCl^{37}; DCl^{35}, and DCl^{37}.

24-8. Use the recurrence relation for the associated Legendre functions to evaluate the integral:

$$\int_0^\pi \int_0^{2\pi} P_{J'}^{|m'|}(\cos\theta)e^{-i|m'|\varphi}xP_J^{|m|}(\cos\theta)e^{i|m|\varphi}\sin\theta\,d\theta\,d\varphi,$$

and thus show that $\Delta J = \pm 1$ for x-polarized radiation.

Chapter Twenty-five

Intermolecular Forces

25-1 Introduction

Two chemically saturated particles, such as two molecules of methane or two atoms of argon, are subject to attractive forces as they approach one another. The *intermolecular forces* are electrical in origin, and, since they are responsible for the phenomena of gas imperfection and liquefaction, they are often called van der Waals forces. The energy of vaporization of a liquid provides a convenient measure of the strength of these forces, since it is the energy required to pull the molecules from the liquid, where they are in close proximity, and bring them into the gas, where they are widely separated. The energy of vaporization is simply related to the heat of vaporization of the liquid at constant pressure:

$$Q_{vap} = \Delta H_{vap} = \Delta E_{vap} + p(\overline{V}_{gas} - \overline{V}_{liq}).$$

Approximately, $\overline{V}_{gas} - \overline{V}_{liq} = \overline{V}_{gas} = RT/p$. So that $\Delta H_{vap} = \Delta E_{vap} + RT$. At the normal boiling point, $\Delta H_{vap} = T_b \Delta S_{vap}$; hence for ΔE_{vap} we obtain $\Delta E_{vap} = T_b(\Delta S_{vap} - R)$. For normal liquids we have the Trouton rule, Section 9–4, $\Delta S_{vap} = 21$ eu; hence,

$$\Delta E_{vap} \approx 19 T_b \text{ cal/mole.} \tag{25-1}$$

Even for substances which do not obey the Trouton rule, the proportionality between ΔE_{vap} and the boiling point is roughly correct. In view of Eq. (25–1) we may take the boiling point of a liquid as a convenient measure of the cohesive energy, which in turn depends on the strength of the intermolecular forces.

To understand the origin of the intermolecular forces we shall need a number of the laws of classical electrostatics. It would be too lengthy a diversion to derive all of these laws here; consequently we shall simply introduce them as they are needed with little apology. Most of the laws we require are based ultimately on Coulomb's law,

$$F = \frac{q_1 q_2}{\epsilon r^2},$$ (25–2)

which relates the force of interaction F between two electrical charges q_1 and q_2, and their distance of separation r. The quantity ϵ is the dielectric constant of the medium, called a dielectric, in which the charges are immersed; in vacuum, $\epsilon = 1$.

Since the intermolecular forces depend on the effect on one molecule of the electrical field produced by another molecule, we begin by looking at the effect of an electrical field on matter in bulk and then at the effect of a field on individual molecules.

25–2 Polarization in a Dielectric

If an electrical field E is applied between two parallel metal plates separated by a fixed distance (a parallel-plate capacitor), one plate acquires a positive charge and the other a negative charge; Fig. 25–1(a). The charge per unit area of the plate is the charge density σ. The field strength is

$$E = 4\pi\sigma/\epsilon.$$ (25–3)

Within the dielectric the application of the field produces a minute shift of negative charge toward the positive plate and a shift of positive charge toward the negative plate; the dielectric is *polarized*. Bound to the surface of the dielectric at the negative plate is a positive charge density, $+\sigma_p$; at the positive plate a negative charge density, $-\sigma_p$, is bound to the surface of the dielectric, Fig. 25–1(b). The quantity σ_p is called the *polarization* of the dielectric.

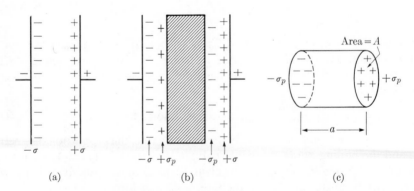

Fig. 25–1 (a) Charged capacitor.   (b) Capacitor with dielectric.
(c) Cylindrical section of dielectric.

Figure 25–1(c) shows a cylindrical element of the dielectric with its axis in the direction of the polarizing field. If the area of each face of the cylinder is A, then the charges on the faces are $+\sigma_p A$ and $-\sigma_p A$. These charges are separated by the length of the cylinder a, so that the cylinder has a dipole moment equal to $(\sigma_p A)a$. The volume of the cylinder is aA; therefore the dipole moment per unit volume is

$$\text{dipole moment/volume} = \sigma_p. \tag{25-4}$$

Therefore, the polarization σ_p in addition to being the charge density on the surface is also equal to the dipole moment per unit volume of the dielectric.

Since $\epsilon = 1$ in vacuum, by Eq. (25–3) the field in vacuum is $E_0 = 4\pi\sigma$. The field E within the dielectric is less than this, since the polarization compensates part of the charge density; the field in the dielectric is a result of the net charge density at the face, so that in the dielectric we have

$$E = 4\pi(\sigma - \sigma_p). \tag{25-5}$$

But, according to Eq. (25–3), the field in the dielectric is $E = 4\pi\sigma/\epsilon$; hence $4\pi\sigma = \epsilon E$ and Eq. (25–5) becomes

$$4\pi\sigma_p = (\epsilon - 1)E. \tag{25-6}$$

Therefore, the polarization is proportional to the field *within* the dielectric; the constant of proportionality depends on ϵ.

To relate E, the field within the dielectric, to E_0, the field in vacuum which is equal to the applied field, we consider the system shown in Fig. 25–2. A sphere of the dielectric is suspended in vacuum between the plates. For this system it may be shown that

$$E = \frac{3}{(\epsilon + 2)} E_0. \tag{25-7}$$

Using this result in Eq. (25–6), we obtain

$$\sigma_p = \frac{3}{4\pi} \frac{(\epsilon - 1)}{(\epsilon + 2)} E_0, \tag{25-8}$$

which relates the polarization, the dipole moment per unit volume, σ_p to the dielectric constant of the material, and the applied field E_0.

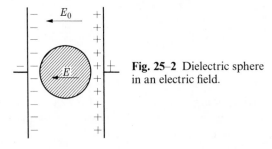

Fig. 25-2 Dielectric sphere in an electric field.

Equation (25–8) is based on purely classical concepts of electrostatics and is in no way dependent on the atomic or molecular structure of the dielectric. The macroscopic property ϵ is easily measurable. Knowing ϵ for the value of the applied field E_0, we can calculate the polarization σ_p.

25–3 Dielectric Polarization and Structure

The dipole moment per unit volume of the dielectric is made up of contributions from all the molecules in the unit volume. If the density of the dielectric is ρ, then the number of moles per cubic centimeter is ρ/M, and the number of molecules per cubic centimeter is $N_0\rho/M$, where N_0 is the Avogadro number, and M is the molecular weight. If m is the average dipole moment per molecule induced by the field, then the dipole moment of 1 cm^3 is $mN_0\rho/M$; thus, we have

$$\sigma_p = mN_0\rho/M. \tag{25–9}$$

Using this result in Eq. (25–8), we obtain

$$m = \frac{3}{4\pi N_0} \frac{(\epsilon - 1)}{(\epsilon + 2)} \left(\frac{M}{\rho}\right) E_0, \tag{25–10}$$

which describes the value of m in terms of the macroscopic properties E_0, M, ρ, and ϵ. Having obtained such a relation, we inquire as to how the dipole moment m is produced in the direction of the field.

If a molecule which has no permanent dipole moment is placed in an electrical field, the electronic cloud will be displaced slightly toward the positive plate. This distorted molecule possesses a dipole moment m, which is proportional to the applied field:

$$m = \alpha_0 E_0. \tag{25–11}$$

The constant of proportionality α_0 is the *distortion polarizability* of the molecule. The polarizability has the dimensions of volume and is measured in cubic centimeters; it is comparable to the volume of the molecule. From Eq. (25–11), we have $\alpha_0 = m/E_0$. The polarizability is the dipole moment produced by an applied field of unit strength.

For any substance whatsoever it can be shown that

$$m = \alpha E_0, \tag{25–12}$$

where α is the polarizability of the substance. If the substance has a permanent dipole moment, then the polarizability is the sum of two terms

$$\alpha = \alpha_0 + \alpha_\mu, \tag{25–13}$$

where α_0 is the distortion polarizability, and α_μ is the *orientation* polarizability. The orientation polarizability arises from the tendency of the permanent dipole moment μ to be oriented in the direction of the applied field.

Using Eq. (25–12) in Eq. (25–10) and rearranging, we obtain

$$\left(\frac{\epsilon-1}{\epsilon+2}\right)\left(\frac{M}{\rho}\right) = \frac{4\pi}{3}N_0\alpha; \tag{25–14}$$

the Clausius-Mosotti equation. The molar polarization† P is defined by

$$P = \left(\frac{\epsilon-1}{\epsilon+2}\right)\left(\frac{M}{\rho}\right). \tag{25–15}$$

The macroscopic quantities in P are easily measured. Then we have

$$P = \frac{4\pi}{3}N_0\alpha. \tag{25–16}$$

If α is a constant characteristic of the molecule, then P is a constant, and Eq. (25–15) is a relation between the dielectric constant and the density of the substance. Furthermore, if α is characteristic of the molecule, then α and the molar polarization P should be independent of the temperature. This is substantiated experimentally for *nonpolar* molecules, those which have no permanent dipole moment.

25–4 Orientation Polarizability

Suppose that a large number of polar molecules each having a permanent dipole moment μ are placed between the plates of a capacitor. In the absence of a field and at reasonably high temperatures, the thermal motions of the molecules will produce a random orientation of the molecules so that there is no net dipole moment in any direction. However, if a field is applied across the plates of the capacitor, the dipole molecules will be oriented in the field, producing a net dipole moment in the direction of the field. The net induced dipole moment divided by the number of molecules is the average dipole moment per molecule in the direction of the field, $\bar{m}$. It can be shown that

$$\bar{m} = \mu^2 E_0/3kT. \tag{25–17}$$

This equation shows that $\bar{m}$ is proportional‡ to the field E_0. The orientation polarizability α_μ is defined by $\bar{m} = \alpha_\mu E_0$; from Eq. (25–17), we obtain

$$\alpha_\mu = \mu^2/3kT. \tag{25–18}$$

At high temperatures $\bar{m}$ and α_μ are much smaller than at low temperatures. At high temperature, the thermal motion is more successful in reducing the orientation in the field.

---

† Note that the *molar polarization* is not the same as the *polarization* in the sense of Section 25–2.

‡ It is clear that $\bar{m}$ cannot continue to be proportional to E_0 if the field is strong. A saturation effect occurs. If all the molecules are completely oriented, increasing the field does not increase $\bar{m}$ any further.

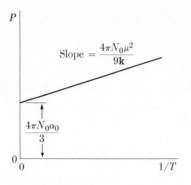

Fig. 25–3 Typical Debye plot of P versus $1/T$.

The total polarizability of any molecule is the sum of the distortion polarizability and the orientation polarizability, Eq. (25–13). Thus we have

$$\alpha = \alpha_0 + \alpha_\mu = \alpha_0 + \mu^2/3\mathbf{k}T. \tag{25–19}$$

Using this result in Eq. (25–14), we obtain the Debye equation,

$$\left(\frac{\epsilon - 1}{\epsilon + 2}\right)\left(\frac{M}{\rho}\right) = \frac{4\pi N_0\alpha_0}{3} + \frac{4\pi N_0\mu^2}{9\mathbf{k}T}, \tag{25–20}$$

which is used to obtain the value of the dipole moment of a molecule from the measured value of the dielectric constant at several temperatures. From the values of ϵ and ρ at several temperatures, the value of the molar polarization (the left-hand side of the equation) is calculated. A plot of the molar polarization against the reciprocal of the temperature should be linear. By Eq. (25–20), the slope is $4\pi N_0\mu^2/9\mathbf{k}$ and the intercept is $4\pi N_0\alpha_0/3$. A typical plot is shown in Fig. 25–3. From the slope and intercept, the dipole moment μ and the distortion polarizability α_0 of the molecule are obtained. A few values of μ are shown in Table 25–1. The unit for the dipole moment of a molecule is the debye; 1 debye $= 1\,\mathrm{D} = 10^{-18}$ esu·cm.

Table 25–1 Dipole moments of molecules

Molecule	μ, D	Molecule	μ, D	Molecule	μ, D
HF	1.91	H_2O	1.85	NH_3	1.47
HCl	1.07	H_2S	0.92	PH_3	0.55
HBr	0.80			AsH_3	0.16
HI	0.42				
CH_3F	1.81	CH_3OCH_3	1.30	C_6H_5Cl	1.70
CH_3OH	1.70	$CH_2{-}CH_2$ $\diagdown O \diagup$	1.90	C_6H_5Br	1.70
CH_3NH_2	1.26			$C_6H_5NO_2$	4.27
o-$C_6H_4Cl_2$	2.50	o-$C_6H_4(NO_2)_2$	6.0		
m-$C_6H_4Cl_2$	1.72	m-$C_6H_4(NO_2)_2$	3.89		
p-$C_6H_4Cl_2$	0	p-$C_6H_4(NO_2)_2$	0		

25–5 Molar Refraction

The dielectric constant is ordinarily measured in an alternating current circuit. The direction of the field across the capacitor changes back and forth with the frequency of the applied potential. If we imagine a single polar molecule between the plates of a capacitor, then if the frequency is not too high, this single molecule will flip back and forth as the field oscillates, always adjusting its orientation to match the direction of the field.

However, the molecule does require a finite time to adjust its orientation. If this time, the *relaxation time*, is very short compared with the time of one cycle of the applied field, then the molecule adjusts itself readily to the different orientations of the field. On the other hand, if the frequency of the applied field is increased, then finally a situation prevails in which the molecule does not have time to change its orientation before the field switches back again. As a result at very high frequencies the molecule is not oriented by the field at all, and the permanent dipole moment ceases to contribute to the molar polarization; only the distortion polarization remains.

At high frequencies, Eq. (25–14), even for molecules with a permanent dipole moment, becomes

$$\left(\frac{\epsilon - 1}{\epsilon + 2}\right)\left(\frac{M}{\rho}\right) = \frac{4\pi N_0 \alpha_0}{3}. \tag{25–21}$$

The distortion polarization remains because even at high frequencies the electron cloud is mobile enough to adjust to the changing field.

An electromagnetic frequency of the order of 10^{14} cycles/sec corresponds to a light wave; then it may be shown that $\epsilon = r^2$, where r is the index of refraction of the light of the frequency in question. Then Eq. (25–21) becomes

$$\left(\frac{r^2 - 1}{r^2 + 2}\right)\left(\frac{M}{\rho}\right) = \frac{4\pi N_0 \alpha_0}{3}, \tag{25–22}$$

where the quantity on the left-hand side is called the molar refraction R; thus

$$R = \left(\frac{r^2 - 1}{r^2 + 2}\right)\left(\frac{M}{\rho}\right) = \frac{4\pi N_0 \alpha_0}{3}. \tag{25–23}$$

The value of R can be calculated from the measured value of the refractive index of the liquid or solid. Equation (25–23) can be combined with Eq. (25–20) to express the dielectric constant at low frequencies:

$$\left(\frac{\epsilon - 1}{\epsilon + 2}\right)\left(\frac{M}{\rho}\right) = R + \frac{4\pi N_0 \mu^2}{9kT}. \tag{25–24}$$

The molar refraction of a substance is approximately the sum of the refractions of the electron groups within it. The molar refraction of NaCl, for example, is the sum of the refractions of the Na^+ ion and the Cl^- ion. A few values of the molar

Table 25-2† Ionic refractions

Ion	R_D, cm³	Ion	R_D, cm³	Gas	R_D, cm³	Ion	R_D, cm³	Ion	R_D, cm³	Ion	R_D, cm³	Ion	R_D, cm³
O^{2-}	7	F^-	2.5	He	0.50	Li^+	0.20	Be^{2+}	0.09	B^{3+}	0.05	C^{4+}	0.03
S^{2-}	15	Cl^-	8.7	Ne	1.00	Na^+	0.50	Mg^{2+}	0.29	Al^{3+}	0.17	Si^{4+}	0.1
				Ar	4.20	K^+	2.2	Ca^{2+}	1.35	Sc^{3+}	1.0	Ti^{4+}	0.7
								Zn^{2+}	0.3				
Se^{2-}	16.3	Br^-	12.2	Kr	6.37	Rb^+	3.6	Sr^{2+}	2.3	Y^{3+}	2.6	Zr^{4+}	2.0
								Cd^{2+}	2.4				
Te^{2-}	24.4	I^-	18.5	Xe	10.42	Cs^+	6.3	Ba^{2+}	4.3	La^{3+}	4.0	Ce^{4+}	3.1
								Hg	5.0				

† By permission from C. P. Smyth, *Dielectric Behavior and Structure*, McGraw-Hill Book Co., Inc., New York, 1955.

refraction appropriate to the D-line of sodium are given in Table 25–2. The refraction of the inert gases can be measured directly. To obtain the refraction of the individual ions from that of their salts, the value of the refraction of at least one ion must be known. The refraction of the fluoride ion has been accurately calculated from the quantum mechanics, and using this value, we can calculate the refractions of the ions Li^+, Na^+, etc., from the refraction of the corresponding fluorides. Table 25–2 is built up in this way.

From the values in Table 25–2 it is apparent that the refraction of a particular group of electrons decreases very much with increase in the nuclear charge. The group of two electrons has a refraction of 0.50 cm^3 in helium, but only has 0.03 cm^3 in C^{4+}. The group of ten in the second row has a high value of 7 cm^3 for O^{2-} which drops to 0.1 in Si^{4+}. Clearly, the contribution of the inner core of two electrons has dropped to a negligible amount in this group of ten; the refraction is essentially the refraction of the outer group of eight electrons. The more tightly the electrons are bound to the central core, the less they are deformed in an applied field and the less contribution they will make to the refraction of the compound. If the electron cloud is large, loose and floppy it is easily deformed in the field; correspondingly, the refraction is large. For this same reason the molar refraction (cm^3) roughly parallels the molar volume (cm^3) of the substance.

The same argument is applied to electron groups in covalent molecules. The refraction of methane is attributed to the refraction of four equivalent electron groups, the pair bonds between the carbon and hydrogen atom. Thus for the carbon–hydrogen bond refraction we can assign $R_{C-H} = \frac{1}{4}R_{CH_4}$. Then for the C—C bond, we derive a value from the refraction of ethane; $R_{C_2H_6} = 6R_{C-H} + R_{C-C}$. This procedure yields $R_{C-H} = 1.70$, $R_{C-C} = 1.21$. Using these values, we can calculate the refraction of any saturated hydrocarbon.

The contribution of double and triple bonds to the refraction is found from the refractions of $H_2C{=}CH_2$ and $HC{\equiv}CH$. Table 21–3 lists a few values of bond

Table 25–3† Refractions of electron groups

Group	R_D, cm^3	Group	R_D, cm^3	Group	R_D, cm^3
H—H	2.08	H⟍Ö⁄H	3.76	C⟍Ö⁄C	2.85
C—H	1.70				
C—C	1.21				
C=C	4.15	C⟍Ö⁄H	3.23	C=Ö	3.42
C≡C	6.03				

†   By permission from C. P. Smyth, *Dielectric Behavior and Structure*, McGraw-Hill Book Co., Inc., New York, 1955.

refractions. Note in comparing the single, double, and triple carbon–carbon bond that the refraction increases with the multiplicity of the bond. The electron pairs in the π bonds are looser than those in the single bond. The values in the table for groups including an oxygen show that the refraction depends on the mode of attachment of the oxygen. The refraction, which includes the two electron pairs on oxygen as well as the bonding pairs, is different in ketones, ethers, and alcohols.

For simple compounds the sum of the group refractions yields the molar refraction of the compound with reasonable accuracy. Difficulties show up in compounds with conjugated double bonds, which have a higher refraction than would be expected. In a conjugated system the electrons in the π bonds are free to move over the entire molecule; consequently they are "looser" and more easily deformed. The additional contribution to the refraction is called "exaltation."

25–6 Intermolecular Forces

Having described the methods used to obtain such properties as polarizability and dipole moment, we return to the problem of intermolecular forces. If two polar molecules have the proper orientation, their positive and negative ends produce a mutual attraction between the molecules. Furthermore, since the field of one polar molecule should induce a dipole moment by distortion of the electron cloud of the other molecule, this effect leads to a mutual attraction of the molecules. It is possible to construct a purely electrostatic theory of intermolecular forces, at least for polar molecules, based on this mutual interaction.

To compute the energy of interaction between two dipoles, consider the approach of two dipoles end to end as shown in Fig. 25–4. Let the charge on the ends of the dipoles be q and the charge separation be a. The dipole at the left, which produces a field E, is fixed at the origin of the coordinate system. The field is, by definition, the force acting on a unit positive charge at the point in question. We may form the second dipole at the distance r from the first by bringing up the two charges $+q$ and $-q$, one at a time. The work required to bring $-q$ from infinity to the position r is the integral of the force acting on the charge $E(-q)$, multiplied by $-dr$, the distance moved: $\int_{\infty}^{r} E(-q)(-dr)$. Similarly, the work required to bring $+q$ from infinity to $r + a$ is $\int_{\infty}^{r+a} E(+q)(-dr)$. The total potential energy W of the dipole at r is the sum of these two integrals:

$$W = \int_{\infty}^{r} Eq\,dr - \int_{\infty}^{r+a} Eq\,dr = -\int_{r}^{\infty} Eq\,dr + \int_{r+a}^{\infty} Eq\,dr,$$

where the change in sign in the second expression is effected by interchanging the

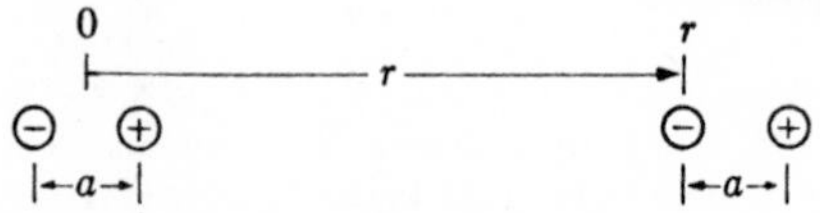

Figure 25–4

limits of integration. The first integral may be written as the sum of two terms so that

$$W = -\left[\int_r^{r+a} Eq\, dr + \int_{r+a}^{\infty} Eq\, dr\right] + \int_{r+a}^{\infty} Eq\, dr = -\int_r^{r+a} Eq\, dr.$$

In the limit as $a \to 0$, the quantity Eq is constant in the range of integration, and we have $W = -Eqa$. But the dipole moment is $m = qa$, so that

$$W = -Em, \tag{25-25}$$

where W is the potential energy of a dipole of moment m in the field E produced by the other dipole.

So far as electrostatics is concerned, Eq. (25-25) is fine for classical dipoles, i.e., disembodied electrical charges at certain fixed distances of separation. The situation with molecules is not so simple. Consider molecule 1, having a permanent dipole moment μ, and a fixed orientation in space. Molecule 2 approaches. The orientation of the dipole axis of molecule 2 relative to molecule 1 is completely random. The field of molecule 1 acts on molecule 2 to induce a dipole moment in molecule 2. The induced moment may result from distortion or orientation of a permanent moment. In either case, the induced moment is in the direction of the field. We write

$$m_2 = \alpha_2 E_1, \tag{25-26}$$

where m_2 is the moment induced in molecule 2 by E_1, the field of molecule 1; the quantity α_2 is the polarizability of molecule 2. [Compare Eq. (25-12).]

To induce the dipole moment, a charge q must be moved through a distance a; the element of work dW' done in this process is the force $E_1 q$ multiplied by the distance moved da. Hence, $dW' = E_1 q\, da$. Since $m_2 = qa$, then $dm_2 = q\, da$ so that

$$W' = \int dW' = \int_0^{m_2} E_1\, dm_2.$$

From Eq. (25-26), $E_1 = m_2/\alpha_2$, so that

$$W' = \int_0^{m_2} \frac{m_2}{\alpha_2} dm_2 = \frac{m_2^2}{2\alpha_2}.$$

This can be expressed in terms of E_1 by using Eq. (25-26) again:

$$W' = \tfrac{1}{2}\alpha_2 E_1^2. \tag{25-27}$$

The total energy W_2 of molecule 2 in the field of molecule 1 is the sum of the potential energy due to its position, which is given by Eq. (25-25), and the energy of distortion or orientation given by Eq. (25-27):

$$W_2 = -E_1 m_2 + \tfrac{1}{2}\alpha_2 E_1^2,$$

which in virtue of Eq. (25-26) becomes

$$W_2 = -\alpha_2 E_1^2 + \tfrac{1}{2}\alpha_2 E_1^2 = -\tfrac{1}{2}\alpha_2 E_1^2.$$

A similar expression, $W_1 = -\frac{1}{2}\alpha_1 E_2^2$, can be written for the energy of molecule 1 in the field of molecule 2. The interaction energy per molecule W_i is the sum

$$W_i = \frac{1}{2}(W_1 + W_2) = -\frac{1}{4}\alpha_1 E_2^2 - \frac{1}{4}\alpha_2 E_2^2.$$

If the molecules are the same kind, then $\alpha_1 = \alpha_2 = \alpha$ and $E_1 = E_2 = E$. Therefore,

$$W_i = -\frac{1}{2}\alpha E^2. \tag{25–28}$$

Since α is positive and E^2 is positive, the interaction energy is negative. The molecules are lower in energy at the distance r than at $r = \infty$ ($E = 0$ at $r = \infty$). This lower energy means that the molecules attract each other because of their mutual influence.

The remaining difficulty is that the field depends on the angle of approach of the two molecules. If E^2 is averaged over all the possible angles of approach, then $\overline{E^2} = 2\mu^2/r^6$. Using this value of $\overline{E^2}$ in Eq. (25–28) yields

$$W_i = -\alpha\mu^2/r^6. \tag{25–29}$$

But, by Eq. (25–19), $\alpha = \alpha_0 + \mu^2/3kT$, so that

$$W_i = -\frac{\mu^2}{r^6}\left(\alpha_0 + \frac{\mu^2}{3kT}\right), \tag{25–30}$$

where the first term represents the attraction resulting from the distortion of the electron cloud of one molecule by the permanent moment of the other molecule, and the second term represents the attraction resulting from the favorable induced orientation of the permanent moment of one molecule by the field of the other. The order of magnitude of W_i is easily estimated: $\mu \approx 10^{-18}$, $\alpha = 10^{-24}$. If we calculate the interaction when the molecules are very close to each other, then $r \approx 10^{-8}$ cm. From Eq. (25–29) we obtain

$$W_i = -\frac{10^{-24}(10^{-36})}{10^{-48}} = -10^{-12} \text{ ergs/molecule} \approx -10 \text{ kcal/mole}.$$

Since this is the correct order of magnitude of energies of vaporization of liquids, it seems that this may be a reasonable way to explain the cohesive energies of liquids.

25–7 The Dispersion Energy

The treatment of intermolecular forces in Section 25–6 presupposed that the molecules possess a permanent dipole moment. We now ask how it is possible for two molecules such as H_2 or CH_4 or argon, which have no permanent dipole moment, to attract one another. This problem was first considered by F. London; the forces producing attraction are sometimes called London forces, sometimes dispersion forces.

To visualize the physical situation consider an atom of an inert gas such as helium or argon. The electron distribution around the positive nucleus is spherical so that there is no net dipole moment. However, the electron distribution is an average over time; Section 19–12. Suppose that the electrons are moving relative to the nucleus in such a way that the time average of the electron positions yields the spherical electron cloud, yet at any instant the atom has a separation of positive and negative charge, a dipole moment. The orientation of the dipole moment vector changes constantly as the motion continues so that the average dipole moment is zero.

If two such atoms are brought near one another, each has a momentary dipole and the electronic motions in the two atoms are coupled by the electrical interaction of the momentary dipoles. The electronic motions in the two atoms synchronize so that the momentary dipoles remain in an attractive orientation, and thus lower the energy of the system. The interaction energy is

$$E_d = -\tfrac{3}{4}h\nu_0\left(\frac{\alpha_0^2}{r^6}\right). \tag{25–31}$$

For many simple molecules the quantity $h\nu_0$ is equal to the ionization energy of the molecule. The polarizability α_0 can be calculated from the molar refraction of the liquid.

The values of α_0 parallel the values of the volume of the molecules. Therefore, the dispersion energy is greater for large than for small molecules. Comparing the large iodine molecule with the small fluorine molecule, we note that iodine is a solid at room temperature, while fluorine is a gas. This implies that the intermolecular forces are larger in iodine than in fluorine. The values of $h\nu_0$ are slightly different, also, but this effect is minor compared with the effect of the larger molecular volume. The dispersion interaction is usually the most important part of the interaction even if the molecules have a dipole moment. For any molecule the interaction energy per pair, E_i, is a sum of terms; Eqs. (25–30), (25–31):

$$E_i = -\frac{2\mu^2\alpha_0}{r^6} - \frac{2\mu^4}{3kTr^6} - \frac{3\alpha_0^2 h\nu_0}{4r^6}. \tag{25–32}$$

25–8 Interaction Energy and the van der Waals "*a*"

The attractive forces discussed so far are inversely proportional to the sixth power of the distance of separation r of the molecules, so we write Eq. (25–32) in the form

$$E_i = -A/r^6, \tag{25–33}$$

where A is a constant of proportionality having a different value for each kind of molecule. The energy E_i is the interaction energy of a pair of molecules separated by a distance r. In a gas, the distances of separation may have many different values. What is the average interaction energy between all the molecules of a gas?

To solve this problem we fix our attention on a molecule at the center of a spherical container of radius R, having a volume $v = 4\pi R^3/3$. If there are N molecules in the container, then the number per cubic centimeter is N/v. How many molecules are at a distance between r and $r + dr$ from the central molecule? The volume of the spherical shell bounded by spheres of radii r and $r + dr$ is $dV_{shell} = 4\pi r^2\, dr$. The number, dN, of molecules in this shell is $dN = (N/v)4\pi r^2\, dr$. The energy of interaction of these molecules with the one at the center is $E_i\, dN$; the average interaction energy of all the molecules with the one at the center is

$$\bar{E}_i = \int E_i\, dN/N.$$

Using Eq. (25–33) for E_i and the value of dN, we obtain

$$\bar{E}_i = -\frac{4\pi A}{v}\int_{\sigma}^{R}\frac{dr}{r^4} = \frac{4\pi A}{3v}\left(\frac{1}{R^3} - \frac{1}{\sigma^3}\right),\qquad(25\text{–}34)$$

where the lower limit σ is the distance of closest approach of the centers of the molecules, the molecular diameter; Fig. 25–5.

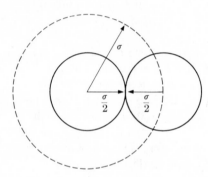

Fig. 25–5 The excluded volume.

In any reasonable situation, the radius R of the container is very much larger than σ, so that Eq. (25–34) reduces to

$$\bar{E}_i = -\frac{4\pi A}{3\sigma^3 v},\qquad(25\text{–}35)$$

which is the average interaction energy *per pair* of molecules in the gas. The total interaction energy is obtained by multiplying by the number of pairs of molecules. If there are N molecules, the first one of a pair may be chosen in N ways, the second member may be chosen in $N - 1$ ways, the total number of pairs is the product of $N(N - 1)$; since N is very large, this is effectively equal to N^2. But this enumeration of the number of pairs counts both the pair between molecules a and b and the pair between molecules b and a as being different; so we divide by 2 and get $\frac{1}{2}N^2$. The total interaction energy is

$$E = \tfrac{1}{2}N^2\bar{E}_i = -\frac{2\pi N^2 A}{3\sigma^3 v}.$$

The energy per mole $\bar{E} = N_0 E/N = -2\pi N_0 N A/3\sigma^3 v$; the volume per mole

$\overline{V} = N_0 v/N$, so that

$$\overline{E} = -\frac{2\pi N_0^2 A}{3\sigma^3 \overline{V}}. \tag{25–36}$$

By differentiation we obtain

$$\left(\frac{\partial \overline{E}}{\partial \overline{V}}\right)_T = \frac{2\pi N_0^2 A}{3\sigma^3 \overline{V}^2}. \tag{25–37}$$

Problem 10–1 required proof that $(\partial \overline{E}/\partial \overline{V})_T = a/\overline{V}^2$ for a van der Waals gas. Comparing this result with Eq. (25–37) shows that the van der Waals constant a is given by

$$a = \frac{2\pi N_0^2 A}{\sigma^3}. \tag{25–38}$$

The form of Eq. (25–37) is in fact a justification of the form of the term $a/\overline{V}^2$ in the van der Waals equation. The van der Waals a is proportional to the coefficient A in the interaction energy. Comparing Eqs. (25–32) and (25–33), we see that A depends on temperature if the molecule has a permanent dipole moment. Thus we expect, correctly, that the van der Waals equation would be improved considerably if a were allowed to depend on temperature.

For the sake of completeness we note that, the van der Waals b is related to the volume of the molecules and therefore to σ. Figure 25–5 shows that the center of a molecule cannot come closer than a distance σ to the center of any other. A molecule thus excludes the volume $v_x = 4\pi\sigma^3/3$. If molecules are added to a container one by one, the volume available to the first molecule is $\overline{V}$, and that available to the second is $\overline{V} - v_x$; if v_i is the volume available to the ith molecule, then $v_i = \overline{V} - (i - 1)v_x$. The average available volume is $\overline{V} - b = (\sum v_i)/N_0$. It is easy to show that $(\sum v_i)/N_0 = \overline{V} - \frac{1}{2}N_0 v_x$, so that

$$b = \tfrac{1}{2}N_0 v_x = \tfrac{2}{3}N_0\pi\sigma^3. \tag{25–39}$$

But the volume of the molecule $v_m = (4\pi/3)(\sigma/2)^3 = v_x/8$, so that

$$b = \tfrac{1}{2}N_0(8v_m) = 4N_0 v_m. \tag{25–40}$$

Since $N_0 v_m$ is the volume of the molecules, b is four times the volume of the molecules. Using Eq. (25–39) to express σ^3 in terms of b, Eq. (25–38) becomes

$$a = \frac{4\pi^2 N_0^3 A}{3b}. \tag{25–41}$$

From the molecular diameter σ, b can be calculated by Eq. (25–39). Then, knowing A we can calculate the constant a using Eq. (25–41).

25–9   Laws of Interaction

Considering the fact that the van der Waals equation is not a particularly good one for the interpretation of the p-V-T data, it seems a hollow victory to be able to calculate the constants a and b. The illustration of the technique involved is the more important thing. Having seen how to proceed from the energy of interaction of two molecules to the macroscopic constants a and b, we gain insight into the refinements and modifications which could be made to yield a more accurate equation of state than the van der Waals equation. Even without that insight it is consoling to be able to calculate van der Waals constants from such seemingly unrelated properties as the refractive index, the dielectric constant, and the ionization potential of the molecule.

Returning to the energy of interaction between two molecules as a function of distance, we see that the energy at large distances decreases as $1/r^6$ until r reaches the value σ. At $r \leq \sigma$, the energy becomes infinitely positive; this is shown by the vertical line in Fig. 25–6(a). This form of the interaction energy results from the supposition that the molecules are "hard spheres" of diameter σ.

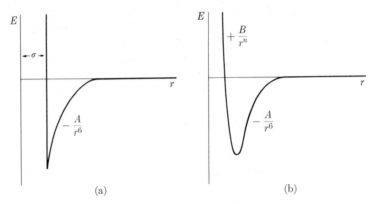

Fig. 25–6 Interaction energy as a function of the distance of separation.
(a) van der Waals potential. (b) Lennard-Jones potential.

Considering the diffuse electron cloud around the molecule, it would be surprising if the molecule behaved as a hard sphere of definite diameter. The repulsion of the molecules should begin smoothly as the electron clouds of the molecules begin to encroach upon each other's domain. The repulsion energy has been given many different mathematical forms. One of the commonest forms is a term proportional to $1/r^n$, where n is a large integer. The total interaction energy would then be written

$$E_i = -\frac{A}{r^6} + \frac{B}{r^n}. \tag{25–42}$$

This form of the interaction law is called the Lennard-Jones potential.[†] In practice the law is most easily handled if $n = 12$; it is then called a 6–12 potential. The shape of the Lennard-Jones potential is shown in Fig. 25–6(b).

[†]   Lennard-Jones is one man, not two!

Other modifications of the interaction energy can be made by adding additional attractive terms in $1/r^8$, for the dipole–quadrupole interaction, and even higher terms for the higher multipole interactions. These higher terms arise because the distance between positive and negative charges in a molecular dipole is not infinitesimal but finite. These higher terms make comparatively small contributions to the interaction energy.

25–10 Comparison of the Contributions to the Interaction Energy

In Section 25–1 it was shown that the boiling point is a qualitative measure of the interaction energy between the molecules of the substance. Three contributions make up the interaction energy:
1. The orientation effect, produced by the mutual action of the permanent dipole moments of the molecules;
2. The distortion effect, produced by the interaction of an induced dipole moment of one molecule with the permanent dipole moment of another molecule;
3. The dispersion effect, produced by the synchronization of the electronic motion in two molecules which results in momentary dipole moments oriented so as to produce an attraction between the molecules.

First we compare the interaction energy, the boiling points, of molecules which have no permanent dipole moment. The interaction energy is due solely to the dispersion effect, and this, by Eq. (25–31), depends on the polarizability α, and $h\nu_0$, which may be thought of as the binding energy of the least tightly bound electron in the molecule. In most molecules, $h\nu_0$ is about 10 eV and does not change very much for different molecules; we will assume it has the same value for all the molecules under discussion. Listed in Table 25–4 are the number of electrons N, the polarizabilities α, and the boiling points T_b, for a number of simple molecules which have no permanent dipole moment. The boiling point increases with the value of α, as we expect. The atoms with more electrons have larger and floppier electron clouds which are more easily deformed in a field; the polarizability is therefore larger, and this is reflected in a larger value of the interaction energy and a higher boiling point.

Table 25–4

Molecule	He	Ne	Ar	Kr	Xe
N	2	10	18	36	54
$\alpha \times 10^{24}$, cm^3	0.203	0.392	1.63	2.46	4.01
T_b, °K	4.216	27.3	87.3	119.9	165.1

As a general rule the more electrons a molecule has, the larger and less tightly held will be the electron cloud. The large loose cloud is easily deformed, so that

Table 25-5

Molecule	H_2	N_2	O_2	CH_4	C_2H_6	C_3H_8
N	2	14	16	10	18	26
$\alpha \times 10^{24}$, cm^3	0.81	1.72	1.55	2.6	4.5	6.4
T_b, °K	20.4	77.3	90.2	111.7	184.5	231

the polarizability, the dispersion energy, and the boiling point are all large. A few more examples of this for molecules which have $\mu = 0$ are listed in Table 25–5. The increase in boiling point of hydrocarbons with increase in molecular weight is, of course, a result of the larger number of electrons, and is not immediately related to the larger mass.

Table 25-6

Molecule	Isobutane	Isobutylene	Trimethyl amine
Formula	$(CH_3)_3CH$	$(CH_3)_2C{=}CH_2$	$(CH_3)_3N$
$\alpha \times 10^{24}$, cm^3	8.36	8.36	8.08
μ, D	0	0.49	0.67
T_b °K	263	267	278

Next we examine the effect of a permanent dipole moment, choosing first a group of small molecules with about the same number of electrons so that α is about the same, as in Table 25–6. The presence of the dipole moment in isobutylene and trimethyl amine results in a slight increase in boiling point. The effect is not very dramatic, but then the dipole moments are not very large. The effect of high dipole moments on moderately sized molecules is illustrated by the compounds listed in Table 25–7. In these compounds, the polarizabilities are roughly the same; the dipole moments are large, and we see a marked effect on the boiling point.

If the molecules are large (have many electrons), the presence or absence of a dipole moment makes little difference in the boiling points; see Table 25–8. The three dichlorobenzenes have roughly the same value of $\alpha = 15 \times 10^{-24}$ cm^3; for

Table 25-7

Molecule	Propane	Dimethyl ether	Ethylene oxide
Formula	$(CH_3)_2CH_2$	$(CH_3)_2O$	C_2H_4O
$\alpha \times 10^{24}$, cm^3	6.4	6.0	5.2
μ, D	0	1.30	1.90
T_b, °K	231	248	284

Table 25–8

	Dichlorobenzene			Dinitrobenzene		
	para	meta	ortho	para	meta	ortho
μ, D	0	1.72	2.50	0	3.89	6.0
T_b, °K	446	445	453	572(subl)	576	592

the three dinitrobenzenes, $\alpha = 19 \times 10^{-24}$ cm³. In these large molecules, the dispersion interaction is the most important part of the cohesive energy. The presence or absence of a dipole moment in the dichlorobenzenes makes little difference in the boiling point. In the case of the dinitrobenzenes, the increase in dipole moment from 0 to 6 D produces only a 20° change in boiling point. Compare this with propane and ethylene oxide, where an increase in μ from 0 to 1.9 D increases the boiling point by 53°.

We can conclude that the larger and more complicated a molecule is, the less the presence or absence of a dipole moment matters to the interaction energy. The dipole moment must be quite large in a large molecule if it is going to affect the interaction energy at all.

25–11　Hydrogen Fluoride, Water, Alcohols, Amines

We return to the discussion of small molecules and consider ethyl alcohol, which has

$$\alpha = 5.2 \times 10^{-24} \text{ cm}^3 \quad \text{and} \quad \mu = 1.70\text{D}.$$

From our experience with dimethyl ether and ethylene oxide, we should expect a boiling point somewhere between the values for those compounds, a value somewhat less than 284°K. The actual boiling point is about 70° higher, 352°K. We can compare HF; $\alpha = 0.8 \times 10^{-24}$ cm³, $\mu = 1.91$ D. Looking at ethylene oxide, which has the same μ but a considerably higher α, we expect that HF should boil at a temperature considerably below 284°K. HF boils at 291°K. Water and ammonia behave in the same way. The properties of the several compounds are shown in Table 25–9. By comparison of any of the compounds on the left of Table 25–9

Table 25–9

Molecule	Neon	Methane	Ammonia	Water	Hydrogen fluoride	Ethylene oxide
Formula	Ne	CH_4	NH_3	H_2O	HF	C_2H_4O
$\alpha \times 10^{24}$, cm³	0.392	2.59	2.34	1.59	0.80	5.2
μ, D	0	0	1.46	1.85	1.91	1.90
T_b, °K	27.3	111.7	240	373	293	284

Table 25–10

Formula	CH_3F	CH_3OH
$\alpha \times 10^{24}$, cm^3	3.84	3.0
μ, D	1.81	1.70
T_b, °K	195	338

with ethylene oxide, all should have boiling points considerably lower than 284°K. Methane and neon fulfill this expectation; ammonia has a somewhat lower boiling point, but not as low as one would expect. Water boils 90° *higher* than ethylene oxide. Since NH_3, HF, and H_2O are very small molecules, it might be argued that this increase in interaction energy results from dipole–dipole interaction at very close distances. This is not so, as one can demonstrate by comparing methyl fluoride and methyl alcohol, which are essentially the same size, as in Table 25–10. Methyl alcohol has a lower value of α and μ and so should have a *lower* boiling point. In fact the boiling point is higher by 143°. A final example is provided by dimethyl ether and ethyl alcohol. These molecules are roughly the same size, as shown in Table 25–11. It is apparent that the boiling point of ethyl alcohol is about 100° too high.

Table 25–11

Formula	CH_3OCH_3	C_2H_5OH
$\alpha \times 10^{24}$, cm^3	6.0	5.2
μ, D	1.30	1.70
T_b, °K	248	352

Most compounds containing OH, NH, or NH_2 groups have higher boiling points than would be predicted on the basis of their dipole moments and polarizabilities. Among the fluorides, only HF has this anomaly; other fluorides behave normally.

25–12 The Hydrogen Bond

The anomalies observed in the boiling points of compounds containing hydroxyl, amino, and imino groups indicate that there exists in these compounds an additional interaction energy over and above the van der Waals interaction energy. The magnitude of this additional energy is comparable to the van der Waals interaction energy, being of the order of 5 to 10 kcal/mole. This additional energy is a result of the formation of a weak bond between the oxygen atoms in two molecules of methanol, for example. A hydrogen atom lies between the two bonded oxygen atoms. This bond is called a *hydrogen bond* and is given the conventional structural representation, $O\cdots H—O$. A hydrogen bond can be formed between any two highly electronegative atoms; this requirement restricts the hydrogen bond to fluorine, oxygen, and nitrogen, so that the following types occur: $F—H\cdots F$, $O—H\cdots O$, $N—H\cdots N$; and, of course, mixed types such as $N—H\cdots O$. The effects of hydrogen bonding with S and Cl are very slight.

Although much has been written in the attempt to explain the stability of the hydrogen bond, its fundamental nature remains somewhat obscure. Because of reluctance to assign two covalent bonds to a hydrogen atom, great emphasis has been placed on interpreting the bond on a purely electrostatic basis, which has been successful in explaining some of the properties of the hydrogen bond. More recently, it has been realized that hydrogen can be connected to more than one atom by bonds which are distinct from the ordinary covalent bond, since they involve three nuclei rather than two, yet seem to have some characteristics of the covalent bond. The theory of the hydrogen bond has been reexamined on this basis. No final pronouncement on the subject can be made at the present time.

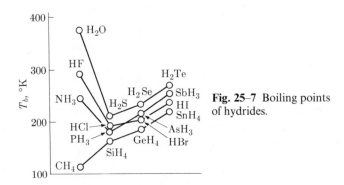

Fig. 25–7 Boiling points of hydrides.

One of the most striking illustrations of the effect of hydrogen bonding on physical properties is shown by the plot of boiling points of the hydrides of the elements in periodic groups IV, V, VI, and VII, shown in Fig. 25–7; the horizontal axis serves only to separate one compound from the next one. The melting points of these compounds exhibit a similar anomaly. The high boiling points of water, ammonia, alcohols, amines, and hydrogen fluoride are a consequence of the fact that these substances are hydrogen bonded in the liquid into polymers; for example, a liquid alcohol can be regarded as a mixture of polymers,

$$\begin{matrix} R & & R & & R \\ | & & | & & | \\ O—H & \cdots & O—H & \cdots & O—H \end{matrix}$$

Once we adopt the view that these kinds of compounds are capable of association, then a number of diverse observations come into focus.

Hydrogen bonding accounts for unusually high melting and boiling points of such compounds as alcohols, sugars, organic acids, and simple inorganic acids such as H_2SO_4, HNO_3, H_3BO_3. A compound such as urea, NH_2—CO—NH_2, is solid at room temperature, while acetone, CH_3—CO—CH_3, having about the same number of electrons, is a volatile liquid. Urea is hydrogen bonded, acetone can form hydrogen bonds only in the enol form, CH_2=$C(OH)CH_3$. Boric acid should be a gas or at worst a volatile liquid at room temperature; BF_3, with the same number of electrons, is a gas. In fact, boric acid is an involatile solid which melts with decomposition at 185°C. The formula, if written correctly as $B(OH)_3$, betrays the possibility of hydrogen bonding.

The unusual values of entropies of vaporization of these associated liquids are understandable. While so-called normal liquids have entropies of vaporization of about 21 eu, the values for these hydrogen-bonded compounds are usually (but not always) higher. This is illustrated by the following data.

Compound	H_2O	NH_3	CH_3OH	CH_3NH_2	HF	CH_3COOH
ΔS_{vap}, eu	26.0	23.3	25.0	23.1	6.1	14.9

In those cases for which ΔS_{vap} is higher than the normal value, the liquid, as a result of being associated, has a higher degree of order than a normal liquid. Therefore, the transition from liquid to gas is attended by an unusually large entropy increase. Substances such as HF and acetic acid, which have unusually low values of ΔS_{vap}, are polymerized even in the vapor state. This polymerization reduces the disorder in the vapor and so lessens the value of ΔS_{vap}. It is known independently of this that HF is highly polymerized in the vapor state. The polymers are zigzag chains (or rings). The chain in HF has the structure

The vapor of acetic acid contains an appreciable amount of the dimer

This accounts for the low value of ΔS_{vap} for this substance.

Hydrogen bonding also affects the appearance of the infrared spectrum. The OH group absorbs at a frequency of 3500 cm^{-1}. If the spectrum of a hydroxyl compound is measured in the vapor, where it is not hydrogen bonded, this absorption shows up as a sharp peak centered at this frequency. If the spectrum of the hydrogen-bonded compound is examined, the peak is greatly broadened and the center is shifted to a lower frequency. The broadening of the absorption peak is characteristic of hydrogen bonding.

Additional evidence of hydrogen bonding in solids is provided by x-ray diffraction studies of crystals. In the crystal of a substance containing hydroxyl groups, certain of the oxygen–oxygen distances are abnormally short, indicating bond formation between these oxygens. There are just enough of these short distances to account for the hydrogen atoms in the hydroxyl groups.

Problems

25–1. By combining the thermodynamic equation of state with the van der Waals equation, it can be shown that $(\partial \bar{E}/\partial \bar{V})_T = a/\bar{V}^2$. Below the critical temperature, the van der Waals equation predicts, approximately, a liquid volume equal to b, and a gas volume equal to RT/p. Assuming that the substance follows the van der Waals equation, what increase in energy attends the isothermal expansion of one mole of a substance from the liquid volume to the gaseous volume?

25–2. The dielectric constant for chlorobenzene is:

t, °C	-50	-20	20
ϵ	7.28	6.30	5.71

Assuming that the density, 1.11 gm/cm^3, does not vary with temperature, estimate the dipole moment of this compound. Molecular weight $= 112.45$.

25–3. From the values in Table 25–3, calculate the molar refraction of butane, propene, and acetone.

25–4. From the value of R_D in Table 25–3, calculate the polarizability of water.

25–5. Compare the magnitude of the average interaction energy between two molecules in the two situations: a liquid having a molar volume of 20 cm^3; a gas having a molar volume of 20,000 cm^3.

25–6. The Lennard-Jones potential $\epsilon = A/r^6 + B/r^n$ can be expressed in terms of ϵ_m, the energy at the minimum, and r_0, the distance of separation at the minimum. Find A and B in terms of r_0, ϵ_m, and n. Write the potential in terms of the new parameters. If σ is the distance of separation when $\epsilon = 0$, find the relation between r_0 and σ.

25–7. Using the results in Problem 25–6, note the simplification in the form of ϵ if $n = 12$. Write ϵ in terms of ϵ_m and r_0 and in terms of ϵ_m and σ if $n = 12$.

25–8. Calculate the average dispersion interaction energy at 5 Å separation between two molecules of (a) neon, $\alpha = 0.392 \times 10^{-24}$ cm^3, $h\nu_0 = 497.2$ kcal; (b) argon, $\alpha = 1.63 \times 10^{-24}$ cm^3, $h\nu_0 = 363.3$ kcal; (c) krypton, $\alpha = 2.46 \times 10^{-24}$ cm^3, $h\nu_0 = 322.7$ kcal; (d) xenon, $\alpha = 4.01 \times 10^{-24}$ cm^3, $h\nu_0 = 279.7$ kcal. (e) Plot the boiling points, given in Section 25–10, as a function of the dispersion energy.

25–9. For argon, $\alpha = 1.63 \times 10^{-24}$ cm^3; for water, $\mu = 1.85 \times 10^{-18}$ esu·cm. Estimate the energy of interaction of a single argon molecule with a single water molecule at the distance of closest approach. Both molecules can be considered as spheres; for argon,

$\sigma = 3.08 \times 10^{-8}$ cm; for water, $\sigma = 2.76 \times 10^{-8}$ cm. [*Note*: Argon forms a solid hydrate, Ar·5H$_2$O, in which the "bond" energy between water and argon is about 10 kcal.]

25-10. According to classical electrostatics, the polarizability of a perfectly conducting sphere is equal to r^3, where r is the radius of the sphere. Using this relation, and the data in Table 25–2, compare the radii of the species in the second row of Table 25–2; O^{2-}, F$^-$, Ne, etc.

Chapter Twenty-six

Structure of Solids and Liquids

26–1 The Structural Distinction Between Solids and Liquids

The word "solid" will be applied only to *crystalline* solids, since it is always possible to distinguish between a crystalline solid and noncrystalline phases such as liquids and amorphous "solids." Structurally, the constituent particles, atoms, molecules, or ions, of a crystalline solid are arranged in an orderly, repetitive pattern in three dimensions. This pattern is such that having observed the pattern in some small region of the crystal, it is possible to predict accurately the positions of particles in any region of the crystal however far it may be removed from the region of observation. The crystal has *long-range order*.

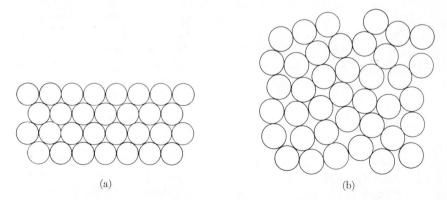

(a)   (b)

Fig. 26–1 Schematic view of structure in a crystal and in a liquid.

The particles composing a liquid are not arranged in such a precise fashion. In a small region of a liquid, there may appear to be a pattern of arrangement. However, if we observe a neighboring region, the pattern will be somewhat different, or if it is nearly the same, it may not be accurately joined to the first region. In terms of the arrangement of the particles, a liquid has *short-range order* but lacks the long-range order which characterizes the solid. Figure 26–1 illustrates the distinction in two dimensions. The difference between solid and liquid is the difference between two ways of putting ball bearings in a box; they may be carefully packed row upon row or may simply be dumped into the box.

The irregular arrangement of particles in the liquid results in the appearance of voids or holes here and there throughout the structure. Note that a *nearly* regular arrangement exists around many of the particles in Fig. 26–1(b). The presence of the holes requires a larger volume, and in most cases the volume of the liquid is larger than that of the solid. The ease of flow is also related to this larger volume of the liquid. Amorphous "solids" have the same structural features as liquids, and are conveniently regarded as extremely viscous liquids.

26–2 An Empirical Classification of Solid Types

There is no unique way of classifying solids. The method chosen depends in great measure on the purpose at hand. For the present we will classify solids according to the type of bond which holds the constituent particles of the solid together.

Four types of bonds are operative in binding individual species into a crystal. On this basis we distinguish four types of solids, (a) metals, (b) ionic crystals, (c) van der Waals crystals, and (d) covalent crystals. The crystals bound by hydrogen bonds are usually classed with van der Waals crystals, since the strength of bonding is of the same order of magnitude in both types. However, since the arrangement of the hydrogen atoms in a molecule permits hydrogen bonds only in special directions, the hydrogen-bonded crystals have some of the features of covalent crystals.

In the first three solid types (metals, ionic crystals, van der Waals crystals), the forces of interaction which hold the particles together do not act in any preferred direction, while in the covalent crystals, the bonds are formed only in special directions because of the directional character of the covalent bond. The principles governing the direction of bond formation in covalent crystals are the same as those governing the covalent bond in individual molecules.

26–3 Geometrical Requirements in the Close-packed Structures

If we ask how atoms or molecules can be arranged in a regular way to build a crystal, it seems that the possibilities might be unlimited. Although there are a great number of possible arrangements, a relatively small number of these recur again and again. One factor which limits the possibilities is the requirement that the arrangement be the most stable one energetically. Some departure from this principle is allowed, but generally different crystalline forms of a single substance do not differ greatly in energy. Energies of transitions between different crystalline forms of a substance are usually only a few tenths of a kilocalorie per mole.

First consider those crystals in which the energy of interaction between the particles does not depend on the direction of approach. As two particles approach, the energy of the system decreases and passes through a minimum at some distance. The two-particle system has its greatest stability at this point. If a third particle is introduced, the energy of the system decreases further. The maximum stability is attained when each particle in the aggregate is surrounded by the greatest possible number of neighbors. In brief, the particles must be as closely packed as possible. If the particles are spheres of the same size, the problem reduces to how to pack as many balls as possible in a given space.

Clearly, the balls will be packed in layers, and each layer must be closely packed. We begin by arranging the layer as shown in Fig. 26–2. In the layer, each sphere has six nearest neighbors. To build the crystal in three dimensions, we stack the layers one on top of the other in a regular way. Two possibilities exist after the second layer is put on. Directing attention to sphere A (Fig. 26–2), suppose that the spheres in the next layer nestle in the notches at the positions marked with dots. Three of these will make contact with A. There is in this second layer a notch over A, but there are also notches over the positions marked with crosses. The third layer may then repeat the arrangement in the first layer or not. If the third layer repeats the first, then there are only two kinds of layers, denoted x and y, and the patterns of layers is $xyxyxy\ldots$ If the third layer does not repeat the first, it is denoted by z and the pattern of layers is $xyzxyzxyz\ldots$† These two arrangements are common ones in metals and in van der Waals crystals composed of effectively spherical molecules such as CH_4, HCl, Ar.

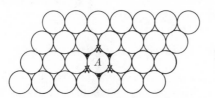

Figure 26–2

The arrangement of close-packed layers in the pattern $xyxy\ldots$ is the hexagonal close-packed structure (hcp); the pattern $xyzxyz\ldots$ is the cubic close-packed (ccp), or face-centered cubic (fcc), structure. In each of these structures, every sphere is in contact with twelve others: six in its own layer, three in the layer above, and three in the layer below. The twelve coordination in these structures is shown by an exploded view in Fig. 26–3(a) and (b), and in a different view in Fig. 26–3(c) and (d). The hcp

---

† If the reader can obtain about twenty or thirty marbles or coins, it is helpful to work out these arrangements.

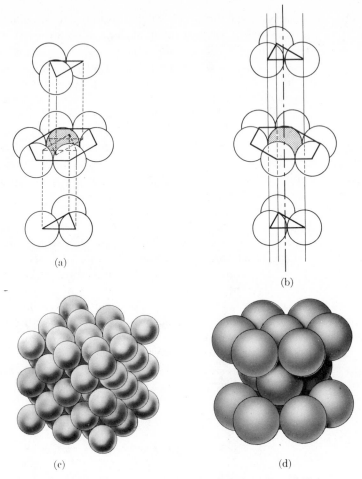

(a)

(b)

(c)

(d)

Fig. 26–3 Close-packed structures. (a) and (c) fcc. (b) and (d) hcp.

and fcc structures are the typical structures encountered in metals. The high co-ordination number (twelve), in these structures results in a crystal of comparatively high density.

Another common arrangement of spheres which occurs in a few metals is the body-centered cubic (bcc), which is built up of layers having the arrangement shown in Fig. 26–4. The second layer fits in the notches of the first and the pattern of layers repeats, $xyxy\ldots.$ In these layers the number of nearest neighbors around any sphere is four, as compared with six in the close-packed layers. In the body-centered struc-ture the overall coordination number is eight; there are four nearest neighbors within the most closely packed layer, two in the layer above, and two in that below. As a result of the less efficient packing, the bcc structure has an inherently lower density than the hcp or the fcc structures.

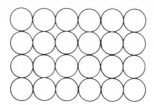

Fig. 26–4 Most closely packed layer in bcc structure.

    The positions of the atom centers in the three structures are shown in a different view in Fig. 26–5. To describe these structures completely requires the specification of the edge length a of the fundamental cube in the face-centered and body-centered cubic arrangements. The hexagonal close-packed structure requires the specification of two lengths, the nearest-neighbor distance a within the close-packed layer, and the distance c between the two repeating layers. If the spheres were truly rigid spheres, geometry would require $c = 1.633a$. Since the particles in a crystal are not truly rigid spheres, this relation is not exactly fulfilled; a and c must be specified separately. In metals having the hcp structure, the relation is nearly fulfilled. Table 26–1 lists a few of the metals which crystallize in the three structures, along with values of the lattice parameters.

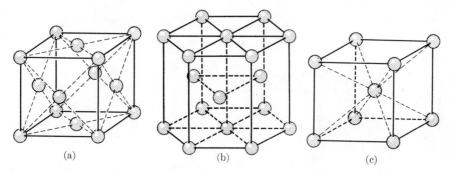

Fig. 26–5 Location of atom centers in fcc, hcp, and bcc structures.

Table 26–1 Crystal structure and lattice constants (Å) for common metals

Face-centered cubic				Hexagonal close-packed			Body-centered cubic	
Metal	a	Metal	a	Metal	a	c	Metal	a
Al	4.05	Pd	3.89	Co	2.51	4.07	Cr	2.88
Cu	3.62	Pt	3.92	Mg	3.21	5.21	Fe	2.87
Au	4.08	Ag	4.09	Ti	2.95	4.68	W	3.16
Pb	4.95	Ni	3.52	Zn	2.66	4.95	Na	4.29

26–4 Packing in Ionic Crystals

The packing of spheres in ionic crystals is complicated by the fact that the ions are positively and negatively charged. Suppose that the electrical charges on the positive and negative ions are equal, though opposite in sign. To build an electrically neutral structure requires that the number of negative ions around each positive ion be the same as the number of positive ions around each negative ion. If the positive and negative ions have the same size, we find that it is not possible to build a layer of alternating positive and negative ions with six positive ions around each negative ion, and vice versa, to yield a total coordination of twelve positive ions around each negative ion. The highest coordination possible, if the structure is to be electrically neutral, is that having the most closely packed layers built in the manner shown in Fig. 26–6. In the layer each positive ion (shaded circles) is surrounded by four negative ions (open circles), and each negative ion by four positive ions. Piling up layers of this kind in the order $xyxy\ldots$ yields a cubic type of structure in which the central particle is an ion of one charge, with eight oppositely charged ions at the cube corners. The structure consists of two interpenetrating simple cubic lattices. The positions of one lattice are occupied by positive ions while those of the other are occupied by negative ions. This is the cesium chloride, CsCl, structure: Fig. 26–7.

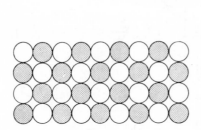

Fig. 26–6 Layer in CsCl structure.

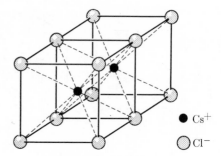

$\bullet$ Cs$^+$

$\bigcirc$ Cl$^-$

Fig. 26–7 CsCl structure.

A curious compromise is reached in many ionic crystals. The crystal NaCl, for example, consists of two interpenetrating close-packed (fcc) lattices. The positions of one lattice are occupied by positive ions, while those of the other are occupied by negative ions. Consider the unit cube of the fcc structure in Fig. 26–8(a). There is a void, or hole, outlined by the octahedron, at the center of the cube. Similar octahedral holes are centered on each edge of the unit cube; Fig. 26–8(b). Each hole is at the center of an octahedron which has atoms at each of the six apices. The centers of the octahedral holes occupy the positions of an fcc lattice which interpenetrates the lattice of atoms. Small foreign atoms, such as H, B, C, N, can occupy these holes. Many carbides, hydrides, borides, and nitrides of the metals are *interstitial* compounds formed in this way.

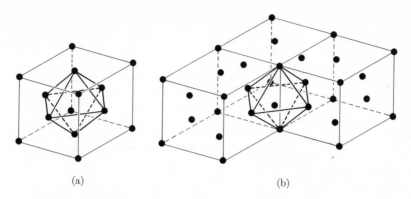

(a)   (b)

Fig. 26–8 Octahedral holes in the fcc structure. (a) Body centered. (b) Edge centered.

If we wish to locate comparatively large particles in these holes, then all the particles of the original lattice must move apart; the structure expands to enlarge the holes to sufficient size. We can view the sodium chloride structure as an fcc arrangement of chloride ions which has expanded sufficiently to permit the sodium ions in the octahedral holes. As a result, neither of these interpenetrating lattices is close packed in the sense of having all the particles in contact as they are in metals, but both have the symmetry of the close-packed fcc lattices. Each sodium ion is in contact with six chloride ions, and each chloride ion is in contact with six sodium ions; 6-6 coordination. Figure 26–9 shows the NaCl structure; the fcc arrangement

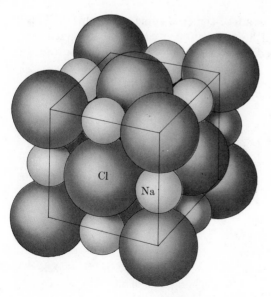

Fig. 26–9 The NaCl structure.

of the negative chloride ions is apparent in Fig. 26–9; there is a sodium ion at the center of the cube.

In addition to the CsCl structure, 8-8 coordination, and the NaCl structure, 6-6 coordination, there are two structures having 4-4 coordination, the zinc blende structure and the wurtzite structure; Fig. 26–10.

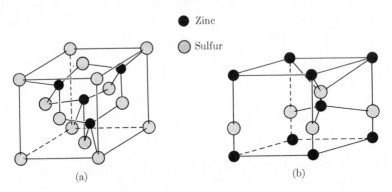

Fig. 26–10 Unit cells in (a) zinc blende and (b) wurtzite.

Unsymmetrical valence types of compounds such as CaF_2 and Na_2O have more complicated structures, since the coordination number of the ion with the larger charge, Ca^{2+} in the case of CaF_2, must be twice that of the ion of lower charge if the crystal is to be electrically neutral. For compounds of the 1-2 valence type, the typical structures are the fluorite (CaF_2) structure and the rutile (TiO_2) structure; Fig. 26–11.

The Ca^{2+} ions in fluorite are in a face-centered cubic arrangement. This lattice has, in addition to the octahedral holes mentioned earlier, holes which are tetrahedrally coordinated. The tetrahedral holes of the fcc structure are occupied by F^- ions in fluorite. Each F^- ion is tetrahedrally coordinated to Ca^{2+} ions. Figure

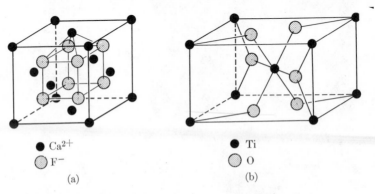

Fig. 26–11 Unit cells in (a) fluorite and (b) rutile.

26–11(a) also shows that the Ca^{2+} ion on the top face is connected to four F^- ions below it; it is similarly connected to four F^- ions (not shown) lying above it. The coordination of the Ca^{2+} ion is eight, and the fluorite structure is described as having 8-4 coordination. Fluorite may be considered as a face-centered cubic lattice of Ca^{2+} ions interpenetrated by a simple cubic lattice of F^- ions.

In rutile, having 6-3 coordination, the octahedral coordination of Ti^{4+} to O^{2-} and the triangular coordination of O^{2-} to Ti^{4+} is evident in Fig. 26–11(b).

We have described structures of metals and ionic crystals in terms of close-packed arrangements of spheres. Clearly, if the particles are not spherical, the close packing must be done in a manner appropriate to the shape of the particle. One could scarcely expect long rod-shaped molecules to pack as spheres would; such particles pack into the crystals as matches in a box.

26–5 The Radius Ratio Rules

A factor which is of great importance in ionic crystals is the difference in size of the positive and negative ions. Pauling has shown how the geometric requirements for close packing of spheres of different sizes can be simply expressed in terms of the radius ratio $\rho = r_s/r_l$, defined as the ratio of the radius of the smaller ion, r_s, to that of the larger ion, r_l.

For simple ionic compounds of symmetrical valence type, the radius ratio rules are:

Value of ρ	$\rho < 0.414$	$0.414 < \rho < 0.732$	$\rho > 0.732$
Coordination	4-4	6-6	8-8
Structure	zinc blende or wurtzite	NaCl	CsCl

These rules enable us to predict the structure of the compound from the relative sizes of the two ions. Applied to many different ionic crystals of different valence types, the rules are quite good. There are at least two reasons for the exceptions to the radius ratio rules: (1) the ions are not rigid spheres; (2) the ions of opposite charge are not in contact.

26–6 Geometrical Requirements in Covalent Crystals

The notable exception to the rule of close packing appears in covalent crystals in which the maximum stability is obtained, not with the greatest possible number of neighbors, but by forming the allowed number of covalent bonds in the proper directions. This requirement is an individual thing so that a generalization of the kind embodied in the radius ratio rules is out of the question for covalent crystals. We cannot build up typical structures with the ease and confidence with which we stacked spheres into layers and layers one upon another. Rather than struggle with this host of individual problems, we will make only a few elementary remarks about the subject.

First of all, comparatively few solids are held together exclusively by covalent bonds. The majority of solids incorporating covalent bonds are bound also by either ionic or van der Waals bonds. The common occurrence is to find distinct molecules held together by covalent bonds and the molecules bound in the crystal by van der Waals bonds. The covalent bonds may hold a complex anion or cation together; the cations and anions are bound in the crystal by ionic bonds.

Only those atoms which form four covalent bonds produce a repetitive three-dimensional structure using only covalent bonds. The diamond structure, Fig. 26–12, is one of several related structures in which only covalent bonds are used to build the solid. The diamond structure is based on a face-centered cubic lattice wherein four out of the eight tetrahedral holes are occupied by carbon atoms. Every atom in this structure is surrounded tetrahedrally by four others. No discrete molecule can be discerned in diamond. The entire crystal is a giant molecule.

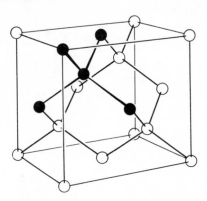

Fig. 26–12 Diamond structure.

Generally the covalent solids have comparatively low densities as a result of the low coordination numbers. This effect is intensified in those crystals in which covalently bound structural units are bound in the crystal by van der Waals forces. The distance between two units held by van der Waals forces is significantly greater than that between units held by covalent, ionic, or metallic bonds; these large distances result in a low density of the solid.

26–7 The Symmetry of Crystals

The symmetry exhibited by a macroscopic crystal is a consequence of the symmetrical arrangement of the units of structure which compose the crystal. To understand the choice of a unit of structure and how a repetitive structure is built from that unit, we examine the problem in two dimensions. Any area-filling repetitive pattern on a plane surface is based on a unit of pattern which may be outlined by a parallelogram. The entire pattern can be generated by translating this parallelogram, the unit of pattern, by definite distances parallel to its edges. We describe the unit of pattern in terms of two vectors of lengths a and b, which form two sides of the parallelogram. Starting at a point and moving any integral multiple of the distance a in the direction of the first vector, we reach an equivalent point in the pattern; similarly, moving an integral multiple of the distance b in the direction of the second vector, we reach an equivalent point in the pattern. The two vectors are the *primitive translations* of the pattern. Figure 26–13 shows a simple pattern of points, and illustrates the fact that the choice of vectors is not unique, either set a, b or a', b' is legitimate.

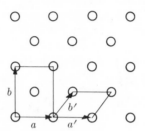

Fig. 26–13 Units of pattern in two dimensions.

In two dimensions there are five possible units of pattern, unit cells, which can build a repetitive pattern by translation in directions parallel to the edges; Fig. 26–14. The unit cell with the 120° angle is interesting because it permits a threefold or sixfold axis of symmetry at a point in the pattern, which is a permissible type of symmetry in two-dimensional patterns.

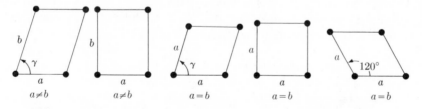

Fig. 26–14 The five units of pattern in two dimensions.

To generate a repetitive pattern in three-dimensions, an additional repetition vector out of the plane of the first two must be added. The three vectors define a parallelepiped. Any repetitive pattern in three dimensions has a parallelepiped as a unit cell. There are seven distinct parallelepipeds, those labeled (P) or (R) in Fig. 26–20, which can generate by translation any repetitive pattern in three dimensions. Crystals are classified into seven *crystal systems* according to the shape (the lengths and inclinations of the vectors) of the unit cell. The lengths of the primitive translation vectors, the axes of the unit cell, are denoted by a, b, and c. The angle between a and b is γ, that between b and c is α, and that between c and a is β. Table 26–2 lists the relations between the lengths and between the angles for the crystal systems. In all cases the edges of the unit cell are parallel to edges or possible edges of the crystal.

Table 26–2 The seven crystal systems

Axes	Angles	System
$a \neq b \neq c$	$\alpha \neq \beta \neq \gamma \neq 90°$	Triclinic
$a \neq b \neq c$	$\gamma \neq \alpha = \beta = 90°$	Monoclinic
$a \neq b \neq c$	$\alpha = \beta = \gamma = 90°$	Orthorhombic
$a = b \neq c$	$\alpha = \beta = 90°, \gamma = 120°$	Hexagonal
$a = b \neq c$	$\alpha = \beta = \gamma = 90°$	Tetragonal
$a = b = c$	$\alpha = \beta = \gamma$	Rhombohedral (trigonal)
$a = b = c$	$\alpha = \beta = \gamma = 90°$	Cubic

·26–8 The Crystal Classes

Having divided crystals into systems according to the possible shapes of the unit cell, a further division can be made according to the combinations of symmetry elements which are compatible with each system. An element of symmetry is an operation which brings the crystal into coincidence with itself. The elements of symmetry are: rotation about an axis, reflection in a plane, inversion through a center of symmetry, and rotation inversion. For example, consider the simple cube shown in Fig. 26–15(a). Rotation through 90° around the vertical axis brings the cube into coincidence with itself. This axis is a fourfold axis of symmetry. (If a rotation of 360°/p around an axis brings the figure into coincidence, then the axis is a p-fold axis of symmetry.) The cube has three fourfold axes of symmetry, which are symbolized by the small squares

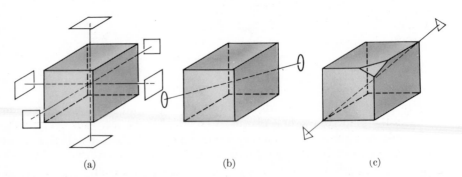

(a)   (b)   (c)

Fig. 26–15 Axes of symmetry of the cube. (a) Fourfold. (b) Twofold. (c) Threefold.

at the ends of the axes. A twofold axis, rotation through 180°, is shown in Fig. 26–15(b), symbolized by ellipses at the ends of the axis. The cube has six twofold axes. The four main diagonals of the cube are threefold axes of symmetry, symbolized by the triangles in Fig. 26–15(c). The threefold axis has been emphasized in Fig. 26–15(c) by truncating one corner of the cube to display the triangle. A plane of symmetry divides any figure into mirror images. The cube has nine planes of symmetry, shown in Fig. 26–16.

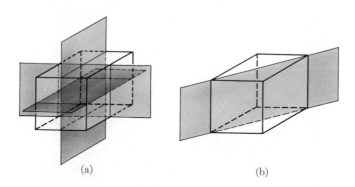

(a)   (b)

Fig. 26–16 Planes of symmetry of the cube. (a) Principal planes (3). (b) Diagonal planes (6).

Finally, the cube has a center of symmetry. Possession of a center of symmetry, center of inversion, means that if any point on the cube is connected to the center by a line, that line produced an equal distance beyond the center will intersect the cube at an equivalent point. More succinctly, a center of symmetry requires that diametrically opposite points in a figure be equivalent. These elements together with rotation–inversion† are the symmetry elements for crystals. The elements of symmetry found in crystals are: (a) center of symmetry; (b) planes of symmetry; (c) 2-, 3-, 4-, and 6-fold axes of symmetry; and (d) 2- and 4-fold axes of rotation–inversion. Of course, every crystal does not have all these elements of symmetry. In fact, there are only 32 possible combinations of these elements of symmetry. These possible combinations divide crystals into 32 *crystal classes*. The class to which a crystal belongs can be determined by the external symmetry of the crystal. The number of crystal classes corresponding to each crystal system are: triclinic, 2; monoclinic, 3; orthorhombic, 3; hexagonal, 7; tetragonal, 7; rhombohedral, 5; cubic, 5.

To illustrate how crystals with the same crystallographic axes (belonging to the same crystal system) can have different combinations of symmetry elements, we

---

† *p*-fold axis of rotation–inversion: rotation through 360°/*p*, followed by inversion through the center.

choose the cubic system with three equal axes at 90°. Rock salt belongs to the cubic system; the crystals have the full symmetry of the cube described above: three four-fold axes, four threefold axes, six twofold axes, nine planes of symmetry and a center of symmetry. The crystal shown in Fig. 26–17(a) also belongs to the cubic system. However, the crystallographic axes have only twofold symmetry; the principal diagonals are threefold axes of symmetry still; there are only six mirror planes. This figure has the symmetry of the tetrahedron; Fig. 26–17(b). The crystal shown in Fig. 26–17(a) belongs to the tetrahedral class of the cubic system rather than the normal class which has the full symmetry of the cube.

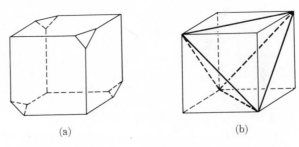

Fig. 26–17 Tetrahedral class of the cubic system. (a) Trun-cated cube. (b) Tetrahedron developed from the cube.

It is possible to determine the symmetry class to which a given crystal belongs by examining a number of specimens of the crystal: measuring interfacial angles, etc. Measurement of optical properties, such as refractive index, which may have different values along different axes, can aid in the determination of symmetry class. It is essential to examine a number of crystals preferably grown under different conditions. The *habit* of the crystal (but not the symmetry class) depends upon how the crystal is grown. Various possible *habits* of NaCl are shown in Fig. 26–18. Both the cube

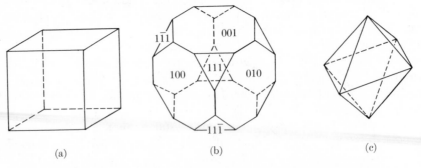

Fig. 26–18 Crystal habit. (a) Cubic NaCl crystal. (b) NaCl crystal showing some development of octahedral faces. (c) NaCl crystal grown in 10% urea solution.

and the octahedron have the same combination of symmetry elements; therefore, a substance such as NaCl may show the faces appropriate to either. Another variation in habit is shown by sodium chlorate, $NaClO_3$. It may grow as a cube, or a tetrahedron, or in a crystal showing a more complicated pattern of faces. The symmetry class is correctly assigned by judging the least symmetrical crystal. The cubic and tetrahedral habits of $NaClO_3$ both exhibit higher symmetry than is proper to the crystal class. The exhibition of a higher symmetry than the true symmetry is a common occurrence. If the crystals are grown quickly, they do not develop all the faces which are proper to the crystal class. The absence of these faces gives the appearance of higher symmetry.

26–9 Symmetry in the Atomic Pattern

The division of crystals into crystal systems and crystal classes is based on the symmetry of the crystal as a finite object, or the symmetry of a single unit cell. In a unit cell all of the corners are equivalent points, since by translation along the axes the entire pattern can be generated; this is shown in Fig. 26–19, where the unit cell is heavily outlined. By translation of the unit cell, the entire pattern of equivalent points, called a *point lattice*, is generated. However, all the possible point lattices cannot be obtained if points are placed only at the corners of the unit cells in the seven systems. It was shown by Bravais that there are seven more unit cells which are required to produce every possible arrangement of equivalent points in space, that is, every point lattice. These fourteen unit cells are the Bravais lattices; Fig. 26–20.

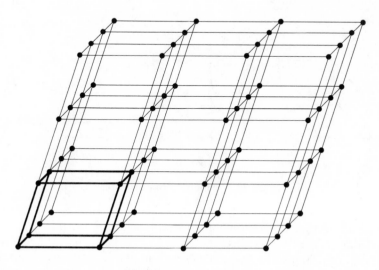

Fig. 26–19 A point lattice.

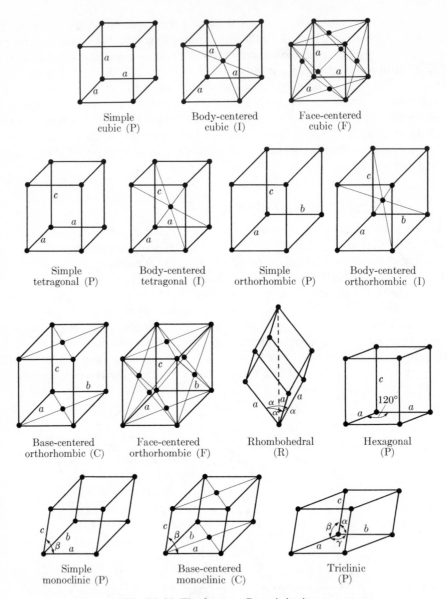

Fig. 26–20 The fourteen Bravais lattices.

In addition to the symmetry within any particular cell, the points in the neighboring cells are related by symmetry to those in that particular cell. Thus we can add symmetry operations which contain an element of translation as well as the other elements appropriate to the finite figure. The addition of translation to the possible symmetry operations greatly increases the number of possible combinations of the symmetry elements. There is a total of 230 possible combinations (space groups); any atomic arrangement in a crystal must have the symmetry corresponding to one of these 230 combinations of symmetry operations. To determine the space group requires a detailed examination of the crystal by x-rays.

The new symmetry elements which are introduced are screw axes, and glide planes. A screw axis in a pattern is exemplified in the structure of selenium which has a threefold screw axis. The chain of selenium atoms winds around the edge of the unit cell. If we imagine a cylinder centered on the edge of a unit cell, Fig. 26–21(a), then a rotation of 120° with a translation of $\frac{1}{3}$ the height of the cell moves atom a to position b, atom b to position c, atom c to a', atom a' to a position in the next unit cell, and so on. Figure 26–21(b) shows the operation of a glide plane. If the upper layer of atoms is moved a distance $a/2$ and then reflected in the plane MM', the lower plane of atoms is generated.

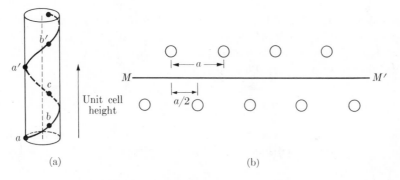

Fig. 26–21 (a) Screw axis. (b) Glide plane.

26–10 The Designation of Crystal Planes and Faces

Knowing that a crystal is built up by the repetition of a unit cell, we can easily explain the development of faces of various kinds; Fig. 26–22(a) illustrates this in two dimensions. The faces F_1 and F_2 are formed by the bottoms and sides of the unit cells. Other faces F_3 and F_4 are possible, formed by the edges of the unit cells. Since the unit cell is of atomic size, we do not see the little steps but see only another face of the crystal. Because the crystal is built in this special way an important relation exists between the axial intercepts of any face and those of any other face. We compare the intercepts on the axes of the face F_3 with those of a possible face F_4. The geometry of these faces is shown in Fig. 26–22(b). The line PL is produced until it intersects the x-axis at a and the y-axis at b. In intercept form the equation of the line is

$$\frac{x}{a} + \frac{y}{b} = 1. \tag{26–1}$$

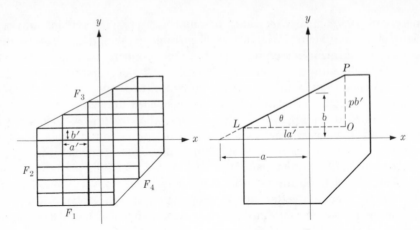

Fig. 26–22 The law of rational intercepts.

Let the width of the unit cell be a', and the height of the unit cell be b'. For the face F_3, suppose OP is pb' and OL is la', where p and l are integers. Then $\tan\theta = pb'/la'$, and also $\tan\theta = b/a$; therefore, $b/a = pb'/la'$, so that $b = (p/l)(a/a')b'$ and Eq. (26–1) becomes, after multiplying by a/a',

$$\frac{x}{a'} + \frac{y}{(p/l)b'} = \frac{a}{a'}. \tag{26–2}$$

But there are other points on F_3: $x = ma'$, $y = nb'$, where m and n are integers. Equation (26–2) must be satisfied at these points. Hence

$$m + \frac{nl}{p} = \frac{a}{a'}; \tag{26–3}$$

the left-hand side of this equation involves only integers; hence a/a' is *rational*, expressible as a ratio of integers. From the earlier equality, $b/b' = (p/l)(a/a')$, it follows that b/b' is rational. Since the face F_3 was not a special one, it follows that the axial intercepts of any face, measured in units of the length of the unit cell, are rational numbers. The argument in three dimensions goes in the same way, except that we deal with the intercepts of planes on the three axes. Since the intercepts of any plane are rational multiples of the length of the unit cell, it follows that the intercepts of two planes are rational multiples of each other. Let the intercepts on the x-axis of two faces be a_1 and a_2; then $a_1 = r_1a'$ and $a_2 = r_2a'$, where r_1 and r_2 are rational. It follows that $a_2 = (r_2/r_1)a_1$. Since r_1 and r_2 are rational a_2 is a rational multiple of a_1. The same argument can be made for the y and z intercepts.

Therefore, if to a given face of a crystal we assign intercepts, a, b, c on the coordinate axes, then the intercepts a_1, b_1, c_1 of any possible face of the crystal are rational multiples of a, b, c. This is a fundamental law of crystallography, the law of rational intercepts.

Instead of describing a given face of the crystal by multiples of standard intercepts, we use the reciprocals of these multiples. That is, in terms of the intercepts a, b, c of the reference face, the intercepts of any face are given by

$$a_1 = a/h, \qquad b_1 = b/k, \qquad c_1 = c/l.$$

The numbers h, k, l are rational numbers or zero. If any of h, k, or l are fractions, the whole set is multiplied by the least common denominator, to yield a set of integers h, k, l. The resulting integers h, k, l are called the Miller indices of the face. Through this process of taking reciprocals and clearing fractions, the law of rational intercepts becomes the law of rational indices. It should be clear that they are one and the same law. The indices of a face describe its orientation relative to the reference face, but do not describe the actual position. The usefulness of the indices in describing the face can be seen from writing the equation of the plane in intercept form:

$$\frac{x}{a'} + \frac{y}{b'} + \frac{z}{c'} = 1.$$

But in terms of the intercepts of the reference plane, $a' = a/h$, etc.; hence

$$\frac{hx}{a} + \frac{ky}{b} + \frac{lz}{c} = 1.$$

If we measure distances in terms of the reference intercepts, x in units of a, y in units of b, etc., then the equation becomes

$$hx' + ky' + lz' = 1, \tag{26–4}$$

where $x' = x/a$, $y' = y/b$, $z' = z/c$.

Consider the cube in Fig. 26–23. The intercepts of the right-hand side of the cube are ∞, 1, ∞. The reciprocals (indices) are 010. The front face of the cube has intercepts 1, ∞, ∞, and indices 100; the left face has intercepts ∞, -1, ∞, and indices $0\bar{1}0$. (The minus sign is written over the number.) The rear face has indices

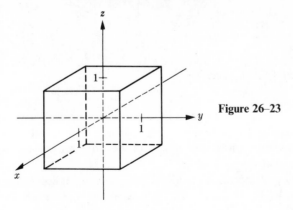

Figure 26–23

$\bar{1}00$. The indices of the top and bottom faces of the cube are 001, and 00$\bar{1}$. Any plane parallel to an axis has an intercept of ∞, and an index of zero for that axis.

The crystal consists of atoms, ions, or molecules; any face of a crystal consists of a layer of atoms, or ions or molecules. The method of describing faces of a crystal can be used to describe the planes of atoms in the crystal. Consider the body-centered cubic cell in Fig. 26–24(a); the intercepts of the shaded planes of atoms are ∞, 1, 1, so the indices are (011). In Fig. 26–24(b), the shaded plane has intercepts ∞, 2, 1, so the reciprocals are 0, $\frac{1}{2}$, 1; clearing the fraction the indices are (012). Figure 26–24(c), (d), and (e) show the (101), the (110), and the (111) planes.

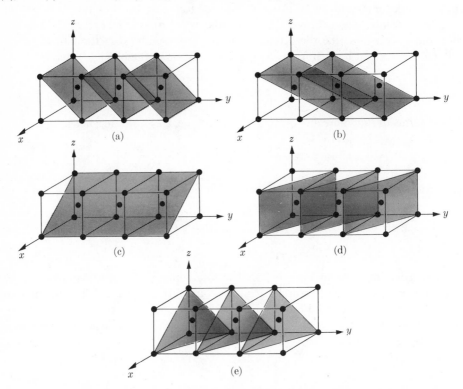

Fig. 26–24 Planes in the bcc lattice. (a) (011) planes. (b) (012) planes. (c) (101) plane. (d) (110) planes. (e) (111) planes.

Figure 26–18(b) showed the development of octahedral faces on a cube. These faces are obtained by truncating the eight corners of the cube. The crystal faces are designated by Miller indices obtained in the usual way. The angle between the 100 and 111 face is always 54° 44′ 8″.

The constancy of interfacial angles is a law of crystallography which has been recognized since the 17th century. In 1669, Nicolaus Steno observed that regardless

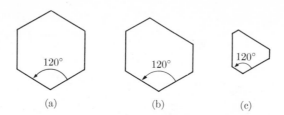

Fig. 26–25 The constancy of interfacial angles. (a) Section of an ideal quartz crystal. (b) and (c) Possible shapes of the crystal section.

of how the real crystal might be distorted from the ideal shape, the angles between two types of face were always constant. For example, a crystal of quartz has hexagonal symmetry, and a section cut perpendicular to the hexagonal axis should have a regular hexagonal shape, Fig. 26–25(a), each interior angle being 120°; in actual fact because the growth of certain faces in the crystal is inhibited by the crowding of neighboring crystals and other influences, the faces of a real crystal develop unequally, and the section of real quartz crystals may have shapes such as those shown in Fig. 26–25(b) and (c). In all the possible distortions, the interfacial angle remains at 120°. This constancy of interfacial angles introduces an enormous simplification into crystallography, for it permits the recognition of the fundamental crystal symmetry from the interfacial angles of imperfect crystals.

26–11 The X-ray Examination of Crystals

The conclusion that x-rays were light rays of very short wavelength and the realization that a crystal consisted of a regular array of planes of atoms prompted Max von Laue in 1912 to suggest that the crystal should behave as a diffraction grating for x-rays if the wavelength were comparable to the spacing in the crystal. This suggestion was confirmed experimentally almost immediately by Friedrich and Knipping. Figure 26–26 shows a typical set-up for the Laue method. The diffraction pattern of spots produced on the photographic plate is called a Laue pattern.

An x-ray beam progressing through a crystal is reflected from every possible plane of atoms in the crystal by the usual law of specular reflection, the incidence angle equal to the reflection angle. Since there are many different planes of atoms all oriented at different angles relative to the incident beam, one might expect that the emergent beam would be completely diffused over all angles. The fact is that the emergent beam appears only at certain particular angles, and thus produces the Laue pattern. This happens because a plane of atoms is not present in the crystal by itself but with an enormous number of similar planes parallel to it. As a general rule the reflected beams from these parallel planes interfere destructively, and there is no emergent beam in most directions.

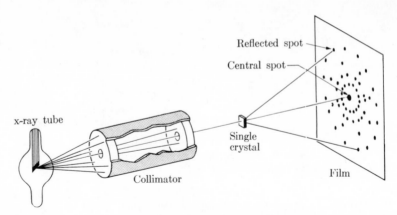

Fig. 26–26 The Laue method.

The condition for the reflected beams from a given set of parallel planes to rein-force each other and produce a spot is easily derived. Figure 26–27 shows a set of planes within a crystal having an interplanar spacing d. The angle between the planes and the direction of the beam is the glancing angle θ. Ray R_1 is reflected specularly by the first plane to yield R_1'. Similarly, ray R_2 is reflected specularly from the second plane to yield R_2'. If the rays R_1' and R_2' are to reinforce one another, they must have the same phase; this condition is met if the extra distance traversed by R_2R_2' is equal to an integral number of wavelengths of the x-ray. The extra distance is $2x$, so that $2x = n\lambda$, where n is an integer. But from the geometry of the situation, $x = d \sin \theta$. Consequently, in terms of the interplanar spacing d, the condition for constructive interference becomes

$$2d \sin \theta = n\lambda, \qquad n = 1, 2, 3, \ldots, \tag{26–5}$$

which is the fundamental law of x-ray crystallography, the Bragg condition, or Bragg's law. It states that for a given wavelength of x-rays, the reflected beam will emerge only at those angles for which the condition is satisfied. This accounts for the Laue pattern of spots. Each spot is produced by a certain set of planes which fulfills the condition. Since similar sets of planes are disposed within the crystal in accordance with the crystal symmetry, the arrangement of the spots in the Laue pattern has the symmetry of the crystal to a certain degree.

The reflected beam makes an angle 2θ with the direction of the incident beam. This is shown by the simple geometry of Fig. 26–27. It is important to note that the Laue method uses "white" radiation containing a wide range of wavelengths; every set of planes can pick out the wavelength which can satisfy the Bragg relation for the particular value of the interplanar spacing, and thus every set of planes can produce a spot in the Laue pattern. If monochromatic radiation were used in the Laue method, unless the Bragg relation were accidentally fulfilled for some set of planes, no pattern would be produced.

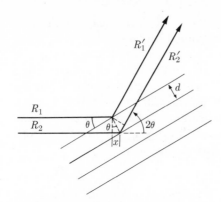

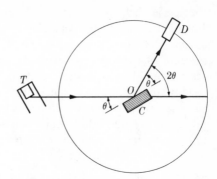

Fig. 26–27 X-ray reflection from a
set of planes.

Fig. 26–28 X-ray diffractometer.

In the Bragg x-ray diffractometer, Fig. 26–28, x-radiation from the tube T falls
upon a crystal C, which is mounted so that it may be rotated, the angle of rotation
being measured on the scale of the instrument. By rotating the crystal it is possible
to bring the coherent scattered beam from each set of planes into the detecting
chamber D. The response of the detector, an ionization chamber or a Geiger counter,
at various angles of rotation can be recorded to produce a pattern, such as that shown
in Fig. 26–29, for a crystal of tungsten. The numbers on the peaks are the indices
of the planes which produce that peak. By measuring the values of θ, the interplanar
spacing can be computed from the Bragg equation.

The fact that some planes have a higher density of atoms than others produces
the variation in intensity of the diffracted beam for different sets of planes, an obvious
feature in Fig. 26–29. The planes of high atomic density scatter x-rays better and
produce the more intense beam. If more than one kind of atom is present in the crystal,

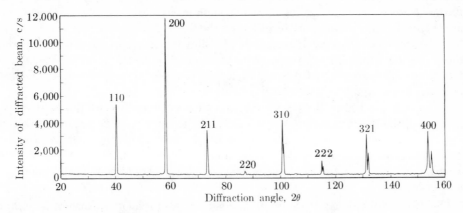

Fig. 26–29 Typical diffraction pattern. Intensity at various angles for tungsten.

the species with the greater number of electrons has the greater scattering power. For light elements, the scattering power is proportional to the number of electrons around the atom. For this reason, it is not possible to deduce anything about the position of hydrogen atoms from x-ray diffraction.† With only one electron, the scattering ability of hydrogen atoms is so small that any effect it produces is over-shadowed by the effect of neighboring atoms which have more electrons. For the same reason it is impossible to distinguish between atoms which differ only by the possession of one additional electron. Carbon and nitrogen, for example, are in-distinguishable because their scattering ability is about the same.

26-12   Debye-Scherrer (Powder) Method

The Bragg method of obtaining interplanar spacings has the disadvantage that mount-ing the crystal on a precise axis is time consuming. The *rotating crystal* method of obtaining a diffraction pattern, one of the most valuable methods for structure determination, also requires precise mounting of the crystal.

The simplest method of obtaining interplanar spacings is the Debye-Scherrer method. A sample of the crystal is ground to a powder and placed in a thin-walled glass tube which is mounted in the x-ray beam. Since many crystals are present, all having different orientations, there are some which will be so oriented as to satisfy the Bragg relation for a given set of planes. Another group will be oriented so that the Bragg relation is satisfied for another set of planes, and so on. In this method a cone of reflected radiation is produced, Fig. 26–30, which is recorded on the film as a curved line. The cone, rather than a single ray which produces a spot, results from the fact that if a given set of planes satisfies the Bragg condition, and that set is rotated about the axis of the incident beam, the Bragg condition will still be satisfied, but the reflected ray will describe a cone. All the orientations about the axis of the beam are present in different particles of the powder. Figure 26–30 shows the conical reflected radiation and the position of the film which results in the line pattern. The film makes a nearly complete circle so that even the rays which are reflected at large angles are recorded on the film.

The distances between the lines on the film are measured accurately. From these and the dimensions of the camera, the diffraction angle 2θ can be calculated for each set of planes. From the Bragg angle θ the interplanar spacings d are calculated from the Bragg equation.

Through the use of the densitometer, a device for measuring the light trans-mission of the developed film, the intensity of the lines on a powder photograph can be measured. If, for example, the material whose structure is under study is a metal, then all the atoms are of the same kind. The most intense line will be reflected from the planes which are most closely packed. These would be the (111) planes in

---

†   The location of hydrogen atoms can be determined in a straightforward manner by nuclear magnetic resonance measurements.

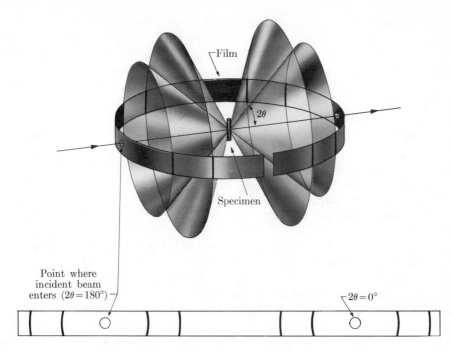

Fig. 26–30 Debye-Scherrer powder method.

an fcc structure and the (001) planes in the hcp structure. Having identified the most closely packed planes and calculated their spacing from the Bragg equation, it is sometimes possible using other simple features of the pattern to identify the structure and establish all of the interplanar spacings by simple geometry.

26–13 Intensities and Structure Determination

Consider the face-centered cubic lattice in Fig. 26–31. The (100) planes are interleaved at just half the spacing by the (200) planes, which contain only face-centered atoms. The reflected rays from this second set of planes are 180° out of phase with

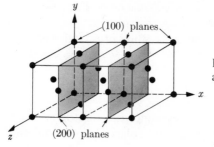

Fig. 26–31 Interleaving of the (200) and (100) planes in the fcc structure.

those from the (100) planes. The two reflections interfere destructively so that a first-order reflection does not appear from the (100) planes in this lattice. (Higher-order reflections appear from both sets, but the intensities are much weaker.) For the same reason the first-order reflection from the (110) plane does not appear, being destroyed by the first-order reflection from (220) planes. The (111) planes are not interleaved in this way so that the first-order reflection comes through loud and clear, especially so because the (111) planes are close packed. The absence of certain lines helps enormously in the assignment of indices, the indexing, of the lines which do appear.

From a complete study of line spacings and intensities in the diffraction pattern it is possible to determine the size and shape of the unit cell and the arrangement of the kinds of atoms within the cell. However, it is only in crystals of high symmetry, which have no more than two kinds of particles, that it is possible to do this in a reasonably straightforward manner, in metals, for example, or in crystals such as NaCl or ZnS.

Symmetrically interleaved layers of atoms of the same kind can completely extinguish certain reflections. Suppose, however, that the interleaved layers are not just halfway between, so that the reflected radiation is neither completely in nor out of phase with that from the first set of layers; add to this the fact that the interleaved layer may contain atoms of different scattering ability. The problem becomes quite complicated, and the solution is accomplished by trial. From the positions of the lines it is possible to establish the principal spacings in the structure, the shape and size of the unit cell, and the space group if the crystal class is known. That much is straightforward. Knowledge of the chemical constitution and density of the crystal leads to the number of atoms of each kind in the unit cell. Then the problem is to fix the positions of the various atoms in the unit cell. This is done by guess, or intuition. The intensities of the lines or spots in the diffraction pattern are calculated from an assumed arrangement of the atoms. The calculated intensities are compared with those observed. This procedure is continued until reasonable agreement is reached. The structure which gives this agreement is assumed to be correct. Because it is not possible to try all possible arrangements of the atoms with all possible parameters and thus exclude every structure but the correct one, the structures of complicated materials are accepted with some reserve.

The calculation of intensities is very tedious; fortunately it can be given to a computer. It is not too much to hope that the computer can make enough trials to eliminate the possibility of assigning an incorrect structure. Even before the use of computers, the structures of thousands of crystals had been worked out by hand calculation. We have seen the fruits of the x-ray study of crystals in the structures described in the earlier parts of this chapter.

26-14 X-ray Diffraction in Liquids

The diffraction pattern of a liquid resembles a powder photograph except that the very sharp lines of the powder photograph are replaced by a few broad bands of reflected radiation. From an analysis of the intensity distribution in these broad bands, it is possible to construct the radial distribution function for particles around a central particle in the liquid. This distribution function is interpreted in terms of the *average* number of atoms surrounding a central atom at the distance corresponding to the peak.

Figure 26–32 shows the radial distribution function, $4\pi r^2 \rho$, in liquid sodium. The upper drawing interprets the peaks in terms of "shells" of atoms around the central atom. At 4 angstroms from the central atom, the average number of atoms in the liquid is 10.6. This number is determined by the shaded area under the curve. The vertical lines show the number of atoms in successive "shells" in solid sodium.

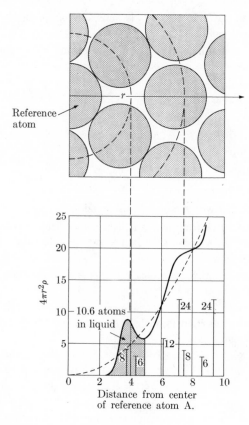

Fig. 26–32 Radial distribution curve in liquid sodium.

Problems

26–1. Using the data in Table 26–1 compute the axial ratio c/a for the metals crystallizing in the hcp system and compare with the ideal value, 1.633.

26–2. Figure 26–5(a) and (c) show the unit cell for the fcc and bcc structures. How many atoms does the unit cell contain in each of these cases? [Note that an atom on a face is shared between two cells, an atom at a corner is shared between a number (how many?) of cells.]

26–3. The hexagonal cell shown in Fig. 26–5(b) consists of three unit cells. How many atoms are in the hexagonal cell shown and how many in the unit cell?

26–4. Referring to Fig. 26–5, which shows the location of the atom centers for the fcc, hcp, and bcc structures, if the edge length of the cube is a, compute, for the fcc and bcc structures, the volume of the cube and the volume of the cube which is actually occupied by the spheres. The spheres are in close contact; keep in mind that a sphere on a face or at a corner is only partially inside the cube. What percent of the space within the cube is empty?

26–5. Figure 26–7 shows two unit cells of CsCl. How many Cs^+ and Cl^- ions in the unit cell?

26–6. Figure 26–9 shows the unit cell of NaCl. How many Na^+ and Cl^- ions in the unit cell? (There is a sodium ion at the center of the cube!)

26–7. Using Fig. 26–8, how many octahedral holes per atom are present in the fcc structure? By sketching the bcc structure decide how many octahedral holes per atom are present.

26–8. Calculate the size sphere which can be accommodated in the octahedral hole of the fcc structure; cube edge $= a$, atom radius $= r_a$.

26–9. How many atoms (or ion pairs) are in the unit cell of

a) diamond, Fig. 26–12;   b) zinc blende, Fig. 26–10(a);   c) wurtzite, Fig. 26–10(b);
d) fluorite, Fig. 26–11(a);   e) rutile, Fig. 26–11(b).

26–10. What are the elements of symmetry of a tetragon?

$$a = b \neq c, \qquad \alpha = \beta = \gamma = 90°.$$

26–11. a) Sketch a cube and label each face with the proper Miller indices.

b) Suppose that every edge of the cube is truncated by a plane perpendicular to the plane containing the edge and center of the cube. Sketch at least two of the faces exposed and find the Miller indices of these faces.

26–12. Using Fig. 26–5(a), sketch the arrangements of atoms in (111) plane, in the (100) plane, in the (011) plane. Which plane is close packed?

26–13. Using x-rays of wavelength $\lambda = 1.790 \times 10^{-8}$ cm, a metal produces a reflection at $2\theta = 47.2°$. If this is a first-order reflection from the (110) planes of a body-centered cubic lattice what is the edge length of the cube?

26–14. The lattice parameter of silver, an fcc structure, is 4.086 Å. An x-ray beam produces a strong reflection from the (111) planes at $2\theta = 38.2°$. What is the wavelength of the x-ray?

26–15. Using x-radiation, $\lambda = 1.542$ Å, a face-centered cubic lattice produces reflections from the (111) and (200) planes. If the density of copper, which is face-centered cubic, is 8.89 gm/cm^3, at what angles will the reflections from copper appear?

Chapter Twenty-seven

The Relation between
Structure and Macroscopic
Properties

27–1 Preliminary Remarks

In the discussions of the kinetic theory of gases and of intermolecular forces, we obtained expressions for properties of matter in bulk in terms of the properties of the individual molecules. In this chapter we will describe the cohesive energy of ionic crystals in terms of the interactions of the ions in the crystals, and some of the properties of metals and covalent crystals in terms of the quantum mechanical picture obtained from the Schrödinger equation. In Chapter 28 the method for calculating the thermodynamic properties of bulk systems from a knowledge of structure will be described.

27–2 Cohesive Energy in Ionic Crystals

A satisfactory theory of the cohesive energy of ionic crystals can be based almost exclusively on Coulomb's law. If two particles i and j, having charges z_i and z_j, are placed a distance r_{ij} apart in vacuum, the energy of interaction between them is

$$E_{ij} = z_i z_j / r_{ij}. \tag{27–1}$$

If z_i and z_j have the same sign, E_{ij} is positive, the particles repel one another, and the system lacks stability. If z_i and z_j are opposite in sign, the energy is negative, and the two are bound together by the energy E_{ij}. The more ions of opposite charge which surround a particular ion, the greater the stability of the structure. We used this result implicitly in arguing for close-packed structures in ionic crystals in Section 26–4.

Consider a crystal, such as NaCl, in which the charges on the ions are equal and opposite in sign, $z_+ = -z_-$. To compute the cohesive energy of a crystal which contains N ions of each sign, N ion pairs, we add the interaction of every ion with all the others. Since the interaction energies depend on the distances between the ions, we must know the geometrical arrangement of the ions in the structure. Fortunately, for symmetrical salts (rock salt, CsCl, zinc blende, and wurtzite), the possible structures are all cubic, so that if we know the distance r between a cation and the nearest anion, then the distance between any two ions in the crystal can be computed from the geometry of the structure. We write for the distance between ions i and j, $r_{ij} = \alpha_{ij}r$ in which α_{ij} is a numerical factor obtained from the geometry of the structure. The expression in Eq. (27–1) becomes

$$E_{ij} = z_i z_j / r\alpha_{ij}.$$

The interaction energy of ion i with all of the others, j, is obtained by summing this expression:

$$E_i = \frac{z_i}{r} \sum_{j \neq i} \frac{z_j}{\alpha_{ij}}.$$

In every term of this sum, $z_j = \pm z_i$; furthermore, each term in the sum is simply a number, determined by geometry, so the sum is a number which we write as $z_i S_i$. Then

$$E_i = z_i^2 S_i / r.$$

Summing the energies of all the ions in the lattice yields the total energy of interaction E_M:

$$E_M = \tfrac{1}{2} \sum_i E_i = \frac{z^2}{r} \sum_i \tfrac{1}{2} S_i,$$

where the factor $\tfrac{1}{2}$ appears because we must count the interaction between any pair of ions only once. The sum is a sum of numbers; it is negative and proportional to N, so we write $\tfrac{1}{2}\sum S_i = -NA$, where A is a numerical factor, the Madelung constant, named after E. Madelung, who first evaluated sums of this type. The total electrostatic energy, the Madelung energy, is

$$E_M = -NAz^2/r. \qquad (27\text{–}2)$$

Table 27–1

Structure	Coordination	A
CsCl	8-8	1.7627
NaCl	6-6	1.7476
Zinc blende	4-4	1.6381
Wurtzite	4-4	1.641

Values of the Madelung constant calculated from the geometry of the symmetrical structures are given in Table 27–1.

The cohesive energy of ionic crystals given by Eq. (27–2) is about 10% too large. This is a consequence of neglecting the repulsion which arises at close distances. Figure 27–1 shows how this comes about. As a function of r, the energy given by Eq. (27–2) follows the dotted curve in the figure. At the equilibrium separation r_0, the depth of the curve is somewhat below that of the minimum in the solid curve, which represents the actual cohesive energy.

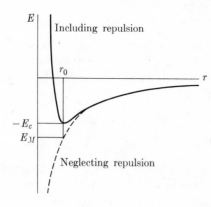

Fig. 27–1 Energy in an ionic crystal as a function of interionic distance.

The repulsion which develops as the ions come in contact is represented, just as with neutral molecules, Section 25–9, by a term b/r^n, where b and n are constants and n is a large power, usually $n = 6$ to 12. This form of the repulsion energy was first introduced by M. Born, and is called the Born repulsion. The cohesive energy is written

$$E = -\frac{NAz^2}{r} + \frac{b}{r^n}. \qquad (27\text{–}3)$$

The two empirical constants b and n are determined by two conditions. First we require that the energy have a minimum value at r_0, the equilibrium separation in the crystal; $(dE/dr)_{r=r_0} = 0$. Differentiating Eq. (27–3), we have

$$\frac{dE}{dr} = \frac{NAz^2}{r^2} - \frac{nb}{r^{n+1}}.$$

Setting this equal to zero at r_0 and solving for b, we obtain

$$b = \frac{NAz^2 r_0^{n-1}}{n}, \qquad (27\text{–}4)$$

which reduces Eq. (27–3) to

$$E = -\frac{NAz^2}{r_0}\left[\frac{r_0}{r} - \frac{1}{n}\left(\frac{r_0}{r}\right)^n\right]. \qquad (27\text{–}5)$$

At $r = r_0$ this becomes the negative of the cohesive energy $-E_c$:

$$-E_c = -\frac{NAz^2}{r_0}\left(1 - \frac{1}{n}\right). \tag{27-6}$$

The cohesive energy E_c is the Madelung energy at $r = r_0$ multiplied by $(1 - 1/n)$. If n is about 10, then the 10% error noted above is explained. Using Eq. (27–6) in Eq. (27–5), we obtain

$$E = -\frac{nE_c}{n-1}\left[\frac{r_0}{r} - \frac{1}{n}\left(\frac{r_0}{r}\right)^n\right]. \tag{27-7}$$

The constant n is determined from the compressibility of the crystal at $0°K$; $\beta = -(1/V_0)(\partial V/\partial p)_T$. At $T = 0°K$ the thermodynamic equation of state, Eq. (10–28), becomes $p = -(\partial E/\partial V)_T$; by differentiating, $(\partial p/\partial V)_T = -(\partial^2 E/\partial V^2)_T$. Combining this result with the definition of β, we obtain

$$\frac{1}{V_0\beta} = \left(\frac{\partial^2 E}{\partial V^2}\right)_T. \tag{27-8}$$

To express the energy in terms of the volume of the crystal, we observe that the volume is proportional to the cube of the interionic distance r, so that r is proportional to the cube root of the volume; therefore, $(r_0/r) = (V_0/V)^{1/3}$, and Eq. (27–7) becomes

$$E = -\frac{nE_c}{n-1}\left[\left(\frac{V_0}{V}\right)^{1/3} - \frac{1}{n}\left(\frac{V_0}{V}\right)^{n/3}\right].$$

Differentiating, we have

$$\frac{dE}{dV} = \frac{nE_c}{3(n-1)}\left[\frac{V_0^{1/3}}{V^{4/3}} - \frac{V_0^{n/3}}{V^{1+n/3}}\right],$$

and

$$\frac{d^2E}{dV^2} = -\frac{nE_c}{9(n-1)}\left[\frac{4V_0^{1/3}}{V^{7/3}} - \frac{(n+3)V_0^{n/3}}{V^{2+n/3}}\right].$$

At $V = V_0$, this becomes $(d^2E/dV^2)_{r_0} = nE_c/9V_0^2$. Using this value and the value of E_c from Eq. (27–6) in Eq. (27–8), we obtain, after solving for n,

$$n = 1 + \frac{9V_0 r_0}{NAz^2\beta}. \tag{27-9}$$

Having determined the value of n from the compressibility and the volume V_0, by Eq. (27–9), the cohesive energy of the crystal can be obtained from Eq. (27–6). A few values for E_c are given in Table 27–2.

Table 27–2

Salt	LiCl	NaCl	KCl	RbCl	CsCl
n	7.0	8.0	9.0	9.5	10.5
E_c, kcal/mole	193.3	180.4	164.4	158.9	148.9

The cohesive energy E_c of an ionic crystal is the energy of the crystal relative to the infinitely separated ions, the energy required for the reaction

$$MX(s) \rightarrow M^+(g) + X^-(g).$$

The energy of this reaction is not directly measurable, and therefore is determined indirectly. The Born-Haber cycle of reactions is used.

Reaction	Energy
$M(s) \rightarrow M(g)$	S = sublimation energy
$M(g) \rightarrow M^+(g) + e^-(g)$	I = ionization energy
$\frac{1}{2}X_2(g) \rightarrow X(g)$	$\frac{1}{2}D = \frac{1}{2}$ the dissociation energy
$e^-(g) + X(g) \rightarrow X^-(g)$	$-E_A$ = minus electron affinity
$X^-(g) + M^+(g) \rightarrow MX(s)$	$-E_c$ = minus cohesive energy.

Summing these yields the formation reaction of MX(s);

$$M(s) + \tfrac{1}{2}X_2(g) \rightarrow MX(s), \qquad \Delta E_f = \text{energy of formation.}$$

Therefore,

$$\Delta E_f = S + I + \tfrac{1}{2}D - E_A - E_c,$$

$$E_c = S + I + \tfrac{1}{2}D - E_A - \Delta E_f. \qquad (27\text{–}10)$$

The values computed for E_c from the experimental values of the quantities on the right of Eq. (27–10), agree with the values predicted by Eq. (27–6) to within about 4% for the alkali halides.

The theory can be refined somewhat by including the van der Waals attraction of the electron clouds of the ions; this is more important for a substance, such as CsI, in which the electron clouds are large and floppy, than for LiF, in which the electron clouds are small and tightly bound. The presence of van der Waals interaction increases the cohesive energy slightly. The only important contribution of quantum mechanics to this problem is the requirement that the zero-point energy of the crystal be included in the calculation. This decreases the calculated value of the cohesive energy by about 0.5 to 1.0%. These additional contributions do not change the values in Table 27–2 by more than 2 or 3%.

Attention should be directed to the magnitude of the cohesive energy in uniunivalent ionic crystals, which ranges from 150 to 200 kcal/mole. This is 10 to 20 times larger than that found in van der Waals crystals. Furthermore, as the ions get larger, the cohesive energy decreases, Eq. (27–6). The extreme values are: LiF, $E_c = 240$ kcal/mole; CsI, $E_c = 136.1$ kcal/mole. The larger ions are simply farther apart in the crystal. Finally, if one considers crystals made up of divalent ions, CaO, BaO, etc., Eq. (27–6) predicts that the cohesive energy should be proportional to the square of the charge so the energies should be roughly four times greater than the energies of 1–1 salts. This is approximately correct; the values for CaO and BaO are 842 kcal and 747 kcal, respectively.

The increase in cohesive energy in the 2–2 salts explains the generally lower solubility of these salts, e.g., the sulfides, as compared with that of the alkali halides. The greater the cohesive energy, the more difficult it is for a solvent to break up the crystal.

27–3 The Electronic Structure of Solids

For an isolated atom the quantum mechanics predicts a set of energy levels of which some, but not all, are occupied by electrons. What happens to this scheme of energy levels if many atoms are packed closely together in a solid? Consider two helium atoms, infinitely far apart; each has two electrons in the $1s$ level. As these two atoms approach, they attract each other slightly, the interaction energy has a shallow minimum at some distance. Since each atom is influenced by the presence of the other, the energy levels on each atom are slightly perturbed. The $1s$ level splits into a set of two levels, which may be thought of as the energy levels for the systems $(He)_2$. Each of these levels can accommodate two electrons; the four electrons of the system fill the two levels. The average energy of the two levels is slightly less than the energy of the $1s$ level of the isolated atom. This slight lowering of the average energy is the cohesive energy, the van der Waals interaction energy, of the system $(He)_2$. If three helium atoms were brought together, the system would have a set of three closely spaced $1s$ levels. In a system of N atoms, the $1s$ levels split into a group of N closely spaced levels called an *energy band*, the $1s$ band. For a collection of N helium atoms, since the $1s$ level is fully occupied in the individual atoms, the $1s$ band is completely occupied. For helium and for any saturated molecule which forms a van der Waals solid, the width of the band, the energy difference between the topmost and lowermost levels in the band, is very small, because of the weak interaction between saturated molecules. To a good approximation, the energy level scheme in a van der Waals solid is much like that in the individual molecules which compose the solid, the filled levels being displaced downward very slightly to account for the cohesive energy of the solid.

Consider a solid which contains N atoms of one kind only. For each energy level in the isolated atom which accommodates two electrons there is in the solid an energy band containing N levels each of which can accommodate two electrons. This energy band has a definite width, a fact which implies that the N levels within the band are very closely spaced. They are so closely spaced that the band may be considered as a continuum of allowed energies; it is often called a quasi-continuous band of levels. Figure 27–2 illustrates schematically the contrast between the energy level systems in an isolated atom and in a solid. The shaded regions in the figure cover the ranges of energy permitted to an electron in the solid, the energy bands, while the spaces between the bands are the values of energy which are not permitted. Figure 27–2 has been drawn so that none of the bands overlap; ordinarily the higher energy bands do overlap.

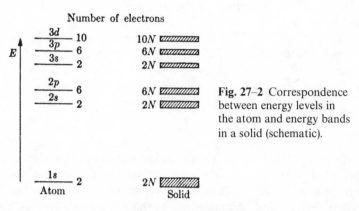

Fig. 27–2 Correspondence between energy levels in the atom and energy bands in a solid (schematic).

Consider metallic sodium. The sodium atom has eleven electrons in the configuration $(1s)^2(2s)^2(2p)^6(3s)^1$. Bringing many sodium atoms together in the crystal scarcely affects the energies of the electrons in the 1s, 2s, and 2p levels, since the electrons in these levels are screened from the influence of the other atoms by the valence electron; the corresponding bands are filled. The levels in the valence shell are very much influenced by the presence of other atoms and split into bands as shown in Fig. 27–3(a). The 3s and 3p bands have been displaced horizontally to illustrate the effect of overlapping bands. The N valence electrons fill the lowest levels available, which results in a partial filling of both the 3s and 3p bands; the filled portion of the bands is indicated by the shading. The overlapping of s and p bands is a characteristic feature of the electronic structure of metals; the merged bands are often designated as an sp band.

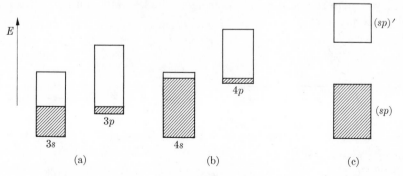

Fig. 27–3 Band structure in solids. (a) Sodium. (b) Calcium. (c) Diamond.

The *s* band can hold $2N$ electrons or 2 electrons/atom. Were it not for the fact that the *p* band overlaps the *s* band, the *s* band would be completely filled in divalent metals such as calcium. As we shall see shortly, if the *s* band were filled and a gap of forbidden energies separated the top of the *s* band from the bottom of the *p* band, then the divalent metals would be insulators. As is shown in Fig. 27–3(b) the *p* and *s* bands in calcium overlap slightly; the shaded area indicates the way in which the two electrons of calcium fill the bands.

Diamond is a crystal with filled bands. The *s* band, which holds 2 electrons/atom, and the *p* band, which holds 6 electrons/atom, interact in diamond to form two distinct bands each of which holds four electrons per atom; these bands are designated *sp* and $(sp)'$ in Fig. 27–3(c). The four electrons per atom in diamond exactly fill the lower band. Diamond with this filled band is an insulator.

27–4   Conductors and Insulators

A crystal with completely filled energy bands is an insulator, and one with partially filled bands is a conductor. The band in a real crystal contains as many levels as atoms in the crystal, but for argument's sake suppose we imagine that the band has only eight levels in it; Fig. 27–4(a). We may suppose that half of these levels are associated with motion of the electrons in the $+x$-direction and half with motion in the $-x$-direction. This is indicated by the arrowheads on the levels. No matter how the band is filled, half of the electrons are in levels corresponding to motion in the $+x$-direction and half in levels corresponding to motion in the $-x$-direction; there is consequently no net motion in one direction and no current flow. If we apply an electric field in the $+x$-direction, the energy of one set of levels is lowered and the energy of the other set is raised; Fig. 27–4(b). If the band is full, then all levels are occupied before and after the application of the field, and there is still no net electronic motion in either direction; the crystal is an insulator. However, if the band is only partly filled, then only the lowest levels are occupied; application of the field rearranges the positions of the levels, and the electrons drop into the lowest set of levels in the presence of the field. In this lowest set of levels, the ones corresponding to motion in the $-x$-direction predominate, so there is a net flow of electrons to the left; a net current flows and the crystal is a conductor.

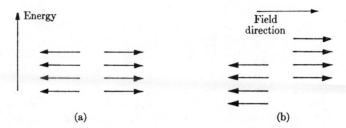

Fig. 27–4 Displacement of energy levels in a band by an electric field. (a) Field off. (b) Field on.

Metals which conduct by electron flow have incompletely filled bands, while insulators such as diamond have completely filled bands. If it is possible to raise electrons from a filled band in an insulator to an empty band of higher energy, then these excited electrons can carry a current. Since the energy gap between the bands is fairly large, this ordinarily cannot be done by an increase in temperature to supply sufficient thermal energy. By using light of high enough frequency it is possible to excite the electrons. The phenomenon is called photoconductivity. Visible light will do this for selenium.

27–5 Ionic Crystals

In the first approximation, the band system of a crystal containing two different kinds of atom may be regarded as a superposition of the band systems of the two individual particles. The band system for sodium chloride is shown in Fig. 27–5. The eight electrons occupy the 3s and 3p bands of the chloride ion, while the 3s band of the sodium ion, which has a higher energy, is vacant. This is a quantum mechanical way of saying that the crystal is made up of sodium ions and chloride ions rather than of atoms of sodium and chlorine. The filled bands are separated from the empty bands by an energy gap so that sodium chloride is an insulator.

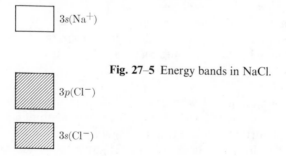

$3s(\text{Na}^+)$

$3p(\text{Cl}^-)$

$3s(\text{Cl}^-)$

Fig. 27–5 Energy bands in NaCl.

27–6 Semiconductors

Semiconductors are solids which exhibit a feeble electrical conductivity which increases with increase in temperature. (The conductivity of metals decreases with increase in temperature.) Semiconductivity appears in insulators which are slightly contaminated with foreign substances, and in compounds, such as Cu_2O and ZnO, which do not contain exactly stoichiometric amounts of metal and nonmetal.

Pure silicon is an insulator, similar to diamond in both crystal structure and electronic structure. The electronic structure in pure silicon can be represented by the filled and empty bands shown in Fig. 27–6(a). Suppose we remove a few of the silicon atoms and replace them by phosphorus atoms, each of which has one more electron than the silicon atom. The energy levels of the phosphorus atoms, impurity levels, are superposed on the band system of the silicon; these levels do not match those in silicon exactly. (Since there are so few phosphorus atoms, the levels are not split into bands.) It is found that the extra electrons introduced by the phosphorus

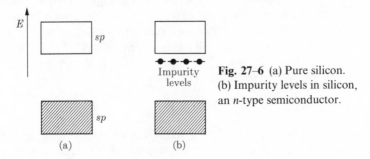

Fig. 27–6 (a) Pure silicon.
(b) Impurity levels in silicon, an *n*-type semiconductor.

atoms occupy the impurity levels shown in Fig. 27–6(b) which are located slightly below the empty band of the silicon lattice. In these levels the electrons are bound to the phosphorus atoms and cannot conduct a current; since the energy gap between these levels and the empty band of silicon, the conduction band, is comparable to **k***T*, the thermal energy, a certain fraction of these electrons are thermally excited to the conduction band in which they can move under the influence of an applied field. At higher temperatures more electrons are excited to the conduction band and the conductivity is larger. If very many phosphorus atoms are introduced in the lattice, the impurity level itself widens into a band which overlaps the conduction band of the silicon; the conductivity then becomes metallic in character. This is an example of *n*-type semiconductivity, so-called because the carriers of the current, the electrons, are *n*egatively charged.

If atoms of aluminum or boron are introduced in the silicon lattice, they also introduce their own system of levels. Since the aluminum atom has one less electron than silicon, the impurity levels are vacant. Figure 27–7(a) shows the position of the impurity levels, which in this case are only slightly above the filled band of the silicon lattice. Electrons from the filled band can be excited thermally to the impurity levels, Fig. 27–7(b), where they are bound to the aluminum atoms to produce the species Al⁻ in the lattice. The holes which are left in the band effectively carry a

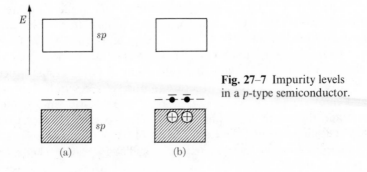

Fig. 27–7 Impurity levels in a *p*-type semiconductor.

positive charge and can move under the influence of an applied field, and thus carry a current. This is an example of p-type semiconductivity, since the carrier is positively charged.

The semiconductivity of nonstoichiometric compounds such as ZnO and Cu_2O can be explained in a similar way. If ZnO loses a little oxygen, it can be considered as ZnO with a few zinc atoms as impurities. The zinc atoms have two more electrons than the zinc ions; therefore, the semiconductivity is n-type. Since the crystal Cu_2O may contain extra oxygen, it may be considered as Cu_2O with some Cu^{2+} ions as impurities. The Cu^{2+} ion has one less electron than the Cu^+ ion, so the conductivity is p-type. Sodium chloride exposed at high temperatures to sodium vapor incorporates sodium atoms as impurities; the impure crystal has n-type semiconductivity. Excess halogen can be introduced into sodium chloride to yield a p-type semiconductor.

Until the late 1940's the study of semiconductivity was a very frustrating occupation. Reproducible measurements were very difficult to obtain. To study the phenomenon in silicon, for example, it is necessary to begin with silicon of a fantastic degree of purity, less than one part per billion of impurity. Accurately controlled amounts of a definite type of impurity are then added. If ordinary silicon is used, the accidental impurities and their concentrations are variable from sample to sample, making the experimental measurements nearly valueless. In the years since 1948 the technique of producing materials of the required degree of purity, the technique of zone refining, has been developed to such an extent that semiconductors with reproducible characteristics are produced with ease on a commercial scale. The transistor, a device made of semiconducting materials, has become a commonplace item.

27–7 Cohesive Energy in Metals

Any detailed calculation of the cohesive energy of metals is quite complicated; however, it is possible from a qualitative examination of the band systems to gain a little insight into the problem. Consider the transition metals which as isolated atoms have partially filled d shells, and as solids have partially filled d bands. The d band can hold 10 electrons/atom. Since the d shell in the atoms is shielded somewhat by the outer electrons, the d band is very narrow compared with the sp band. Figure 27–8(b) shows the relative widths and the filling of the d and sp bands in copper. The d band is completely filled. In nickel, which has one less electron per atom, the d band is only partially filled; Fig. 27–8(a). The lower cohesive energy of copper compared with that of nickel is a result of the higher average energy of the electrons in the sp band of copper. Zinc has one more electron than copper; adding this electron to the sp band fills it to a much higher level, Fig. 27–8(c), resulting in a marked decrease in the cohesive energy. The cohesive energies are (kcal/mole): nickel, 101.61; copper, 81.52; zinc, 31.19. The lower the energy of the electrons in the metal, the more stable the system and the greater the cohesive energy. It is evident from Fig. 27–8 that a partially empty d band in a metal is an indication of a large cohesive energy, since the

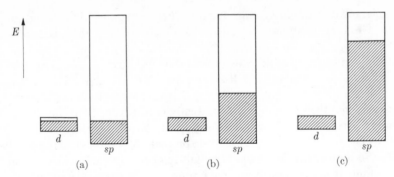

Fig. 27-8 Effect of the d band on the electronic energy. (a) Nickel. (b) Copper. (c) Zinc.

average energy of the electrons is low. Addition of an electron to the sp band, as in going from copper to zinc, increases the average electronic energy rapidly, since the very wide sp band accommodates only 4 electrons/atom, while the very narrow d band accommodates 10.

Problems

27-1. Consider the following arrangement of ions:

a) $+ \quad -$

b) $+ \quad - \quad + \quad -$

c) $+ \quad - \quad + \quad - \quad + \quad -$

d) $+ \quad -$
 $\quad - \quad +$

The charge on the positive and negative ions is $+e$ and $-e$ respectively; the spacing in the linear arrays is r between any two neighbors. Compute the Madelung constant for these arrangements of ions.

27-2. The ion radii for Na^+ and Cl^- are 0.95 Å and 1.81 Å.. Calculate the cohesive energy in kcal per mole neglecting repulsion; $A = 1.7476$. Calculate the cohesive energy if $n = 8.0$ ($e = 4.80 \times 10^{-10}$ esu).

27-3. Using the Madelung constants in Table 27-1, compare the cohesive energy of RbCl in the NaCl structure and in the CsCl structure. The radii are $r_+ = 1.48$ Å, $r_- = 1.81$ Å, and are assumed to be the same in both structures.

27-4. a) Arrange the alkali metal fluorides in order of increasing cohesive energy.

b) Arrange the potassium halides in order of increasing cohesive energy.

27-5. What is the approximate ratio of cohesive energies of NaF and MgO?

27-6. The density of NaCl is 2.165 gm/cm³. Calculate the interionic, Na^+–Cl^-, distance. If $n = 8.0$, calculate the compressibility of solid NaCl.

Chapter Twenty-eight

Structure and Thermodynamic Properties

28–1 The Energy of a System

The energy of an individual atom or molecule can be computed from the quantum mechanics. In a collection of a large number of molecules there is an energy distribution; some molecules have more energy and some less. The *average* energy of the collection of molecules is identified with the thermodynamic energy of the system. It is our aim to discover the relation between the properties of the individual molecule which are obtained from the Schrödinger equation and the thermodynamic properties of the bulk system, which contains very many individual molecules.

Consider a system of fixed volume V, which contains a very large number N of molecules. Since the energies of the individual molecules have discrete values, the possible energies of the system have discrete values $E_1, E_2, E_3, \ldots, E_i$. These values of the energy are found by solution of the Schrödinger equation. We specify that the temperature is constant, since the system is immersed in a heat reservoir at constant temperature. Since the system exchanges energy with the reservoir, if we make a number of observations of the system, we will find that it is in a different quantum state, has a different E_i, in each observation. The thermodynamic energy is the average of the energies exhibited in a large number of observations. If we wish, instead of observing one system a large number of times, we can construct a large number of identical systems, immerse them in the same temperature reservoir, and make one observation on each system. Each is found to be in a different quantum state; the energy is obtained by averaging over all the systems.

Consider a collection of a large number, $\mathbf{N}$, of identical systems, an *ensemble*. Every system in the ensemble has one of the energy values, so we may write the energy distribution as follows:

Energy: E_1 E_2 E_3 ... E_i ...

Number of systems: n_1 n_2 n_3 ... n_i ...

The probability of finding a system with the energy E_i is $P_i = n_i/N$. This probability depends on the energy E_i, so we write

$$P_i = f(E_i). \tag{28-1}$$

Similarly, the probability of finding a system with energy E_j is

$$P_j = f(E_j). \tag{28-2}$$

Suppose that we choose two systems from the ensemble; the probability P_{ij} that one has E_i and the other has E_j is the product of the individual probabilities,

$$P_{ij} = P_i P_j = f(E_i)f(E_j). \tag{28-3}$$

There is another way to choose two systems from the ensemble. Suppose that we pair off the systems randomly, to form $N/2$ paired systems. The probability that a pair has energy $E_i + E_j$ is also P_{ij} and must be the same function of the energy of the paired system, as the P_i is of the energy of the single system; P_{ij} differs at most by a multiplicative constant B, since the total number of systems involved is different. Therefore,

$$P_{ij} = Bf(E_i + E_j). \tag{28-4}$$

Combining this with the result in Eq. (28–3), we obtain the functional equation

$$f(E_i)f(E_j) = Bf(E_i + E_j). \tag{28-5}$$

We have met a similar equation, Eq. (4–27), in the kinetic theory of gases. Equation (28–5) is satisfied if $f(E_i)$ has the form

$$f(E_i) = Be^{-\beta E_i}, \tag{28-6}$$

where β is a positive constant, and the negative sign in the exponential has been chosen to avoid predicting an infinite probability of finding systems with infinite energy. The constant β must be the same for all systems; otherwise the functional relation, Eq. (28–5), would not be fulfilled. The property which is common to all the systems is the temperature, so without further argument, we set

$$\beta = 1/kT, \tag{28-7}$$

where k is the Boltzmann constant. The relation in Eq. (28–7) can be proven rigorously, of course, but to avoid rewriting many equations, we will not undertake the proof here.

The probability becomes finally

$$P_i = Be^{-E_i/kT}. \tag{28-8}$$

The constant B is determined by the condition that the sum of the probabilities over all possible energy states is unity:

$$\sum_i P_i = 1, \tag{28-9}$$

so that

$$B \sum_i e^{-E_i/kT} = 1. \tag{28-10}$$

The summation in Eq. (28–10) is called the *partition function*, or the *state sum*, and is given the symbol Q:

$$Q = \sum_i e^{-E_i/kT}. \tag{28-11}$$

Thus, $B = 1/Q$, and

$$P_i = \frac{e^{-E_i/kT}}{Q}. \tag{28-12}$$

Knowing the probability of finding the system with energy E_i, we can calculate the thermodynamic energy E of the system, which is the average energy of the ensemble:

$$E = \frac{\sum_i \mathbf{n}_i E_i}{\mathbf{N}}.$$

Since $\mathbf{n}_i/\mathbf{N} = P_i$, this becomes

$$E = \sum_i P_i E_i. \tag{28-13}$$

By the same reasoning any function of the energy $Y(E_i)$ has the average value Y given by

$$Y = \sum_i P_i Y(E_i). \tag{28-14}$$

The argument assumes that the probabilities of choosing one system with energy E_i and another with E_j are independent; this leads to a distribution function P_i which is of the Maxwell-Boltzmann type. The independence of the probabilities implies that the distribution is a random one.

28-2 Definition of the Entropy

In the ensemble the systems are distributed over the various quantum states; every possible way of arranging the systems in the quantum states is called a *complexion* of the ensemble. The number of complexions is denoted by $\mathbf{\Omega}$; then the entropy of the ensemble is defined, as in Section 9–14, by

$$\mathbf{S} = \mathbf{k} \ln \mathbf{\Omega}. \tag{28-15}$$

The entropy of the system is the entropy of the ensemble divided by the number of systems **N**, so that

$$S = \frac{\mathbf{S}}{\mathbf{N}} = \frac{k \ln \Omega}{\mathbf{N}}. \tag{28–16}$$

We regard the quantum states with energies E_i as boxes and the systems as balls to be distributed among the boxes. The total number of distinguishable ways of arranging the balls in the boxes (the systems in the quantum states) is the number of complexions Ω of the ensemble. This number is given by Eq. (9–70):

$$\Omega = \frac{\mathbf{N}!}{\mathbf{n}_1! \mathbf{n}_2! \mathbf{n}_3! \ldots}. \tag{28–17}$$

To compute S we first compute $\ln \Omega$.

$$\ln \Omega = \ln \mathbf{N}! - \sum_i \ln \mathbf{n}_i!$$

If **N** is large, the Stirling formula yields $\ln \mathbf{N}! = \mathbf{N} \ln \mathbf{N} - \mathbf{N}$. Then

$$\ln \Omega = \mathbf{N} \ln \mathbf{N} - \mathbf{N} - \sum_i \mathbf{n}_i \ln \mathbf{n}_i + \sum_i \mathbf{n}_i.$$

Since $\sum_i \mathbf{n}_i = \mathbf{N}$, and $\mathbf{n}_i = P_i \mathbf{N}$, this reduces to

$$\ln \Omega = \mathbf{N} \ln \mathbf{N} - \mathbf{N} \sum_i P_i \ln (\mathbf{N} P_i).$$

Expanding $\ln (\mathbf{N} P_i)$ and using the fact that $\sum_i P_i = 1$, this becomes

$$\ln \Omega = -\mathbf{N} \sum_i P_i \ln P_i. \tag{28–18}$$

Using this result in Eq. (28–16), we obtain for the entropy of the system

$$S = -k \sum_i P_i \ln P_i. \tag{28–19}$$

Equation (28–19) expresses the dependence of the entropy on the P_i. It is important to observe that P_i is the fraction of the systems in the state with energy E_i, so that the form of the sum in Eq. (28–19) has the appearance of an entropy of mixing. The systems in the ensemble are "mixed," or spread over the possible energy states of the system. It is this "mixing" or spreading which gives rise to the property of a system which we call the entropy.

28–3 The Thermodynamic Functions in Terms of the Partition Function

Equations (28–13) and (28–19) relate the energy and entropy to the P_i. From these equations, the relation between P_i and Q, Eq. (28–12), and the definition of Q, Eq. (28–11), all of the thermodynamic functions can be expressed in terms of the partition function Q and its derivatives. We begin by differentiating Eq. (28–11) with respect to temperature:

$$\left(\frac{\partial Q}{\partial T}\right)_V = \frac{1}{kT^2}\sum_i E_i e^{-E_i/kT}. \tag{28–20}$$

Since the E_i are obtained ultimately from the Schrödinger equation, they do not depend on temperature; they may, however, depend on the volume, so the derivative is a partial derivative. Using Eq. (28–12), the exponential in the sum in Eq. (28–20) may be replaced by QP_i, which brings the equation to the form

$$kT^2\left(\frac{\partial Q}{\partial T}\right)_V = Q\sum_i P_i E_i.$$

By comparison with Eq. (28–13) this summation is equal to the energy, so we have

$$E = \frac{kT^2}{Q}\left(\frac{\partial Q}{\partial T}\right)_V = kT^2\left(\frac{\partial \ln Q}{\partial T}\right)_V, \tag{28–21}$$

which relates the energy to the partition function.

To obtain the entropy, we calculate $\ln P_i$ using Eq. (28–12):

$$\ln P_i = -\frac{E_i}{kT} - \ln Q.$$

Putting this expression in Eq. (28–19) for the entropy yields

$$S = -k\left[-\frac{1}{kT}\sum_i P_i E_i - \ln Q \sum_i P_i\right].$$

Using Eqs. (28–9) and (28–13) this becomes

$$S = \frac{E}{T} + k\ln Q.$$

Insertion of the value of E from Eq. (28–21) yields

$$S = k\ln Q + kT\left(\frac{\partial \ln Q}{\partial T}\right)_V, \tag{28–22}$$

which expresses the entropy in terms of the partition function. Since all of the other thermodynamic functions are simply related to S, E, T, and V, it is an easy matter to calculate them. For example, the Helmholtz function $A = E - TS$. Using the values for E and S in terms of Q, we obtain for A,

$$A = -kT\ln Q. \tag{28–23}$$

From the fundamental equation, Eq. (10–21), $p = -(\partial A/\partial V)_T$. Differentiating Eq. (28–23), we obtain for the pressure,

$$p = kT\left(\frac{\partial \ln Q}{\partial V}\right)_T. \tag{28–24}$$

Then the values of H and G follow immediately from the definitions:

$$H = kT\left[T\left(\frac{\partial \ln Q}{\partial T}\right)_V + V\left(\frac{\partial \ln Q}{\partial V}\right)_T\right],$$
(28–25)

$$G = -kT\left[\ln Q - V\left(\frac{\partial \ln Q}{\partial V}\right)_T\right].$$
(28–26)

Finally by differentiating Eq. (28–21), we obtain the heat capacity C_v:

$$C_v = kT\left[2\left(\frac{\partial \ln Q}{\partial T}\right)_V + T\left(\frac{\partial^2 \ln Q}{\partial T^2}\right)_V\right].$$
(28–27)

In a certain sense we have solved the problem of obtaining thermodynamic functions from the properties of molecules. These functions have been related to Q, which, by its definition, is related to the energy levels of the system, which are in turn related to the energy levels of the molecules in the system. To make these expressions useful, we must express the partition function in terms of the energies of the molecules.

28-4 The Molecular Partition Function

Consider the quantum state of the system which has the energy E_i. This energy is composed of the sum of the energies of the molecules $\epsilon_1, \epsilon_2, \ldots$ plus any interaction energy W between the molecules:

$$E_i = \epsilon_1 + \epsilon_2 + \cdots + W.$$
(28–28)

For the present we assume that the particles do not interact (ideal gas) and set $W = 0$. Each energy ϵ_i corresponds to one of the allowed energy levels of the molecule. Because the energy E_i has the form given by Eq. (28–28), it is possible to write the partition function as a product of partition functions of the individual molecules q. The final form is, for indistinguishable molecules,

$$Q = \frac{1}{N!}q^N,$$
(28–29)

where N is the number of molecules in the system and

$$q = \sum_i e^{-\epsilon_i/kT}.$$
(28–30)

The sum in Eq. (28–30) is over all the energy levels of the *molecule*, so q is the *molecular partition function*. If two kinds of molecules are present, N_a of A and N_b of B, then

$$Q = \frac{q_a^{N_a}q_b^{N_b}}{N_a!N_b!}.$$
(28–31)

No attempt will be made to justify these equations except to say that if the Q were written simply as q^N, too many terms would be included; division by $N!$ is required to yield the correct result.

Since only $\ln Q$ appears in the formulas for the thermodynamic functions, we find, using the Stirling formula, from Eq. (28–29)

$$\ln Q = N \ln q - N \ln N + N. \qquad (28\text{–}32)$$

Using the expression in Eq. (28–32), we can express all the thermodynamic functions in terms of $\ln q$ instead of $\ln Q$.

28–5 The Chemical Potential

The value of the chemical potential in a mixture is calculated using the relation $\mu_a = (\partial A/\partial N_a)_{T,V}$. From Eq. (28–31) we have

$$\ln Q = N_a \ln q_a + N_b \ln q_b - N_a \ln N_a + N_a - N_b \ln N_b + N_b.$$

Differentiating with respect to N_a, this becomes

$$\left(\frac{\partial \ln Q}{\partial N_a}\right)_{T,V} = \ln q_a - 1 - \ln N_a + 1 = \ln\left(\frac{q_a}{N_a}\right).$$

By differentiating Eq. (28–23), we obtain

$$\left(\frac{\partial A}{\partial N_a}\right)_{T,V} = -\mathbf{k}T\left(\frac{\partial \ln Q}{\partial N_a}\right)_{T,V} = \mu_a.$$

Thus,

$$\mu_a = -\mathbf{k}T \ln\left(\frac{q_a}{N_a}\right), \qquad (28\text{–}33)$$

which expresses the chemical potential of a gas (indistinguishable molecules) in terms of the molecular partition function per molecule q_a/N_a, a result which is useful for the discussion of chemical equilibria. If we were dealing with a solid in which the molecules are locked in place and therefore are distinguishable, the factors $N_a!$ and $N_b!$ do not appear in Eq. (28–31), so we have the simpler result

$$\mu = -\mathbf{k}T \ln[q e^{-W/N\mathbf{k}T}], \qquad (28\text{–}34)$$

where the interaction energy W appears, since W is not zero in a solid.

28–6 Application to Translational Degrees of Freedom

The application of the formulas is simplest if the molecules possess energy in only one form. Therefore, we consider a system such as a monatomic gas which has only translational energy. (We ignore any contribution of the internal electronic energy of the atom to the properties of the system. In rigorous treatments it must be included.)

The energy of translation ϵ_t is made up of the energies in each component of the motion, so we write

$$\epsilon_t = \epsilon_x + \epsilon_y + \epsilon_z.$$

Again, because these energies are additive, the translational partition function q_t factors into a product:

$$q_t = q_x q_y q_z. \tag{28–35}$$

The energy levels for translation are the energy levels for a particle in a box, Section 21–3. If the width of the box in the x-direction is a, then the permitted values of the kinetic energy from the Schrödinger equation are

$$\epsilon_x = \frac{h^2 n^2}{8ma^2}, \qquad n = 1, 2, 3, \ldots. \tag{28–36}$$

It was shown in Section 21–4 that the spacing between levels in a box of macroscopic dimensions is extremely small, too small to distinguish the levels observationally. Therefore, we choose a new set of distinct levels, ϵ_i, but separated by an energy $d\epsilon$. Let there be g_i levels in the energy range $d\epsilon$ between ϵ_i and ϵ_{i+1}. All of these levels in this energy range will be assigned the single energy value ϵ_i. Then the terms in the partition function group into sets, and the partition function can be written as

$$q_x = \sum_i g_i e^{-\epsilon_i/kT}, \tag{28–37}$$

since g_i terms containing the single exponential $e^{-\epsilon_i/kT}$ appear when the sum is made over all levels.

To compute g_i we compute the spacing between levels,

$$\epsilon_{n+1} - \epsilon_n = (2n + 1)\frac{h^2}{8ma^2}.$$

If n is very large, then $2n + 1 = 2n = (4a/h)(2m\epsilon)^{1/2}$, where the last form is obtained by solving Eq. (28–36) for $2n$ (the subscript x on ϵ has been dropped). The value of the spacing becomes

$$\epsilon_{n+1} - \epsilon_n = (h/a)(\epsilon/2m)^{1/2}.$$

The number of levels in the range $d\epsilon$ is g and is obtained by dividing the range by the spacing between levels:

$$g = (a/h)(2m/\epsilon)^{1/2} \, d\epsilon.$$

Putting this value of g in the partition function and changing the summation to integration from $\epsilon = 0$ to $\epsilon = \infty$, Eq. (28–37) becomes

$$q_x = \int_0^\infty \frac{a}{h}\left(\frac{2m}{\epsilon}\right)^{1/2} e^{-\epsilon/kT} \, d\epsilon. \tag{28–38}$$

By changing variables to $y^2 = \epsilon/kT$, then $d\epsilon = 2kT\,y\,dy$, and the integral in Eq. (28–38) becomes

$$q_x = \frac{2a}{h}(2mkT)^{1/2}\int_0^\infty e^{-y^2}\,dy.$$

The integral has the value $\sqrt{\pi}/2$; hence

$$q_x = \frac{a}{h}(2\pi mkT)^{1/2}.$$

Similarly, if b and c are the widths of the box in the y- and z-directions, we get

$$q_y = \frac{b}{h}(2\pi mkT)^{1/2}, \qquad q_z = \frac{c}{h}(2\pi mkT)^{1/2}.$$

The translational partition function q_t is the product of these, by Eq. (28–35); we obtain

$$q_t = \left(\frac{2\pi mkT}{h^2}\right)^{3/2} V, \tag{28–39}$$

where the product of the dimensions of the box abc has been replaced by the volume V. Using Eq. (28–32), we obtain for Q

$$\ln Q = N \ln q_t - N \ln N + N. \tag{28–40}$$

The derivatives of $\ln Q$ are

$$\left(\frac{\partial \ln Q}{\partial T}\right)_V = \frac{3N}{2T} \quad \text{and} \quad \left(\frac{\partial \ln Q}{\partial V}\right)_T = \frac{N}{V}.$$

The reader may easily verify that the energy of the system is $\frac{3}{2}NkT$ and that $p = NkT/V$, which are the values we expect for a monatomic gas in which the interaction energy is zero. From the values of the derivatives and $\ln Q$ itself, any of the thermodynamic quantities can be calculated using the formulas in Section 28–3.

28–7 Partition Function of the Harmonic Oscillator

If the particles composing the system have only vibrational motion, the energies permitted are given by

$$\epsilon_s = (s + \tfrac{1}{2})h\nu, \qquad s = 0, 1, 2, \ldots,$$

where ν is the frequency of the oscillator. Using this value for the energy, the vibrational molecular partition function becomes

$$q_v = \sum e^{-\epsilon_s/kT} = \sum_{s=0}^\infty e^{-(s+1/2)h\nu/kT},$$

where the sum is over all the integral values of s from zero to infinity. To simplify, let $y = e^{-hv/kT}$, then

$$q_v = y^{1/2} \sum_{s=0}^{\infty} y^s = y^{1/2}(1 + y + y^2 + \cdots).$$

Since $1/(1 - y) = 1 + y + y^2 + y^3 + \cdots$, we see that q_v becomes

$$q_v = \frac{y^{1/2}}{1 - y} = \frac{e^{-hv/2kT}}{1 - e^{-hv/kT}}. \tag{28–41}$$

It is customary to define a characteristic temperature for the oscillator, $\theta = hv/k$. Then

$$q_v = \frac{e^{-\theta/2T}}{1 - e^{-\theta/T}}. \tag{28–42}$$

If the temperature is either very high or very low, this equation takes on a simpler form.

Case I. T is very low; $\theta/T \gg 1$. Then, $e^{-\theta/T}$ is negligible compared with unity in the denominator of Eq. (28–42), and we have

$$q_v = e^{-\theta/2T} \qquad \text{(low temperature)}. \tag{28–43}$$

Case II. T is very high; $\theta/T \ll 1$. Then the exponential in the denominator can be expanded $e^{-\theta/T} = 1 - \theta/T$, and we have

$$q_v = \frac{T}{\theta} e^{-\theta/2T} \quad \text{(high temperature)}. \tag{28–44}$$

28–8 The Monatomic Solid

The monatomic solid has only vibrational motion. The partition function can be written as a product of the partition functions of the atoms composing the solid and an exponential factor which includes the interaction energy W of the atoms in the solid:

$$Q = e^{-W/kT} q_1 q_2 q_3 \cdots q_N. \tag{28–45}$$

Since each particle in the solid has three vibrational degrees of freedom, the partition function for each atom is a product of three vibrational partition functions, so that Q contains a product of $3N$ vibrational partition functions. Thus

$$Q = e^{-W/kT} q_{v_1} q_{v_2} q_{v_3} \cdots q_{v_{3N}}. \tag{28–46}$$

Each partition function contains a frequency, so that there is a total of $3N$ frequencies and, correspondingly, $3N$ values of θ which are involved. We cannot do anything further at low temperatures without knowing something more about the frequencies.

At very high temperatures we can do a little bit. Taking the logarithm of Q from Eq. (28–46), we obtain

$$\ln Q = -\frac{W}{\mathbf{k}T} + \ln q_{v_1} + \ln q_{v_2} + \cdots + \ln q_{v_3N}. \tag{28–47}$$

But from Eq. (28–44) at high temperatures we have

$$\ln q_v = -\frac{\theta}{2T} - \ln \theta + \ln T,$$

and the temperature derivative

$$\left(\frac{\partial \ln q_v}{\partial T}\right)_V = \frac{\theta}{2T^2} + \frac{1}{T}, \tag{28–48}$$

so that the energy in the individual vibration is

$$E_v = \tfrac{1}{2}\mathbf{k}\theta + \mathbf{k}T = \tfrac{1}{2}hv + \mathbf{k}T. \tag{28–49}$$

It is apparent from the form of Eq. (28–47) that the total energy of the solid is made up of a sum of the contributions from each vibration, E_v, and a contribution from the term $-W/\mathbf{k}T$. This last term contributes W, so we have for the energy of the solid,

$$E = W + \sum_{i=1}^{3N} E_{v_i}.$$

From Eq. (28–49) this becomes

$$E = W + \sum_{i=1}^{3N} \tfrac{1}{2}hv_i + \sum_{i=1}^{3N} \mathbf{k}T.$$

The summation in the second member on the right is the sum of the zero-point energies of all the oscillators. The third member is a sum of $3N$ terms each of value $\mathbf{k}T$, so that it is equal to $3N\mathbf{k}T$. Since the first two members are constant, we combine them in a single term E_0. The final result at high temperatures is

$$E = E_0 + 3N\mathbf{k}T. \tag{28–50}$$

The heat capacity of the solid is, by differentiation,

$$C_v = 3N\mathbf{k}. \tag{28–51}$$

The value of the heat capacity in Eq. (28–51) does not depend on any assumption about the frequencies in the solid. This should be the value of the heat capacity of any monatomic solid if the temperature is sufficiently high. If we deal with one mole of the solid, then $N = N_0$, and $N_0\mathbf{k} = R$. For one mole, $C_v = 3R \approx 6.0$ cal·deg^{-1}·mole^{-1}. This result is the law of Dulong and Petit, recognized for nearly a century and a half. In modern terms the Dulong and Petit law says that the molar heat capacity (constant pressure) of an element should be 6.4 cal. Since the value of C_p is somewhat larger than C_v, the discrepancy is understandable.

The result in Eq. (28–51) is confirmed at high temperatures for many solids. At ordinary temperatures the heat capacity is often less than the ideal value $3R$. In diamond at room temperature the heat capacity is only 1.45 cal·deg^{-1}·mole^{-1}, indicating that the vibrations are not fully excited; temperatures of the order of 2000 to $3000°K$ are required before diamond has the high-temperature value. A crystal such as NaCl has $2N_0$ atoms per mole and should therefore have $C_v = 6R = 12.0$ cal·deg^{-1}·mole^{-1}. For NaCl at $25°C$, the value of $C_p = 11.88$ cal·deg^{-1}·mole^{-1}. Ferric oxide, Fe_2O_3, with $5N_0$ atoms per mole should have $C_v = 15R = 30$ cal·deg^{-1}·mole^{-1}. The $25°C$ value of $C_p = 25.0$ cal·deg^{-1}·mole^{-1}.

A salt such as $NaNO_3$ has a vibrational heat capacity of $6R$ contributed by vibrations of Na^+ and NO_3^- in the solid and an additional contribution from the vibrations within the nitrate ion, which are partly but not fully excited. Some metals, notably the transition metals, exhibit values of C_v greater than $3R$ at high temperatures; this extra contribution comes from the heat capacity of the "electron gas" in the metal.

To discuss the heat capacity at intermediate and low temperatures requires some additional assumption about the $3N$ frequencies. The simplest approach is that of Einstein who assumed that all the frequencies have the same value, v_E. The partition function then takes the form

$$Q = e^{-W/kT} q_v^{3N}, \tag{28–52}$$

where q_v has the form given by Eq. (28–41) with $v = v_E$. The Einstein model agrees well with experimental values of C_v at intermediate and high temperatures, but predicts values which are too low at low temperatures.

The Debye theory assumes that there is a continuous distribution of frequencies from $v = 0$ to a certain maximum value $v = v_D$. The final expression obtained for the heat capacity is complicated, but succeeds in interpreting the heat capacity of many solids over the entire temperature range rather more accurately than the Einstein expression. At low temperatures, the Debye theory yields the simple result

$$C_v = \frac{12\pi^4}{5} Nk \left(\frac{T}{\theta_D} \right)^3, \tag{28–53}$$

where $\theta_D = hv_D/k$. This is the famous Debye "T-cubed" law for the heat capacity of a solid. At temperatures near the absolute zero the great majority of solids follow this law quite accurately.

28–9 General Expressions for the Partition Function

In general the molecules composing a system will possess energy in several ways: in translation, rotation, and vibration. In a diatomic gas, for example, there are three degrees of translational freedom, two rotational degrees of freedom, and one vibrational degree of freedom. If the energies in these various degrees of freedom are additive, the molecular partition function factors into a product of partition functions for the various degrees of freedom. For the diatomic molecule,

$$q = q_t^3 q_r^2 q_v, \tag{28–54}$$

where q_t is the partition function for a single translational degree of freedom. The rotational partition function, which we will not discuss in detail, has the following value for diatomic molecules:

$$q_r^2 = \frac{8\pi^2 I k T}{\sigma h^2}, \tag{28–55}$$

where $\sigma = 2$ for homonuclear molecules and $\sigma = 1$ for heteronuclear diatomic molecules; I is the moment of inertia.

28–10 Equilibrium Constants

Consider the equilibrium in the ideal gaseous reaction

$$A + 2B \rightleftharpoons 3C.$$

The equilibrium condition is

$$\mu_a + 2\mu_b = 3\mu_c.$$

Since all the substances are ideal gases, the chemical potentials will all have the form given by Eq. (28–33); using this form in the equilibrium condition, we obtain, after dividing through by $-kT$,

$$\ln\left(\frac{q_n}{N_a}\right) + \ln\left(\frac{q_b}{N_b}\right)^2 = \ln\left(\frac{q_c}{N_c}\right)^3,$$

which can be rearranged to the form

$$\frac{N_c^3}{N_a N_b^2} = \frac{q_c^3}{q_a q_b^2}.$$

Dividing numerator and denominator on both sides by V^3, we obtain

$$K_c' = \frac{c_C^3}{c_A c_B^2} = \frac{(q_c/V)^3}{(q_a/V)(q_b/V)^2}. \tag{28–56}$$

The ratio of concentrations on the left-hand side of Eq. (28–56) (concentrations in molecules/cm^3) is the equilibrium constant of the reaction. The proper quotient of molecular partition functions per unit volume (q/V) is equal to the equilibrium constant in this special concentration unit. Equation (28–56) is an important link between the quantum mechanics and chemistry. Knowing the energy levels of the molecules, we can calculate the molecular partition functions and then by Eq. (28–56) the equilibrium constant for the chemical reaction is obtained.

28–11   Conclusion

We have explored some of the simpler aspects of statistical thermodynamics, a very powerful theoretical tool. If the energy levels of the molecules composing the system can be obtained by solution of the Schrödinger equation, the partition function can be calculated; then any thermodynamic property can be evaluated. One of the great virtues of statistical thermodynamics is its ability to reveal general laws. For example, we reached the conclusion that all monatomic solids should have the same heat capacity at high temperatures. Restrictions on the laws are made apparent; for example, the heat capacity of a monatomic solid at low temperatures depends very much on what is assumed about the frequencies in the solid.

A number of objections may be raised at this point. Difficulties in solving the Schrödinger equation, approximations which must be made in many of the mathematical steps, and so on. These are legitimate objections, but we have concentrated here in presenting the more theoretical side of the statistical thermodynamics. The actual values of the energies may be known from experiment! Analysis of the spectrum of a molecule will give us all the information about energy levels that we need. We simply insert the experimental values of the energies into the exponentials of the partition function, add all the exponentials together, and by brute force evaluate any thermodynamic quantity which happens to be of interest. This is somewhat laborious but very practical. The most difficult part is obtaining and analyzing the spectral data in terms of energy levels. To obtain the heat capacity of hydrogen at 2000°K by a calorimetric method would be a nasty job; to study the spectrum and calculate the heat capacity from the spectral data using partition functions is very much easier and yields a much more accurate result. In this connection, it should be said that a very large proportion of the tabulated thermodynamic data is obtained from spectral data.

A final word should be said about the entropy. Although we can consider that the energy of a system is the sum of the energies of the individual molecules, the entropy of a system is not the sum of the entropies of individual molecules. The entropy is defined in terms of the complexions of a very large number of systems in an ensemble. The entropy of a molecule has no real meaning. We can divide the entropy of a system by the number of molecules and talk about an entropy per molecule, but in the final analysis this is a fiction. Entropy and temperature have meaning only for matter in bulk which is composed of a very large number of individual particles. Systems containing only a few molecules need not obey the second law of thermodynamics. The number of molecules must be very large before probabilities become actualities with a negligible chance of observing a deviation.

Problems

28–1.   Compute the translational entropy of one mole of a monatomic gas at 1 atm = 1.013 × 10^6 dynes/cm^2 as a function of M and T from the partition function. Evaluate the expression for argon at 298°K.

28–2. Using the partition function, show that for a monatomic gas, $E = \frac{3}{2}NkT$ and that $p = NkT/V$.

28–3. a) Using the complete expression for the vibrational partition function, Eq. (28–42), derive the expression for C_v as a function of θ/T.

b) Using the expression in Eq. (28–44), compute C_v^∞, the heat capacity at infinite temperature.

c) Calculate the values of C_v/C_v^∞ for $\theta/T = 0, 0.5, 1.0, 1.5, 2.0, 3.0, 4.0, 5.0, 6.0$. Plot these values against θ/T.

28–4. The bending vibration in CO_2 has a frequency of 1.95×10^{13} sec^{-1}.

a) Compute the characteristic temperature θ for this vibration.

b) What contribution does this vibration make to the heat capacity of CO_2 at 300°K?

c) This vibration is doubly degenerate, that is, the CO_2 molecule has two vibrations of this frequency. How does this affect the heat capacity?

d) The stretching vibrations have much higher frequencies. What contribution does the frequency, the asymmetrical stretch, at 6.9×10^{13} sec^{-1} make to the heat capacity at 300°K?

28–5. The chemical potentials of an ideal monatomic gas and a monatomic solid are given by

$$\mu_{\text{gas}} = -kT \ln (q_t/N)$$
$$\mu_{\text{solid}} = -kT \ln (q_v^3 e^{-W/NkT}),$$

if we assume that the frequencies in the solid are all the same.

a) Derive an expression for the equilibrium vapor pressure of a monatomic solid. Use the high-temperature value for q_v; $q_v = T/\theta$.

b) By differentiating and comparing with the Gibbs-Helmholtz equation, compute the value of the enthalpy of vaporization.

28–6. a) For any ideal gas, $\mu = -kT \ln (q/N)$. How does the possession of rotational and vibrational degrees of freedom in addition to translational degrees of freedom affect the value of the chemical potential?

b) Other things being equal (!), which crystal will have the higher vapor pressure at a specified temperature, a crystal of a monatomic substance or a crystal of a diatomic substance? Assume the diatomic molecules do not rotate in the solid.

Chapter Twenty-nine

Transport Properties

29–1 Introductory Remarks

We turn our attention at this point to the changes in the properties of a system with time. Thermodynamics describes the change in properties of a system in a change in state, but provides no information about the time which is required to effect the change. On a practical level, the time interval required for any particular change in state is of utmost importance. There is little consolation in knowing that a certain reaction can occur naturally if a million years is required for the transformation. Hydrogen and oxygen if left to themselves do not form water within any practical length of time. If a trace of platinum is added to the vessel, the conversion to water is complete within a few microseconds.

Much of the importance of thermodynamics lies in the fact that the thermodynamic predictions do not depend on the detailed way in which a system is transformed from an initial to a final state. The time interval required for the transformation depends very much on these details. A certain free-energy difference exists between the initial state, $H_2 + \frac{1}{2}O_2$, and the final state, H_2O. This free-energy difference has nothing to do with the presence or absence of a bit of platinum. As we have seen, the presence of a bit of platinum changes the time interval required by an enormous factor. The discussion of the rates of transformations requires in every case the postulation of some model of the structure of the system, while thermodynamics needs no model. By testing the rate predicted by a certain model against the experimental data, we can judge the adequacy of the model.

At first we look at the purely empirical laws which are observed to govern the rates of various processes. Later, by application of knowledge from thermodynamics and (mainly) structure, we attempt to interpret these laws in terms of the constitution of the system and the fundamental properties of the atoms and molecules which compose it.

29–2 Transport Properties

There is a particularly simple group of processes, transport processes, in which some physical quantity such as mass or energy or momentum or electrical charge is transported from one region of a system to another. Consider a metal bar connecting two heat reservoirs at different temperatures. Heat flows through the bar from the high-temperature reservoir to the low-temperature reservoir; the heat flow is the manifestation of the transport of energy through the bar. The energy flow is easily measurable. Another example is the transport of electrical charge through a conductor by the application of an electrical potential difference between the ends of the conductor. Mass is transported in the flow of a fluid through a pipe resulting from a pressure difference between the ends of the pipe. Diffusion is the mass transport which occurs in a mixture if a concentration gradient is present. The phenomenon of viscosity, the resistance to flow exhibited by fluids, results from the transport of momentum in a direction perpendicular to the direction of flow.

In all cases the *flow*, the amount of the physical quantity transported in unit time through a unit of area perpendicular to the direction of flow, is proportional to the gradient of some other physical property such as temperature, pressure, or electrical potential. Choosing the z-axis as the direction of flow, the general law for transport is

$$J_z = -B\frac{\partial Y}{\partial z}, \tag{29–1}$$

where J_z is the flow, the amount of the quantity transported per cm^2 per sec, $-B$ is the proportionality constant, and $(\partial Y/\partial z)$ is the gradient of Y in the direction of flow; Y may be any of the quantities temperature, electrical potential, pressure, etc. Since flow occurs in a particular direction, it is a vector quantity; Eq. (29–1) describes the z-component of the vector. We shall not require the more general three-dimensional equations. For the examples mentioned above we have the individual equations for flow in the z-direction:

$$\text{Heat flow:} \qquad J_z = -\kappa_T\frac{\partial T}{\partial z} \quad \text{(Fourier's law)} \qquad (29–2)$$

$$\text{Electrical current:} \quad J_z = -\kappa\frac{\partial \mathscr{V}}{\partial z} \quad \text{(Ohm's law)} \qquad (29–3)$$

$$\text{Fluid flow:} \qquad J_z = -C\frac{\partial p}{\partial z} \quad \text{(Poiseuille's law)} \qquad (29–4)$$

$$\text{Diffusion:} \qquad J_z = -D\frac{\partial c}{\partial z} \quad \text{(Fick's law)} \qquad (29–5)$$

In these equations, κ_T is the thermal conductivity coefficient, κ is the electrical conductivity, C is a frictional coefficient related to the viscosity, and D is the diffusion coefficient.

These laws are well established experimentally. They were proposed initially as empirical laws, generalizations from experiment. It is our aim now to give these laws an interpretation in terms of the structure of the substance.

29–3 The General Equation for Transport

If any physical quantity is transported, the amount transported through unit area in unit time is the number of molecules passing through the unit area in unit time multiplied by the amount of the physical quantity carried by each molecule. For any transport

$$j = n'q, \tag{29–6}$$

where j is the flow per $cm^2 \cdot sec$, n' is the number of carriers passing through $1\ cm^2$ in 1 sec, and q is the amount of the physical quantity possessed by each carrier. By calculating n' and q, we obtain the value of j. We begin with n'.

How many molecules pass the base, $1\ cm^2$, of the parallelepiped in Fig. 29–1 in unit time? If all the molecules were moving downward with an average velocity $\bar{c}$, then each travels a distance $\bar{c}\ dt$ in the time interval dt. Therefore, all the molecules in the parallelepiped of height $\bar{c}\ dt$ will pass the bottom face in the interval dt. The volume of the parallelepiped is $\bar{c}\ dt\ cm^3$; if n is the number of molecules per cubic centimeter, then the number crossing the base in dt is $n\bar{c}\ dt$. In unit time the number crossing $1\ cm^2$ area is

$$n' = n\bar{c}. \tag{29–7}$$

The expression for the flow, Eq. (29–6) becomes

$$j = n\bar{c}q, \tag{29–8}$$

which is applicable to any transport process; the flow is equal to the product of

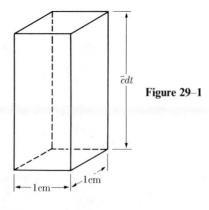

$\bar{c}dt$

Figure 29–1

$\longleftarrow 1\,cm \longrightarrow$ $1\,cm$

the number of carriers per cubic centimeter, the average velocity in the direction of the flow, and the amount of the physical quantity carried by each.

29–4 Thermal Conductivity in a Gas

Suppose that two large metal plates parallel to the xy-plane and separated by a distance Z are at temperatures T_1 and T_2, the hotter plate (T_2) being the upper one. After some time a steady state will be established in which there is a downward flow of heat at a constant rate. This flow of heat results from the fact that the molecules at the upper levels have a greater thermal energy than those at the lower levels; the molecules moving downward carry more energy than do those moving upward.

To calculate the net energy flow in unit time through 1 cm^2 parallel to the xy-plane, we imagine a large number of horizontal layers in the gas, each successive layer being at a slightly higher temperature than the one below it. The change in temperature with height is

$$\frac{\partial T}{\partial z} = \frac{\Delta T}{\Delta z} = \frac{T_2 - T_1}{Z - 0}, \tag{29–9}$$

if the lower plate lies at the position $z = 0$, the upper one at $z = Z$. The gradient $(\partial T/\partial z)$ is constant, so at any height z the temperature is

$$T = T_1 + \left(\frac{\partial T}{\partial z}\right)z. \tag{29–10}$$

If the gas is monatomic with an average thermal energy $\bar{\epsilon} = \frac{3}{2}kT$, then the average energy of the molecules at the height z is

$$\bar{\epsilon} = \tfrac{3}{2}kT = \tfrac{3}{2}k\left[T_1 + \left(\frac{\partial T}{\partial z}\right)z\right]. \tag{29–11}$$

To calculate the heat flow, we consider an area 1 cm^2 in a horizontal plane at the height z; Fig. 29–2. The energy carried by a molecule as it passes through the plane

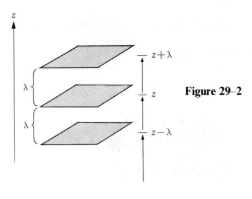

Figure 29–2

depends on the temperature of the layer of gas at which the molecule had its last opportunity to adjust its energy. This last adjustment occurred during the last collision with another molecule. Suppose that, on the average, the molecules have traveled a distance λ since their last collision. If the surface of interest lies at a height z, the molecules going down made their last collision at a height $z + \lambda$, while those going up made their last collision at a height $z - \lambda$; Fig. 29–2. The molecules carry an amount of energy appropriate to the height where the last collision occurred. The downward flow of energy is, by Eqs. (29–8) and (29–11)

$$\epsilon\downarrow = \tfrac{1}{6}(n\bar{c})_{z+\lambda}\tfrac{3}{2}\mathbf{k}\left[T_1 + \left(\frac{\partial T}{\partial z}\right)(z + \lambda)\right],$$

while the upward flow is given by

$$\epsilon\uparrow = \tfrac{1}{6}(n\bar{c})_{z-\lambda}\tfrac{3}{2}\mathbf{k}\left[T_1 + \left(\frac{\partial T}{\partial z}\right)(z - \lambda)\right].$$

The factor $\tfrac{1}{6}$ appears since, on the average, only $\tfrac{1}{6}$ of the molecules are going down and only $\tfrac{1}{6}$ are going up. The net flow upward is denoted by J_ϵ and is

$$J_\epsilon = \epsilon\uparrow - \epsilon\downarrow. \tag{29–12}$$

Before writing out the equation in detail, we should note that if the gas is not to have net motion through the surface we require that the number of molecules going up in unit time must equal the number going down, so that

$$\tfrac{1}{6}(n\bar{c})_{z+\lambda} = \tfrac{1}{6}(n\bar{c})_{z-\lambda}, \tag{29–13}$$

which means that $n\bar{c}$ has the same value at every height.† Introducing the expressions for $\epsilon\uparrow$ and $\epsilon\downarrow$ into Eq. (29–12) and using Eq. (29–13), we obtain

$$J_\epsilon = \tfrac{1}{6}n\bar{c}\tfrac{3}{2}\mathbf{k}\left(\frac{\partial T}{\partial z}\right)[z - \lambda - (z + \lambda)] = -\tfrac{1}{2}n\bar{c}\mathbf{k}\lambda\left(\frac{\partial T}{\partial z}\right).$$

Comparing this result with the empirical law, Eq. (29–2), for thermal conductivity, $J_\epsilon = -\kappa_T(\partial T/\partial z)$, we obtain

$$\kappa_T = \tfrac{1}{2}n\bar{c}\mathbf{k}\lambda = \tfrac{1}{3}(n/N_0)C_v\bar{c}\lambda, \tag{29–14}$$

where $C_v = \tfrac{3}{2}\mathbf{k}N_0$ has been used. The factor (n/N_0) is the concentration in moles per cubic centimeter. From Eq. (4–58) the average velocity is

$$\bar{c} = \sqrt{8\mathbf{k}T/\pi m}. \tag{29–15}$$

It is therefore possible to calculate κ_T, the coefficient of thermal conductivity, if we can calculate λ, the average distance traveled by a molecule since its last collision.

---

† This is not actually correct, but to do the derivation without this assumption complicates matters considerably.

Or, looking at matters from the bright side, we can evaluate λ if the value of κ_T has been measured.

Unfortunately, there are several things about the above derivation which can be criticized. Both $\bar{c}$ and n contain the temperature T, yet the temperature is different at different positions. Since the simple law of heat conduction is only correct if $T_2 - T_1$ is small compared with *either of the two temperatures*, it is sufficient to use the average temperature in computing n and $\bar{c}$. A more serious objection is that we use quantities such as n and $\bar{c}$ derived from the equilibrium distribution function and apply them to a nonequilibrium situation. The fact of the matter is that if a nonequilibrium distribution is used, the mathematical complication introduced is enormous. Happily, the result of the more accurate treatment is not substantially different but only changes the numerical constant $\frac{1}{3}$ in Eq. (29–14), assuming the absence of attractive forces. The distance λ has been introduced in a somewhat arbitrary way. To understand Eq. (29–14) we must have a more definite idea about λ.

29–5   Collisions in a Gas; the Mean Free Path

The *mean free path* λ of the molecule by definition is the average distance traveled between collisions. In one second, a molecule travels ($\bar{c} \times 1$ sec) cm and makes Z_1 collisions. Dividing the distance traveled by the number of collisions, we obtain the distance traveled between collisions:

$$\lambda = \bar{c}/Z_1. \qquad (29\text{–}16)$$

To calculate λ we calculate Z_1.

Let σ be the diameter of the molecule and consider a cylinder, Fig. 29–3, of radius σ and height $\bar{c}$. In one second, the molecule travels a distance $\bar{c}$ and sweeps out the cylinder; it collides with all the molecules within the cylinder.[†] The number of molecules in the cylinder is $\pi\sigma^2\bar{c}n$; this is the number of collisions made by one molecule in one second. The formula $Z_1 = \pi\sigma^2\bar{c}n$ must be multiplied by the factor $\sqrt{2}$ to take into account the fact that it is the average velocity along the line of centers of two molecules which matters and not the average velocity of a molecule. Consider two molecules which have their velocity vectors in the orientations shown in Fig. 29–4. For molecules moving in the same direction with the same velocity, the relative velocity of approach is zero. In the second case, where they approach head-on, the relative velocity of approach is $2\bar{c}$. If they approach at $90°$, the relative velocity of approach is the sum of the velocity components along the line joining the centers; this is $\frac{1}{2}\sqrt{2}\bar{c} + \frac{1}{2}\sqrt{2}\bar{c} = \sqrt{2}\bar{c}$. The third situation represents the average situation, so we write more exactly

$$Z_1 = \sqrt{2}\pi\sigma^2\bar{c}n. \qquad (29\text{–}17)$$

---

†   The fact that collisions result in the molecule following a zig-zag path does not matter; the volume swept out is the same.

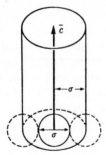

Fig. 29-3 Volume swept out by a molecule in 1 sec.

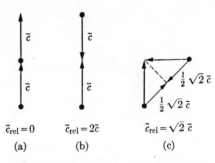

$\bar{c}_{rel} = 0$ $\bar{c}_{rel} = 2\bar{c}$ $\bar{c}_{rel} = \sqrt{2}\,\bar{c}$

(a)   (b)   (c)

Fig. 29-4 Relative velocity $\bar{c}_{rel}$ along the line of centers.

At 1 atm pressure and ordinary temperature, $n \approx 10^{19}$ cm^{-3}, $\sigma \approx 10^{-8}$ cm, $\bar{c} \approx 10^4$ cm/sec; the other numerical factors produce another factor of 10, so that $Z_1 \approx 10^8$ sec^{-1}, 100 million collisions every second. By combining Eqs. (29–16) and (29–17), the mean free path is

$$\lambda = \frac{1}{\sqrt{2}\pi\sigma^2 n}. \tag{29–18}$$

At 1 atm pressure, $\lambda \approx 10^{-4}$ cm. The mean free path depends on $1/n$ and is proportional to $1/p$ by the gas law $1/n = RT/N_0 p$. The lower the pressure, the fewer collisions in unit time and the longer will be the mean free path.

If there are n molecules/cm^3 and each makes Z_1 collisions/sec, the total number of collisions per cm^3 in 1 sec is

$$Z_{11} = \tfrac{1}{2}Z_1 n = \tfrac{1}{2}\sqrt{2}\pi\sigma^2 \bar{c}n^2. \tag{29–19}$$

The factor $\tfrac{1}{2}$ is introduced because a simple multiplication of Z_1 by n would count every collision twice. Without giving a detailed proof, the number of collisions·cm^{-3}·sec^{-1} between unlike molecules in a mixture is

$$Z_{12} = \pi\sigma_{12}^2\sqrt{\frac{8kT}{\pi\mu}}n_1 n_2, \tag{29–20}$$

where n_1 and n_2 are the numbers of molecules per cubic centimeter of kind 1 and kind 2, σ_{12} is the average of the diameters of the two kinds of molecules, and μ is the reduced mass, $1/\mu = 1/m_1 + 1/m_2$. These values for collision numbers will be useful later in the calculation of the rates of chemical reactions. A chemical reaction between two molecules can occur only when the molecules collide.

29–6   Final Expression for the Thermal Conductivity

Having obtained an expression for the mean free path λ, we write the formula for the coefficient of thermal conductivity by combining Eqs. (29–18) and (29–14);

$$\kappa_T = \frac{\bar{c}C_v}{3\sqrt{2\pi N_0 \sigma^2}}. \tag{29–21}$$

This equation leads to the interesting conclusion that the thermal conductivity is independent of the pressure. This lack of dependence on pressure is a result of two compensating effects. By Eq. (29–14), κ_T is proportional to n and to λ; but λ is inversely proportional to n so that the product $n\lambda$ is independent of pressure. At lower pressures fewer molecules cross the surface in one second, but they come from a larger distance (λ is larger at lower p) and so carry a proportionately greater excess energy. Experiment confirms that κ_T is independent of pressure.

If C_v is independent of temperature, then everything on the right of Eq. (29–21) is constant except $\bar{c}$, which is proportional to $T^{1/2}$. Therefore, κ_T should increase as $T^{1/2}$. This is also confirmed experimentally.

In this derivation of the expression for κ_T, we have assumed that the pressure is high enough so that λ is much smaller than the distance separating the two plates. At very low pressures where λ is much larger than the distance between the plates, the molecule bounces back and forth between the plates and only rarely collides with another gas molecule. In this case the mean free path does not enter the calculation, and the value of κ_T depends on the separation of the plates. At these low pressures the thermal conductivity is proportional to the pressure, since it must be proportional to n, and λ does not appear in the formula to compensate for the pressure dependence of n.

29–7 Viscosity

The formula for the viscosity coefficient of a gas can be derived in a way similar to that used for heat conduction. We imagine two very large parallel flat plates, one lying in the xy-plane, the other at a distance Z above the xy-plane. We keep the lower plate stationary and pull the upper plate in the $+x$-direction with a velocity U. The viscosity of the gas exerts a drag on the moving plate. To keep the plate in uniform motion, a force must be applied to balance the viscous drag. Looking at the situation in another way, if the upper plate moves with a velocity U, the viscous force will tend to set the lower plate in motion. A force must be applied to the lower plate to keep it in place.

Again we suppose that the gas between the plates is made up of a series of horizontal layers. The layer next to the lower plate is immobile; as we move upward, each successive layer has a slightly larger component of velocity in the x-direction, the topmost layer at the height Z having the velocity U. This type of flow, in which there is a regular gradation of velocity in passing from one layer to the next, is called *laminar* flow. The layer at the height z has a velocity in the x-direction given by u_z:

$$u_z = \frac{\partial u}{\partial z}z. \tag{29–22}$$

At $z = Z, u = U$, so that

$$\frac{\partial u}{\partial z} = \frac{U}{Z}. \tag{29-23}$$

If we observe a layer at the height z, molecules enter this layer from the neighboring layers. The molecules from the upper layers will bring extra x-momentum to this layer, while those which come from below are deficient in x-momentum. There is therefore a net downward flow of x-momentum through the layer. Now we compute the rate of this flow through 1 cm^2 of the layer at the height z; Fig. 29–5.

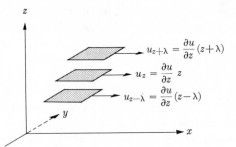

Fig. 29–5 Velocity of layers in a flowing gas.

The number of molecules passing downwards through 1 cm^2/sec is, by Eq. (29–7), $\frac{1}{6}n\bar{c}$, and as many come upward as come downward. The molecules which pass downward through the layer at z carry x-momentum appropriate to the layer in which they made their last collision, the layer at height $z + \lambda$. This x-momentum is

$$mu_{z+\lambda} = m\left(\frac{\partial u}{\partial z}\right)(z + \lambda).$$

So the amount of momentum coming down through 1 cm^2 in 1 sec is

$$(mu)\!\downarrow = \tfrac{1}{6}n\bar{c}m\left(\frac{\partial u}{\partial z}\right)(z + \lambda).$$

Similarly, the amount coming up is

$$(mu)\!\uparrow = \tfrac{1}{6}n\bar{c}m\left(\frac{\partial u}{\partial z}\right)(z - \lambda),$$

since the molecules coming up adjusted their momentum in the layer at $z - \lambda$. The net downward flow of x-momentum is

$$(mu)\!\downarrow - (mu)\!\uparrow = \tfrac{1}{3}n\bar{c}m\lambda\frac{\partial u}{\partial z}.$$

Since this quantity is independent of z, it must also be equal to the net x-momentum transferred in 1 sec to 1 cm^2 of the lower plate. Since the momentum transfer in unit time is the force, the force acting in the x-direction on 1 cm^2 of the lower plate is

$$f_x = \tfrac{1}{3}n\bar{c}m\lambda\frac{\partial u}{\partial z}. \qquad (29\text{--}24)$$

To hold this plate stationary, we must apply an equal and opposite force f_{-x}, such that $f_x + f_{-x} = 0$. The viscosity coefficient is defined by

$$f_{-x} = -\eta\frac{\partial u}{\partial z}. \qquad (29\text{--}25)$$

The viscosity coefficient is the force which must be applied to hold the lower plate stationary if the velocity gradient $\partial u/\partial z$ is unity and the area of the plate is 1 cm^2. Comparing Eqs. (29–24) and (29–25), we see that

$$\eta = \tfrac{1}{3}n\bar{c}m\lambda. \qquad (29\text{--}26)$$

If the density of the gas is ρ, then $\rho = nm$, and

$$\eta = \tfrac{1}{3}\rho\bar{c}\lambda. \qquad (29\text{--}27)$$

Again, the numerical factor $\tfrac{1}{3}$ is not quite correct, since the flow of gas produces a nonequilibrium situation. For elastic spheres, the factor should be $\tfrac{1}{2}$. The unit of the viscosity coefficient is the *poise*; 1 poise = 1 gm·cm^{-1}·sec^{-1}.

The coefficient of viscosity depends on the product $n\lambda$ and so is independent of pressure; Eq. (29–26). This rather surprising result, along with the similar result for thermal conductivity, was one of the great initial triumphs of the kinetic theory of gases. It seems as though the viscosity of a gas, which is a measure of its resistance to flow, ought to be greater at high pressures than at low. This contrary prediction of kinetic theory and the subsequent experimental verification gave great impetus to the further development of the theory.

Comparison of Eqs. (29–26) and (29–14) shows that since $M = N_0 m$,

$$\frac{\kappa_T}{\eta} = \frac{C_v}{M}. \qquad (29\text{--}28)$$

This ratio is the heat capacity per gram. More accurate theory, as well as experiment, shows that for monatomic gases

$$\frac{\kappa_T}{\eta} = 2.5\frac{C_v}{M}, \qquad (29\text{--}29)$$

is more nearly correct.

29–8 Molecular Diameters

Using $\bar{c} = \sqrt{8\mathbf{k}T/\pi m}$ and λ from Eq. (29–18) in Eq. (29–26), we obtain

$$\eta = \frac{2}{3}\frac{\sqrt{mkT}}{\pi^{3/2}\sigma^2} = \frac{2}{3}\frac{\sqrt{MRT}}{\pi^{3/2}N_0\sigma^2}, \qquad (29\text{--}30)$$

which expresses η in terms of M, T, and the quantity $N_0\sigma^2$. If we know N_0, the value of the molecular diameter can be calculated from the measured values of η.

Alternatively, if another expression involving N_0 and σ is available, values of both N_0 and σ can be determined. In Section 25–8, we related N_0 and σ to the van der Waals b. From Eq. (25–39), we have

$$b = \tfrac{2}{3}\pi N_0 \sigma^3. \tag{29–31}$$

Eliminating N_0 between Eqs. (29–30) and (29–31) and solving for σ, we have

$$\sigma = \tfrac{9}{4}\eta b \sqrt{\frac{\pi}{MRT}}. \tag{29–32}$$

Using this result in Eq. (29–30), we obtain

$$N_0 = \left(\frac{32}{243\pi b^2}\right)\left(\frac{MRT}{\pi\eta^2}\right)^{3/2}. \tag{29–33}$$

It was from Eqs. (29–32) and (29–33) that the first concrete estimates of N_0 and σ were obtained. In Table 29–1 we list values of σ, calculated from η and the currently accepted value of N_0 using Eq. (29–30).

Table 29–1

Gas	t, °C	$\eta \times 10^6$, dyne·sec cm^{-2}	σ, Å
NH_3	20	98.2	3.64
CO_2	20	148.0	3.76
Ar	20	221.7	2.98
C_2H_4	20	100.8	4.07
CH_4	20	108.7	3.41

29–9 Diffusion

If the concentration is not uniform in a mixture of two gases, the gases diffuse into one another until the composition is uniform. The derivation of the diffusion coefficient in such a situation is lengthy and somewhat complicated, since each gas has a different value for $\bar{c}$ and for λ. To simplify matters we treat the case of a single gas so that there is only one value of $\bar{c}$ and of λ. The result obtained is very nearly correct for the diffusion of one isotope into another. To define the problem suppose that some of the molecules of the gas are painted red; the gas is confined in a vertical tube and the number n' of red molecules per cubic centimeter is greater at one end than at the other; then the number n of unpainted molecules per cubic centimeter must also vary from one end to the other if the total pressure is to be uniform throughout the tube. For each species we write for the number at the height z:

$$n = n_0 + \frac{\partial n}{\partial z}z, \qquad n' = n_0' + \frac{\partial n'}{\partial z}z,$$

where n_0 and n_0' are the numbers per cubic centimeter at $z = 0$.

Consider a horizontal area of 1 cm^2 at the height z. The number of red molecules passing downward through this area per second is

$$n' \downarrow = \tfrac{1}{6}\bar{c}n'_{z+\lambda} = \tfrac{1}{6}\bar{c}\left[n_0' + \frac{\partial n'}{\partial z}(z + \lambda) \right],$$

since the molecules originate in the layer at $z + \lambda$. Similarly the number coming up from below is

$$n' \uparrow = \tfrac{1}{6}\bar{c}n_{z-\lambda} = \tfrac{1}{6}\bar{c}\left[n_0' + \frac{\partial n'}{\partial z}(z - \lambda) \right].$$

The net flow upward is

$$n' \uparrow - n' \downarrow = -\tfrac{1}{3}\bar{c}\lambda\frac{\partial n'}{\partial z}.$$

By the law for diffusion, Eq. (29–5), the upward flow is $-D'(\partial n'/\partial z)$, where D' is the diffusion coefficient of the red molecules. Thus we have $D' = \tfrac{1}{3}\bar{c}\lambda$; but the red molecules differ from the others only by a coat of paint, so that

$$D = \tfrac{1}{3}\bar{c}\lambda. \tag{29–34}$$

(The numerical factor $\tfrac{1}{3}$ is wrong as usual!) Since $\bar{c}$ is inversely proportional to $M^{1/2}$, we can understand *Graham's law* which states that the rate of diffusion of a gas is inversely proportional to the square root of the molecular weight.

Since the mean free path is inversely proportional to the pressure, the diffusion coefficient decreases with increase in pressure. The molecules have to fight their way through the swarm of other molecules by making many collisions. At high pressures they make many more collisions, and their progress in any given direction is slowed. This refutes an early objection to the kinetic theory, which was that the high molecular velocities predicted by kinetic theory were obviously ridiculous since, if the molecules moved that quickly, the smell of a gas such as NH_3 or H_2S released in one corner of a room should be noticed instantly everywhere in the room, while in fact it takes some time before the odor is detected in another part of the room. In answer it was pointed out that at ordinary pressures the gas molecule makes many collisions and the path of a molecule is a fantastic zig-zag with little net motion in any particular direction in spite of the high velocity.

An important application of the concept of the mean free path and its relation to diffusion was made by Irving Langmuir. In the ordinary incandescent lamp, the passage of the current heats a tungsten filament white hot. To prevent oxidation of the filament and immediate burnout, the bulb must be evacuated. However, if the pressure is reduced too far, the mean free path of the tungsten atom becomes large compared with the size of the bulb. Tungsten atoms which are boiled off the filament can go directly to the glass wall without an intervening collision with a gas molecule. The atoms condense on the glass wall, blacken the bulb, and weaken the filament which soon breaks. Langmuir produced argon under a few centimeters pressure. This reduces the mean free path to something less than the diameter of the filament. In this situation, a tungsten atom which has been boiled off travels only a short distance before it hits a gas molecule. A likely result of this collision is that the tungsten atom is reflected back onto the filament. In any event the tungsten atoms must leave the region of the filament by diffusion through the argon which is slow. The presence of argon in the bulb lengthens bulb life enormously.

29–10 Summary of Transport Properties in a Gas

Kinetic theory interprets the phenomenological laws of transport in gases on the basis of a single mechanism, and expresses the values of κ_T, η, and D in terms of the mean free path, the density, and the average velocity of the molecules. The equations are

$$\kappa_T = \tfrac{1}{3}(n/N_0)C_v\,\bar{c}\lambda, \qquad \eta = \tfrac{1}{3}nm\bar{c}\lambda, \qquad D = \tfrac{1}{3}\bar{c}\lambda.$$

Since all of these depend on λ, they are sometimes called *free path* phenomena.

29–11 The Nonsteady State

In the preceding sections we assumed that the flow was in a steady state, where the amount of a quantity flowing into any volume element is just balanced by an equal amount of the quantity flowing out in the same time interval. For diffusion this means that the concentration in any volume element is independent of time, $\partial c/\partial t = 0$. For thermal conductivity it means that energy does not accumulate in any volume element, or that $\partial T/\partial t = 0$.

To treat diffusion in the nonsteady state, we consider the situation shown in Fig. 29–6. Molecules diffuse in the $+x$-direction through two elements of area, each being 1 cm^2, perpendicular to the x-axis and located at x and $x + \Delta x$. The flow through the element at x is J_x, that through the element at $x + \Delta x$ is $J_{x+\Delta x}$. The element enclosed by the parallelepiped has a volume equal to 1 cm$^2\cdot\Delta x$ cm $= \Delta x$ cm^3.

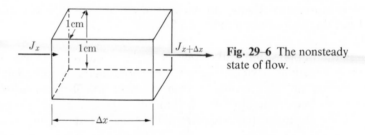

Fig. 29–6 The nonsteady state of flow.

In the time dt, the number of moles entering the volume element from the left is $J_x\, dt$, while the number leaving at $x + \Delta x$ is $J_{x+\Delta x}\, dt$. If the increase in the number of moles in the volume element in the interval dt is ΔN, the excess of what flows in at x over what flows out at $x + \Delta x$, then $\Delta N = J_x\, dt - J_{x+\Delta x}\, dt$. But $J_{x+\Delta x} = J_x + (\partial J_x/\partial x)\, \Delta x$, so that

$$\Delta N = -\frac{\partial J_x}{\partial x}\Delta x\, dt.$$

The increase in concentration in the volume element is $dc = \Delta N/\Delta x$, so that

$$dc = -\frac{\partial J_x}{\partial x}\, dt,$$

or

$$\frac{\partial c}{\partial t} = -\frac{\partial J_x}{\partial x}. \qquad (29\text{–}35)$$

By Eq. (29–5), $J_x = -D(\partial c/\partial x)$. Using this in Eq. (29–35), we obtain

$$\frac{\partial c}{\partial t} = \frac{\partial}{\partial x}\left(D\frac{\partial c}{\partial x}\right),$$

or, if D is independent of x,

$$\frac{\partial c}{\partial t} = D\frac{\partial^2 c}{\partial x^2}. \qquad (29\text{–}36)$$

The solution of Eq. (29–36), Fick's second law of diffusion, under a specified set of conditions, yields the concentration as a function of x and t. From Eq. (29–36) we discover that the condition for the steady state, $\partial c/\partial t = 0$, implies that $\partial^2 c/\partial x^2 = 0$ or $\partial c/\partial x = $ constant. Therefore, in the steady state the concentration varies linearly with the coordinate.

By a similar argument using the equation for heat conduction, we obtain

$$\frac{\partial T}{\partial t} = \frac{\kappa_T}{\rho c_v}\frac{\partial^2 T}{\partial x^2}, \qquad (29\text{–}37)$$

where the factor ρc_v appears when the energy increment in the volume element is converted to a temperature increment; c_v is the heat capacity per gram and ρ is the density. Equations (29–36) and (29–37) are formally the same. Solving a problem in diffusion solves an analogous problem in heat conduction.

29–12 The Poiseuille Formula

The rate at which a fluid flows through a tube depends on the dimensions (radius and length) of the tube, the viscosity of the fluid, and the pressure drop between the ends of the tube. To discover the relation between these quantities, we first calculate the volume passing any point in a circular tube in unit time.

In narrow circular tubes the flow is laminar so that the cylindrical sheath at the boundary of the tube is stationary; as we move to the center each successive cylindrical sheath moves with a slightly larger velocity. Suppose the tube lies with its length along the x-axis. Then consider the sheath shown in Fig. 29–7, having an inner radius r and an outer radius $r + dr$. If the velocity of this sheath in the x-direction is v cm/sec, then in 1 sec the sheath moves v cm and carries all the fluid in it past a given point. The volume passing any point in unit time is $2\pi rv\, dr$. The total volume passing any point in unit time is V, and is the sum of the contribution of every sheath in the tube. Therefore,

$$V = \int_0^a 2\pi rv\, dr, \tag{29–38}$$

where a is the radius of the tube. To obtain V, the volume delivered by the tube in unit time, v must be known as a function of r.

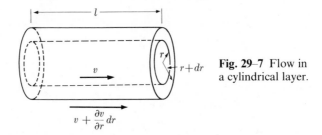

Fig. 29–7 Flow in a cylindrical layer.

The relation between v and r is obtained by balancing the forces due to the pressure difference and the viscosity. Let the pressure on the left end of the tube be p_1 and that on the right end be p_2. The force acting on the left end of the sheath is $p_1 2\pi r\, dr$, that on the right end is $p_2 2\pi r\, dr$. The net force in the $+x$-direction, f_x, due to the pressure difference is

$$f_x = (p_1 - p_2)2\pi r\, dr. \tag{29–39}$$

Each square centimeter of the inner surface of the sheath is subject to a viscous force in the $+x$-direction equal to $-\eta(\partial v/\partial r)$. If the area of the inner surface is $S = 2\pi rl$, then the total force acting on the inner surface is $-\eta S(\partial v/\partial r)$. This inner surface is being pulled along by the faster moving interior cylinder. The outer surface of the sheath is *retarded* by the slower moving fluid outside the sheath, the force in the x-direction on the outer surface being

$$\eta S\frac{\partial v}{\partial r} + d\left(\eta S\frac{\partial v}{\partial r}\right).$$

The net viscous force is the sum of the forces on the inner and outer surfaces, f'_x:

$$f'_x = d\left(\eta S\frac{\partial v}{\partial r}\right). \tag{29–40}$$

For balance, the sum of the forces in the $+x$-direction due to pressure difference and viscous forces must be zero; $f_x + f'_x = 0$. Using Eqs. (29–39) and (29–40) and rearranging, this becomes

$$d\left(\eta S \frac{\partial v}{\partial r}\right) = -2\pi r(p_1 - p_2)\, dr,$$

which integrates immediately to

$$\eta S \frac{\partial v}{\partial r} = -\pi(p_1 - p_2)r^2 + A,$$

where A is the integration constant. Using the value of $S = 2\pi r l$, this becomes

$$\frac{\partial v}{\partial r} = -\frac{(p_1 - p_2)r}{2\eta l} + \frac{A}{2\pi \eta l r}.$$

Integrating again, we obtain

$$v = -\frac{(p_1 - p_2)r^2}{4\eta l} + \frac{A}{2\pi \eta l}\ln r + B, \tag{29-41}$$

where B is another integration constant. Now the velocity must be finite at $r = 0$, and this is not possible if the logarithmic term appears in Eq. (29–41); therefore it must be that $A = 0$. Then

$$v = -\frac{(p_1 - p_2)r^2}{4\eta l} + B.$$

At the radius of the tube, $r = a$, the velocity of the fluid is zero, so we have

$$0 = -\frac{(p_1 - p_2)a^2}{4\eta l} + B.$$

Using this value of B, the velocity becomes

$$v = \frac{(p_1 - p_2)(a^2 - r^2)}{4\eta l}, \tag{29-42}$$

which expresses the velocity as a function of r, which is required to evaluate the volume delivered in unit time. Using this value of v in the integral of Eq. (29–38), we obtain

$$V = \frac{\pi(p_1 - p_2)}{2\eta l}\int_0^a (a^2 - r^2)r\, dr = \frac{\pi a^4(p_1 - p_2)}{8\eta l}, \tag{29-43}$$

which is Poiseuille's formula; it has been verified quite accurately for fluid flow through tubes for which $a \ll l$. Knowing the radius of the tube, the length, and the pressure difference, we can calculate the value of η from the measured volume of liquid discharged in unit time. Conversely, if η is known the radius of the tube can be calculated from the volume discharged; this is useful for measuring the average cross section of a fine capillary tube.

Since the pressure gradient $\partial p/\partial x = (p_2 - p_1)/l$, Eq. (29–43) can be written in the form

$$V = -\frac{\pi a^4}{8\eta}\frac{\partial p}{\partial x}, \qquad (29\text{–}44)$$

which is again Poiseuille's law; compare with Eq. (29–4).

29–13 The Viscosimeter

The viscosimeter is an instrument for determining viscosity by measuring the time required for a fixed volume of a liquid to flow through a capillary tube, the efflux time. A simple viscosimeter is shown in Fig. 29–8. Two bulbs are connected by a length of capillary tubing. The liquid is forced into the left-hand limb until it rises above the mark at a. It is then allowed to flow into the lower bulb. The time required for the liquid level to drop from a to b is measured. This is the time required for a fixed volume of liquid to flow through the capillary. The pressure difference varies with time during efflux but is proportional to the density ρ of the liquid. So if two different liquids are compared in the same viscosimeter, we have

$$\frac{\eta_1}{\eta_2} = \frac{\rho_1 t_1}{\rho_2 t_2}. \qquad (29\text{–}45)$$

This is a convenient method for measuring the viscosity of one liquid relative to that of another.

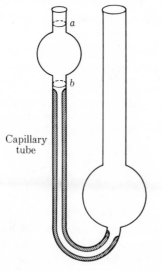

Capillary
tube

Fig. 29–8 Simple viscosimeter.

The temperature dependence of the viscosity is quite different for liquids and for gases. The simple derivation for gases predicts a proportionality of η to $\sqrt{T}$, and this is observed experimentally. In contrast, the viscosity of liquids *decreases* with increase in temperature. An equation, first proposed empirically and later given a theoretical foundation, represents the data reasonably well:

$$\ln \eta = a + \frac{E}{RT},$$

where a and E are constants. This may be written in the alternative form

$$\eta = Ae^{E/RT}; \tag{29-46}$$

E is called the *activation* energy for flow. The significance of Eq. (29-46) will be more fully appreciated after a study of the rates of chemical reactions.

Problems

29-1. If the molecular diameter of H_2 is 1.9×10^{-8} cm, calculate the number of collisions made by a hydrogen molecule in 1 sec.

a) if $T = 300°$ and $p = 1$ atm,

b) if $T = 500°$ and $p = 1$ atm,

c) if $T = 300°$ and $p = 10^{-4}$ atm.

d) Compute the total number of collisions per sec occurring in 1 cm^3 for each of the gases in (a), (b), (c).

29-2. The molecular diameter of N_2 is about 2.7 Å.

a) Compute the mean free path of N_2 at 300°K and 1 atm, 0.1 atm, 0.01 atm.

b) A reasonably good vacuum system achieves a pressure of about 10^{-9} atm. What is the mean free path at this pressure?

c) If the diameter of the evacuated tube ($p = 10^{-9}$ atm) is 5 cm, how many times does the molecule strike the walls between two successive collisions with other gas molecules?

29-3. Suppose there are 20 couples on a dance floor which is 50 ft × 50 ft. If the diameter of each couple is 2 ft and the velocity is 2 ft/sec, derive formulas for, and then calculate the mean free path, the number of collisions per minute made by each couple, and the total number of collisions per minute. (Assume the motion is chaotic.)

29-4. The viscosity coefficient of methane at 0°C is 102.6×10^{-6} poise. (1 poise = 1 gm·cm^{-1}·sec^{-1}.) Calculate the molecular diameter.

29-5. Compare the thermal conductivities of O_2 and H_2; ignore the difference in molecular diameter. Both have $C_v = \frac{5}{2}R$.

29-6. Two parallel plates 0.5 cm apart are maintained at 298°K and 301°K. The space between the two plates is filled with H_2, $\sigma = 1.9$ Å, $C_v = \frac{5}{2}R$. Calculate the flow of heat between the two plates in cal·cm^{-2}·sec^{-1}.

29–7. Ethane has a molecular weight of 30, compared with 28 for N_2 and 32 for O_2. The molecular diameter is not greatly different from that of oxygen or nitrogen. The thermal conductivity of ethane is significantly larger than that of O_2 or N_2. Explain.

29–8. One end of a capillary tube 10 cm in length and 1 mm in diameter is connected to a water supply which has a pressure of 2 atm. The viscosity coefficient of water is 0.01 poise. How much water does the tube deliver in 1 sec? (1 atm $= 10^6$ dynes/cm^2.)

29–9. At 25°C, H_2 has a viscosity of 88×10^{-6} poise. What diameter tube must be used if a 1 m length under a pressure difference of 0.3 atm is to deliver 1 liter/min?

29–10. Consider the flow through a cylindrical sheath of inner radius a and outer radius b. The flow per second is given by Eq. (29–38), where the limits of integration are a and b. The velocity is given by Eq. (29–41), but in this case $A \neq 0$. The constants A and B are determined by the conditions that $v = 0$ at $r = a$ and $v = 0$ at $r = b$. Derive the formula corresponding to Poiseuille's equation for this case.

29–11. The thermal conductivity of silver is about $1.0 \, cal \cdot deg^{-1} \cdot cm^{-1} \cdot sec^{-1}$. Calculate the heat flow per second through a silver disc 0.1 cm in thickness and having 2 cm^2 area if the temperature difference between the two sides of the disc is 10°.

29–12. The densities of acetone and water at 20°C are 0.792 and 0.9982 gm/cm^3, respectively. The viscosity of water is 1.0050×10^{-2} poise at 20°C. If water requires 120.5 sec to run between the marks on a viscosimeter and acetone requires 49.5 sec, what is the viscosity of acetone?

29–13. The viscosities of acetone are

t, °C	-60	-30	0	30
$\eta \times 10^2$, poise	0.932	0.575	0.399	0.295

By plotting $\ln \eta$ versus $1/T$, determine the value of E in Eq. (29–46).

Chapter Thirty

Electrical Conduction

30–1 Electrical Transport

The quantity of electrical charge which passes any point in a conductor in unit time is the *current*. The current passing through an area of 1 cm^2 perpendicular to the direction of the current is the current density j (A/cm^2). By the general law of transport, the current density is proportional to the potential gradient, Eq. (29–3),

$$j = -\kappa \frac{\partial \mathcal{V}}{\partial x} \tag{30–1}$$

if the direction of flow is along the x-axis. The constant of proportionality κ is the *conductivity* of the substance. The electrical field E is defined by $E = -\partial \mathcal{V}/\partial x$, so Eq. (30–1) can be written in the form

$$j = \kappa E. \tag{30–2}$$

Equations (30–1) and (30–2) are expressions of Ohm's law.

To transform Ohm's law into a more familiar form, we consider a conductor of length l' and cross-sectional area A. If the emf across the ends is $\mathcal{E} = \mathcal{V}_2 - \mathcal{V}_1$, then $E = (\mathcal{V}_2 - \mathcal{V}_1)/l' = \mathcal{E}/l'$. The current I carried by the conductor is related to the current density by $I = jA$. Using these expressions for E and j in Eq. (30–2), we obtain

$$I = \kappa A \mathcal{E}/l'. \tag{30–3}$$

We define the *conductance* $L = \kappa A/l'$. Then

$$I = L\mathcal{E}. \tag{30–4}$$

The *resistance* R of the conductor is defined by $R = 1/L = l'/\kappa A = \rho l'/A$, where the *resistivity* $\rho = 1/\kappa$. This definition brings Ohm's law, Eq. (30–4), into its familiar form

$$\mathscr{E} = IR. \qquad (30\text{–}5)$$

By putting the definition of the resistivity into Eq. (30–2), we obtain an analogue of Eq. (30–5):

$$E = j\rho. \qquad (30\text{–}6)$$

Ordinarily we shall use Ohm's law in the form of Eq. (30–2) or Eq. (30–6). This is convenient since κ and ρ are properties of the material composing the conductor and do not depend on its geometry. The resistance depends on the geometry of the conductor through the relation

$$R = \rho\frac{l'}{A}. \qquad (30\text{–}7)$$

Lengthening the conductor increases its resistance, while thickening it decreases its resistance. Since R is measured in ohms, ρ has the units ohm·cm. The conductivity κ has the units $\text{ohm}^{-1}\cdot\text{cm}^{-1}$, sometimes written $\text{mho}\cdot\text{cm}^{-1}$.

30–2 Conduction in Metals

The current in metals is carried entirely by the electrons, each of which carries a negative charge e. Using Eq. (29–8) for the flow, the product of the number of electrons per cubic centimeter, their average velocity in the direction of the flow, and their charge, we obtain

$$j = nve. \qquad (30\text{–}8)$$

Combining this result with Eq. (30–2), the expression for the conductivity becomes

$$\kappa = nve/E. \qquad (30\text{–}9)$$

Ohm's law requires that κ be a constant; it must be independent of the field E. Therefore one of the quantities in the numerator of Eq. (30–9) must be proportional to E to compensate for the presence of E in the denominator. Obviously the charge e on the electron does not depend on the field. The number of electrons per cubic centimeter could conceivably depend on the field, but it can be shown that such a dependence would not be a simple proportionality. It must be that the velocity of the carrier is proportional to the field and that the numbers of carriers is independent of the field. This is the condition which must be satisfied if any conductor is to obey Ohm's law. Therefore we write

$$v = lE. \qquad (30\text{–}10)$$

The constant of proportionality l is called the *mobility*, which is the velocity acquired by a carrier in a field of unit strength; $l = v/E$. Since the field has the units volt·cm^{-1}, and the velocity is in cm·sec^{-1}, the mobility has the units cm^2·volt^{-1}·sec^{-1}.

From the requirement that the velocity must be proportional to the field, we conclude that the main force of retardation of the carrier is due to friction. If the charge on the carrier is q, then the force due to the electrical field is qE, which must be balanced by the inertial force $ma = m(dv/dt)$, and the frictional force Bv, which is proportional to the velocity. Thus

$$qE = m\frac{dv}{dt} + Bv,$$

where B is a constant. From this equation it is clear that if the velocity is to be proportional to E, the first term, the inertial force, must be negligibly small in comparison with the second, the frictional retardation, so that we have

$$qE = Bv. \tag{30–11}$$

In a metal the frictional force arises from the scattering of the electrons by collisions with the metal ions in the lattice.

In terms of the mobility, the expression in Eq. (30–9) for the conductivity becomes

$$\kappa = nle. \tag{30–12}$$

From a measurement of the resistance of a metal, the resistivity and the conductivity can be determined. Since we know the value of e, the measurement yields a value of the product nl. To determine n and l individually requires an independent measurement of some other quantity which depends on one or both of these quantities.

30–3 The Hall Effect

Consider the following experiment: a current having a current density j is passed through a metal strip in the x-direction; simultaneously a magnetic field H is applied in the z-direction. Two probes A and A' are placed on opposite sides of the strip; Fig. 30–1. The magnetic field, indicated by the dashed circle in Fig. 30–1, deflects the electron stream in the metal with the result that an electrical field E_y develops

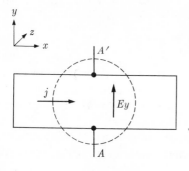

Fig. 30–1 The Hall effect.

across the width of the strip and produces a potential difference $\mathscr{V}_H$, the Hall potential, between the two probes A and A'.

If v is the velocity of the electrons in the x-direction, the force acting in the y-direction due to the magnetic field is Hev; this force is balanced by the force from the electrical field in the y-direction, which is eE_y. Thus we have

$$eE_y = Hev \quad \text{or} \quad E_y = Hv.$$

Using the value of v from Eq. (30–8), this becomes

$$E_y = Hj/ne.$$

The Hall potential is $\mathscr{V}_H = E_y w$, where w is the width of the strip. So that

$$\mathscr{V}_H = \frac{wHj}{ne} = R_H wHj, \tag{30–13}$$

where $R_H = 1/ne$ is the Hall coefficient. Measurement of w, j, H, and $\mathscr{V}_H$ suffices to determine the value of R_H. This determines n, since from the definition,

$$n = 1/eR_H. \tag{30–14}$$

Combining Eq. (30–14) with Eq. (30–12), we obtain the mobility of the electrons:

$$l = \kappa R_H. \tag{30–15}$$

By measuring the conductivity and the Hall coefficient, it is possible to obtain values of the mobility and the number of carriers per cm^3. Table 30–1 lists values of l, n, and the number of carriers per atom which contribute to the conductivity for several metals.

Table 30–1

Metal	Cu	Ag	Au	Li	Na	Zn†	Cd†
l, cm$^2 \cdot$volt$^{-1} \cdot$sec^{-1}	35	56	30	19	48	5.8	7.9
$n \times 10^{-22}$, cm^{-3}	11	7.5	8.7	3.7	2.4	19	10
Electrons/atom	1.4	1.3	1.5	0.79	0.94	3.0	2.2

†   Sign of R_H indicates carrier is positively charged.

The values of the mobility are interesting because they are so small. This emphasizes that it is the frictional resistance which retards the motion. An electron moving in free space subject only to inertial retardation would have a mobility about one million times larger than the mobilities in metals.

Another point of interest is that not all but only about one electron per atom is free to carry the current. Only the electrons in levels near the top of the filled part of the partially filled band are "free" to move under the application of a field. As we saw in Section 27–4, to carry a current the electrons must be able to shift from one set of levels to another set; vacant levels must be available which are not very much different in energy. Vacant levels are available only near the top of the filled part of a partially filled band and so only these electrons contribute to the conductivity.

Finally, there is the curious result that if the electrons which carry the current are in levels near the top of a band, then the field pushes the electrons in the *wrong* direction; wrong in the sense that they are accelerated in the direction opposite to the usual one. These electrons behave as if they were positively charged. This happens with zinc and with cadmium as well as a number of other metals. The effect is detected in the Hall experiment; the Hall potential for these metals has the opposite sign when compared with a metal such as copper.

Measurement of the magnitude and sign of the Hall potential in semiconductors enables us to distinguish experimentally between *p*- and *n*-type semiconductors, and to determine, knowing κ, the number and mobility of the carriers.

30–4 The Electrical Current in Ionic Solutions

The passage of an electrical current in an ionic solution is a more complex event than the passage of a current through a metal. In the metal, the nearly weightless electrons carry all the current. In the ionic solution, the current is carried by the motion of massive positive and negative ions. Consequently, *the passage of a current is accompanied by a transport of matter.* The positive and negative ions do not carry equal portions of the current, so that a concentration gradient develops in the solution. Furthermore, transfer of the electrical charge through the solution–electrode interface is accompanied by a chemical reaction (electrolysis) at each electrode. For clarity we must keep the phenomena in the body of the solution separate from the phenomena at the electrodes. We begin by dealing briefly with the phenomena at the electrodes (electrolysis), and then we will describe the occurrences in the body of the solution which are our main concern in this chapter.

If a direct current is passed between two electrodes in an electrolytic solution, a chemical reaction, electrolysis, occurs at the electrodes. After a study of various types of electrolytic reactions, Faraday (1834) discovered two simple and fundamental rules of behavior, now called Faraday's laws of electrolysis. Faraday's first law states that the amount of chemical reaction which occurs at any electrode is proportional to the quantity Q of electricity passed. Since Q is the product of the current and the time, $Q = It$, we write

$$W = ZIt, \tag{30–16}$$

where W is the weight of any product of the electrolysis and Z, the electrochemical equivalent of that product, is the constant of proportionality. The second law states that the passage of a fixed quantity of electricity produces amounts of two different substances in proportion to their chemical equivalent weights. Faraday's experiments showed that these rules were followed with great accuracy. So far as we know these laws are *exact*. One equivalent of chemical reaction is produced by the passage of one Faraday, $\mathscr{F} = 96,490$ coulomb/equivalent. The faraday is the magnitude of the charge on an Avogadro number of electrons; $\mathscr{F} = N_0 e$.

30–5 The Measurement of Conductivity in Electrolytic Solutions

A simple conductivity cell is shown in Fig. 30–2. Two platinum electrodes are sealed in the ends of the cell. These are usually coated with a deposit of finely divided platinum, platinum black, to eliminate some of the effects of electrolysis. The cell is filled with the solution, and the resistance is measured by placing the cell in one arm of the alternating current version of a Wheatstone bridge. The frequency of current ordinarily used is about 1000 c/s.

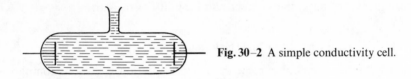

Fig. 30–2 A simple conductivity cell.

From Eq. (30–7) the resistance is

$$R = \rho \frac{l'}{A} = \frac{1}{\kappa} \frac{l'}{A},$$

since $\rho = 1/\kappa$. For κ we obtain

$$\kappa = l'/RA. \tag{30–17}$$

The cell constant $K = l'/A$ depends on the geometry of the cell; it can be determined for cells of simple design by measuring the distance l' between the electrodes and the area A of the electrodes. More conveniently the cell constant is determined indirectly by measuring the resistance of the cell when it contains a solution of

Table 30–2† Conductivity of KCl solutions (ohm^{-1}·cm^{-1})

gm KCl/kgm soln (vac)	Concentration, D	0°C	18°C	25°C
71.1352	1	0.06517$_6$	0.09783$_8$	0.11134$_2$
7.41913	0.1	0.007137$_9$	0.011166$_7$	0.012856$_0$
0.745263	0.01	0.0007736$_4$	0.0012205$_2$	0.0014087$_7$

† G. Jones and B. C. Bradshaw, *J. Amer. Chem. Soc.* **55**, 1780 (1933).

accurately known conductivity. Solutions of potassium chloride are commonly used for this purpose. Some values of κ for KCl solutions are given in Table 30–2. The concentrations are expressed in *demal* units (D). A one demal solution (1 D) contains 1 gram-mole of salt in one cubic decimeter of solution at $0°C$.

If R_s is the resistance of the cell containing a solution of known conductivity κ_s, then

$$K = l'/A = \kappa_s R_s, \tag{30–18}$$

so that by Eq. (30–17)

$$\kappa = \kappa_s \left(\frac{R_s}{R} \right). \tag{30–19}$$

In precision work great care must be taken to eliminate effects due to electrolysis and those due to variation in temperature. Controlling the temperature is a particularly difficult problem because of the heating effect of the current. Water of extreme purity (conductivity water) must be used, since stray impurities in the water can produce sensible variations in the value of the conductivity of the solution. The contribution of water itself to the conductivity must be subtracted from the measured value for the solution.

30–6 The Migration of Ions

Kohlrausch established that electrolytic solutions obeyed Ohm's law accurately once the effect of the electrolysis products was eliminated by using high-frequency alternating current. Kohlrausch also showed from the experimental data that the conductivity of a solution could be composed of separate contributions from each ion; Kohlrausch's *law of the independent migration of ions*.

Suppose that n_+ and n_- are the number per cm^3 of positive and negative ions, respectively. Let their velocities be v_+ and v_-, and their charges $z_+ e$ and $z_- e$. Then, by the fundamental law of transport, Eq. (30–8), the current density is

$$j = n_+ v_+ z_+ e + n_- v_- z_- e. \tag{30–20}$$

(Note that both the velocity and the charge of the negative ion are opposite in sign to those of the positive ion. However, the product $v_- z_- e$ has the same sign as that product for the positive ion, so the terms add together in Eq. (30–20). For convenience we will take all the quantities positively, since this will not affect the final result.) Physically, Eq. (30–20) says that the effects of positive ions moving in one direction and negative ions moving in the other add up to produce the total flow of charge.

In spite of the passage of the current, the solution must remain electrically neutral; thus, the positive charge in $1\,cm^3$, $n_+ z_+ e$, must be equal to the negative charge in $1\,cm^3$, $n_- z_- e$. If c^* is the number of equivalents per cm^3, then the total positive or negative charge in $1\,cm^3$ is $c^* \mathscr{F}$; since $\mathscr{F} =$ the charge per equivalent. Therefore we have

$$c^* \mathscr{F} = n_+ z_+ e = n_- z_- e. \tag{30–21}$$

In view of this relation, Eq. (30–20) reduces to

$$j = c^* \mathscr{F}(v_+ + v_-).$$ (30–22)

Introducing the mobilities of the ions defined by Eq. (30–10), $v_+ = l_+ E$, and $v_- = l_- E$, brings Eq. (30–22) to the form

$$j = c^* \mathscr{F}(l_+ + l_-)E.$$ (30–23)

Comparison of this equation with Ohm's law in the form of Eq. (30–2) yields the result

$$\kappa = c^* \mathscr{F}(l_+ + l_-).$$ (30–24)

If the mobilities are independent of the concentration, Eq. (30–24) says that the conductivity is proportional to the concentration c^* of the electrolyte.

To bring the data to a consistent basis, Kohlrausch defined the equivalent conductivity Λ:

$$\Lambda = \frac{\kappa}{c^*}.$$ (30–25)

The equivalent conductivity is the conductivity if one equivalent of electrolyte were present in every cubic centimeter of the solution. Combining Eqs. (30–24) and (30–25), we obtain the result

$$\Lambda = \mathscr{F}(l_+ + l_-).$$ (30–26)

The equivalent conductivities of the individual ions λ_+ and λ_- are defined by

$$\lambda_+ = l_+ \mathscr{F} \qquad \lambda_- = l_- \mathscr{F}.$$ (30–27)

These definitions bring Eq. (30–26) to the form

$$\Lambda = \lambda_+ + \lambda_-.$$ (30–28)

Equation (30–28) expresses the equivalent conductivity as the sum of contributions from each kind of ion present; this is Kohlrausch's law; it is strictly correct only if the electrolytic solution is infinitely dilute, $c^* = 0$. This is not surprising, since the electrically charged ions should exert a mutual influence on each other, especially if they are present in appreciable concentration. Thus, if Λ^0 is the equivalent conductivity at infinite dilution, then the expression for Kohlrausch's law is

$$\Lambda^0 = \lambda_+^0 + \lambda_-^0.$$ (30–29)

30–7 The Determination of Λ^0

Kohlrausch found that the equivalent conductivity depends on the concentration of the electrolyte, and that in dilute solutions of strong electrolytes this dependence could be expressed by the equation

$$\Lambda = \Lambda^0 - b\sqrt{c},$$ (30–30)

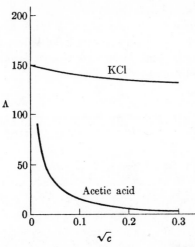

Fig. 30–3 Equivalent conductivity of strong and weak electrolytes (c in equivalents/liter).

where Λ^0 and b are constants. Plotting the value of Λ against the square root of the concentration, yields a straight line at low concentrations. The line can be extrapolated to $c = 0$ to yield Λ^0, the value of Λ at infinite dilution.

The equivalent conductivity of weak electrolytes falls off much more rapidly with increasing concentration than Eq. (30–30) predicts. The comparative behavior of KCl and acetic acid is shown schematically in Fig. 30–3. Arrhenius suggested that the degree of dissociation of an electrolyte was related to the equivalent conductivity by

$$\alpha = \Lambda/\Lambda^0. \tag{30–31}$$

Ostwald used this relation in conjunction with the law of mass action to explain the variation of the equivalent conductivity of weak electrolytes with concentration. Consider the dissociation of acetic acid:

$$\text{HAc} \rightleftharpoons \text{H}^+ + \text{Ac}^- ;$$

if α is the degree of dissociation, then $c_{\text{H}^+} = c_{\text{Ac}^-} = \alpha c$, and $c_{\text{HAc}} = (1 - \alpha)c$. The equilibrium constant is

$$K = \frac{\alpha^2 c}{1 - \alpha}.$$

Using $\alpha = \Lambda/\Lambda^0$, we obtain

$$K = \frac{c\Lambda^2}{\Lambda^0(\Lambda^0 - \Lambda)} \tag{30–32}$$

as the relation between Λ and c; the *Ostwald dilution law*. Using the values of Λ at various concentrations and the value of Λ^0, it is found that the right-hand side of Eq. (30–32) is very nearly constant, and in fact, this is a reasonable way to determine the value of the dissociation constant of a weak electrolyte.

To use Eq. (30–32) in the way described above, the value of Λ^0 must be known. The extrapolation used for strong electrolytes is useless for a weak electrolyte. Because of the steepness of the curve near $c = 0$, any extrapolated value would be subject to gross errors. To obtain Λ^0 for a weak electrolyte we use Kohlrausch's law. Using acetic acid as an example, we have at infinite dilution,

$$\Lambda^0_{HAc} = \lambda^0_{H^+} + \lambda^0_{Ac^-}.$$

To each side of this equation we add the Λ^0 of the salt of a strong acid and strong base, e.g., NaCl

$$\Lambda^0_{HAc} + \Lambda^0_{NaCl} = \lambda^0_{H^+} + \lambda^0_{Cl^-} + \lambda^0_{Na^+} + \lambda^0_{Ac^-},$$

which can be written in the form

$$\Lambda^0_{HAc} + \Lambda^0_{NaCl} = \Lambda^0_{HCl} + \Lambda^0_{NaAc};$$

hence

$$\Lambda^0_{HAc} = \Lambda^0_{HCl} + \Lambda^0_{NaAc} - \Lambda^0_{NaCl}. \tag{30–33}$$

The equivalent conductivities on the right-hand side can all be obtained by the extrapolation of a Λ versus $\sqrt{c}$ plot, since the substances involved are all strong electrolytes.

After the Arrhenius theory was first proposed, an attempt was made to fit all conductance data to the Ostwald dilution law. It soon became apparent that many substances did not conform to this law. These substances are strong electrolytes; they are completely dissociated into ions. The discussion of the dependence of the equivalent conductivity of strong electrolytes on concentration is based on the ideas contained in the Debye-Hückel theory.

30–8 The Onsager Equation

If we imagine a single ion immersed in a fluid and subjected to an electrical field, then the only retardation the ion experiences is that due to the viscosity of the fluid. If the ion is a sphere of radius r, then the frictional force opposing its motion is given by Stokes' law:

$$f = 6\pi\eta r v, \tag{30–34}$$

where η is the viscosity of the medium, and v is the velocity of the ion. We balance this by the electrical force acting on the ion, zeE:

$$zeE = 6\pi\eta r v.$$

From this equation we obtain the mobility,†

$$l = \frac{v}{E} = \frac{ze}{6\pi\eta r(300)}.$$

---

†   The factor 300 appears because of the conversion of esvolts to volts.

The equivalent conductivity of an electrolyte is given by $\Lambda^0 = \mathscr{F}(l_+ + l_-)$, so that

$$\Lambda^0 = \frac{\mathscr{F}e}{6\pi\eta(300)}\left(\frac{z_+}{r_+} + \frac{z_-}{r_-}\right). \tag{30–35}$$

The only quantity on the right-hand side which depends on the medium is η, so that we should have the relation

$$\Lambda^0\eta = \text{constant}, \tag{30–36}$$

which is Walden's rule. Comparison of the $\Lambda^0\eta$ product for a specified electrolyte in several different solvents shows that the agreement is not very good except for electrolytes with very large ions. The difficulty results from the fact that the ions are solvated. An ion is attached to molecules of solvent which are carried along with the ion as it moves. The effective radius of the ion is therefore larger than its crystallographic radius and is different in each solvent. The amount of solvent held to the ion is less with larger ions (since the electrical field due to the ion itself is smaller), so that the effective radius is more nearly the same in various solvents; consequently, Walden's rule is more accurate for large ions. If water is included in the solvents under comparison, the $\Lambda^0\eta$ product in water is usually quite different than for the others, indicating more marked solvation in water. If H_2O and D_2O are compared, the $\Lambda^0\eta$ products are very nearly equal.

If the solution of electrolyte is not infinitely dilute, then the ion is retarded in its motion because of the electrical attraction between ions of opposite sign, *asymmetry effect*, and because the positive and negative ions are moving in opposite directions each carrying some solvent, *electrophoretic effect*. Both of these effects are intensified as the concentration of the electrolyte increases so that the retarding forces increase and the conductivity decreases.

The Debye-Hückel theory of ionic solutions provides the concept of an ionic atmosphere surrounding each ion. In the absence of an applied field, this atmosphere can be imagined as a sphere of opposite charge whose radius $r_a = 1/\kappa'$, the Debye length. In the absence of a field, Fig. 30–4(a), the atmosphere is symmetrically disposed about the ion, so that it exerts no net force on the ion. In the presence of a field, Fig. 30–4(b), as the ion moves in one direction the atmosphere does not have time to adjust itself to remain spherically disposed about the ion, and it lags behind. As a result, the ion is retarded in its motion by the atmosphere, which cannot keep

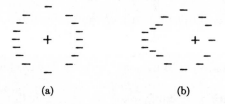

(a)   (b)

Fig. 30–4 Asymmetry effect. (a) Field off. (b) Field on.

up. The effect of the ionic atmosphere is less when $1/\kappa'$ is large, that is, when the atmosphere is far away; less in solvents of high dielectric constant, because the force between ions is reduced by high dielectric constant; and less when $\mathbf{k}T$ is large, since increase in temperature yields a less coherent atmosphere. The asymmetry effect reduces the value of Λ by a term of the form (for uni-univalent electrolytes), $B\Lambda^0\sqrt{c}$. The constant $B = 8.20 \times 10^5/(\epsilon T)^{3/2}$, where ϵ is the dielectric constant of the solvent.

The electrophoretic effect arises from the motion of the atmosphere in the direction opposite to that of the ion. Both the atmosphere and the ion pull solvent with them and each is, in effect, swimming upstream against the solvent pulled along by the motion of the other. This retardation is less in very viscous solvents because the motion of both the atmosphere and the ion is slowed down. The expression for the electrophoretic retardation has the form, for uni-univalent electrolytes, $A\sqrt{c}$, where $A = 82.4/(\epsilon T)^{1/2}\eta$, where η is the viscosity coefficient of the solvent. When written out in detail the electrophoretic retardation resembles Eq. (30–35); it has the form

$$\frac{\mathscr{F}e}{300(6\pi\eta)}\left(\frac{z_+}{r_a} + \frac{z_-}{r_a}\right) = \frac{\mathscr{F}e(z_+ + z_-)\kappa'}{300(6\pi\eta)},$$

since the radius of the atmosphere is $r_a = 1/\kappa'$. If we subtract this term from Λ^0 to obtain Λ, and use Eq. (30–35) for Λ^0, we get

$$\Lambda = \frac{\mathscr{F}e}{6\pi\eta(300)}\left(\frac{z_+}{r_+} - \frac{z_+}{r_a} + \frac{z_-}{r_-} - \frac{z_-}{r_a}\right).$$

This equation implies that the effect of the motion of the atmosphere is to increase the effective radius of each ion to

$$(r_+)_{\text{eff}} = r_+\left(1 + \frac{r_+}{r_a - r_+}\right) \quad \text{and} \quad (r_-)_{\text{eff}} = r_-\left(1 + \frac{r_-}{r_a - r_-}\right).$$

As the radius of the atmosphere decreases, the effective radius of the ion increases, and the motion of the ion is slowed.

The final expression for Λ which includes both the asymmetry effect and the electrophoretic effect is (for uni-univalent electrolytes)

$$\Lambda = \Lambda^0 - \left[\frac{82.4}{(\epsilon T)^{1/2}\eta} + \frac{8.20 \times 10^5\Lambda^0}{(\epsilon T)^{3/2}}\right]\sqrt{c}, \qquad (30\text{–}37)$$

which is the Onsager equation; it is usually abbreviated to

$$\Lambda = \Lambda^0 - (A + B\Lambda^0)\sqrt{c}, \qquad (30\text{–}38)$$

where c is the concentration in equivalents per liter. The test of Eq. (30–37) is whether the limiting slope of a plot of experimental values of Λ versus $\sqrt{c}$ has the value predicted by the equation. A comparison of the data with the values predicted by the Onsager equation is shown for several salts in water in Fig. 30–5. The agreement

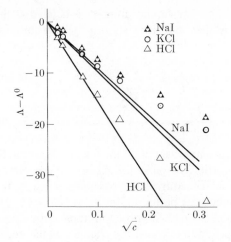

Fig. 30–5 Test of the Onsager equation. The lines are the limiting slopes.

is usually excellent in very dilute solutions up to about 0.02 molar. In more concentrated solutions the conductivity is usually higher than one would predict from the Onsager equation.

30–9 Conductance at High Fields and High Frequencies

The concept of the ion atmosphere is further substantiated by the Wien effect and the Debye-Falkenhagen effect. In very high fields, $E > 10^5$ V/cm, an increase in conductivity is observed (Wien effect). This increase in conductivity results from the fact that a finite time (the relaxation time) is required for the atmosphere to form about an ion. In very high fields the ion moves so quickly that it effectively loses its atmosphere; the atmosphere does not have time to form and so cannot slow the ion. The asymmetry effect disappears and the conductance increases.

For the same reason the conductivity increases at high frequencies, 3×10^6 c/s (Debye-Falkenhagen effect). The ion changes its direction of motion so quickly that the more sluggish atmosphere cannot adjust and follow the motion of the ion. The ion moves as if it had no atmosphere, and the conductivity increases. At high frequencies both the asymmetry and electrophoretic effects are absent.

30–10 Transference Numbers

The measurement of the conductivity yields a value of the sum of the positive and negative ion conductivities. To obtain the individual values of the ion conductivities, an additional independent measurement is necessary. Even before Kohlrausch demonstrated the law of independent migration of ions it was commonly supposed that each ion contributed to the flow of current. In 1853 Hittorf devised a method to measure the contribution of the individual ions. The principle of the Hittorf method is best illustrated by an example. The transference number is defined as the fraction of the current which is carried by that ion. In a solution which contains only one electrolyte,

$$t_+ = \frac{\lambda_+}{\Lambda} = \frac{\lambda_+}{\lambda_+ + \lambda_-} = \frac{l_+}{l_+ + l_-} = \frac{v_+}{v_+ + v_-}, \tag{30–39}$$

$$t_- = \frac{\lambda_-}{\Lambda} = \frac{\lambda_-}{\lambda_+ + \lambda_-} = \frac{l_-}{l_+ + l_-} = \frac{v_-}{v_+ + v_-}. \tag{30–40}$$

It is apparent that the sum of the transference numbers of all the ions in the solution must equal unity. If one equivalent of electricity is passed, the positive ion carries t_+ equivalents and the negative ion carries t_- equivalents past any plane in the body of the solution perpendicular to the current path.

Consider the electrolysis cell shown in Fig. 30–6. Suppose that the solution contains copper sulfate and the anode is copper. We examine the changes which occur

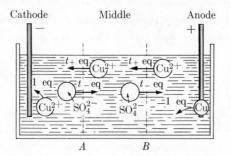

Fig. 30–6. Transference in $CuSO_4$ with copper electrodes.

Table 30–3

Cathode compartment	Middle compartment	Anode compartment
1 eq Cu^{2+} plates out on cathode t_+ eq Cu^{2+} migrate in t_- eq SO_4^{2-} migrate out	t_+ eq of Cu^{2+} migrate out at A t_+ eq of Cu^{2+} migrate in at B t_- eq of SO_4^{2-} migrate in at A t_- eq of SO_4^{2-} migrate out at B	1 eq Cu^{2+} dissolves from anode t_+ eq Cu^{2+} migrate out t_- eq SO_4^{2-} migrate in
Net change $(\Delta n_{Cu^{2+}})_c = (t_+ - 1)\,\text{eq}$ $\quad\quad = -t_-\,\text{eq}$ $(\Delta n_{SO_4^{2-}})_c = -t_-\,\text{eq}$	Net change $\Delta n_{Cu^{2+}} = 0$ $\Delta n_{SO_4^{2-}} = 0$	Net change $(\Delta n_{Cu^{2+}})_a = (1 - t_+)\,\text{eq}$ $\quad\quad = t_-\,\text{eq}$ $(\Delta n_{SO_4^{2-}})_a = t_-\,\text{eq}$

in each compartment if *one equivalent* of electricity passes; see Table 30–3. If a quantity of electricity Q passes, this is $Q/\mathscr{F}$ equivalents, so all of the changes are multiplied by $Q/\mathscr{F}$. In this experiment, the amount of $CuSO_4$ in the cathode compartment decreases by $t_-(Q/\mathscr{F})$ equivalents, while in the anode compartment the number of equivalents increases by $t_-(Q/\mathscr{F})$. The concentration in the middle part of the cell is unchanged by the passage of the current. By arranging the apparatus properly, the boundaries indicated at A and B in Fig. 30–6 can be replaced by stopcocks (Fig. 30–7), so that the three portions of the solution can be drawn off separately after the experiment. The weight and concentration of electrolyte in each portion is measured after the experiment. Knowing the original concentration, we can calculate the changes in number of equivalents in each compartment. Analysis of the middle compartment is used as a check to determine if any interfering effects have occurred. The changes in numbers of equivalents in the compartments can be related to the transference numbers of the ions by a procedure such as the one given above. It is not possible to write a general formula relating the changes to the transference numbers, since what happens in every case depends on the chemical effect produced by the electrode reactions. The changes must be figured out using the above method for each combination of electrodes.

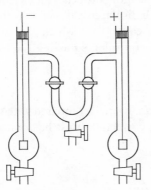

Fig. 30–7 The Hittorf cell.

The Hittorf experiment is subject to many difficulties in practice. The development of a concentration gradient by the flow of current results in diffusion of the electrolyte from the more concentrated to the less concentrated regions. This tends to undo the effect which is to be measured; to minimize diffusion the experiment must not be carried over too long a time. On the other hand, if the time is short, the concentration changes are small because a small current must be used. If large currents are used, heating effects occur unevenly and produce convection in the solution; this mixes the solution up again. In addition to all this, density differences which develop with the concentration differences between the parts of the solution may produce convection also. In spite of all these difficulties, reasonably good measurements of the transference numbers can be made using the Hittorf method.

A difficulty in interpretation arises because the ions are solvated, and in their motion they carry solvent with them from one compartment to another. We shall return to this problem in Section 30–12.

30–11 The Moving-boundary Method

A method for the measurement of transference numbers which has been brought to a high state of perfection is the moving-boundary method. A schematic diagram of the apparatus is shown in Fig. 30–8. A tube has two electrodes fixed at the ends and contains two solutions having an ion in common, one of a compound M′A and another of a compound MA. The system is arranged so that the boundary between the solutions is reasonably sharp; the position of the boundary is visible because of a difference in refractive index of the solutions, or in some cases a difference in color. To avoid mixing and destruction of the boundary, the denser solution is placed beneath the less dense. Suppose that the boundary between the two solutions is initially at b, and that a current is passed for a time t, in which $Q/\mathscr{F}$ equivalents are passed. The M^+ ion carries $t_+(Q/\mathscr{F})$ equivalents past the plane at b. The boundary must move up far enough (to b') so that $t_+(Q/\mathscr{F})$ equivalents may be accommodated in the volume between b and b'. If l is the length between b and b', and a is the cross-sectional area of the tube, then the volume displaced is la. If c^* is the concentration of MA in equivalents/cm^3, the number of equivalents which can be contained in la is c^*la; but this is simply the number of equivalents passing the plane at b. Thus $c^*la = t_+(Q/\mathscr{F})$, so that

$$t_+ = c^*la\mathscr{F}/Q, \tag{30–41}$$

which assumes that the volume displaced, la, is small compared with the total volume of the solution of MA; in precise work a correction must be applied.

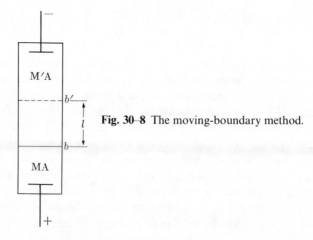

Fig. 30–8 The moving-boundary method.

The moving-boundary method yields more accurate data on transference numbers than does the Hittorf method. Experimentally it is easier to handle. The difficulties lie in the establishment of a sharp boundary, the necessity of avoiding convection currents, and excessive heating by the current. However, once the boundary is established, the flow of current sharpens the boundary, making this a minor difficulty. The relative concentrations of the two solutes is important in maintaining a sharp boundary. The faster moving ion, M′ in this example, does not lead by more than a few angstroms, since this develops a potential difference in such a sense as to slow it down; in the steady state the two ions move with the same velocity, but M′ is always a little bit ahead of M.

The measurements of the transference number are made over a range of concentration of electrolyte; the plot of t versus $\sqrt{c}$ is linear in dilute solution and can be extrapolated to $c = 0$ to obtain the value of the transference number at infinite dilution t^0.

30–12 Equivalent Ionic Conductivities

Once measurements of transference numbers have been made, it is possible to calculate the value of the equivalent ionic conductivities of the ions, using the relations

$$\lambda_+^0 = t_+^0 \Lambda^0, \qquad \lambda_-^0 = (1 - t_+^0)\Lambda^0.$$

Values of λ_+^0 and λ_-^0 for a number of ions are given in Table 30–4.

It is interesting to compare the conductivities of the alkali metal ions:

Ion	Li^+	Na^+	K^+	Rb^+	Cs^+
λ_+^0	38.66	50.11	73.52	77.8	77.3

If we insist on the Stokes' law interpretation of these values, in analogy to Eq. (30–35) we would write

$$\lambda_+^0 = \frac{\mathscr{F}ez_+}{6\pi\eta(300)r_+}. \tag{30–42}$$

We would be forced to conclude that the radius of lithium ion is *larger* than that of cesium ion. Since the crystallographic radius of lithium is much smaller than that of cesium, this indicates a difficulty with the Stokes' law interpretation. However, we can argue that the lithium ion is large because it carries a load of water molecules with it, while the cesium ion, which has a relatively weak field to hold the water molecules to it, carries very little water. This is, in fact, correct although it does not justify the use of Stokes' law.

The transport of water by the ions was first measured by Washburn. Using the Hittorf method, a reference substance such as sugar or urea is added to the solution.

Table 30–4† Ion conductances at infinite dilution at 25°C

Cation	λ_+^0	Anion	λ_-^0
H^+	349.8	OH^-	197.8
Li^+	38.66	Cl^-	76.35
Na^+	50.11	Br^-	78.20
K^+	73.52	I^-	76.9
Rb^+	77.8	NO_3^-	71.44
Cs^+	77.3	ClO_3^-	64.6
Ag^+	61.92	BrO_3^-	55.8
Tl^+	74.7	IO_3^-	40.5
NH_4^+	73.4	ClO_4^-	67.3
$(CH_3)_4N^+$	45.0	IO_4^-	54.5
$\frac{1}{2}Mg^{2+}$	53.06	HCO_3^-	44.5
$\frac{1}{2}Ca^{2+}$	59.50	Acetate$^-$	40.9
$\frac{1}{2}Sr^{2+}$	59.46	Benzoate$^-$	32.3
$\frac{1}{2}Ba^{2+}$	63.64	Picrate$^-$	30.4
$\frac{1}{2}Cu^{2+}$	54	$\frac{1}{2}C_2O_4^{2-}$	24.0
$\frac{1}{2}Zn^{2+}$	53	$\frac{1}{2}SO_4^{2-}$	80.0

† By permission from H. S. Harned and B. B. Owen, *The Physical Chemistry of Electrolytic Solutions.* 3rd ed. Reinhold Publishing Corp., New York, 1958.

Presumably the reference substance does not move in the field, and the transport of the solvent can be calculated from the analysis of the solution in the three compartments. If a value is assumed for the number of water molecules attached to one ion, a value for the number attached to the other ion can be calculated. Presently other methods for evaluation of hydration numbers are preferred from the partial molar volume of the salt in the solution, for example. The different methods are internally consistent but often do not agree well with each other. It is generally agreed that the negative ions are not hydrated. Then the hydration numbers are, approximately: Li^+, 6; Na^+, 4; K^+, 2; Rb^+, 1. (These values have been rounded to integers.)

The data in Table 30–4 also show that the equivalent conductivities of the hydrogen ion and the hydroxyl ion are much larger than those of other ions. The very large values of the equivalent ionic conductivity observed for H^+ and OH^- have been explained on the basis of a proton jump from one species to another. For conduction by H^+ ion, we have the scheme shown in Fig. 30–9. A proton is transferred from the H_3O^+ ion to an adjacent water molecule, thereby converting the water molecule to an H_3O^+ ion. The process is repeated, the newly formed H_3O^+ ion handing on a proton to the next water molecule, and so on. The occurrence of this process leaves the water molecules in an unfavorable orientation; for the process to happen again, they must rotate through 90°. The initial stage is shown in

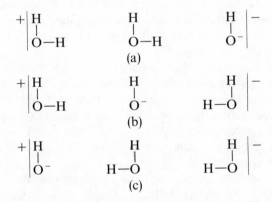

Fig. 30–9 Mechanism of conduction for hydrogen ion.

Fig. 30–9(a), an intermediate stage in Fig. 30–9(b), and the final stage in Fig. 30–9(c). The analogous process for the hydroxyl ion is shown in Fig. 30–10.

The process of proton transfer results in a more rapid transfer of positive charge from one region of the solution to another than would be possible if the ion H_3O^+ has to push its way through the solution as other ions must. For this reason also the conductivities of H^+ and OH^- ions are not related to the viscosity of the solution.

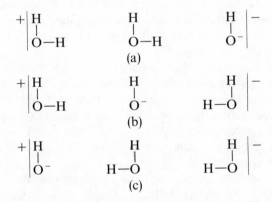

Fig. 30–10 Mechanism of conduction for hydroxyl ion.

30–13 Temperature Dependence of the Ionic Equivalent Conductivities

For most ions in water solution, the value of λ^0 increases with temperature by about 2% per degree. This is expressed by

$$\lambda^0 = \lambda^0_{25^\circ}[1 + 0.02(t - 25)],$$

where t is in °C. The temperature dependence results from the decrease in viscosity of water with temperature, which amounts to about 2% per degree. In this regard, the equivalent conductivities of H^+ and OH^- have temperature coefficients which are peculiar, being about 0.014 for H^+ and 0.016 for OH^-; this is a result of the difference in the conduction mechanism. The decrease in λ^0 with increase in pressure is mainly a consequence of the increase in viscosity with pressure.

30–14 Conductance in Nonaqueous Solvents

The principles governing conductivity in nonaqueous solvents are the same as in aqueous solutions, of course. However, in solvents having low dielectric constants, there is a lessening of the degree of ionization of many substances. Electrolytes which are completely dissociated in water may be only partially dissociated in a low dielectric constant solvent. Hydrochloric acid is completely dissociated in water; HCl is a "strong" acid. In ethyl alcohol, however, HCl is a "half-strong" acid, with a dissociation constant of about 1.5×10^{-2}.

Suppose we compare the energy of interaction of two ions having charges $+ze$ and $-ze$ at a distance r in a medium of dielectric constant ϵ. This energy is

$$V = -z^2 e^2 / \epsilon r. \tag{30–43}$$

If ϵ is large (in H_2O, $\epsilon = 80$), then the ions must come rather close together before the energy of interaction becomes appreciable. If we choose ethyl alcohol ($\epsilon = 24$), then at the same distance of approach, the interaction energy will be $\frac{80}{24} = 3.3$ times greater; or, put in another way, the energy of interaction becomes appreciable at a distance 3.3 times greater than in water. As a result, since most solvents have much lower dielectric constants than water, the effects due to ionic interaction are much larger than in water.

The large ionic interaction often renders the Onsager equation useless (it is still presumably correct) for the extrapolation to obtain Λ^0. The solutions for which the Onsager relation is valid are so dilute it is not possible to obtain reliable measurements of their conductivity. In these cases, special methods of obtaining Λ^0 must be used. If the electrolyte is weakly dissociated, then the Λ^0 can be obtained by application of the Ostwald dilution law, modifying this in precise work to correct for the interionic forces.

In solvents of low dielectric constant, ion association occurs. The appearance of ion pairs A^+B^- and ion triplets $A^+B^-A^+$ and $B^-A^+B^-$ results in a very rapid variation of conductivity with concentration.

30–15 Determination of the Ion Product of Water

The ion product of water is $K_w = a_{H^+} a_{OH^-}$. Since in pure water the concentrations of the ions are exceedingly small, we may set the activities equal to the concentrations of the species present; so $K_w = c_{H^+} c_{OH^-}$. In pure water, $c_{H^+} = c_{OH^-} = K_w^{1/2}$.

The conductivity of pure water κ_w is related to the molar concentrations by the equation

$$\kappa_w = \left(\frac{c_{H^+}}{1000}\right)\lambda_{H^+} + \left(\frac{c_{OH^-}}{1000}\right)\lambda_{OH^-},$$

which becomes

$$\kappa_w = \frac{K_w^{1/2}}{1000}(\lambda_{H^+} + \lambda_{OH^-}).$$

The ions are present in such low concentration that the equivalent ionic conductivities may be taken as the ones at infinite dilution, Table 30–4, so that $\lambda_{H^+} + \lambda_{OH^-} = 547.6$, and we obtain for K_w

$$K_w = \left(\frac{1000\kappa_w}{547.6}\right)^2.$$

The value of κ_w at 25° obtained by Kohlrausch and Heydweiller (1894) is 0.062×10^{-6} ohm^{-1}·cm^{-1}. Using this value, we obtain

$$K_w = \left(\frac{620}{547.6} \times 10^{-7}\right)^2 = 1.28 \times 10^{-14}.$$

This value is somewhat high. The best values of K_w are obtained from emf measurements of electrochemical cells, and these agree well with the best values from conductivity measurements. At 25°C the most reliable value of K_w is 1.008×10^{-14}. Values of K_w at several temperatures are given in Table 30–5. The variation with temperature should be noted.

Table 30–5†   The ion product of water

t, °C	0	10	20	25	30	40	50	60
$K_w \times 10^{14}$	0.1139	0.2920	0.6809	1.008	1.469	2.919	5.474	9.614

†   By permission from H. S. Harned and B. B. Owen, *The Physical Chemistry of Electrolytic Solutions*, 3rd ed. Reinhold Publishing Corp., New York, 1958.

30–16   Determination of Solubility Products

Another application of conductance measurements is in the determination of the solubility of a slightly soluble salt. For example, a saturated solution of silver chloride in water has a conductivity which is given by

$$\kappa = c_{Ag^+}^* \lambda_{Ag^+} + c_{Cl^-}^* \lambda_{Cl^-} + c_{H^+}^* \lambda_{H^+} + c_{OH^-}^* \lambda_{OH^-}.$$

If the salt dissolves to only a small extent, then the ionization of water will be un-affected by the presence of the salt, and the last two terms in the equation are simply the conductivity of the water. Therefore,

$$\kappa - \kappa_w = c_{Ag^+}^* \lambda_{Ag^+} + c_{Cl^-}^* \lambda_{Cl^-}.$$

If s is the solubility in equivalents per liter, then $s = 1000c_{Ag^+}^* = 1000c_{Cl^-}^*$. Then

$$\kappa - \kappa_w = \frac{s}{1000}(\lambda_{Ag^+} + \lambda_{Cl^-}).$$

If the solution is very dilute, the values of λ^0 may be used from Table 30–4; then

$$s = \frac{1000(\kappa - \kappa_w)}{\Lambda_{AgCl}^0}.$$

The solubility product constant is given by $K_{sp} = a_{Ag^+}a_{Cl^-}$. If the solution is dilute enough to regard the activity coefficients as unity, then $K_{sp} = s^2$. In the case of silver chloride $\kappa - \kappa_w = 1.802 \times 10^{-6}$, so that

$$K_{sp} = \left[\frac{1000(1.802)(10^{-6})}{138.27}\right]^2 = 1.70 \times 10^{-10}.$$

This value is in excellent agreement with that obtained from emf measurements.

30–17 Conductometric Titrations

The variation of the conductance of a solution during a titration can serve as a useful method of following the course of the reaction. Consider a solution of a strong acid, HA, to which a solution of a strong base, MOH, is added. The reaction

$$H^+ + OH^- \rightarrow H_2O$$

occurs. For each equivalent of MOH added, one equivalent of hydrogen ion is removed. Effectively, the faster-moving H^+ ion is replaced by the slower-moving M^+ ion, and the conductance of the solution falls. This continues until the equivalence point is reached, at which we have a solution of the salt MA. If more base is added, the conductance of the solution increases, since more ions are being added and the reaction no longer removes an appreciable number of them. Consequently, in the titration of a strong acid with a strong base, the conductance has a minimum at the equivalence point. This minimum can be used instead of an indicator dye to deter-mine the endpoint of the titration. A schematic plot of the conductance of the solution against the number of milliliters of base added is shown in Fig. 30–11. This technique is applicable to any titration which involves a sharp change in conductivity at the equivalence point.

Consider the titration of a silver nitrate solution with sodium chloride. In the precipitation reaction $Ag^+ + Cl^- \rightarrow AgCl$, the sodium ion replaces the silver ion in solution. This in itself produces little change in conductance, so that the plot

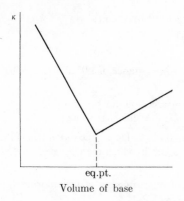

Fig. 30–11 Conductometric titration of a strong acid with a strong base.

eq.pt.

Volume of base

of conductance versus number of milliliters of titrant is nearly horizontal. However, after the equivalence point is passed, the conductance increases sharply because of the additional ions. The endpoint can be determined easily.

The equation for the conductance is simple and is given in the case of the acid–base titration by

$$\kappa = \tfrac{1}{1000}(c_{H^+}\lambda_{H^+} + c_{A^-}\lambda_{A^-} + c_{M^+}\lambda_{M^+} + c_{OH^-}\lambda_{OH^-}), \qquad (30\text{–}44)$$

where the concentrations are in equivalents per liter. Using this equation and knowing the concentrations of the acid (HA) and base (MOH) solutions, it is easy to calculate the conductance of the solution as a function of the volume of base added.

Problems

30–1. A metal wire carries a current of 1 A. How many electrons pass a point in the wire in 1 sec?

30–2. An emf of 100 V is applied across a wire which is 2 m long and has a diameter of 0.05 cm. If the current is 25 A, calculate

a) the resistance and conductance of the wire,

b) the field strength,

c) the current density,

d) the resistivity and conductivity of the wire.

30–3. A solution of sulfuric acid is electrolysed using a current of 0.10 A for three hours. How many cm^3 (at STP) of hydrogen and oxygen are produced?

30–4. Potassium chlorate is prepared by the electrolysis of KCl in basic solution:

$$6OH^- + Cl^- \rightarrow ClO_3^- + 3H_2O + 6e^-$$

If only 60% of the current is utilized in this reaction, what time will be required to produce 10 cm of KClO$_3$ using a current of 2 A?

30–5. A 0.01 D solution of KCl has a conductivity of 0.0014088 ohm^{-1}·cm^{-1}. A cell filled with this solution has a resistance of 4.2156 ohms.

a) What is the cell constant?

b) The same cell filled with a solution of HCl has a resistance of 1.0326 ohms. What is the conductivity of the HCl solution?

30–6. The equivalent conductivities at infinite dilution are: sodium benzoate, 82.4; hydrochloric acid, 426.2; sodium chloride, 126.5. Calculate Λ^0 for benzoic acid.

30–7. For H$^+$ and Na$^+$ the values of λ^0_+ are 349.8 and 50.11. Calculate the mobilities of these ions and the velocities of the ions if they are in a cell in which the electrodes are 5 cm apart and to which a potential of 2 V is applied.

30–8. From the data in Table 30–4, calculate the transference number of the chloride ion in each of the infinitely dilute solutions: HCl, NaCl, KCl, NH$_4$Cl, CaCl$_2$.

30–9. The conductivity of any solution is expressed by $\kappa = (1/1000)\Sigma_i c_i \lambda_i$, where c_i is the concentration (equivalents/liter) of the ion, and λ_i is its equivalent conductivity; the sum is taken over all the ions in the solution. Calculate the fraction of the current carried by each ion, the transference number, in a solution which is 0.1 molar in Na$_2$SO$_4$ and 0.01 molar in H$_2$SO$_4$. (Values of λ in Table 30–4.)

30–10. What must be the ratio of concentrations of HCl and NaCl in a solution if the transference number of the hydrogen ion is 0.5? (Data in Table 30–4.)

30–11. The equivalent conductivity of acetic acid at different concentrations is:

Λ	49.50	35.67	25.60
c, equivalents/liter	9.88×10^{-4}	19.76×10^{-4}	39.52×10^{-4}

Using the value of Λ^0 calculated from Table 30–4, compute

a) the degree of dissociation at each concentration,

b) the value of the dissociation constant.

30–12. a) Relate the changes in concentration in the Hittorf cell to the transference number of the positive ion and the quantity of electricity passed if the cell is filled with hydrochloric acid and both electrodes are silver–silver chloride electrodes.

b) How is the relation affected if the cathode is replaced by a platinum electrode so that H$_2$ is evolved?

c) If the anode is replaced by a platinum electrode and oxygen is evolved? [*Note:* Using an electrode which allows gas evolution would be a very bad way to do a Hittorf experiment! Why?]

30–13. A Hittorf cell fitted with silver–silver chloride electrodes is filled with HCl solution which contains 0.3856×10^{-3} gm HCl/gm water. A current of 2.00 mA is passed for exactly 3 hours. The solutions are withdrawn, weighed, and analyzed. The total weight of the cathode solution is 51.7436 gm; it contains 0.0267 gm of HCl. The anode solution weighs 52.0461 gm and contains 0.0133 gm of HCl. What is the transference number of the hydrogen ion?

30–14. A moving boundary experiment is done with 0.01 molar LiCl. In a tube having a cross-sectional area of 0.125 cm², the boundary moves 7.3 cm in 1490 sec using a current of 1.80×10^{-3} A. Calculate t_+.

30–15. The conductivity of a saturated solution of $BaSO_4$ has the value 3.48×10^{-6} ohm⁻¹·cm⁻¹. The conductivity of pure water is 0.50×10^{-6} ohm⁻¹·cm⁻¹. Compute the solubility product of $BaSO_4$.

30–16. a) Suppose that an acid having a concentration c_a (equivalents/liter) is titrated with a base having a concentration c_b (equivalents/liter). If v_0 is the volume (in cm³) of the acid, and v the volume of base added at any stage of the titration, show that the conductivity before the equivalence point is reached is given as a function of v by

$$\kappa = \left(\frac{v_0}{v_0 + v}\right)\left[\kappa_a - \left(\frac{c_b}{1000}\right)\left(\frac{v}{v_0}\right)(\lambda_{H^+} - \lambda_{M^+})\right],$$

where κ_a is the conductivity of the acid solution before any base has been added. Assume that the values of λ do not change with the volume of the solution. (Before the equivalence point is reached, the concentration of OH⁻ is negligible compared with that of H⁺; this situation is reversed after the equivalence point.)

b) Show that after the equivalence point is passed

$$\kappa = \left(\frac{v_0 + v_e}{v_0 + v}\right)\kappa_e + \left(\frac{v - v_e}{v_0 + v}\right)\kappa_b,$$

where κ_b is the conductivity of the basic solution, κ_e that of the solution at the equivalence point, and v_e the volume of base added at the equivalence point.

30–17. The crystallographic radii of Na⁺ and Cl⁻ are 0.95 Å and 1.81 Å. Estimate the ion conductivities using Stokes' law and compare with the values in Table 30–4. ($\eta = 0.89 \times 10^{-2}$ poise.)

30–18. At 25°C the constants in the Onsager equation are $A = 60.19$ and $B = 0.2289$ if c is expressed in equivalents/liter. From the data in Table 30–4 compute the equivalent conductivity for 0.01 molar HCl, 0.01 molar KCl, and 0.01 molar LiCl.

Chapter Thirty-one

Chemical Kinetics

I. Empirical Laws and Mechanism

31–1 Introduction

The rates of chemical reactions form the subject matter of chemical kinetics. Experimentally it is found that the rate of a chemical reaction is dependent on the temperature, pressure, and the concentrations of the species involved. The presence of a catalyst can increase the rate by many powers of ten. From the study of the rate of a reaction and its dependence on all these factors, much can be learned about the detailed steps by which the reactants are transformed to products.

Reactions can be classified kinetically as homogeneous reactions or heterogeneous reactions. A homogeneous reaction occurs entirely in one phase, while a heterogeneous reaction occurs, at least in part, in more than one phase. The most usual type of heterogeneous reaction has a rate which depends on the area of a surface which is exposed to the reaction mixture. This surface may be the interior wall of the reaction vessel or it may be the surface of a solid catalyst. In any kinetic study at some stage it is necessary to find out if the reaction is influenced by the walls of the vessel. If the vessel is made of glass, it is usually packed with glass wool, or beads, or many fine glass tubes so as to increase the exposed area. Any effect on the rate of the reaction is noted. If the reaction is strictly homogeneous, the rate will not be affected by packing the vessel in this way. In this chapter the discussion will be restricted almost entirely to homogeneous reactions.

31–2 Rate Measurements

In the course of a chemical reaction the concentrations of all the species present change with time, and so the properties of the system change. The rate of the reaction is measured by measuring the value of any convenient property which can be related to the composition of the system as a function of time. The property chosen should be reasonably easy to measure; it should change sufficiently in the course of the reaction to permit an accurate distinction to be made between the various compositions of the system as time passes. The property chosen depends on the individual reaction. In one of the first quantitative studies of reaction rates, Wilhelmy (1850) measured the rate of inversion of sucrose by measuring the change with time of the angle of rotation of a beam of plane-polarized light which passed through the sugar solution.

The number of methods of following a reaction with time is very large. Just to mention a few: changes in pressure, changes in pH, changes in refractive index, changes in optical density at one or more wavelengths, changes in thermal conductivity, changes in volume, changes in electrical resistance. If physical methods such as these can be applied, they are usually more convenient than a chemical method.

Since the rate of the majority of chemical reactions is very sensitive to temperature, the reaction vessel must be kept in a thermostat so that the temperature is accurately maintained at a constant value. In some cases it is necessary to control the pressure.

No matter what property we choose to measure, the data can ultimately be translated into a variation of the concentration of a reactant or a product with time. Figure 31–1 shows this variation schematically for a reactant and for a product. The concentration of any reactant decreases from its initial to the equilibrium value and the concentration of any product increases from its initial value (usually zero) to the equilibrium value.

Our task now is to describe the curves in Fig. 31–1 more accurately. We begin by describing the various rate laws which have been found experimentally. Later we shall interpret these laws in terms of the molecular processes involved.

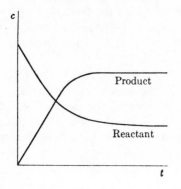

Fig. 31–1 Variation of concentration with time.

31–3 Rate Laws

Consider a chemical reaction which may be written

$$2A \rightarrow B + C.$$

The rate of this reaction may be expressed in terms of the time rate of change of concentration of any of the substances involved. The rate may be written as the decrease in concentration of A with time, $-(dc_A/dt)$, or as the increase in concentration of either B or C with time, (dc_B/dt) or (dc_C/dt). These expressions of the rate are related to one another through the stoichiometry of the reaction. Since one molecule each of B and C is produced in the overall reaction, the rate of increase of c_B and c_C must be the same. However, since two molecules of A disappear for each molecule of B or C produced, the rate of decrease of c_A must be twice the rate of increase of c_C or c_B, so that we have

$$\frac{-dc_A}{dt} = 2\frac{dc_B}{dt} = 2\frac{dc_C}{dt}. \tag{31–1}$$

Because of this relation it does not matter which derivative we choose to express the rate; any of them is satisfactory.

In general the rate of any reaction is a function of the concentrations of all the reactants and products. Applying this to the reaction above, we have

$$\frac{-dc_A}{dt} = f(c_A, c_B, c_C), \tag{31–2}$$

which is the *rate law* for the reaction. In many cases, the rate law has the simple form

$$\frac{-dc_A}{dt} = kc_A^\alpha c_B^\beta c_C^\gamma, \tag{31–3}$$

where k, α, β, and γ are constants. The constant k is the *rate constant* of the reaction, or *the specific rate* of the reaction, since k is the rate if all the concentrations were unity. The constant α is the *reaction order* with respect to A, β is the reaction order with respect to B, and γ is the reaction order with respect to C. The overall reaction order is the sum $\alpha + \beta + \gamma$.

It cannot be emphasized too strongly that the order of the reaction with respect to a given substance has *no relation whatsoever* to the stoichiometric coefficient of that substance in the chemical equation. For example, in the chemical equation above, the coefficient of A is 2. We cannot infer from this that $\alpha = 2$. (*Elementary reactions*, reactions which take place in a single step, are excepted from this statement.) The order of the reaction governs the mathematical form of the rate law, and therefore the variation in concentration of all the species with time. The determination of the order of the reaction with respect to the various substances taking part is one of the first objectives of a kinetic investigation.

There is a certain ambiguity in the definition of the rate constant of a reaction. Consider the reaction

$$Br_2 \rightarrow 2Br.$$

The rate constant of the reaction may be defined in at least two ways:

$$\frac{-dc_{Br_2}}{dt} = kc_{Br_2} \quad \text{or} \quad \frac{dc_{Br}}{dt} = k'c_{Br_2}.$$

From the stoichiometry of the reaction, the rate of appearance of bromine atoms is twice the rate of disappearance of bromine molecules; that is,

$$\frac{dc_{Br}}{dt} = 2\left(-\frac{dc_{Br_2}}{dt}\right),$$

so we conclude that

$$k' = 2k.$$

Which is the rate constant of the reaction, k or k'? There seems to be no general agreement or accepted convention on this point. So long as one is careful to specify with respect to which species in the reaction the constant is defined, and so long as one is consistent after that, the difficulty disappears. One attractive possibility would be to write the reaction, then define the rate constant in terms of the advancement per unit volume, $d(\xi/V)/dt$. For the example here, we would have

$$\frac{d(\xi/V)}{dt} = kc_{Br_2}.$$

Then, since $c_{Br_2} = c^0_{Br_2} - (\xi/V)$, we would have

$$\frac{dc_{Br_2}}{dt} = \frac{(-1)d(\xi/V)}{dt} = -kc_{Br_2}.$$

On the other hand, for bromine atoms we would have $c^0_{Br} = c^0_{Br} + 2(\xi/V)$, and therefore

$$\frac{dc_{Br}}{dt} = \frac{2d(\xi/V)}{dt} = 2kc_{Br_2}.$$

Using this method, the rate constant is completely defined once the balanced chemical equation is written down.

31–4 First-order Reactions

Consider a reaction of the type,

$$A \rightarrow \text{Products.}$$

Suppose that the reaction is first order with respect to A, and that the products are not involved in the rate law. Then, if c is the concentration of A, the rate law is

$$\frac{-dc}{dt} = kc. \tag{31–4}$$

Rearranging, we have

$$dc/c = -k\, dt.$$

Integrating, we obtain

$$\ln c = -kt + C,$$

where C is the constant of integration. If initially, $c = a$ when $t = 0$, then $\ln a = C$, and the equation becomes, after combining the logarithmic terms,

$$\ln \frac{c}{a} = -kt \tag{31–5}$$

$$c = ae^{-kt}. \tag{31–6}$$

The concentration of A decreases exponentially with time. According to Eq. (31–5), the plot of $\ln(c/a)$ versus t should be linear if the reaction is first order. The data are plotted in this way; if the points lie on a straight line, the reaction is first order and the rate constant is equal to the slope of the line.

The half-life τ of the reaction is the time required for the concentration of A to reach one-half of its initial value. Therefore, when $t = \tau$, $c = \frac{1}{2}a$. Putting these values into Eq. (31–5), we obtain $\ln \frac{1}{2} = -k\tau$, so that

$$\tau = \frac{\ln 2}{k} = \frac{0.693}{k}. \tag{31–7}$$

One of the ways of evaluating the rate constant of a reaction is to determine the half-life for various initial concentrations of the reactant A. If the half-life is independent of the initial concentration, then the reaction is first order, and the rate constant is calculated using Eq. (31–7). It is only for first-order reactions that the half-life is independent of the initial concentration.

The decomposition of N_2O_5 is an example of a first-order reaction. The stoichiometry is represented by

$$2N_2O_5 \rightarrow 4NO_2 + O_2,$$

and the rate law is

$$\frac{-dc_{N_2O_5}}{dt} = kc_{N_2O_5}.$$

At 25°C the rate constant is 3.38×10^{-5} sec^{-1}. Note the absence of any relation between the order of the reaction and the stoichiometric coefficient of N_2O_5 in the chemical equation.

31–5 Radioactive Decay

The radioactive decay of an unstable nucleus is an important example of a process which follows a first-order rate law. Choosing Cu^{64} as an example, we have the transformation

$$Cu^{64} \rightarrow Zn^{64} + \beta^-, \qquad \tau = 12.8\,hr.$$

The emission of a β-particle occurs with the formation of a stable isotope of zinc. The probability of this occurrence in the time interval dt is simply proportional to dt. Therefore

$$\frac{-dN}{N} = \lambda\,dt, \tag{31–8}$$

where $-dN$ is the number of copper nuclei which disintegrate in the interval dt. Equation (31–8) is a first-order law, and can be integrated to the form

$$N = N_0 e^{-\lambda t}, \tag{31–9}$$

N_0 being the number of Cu^{64} nuclei present at $t = 0$, N the number at any time t. The constant λ is the *decay constant* and is related to the half-life by

$$\lambda = \frac{\ln 2}{\tau}. \tag{31–10}$$

In contrast to the rate constant of a chemical reaction, the decay constant λ is completely independent of any external influence such as temperature or pressure. Using the value of λ from Eq. (31–10) in Eq. (31–9), we obtain, since $\exp(\ln 2) = 2$,

$$N = N_0 e^{-(\ln 2)t/\tau} = N_0(2)^{-t/\tau} = N_0(\tfrac{1}{2})^{t/\tau}. \tag{31–11}$$

From Eq. (31–11) it is clear that after the elapse of a period equal to two half-lives, $(\tfrac{1}{2})^2 = \tfrac{1}{4}$ of the substance remains. After three half-lives have elapsed $\tfrac{1}{8}$ remains, after 4 half-lives, $\tfrac{1}{16}$, and so on. The mathematics is the same as that of the barometric distribution, Section 2–10.

31–6 Second-order Reactions

We return to the reaction

$$A \rightarrow Products,$$

now supposing that the reaction is second order. If c is the concentration of A at any time, then the rate law is

$$\frac{-dc}{dt} = kc^2. \tag{31–12}$$

Separating variables, we have

$$\frac{-dc}{c^2} = k\,dt.$$

Integrating, we obtain

$$1/c = kt + C,$$

where the constant C is evaluated by requiring that at $t = 0$, $c = a$. Then $C = 1/a$, so that

$$\frac{1}{c} = \frac{1}{a} + kt, \tag{31–13}$$

which is the integrated rate law for a second-order reaction. The reciprocal of the concentration of A is plotted against t. If the data fall on a straight line, this is evidence that the reaction is second order. The slope of the line is the rate constant.

The half-line is defined as before. When $t = \tau$, $c = a/2$. Using these values in Eq. (31–13), we obtain

$$\tau = \frac{1}{ka}. \tag{31–14}$$

For a second-order reaction, the half-life depends on the initial concentration of the reactant. If the initial concentration is doubled, the time required for half of A to react will be reduced by one-half. If the half-life for various initial concentrations is plotted against $1/a$, the rate constant is the reciprocal of the slope of the line.

Consider a reaction of the type

$$A + rB \rightarrow \text{Products,}$$

where r is the number of moles of B which react with one mole of A. If at any time c_A is the concentration of A and c_B is that of B, then assuming that the reaction is first order with respect to both A and B, the overall order is second and the rate law can be written

$$\frac{-dc_A}{dt} = kc_A c_B. \tag{31–15}$$

Since B disappears r times faster than A, we could also write

$$\frac{-dc_B}{dt} = rkc_A c_B. \tag{31–16}$$

To get Eq. (31–15) into an integrable form, c_A and c_B must be expressed in terms of a single variable. If a and b are the initial concentrations of A and B (moles/liter), and x is the number of moles per liter of A which have reacted at t, then at t, $c_A = a - x$, and $c_B = b - rx$. Using these values, the rate law becomes

$$\frac{dx}{dt} = k(a - x)(b - rx). \tag{31–17}$$

Separating variables, we obtain

$$\frac{dx}{(a - x)(b - rx)} = k \, dt.$$

Using the method of partial fractions,† the left-hand side of this equation can be written in the form

$$\frac{1}{b - ra}\left[\frac{1}{a - x} - \frac{1}{(b/r) - x}\right] dx = k\,dt.$$

Integrating, we have

$$\frac{1}{b - ra}\left[-\ln(a - x) + \ln\left(\frac{b}{r} - x\right)\right] = kt + C.$$

At $t = 0$, $x = 0$ so that

$$C = \frac{1}{b - ra}\left(-\ln a + \ln\frac{b}{r}\right).$$

This value of C brings the integrated rate law to the form

$$\frac{1}{b - ra}\left\{\ln\left(\frac{a}{a - x}\right) - \ln\left[\frac{(b/r)}{(b/r) - x}\right]\right\} = kt. \tag{31–18}$$

The quantities on the left-hand side are all known from the experimental data, so the left-hand side can be plotted against t to yield a straight line with a slope equal to k. Equation (31–18) is not correct if A and B are mixed in the stoichiometric proportions; that is, if $b = ra$. In this case $c_B = rc_A$ at all times so that the differential equation, (31–15), becomes

$$\frac{-dc_A}{dt} = krc_A^2,$$

which has the form of Eq. (31–12). Therefore, the integrated equation has the same form as Eq. (31–13),

$$\frac{1}{c_A} = \frac{1}{a} + (kr)t.$$

31–7 Higher-order Reactions

Reactions of order higher than second are occasionally important. A third-order rate law may have any of the forms

$$\frac{-dc}{dt} = kc^3, \qquad \frac{-dc_A}{dt} = kc_A^2 c_B, \qquad \text{or} \qquad \frac{-dc_A}{dt} = kc_A c_B^2,$$

and so on. These equations can either be integrated directly or integrated after expressing all the concentrations in terms of a single variable, as in the preceding

---

†   If the method of partial fractions is unfamiliar, an elementary calculus text should be consulted.

example. The procedure is straightforward, but the results are not of sufficiently general interest to be included in detail here. The most common third-order reactions are several which involve nitric oxide; for example,

$$2NO + O_2 \rightarrow 2NO_2, \qquad \frac{-dc_{NO}}{dt} = kc_{NO}^2 c_{O_2},$$

$$2NO + Cl_2 \rightarrow 2NOCl, \qquad \frac{-dc_{NO}}{dt} = kc_{NO}^2 c_{Cl_2}.$$

If a reaction is of nth order with respect to a particular reactant, then the rate law is

$$\frac{-dc}{dt} = kc^n.$$

Taking the logarithm of both sides of this equation, we obtain

$$\log_{10}\left(\frac{-dc}{dt}\right) = \log_{10} k + n \log_{10} c, \qquad (31\text{--}19)$$

which can be used in the following way. A plot of c versus t is constructed from the data. The slope of the curve dc/dt is measured at several different values of t; the corresponding value of c is read from the plot. The logarithm of $(-dc/dt)$ is then plotted against $\log_{10} c$. The slope of the line is the order n of the reaction.

Equation (31–19) can be used in another way. The initial slope of the curve of c versus t is measured for several different initial concentrations. Then the logarithm of the initial slope is plotted against the logarithm of the initial concentration. The slope of this plot, according to Eq. (31–19), is the order of the reaction.

Since a reaction may be proportional to different powers of the concentrations of the several reactants, it is necessary to determine the dependence of the rate on each of these concentrations. If, for example, the rate is $kc_A^\alpha c_B^\beta x_C^\gamma$, then if B and C are present in large excess, while the concentration of A is very small, the concentrations of B and C will remain effectively constant throughout the reaction. The rate will then be proportional only to c_A^α. By altering the initial concentration of A, the order α can be determined. The procedure is repeated by having A and C present in excess to determine β, and so on. This is the isolation method for determination of the order of a reaction.

31-8 Relaxation Methods

Since 1953, Manfred Eigen and his colleagues have invented and developed several powerful techniques for the measurement of the rates of very fast reactions, reactions which are effectively complete within a time period less than about 10 μsec. These techniques are appropriately called relaxation techniques. If we attempt to measure the rate of a very fast reaction by traditional methods, it is clear that the time required to mix the reactants will be a limiting factor. Many devices have been designed to produce rapid mixing of reactants. The best of these cannot mix two solutions in a time shorter than a few hundred microseconds. Any method which requires mixing of the reactants cannot succeed with reactions which take place in times shorter than the mixing time. The relaxation methods avoid the mixing problem completely.

Suppose that in the chemical reaction of interest we are able to monitor the concentration of a colored species by passing light of the appropriate frequency through the mixture and observing the intensity of the transmitted beam. Consider a chemical reaction at equilibrium and suppose that the species we are monitoring has the concentration c, Fig. 31–2. Suppose that at time t_0 one of the parameters on which the equilibrium depends (e.g., temperature) is instantaneously brought to some new value. Then the concentration of the species we are observing must achieve some new equilibrium value, $\bar{c}$. Since chemical reactions occur at a finite rate, the concentration of the species will not change instantaneously to the new value, but will follow the course indicated by the dashed curve in Fig. 31–2. The system, having been perturbed from its old equilibrium position, *relaxes* to its new equilibrium condition. As we shall show below, if the difference in concentration between the two states is not too large, then the curve in Fig. 31–2 is a simple exponential function, characterized by a single constant, the relaxation time τ. The relaxation time is the time required for the difference in concentration between the two states to decay to $1/e$ of its initial value.

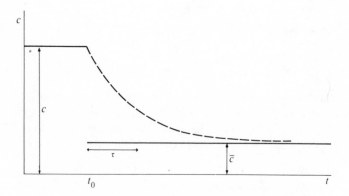

Fig. 31–2 Concentration change after impulse.

The apparatus for the "temperature-jump" method is shown in scheme in Fig. 31–3. A high-voltage power supply charges a capacitor, C. At a certain voltage the spark gap, G, breaks down and the capacitor discharges, sending a heavy current through the cell which contains the reactive system at equilibrium in a conductive aqueous solution. The passage of the current raises the temperature of the system about 10°C in a few microseconds. In the following time interval the concentration

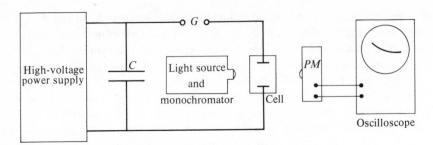

Fig. 31–3 Temperature-jump apparatus.

of the absorbing species adjusts to the equilibrium value appropriate to the higher temperature. This changes the intensity of the light beam emerging from the cell into the detecting photomultiplier tube, *PM*. The output of the photomultiplier tube is displayed on the vertical axis of an oscilloscope; the horizontal sweep of the oscilloscope is triggered by the spark discharge. In this way the concentration versus time curve is displayed on the oscilloscope screen.

The relaxation methods have the advantage that the mathematical interpretation is exceptionally simple. This simplicity is a consequence of arranging matters so that the displacement from the original equilibrium position is small.

Consider the elementary reaction

$$A + B \rightleftharpoons C.$$

The rate equation for this reaction can be written

$$\frac{dc_C}{dt} = k_f c_A c_B + (-k_r c_C), \tag{31–20}$$

in which we have expressed the net rate as the *sum* of the forward rate ($k_f c_A c_B$) and the reverse rate ($-k_r c_C$). It is convenient for graphical representation to give the reverse rate a negative sign here. The mole numbers, and therefore the concentrations, of each species can be expressed in terms of the advancement of the reaction, ξ (Section 11–7).

$$n_A = n_A^0 - \xi, \qquad n_B = n_B^0 - \xi, \qquad n_C = n_C^0 + \xi.$$
$$c_A = c_A^0 - (\xi/V), \qquad c_B = c_B^0 - (\xi/V), \qquad c_C = c_C^0 + (\xi/V).$$

The mole numbers, n^0, and the concentrations, c^0, are reference values of these quantities at $\xi = 0$; and V is the volume. Using these values for the concentrations, Eq. (31–20) becomes

$$\frac{1}{V}\frac{d\xi}{dt} = k_f [c_A^0 - (\xi/V)][c_B^0 - (\xi/V)] + \{-k_r [c_C^0 + (\xi/V)]\}.$$

From this equation it is apparent that the forward rate is a quadratic function of ξ,

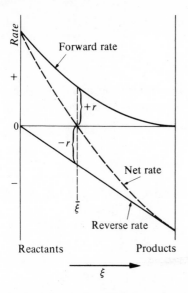

Fig. 31–4 Forward, reverse, and net rates of reaction.

while the reverse rate is a linear function of ξ; these rates are shown as functions of ξ in Fig. 31–4. The sum of the two functions is the net rate, indicated by the dashed line in Fig. 31–4.

At $\bar{\xi}$, the equilibrium value of the advancement, the net rate is zero, and we have

$$r = k_f \bar{c}_A \bar{c}_B = k_r \bar{c}_C \qquad (31\text{--}21)$$

in which the bar over the concentration indicates the equilibrium value. The exchange rate, r, is the rate of either the forward or the reverse reaction (without the minus sign) at equilibrium. Equation (31–21) can be rearranged to the form

$$K = \frac{k_f}{k_r} = \frac{\bar{c}_C}{\bar{c}_A \bar{c}_B}, \qquad (31\text{--}22)$$

which is the equilibrium relation for the elementary reaction; K is the equilibrium constant.

Although the detailed shapes of the curves will depend on the order of the reaction, the forward rate, the reverse rate, and the net rate of any elementary reaction will be related in the general way indicated in Fig. 31–4. Most importantly, it is apparent that the net rate can be approximated by a straight line over a narrow range near the equilibrium position. Let the net rate, $(1/V)(d\xi/dt) = v$. Then we expand v in a Taylor series about the equilibrium value of ξ:

$$v = v_{\xi = \bar{\xi}} + \left(\frac{dv}{d\xi}\right)_{\xi = \bar{\xi}} (\xi - \bar{\xi}).$$

However, $v_{\xi = \bar{\xi}}$ is the net rate at equilibrium, which is zero. Introducing the definition of v and multiplying by the volume, the equation becomes

$$\frac{d\xi}{dt} = V\left(\frac{dv}{d\xi}\right)_{\xi=\bar{\xi}} (\xi - \bar{\xi}).$$ (31–23)

We note that $V(dv/d\xi)_{\xi=\bar{\xi}}$ has the dimensions of a reciprocal time, and depends only upon $\bar{\xi}$, that is, only upon equilibrium values of concentration, not upon ξ or t. We define the constant τ, the relaxation time, by

$$\frac{1}{\tau} = -V\left(\frac{dv}{d\xi}\right)_{\xi=\bar{\xi}}.$$ (31–24)

The minus sign is introduced to compensate the negative sign of the derivative; see Fig. 31–4.

The introduction of τ brings the rate equation, Eq. (31–23), to the form

$$\frac{d\xi}{dt} = -\frac{1}{\tau}(\xi - \bar{\xi}),$$ (31–25)

in which τ is independent of ξ or t. This equation has the form of a first-order law and integrates immediately to

$$\xi - \bar{\xi} = (\xi - \bar{\xi})_0 e^{-t/\tau},$$ (31–26)

in which $(\xi - \bar{\xi})_0$ is the initial displacement (at $t = 0$) from equilibrium. Since the displacement of the concentration of any species from its equilibrium value is $\Delta c_i = c_i - \bar{c}_i$, and since $c_i = c_i^0 + (v_i/V)\xi$, where v_i is the stoichiometric coefficient of the species in the reaction, we obtain immediately $\Delta c_i = (v_i/V)(\xi - \bar{\xi})$. Thus, the displacement of the concentration of any species from the equilibrium value is proportional to the displacement of the advancement. Consequently, the time dependence of the concentration of any species is given by the same relation as in Eq. (31–26).

$$(c_i - \bar{c}_i) = (c_i - \bar{c}_i)_0 e^{-t/\tau}.$$ (31–27)

The pattern which appears on the oscilloscope screen in the temperature-jump experiment is therefore a simple exponential one provided only one reaction is involved. The value of τ can be obtained by measuring the horizontal distance (time axis) required for the value of the vertical displacement to fall to $1/e = 0.3679$ of its initial value; Fig. 31–5.

It must be emphasized that Eqs. (31–23) through (31–27) are quite general; they do not depend on the order of the reaction and most particularly they do not depend upon the example we chose for illustration. Equation (31–27) is a typical example of a relaxation law. It implies that any small perturbation from equilibrium in a chemical system disappears exponentially with time. If there are several elementary steps in the mechanism of a reaction then there will be several relaxation times.

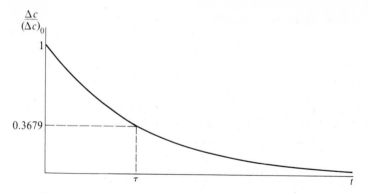

Fig. 31–5 The relaxation time.

We can evaluate the relaxation time for the example above by evaluating the derivative, $dv/d\xi$, at $\xi = \bar{\xi}$. Since $v = k_f c_A c_B - k_r c_C$, then

$$\frac{dv}{d\xi} = k_f c_A (dc_B/d\xi) + k_f c_B (dc_A/d\xi) - k_r (dc_C/d\xi).$$

But $dc_A/d\xi = -1/V = dc_B/d\xi$, and $dc_C/d\xi = 1/V$. Thus, at $\xi = \bar{\xi}$, this becomes

$$\left(\frac{dv}{d\xi}\right)_{\xi = \bar{\xi}} = -(1/V)k_f(\bar{c}_A + \bar{c}_B) + k_r.$$

Then, by the definition of τ, Eq. (31–24),

$$\frac{1}{\tau} = k_f(\bar{c}_A + \bar{c}_B) + k_r. \qquad (31\text{–}28)$$

By making measurements on the system with different values of the equilibrium concentrations, we can evaluate both k_f and k_r. Knowledge of the equilibrium constant, in view of Eq. (31–22), provides additional information about k_f and k_r.

The relaxation method is not restricted to the study of very fast reactions. With appropriate choices of sensing and recording devices, it could be used to study the rate of any reaction. The value of the relaxation technique for the study of very fast reactions lies in the fact that ordinarily it is the only technique which is available for these reactions.

A few rate constants which have been measured by relaxation techniques are given in Table 31–1. It should be noted that the rate constant k_f for the combination of two oppositely charged ions is very large. This process is always very fast since it is only limited by the rate at which the two ions can diffuse through the medium and get close enough to each other to combine. It should be mentioned that the reaction, $H^+ + OH^- \rightarrow H_2O$ has the largest second-order rate constant which is known.

Table 31–1† Rate constants of some very rapid reactions (at 25°C)

Reaction	k_f (M^{-1}-sec^{-1})	k_r (sec^{-1})
$H^+ + OH^- \rightleftharpoons H_2O$	1.4×10^{11}	2.5×10^{-5}
$H^+ + F^- \rightleftharpoons HF$	1.0×10^{11}	7×10^7
$H^+ + HCO_3^- \rightleftharpoons H_2CO_3$	4.7×10^{10}	$\sim 8 \times 10^6$
$OH^- + NH_4^+ \rightleftharpoons NH_3 + H_2O$ (22°C)	3.4×10^{10}	6×10^5
$OH^- + HCO_3^- \rightleftharpoons CO_3^{--} + H_2O$ (20°C)	$\sim 6.0 \times 10^9$	—

†   M. Eigen and L. DeMaeyer in *Techniques of Organic Chemistry* S. L. Friess, E. S. Lewis and A. Weissburger, eds., Vol. VIII, Part II. Interscience Publishers, New York, 1963.

31–9 The Dependence of Rate of Reaction on Temperature

With very few exceptions the rate of reaction increases, often very sharply, with increase in temperature. The relation between the rate constant k and temperature was first proposed by Arrhenius:

$$k = Ae^{-E^*/RT}. \tag{31–29}$$

The constant A is called the *frequency factor*, or pre-exponential factor; E^* is the *activation energy*. Converting Eq. (31–29) to logarithmic form, we have

$$\log_{10} k = \log_{10} A - \frac{E^*}{2.303\,RT}; \tag{31–30}$$

it is apparent that by determining the value of k at several temperatures, the plot of $\log_{10} k$ versus $1/T$ will yield the activation energy from the slope of the curve and the frequency factor from the intercept. Although the frequency factor may depend slightly on the temperature, unless the temperature range is very great, this effect can ordinarily be ignored. The determination of the activation energy is an important objective of any kinetic investigation.

The justification of the Arrhenius equation on theoretical grounds will be discussed in the next chapter. We can give a qualitative idea of the meaning of the equation for a reaction which occurs upon the collision of two molecules. In such a case, the reaction rate should be proportional to Z, the number of collisions per second. Furthermore, if we assume that not all collisions, but only those collisions involving an energy greater than some critical value E^*, are effective, then the rate of the reaction will have the form

$$\text{rate} = Ze^{-E^*/RT}, \tag{31–31}$$

since the fraction of collisions having energies greater than E^* is $\exp(-E^*/RT)$ so long as $E^* \gg RT$. The form of Eq. (31–31) is that required to yield the Arrhenius equation for the rate constant in this case.

31–10 Mechanism

It was pointed out in Section 31–3 that the exponents of the concentrations in the rate law in general do not bear any relation to the stoichiometric coefficients in the balanced chemical equation. This is so because the overall chemical equation yields no information about the *mechanism* of the reaction. By the mechanism of a reaction we mean the detailed way by which the reactants are converted into products. The rate at which equilibrium is attained in a system depends on the mechanism of the process while the equilibrium state itself is independent of the mechanism and depends only on the relative free energies. From a study of the position of equilibrium, values of changes of free energy, entropy, and enthalpy can be obtained. From a study of the rate of reaction under various conditions, information about the mechanism can be gained. The kinetic study is generally complicated and often requires a great deal of ingenuity in the interpretation of the data simply because it is as likely as not that the mechanism is complicated. Also it often happens that from kinetic data alone it is not possible to decide which of several reasonable mechanisms is the actual mechanism of the reaction. All too often it is not possible to distinguish on any basis which of, let us say, two mechanisms is the actual one. We may be reduced to saying that one seems more plausible than the other.

The attack on the problem of mechanism in a chemical reaction begins with the resolution of the reaction into a postulated sequence of *elementary* reactions. An elementary reaction is one which occurs in a single act. As an example, consider the reaction

$$H_2 + I_2 \rightarrow HI + HI.$$

As a hydrogen molecule and an iodine molecule collide, we may assume that they momentarily have the configuration

$$
\begin{array}{c}
H\cdots I \\
|\quad | \\
H\cdots I
\end{array}
$$

and that this complex can then dissociate into two molecules of HI. The sequence of events is illustrated as follows:

$$
\begin{array}{ccccc}
\begin{array}{c} H \\ | \\ H \end{array} & + & \begin{array}{c} I \\ | \\ I \end{array} & \rightarrow & \begin{array}{c} H\cdots I \\ |\quad | \\ H\cdots I \end{array} & \rightarrow & \begin{array}{c} HI \\ + \\ HI \end{array}
\end{array}
$$

approaching   collision   separating

Thus, in this single act of collision the reactants disappear and the products are formed. The reverse of this reaction is also an elementary reaction, the collision of two molecules of HI to form H_2 and I_2.

An elementary reaction which involves two molecules, such as the one above, is a *bimolecular* reaction. A *unimolecular* reaction is an elementary reaction which involves only one molecule; e.g., the dissociation of a molecule such as HO_2:

$$\overset{\cdot}{H}O_2 \rightarrow H + O_2.$$

In a single act the HO_2 molecule simply falls apart into two fragments. The reverse reaction,

$$H + O_2 \rightarrow HO_2,$$

is an elementary reaction, and since it involves two molecules, is a bimolecular reaction. Only elementary reactions can be characterized by their *molecularity*; the adjectives "unimolecular" and "bimolecular" do not have meaning for complex reactions which involve a sequence of many elementary steps.

The rate laws for elementary reactions can be written down immediately. Under any prescribed set of conditions, the probability of a molecule A falling into fragments in unit time is a constant. So for the unimolecular elementary reaction

$$A \rightarrow \text{fragments,}$$

the rate law is

$$\frac{-dc_A}{dt} = kc_A. \tag{31-32}$$

Since the probability of falling apart in unit time is constant, the greater the number of molecules present, the greater will be the rate of disappearance; hence the rate law, Eq. (31–32).

For a bimolecular reaction, the rate depends on the number of collisions in unit time; in Section 29–5 it was shown that the number of collisions between like molecules is proportional to the square of the concentration; therefore, for a bimolecular elementary reaction of the type

$$2A \rightarrow \text{Products,}$$

the rate law is

$$\frac{-dc_A}{dt} = kc_A^2. \tag{31-33}$$

Similarly, the number of collisions per second between unlike molecules is proportional to the product of the concentrations of the two kinds of molecules; hence for the bimolecular elementary reaction of the type

$$A + B \rightarrow \text{Products,}$$

the rate law is

$$\frac{-dc_A}{dt} = \frac{-dc_B}{dt} = kc_A c_B. \tag{31-34}$$

A termolecular reaction is an elementary reaction which involves the simultaneous collision of three molecules, e.g.,

$$A + B + C \rightarrow \text{Products,}$$

$$\frac{-dc_A}{dt} = kc_A c_B c_C. \tag{31-35}$$

The frequency of occurrence of three-body collisions is very much smaller than that of two-body collisions. Consequently, if a termolecular step is essential to the progress of the reaction, the reaction is very slow.

Examination of Eqs. (31–32), (31–33), and (31–34) shows that for elementary reactions the order of the reaction can be inferred from the stoichiometric coefficients. This is true *only for elementary reactions*.

31–11 Opposing Reactions; the Hydrogen–Iodine Reaction

The gas phase reaction of hydrogen with iodine, investigated extensively by Bodenstein in the 1890's, is kinetically simple and provides a classic example of opposing second-order reactions. Between 300 and 500°C the reaction proceeds at rates which are conveniently measurable. The mechanism is simple, consisting of one elementary reaction and its reverse†

$$H_2 + I_2 \underset{k_{-1}}{\overset{k_1}{\rightleftharpoons}} 2HI.$$

The net rate of formation of HI is the rate of the forward reaction less the rate of the reverse action. Since both are elementary reactions, we have

$$\frac{dc_{HI}}{dt} = k_1 c_{H_2} c_{I_2} - k_{-1} c_{HI}^2.$$

If, at time t, x moles/liter of HI have been formed, then

$$c_{H_2} = a - \tfrac{1}{2}x, \qquad c_{I_2} = b - \tfrac{1}{2}x, \qquad c_{HI} = x,$$

where a and b are the initial concentrations of H_2 and I_2. Using these values of the concentrations, the rate law is

$$\frac{dx}{dt} = k_1(a - \tfrac{1}{2}x)(b - \tfrac{1}{2}x) - k_{-1}x^2. \qquad (31–36)$$

At equilibrium $(dx/dt) = 0$, and Eq. (31–36) can be written in the form

$$\frac{x_e^2}{(a - \tfrac{1}{2}x_e)(b - \tfrac{1}{2}x_e)} = \frac{k_1}{k_{-1}},$$

where x_e is the equilibrium value of x. The left-hand side of this equation is the proper quotient of equilibrium concentrations, the equilibrium constant; therefore,

$$K = k_1/k_{-1}. \qquad (31–37)$$

The equilibrium constant of an *elementary* reaction is equal to the ratio of the rate constants of the forward and reverse reactions. This relation is correct *only* for elementary reactions.

---

† It is customary to write the rate constant for the forward reaction over the arrow, and that for the reverse reaction under the arrow.

Using Eq. (31–37), which yields $k_{-1} = (1/K)k_1$, the rate law, after multiplying out, becomes

$$\frac{dx}{dt} = \frac{k_1}{4}\left(1 - \frac{4}{K}\right)\left[x^2 - \frac{2(a + b)x}{1 - (4/K)} + \frac{4ab}{1 - (4/K)}\right].$$

The expression in the bracket can be written $(x - x_1)(x - x_2)$, where x_1 and x_2 are the roots of the expression. Then

$$x_1 = \frac{a + b + m}{1 - (4/K)}, \quad x_2 = \frac{a + b - m}{1 - (4/K)}, \quad m = \sqrt{(a + b)^2 - 4ab(1 - 4/K)}.$$

Note that m as well as x_1 and x_2 are known quantities computed from the values of a, b, and K. The rate equation becomes

$$\frac{dx}{(x - x_1)(x - x_2)} = \frac{k_1}{4}\left(1 - \frac{4}{K}\right) dt.$$

By the partial-fraction method we write

$$\left(\frac{1}{x - x_1} - \frac{1}{x - x_2}\right) dx = \frac{k_1}{4}\left(1 - \frac{4}{K}\right)(x_1 - x_2)\, dt.$$

Integrating and using $x_1 - x_2 = 2m/(1 - 4/K)$, this reduces to

$$\ln\left(\frac{x - x_1}{x - x_2}\right) = \tfrac{1}{2}k_1 mt + C.$$

Since $x = 0$ at $t = 0$, we find $C = \ln(x_1/x_2)$, so that

$$\ln\left(\frac{x_1 - x}{x_1}\right)\left(\frac{x_2}{x_2 - x}\right) = \tfrac{1}{2}k_1 mt. \tag{31–38}$$

Knowing K, a, and b, as well as x as a function of t, we can plot the left-hand side of Eq. (31–38) against t to obtain the value of the rate constant k_1 from the slope. For comparison with the ordinary rate expression without the reverse reaction, we write Eq. (31–18) in the form $(r = 1)$

$$\ln\left(\frac{b - x}{b}\right)\left(\frac{a}{a - x}\right) = k(b - a)t.$$

The similarity in form of the two expressions is apparent. Using Eq. (31–38), Bodenstein obtained satisfactory values of the rate constant at several temperatures.

For many years the hydrogen–iodine reaction has been the traditional example of opposing second-order reactions. Recent work by J. H. Sullivan indicates that the mechanism is not as simple as we have assumed here; in fact, the mechanism now seems to be obscure. For a discussion and references see R. M. Noyes, *J. Chem. Phys.* **48**, 323 (1968).

31–12 Consecutive Reactions

When it is necessary for a reaction to proceed through several successive elementary steps before the product is formed, the rate of the reaction is determined by the rates of all these steps. If it should happen that one of these reactions is very much slower than any of the others, then the rate will depend on the rate of this single slowest step. The slow step is the *rate-determining* step. The situation is analogous to water flowing through a series of pipes of different diameters. The rate of delivery of the water will depend on the rate at which it can pass through the narrowest pipe. An apt illustration of this feature of consecutive reactions is offered by the Lindemann mechanism of activation for unimolecular decompositions.

31–13 Unimolecular Decompositions; Lindemann Mechanism

Before 1922 the existence of unimolecular decompositions posed a severe problem in interpretation. The unimolecular elementary step consists in the breaking of a molecule into fragments:

$$A \rightarrow \text{Fragments}.$$

If this occurs, it does so because the energy content of the molecule is too large. Too much energy somehow gets into a particular vibrational degree of freedom; this vibration then produces dissociation of the molecule into fragments.

The molecules which have this excess amount of energy decompose. If the decomposition is to continue, other molecules must gain an excessive amount of energy. The problem arises as to just how the molecules acquire this extra amount of energy. In 1919, Perrin suggested that this energy was supplied by radiation, that is, by the absorption of light. This radiation hypothesis implies that in the absence of light the reaction will not occur. Immediate experimental tests of this hypothesis proved it wrong, and the puzzle remained. It appeared that the molecules could not gain the needed energy by collisions with other molecules, since the collision rate depends on the square of the concentration; this would make the reaction second order, whereas it is observed to be first order.

In 1922, Lindemann proposed a mechanism by which the molecules could be activated by collision and yet the reaction could, nonetheless, be first order. The activation of the molecule is by collision

$$A + A \xrightarrow{k_1} A^* + A,$$

where A is a normal molecule, and A^* an activated molecule. The collision between two normal A molecules produces an activated molecule A^* which has an excess energy in the various vibrational degrees of freedom; the remaining molecule is deficient in energy.

Once the activated molecule is formed, it may suffer either of two fates: it may be deactivated by collision,

$$A^* + A \xrightarrow{k_{-1}} A + A,$$

or it may decompose into products,

$$A^* \xrightarrow{k_2} \text{Products}.$$

The rate of disappearance of A is the rate of the last reaction:

$$\frac{-dc_A}{dt} = k_2 c_{A^*}. \tag{31–39}$$

With this equation we are faced with the problem of expressing the concentration of an active species in terms of the concentration of normal species. We assume that once the reaction starts, a steady state is reached in which the concentration of the activated molecules does not change very much, so that $(dc_{A^*}/dt) = 0$. This is the *steady-state approximation*. Since A* is formed in the first reaction and removed in the others, we have

$$\frac{dc_{A^*}}{dt} = 0 = k_1 c_A^2 - k_{-1} c_A c_{A^*} - k_2 c_{A^*}.$$

Using this equation we can express c_{A^*} in terms of c_A, the concentration of the normal molecules,

$$c_{A^*} = \frac{k_1 c_A^2}{k_{-1} c_A + k_2}.$$

This value of c_{A^*} brings the rate law, Eq. (31–38), to the form

$$\frac{-dc_A}{dt} = \frac{k_2 k_1 c_A^2}{k_{-1} c_A + k_2}. \tag{31–40}$$

There are two important limiting forms of Eq. (31–40).

Case I. $k_{-1} c_A \ll k_2$. Suppose that the rate of decomposition, $k_2 c_{A^*}$, is extremely fast, so fast that as soon as the activated molecule is formed it falls apart. Then there is no time for a deactivating collision to occur, and the rate of deactivation is very small compared with the rate of decomposition. Then $k_{-1} c_A c_{A^*} \ll k_2 c_{A^*}$, or $k_{-1} c_A \ll k_2$. Hence the denominator $k_{-1} c_A + k_2 \approx k_2$, and Eq. (31–39) becomes

$$\frac{-dc_A}{dt} = k_1 c_A^2. \tag{31–41}$$

The rate of the reaction is equal to the rate at which the activated molecules are formed, since the activated molecule decomposes immediately. The kinetics are second order, since the collision is a second-order process.

Case II. $k_{-1} c_A \gg k_2$. If after activation there is an appreciable time lag before the molecule falls apart, then there is opportunity for the activated molecule to make a number of collisions which may deactivate it. If the time lag is long, then the rate of deactivation, $k_{-1} c_A c_{A^*}$, is very much greater than the rate of decomposition, $k_2 c_{A^*}$. This means that $k_{-1} c_A \gg k_2$, and $k_{-1} c_A + k_2 \approx k_{-1} c_A$. This brings

Eq. (31–40) to the form

$$\frac{-dc_A}{dt} = k_2\left(\frac{k_1}{k_{-1}}\right)c_A, \tag{31–42}$$

and the rate law is first order. The usual fate of an activated molecule is deactivation by collision. A very small fraction of the activated molecules decompose to yield products.

In a gas-phase reaction, high pressures increase the number of collisions so that $k_{-1}c_A$ is large and the rate is first order. The supply of activated molecules is adequate, and the rate at which they fall apart limits the rate of the reaction. At lower pressures the number of collisions decreases, $k_{-1}c_A$ is small, and the rate is second order. The rate then depends on the rate at which activated molecules are produced by collisions.

The apparent first-order rate "constant" decreases at low pressures. Physically the decrease in value of the rate constant at lower pressures is a result of the decrease in number of activating collisions. If the pressure is increased by addition of an inert gas, the rate constant increases again in value, showing that the molecules can be activated by collision with a molecule of an inert gas as well as by collision with one of their own kind. Several first-order reactions have been investigated over a sufficiently wide range of pressure to confirm the general form of Eq. (31–40). The Lindemann mechanism is accepted as the mechanism of activation of the molecule.

31–14 Complex Reactions. The Hydrogen–Bromine Reaction

The kinetic law for the hydrogen–bromine reaction is considerably more complicated than that for the hydrogen–iodine reaction. The stoichiometry is the same,

$$H_2 + Br_2 \rightarrow 2HBr,$$

but the rate law established by M. Bodenstein and S. C. Lind in 1906 is expressed by the equation

$$\frac{d(HBr)}{dt} = \frac{k(H_2)(Br_2)^{1/2}}{1 + m(HBr)/(Br_2)}, \tag{31–43}$$

where k and m are constants, and we have used parentheses to indicate the concentration of the species. The appearance of the term $(HBr)/(Br_2)$ in the denominator implies that the presence of the product decreases the rate of the reaction; the product acts as an inhibitor. However, the inhibition is less if the concentration of bromine is high.

The expression in Eq. (31–43) was not explained until 1919, when J. A. Christiansen, K. F. Herzfeld, and M. Polanyi independently proposed the correct mechanism. The mechanism consists of five elementary reactions:

(1) $$Br_2 \xrightarrow{k_1} 2Br,$$

(2) $$Br + H_2 \xrightarrow{k_2} HBr + H,$$

(3)
$$H + Br_2 \xrightarrow{k_3} HBr + Br,$$

(4)
$$H + HBr \xrightarrow{k_4} H_2 + Br,$$

(5)
$$Br + Br \xrightarrow{k_5} Br_2.$$

The HBr is formed in reactions (2) and (3) and removed in reaction (4), so we have for the rate of formation of HBr

$$\frac{d(\mathrm{HBr})}{dt} = k_2(\mathrm{H_2})(\mathrm{Br}) + k_3(\mathrm{H})(\mathrm{Br_2}) - k_4(\mathrm{H})(\mathrm{HBr}). \tag{31-44}$$

The difficulty with this expression is that it involves the concentrations of H atoms and Br atoms; the concentrations of these atoms are not readily measurable, so the equation is useless unless we can express the concentrations of the atoms in terms of the concentrations of the molecules, H_2, Br_2, and HBr. Since the atom concentrations are, in any case, very small, it is assumed that a steady state is reached in which the concentration of the atoms does not change with time; the atoms are removed at the same rate as they are formed. From the elementary reactions, the rates of formation of bromine atoms and of hydrogen atoms are

$$\frac{d(\mathrm{Br})}{dt} = k_1(\mathrm{Br_2}) - k_2(\mathrm{Br})(\mathrm{H_2}) + k_3(\mathrm{H})(\mathrm{Br_2}) + k_4(\mathrm{H})(\mathrm{HBr}) - k_5(\mathrm{Br})^2,$$

$$\frac{d(\mathrm{H})}{dt} = k_2(\mathrm{Br})(\mathrm{H_2}) - k_3(\mathrm{H})(\mathrm{Br_2}) - k_4(\mathrm{H})(\mathrm{HBr}).$$

The steady-state conditions are $d(\mathrm{Br})/dt = 0$ and $d(\mathrm{H})/dt = 0$, so these equations become

$$0 = k_1(\mathrm{Br_2}) - k_2(\mathrm{Br})(\mathrm{H_2}) + k_3(\mathrm{H})(\mathrm{Br_2}) + k_4(\mathrm{H})(\mathrm{HBr}) - k_5(\mathrm{Br})^2,$$

$$0 = k_2(\mathrm{Br})(\mathrm{H_2}) - k_3(\mathrm{H})(\mathrm{Br_2}) - k_4(\mathrm{H})(\mathrm{HBr}).$$

By adding these two equations, we obtain $0 = k_1(\mathrm{Br_2}) - k_5(\mathrm{Br})^2$, which yields

$$(\mathrm{Br}) = \left(\frac{k_1}{k_5}\right)^{1/2} (\mathrm{Br_2})^{1/2}. \tag{31-45}$$

From the second equation,

$$(\mathrm{H}) = \frac{k_2(k_1/k_5)^{1/2}(\mathrm{H_2})(\mathrm{Br_2})^{1/2}}{k_3(\mathrm{Br_2}) + k_4(\mathrm{HBr})}. \tag{31-46}$$

By using these values for (Br) and (H) in Eq. (31–44), we obtain, after collecting terms and dividing numerator and denominator by $k_3(\mathrm{Br_2})$,

$$\frac{d(\mathrm{HBr})}{dt} = \frac{2k_2(k_1/k_5)^{1/2}(\mathrm{H_2})(\mathrm{Br_2})^{1/2}}{1 + (k_4/k_3)(\mathrm{HBr})/(\mathrm{Br_2})}. \tag{31-47}$$

This equation has the same form as Eq. (31–43), the empirical equation of Bodenstein and Lind. (The integrated form of this equation has no particular utility.)

There are several points of interest in this mechanism. First of all, the reaction is initiated by the dissociation of a bromine molecule into atoms. Once bromine atoms are formed, a single bromine atom can produce a large number of molecules of HBr through the sequence of reactions (2) and (3). These reactions form a *chain* in which an active species such as a Br or H atom is consumed, product is formed, and the active species regenerated. These reactions are *chain-propagating* reactions. The reaction (4) propagates the chain in the sense that the active species (H) is replaced by another active species (Br), but the product (HBr) is removed by this reaction thus decreasing the net rate of formation of HBr. Reaction (4) is an example of an *inhibiting* reaction. The final reaction (5) removes active species and therefore is a *chain-terminating* reaction.

Comparing reactions (3) and (4) it is apparent that Br_2 and HBr are competing for the H atoms; the success of HBr in this competition will determine the extent of the inhibition. The success of the competition is determined by the relative rates of reactions (4) and (3):

$$\frac{(\text{rate})_4}{(\text{rate})_3} = \frac{k_4(\text{H})(\text{HBr})}{k_3(\text{H})(\text{Br}_2)} = \frac{k_4(\text{HBr})}{k_3(\text{Br}_2)}.$$

This accounts for the form of the second term in the denominator of Eq. (31–47).

Since $(\text{HBr}) = 0$ at $t = 0$, the initial rate of formation of HBr is given by

$$\left[\frac{d(\text{HBr})}{dt}\right]_0 = 2k_2\left(\frac{k_1}{k_5}\right)^{1/2}(\text{H}_2)_0(\text{Br}_2)_0^{1/2}.$$

By plotting (HBr) versus t, the limiting value of the slope $[d(\text{HBr})/dt]_0$ at $t = 0$ is obtained. By doing this for several different values of the initial concentrations $(\text{H}_2)_0$ and $(\text{Br}_2)_0$, the constant $k = 2k_2(k_1/k_5)^{1/2}$ is determined.

31–15 Free-radical Mechanisms

In 1929, F. Paneth and W. Hofeditz detected the presence of free methyl radicals from the thermal decomposition of lead tetramethyl. The apparatus used is shown in Fig. 31–6. Lead tetramethyl is a volatile liquid. After evacuating the apparatus, a stream of H_2 under about 1 mm pressure is passed over the liquid where it entrains the vapor of $Pb(CH_3)_4$ and carries it through the tube. The gases are removed by a high-speed vacuum pump at the other end. The furnace is at position M. After a short period, a lead mirror deposits in the tube at M, formed by the decomposition of the $Pb(CH_3)_4$. If the furnace is moved upstream to position M', a new mirror forms at M', while the original mirror at M slowly disappears.

The explanation of the phenomenon lies in the fact that $Pb(CH_3)_4$ decomposes on heating to form lead and free methyl radicals:

$$Pb(CH_3)_4 \rightarrow Pb + 4CH_3.$$

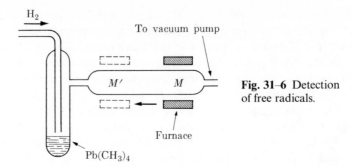

Fig. 31-6 Detection of free radicals.

The lead deposits as a mirror on the wall of the tube. The methyl radicals are swept down the tube mixed in the stream of carrier gas. If the radicals find a mirror downstream, as at M, they can remove it by the reaction

$$Pb + 4CH_3 \rightarrow Pb(CH_3)_4.$$

Following the discovery by Paneth, the technique was extensively developed especially by F. O. Rice and his co-workers.

In 1934, F. O. Rice and K. F. Herzfeld were able to show that the kinetic laws observed for many organic reactions could be interpreted on the basis of mechanisms involving free radicals. They showed that although the mechanism might be complex, the kinetic law could be quite simple. The mechanisms proposed were also capable of explaining the products formed in the reaction.

For illustration, the Rice-Herzfeld mechanism for the decomposition of ethane is

(1) $$C_2H_6 \xrightarrow{k_1} 2CH_3,$$

(2) $$CH_3 + C_2H_6 \xrightarrow{k_2} CH_4 + C_2H_5,$$

(3) $$C_2H_5 \xrightarrow{k_3} C_2H_4 + H,$$

(4) $$H + C_2H_6 \xrightarrow{k_4} H_2 + C_2H_5,$$

(5) $$H + C_2H_5 \xrightarrow{k_5} C_2H_6.$$

The reactions (1) and (2) are required for initiation, (3) and (4) constitute the chain, and (5) is the termination step. The principal products are those which are formed in the chain, so that the overall reaction can be written

$$C_2H_6 \rightarrow C_2H_4 + H_2.$$

A very minor amount of CH_4 is produced.

The rate of disappearance of C_2H_6 is

$$\frac{-d(C_2H_6)}{dt} = k_1(C_2H_6) + k_2(CH_3)(C_2H_6) + k_4(C_2H_6)(H) - k_5(H)(C_2H_5). \quad (31\text{--}48)$$

The steady-state conditions are: for CH_3,

$$0 = \frac{d(CH_3)}{dt} = 2k_1(C_2H_6) - k_2(C_2H_6)(CH_3);$$

for C_2H_5,

$$0 = \frac{d(C_2H_5)}{dt} = k_2(CH_3)(C_2H_6) - k_3(C_2H_5) + k_4(H)(C_2H_6) - k_5(H)(C_2H_5);$$

for H,

$$0 = \frac{d(H)}{dt} = k_3(C_2H_5) - k_4(H)(C_2H_6) - k_5(H)(C_2H_5).$$

Solution of the first equation yields

$$(CH_3) = \left(\frac{2k_1}{k_2}\right). \tag{31–49}$$

Addition of the three equations yields $0 = 2k_1(C_2H_6) - 2k_5(H)(C_2H_5)$, or

$$(H) = \left(\frac{k_1}{k_5}\right)\frac{(C_2H_6)}{(C_2H_5)}.$$

Using this result in the last steady-state equation yields

$$(C_2H_5)^2 - \left(\frac{k_1}{k_3}\right)(C_2H_6)(C_2H_5) - \left(\frac{k_1k_4}{k_3k_5}\right)(C_2H_6)^2 = 0,$$

which must be solved for (C_2H_5):

$$(C_2H_5) = (C_2H_6)\left[\frac{k_1}{2k_3} + \sqrt{\left(\frac{k_1}{2k_3}\right)^2 + \left(\frac{k_1k_4}{k_3k_5}\right)}\right].$$

Since k_1, the rate constant for the initiation step, is very small, the higher powers of it are negligible; then we have

$$(C_2H_5) = \left(\frac{k_1k_4}{k_3k_5}\right)^{1/2}(C_6H_6). \tag{31–50}$$

Then the value of (H) is

$$(H) = \left(\frac{k_1k_3}{k_4k_5}\right)^{1/2}. \tag{31–51}$$

Using the values of (CH_3), (C_2H_5), and (H) from Eqs. (31–49), (31–50), and (31–51) in Eq. (31–48), we obtain, after collecting terms,

$$\frac{-d(C_2H_6)}{dt} = \left[2k_1 + \left(\frac{k_1k_3k_4}{k_5}\right)^{1/2}\right](C_2H_6),$$

or, neglecting the higher power of k_1,

$$\frac{-d(C_2H_6)}{dt} = \left(\frac{k_1 k_3 k_4}{k_5}\right)^{1/2} (C_2H_6). \tag{31-52}$$

Equation (31–52) is the rate law. In spite of the complexity of the mechanism, the reaction is a first-order reaction. The rate constant is a composite of the rate constants of the individual elementary steps.

The Rice-Herzfeld mechanisms usually yield simple rate laws; the reaction orders predicted for various reactions are $\frac{1}{2}$, 1, $\frac{3}{2}$, and 2.

The rate of decomposition of organic compounds can often be increased by the addition of compounds such as $Pb(CH_3)_4$ or $Hg(CH_3)_2$, which introduce free radicals into the system. These compounds are said to *sensitize* the decomposition of the organic compound. In contrary fashion a compound such as nitric oxide combines with free radicals to remove them from the system. This inhibits the reaction by breaking the chains.

31–16 The Temperature Dependence of the Rate Constant for a Complex Reaction

The rate constant of any chemical reaction depends on temperature through the Arrhenius equation, Eq. (31–29). For a complex reaction such as the thermal decomposition of ethane, in which, by Eq. (31–52),

$$k = \left(\frac{k_1 k_3 k_4}{k_5}\right)^{1/2},$$

the rate constant for each elementary reaction can be replaced by its value from the Arrhenius equation; $k_1 = A_1 \exp(-E_1^*/RT)$, and so on. Then

$$k = \left(\frac{A_1 A_3 A_4}{A_5}\right)^{1/2} e^{-(1/2)(E_1^* + E_3^* + E_4^* - E_5^*)/RT}.$$

This is equivalent to the Arrhenius equation for the complex reaction

$$k = Ae^{-E^*/RT},$$

so that, by comparison, we have

$$A = \left(\frac{A_1 A_3 A_4}{A_5}\right)^{1/2}, \tag{31-53}$$

and

$$E^* = \tfrac{1}{2}(E_1^* + E_3^* + E_4^* - E_5^*). \tag{31-54}$$

Therefore if we know the values of A and E^* for each elementary step, the values of A and E^* can be calculated for the reaction. For the ethane decomposition, $E_1^* = 84$ kcal, $E_3^* = 40$ kcal, $E_4^* = 7$ kcal, and $E_5^* = 0$. The activation energy for the reaction should be

$$E^* = \tfrac{1}{2}(84 + 40 + 7 - 0) = 66 \text{ kcal.}$$

The experimental values found for the activation energy are about 69 to 70 kcal. The agreement between the experimental value and that predicted by the mechanism is quite reasonable.

31–17 Branching Chains; Explosions

A highly exothermic reaction which goes at a rate which intrinsically is only moderate may, nonetheless, explode. If the heat liberated is not dissipated, the temperature rises rapidly and the rate increases very rapidly. The ultimate result is a *thermal* explosion.

Another type of explosion is due to *chain branching*. In the treatment of chain reactions we employed the steady-state assumption, and balanced the rate of production of active species against their rate of destruction. In the cases described so far, this treatment yielded values for the concentration of radicals which were finite and small in all circumstances. Two things are clear about the steady-state assumption. First, it is obvious that it cannot be *exactly* correct, and second, it must be very nearly correct. If it were not very nearly correct, then the concentration of active species would change appreciably as time passed. If the concentration of active species decreased appreciably, the reaction would slow down and come to a halt before reaching the equilibrium position. On the other hand, if the concentration of active species increased appreciably with time, the rate of the reaction would increase very rapidly. This in turn would further increase the concentration of active species. The reaction would go at an explosive rate. In fact, explosions do occur if active species such as atoms or radicals are produced more quickly than they can be removed.

If in some elementary reaction an active species reacts to produce more than one active species, then the chain is said to branch. For example,

$$H + O_2 \rightarrow OH + O.$$

In this reaction the H atom is destroyed, but two active species, OH and O, which can propagate the chain, are generated. Since one active species produces two, there are circumstances in which the destruction cannot keep up with the production. The concentration of radicals increases rapidly, thus producing an explosion.

The mechanism of the hydrogen–oxygen reaction is probably not fully understood even today. Much of the modern work has been done by C. N. Hinschelwood and his co-workers. The steps in the chain reaction are

(1) $H_2 \rightarrow 2H$ Initiation,

(2) $H + O_2 \rightarrow OH + O$ $\Big\}$

(3) $O + H_2 \rightarrow OH + H$ Branching,

(4) $OH + H_2 \rightarrow H_2O + H$ Propagation.

The reactions which multiply radicals or atoms must be balanced by processes which destroy them. At very low pressures the radicals diffuse quickly to the walls of the vessel and are destroyed at the surface. The destruction reactions can be written

$$H \rightarrow \text{destruction at the surface,}$$

$$OH \rightarrow \text{destruction at the surface,}$$

$$O \rightarrow \text{destruction at the surface.}$$

If the pressure is low, the radicals reach the surface quickly and are destroyed. The production rate and destruction rate can balance and the reaction goes smoothly. The rate of these destruction reactions depends very much on the size and shape of the reaction vessel, of course.

As the pressure is increased, the branching rate and propagation rate increase, but the higher pressure slows the rate of diffusion of the radicals to the surface so the destruction rate falls. Above a certain critical pressure, the lower explosion limit, it is not possible to maintain a steady concentration of atoms and radicals; the concentration of active species increases rapidly with time, which increases the rate of the reaction enormously. The system explodes; the lower explosion limit depends on the size and shape of the containing vessel.

At higher pressures, three-body collisions which can remove the radicals become more frequent. The reaction,

$$H + O_2 + M \rightarrow HO_2,$$

where M is O_2 or H_2 or a foreign gas, competes with the branching reactions. Since the species HO_2 does not contribute to the reaction, radicals are effectively removed and at high enough pressures a balance between radical production and destruction can be established. Above a second critical pressure, the upper explosion limit, the reaction goes smoothly rather than explosively. There is a third explosion limit at high pressures above which the reaction again goes explosively.

The rate of the reaction as a function of pressure is shown schematically in Fig. 31–7. The rate is very slow at pressures below p_1, the lower explosion limit. Between p_1 and p_2 the reaction velocity is infinite, explosive. Above p_2, the upper explosion

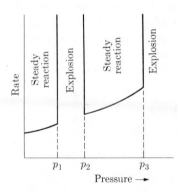

Fig. 31–7 Explosion limits.

limit, the reaction goes smoothly, the rate increasing with pressure. Above p_3, the third explosion limit, the reaction is explosive.

The explosion limits depend on temperature. Below about 460°C explosion does not occur in the low-pressure region.

31–18 Nuclear Fission; the Nuclear Reactor and the "Atomic" Bomb

The explosion of the "atomic" bomb depends on the same general kinetic principles as the $H_2 + O_2$ explosion. The situation in the bomb is somewhat simpler.

If the nucleus of U^{235} absorbs a thermal neutron, the nucleus splits into two fragments of unequal mass and releases several neutrons. If one adds the rest masses of the products and compares this sum with the rest masses of the original U^{235} and the neutron, there is a discrepancy. The products have less mass than do the reactants. The difference in mass Δm is equivalent to an amount of energy by the Einstein equation $E = (\Delta m)c^2$, where c is the velocity of light. This is the energy which is released in the reaction. Only a small fraction ($< 1\%$) of the total mass is converted to energy, but the equivalence factor c^2 is so large that the energy released is enormous.

The fission reaction can be written

$$ n + U^{235} \to X + Y + \alpha n. $$

The atoms X and Y are the fission products, α is the number of neutrons released, and is, on the average, between 2 and 3. This is the same type of chain branching reaction as was encountered in the hydrogen–oxygen reaction. Here the action of one neutron can produce several. If the size and shape of the uranium is such that most of the neutrons escape before they hit another uranium nucleus, then the reaction cannot sustain itself. However, in a large chunk of U^{235}, the neutrons hit other uranium nuclei before escape is possible, and the number of neutrons multiplies rapidly, thus producing an explosive reaction. The awe-inspiring appearance of the explosion of the bomb results from the enormous amount of energy which is released, this energy being, gram for gram, some 10 to 50 million times greater than that released in any chemical reaction.

The fission reaction occurs in a controlled way in the nuclear pile. Here rods of ordinary U^{238} which has been enriched with U^{235} are built into a structure with a moderator such as graphite or D_2O. The neutrons which are emitted at high speeds from the fission of U^{235} are slowed to thermal speeds by the moderator. The thermal neutrons suffer three important fates: some continue the chain to produce the fission of more U^{235}, others are captured by U^{238}, some are absorbed by the control rods of the reactor. The neutron flux in the reactor is monitored constantly. Moving the absorbing control rods into or out of the pile reduces or increases the neutron flux. In this way sufficient neutrons are permitted to maintain the chain reaction at a smooth rate, but enough are absorbed to prevent an explosion.

The U^{238} absorbs a thermal neutron and by radioactive decay yields neptunium and plutonium. The sequence is

$$_{92}U^{238} + {_0}n^1 \rightarrow {_{92}}U^{239} \xrightarrow[t_{1/2} = 23\,min]{} {_{93}}Np^{239} + \beta^-,$$

$$_{93}Np^{239} \xrightarrow[t_{1/2} = 2.3\,day]{} {_{94}}Pu^{239} + \beta^-,$$

$$_{94}Pu^{239} \xrightarrow[t_{1/2} = 24,000\,yr]{} {_{92}}U^{235} + \alpha.$$

The plutonium produced can contribute to the chain reaction since it is fissionable by thermal neutrons.

31–19 Reactions in Solution

The empirical rate laws found for reactions in solution are the same as those for reactions in the gas phase. An intriguing fact about reactions which can be studied in both solution and the gas phase is that quite often the mechanism is the same, and the rate constant has the same value in both situations. This indicates that in such reactions the solvent plays no part, but serves only as a medium to separate the reactants and products. It is worthwhile to mention that reactions in solution may well be faster than in the gas phase because of our tendency to use comparatively concentrated solutions. For example, in a gas at 1 atm pressure, the molar concentration is about 10^{-4} mole/liter. In making up solutions, our first tendency would be to make up a 0.1 or 0.01 molar solution. The reaction would go faster in solution simply because of the increased concentration, not because of a different rate constant. In those cases in which the solvent does not affect the rate constant, it is found that the frequency factors and activation energy have essentially the same values in solution as in the gas phase.

31–20 Catalysis

A catalyst is a substance which increases the rate of a reaction and can itself be recovered unchanged at the end of the reaction. If a substance slows a reaction, it is called an inhibitor or a negative catalyst.

As we have seen, the rate of a reaction is determined by rates of the several reactions in the mechanism. The general function of a catalyst is simply to provide an additional mechanism by which reactants can be converted to products. This alternative mechanism has a lower activation energy than mechanism in the absence of a catalyst, so that the catalysed reaction is faster. Consider reactants A going to products B by an uncatalysed mechanism at a rate v_0; Fig. 31–8(a). If an additional mechanism is provided by a catalyst, Fig. 31–8(b), so that B is formed at a rate v_c by the catalytic mechanism, then the total rate of formation of B is the sum of the rates of formation by each path.

For a catalysed reaction, we have

$$v = v_0 + v_c. \tag{31–55}$$

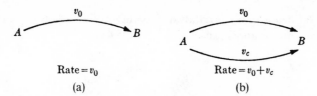

Fig. 31–8 (a) Uncatalysed reaction. (b) Catalysed reaction.

It often happens that in the absence of a catalyst the reaction is immeasurably slow, $v_0 = 0$; then, $v = v_c$. The rate v_c is usually proportional to the concentration of the catalyst. The analogy to an electrical network of parallel resistances, Fig. 31–9(a), or to parallel pipes carrying a fluid, Fig. 31–9(b), is apparent. In each case the flow through the network is the sum of that passing through each branch.

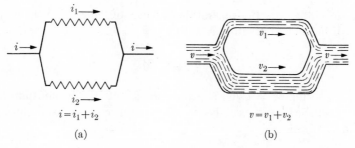

Fig. 31–9 Electrical and hydraulic analogues of catalysed reaction.

For a catalyst to function in this way, the catalyst must enter into chemical combination either with one or more of the reactants or at least with one of the intermediate species involved. Since it must be regenerated after a sequence of reactions, the catalyst is free to act again and again. As a result, a little catalyst produces a great deal of reaction, just as a minute concentration of radicals in a chain reaction produces a lot of the product.

The action of inhibitors is not so simply described, since they may act in a number of different ways. An inhibitor may slow a radical chain reaction by combining with the radicals; nitric oxide functions in this way. In other cases, the inhibitor is consumed by combination with one of the reactants and only delays the reaction until it is used up. Some inhibitors may simply "poison" a trace of catalyst whose presence is unsuspected.

For the present, we restrict our attention to homogeneous catalysis and defer discussion of catalysis by surfaces. The sequence of reactions,

(1) $$S + C \underset{k_{-1}}{\overset{k_1}{\rightleftharpoons}} X,$$

(2) $$X \xrightarrow{k_2} P + C,$$

illustrates the simple type of mechanism by which a catalyst may act. The reactant

S is called the *substrate*, C is the catalyst, P is the product, and X is an intermediate compound. The rate of appearance of the product is

$$\frac{d(P)}{dt} = k_2(X).$$

The steady-state condition for the intermediate is

$$\frac{d(X)}{dt} = 0 = k_1(S)(C) - k_{-1}(X) - k_2(X), \tag{31–56}$$

so that $(X) = k_1(S)(C)/(k_{-1} + k_2)$. The rate is

$$\frac{d(P)}{dt} = \frac{k_2 k_1(S)(C)}{k_{-1} + k_2}. \tag{31–57}$$

This expression illustrates the general dependence of the rate on the concentration of catalyst. If the rate of reaction is written

$$\frac{d(P)}{dt} = k'(S), \tag{31–58}$$

then the rate constant $k' = k_C(C)$ is proportional to (C). The constant k_C is called the catalytic coefficient for the catalyst C. The complete rate constant for the reaction is written in the form

$$k = k_0 + k_C(C),$$

k_0 being the rate constant for the uncatalysed reaction.

If one attempts to apply the rate expression in Eq. (31–57) to express the rate of the reaction throughout the entire course of the reaction, the rate expression is very cumbersome. We therefore consider the complications which arise only as they affect the initial rate of the reaction at $t = 0$. Using subscripts zero for initial values, we have

$$(C) = (C)_0 - (X)$$

and

$$(S) = (S)_0 - (X) - (P) = (S)_0 - (X),$$

since initially $(P) = 0$. Using these values in the steady-state equation, Eq. (31–56), we have

$$0 = k_1[(S)_0 - (X)][(C)_0 - (X)] - k_{-1}(X) - k_2(X).$$

Since the concentration of X is limited by either $(S)_0$ or $(C)_0$, and matters are arranged so that one of these is always small compared with the other, the term in $(X)^2$ is always negligible and the equation can be solved for (X):

$$(X) = \frac{k_1(S)_0(C)_0}{k_1(S)_0 + k_1(C)_0 + k_{-1} + k_2}.$$

The initial rate of the reaction is obtained by using this value of (X) in the rate law:

$$\left[\frac{d(P)}{dt}\right]_0 = \frac{k_2(S)_0(C)_0}{(S)_0 + (C)_0 + K_m + (k_2/k_1)},\tag{31–59}$$

where $K_m = (k_{-1}/k_1)$. Two limiting cases of Eq. (31–59) are important.

Case I. The catalyst concentration is small compared with that of the substrate: $(C)_0 \ll (S)_0$. In this case, $(C)_0$ is dropped from the denominator and we have

$$\left[\frac{d(P)}{dt}\right]_0 = \frac{k_2(S)_0(C)_0}{(S)_0 + K_m + (k_2/k_1)}.\tag{31–60}$$

Then if (k_2/k_1) is small compared with the other terms, we have

$$\left[\frac{d(P)}{dt}\right]_0 = \frac{k_2(S)_0(C)_0}{(S)_0 + K_m},\tag{31–61}$$

which shows that if $(C)_0 \ll (S)_0$, the initial rate is proportional to the catalyst concentration. This rate law is applicable to many enzyme-catalysed reactions; for enzyme reactions, Eq. (31–61) is the Michaelis-Menten law and K_m is the Michaelis constant. This same form of rate law is also applicable to many surface reactions such as the decomposition of compounds on surfaces.

The dependence of the initial rate on the initial concentration of substrate is interesting. If $(S)_0$ is very small, then

$$(S)_0 + K_m \approx K_m,$$

and the rate is first order in $(S)_0$:

$$\left[\frac{d(P)}{dt}\right]_0 = \frac{k_2}{K_m}(S)_0(C)_0.\tag{31–62}$$

However, if $(S)_0$ is very large,

$$(S)_0 + K_m \approx (S)_0.$$

Then the rate is zero order in $(S)_0$:

$$\left[\frac{d(P)}{dt}\right]_0 = k_2(C)_0.\tag{31–63}$$

The initial rate as a function of $(S)_0$ is shown in Fig. 31–10. The limiting value of the rate is a result of the limited amount of catalyst present. The catalyst is needed to produce the reactive compound X. As soon as the concentration of S reaches the point where essentially all of the catalyst is found in the complex X, then further increase in (S) produces no change in the initial rate. It is apparent that by measuring initial rates of reaction with various initial concentrations of S, it is possible to determine the values of k_2 and K_m.

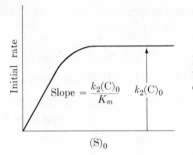

Fig. 31–10 Initial rate versus initial concentration of substrate.

Case II. The substrate concentration is small compared with that of the catalyst: $(S)_0 \ll (C)_0$. In this case, Eq. (31–60) becomes

$$\left[\frac{d(P)}{dt}\right]_0 = \frac{k_2(S)_0(C)_0}{(C)_0 + K_m + (k_2/k_1)}. \tag{31–64}$$

The reaction is always first order in $(S)_0$ in this case, but may be first order or zero order in $(C)_0$, depending on the value of $(C)_0$. This case is not usually as convenient experimentally as Case I.

31–21 Acid–base Catalysis

There are many chemical reactions which are catalysed by acids or bases, or by both. The most common acid catalyst in water solution is the hydronium ion and the most common base is hydroxyl ion. However, some reactions are catalysed by any acid or by any base. If any acid catalyses the reaction, the reaction is said to be subject to *general* acid catalysis. Similarly, *general* base catalysis refers to catalysis by any base. If only certain acids or bases are effective, the phenomenon is called *specific* acid or base catalysis.

A classical example of specific acid–base catalysis is the hydrolysis of esters. The hydrolysis is catalysed by H_3O^+ and OH^- but not by other acids or bases. The rate of hydrolysis in the absence of acid or base is extremely slow.

The mechanism of acid hydrolysis of an ester may be illustrated as follows:

The base-catalysed reaction has the mechanism

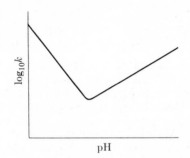

The rate of the reaction is given by

$$\frac{-d(\text{RCOOR}')}{dt} = [k_{\text{H}^+}(\text{H}^+) + k_{\text{OH}^-}(\text{OH}^-)](\text{RCOOR}'). \qquad (31\text{–}65)$$

The concentration of water does not appear in the rate law, since it is effectively constant during the course of the reaction in aqueous solution. Because of the relation $(\text{H}^+)(\text{OH}^-) = K_w$, the rate constant $k = k_{\text{H}^+}(\text{H}^+) + k_{\text{OH}^-}(\text{OH}^-)$ has a minimum at a pH which depends on K_w, k_{H^+}, and k_{OH^-}. The dependence of $\log_{10} k$ on pH is shown schematically in Fig. 31–11.

Fig. 31–11 Rate constant versus pH for a reaction catalysed by H^+ and OH^-.

Problems

31–1. a) Consider a reaction, A → Products, which is one-half order with respect to A. Integrate the rate equation and decide what function should be plotted from the data to determine the rate constant.

b) Repeat the calculation in (a) for a reaction which is three-halves order and nth order.

c) Derive the relation between the half-life, the rate constant, and the initial concentration of A for an nth-order reaction.

31–2. A certain reaction is first order; after 540 sec, 32.5% of the reactant remains.

a) Calculate the rate constant for the reaction.

b) How long would be required for 25% of the reactant to be decomposed?

31–3. The half-life of a first-order reaction is 30 min.

a) Calculate the specific rate constant of the reaction.

b) What fraction of the reactant remains after 70 min?

31–4. Copper 64 emits a β-particle. The half-life is 12.8 hr. At the time you received a sample of this radioactive isotope it had a certain initial activity (disintegrations/min). To do the experiment you have in mind, you have calculated that the activity must not go below 2% of the initial value. How much time do you have to complete your experiment?

31–5. A substance decomposes according to a second-order rate law. If the rate constant is 6.8×10^{-4} liter·mole^{-1}·sec^{-1}, calculate the half-life of the substance

a) if the initial concentration is 0.05 mole/liter;

b) if it is 0.01 mole/liter.

31–6. The decomposition of HI is an elementary reaction,

$$2HI \underset{k_{-1}}{\overset{k_1}{\rightleftharpoons}} H_2 + I_2.$$

The rate of the opposing reaction must be included in the rate expression. Integrate the rate equation under the conditions that the initial concentrations of H_2 and I_2 are zero and that of HI is a.

31–7. In Section 31–6 we reduced Eq. (31–15) to Eq. (31–17). Show that with the same notation Eq. (31–16) also reduces to Eq. (31–17).

31–8. Consider the opposing reactions,

$$A \underset{k_{-1}}{\overset{k_1}{\rightleftharpoons}} B,$$

both of which are first order. If the initial concentration of A is a, that of B is zero and if x moles/liter of A have reacted at time t, integrate the rate expression. Express k_{-1} in terms of the equilibrium constant K, and arrange the result in a form which resembles that for a first-order reaction in which the opposing reaction does not appear.

31–9. Near room temperature, 300°K, an old chemical rule of thumb is that the rate of a reaction doubles if the temperature is increased by 10°. Assuming that it is the rate constant that doubles, calculate the value the activation energy must have if this rule is to hold exactly.

31–10. For the reaction of hydrogen with iodine, the rate constant is 2.45×10^{-4} liter·mole^{-1}· sec^{-1} at 302°C and 0.950 at 508°C.

a) Calculate the activation energy and the frequency factor for this reaction.

b) What is the value of the rate constant at 400°C?

31–11. In the Lindemann mechanism, the "apparent" first-order rate constant, $k_{app} = k_2 k_1 c / (k_{-1} c + k_2)$. At low concentrations, the value of k_{app} decreases. If, when the concentration is 10^{-5} mole/liter, the value of k_{app} reaches 90% of its limiting value at $c = \infty$, what is the ratio of k_2/k_{-1}?

31–12. Using the steady-state treatment, develop the rate expression for the hypothetical mechanisms of formation of HBr.

a)
$$Br_2 \overset{k_1}{\rightarrow} 2Br,$$
$$Br + H_2 \overset{k_2}{\rightarrow} HBr + H.$$

b)
$$Br_2 \xrightarrow{k_1} 2Br,$$
$$Br + H_2 \xrightarrow{k_2} HBr + H,$$
$$Br + HBr \xrightarrow{k_3} Br_2 + H.$$

(Note that these are not chain mechanisms.)

31–13. The Rice-Herzfeld mechanism for the thermal decomposition of acetaldehyde is

(1)
$$CH_3CHO \xrightarrow{k_1} CH_3 + CHO,$$

(2)
$$CH_3 + CH_3CHO \xrightarrow{k_2} CH_4 + CH_2CHO,$$

(3)
$$CH_2CHO \xrightarrow{k_3} CO + CH_3,$$

(4)
$$CH_3 + CH_3 \xrightarrow{k_4} C_2H_6.$$

Using the steady-state treatment, obtain the rate of formation of CH_4.

31–14. The activation energies for the elementary reactions in Problem 31–13 are $E_1^* = 76$ kcal, $E_2^* = 10$ kcal, $E_3^* = 18$ kcal, and $E_4^* = 0$. Calculate the overall activation energy for the formation of methane.

31–15. The reaction between iodine and acetone,

$$CH_3COCH_3 + I_2 \rightarrow CH_3COCH_2I + HI,$$

is catalysed by hydrogen ion. The catalytic coefficient is $k_{H^+} = 4.48 \times 10^{-4}$. Calculate the rate constant in 0.05 and in 0.10 molar acid solution.

Chapter Thirty-two

Chemical Kinetics

II. Theoretical Aspects

32–1 Introduction

The ultimate goal of theoretical chemical kinetics is the calculation of the rate of any reaction from a knowledge of the fundamental properties of the reacting molecules; properties such as the masses, diameters, moments of inertia, vibrational frequencies, binding energies, and so on. At present this problem must be regarded as incompletely solved from the practical standpoint. Two approaches will be described here: the collision theory and the theory of absolute reaction rates. The collision theory is intuitively appealing and can be expressed in very simple terms. The theory of absolute reaction rates is more elegant. Neither theory is able to account for the magnitude of the activation energy except by approximations of questionable validity. The accurate calculation of activation energies from theory is a problem of extreme complexity and has been done for only a few very simple systems.

If we succeed in calculating the rate constant k for a reaction, we will have an interpretation of the Arrhenius equation,

$$k = Ae^{-E^*/RT}. \tag{32–1}$$

We begin by looking a little more closely into the meaning of the activation energy of a reaction.

32–2 The Activation Energy

The expression in Eq. (32–1) is reminiscent of the form of the equation for the equilibrium constant of a reaction. Since

$$d(\ln K)/dT = \Delta H^0/RT^2,$$

we have after integrating,

$$\ln K = -\frac{\Delta H^0}{RT} + \ln K_\infty, \tag{32-2}$$

where $\ln K_\infty$ is the integration constant. For an elementary reaction, $K = k_f/k_r$ and $K_\infty = (k_f)_\infty/(k_r)_\infty$. Furthermore, $\Delta H^0 = H_P^0 - H_R^0$, where H_P^0 and H_R^0 are the total enthalpies of the products and the reactants, respectively. Using these values in Eq. (32–2) and rearranging, we have

$$\ln\frac{k_f}{(k_f)_\infty} - \frac{H_R^0}{RT} = \ln\frac{k_r}{(k_r)_\infty} - \frac{H_P^0}{RT}. \tag{32-3}$$

The rate constant for the forward reaction presumably depends only on the properties of the reactants, while that of the reverse reaction depends only on the properties of the products. The left-hand side of Eq. (32–3) apparently depends only on the properties of reactants, while the right depends only on products. Each side must therefore be equal to a constant, which may be written $-H^*/RT$; then

$$\ln\frac{k_f}{(k_f)_\infty} = -\frac{H^* - H_R^0}{RT} \quad \text{and} \quad \ln\frac{k_r}{(k_r)_\infty} = -\frac{H^* - H_P^0}{RT}.$$

So that

$$k_f = (k_f)_\infty e^{-(H^* - H_R^0)/RT} \quad \text{and} \quad k_r = (k_r)_\infty e^{-(H^* - H_P^0)/RT}.$$

This argument can rationalize the form of the Arrhenius equation for the rate constants of any elementary reaction in either direction. The quantity $H^* - H_R^0$ is the energy quantity which the Arrhenius equation writes as E_f^*. Since experimentally it is observed that E_f^* is positive, it follows that $H^* - H_R^0$ is positive, and that $H^* > H_R^0$. Similar argument shows that H^* is also greater than H_P^0.

The variation in enthalpy through the course of the elementary step as reactants are converted to products is shown in Fig. 32–1. According to this view of the situation, an energy barrier separates the reactant state from the product state. The

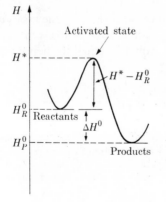

Fig. 32–1 Variation of enthalpy in a reaction.

reactants upon collision must have sufficient energy to surmount this barrier if products are to be formed. The height of this barrier is $H^* - H_R^0$; this is the activation energy† for the reaction in the forward direction E_f^*. Reactants which, upon collision, do not have sufficient energy to surmount the barrier will remain as reactants.

Viewed from the product side, the height of the barrier is $H^* - H_P^0$. This is the activation energy for the reverse reaction E_r^*. The relation between the two activation energies is obtained very simply. We write

$$H^* - H_P^0 = H^* - H_R^0 + H_R^0 - H_P^0 = H^* - H_R^0 - \Delta H^0.$$

Thus,

$$E_r^* = E_f^* - \Delta H^0, \tag{32–4}$$

which is the general relation between the activation energies and the energy change in the reaction. If the activation energy for the reaction in the forward direction is known, that for the reverse reaction can be calculated directly from Eq. (32–4) if ΔH^0 is known.

32–3 The Collision Theory of Reaction Rates

In its simplest form the collision theory is applicable only to bimolecular elementary reactions. With additional assumptions it can be applied to first-order reactions, and with some elaboration it is applicable to termolecular elementary reactions. As an example, we choose an elementary reaction of the type

$$A + B \rightarrow C + D.$$

It is obvious that this reaction cannot occur more often than the number of times the molecules A and B collide. The number of collisions between molecules A and B in 1 cm³ in 1 sec is given by Eq. (29–20):

$$Z_{AB} = \left(\frac{\sigma_A + \sigma_B}{2}\right)^2 \sqrt{\frac{8\pi(m_A + m_B)kT}{m_A m_B}} n_A n_B,$$

where σ_A and σ_B are the molecular diameters, m_A and m_B the molecular masses, and n_A and n_B the number of molecules of A and B per cubic centimeter. If reaction occurred with every collision, then this would be equal to the rate of disappearance of either A or B per cubic centimeter:‡

$$-\frac{dn_A}{dt} = -\frac{dn_B}{dt} = \left(\frac{\sigma_A + \sigma_B}{2}\right)^2 \sqrt{\frac{8\pi(m_A + m_B)kT}{m_A m_B}} n_A n_B.$$

---

† There is a distinction between activation energy and activation enthalpy; however, the relation between them depends on the type of reaction in question. The term "activation energy" will be used here to describe whichever one we are interested in at the moment.

‡ The concentrations are in molecules/cm³ here; we could convert this to concentrations in moles/liter, c, by using $n = (N_0/1000)c$.

Every collision does not, in fact, result in the reaction of A and B, but only those collisions in which the energy of the colliding molecules exceeds E^*. The fraction of collisions in which the energy exceeds E^* is proportional to $\exp(-E^*/RT)$ so that the rate of the reaction is

$$-\frac{dn_A}{dt} = \left(\frac{\sigma_A + \sigma_B}{2}\right)^2 \sqrt{\frac{8\pi(m_A + m_B)kT}{m_A m_B}} e^{-E^*/RT} n_A n_B. \tag{32-5}$$

The empirical law for the rate of the elementary reaction is $-dn_A/dt = kn_A n_B$, so for the rate constant we obtain

$$k = \left(\frac{\sigma_A + \sigma_B}{2}\right)^2 \sqrt{\frac{8\pi(m_A + m_B)kT}{m_A m_B}} e^{-E^*/RT}, \tag{32-6}$$

$$k = Z' e^{-E^*/RT}, \tag{32-7}$$

where $Z' = Z_{AB}/n_A n_B$.

The Arrhenius equation has the same form as Eq. (32–6), so the collision theory predicts for the frequency factor

$$A = Z' = \left(\frac{\sigma_A + \sigma_B}{2}\right)^2 \sqrt{\frac{8\pi(m_A + m_B)kT}{m_A m_B}}. \tag{32-8}$$

Strictly speaking, A should be independent of temperature. However, the square-root dependence in Eq. (32–8) is rather slight, so a weak dependence on temperature is not really a difficulty. The order of magnitude of A can be readily estimated. The value of the radical in Eq. (32–8) is a molecular speed which, at ordinary temperatures, is about 4×10^4 cm/sec. The value of σ is of the order of 10^{-8}, so we have

$$A = (10^{-8})^2(4 \times 10^{+4}) = 4 \times 10^{-12} \text{ cm}^3 \cdot \text{molecule}^{-1} \cdot \text{sec}^{-1}.$$

If the concentration unit is moles per liter, this must be multiplied by $(N_0/1000) = 6 \times 10^{20}$, so that $A = 2 \times 10^9$ liter·mole^{-1}·sec^{-1}. The order of magnitude of the frequency factor for bimolecular reactions is 10^9 to 10^{10} if the concentration units are in moles per liter and the temperature is around 300°K. The frequency factor for reactions involving a light molecule such as H_2 are larger, about 10^{11}.

The collision theory predicts the value of the rate constant satisfactorily for reactions which involve relatively simple molecules if the activation energy is known. Difficulties are encountered with reactions between complicated molecules. The rates tend to be smaller than the collision theory predicts, in many cases by a factor of 10^5 or more. To account for this, an additional factor P, called the *probability factor* or the *steric factor*, is inserted in the expression for k:

$$k = PZ' e^{-E^*/RT}. \tag{32-9}$$

The idea behind this is that even those collisions having the requisite energy may not produce reaction; originally it was supposed that the molecules had to collide in a particular configuration, hence the name *steric factor*. This idea has some validity, especially since the low rates of reaction are usually observed with complex molecules. Presumably two complex molecules will have less chance of colliding in the correct orientation for reaction than will two simple molecules. It will be seen shortly that the probability factor receives a more acceptable interpretation in terms of the entropy of activation of a reaction. In particular, the collision theory offers no explanation for abnormally fast reactions in which P would have to be greater than unity.

32–4 Termolecular Reactions

The problem of termolecular reactions can be treated by collision theory also. A number of such reactions are known; reactions of NO with H_2, O_2, Cl_2 are famous examples. Choosing the reaction with oxygen

$$2NO + O_2 \rightarrow 2NO_2,$$

the rate of the reaction is

$$\frac{-d(O_2)}{dt} = k(NO)^2(O_2).$$

Apparently the reaction as written is elementary and involves the simultaneous collision of two molecules of NO with one molecule of O_2. A remarkable feature of this reaction is the fact that the rate of the reaction *decreases* with *increase* in temperature. This behavior is exhibited by only a very few reactions.

A difficulty in the treatment by collision theory arises as soon as one attempts to define a triple collision. If the molecules are hard spheres, the time in which they are in contact is zero. The probability of being hit by a third molecule during the collision is therefore zero. A finite time interval must be specified for the collision of two molecules if a third is to collide with the two. The time interval is arbitrary; a common specification is that the molecules are in collision so long as the distance between them is less than the molecular diameter. With this specification it may be shown that, approximately, $Z_3/Z_2 = \sigma/\lambda$, where Z_3 and Z_2 are the numbers of triple collisions and binary collisions per cubic centimeter per second, σ is the molecular diameter, and λ is the mean free path. Since λ is inversely proportional to the number of molecules per cubic centimeter, it follows that the number of triple collisions increases as the cube of the number of molecules per cubic centimeter. At ordinary pressures $\lambda \approx 10^{-4}$ cm so that $Z_3 \approx Z_2(10^{-8}/10^{-4}) = 10^{-4}Z_2$. Roughly speaking, there is one triple collision for every 10,000 ordinary collisions. Therefore, reactions requiring triple collisions are slower, other things being equal, than those involving binary collisions. The rate constant calculated on the basis of triple collisions is much larger than the experimental value, indicating that the probability factor is quite small.

An alternative mechanism has been proposed for these reactions. The equilibrium

$$NO + X_2 \rightleftharpoons NOX_2$$

is assumed to be established very rapidly. Then

$$(NOX_2) = K(NO)(X_2),$$

where K is the equilibrium constant. The slow reaction follows:

$$NOX_2 + NO \rightarrow 2NOX.$$

The rate of this reaction is

$$rate = K(NO)(NOX_2) = kK(NO)^2(X_2).$$

This mechanism accounts for the rate law. It is apparent that the equilibrium

$$2NO \rightleftharpoons N_2O_2$$

followed by the slow step

$$N_2O_2 + X_2 \rightarrow 2NOX$$

would also account for the empirical rate law. The difference between the triple collision viewpoint and these mechanisms is not very great. The molecules NOX_2 or N_2O_2 can be thought of as two molecules which are involved in a "sticky" collision. The equilibrium assumption explains the negative temperature coefficient, since it implies that at higher temperatures more double molecules, NOX_2 or N_2O_2, are dissociated; the lower concentration of double molecules results in a lower rate. This implies that the activation energy is negative. This explanation has its problems, since in the reaction of NO with O_2, the activation energy is zero, the decrease in rate constant with temperature being due to the frequency factor which is inversely proportional to T^3.

32–5 Unimolecular Reactions

Collision theory does not deal directly with unimolecular reactions but touches on the subject through the Lindemann mechanism. Once the molecule has been provided with sufficient energy by collision, the problem is to calculate the rate constant for the unimolecular decomposition

$$A^* \rightarrow Products.$$

The theory of this type of decomposition has been developed by O. K. Rice, H. C. Ramsberger, and L. S. Kassel; more recently, N. B. Slater has treated the problem in more exact and elegant terms. The treatment is based on the supposition that if too much energy gets into a particular mode of vibration, then vibration of the molecule in this mode leads to dissociation of the molecule.

The Rice-Ramsberger-Kassel approach assumes that the activated molecule has a certain amount of vibrational energy spread among the various vibrational degrees of freedom of the molecule. Then the probability of one particular mode of vibration acquiring so much of this energy that the vibration leads to dissociation into fragments is calculated.

We assume that there are s vibrational degrees of freedom and that the molecule has j quanta of energy distributed in the s degrees of freedom. Let N_j be the number of ways of distributing the j quanta in the s degrees of freedom. Let N_m be the number of ways of distributing the j quanta in the s degrees, so that a particular degree of freedom has m quanta. Then the probability that that particular degree of freedom has m quanta is N_m/N_j. If j and $j - m$ are large compared with s, it can be shown† that, approximately,

$$\frac{N_m}{N_j} = \left(\frac{j - m}{j}\right)^{s-1}. \tag{32-10}$$

Since j is the total number of quanta, it is proportional to the total vibrational energy of the molecule E; the number m is proportional to E_c, the critical minimum energy required for dissociation to occur. Therefore, the probability of the particular degree of freedom having the critical energy can be written

$$\frac{N_m}{N_j} = \left(\frac{E - E_c}{E}\right)^{s-1}. \tag{32-11}$$

The rate of dissociation of the molecule is proportional to this probability, so that the rate constant is given by

$$k = k'\left(\frac{E - E_c}{E}\right)^{s-1}, \tag{32-12}$$

where k' is a constant. Since E may have any value from E_c to infinity, the rate constant must be averaged, using a Boltzmann distribution, over all values of E from E_c to infinity. The integral is evaluated graphically for particular values of s, and E_c. Reasonable agreement with experiment is obtained using values of s comparable to the number of vibrational degrees of freedom in the molecule.

Thus we see that for unimolecular reactions, as well as for others, it is required that the molecule must have at least a critical minimum energy for reaction to occur. The interpretation of the Arrhenius equation for unimolecular reactions is more complex, however. The pre-exponential factor A is a function of the number of degrees of vibrational freedom s, as well as E_c and T. It should be observed that Eq. (32-11) implies that the higher the energy E in the vibrational modes and the greater the number s of these modes, the greater the probability that the molecule will have the required m quanta in the critical vibration. A related fact is that the rate of activation may be larger than the collision rate predicted by the Lindemann mechanism. Some molecules may be "self-activated" in the sense that quanta of vibrational energy which are spread over the various modes of vibration may flow into the critical mode and supply it with the critical energy. This process enhances the rate of activation.

---

†   The proof is elementary but is too lengthy to be included here.

32–6 Irreversible Thermodynamics

Considerable effort has been expended in the attempt to develop a general theory of reaction rates through some extension of thermodynamics or statistical mechanics. Since neither of these sciences can, by themselves, yield any information about rates of reactions, some additional assumptions or postulates must be introduced. An important method of treating systems which are not in equilibrium has acquired the title of *irreversible thermodynamics*. Irreversible thermodynamics can be applied to those systems which are "not too far" from equilibrium. The theory is based on the thermodynamic principle that in every irreversible process, every process proceeding at a finite rate, entropy is created. This principle is used together with the fact that the entropy of an isolated system is a maximum at equilibrium, and with the principle of microscopic reversibility.† The additional assumption involved is that systems which are slightly removed from equilibrium may be described statistically in much the same way as systems in equilibrium.

An outstanding success of the theory has been the general derivation of the relations between certain pairs of rate constants in transport processes, the Onsager reciprocal relations. Although these relations were known before, the derivations were individualized and in certain cases the validity of the derivation was suspect. The theory is not applicable to the data obtained from the usual type of investigation in chemical kinetics in which the system is far removed from equilibrium. Investigations specifically designed to test the theory have supported its conclusions. An interesting aspect of the theory is that it requires certain relations between the rate constants of coupled reactions in systems which are "not too far" from equilibrium.

Central to the thermodynamic discussion of irreversible processes is the concept of entropy production. Consider the Clausius inequality, $dS \geq dQ/T$, which we can rearrange to the form

$$dS - \frac{dQ}{T} \geq 0.$$

The quantity on the left is greater than or equal to zero, so we may write

$$dS - \frac{dQ}{T} = d\sigma, \qquad (32\text{--}13)$$

if we insist that $d\sigma$ be either zero or positive.

If we suppose that the system is in contact with a reservoir at T, and a quantity of heat dQ flows into the system, then a quantity, $-dQ$, flows into the reservoir. If the quantity, $-dQ$, is transferred reversibly to the reservoir, then the entropy change of the reservoir is $dS_{\text{res}} = -dQ/T$, so that Eq. (32–13) can be written

$$dS + dS_{\text{res}} = d\sigma.$$

---

† Principle of microscopic reversibility: at equilibrium, any molecular process occurs at the same rate as the reverse of that process.

The quantity $d\sigma$ is the entropy increase of the system plus that of the surroundings (the reservoir); $d\sigma$ is called the *entropy production* of the process. For any irreversible transformation, the entropy production is positive, while for a reversible transformation the entropy production is zero.

We may write Eq. (32–13) in the form

$$T\,d\sigma = T\,dS - dQ. \tag{32–14}$$

If we apply this equation to a transformation at constant T and p, we have $dQ_p = dH$, and $T\,dS = d(TS)$, so that $T\,dS - dQ_p = d(TS) - dH = -d(H - TS)$; then

$$T\,d\sigma = -dG. \tag{32–15}$$

For a chemical reaction at constant T and p, we have $dG = (\partial G/\partial\xi)_{T,p}\,d\xi$, and therefore,

$$T\,d\sigma = -\left(\frac{\partial G}{\partial\xi}\right)_{T,p}d\xi. \tag{32–16}$$

DeDonder has introduced **A**, the affinity of the reaction, for the quantity, $-(\partial G/\partial\xi)_{T,p}$. Thus,

$$\mathbf{A} \equiv -\left(\frac{\partial G}{\partial\xi}\right)_{T,p}. \tag{32–17}$$

Combining this definition with Eq. (32–16) yields

$$T\,d\sigma = \mathbf{A}\,d\xi. \tag{32–18}$$

Note that for the spontaneous direction of a reaction, $(\partial G/\partial\xi)_{T,p}$ is negative, so that the affinity is positive. Dividing by dt, we obtain the *rate* of entropy production, $d\sigma/dt$.

$$T\frac{d\sigma}{dt} = \mathbf{A}\frac{d\xi}{dt}. \tag{32–19}$$

Since the rate of entropy production by the second law must always be positive or zero, it follows from Eq. (32–19) that the product of the affinity and the rate of reaction, $d\xi/dt$, must always be positive or zero. This result

$$\mathbf{A}\frac{d\xi}{dt} \geq 0 \tag{32–20}$$

is known as DeDonder's inequality.

There is an important general relation which can be obtained with relative ease by combining a rate equation with a thermodynamic equation. Consider the reaction

$$A + B \rightleftharpoons C.$$

We write the rate equation as in Section 31–8:

$$\frac{1}{V}\frac{d\xi}{dt} = k_f c_A c_B - k_r c_C$$

$$\frac{1}{V}\frac{d\xi}{dt} = k_f c_A c_B \left[1 - \frac{k_r c_C}{k_f c_A c_B} \right].$$

We recognize that $c_C/c_A c_B = Q$, the proper quotient of concentrations for the reaction, and $k_f/k_r = K$, the equilibrium constant for the reaction. Then we have

$$\frac{1}{V}\frac{d\xi}{dt} = k_f c_A c_B \left(1 - \frac{Q}{K} \right). \tag{32–21}$$

Near equilibrium, the quantity $k_f c_A c_B$ approaches $r = k_f \bar{c}_A \bar{c}_B$, the exchange rate of the reaction. Then

$$\frac{1}{V}\frac{d\xi}{dt} = r \left(1 - \frac{Q}{K} \right).$$

However, we have the relations

$$\left(\frac{\partial G}{\partial \xi} \right)_{T,p} = \Delta G^0 + RT \ln Q \qquad \text{and} \qquad \Delta G^0 = -RT \ln K,$$

which combine to yield

$$\left(\frac{\partial G}{\partial \xi} \right)_{T,p} = RT \ln \frac{Q}{K} = -\mathbf{A},$$

or

$$\frac{Q}{K} = e^{-\mathbf{A}/RT}.$$

Since $\mathbf{A}$ is very small near equilibrium, we can expand the exponential in series to obtain: $Q/K = 1 - \mathbf{A}/RT + \cdots$. This brings the rate equation to the form

$$\frac{d\xi}{dt} = Vr\frac{\mathbf{A}}{RT}. \tag{32–22}$$

Equation (32–22) expresses the important result that the rate of a reaction near equilibrium is proportional to the affinity of the reaction.

Equation (32–22) is a chemical example of a linear law analogous to those mentioned in Section 29–2. In each of those cases, a flow, such as a heat flow, an electrical current, a fluid flow, or a diffusive flow, was proportional to a driving force such as a temperature gradient, an electrical. potential gradient, a pressure gradient or a concentration gradient. In the chemical case, Eq. (32–22), the "flow" is the rate of the reaction, while the driving force is the affinity of the reaction divided by T.

If we combine the result of Eq. (32–22) with Eq. (32–19) for the rate of entropy production we obtain

$$\frac{d\sigma}{dt} = R(Vr)\left(\frac{\mathbf{A}}{RT}\right)^2 = \frac{R}{Vr}\left(\frac{\partial \xi}{\partial t}\right)^2.$$

(32–23)

This shows the positive character of $d\sigma/dt$ since it is proportional to the square of the affinity or to the square of the reaction rate.

The two equations, Eqs. (32–22) and (32–23) are typical of the application of thermodynamics to irreversible processes. One obtains, or assumes, a linear rate law such as the one in Eq. (32–22) in which the flow is proportional to the driving force; and one obtains a quadratic law for the entropy production in which, as in Eq. (32–23), the rate of entropy production is proportional to the square of the driving force.

32–7 The Theory of Absolute Reaction Rates

The theory of absolute reaction rates, which is based on statistical mechanics, was developed in full generality by H. Eyring in 1935, although there was a foreshadowing of it in kinetic theory investigations as early as 1915. A simplified development of the equations will be given here. In this theory, we have a postulate of "equilibrium" away from equilibrium, applied more broadly here than in the irreversible thermodynamics.

The fundamental postulate of the theory of absolute reaction rates is that the reactants are always in equilibrium with activated complexes. The activated complex is that configuration of the atoms which corresponds energetically to the top of the energy barrier separating the reactants from the products, Fig. 32–2. The equilibrium is written

$$A + B \rightleftharpoons M^{\neq},$$

and the equilibrium constant is

$$K_{\neq} = \frac{c^{\neq}}{c_A c_B}.$$

(32–24)

The concentration of activated complexes is

$$c^{\neq} = K_{\neq} c_A c_B.$$

(32–25)

Knowing the concentration of activated complexes, the problem resolves into the calculation of the rate at which these complexes decompose into products; that is, we must calculate the rate of the reaction

$$M^{\neq} \rightarrow \text{Products.}$$

The activated complex is an aggregate of atoms which may be thought of as being similar to an ordinary molecule except that it has one special vibration with respect to which it is unstable. This vibration leads to dissociation of the complex into

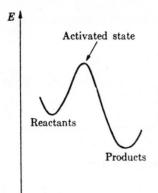

Fig. 32–2 Energy variation in the transformation from reactants to products.

products. If the frequency of this vibration is v, then the rate at which products are formed is

$$\text{rate} = vc^{\neq} \text{ (molecules·cm}^{-3}\text{·sec}^{-1}). \tag{32–26}$$

Using Eq. (32–25) this becomes

$$\text{rate} = vK_{\neq}c_A c_B. \tag{32–27}$$

But, the elementary reaction, $A + B \rightarrow$ Products, has the rate,

$$\text{rate} = kc_A c_B. \tag{32–28}$$

Comparing Eqs. (32–27) and (32–28), we find that the rate constant is given by

$$k = vK_{\neq}. \tag{32–29}$$

A review of the steps involved in deriving Eq. (32–29) shows that it is not restricted by the choice of two reactants, but is correct for any elementary reaction.

The values of v and $K_{\neq}$ can be calculated if we write the equilibrium constant in terms of molecular partition functions per unit volume, f' (Section 28–10). Then

$$K_{\neq} = f'_{\neq}/f'_A f'_B. \tag{32–30}$$

Any molecular partition function can be written in the form $f' = fe^{-\epsilon_0/kT}$. The function f is the partition function evaluated using energies relative to the zero-point energy ϵ_0 of the molecule. Then Eq. (32–30) becomes

$$K_{\neq} = \frac{f_{\neq}}{f_A f_B} e^{-(\epsilon_{0\neq} - \epsilon_{0A} - \epsilon_{0B})/kT} = \frac{f_{\neq}}{f_A f_B} e^{-E_0/RT}. \tag{32–31}$$

The activation energy E_0 is defined as the difference in zero-point energies between the activated complex and the reactants: $E_0 = N_0(\epsilon_{0\neq} - \epsilon_{0A} - \epsilon_{0B})$.

As we have seen in Section 28–9, the partition function can be written as a product of partition functions for translation, rotation, and vibration. We direct our attention to that particular vibration which dissociates the activated complex into products, and factor that vibrational partition function out of $f_{\neq}$. Let

$$f_{\neq} = f_v f^{\neq},\tag{32--32}$$

where $f^{\neq}$ is what remains of $f_{\neq}$ after f_v has been factored out. If the frequency of this vibration is v, then by Eq. (28--44),

$$f_v = \frac{\mathbf{k}T}{hv}e^{-hv/2\mathbf{k}T},$$

if v is small and $hv/\mathbf{k}T \ll 1$. Since the exponential is about equal to unity, $f_v = \mathbf{k}T/hv$ and Eq. 32--32 becomes

$$f_{\neq} = \frac{\mathbf{k}T}{hv}f^{\neq}.\tag{32--33}$$

Using this value of $f_{\neq}$ in Eq. (32--31), we obtain

$$K_{\neq} = \frac{\mathbf{k}T}{hv}\frac{f^{\neq}}{f_A f_B}e^{-E_0/RT}.\tag{32--34}$$

Define $K^{\neq}$ by

$$K^{\neq} = \frac{f^{\neq}}{f_A f_B}e^{-E_0/RT};\tag{32--35}$$

then $K_{\neq} = (\mathbf{k}T/(hv)K^{\neq}$. Using this value in Eq. (32--29) yields, for the rate constant,

$$k = \frac{\mathbf{k}T}{h}K^{\neq},\tag{32--36}$$

which is the Eyring equation for the rate constant of a reaction. The value of $K^{\neq}$ can be calculated from the partition functions of the reactants and the activated complex using an equation having the form of Eq. (32--35). If we use Eq. (32--35) in Eq. (32--36), we obtain

$$k = \frac{\mathbf{k}T}{h}\frac{f^{\neq}}{f_A f_B}e^{-E_0/RT}.\tag{32--37}$$

Comparing this result with the Arrhenius equation, we see that the frequency factor A is given by

$$A = \frac{\mathbf{k}T}{h}\frac{f^{\neq}}{f_A f_B}.\tag{32--38}$$

The expression in Eq. (32--38) is interesting because the partition functions depend on the translational degrees of freedom of the molecules and the activated complex and on the internal degrees of freedom as well. The collision theory cannot take any account of the internal degrees of freedom without becoming incredibly complicated mathematically. The Eyring theory includes the internal degrees of freedom in a very simple way.

Two things are required to calculate the rate constant by Eq. (32–37). First, the activated complex must be specified sufficiently so that $f^{\neq}$ can be calculated; this implies knowing the size and shape so that the moments of inertia can be calculated. The calculation of the vibrational frequencies can be done quantum mechanically, but is quite complicated. Secondly, E_0 must be known. The calculation of E_0 from the quantum mechanics is quite complicated unless drastic approximations are made. This procedure has been carried out in full detail for a number of reactions, particularly for reactions involving hydrogen atoms and hydrogen molecules. Considering the approximations involved, the results are very good.

It is relatively easy to obtain a rough estimate of the order of magnitude of the frequency factor using Eq. (32–38). Consider the reaction

$$H_2 + I_2 \rightarrow 2HI.$$

The frequency factor is

$$A = \frac{\mathbf{k}T}{h} \frac{f^{\neq}}{f_{H_2} f_{I_2}}.$$

The partition function f_{H_2} can be written as a product of partition functions for three translational, two rotational, and one vibrational degree of freedom:

$$f_{I_2} = f_{H_2} = f_t^3 f_r^2 f_v.$$

We use the same value for f_{I_2} since we wish to make only a calculation of the order of magnitude. The complex $(HI)_2$ has three translational, three rotational, and five vibrational degrees (one vibrational degree was removed in the early part of the derivation); thus we write $f^{\neq} = f_t^3 f_r^3 f_v^5$. Using these values, we find for A,

$$A = \frac{\mathbf{k}T}{h} \frac{f_t^3 f_r^3 f_v^5}{f_t^3 f_r^2 f_v f_t^3 f_r^2 f_v} = \frac{\mathbf{k}T}{h} \left(\frac{f_v^3}{f_t^3 f_r} \right).$$

At ordinary temperatures the usual orders of magnitude of these quantities are $f_t \approx 10^8$, $f_r \approx 10$, $f_v \approx 1$, and $(\mathbf{k}T/h) \approx 10^{13}$. Using these values, we obtain

$$A = \frac{10^{13}}{(10^8)^3 10} = 10^{-12}.$$

To convert this value in $cm^3 \cdot molecule^{-1} \cdot sec^{-1}$ to $liter \cdot mole^{-1} \cdot sec^{-1}$, we multiply by $(N_0/1000) \approx 10^{21}$; so that $A \approx 10^9$ $liter \cdot mole^{-1} \cdot sec^{-1}$. This value of the frequency factor usually agrees roughly with the values A found for bimolecular reactions which are usually between 10^9 and 10^{11}. Considering the very approximate values used for the partition functions, the agreement is good.

32–8 Comparison of the Collision Theory with the Absolute Reaction Rate Theory

For bimolecular reactions, the comparison of collision theory with absolute reaction rate theory is rather easily done using the results of the last section. Consider the bimolecular reaction between two polyatomic molecules A and B to yield a complex

$$A + B \rightarrow (AB)^{\neq}.$$

If n_A and n_B are the number of atoms in A and B and if both molecules are nonlinear, then

$$f_A = f_t^3 f_r^3 f_v^{3n_A - 6} \qquad f_B = f_t^3 f_r^3 f_v^{3n_B - 6} \qquad f^{\neq} = f_t^3 f_r^3 f_v^{3(n_A + n_B) - 7},$$

since the complex contains $n_A + n_B$ atoms. Using these values in Eq. (32–37), we obtain

$$k = \frac{\mathbf{k}T}{h} \left(\frac{f_t^3 f_r^3 f_v^{3(n_A + n_B) - 7}}{f_t^3 f_r^3 f_v^{3n_A - 6} f_t^3 f_r^3 f_v^{3n_B - 6}} \right) e^{-E_0/RT} = \frac{\mathbf{k}T}{h} \frac{f_v^5}{f_t^3 f_r^3} e^{-E_0/RT}. \tag{32–39}$$

Now the Eyring equation yields the same result as the collision theory if we treat A and B as if they were atoms and $(AB)^{\neq}$ as if it were diatomic; then we would have $f_A = f_B = f_t^3$ and $f^{\neq} = f_t^3 f_r^2$, and

$$k_{\text{collis}} = \frac{\mathbf{k}T}{h} \frac{f_t^3 f_r^2}{f_t^3 f_t^3} e^{-E_0/RT} = \frac{\mathbf{k}T}{h} \frac{f_r^2}{f_t^3} e^{-E_0/RT}. \tag{32–40}$$

Comparing Eqs. (32–39) and (32–40), we obtain

$$k = \left(\frac{f_v}{f_r} \right)^5 k_{\text{collis}}. \tag{32–41}$$

The probability factor P, which must be introduced arbitrarily in the collision theory, is according to Eq. (32–41),

$$P = \left(\frac{f_v}{f_r} \right)^5. \tag{32–42}$$

If $f_v = 1$ and $f_r = 10$, then $P = 10^{-5}$, which is a not uncommon value of P. This equation gives us a little insight into the effect of the internal degrees of freedom which make reactions between polyatomic molecules very much slower than those between simple molecules.

The great advantage of the theory of absolute reaction rates is that the postulated equilibrium between activated complex and reactants evades entirely the question of just how the complex is formed; the theory resembles thermodynamics in this regard. For the same reason a trace of dissatisfaction with the theory can be voiced, for the object of kinetics is to look into the details of how the reaction goes. The great simplicity introduced by the equilibrium assumption enables us to suppress this feeling of dissatisfaction to some extent. The collision theory could take into account the presence of internal degrees of freedom in the molecule. However, the mathematical treatment using this approach would be intolerably complex.

32–9 Free Energy and Entropy of Activation

According to Eq. (32–36), the Eyring equation, the rate constant is

$$k = \frac{kT}{h} K^{\neq}.$$

The equilibrium constant $K^{\neq}$ can be written in terms of a standard† free energy of activation $\Delta G^{\neq}$:

$$K^{\neq} = e^{-\Delta G^{\neq}/RT}. \tag{32–43}$$

Then we obtain for the rate constant

$$k = \frac{kT}{h} e^{-\Delta G^{\neq}/RT}, \tag{32–44}$$

which emphasizes the free energy of activation as the fundamental quantity, rather than the energy of activation. The free energy of activation can be written $\Delta G^{\neq} = \Delta H^{\neq} - T\Delta S^{\neq}$, so that

$$k = \frac{kT}{h} e^{\Delta S^{\neq}/R} e^{-\Delta H^{\neq}/RT}, \tag{32–45}$$

which resembles the Arrhenius equation, except that $\Delta H^{\neq}$ appears instead of E^*. The quantity $\Delta H^{\neq}$ is often called the heat of activation. The frequency factor A, according to Eq. (32–45), is

$$A = \frac{kT}{h} e^{\Delta S^{\neq}/R}. \tag{32–46}$$

A negative entropy of activation will result in a quite low frequency factor, while a positive entropy of activation will raise its value. The probability factor introduced in the collision theory can be interpreted in terms of the entropy of activation. We write

$$PZ' = \frac{kT}{h} e^{\Delta S^{\neq}/R}. \tag{32–47}$$

The collision frequency Z' is readily calculated so that from the value of P the value of $\Delta S^{\neq}$, the entropy of activation, can be calculated.

If the activated complex resembles the products more than the reactant, then the entropy of activation may be nearly equal to the ΔS^0 of the overall reaction. The values of $\Delta S^{\neq}$ and ΔS^0 do seem to parallel one another in many cases, although it is rare that they are equal.

---

† The usual superscript zero to designate a standard free energy is omitted on the symbol $\Delta G^{\neq}$ to avoid a cumbersome symbol.

32–10 Reactions in Solution

In Section 31–19 we mentioned the fact that the rate constant for a reaction in solution is often very nearly the same as that for the same reaction in the gas phase. A rate constant in solution which is very much different from that in the gas indicates a comparatively strong interaction between the solvent and the reactants or the activated complex.

The reason for the equality between the rate of reaction in the gas and that in the solution can be explained rather simply in terms of the collision theory. Suppose a reaction requires the collision of two molecules in a pure gas A

$$A + A \rightarrow Products.$$

The number of collisions·cm^{-3}·sec^{-1} can be written as

$$Z_{11} = Zn_A^2. \tag{32–48}$$

Now if foreign molecules B are introduced, there will be collisions between A and B and collisions between B and B. This fact does not change the number of collisions between A and A, which is still given by Eq. (32–48). Thus, even if all the intervening space in the gas is filled with foreign molecules B, the rate constant should not change. On this basis, the rate constant in solution should have the same value as in the gas. The argument is correct only for reasonably ideal solutions. Nonideality in the solution implies solvation effects of the type alluded to in the preceding paragraph.

Consider the reaction in solution:

$$A + B \rightleftharpoons M^{\neq} \rightarrow Products.$$

The rate of this reaction is given by Eq. (32–26),

$$rate = vc^{\neq}.$$

But in solution,

$$K_{\neq} = \frac{a^{\neq}}{a_A a_B} = \frac{\gamma^{\neq} c^{\neq}}{\gamma_A \gamma_B c_A c_B}, \tag{32–49}$$

where the a's are the activities and the γ's are activity coefficients. Then the rate becomes

$$rate = vK_{\neq}\left(\frac{\gamma_A \gamma_B}{\gamma^{\neq}}\right)c_A c_B,$$

and the rate constant is

$$k = vK_{\neq}\left(\frac{\gamma_A \gamma_B}{\gamma^{\neq}}\right).$$

Comparison of Eq. (32–29) and (32–36) shows that $vK_{\neq}$ may be replaced by $(\mathbf{k}T/h)K^{\neq}$, so we have

$$k = \frac{\mathbf{k}T}{h} K^{\neq} \left(\frac{\gamma_A \gamma_B}{\gamma^{\neq}} \right), \tag{32–50}$$

which is the equation we must use to discuss reactions in solution.

The definition of the activity coefficient depends on the choice of the reference state in which $\gamma = 1$. If we wish to compare the rate of reaction with the rate in the gas phase, then we will choose the reference state of unit activity coefficient as the ideal gas. This reduces Eq. (32–50) to

$$k_g = \frac{\mathbf{k}T}{h} K^{\neq}, \tag{32–51}$$

where k_g is the rate constant for the reaction in the gas phase. Then in solution

$$k = k_g \left(\frac{\gamma_A \gamma_B}{\gamma^{\neq}} \right)_g. \tag{32–52}$$

In Eq. (32–52) the subscript g on the activity coefficient ratio indicates the choice of reference state. The deviation of the value of the rate constant in the solution from that in the gas depends on this ratio of activity coefficients. If the solvent lowers the free energy of the reactants more than it does that of the activated complex (if the reactants are strongly solvated), then γ_A and γ_B will be small, while $\gamma^{\neq}$ will not be so small. The rate constant in this case will be smaller in solution than in the gas. Conversely, if the activated complex is strongly solvated while the reactants are not, the rate constant will be larger in solution than in the gas.

If the reaction is one which does not take place in the gas phase, then it is more useful to choose the infinitely dilute solution as the reference state of unit activity coefficient. If the rate constant in infinitely dilute solution is k_0, since the γ's are unity, we have

$$k_0 = \frac{\mathbf{k}T}{h} K^{\neq}$$

and

$$k = k_0 \left(\frac{\gamma_A \gamma_B}{\gamma^{\neq}} \right)_0. \tag{32–53}$$

Here the subscript zero on the activity coefficient ratio indicates the infinitely dilute solution as the reference state.

32–11 Ionic Reactions; Salt Effects

The majority of reactions between ions in solution, particularly between simple ions of opposite charge, occur so extremely rapidly that until very recently it was impossible to measure the rates of these reactions. Relaxation techniques such as those described in Section 31–8 are now used to determine the rate of reactions such as $H_3O^+ + OH^- \rightarrow 2H_2O$. The rate constant of this particular reaction is 1.4×10^{11} liter·mole^{-1}·sec^{-1}.

There are some reactions between ions and neutral molecules which progress slowly enough so that ordinary methods of measurement can be used. The rate constants of these reactions depend on the ionic strength of the solution. Equation (32–53) was first derived by Bronsted and Bjerrum before the theory of absolute reaction rates was developed; applied to ionic reactions, Eq. (32–53) is called the Bronsted-Bjerrum equation. By combining Eq. (32–53) with the Debye-Hückel limiting law for ionic activity coefficients, the dependence of the rate constant on the ionic strength can be deduced. Writing Eq. (32–53) in logarithmic form, we have

$$\log_{10} k = \log_{10} k_0 + \log_{10} \gamma_A + \log_{10} \gamma_B - \log_{10} \gamma^{\neq}. \tag{32–54}$$

The value of $\ln \gamma$ is given by Eq. (16–74); by comparing Eqs. (16–74) and (16–78) we find that for a single ion the Debye-Hückel limiting law is

$$\log_{10} \gamma_i = -A z_i^2 \sqrt{\mu}, \tag{32–55}$$

where A is a constant; in water at 25°C, $A = 0.50$. Using the limiting law in Eq. (32–54), realizing that $z^{\neq} = z_A + z_B$, it becomes

$$\log_{10} k = \log_{10} k_0 - A[z_A^2 + z_B^2 - (z_A + z_B)^2]\sqrt{\mu} = \log_{10} k_0 + 2A z_A z_B \sqrt{\mu}.$$

Using $A = 0.50$, we have

$$\log_{10} k = \log_{10} k_0 + z_A z_B \sqrt{\mu}. \tag{32–56}$$

A plot of $\log_{10} k$ against the square root of the ionic strength should yield, in dilute solution, a straight line with a slope equal to $z_A z_B$.

If the ions have like signs, $z_A z_B$ is positive and the rate constant increases with increase in ionic strength. If the ions are oppositely charged, the rate constant decreases with increase in ionic strength. Equation (32–56) is a description of the primary kinetic salt effect or, more simply, the primary salt effect. Figure 32–3 shows a verification of this equation by LaMer. The agreement is eminently satisfactory. The reactions for Fig. 32–3 are

$$
\begin{aligned}
&\text{I.} \quad &&Co(NH_3)_5Br^{2+} + Hg^{2+}, \quad &&z_A z_B = 4, \\
&\text{II.} \quad &&S_2O_8^{2-} + I^-, \quad &&z_A z_B = 2, \\
&\text{III.} \quad &&NO_2NCO_2C_2H_5^- + I^-, \quad &&z_A z_B = 1, \\
&\text{IV.} \quad &&C_{12}H_{22}O_{11} + OH^-, \quad &&z_A z_B = 0, \\
&\text{V.} \quad &&H_2O_2 + H^+ + Br^-, \quad &&z_A z_B = -1, \\
&\text{VI.} \quad &&Co(NH_3)_5Br^{2+} + OH^-, \quad &&z_A z_B = -2.
\end{aligned}
$$

The physical reason behind the behavior in the cases of like and unlike charges on the ions is a result of the relative net charge on the complex. The value of the activity coefficient goes down exponentially with z^2. If both ions have the same sign,

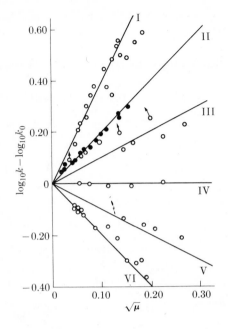

Fig 32–3 Primary salt effect. [Redrawn by permission from V. K. LeMer, *Chem. Revs.* **10**, 179 (1932).]

the complex has a net charge which is high compared with either one. This makes $\gamma^{\neq}$ very small, and the ratio $(\gamma_A\gamma_B/\gamma^{\neq})_0$ is very large. If the ions differ in sign, the net charge on the complex is less than that on either ion, $\gamma^{\neq}$ is then much larger than γ_A and γ_B, and the net result is that the ratio $(\gamma_A\gamma_B/\gamma^{\neq})_0$ and the reaction rate are small. If one species is uncharged (B for example), then A and the complex have the same charge and the ratio $\gamma_A/\gamma^{\neq}$ is unity and independent of $\sqrt{\mu}$. The value of γ_B is not much affected by changes in μ because B is a neutral molecule.

Problems

32–1. Using the expression from the collision theory, compute the rate constant for the reaction $H_2 + I_2 \rightarrow 2HI$, at 700°K. Use $\sigma_A + \sigma_B = 2 \times 10^{-8}$ cm and $E^* = 40$ kcal. Compare with the experimental value of 6.42×10^{-2} liter·mole^{-1}·sec^{-1}.

32–2. If the activation energy for the reaction $H_2 + I_2 \rightarrow 2HI$ is 40 kcal, and the ΔE for the reaction is 6.7 kcal, what is the activation energy for the decomposition of HI?

32–3. If the molecules of a gas have a diameter of 2×10^{-8} cm, calculate the number of triple collisions compared with the number of binary collisions in the gas at 300°K and 0.1, 1, 10, 100 atm pressure. What would the values be at 600°K?

32–4. Suppose that a molecule which decomposes unimolecularly has four vibrational degrees of freedom. If 30 quanta of energy are distributed among these degrees of freedom, what is the probability that 10 quanta will be found in a particular degree of freedom? What is the probability that 20 quanta will be in a particular degree of freedom?

32–5. Estimate the value of the frequency factor for the reaction between an atom and a diatomic molecule, $A + BC \rightarrow AB + C$, using the values of the partition functions given in Section 32–7.

32–6. Predict the effect of increase in ionic strength on the rate constant for each of the following reactions.

a) $Pr(NH_3)_3Cl^+ + NO_2^-$ b) $PtCl_4^{2-} + OH^-$ c) $Pt(NH_3)_2Cl_2 + OH^-$

Chapter Thirty-three

Chemical Kinetics

III. Heterogeneous Reactions, Electrolysis, Photochemistry

33–1 Heterogeneous Reactions

Very early in the development of the art of chemistry finely divided powders of various sorts were recognized as catalysts for many reactions. Only relatively recently have the details of the mechanism of reactions on surfaces been elucidated. For a long time it was thought that the function of the surface was simply to concentrate the reactants on the surface; the increased rate was attributed to the increase in "concentration." It can be shown that this certainly is not correct for the great majority of reactions. Calculation shows that for a concentration effect of this type to produce the increases in rate ordinarily observed would require surface areas per gram of catalyst which are impossible to attain.

In the majority of cases the increased rate of reaction on a surface is the result of the surface reaction having a lower activation energy than that of the homogeneous reaction. At ordinary temperatures, each kilocalorie difference between the activation energies means a factor of five in the rate. The mode of action of the surface therefore is the same as that of any catalyst (Section 31–19) in its provision of an alternative path of lower activation energy for the reaction.

33–2 Steps in the Mechanism of Surface Reactions

For a reaction to occur on a surface the following sequence of steps is required.

1. Diffusion of reactants to the surface.
2. Adsorption of the reactants on the surface.
3. Reaction on the surface.

4. Desorption of products.
5. Diffusion of products from the surface.

Any one or a combination of these may be slow and therefore be rate determining.

In gaseous reactions the diffusion steps (1) and (5) are very fast and are rarely, if ever, rate determining. For very fast reactions in solution the rate may be limited by diffusion to or from the surface of the catalyst. If diffusion is the slow step, then the concentration c' of the diffusing species at the surface will differ from the concentration c in the bulk. In Fig. 33–1, the concentration is plotted as a function of the distance from the surface. This curve is conveniently approximated by the two dashed lines. The distance δ is the thickness of the diffusion layer. This approximation was introduced by Nernst, and the layer in which the concentration differs appreciably from that in the bulk is called the Nernst diffusion layer. The concentration gradient across the diffusion layer is given by $(c - c')/\delta$, so that the rate of transport per square centimeter of the surface is $-D(c - c')/\delta$ moles·cm^{-2}·sec^{-1} if D is the diffusion coefficient. This approximation is a simple correction to the kinetic equations if diffusion is slow enough to matter. The rate of diffusion can be enhanced considerably by vigorous stirring, which thins the diffusion layer. The thickness δ in a well-stirred solution is of the order of 0.001 cm. In less well-stirred solutions the thickness is of the order of 0.005 to 0.010 cm.

It is more commonly observed that the rate of reaction is determined by step (2), or by a combination of steps (3) and (4). We consider these cases in order.

33–3 Simple Decompositions on Surfaces

In the case of the simple decomposition of a molecule on a surface, the process can be represented as a chemical reaction between the reactant A and a vacant site S on the surface. After adsorption, the molecule A may desorb unchanged or may decompose to products. The elementary steps are written

$$\text{Adsorption:} \quad A + S \xrightarrow{k_1} AS,$$

$$\text{Desorption:} \quad AS \xrightarrow{k_{-1}} A + S,$$

$$\text{Decomposition:} \quad AS \xrightarrow{k_2} \text{Products.}$$

If v is the rate of reaction per square centimeter of surface, then

$$v = k_2 c_{AS}, \tag{33–1}$$

where c_{AS} is the concentration (moles/cm^2) of A on the surface.

Let c_s be the total concentration of surface sites per square centimeter and let θ be the fraction of these sites which are covered by A. Then $c_{AS} = c_s\theta$, and $c_s(1 - \theta) = c_S$, the concentration of vacant sites on the surface. Since c_s is a constant, it can be incorporated into the rate constants. Then the rate of the reaction

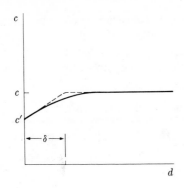

Fig. 33–1 The Nernst diffusion layer.

can be written

$$v = k_2\theta. \tag{33-2}$$

The value of θ is obtained by applying the steady-state condition to the rate of formation of AS:

$$\frac{dc_{AS}}{dt} = 0 = k_1 c_a(1 - \theta) - k_{-1}\theta - k_2\theta, \tag{33-3}$$

where c_a is the concentration of the reactant A either in the gas or in solution. From Eq. (33–3) we obtain

$$\theta = \frac{k_1 c_a}{k_1 c_a + k_{-1} + k_2}. \tag{33-4}$$

This value of θ in the rate law, Eq. (33–2), yields

$$v = \frac{k_2 k_1 c_a}{k_1 c_a + k_{-1} + k_2}. \tag{33-5}$$

If Eq. (33–5) must be considered in full, then it is convenient to invert it:

$$\frac{1}{v} = \frac{1}{k_2} + \frac{k_{-1} + k_2}{k_2 k_1 c_a}. \tag{33-6}$$

A plot of $1/v$ versus $1/c_a$ yields $1/k_2$ as the intercept and $(k_{-1} + k_2)/k_2 k_1$ as the slope. Usually it is more convenient to consider the limiting cases of Eq. (33–5).

Case I. The rate of decomposition is very large compared with the rates of adsorption and desorption. In this case, $k_2 \gg k_1 c_a + k_{-1}$, and the denominator in Eq. (33–5) is equal to k_2; then the rate is given by

$$v = k_1 c_a. \tag{33-7}$$

This is simply the rate of adsorption. Physically, the assumption that k_2 is large implies that an adsorbed molecule decomposes immediately, so that the rate of decomposition depends on how quickly the molecules can be adsorbed. From Eq. (33–4) and the assumption that $k_2 \gg k_1 c_a$, it follows that $\theta \ll 1$. The surface is sparsely covered with the reactant. The reaction is first order in the concentration of the reactant A. This situation is realized in the decomposition of N_2O on gold, and of HI on platinum.

Case II. The rate of decomposition is very small compared with the rate of adsorption and desorption. In this case, k_2 is very small so that the denominator of Eqs. (33–4) and (33–5) is $k_1 c_a + k_{-1}$. Introducing the adsorption equilibrium constant $K = k_1/k_{-1}$, Eq. (33–4) becomes

$$\theta = \frac{Kc_a}{Kc_a + 1}, \tag{33–8}$$

which is the adsorption isotherm. The occurrence of the decomposition does not affect the adsorption equilibrium at all. The rate becomes

$$v = \frac{k_2 Kc_a}{Kc_a + 1}. \tag{33–9}$$

In this case both the surface coverage θ and the rate depend on the concentration c_a. At low concentrations, $Kc_a \ll 1$, and $\theta \approx Kc_a$; the coverage is small. Then

$$v = k_2 Kc_a, \tag{33–10}$$

and the reaction is first order in the concentration of the reactant. At high concentrations, $Kc_a \gg 1$, and $\theta \approx 1$; the surface is nearly fully covered with A. Then

$$v = k_2, \tag{33–11}$$

and the reaction is zero order. Since the surface coverage ceases to vary significantly with the concentration of A at high concentrations, the reaction rate becomes independent of the concentration of A. The decomposition of HI on gold and of NH_3 on molybdenum are zero order at high pressures of HI and NH_3.

The typical variation in the rate of reaction as a function of the concentration of the reactant is shown in Fig. 33–2. This figure should be compared with Fig. 31–6,

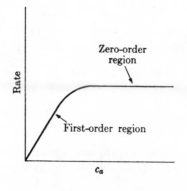

Fig. 33–2 Rate of a surface reaction as a function of reactant concentration.

which shows the same type of behavior for a homogeneous catalyst. The similarity in form between Eq. (33–5) and Eq. (31–59) for homogeneous catalysis should also be noted.

33–4 Bimolecular Reactions on Surfaces

Two molecules A and B can react on a surface if they occupy neighboring sites of the surface. Let θ_a and θ_b be the fractions of the surface sites covered by A and B, respectively, and let θ_v be the fraction of sites which are vacant; $\theta_v = 1 - \theta_a - \theta_b$. We represent the reaction by

$$AS + BS \xrightarrow{k} Products.$$

The rate v is

$$v = k\theta_a\theta_b. \tag{33–12}$$

To evaluate θ_a and θ_b we consider the two adsorption reactions

$$A + S \underset{k_{-1}}{\overset{k_1}{\rightleftarrows}} AS \quad \text{and} \quad B + S \underset{k_{-2}}{\overset{k_2}{\rightleftarrows}} BS.$$

The steady-state equations are

$$\frac{dc_{AS}}{dt} = 0 = k_1 c_a \theta_v - k_{-1}\theta_a - k\theta_a\theta_b,$$

$$\frac{dc_{BS}}{dt} = 0 = k_2 c_b \theta_v - k_{-2}\theta_b - k\theta_a\theta_b.$$

Since $\theta_v = 1 - \theta_a - \theta_b$, these two equations can be solved for θ_a and θ_b. We shall consider only the case for which k is very small; setting $k = 0$, these equations reduce to

$$\theta_a = K_1 c_a \theta_v \qquad \theta_b = K_2 c_b \theta_v, \tag{33–13}$$

where $K_1 = k_1/k_{-1}$ and $K_2 = k_2/k_{-2}$. Using these values of θ_a and θ_b in $\theta_v = 1 - \theta_a - \theta_b$, we obtain

$$\theta_v = 1 - K_1 c_a \theta_v - K_2 c_b \theta_v,$$

so that

$$\theta_v = \frac{1}{1 + K_1 c_a + K_2 c_b}. \tag{33–14}$$

This value of θ_v brings Eqs. (33–13) to the form

$$\theta_a = \frac{K_1 c_a}{1 + K_1 c_a + K_2 c_b}, \qquad \theta_b = \frac{K_2 c_b}{1 + K_1 c_a + K_2 c_b}. \tag{33–15}$$

These values used in Eq. (33–12) yield the rate law

$$v = \frac{kK_1 K_2 c_a c_b}{(1 + K_1 c_a + K_2 c_b)^2},$$ (33–16)

which has some unusual characteristics. We examine each case separately.

Case I. Both A and B are weakly adsorbed; the surface is sparsely covered. In this case, $K_1 c_a \ll 1$ and $K_2 c_b \ll 1$. The denominator of Eq. (33–16) is about equal to unity and the rate law is

$$v = kK_1 K_2 c_a c_b.$$ (33–17)

The reaction is second order overall; and is first order with respect to both A and B.

Case II. One reactant, A, more strongly adsorbed than the other. In this case, $K_1 c_a \gg K_2 c_b$; the denominator is about equal to $1 + K_1 c_a$, and Eq. (33–16) takes the form

$$v = \frac{kK_1 K_2 c_a c_b}{(1 + K_1 c_a)^2}.$$ (33–18)

The rate is first order with respect to the less strongly adsorbed reactant; the dependence of the rate on the concentration of the more strongly adsorbed reactant is more complicated.

Case III. One reactant very strongly absorbed. If A is very strongly adsorbed, we have the same situation as in Case II but with the additional condition that $K_1 c_a \gg 1$, so that the denominator of Eq. (33–18) is $(K_1 c_a)^2$. Then the rate is

$$v = \frac{kK_2 c_b}{K_1 c_a}.$$ (33–19)

The rate of the reaction is inversely proportional to the concentration of the strongly adsorbed species. This is an example of inhibition, or poisoning. In this case one of the reactants itself inhibits the reaction. The reaction between ethylene and hydrogen on copper is of this type. At low temperatures the rate is given by

$$v = k\frac{p_{H_2}}{p_{C_2H_4}},$$

the ethylene being strongly adsorbed. At higher temperatures the ethylene is less strongly adsorbed, the surface is sparsely covered, and the rate expression reduces to that given by Eq. (33–17):

$$v = k' p_{H_2} p_{C_2H_4}.$$

It is generally true that if one substance is strongly adsorbed on the surface, whether it be reactant, product, or a foreign material, the rate is inversely proportional to the concentration of the strongly adsorbed substance which acts as a poison for the reaction.

33-5   The Role of the Surface in Catalysis

In homogeneous catalysis the catalyst combines chemically with one of the reactants to form a compound which reacts readily to form products. The same is true of a surface acting as a catalyst. One or more of the reactants are chemisorbed on the surface; this is equivalent to the formation of the chemical intermediate in the homogeneous case. In both cases, the effect of the catalyst is to provide an alternative path of lower activation energy. This lower energy is the principal reason for the increased rate of reaction. Figure 33–3 shows schematically the energy variation as the reactants pass to products. It is apparent from the figure that if the activation energy for the forward reaction is lowered, then that for the reverse reaction is lowered by the same amount. The catalyst therefore increases the rate of the forward *and the reverse reaction* by the same factor.

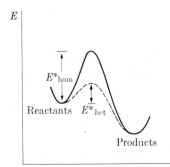

Fig. 33–3 Energy surfaces for homogeneous and heterogeneous reactions.

Table 33–1 lists a few values of the activation energies for various reactions on surfaces, E_{het}, and the corresponding values for the homogeneous, uncatalysed reaction, E_{hom}.

Table 33–1†   Activation energies for heterogeneous and homogeneous reactions

Decomposition of	Surface	E_{het}, kcal	E_{hom}, kcal
HI	Au	25.0	44.0
	Pt	14.0	
N_2O	Au	29.0	58.5
	Pt	32.5	
NH_3	W	39.0	80
	Os	47.0	
	Mo	32–42	
CH_4	Pt	55–60	80

†   By permission from K. J. Laidler, *Chemical Kinetics*. McGraw-Hill Book Co., Inc., New York, 1950.

An important fact about surface reactions is that the surface sites on a catalyst differ in their ability to adsorb the reactant molecules. This is demonstrated by the action of catalytic poisons. In the preceding section, the effect of strong adsorption of one reactant was to inhibit reaction or poison the catalyst. Foreign molecules which do not take part in the reaction can also poison the surface if they are strongly adsorbed. The algebraic effect on the rate equation is to make the rate inversely proportional to some power, usually the first power, of the concentration of the poison.

It has been shown that the amount of poison required to stop the reaction is ordinarily significantly smaller than the amount needed to form a monolayer of poison on the surface. This observation led H. S. Taylor to postulate that the adsorption and subsequent reaction takes place preferentially on certain parts of the surface, which he called "active centers." The active centers may constitute only a small fraction of the total number of surface sites. If these active centers are covered by molecules of the poison, the reaction is unable to proceed except at an extremely slow rate.

Imagine the appearance of a surface on the atomic scale. There are cracks, hills and valleys, boundaries between individual grains, different crystal faces exposed, edges, points, and so on. It is not surprising that adsorption takes place more easily in some places than in others. The chemical kinetic consequences of this lack of uniformity in the surface have been explored extensively in the last few years, both from the theoretical and the experimental standpoints.

The chemical nature of the surface determines its ability to act as a catalyst for a particular type of reaction. For illustration, two reactions of an alcohol can be considered. On metals of the platinum group such as Ni, Pd, Pt, the alcohol is dehydrogenated. Consider CH_3CH_2OH as an example:

$$CH_3CH_2OH \rightarrow CH_3CHO + H_2.$$

On a surface such as alumina, dehydration occurs:

$$CH_3CH_2OH \rightarrow CH_2CH_2 + H_2O.$$

In the two cases the mode of attachment is different.

Nickel has a strong affinity for hydrogen so that on nickel the attachment is presumably to the hydrogen atoms:

On alumina, there are hydroxyl groups at the surface as well as oxide groups. The surface could be imagined as having the configuration

$$
\begin{array}{c}
\quad\quad H \\
\quad\quad \backslash \\
O \quad\quad O \\
| \quad\quad\quad | \\
Al \quad\quad Al \\
\backslash \quad\quad / \\
O
\end{array}
$$

Then the attachment of the alcohol could be

$$
\begin{array}{cc}
\begin{array}{c}
H \quad H \\
| \quad\; | \\
H{-}C{-}C{-}H \\
| \quad\; | \\
H \quad O{-}H \\
\vdots \quad\; \vdots \\
\quad\;\; H \\
\quad\;\; \backslash \\
O \quad\quad O \\
| \quad\quad\quad | \\
Al \quad\quad Al \\
\backslash \quad\quad / \\
O
\end{array}
&
\begin{array}{c}
H \quad H \\
| \quad\; | \\
H{-}C{=}C{-}H \\
\quad O{-}H \\
H \quad\; H \\
\backslash \quad\quad / \\
O \quad\quad O \\
| \quad\quad\quad | \\
Al \quad\quad Al \\
\backslash \quad\quad / \\
O
\end{array}
\end{array}
$$

$\rightarrow$ $\rightarrow H_2O$

After desorption of the water molecule the surface is left unchanged.†

33–6 Electrolysis and Polarization

Electrolysis refers to the chemical reaction or reactions which accompany the passage of a current through an electrolytic solution. An electrochemical cell through which a current is passing is said to be *polarized*. Polarization refers to any or all of the phenomena associated with the passage of a current through a cell.

By Faraday's law, the number of equivalents N of chemical reaction which result from the passage of a current I is given by

$$N = It/\mathscr{F}, \tag{33–20}$$

where $\mathscr{F} = 96{,}500$ coulombs/equivalent, and t is the time. The rate of the reaction‡ is $dN/dt = I/\mathscr{F}$, so that

$$I = \mathscr{F}\, dN/dt. \tag{33–21}$$

According to Eq. (33–21) the current is directly proportional to the rate of reaction in equivalents per second, so that the rate of an electrolytic reaction is ordinarily expressed in amperes or in amperes per square centimeter of electrode area.

---

†   These diagrams are intended to represent plausible suppositions about the surface structure and the mode of attachment of the molecule, nothing more.

‡   Note that the rate of an electrolytic reaction is expressed as the number of equivalents of reaction which take place in unit time, not as a change in concentration with time.

The study of electrode reactions is unique in the sense that within limits the rate of the reaction can be controlled by simply increasing or decreasing the current through the cell. The electrolysis reaction also differs from other chemical reactions in that "half" of it occurs at one electrode and the other "half" occurs at the second electrode, which may be spatially distant from the first. The electrolysis of water,

$$H_2O \rightarrow H_2 + \tfrac{1}{2}O_2,$$

can be broken down into two "half" reactions:

$$\text{At the cathode:} \quad 2H^+ + 2e^- \rightarrow H_2,$$

$$\text{At the anode:} \quad H_2O \rightarrow \tfrac{1}{2}O_2 + 2H^+ + 2e^-.$$

Each of these reactions is proceeding at the same rate I, the current being passed. If the area of the cathode is A_c and that of the anode is A_a, then the rate of the cathodic reaction per square centimeter of cathode is $i_c = I/A_c$, and that of the anodic reaction per square centimeter of anode is $i_a = I/A_a$. The i_c and i_a are anodic and cathodic *current densities*, and these are the quantities of significance, since they are independent of the size of electrode used. The current density at either electrode depends on the concentrations of reactants and products near the electrode, just as any reaction rate depends on concentrations. In addition, the current density depends on the potential of the electrode. The phenomena associated with electrolysis are properly linked with the kinetics of reactions on surfaces. Because of great experimental difficulties, particularly the problem of controlling impurities in liquid solutions, the study of electrode kinetics has become reasonably scientific only relatively recently. Some earlier work is excellent, but much of the earlier work is erratic.

33-7 Polarization at an Electrode

Rather than describe the electrolysis of any solution with any two electrodes, we begin by considering a single reversible electrode at equilibrium and then ask what happens if we pass a current into the electrode.

Consider a hydrogen electrode in equilibrium with H^+ ion at a concentration c and hydrogen gas at a pressure p. The equilibrium potential of this electrode is denoted by $\mathscr{V}_0$. The equilibrium is $2H^+ + 2e^- \rightleftharpoons H_2$. If the potential of the electrode is increased, made more positive, this equilibrium will be disturbed. The reaction from right to left will predominate, H_2 will be oxidized, and a positive current will flow into the solution. If the potential of the electrode is lowered (made more negative), the equilibrium will be disturbed. The reaction from left to right will predominate, H_2 will be liberated, and a positive current will flow into the electrode or a negative current will flow into the solution. The current which flows to the electrode, therefore, depends on the departure of the potential from the equilibrium value $\mathscr{V} - \mathscr{V}_0$. This difference between the applied potential $\mathscr{V}$ and the equilibrium potential $\mathscr{V}_0$ is the overpotential, or overvoltage, of the electrode η:

$$\eta = \mathscr{V} - \mathscr{V}_0. \tag{33-22}$$

If the overpotential is small, then the current density at the electrode is proportional to η:

$$i = \frac{i_0 \mathscr{F} \eta}{RT},\qquad(33\text{–}23)$$

which shows that the sign of i (the direction of the current flow) depends on the sign of η. The constant i_0 is the analogue of the rate constant for an ordinary reaction; the i_0 is the *exchange current* of the electrode reaction. The value of i_0 depends on the concentration of H^+ ion, on the pressure of the hydrogen gas, and especially on the electrode surface. For the hydrogen evolution reaction on platinum, for example, $i_0 \approx 10^{-2}\,A/cm^2$, while on mercury, $i_0 \approx 10^{-14}\,A/cm^2$. It is, in fact, these values of i_0 which allow us to use platinum as the electron collector for a reversible hydrogen electrode and forbid us to use mercury for this purpose.

In the kinetic sense there is a gradation between what are called reversible electrodes and irreversible electrodes. The reversible potential of an electrode is measured by balancing a cell in a potentiometer circuit. This involves detecting the point at which the current flow to the electrode is zero. Suppose the galvanometer registers "zero current" for any value of current between i' and $-i'$. Among other factors the magnitude of i' depends on the design of the galvanometer. The characteristics of the "very reversible" electrode and the "very irreversible" electrode are shown in Fig. 33–4. The balance will be observed for the reversible electrode anywhere in the region between $\mathscr{V}_2$ and $\mathscr{V}_1$. This uncertainty in the measurement of $\mathscr{V}_0$ is very small.

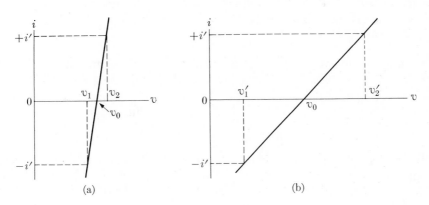

Fig. 33–4 Current–potential relation at (a) reversible and (b) irreversible electrodes.

For the irreversible electrode, Fig. 33–4(b), the null point is registered anywhere in the wide range of potential between $\mathscr{V}'_1$ and $\mathscr{V}'_2$. The slope of the curve is very small; that is, i_0 is very small. An electrode with a large i_0 is therefore "more reversible" than one with a small i_0.

33–8 Measurement of Overvoltage

Before considering the theoretical ideas which relate the current to the overvoltage, it is well to understand the principle of the measurement of overvoltage. A cell is shown schematically in Fig. 33–5. A measured current is passed between the two electrodes A and B. The reference electrode R is the same kind of electrode as B. Matters are arranged so that the same electrode equilibrium is established at both B and R. When $i = 0$, B and R both have the same potential. When the current passes into B, this electrode has a potential measured on the potentiometer P which is different than that of R, which carries no current. This difference in potential is the measured overvoltage, $\eta_m = \mathscr{V}_B - \mathscr{V}_R$. The value of η_m is measured for various values of the current density.

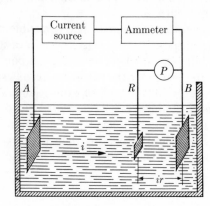

Fig. 33–5 Cell for the measurement of overvoltage.

As the experiment stands, the measured value contains an ohmic component from the ir drop between R and B, a concentration component resulting from concentration changes in the vicinity of the electrode, and a component, denoted by η, which is related to the rate constant of the reaction. Thus,

$$\eta_m = \eta_{\text{ohmic}} + \eta_{\text{conc}} + \eta.$$

There are methods for measuring η_{ohmic} separately; η_{conc} can usually be reduced to a negligible value by vigorous stirring. Thus, from η_m the value of η can be found as a function of the current density. This η, which is related to the rate constant of the reaction, is often called the *activation overvoltage*.

33–9 The Current–Potential Relation

One of the earliest relations between the current and potential was proposed by Tafel in 1913. At the hydrogen electrode, the equilibrium

$$2\text{H}^+ + 2\text{e}^- \rightleftharpoons \text{H}_2$$

is established. If the potential of the electrode is made slightly more negative than the equilibrium value, H_2 will be evolved at a finite rate. Suppose that the elementary reactions at the electrode are

$$(1) \qquad H^+ + e^- \rightleftharpoons H \qquad \text{(fast)},$$

$$(2) \qquad H + H \xrightarrow{k} H_2 \qquad \text{(slow)}.$$

If the second reaction is very slow, then the equilibrium in reaction (1) is undisturbed. The Nernst equation for reaction (1) is

$$\mathscr{V} = \mathscr{V}^0 - \frac{RT}{\mathscr{F}} \ln \frac{c_H}{c_{H^+}}, \tag{33–24}$$

where c_H is the concentration of H atoms per square centimeter of electrode surface, c_{H^+} is the H^+ ion concentration in the solution. The rate of reaction (2) per square centimeter of the electrode surface is the current density i and this is proportional to c_H^2:

$$i = k c_H^2,$$

where k is the rate constant of reaction (2). Then, $c_H = \sqrt{i/k}$. Using this value of c_H in Eq. (33–24), we obtain

$$\mathscr{V} = \mathscr{V}^0 - \frac{2.303RT}{\mathscr{F}} \log_{10} i^{1/2} + \frac{RT}{\mathscr{F}} \ln k^{1/2} + \frac{RT}{\mathscr{F}} \ln c_{H^+}.$$

If the equilibrium value of the potential is $\mathscr{V}_0$, then $\eta = \mathscr{V} - \mathscr{V}_0$, so that

$$\eta = \mathscr{V}^0 - \mathscr{V}_0 + \frac{RT}{\mathscr{F}} \ln k^{1/2} + \frac{RT}{\mathscr{F}} \ln c_{H^+} - \frac{2.303RT}{2\mathscr{F}} \log_{10} i. \tag{33–25}$$

We symbolize the first four terms in Eq. (33–25) by $-a$, and define

$$b = \frac{2.303RT}{2\mathscr{F}}. \tag{33–26}$$

Then Eq. (33–25) can be written

$$-\eta = a + b \log_{10} i, \tag{33–27}$$

which is the Tafel equation; it says that $-\eta$ should be a linear function of $\log_{10} i$. The slope of the plot of $-\eta$ versus $\log_{10} i$ is the Tafel b. At 25°C, $b = 0.05916/2 = 0.0296 \approx 0.030$. Platinum exhibits this value of b over a short range of values of $\log_{10} i$.

If $\eta = 0$, the value of i is the rate at which reaction (2) proceeds at equilibrium;[†] this is called the exchange rate, or in this case the exchange current i_0. Thus, setting

---

† At equilibrium the reverse of reaction (2) goes on at an equal rate, of course.

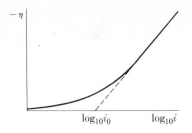

Fig. 33–6 Tafel plot for cathodic overvoltage.

$\eta = 0$ in Eq. (33–27), we obtain $a = -b \log_{10} i_0$, so that

$$-\eta = -b \log_{10} i_0 + b \log_{10} i. \tag{33–28}$$

Figure 33–6 shows a Tafel plot of $-\eta$ versus $\log_{10} i$. The departure from linearity at low values of $-\eta$ is a consequence of ignoring the rate of the reverse of reaction (2) in the derivation.

Although the value of the Tafel b found experimentally is not always the 0.030 predicted by this simple derivation, the Tafel plot is commonly used for the display and discussion of the overvoltage–current density relation. The value of b depends on the slow step in the mechanism of the reaction. If a reaction other than (2) is slow, then the value of b will not be 0.030. The Tafel slope is a criterion of the mechanism of the reaction; however, by itself it is not conclusive.

33–10 General Consequences of the Current–Potential Relation

Rather than discuss the mechanisms of electrode reactions in any detail, some general implications of the current–potential relation will be described.

We ask if it is possible to plate zinc onto a platinum electrode. Consider a solution containing HCl and $ZnCl_2$ in which $a_{H^+} = 1$ and $a_{Zn^{2+}} = 1$. Suppose we electrolyse this solution using a platinum cathode. As soon as hydrogen starts to evolve at the cathode, it behaves as a hydrogen electrode. Since the pressure of H_2 in the vicinity is very close to 1 atm and $a_{H^+} = 1$, the potential of the electrode on the conventional scale is zero if no current flows, and slightly less than zero if current is flowing. The dependence of the potential of the electrode as a function of current density is shown in Fig. 33–7(a). Since the hydrogen overvoltage on platinum is very small, the potential decreases very slowly as the current density is increased. If zinc is to deposit on the electrode, the potential must be more negative than the value of the reversible potential of the Zn^{2+}-Zn couple which is -0.763 V in this situation. It is clear from the figure that a very high current density will be required to bring the potential to a value below -0.763 V and thus permit the deposition of zinc. The current density required is so large that as a practical matter zinc cannot be plated onto a platinum surface.

The electrolysis of this solution proceeds quite differently if a lead cathode is used. The hydrogen overpotential on lead is much larger than that on platinum at

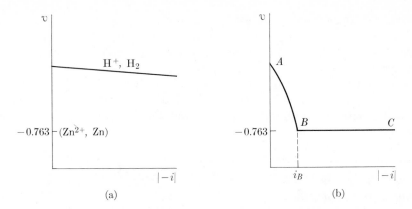

Fig. 33–7 Current–potential curves for the deposition of zinc on (a) a platinum electrode and (b) a lead electrode.

every current density. The current–potential relation is shown schematically for the system with the lead electrode in Fig. 33–7(b). Only a very small current density (less than $1\,\mathrm{mA/cm^2}$) is required to bring the potential down far enough so that zinc will deposit. After this value is reached, the potential does not drop very much with increase in current density because the overvoltage for deposition of zinc on a lead (or zinc) surface is very small. (After the lead is coated with zinc, of course, the electrode is a zinc electrode. It happens that the hydrogen overvoltage on zinc is also quite high, so the shape of the curve is essentially the same as on lead.) Figure 33–7(b) can be interpreted as follows. In the region from A to B, all of the current goes into hydrogen evolution. At B, zinc deposition commences. At any point beyond B, both hydrogen evolution and metal deposition occur. The rate of hydrogen evolution is i_B, and this rate remains nearly constant in the region from B to C, since the potential is effectively constant in this range. The rate of metal deposition, i_M, is therefore

$$i_M = i - i_B \tag{33–29}$$

and increases as i increases. The fraction of the current used in metal deposition is $i_M/i = 1 - i_B/i$. The ratio i_M/i is the current efficiency for metal deposition. Since i_B is very small, the current efficiency is nearly unity at high values of i.

A very active metal such as sodium cannot be plated from aqueous solutions except under special circumstances. The reversible potential for the reduction of Na^+ is $-2.714\,\mathrm{V}$. Even with a lead cathode an enormous current density would be required to bring the cathode below this potential, and then the current efficiency for sodium deposition would be exceedingly small. Sodium can be plated into mercury, which has a high hydrogen overpotential, if a highly alkaline solution is used. Three factors are influential in the process.
1. The alkaline solution, which brings the potential at which hydrogen is deposited closer to the potential for sodium deposition.

2. The high hydrogen overvoltage on mercury.
3. The fact that metallic sodium will dissolve in mercury; this brings the sodium deposition potential nearer the hydrogen value and also keeps the sodium which has been deposited from reacting with water.

High current densities are required and the current efficiency is very low.

It is worthwhile mentioning that charging the lead storage battery would not be possible if it were not for a high hydrogen overvoltage on the negative plate which permits the reaction

$$PbSO_4 + 2e^- \rightarrow Pb + SO_4^{2-}$$

to occur with high efficiency. A high oxygen overvoltage on the positive plate, PbO_2, is required so that the reaction

$$PbSO_4 + 2H_2O \rightarrow PbO_2 + SO_4^{2-} + 4H^+ + 2e^-$$

can occur with high efficiency. If these overvoltages were not large, it would not be possible to polarize the plates to the potentials required for the charging reactions to occur, and passage of the charging current would only decompose the water into hydrogen and oxygen.

Corrosion reactions are another group of reactions which depend critically on the presence or absence of a significant hydrogen overvoltage on the surface.

33–11 Photochemistry

In its broadest sense, photochemistry embraces any chemical effect associated with the emission or absorption of radiation of any sort. This definition includes such diverse phenomena as flames, fluorescence and phosphorescence, photographic reactions, luminescent chemical reactions of all sorts, catalysis of chemical reactions by light, and chemical effects produced by the passage of high-energy nuclear particles through a system.

A division of this broad field can be made immediately on the basis of the energy of the light involved. If a molecule absorbs a photon of frequency v, the energy absorbed is $\epsilon = hv$. The energy of an Avogadro number of quanta, N_0hv, is called an *einstein*. For light in the visible region of the spectrum, let us say a wavelength of $6000 \text{ Å} = 6 \times 10^{-5} \text{ cm}$, the frequency is $v = c/\lambda = 3 \times 10^{10}/(6 \times 10^{-5}) = 5 \times 10^{14} \text{ sec}^{-1}$. The energy of the quantum is $\epsilon = hv = 6.6 \times 10^{-27} (5 \times 10^{14}) = 3.3 \times 10^{-12}$ ergs ≈ 2 eV. The far ultraviolet reaches to about 2000 Å and the energy of such light would be three times as large, about 6 eV. The energy involved in chemical reactions is of this same order of magnitude; therefore, it is not surprising that light in the visible and ultraviolet regions exerts an effect on the course of chemical reactions. The influence of visible and ultraviolet light on chemical reactions is the subject of photochemistry.

Nuclear particles possess energies of the order of a million electron volts, 1 MeV. The effects produced by the passage of such energetic particles through a chemical system are usually, though not always, more drastic than those produced by a photon of ordinary wavelength. These phenomena are incorporated under the title of radiation chemistry. Since many of the effects produced by nuclear radiation are observed only because of the enormous energy involved, they do not have any implication for the rest of chemistry. For this reason our attention in the following paragraphs will be restricted exclusively to photochemistry.

33–12 The Absorption of Light

The influence of light on chemical systems may be trivial or profound. If the light quanta are not energetic enough to produce a profound effect such as the dissociation of a molecule, the energy may simply be degraded into thermal energy. This latter effect may be regarded as trivial in a photochemical sense, since the same result could be achieved by raising the temperature by any means whatsoever.

Any effect of light, whether trivial or profound, can be produced only by light which is absorbed by the system in question. This fact, which today seems obvious, was first recognized at the beginning of the 19th century by Grotthuss and Draper, and is called the law of Grotthuss and Draper. We begin by examining the laws of absorption.

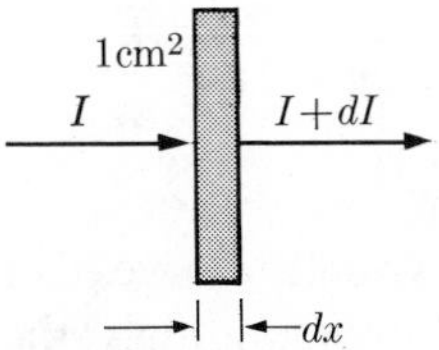

Figure 33–8

Consider a beam of monochromatic light passing through an absorber of thickness dx. Let I be the intensity of the incident beam and $I + dI$ be the intensity of the emergent beam; Fig. 33–8. The intensity of the beam may be defined as the number of quanta of light passing unit area of a plane perpendicular to the direction of the beam in unit time. Let this number be N. Let dN be the number of quanta absorbed in the thickness dx. Then the probability of absorption in the thickness dx is dN/N; and this is proportional to the number of absorbing molecules in the layer, which is $c\,dv$, where c is the concentration in moles per liter, and dv is the volume of the layer. Since $dv = 1 \cdot dx$, we have

$$dN/N = \alpha c\,dx,$$

where α is a constant. Since by definition, $I = N$ and $I + dI = N - dN$, we have $dN = -dI$, and, therefore,

$$-dI/I = \alpha c\,dx. \tag{33–30}$$

This equation says that the *relative* decrease in intensity of the beam is proportional to the number of absorbing molecules in the slab of material. If there are several kinds of molecules present, each with a different ability to absorb light of the frequency in question, then

$$-dI/I = (\alpha_1 c_1 + \alpha_2 c_2 + \cdots) dx. \tag{33–31}$$

The constants $\alpha_1, \alpha_2, \ldots$ are the molar absorption coefficients of the substances in question. For a given wavelength of light each substance has a characteristic value of α. For any particular substance, α is a function of the wavelength. If a substance is transparent at a particular wavelength, then all the light goes through, and $\alpha = 0$. If at a particular wavelength all of the substances are transparent except one, then Eq. (33–31) reduces to Eq. (33–30). Integration of Eq. (33–30) yields

$$\ln I = \alpha c x + C,$$

where I is the intensity of the emergent beam after passage through a layer of thickness x. When $x = 0$, $I = I_0$, the intensity of the incident light. Therefore, $\ln I_0 = C$, and $\ln (I/I_0) = -\alpha c x$. Then

$$I = I_0 e^{-\alpha c x}. \tag{33–32}$$

The absorbed intensity is defined by

$$I_{abs} = I_0 - I. \tag{33–33}$$

Using Eq. (33–33) in Eq. (33–32), we obtain

$$I_{abs} = I_0(1 - e^{-\alpha c x}). \tag{33–34}$$

If for the wavelength in question, α is extremely large, then $e^{-\alpha c x} \approx 0$ and $I_{abs} = I_0$. All the incident light I_0 is absorbed. Correspondingly, the intensity of the emergent beam is zero. This accounts for the familiar observation that a solution, such as copper sulfate, which absorbs in the red end of the spectrum and is transparent in the blue end has a blue color.

Equation (33–32) can be written in either of the forms,

$$I = I_0 10^{-0.4343\alpha c x}, \qquad I = I_0 10^{-\epsilon c x}. \tag{33–35}$$

The constant ϵ is the *molar extinction coefficient*. Either of Eqs. (33–32) or (33–35) is Beer's law. If we are dealing with a pure liquid, then the concentration c is a constant and Eq. (33–35) becomes

$$I = I_0 10^{-kx}, \tag{33–36}$$

where k is the extinction coefficient of the liquid. Equation (33–36) is Lambert's law.

Beer's law is the basic equation for the various colorimetric methods of analysis. By measuring the light transmitted by a solution of absorbing molecules, the concentration of those molecules can be determined from Eq. (33–35); the extinction coefficient is measured by measuring the transmission of solutions of known concentration in the same cell and plotting $\log_{10}(I/I_0)$ against c. The slope of the curve is the value of the constant ϵx. In concentrated solutions, ϵ may depend slightly on concentration; this complication is easily dealt with, since the calibration plot shows it up immediately.

33–13 The Stark-Einstein Law of Photochemical Equivalence

The Stark-Einstein law of photochemical equivalence is in a sense simply a quantum mechanical statement of the Grotthuss-Draper law. The Stark-Einstein law states that each molecule which takes part in the photochemical reaction absorbs one quantum of the light which induces the reaction.†

If we define the primary act of the photochemical reaction as the absorption of the quantum, then the quantum efficiency for the primary act is, by the Stark-Einstein law, equal to unity. For each quantum absorbed, one primary act occurs. For any substance X taking part in a photochemical reaction, the quantum efficiency or quantum yield for the formation (or decomposition) of X is φ_X and is defined by

$$\varphi_X = \frac{\text{no. of molecules of X formed (or decomposed)}}{\text{no. of quanta absorbed}}. \tag{33–37}$$

More conveniently, if we measure the rate of formation of X in molecules per second, dn_X/dt, then the quantum yield is

$$\varphi_X = \frac{dn_X/dt}{\text{no. of quanta absorbed/sec}}.$$

The number of quanta absorbed per second is the absorbed intensity, so that

$$\varphi_X = \frac{dn_X/dt}{I_{\text{abs}}}. \tag{33–38}$$

It is clear that to determine the quantum yield of the reaction it is necessary to measure the rate of reaction and the amount of radiation absorbed. The rate of the reaction is measured in any convenient way. Figure 33–9 shows a typical arrangement for measuring the absorbed intensity. The reacting system is confined to a cell. The intensity of the transmitted beam is measured with the reaction cell empty and

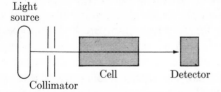

Light source

Collimator   Cell   Detector

Fig. 33–9 Schematic diagram of apparatus for measurement of light intensity.

---

† The Stark-Einstein law (1905) is another example of the break with classical physics. One molecule absorbs the entire quantum; the energy of the light beam is not spread continuously over a number of molecules.

with the cell filled with the reaction mixture. The detector is most frequently a thermopile, which is a junction of dissimilar metals covered with a blackened metal foil. All the radiation is absorbed on the blackened metal and the energy of the radiation is converted to a temperature increase; the temperature increase is converted to a difference in emf of the thermocouple. The device must be calibrated against a standard light source. It has the advantage of being usable for light of any frequency. Photoelectric cells are convenient detectors, but the response varies with frequency and they must be calibrated for each frequency.

Chemical actinometers are often used for the determination of the total number of quanta absorbed in a measured period of time; from this measurement the intensity can be computed. The chemical actinometer utilizes a chemical reaction which has been accurately investigated in its photochemical behavior. The decomposition of oxalic acid in the presence of uranyl sulfate is a common actinometric reaction. The light may have any wavelength between 2500 and 4500 Å. The absorption of the light quantum activates the uranyl ion, which transfers its energy to a molecule of oxalic acid, thereupon the oxalic acid decomposes. The reaction may be written

$$UO_2^{2+} + h\nu \rightarrow (UO_2^{2+})^*$$

$$(UO_2^{2+})^* + H_2C_2O_4 \rightarrow UO_2^{2+} + H_2O + CO_2 + CO.$$

The quantum yield depends on the frequency of the light, varying between about 0.5 and 0.6. After a measured period of time, the light is turned off and the residual oxalic acid determined by titration with $KMnO_4$. From these data and knowing the original amount of oxalic acid, the intensity is calculated using the value of the quantum yield appropriate to the frequency of light used.

33–14 Photochemical Processes

The spectrum of an atom consists of a series of lines which converge at a definite short wavelength, beyond which the spectrum is continuous. The lines are produced by transition of the atom from one quantum state to another. The continuum is a consequence of the transition from a quantum state of the atom to a state in which an electron and an ion are present. Since the translational energies of the separated electron and ion are not quantized, any energy above a certain minimum value can be absorbed by the system; this accounts for the continuum. Similarly, the spectrum of a molecule consists of a series of bands, and beyond a certain short wavelength, a continuum. The bands result from transitions between states of the bound molecule; the continuum is a consequence of absorbing sufficient energy to dissociate the molecule into two or more fragments. With this reminder of the spectral structure, we can consider the fate of the energy of a quantum of absorbed light.

Suppose that an atom in its ground state is irradiated with light having a frequency corresponding to an absorption frequency of the atom. The atom may absorb a quantum and be raised to an excited state. The process is represented by

$$A + h\nu \rightarrow A^*.$$

The excited atom A* may suffer one of several fates.

    1. Emission of resonance radiation. Atom A* may simply emit the quantum which it absorbed and return to the ground state

$$A^* \rightarrow A + h\nu.$$

In this event, the light emitted is called *resonance radiation,* since it has the same frequency as the incident light.

    2. Fluorescence. The excited atom may make several transitions to lower energy states, emitting light of different frequency for each transition. For example,

$$A^* \rightarrow A^{*\prime} + h\nu',$$

$$A^{*\prime} \rightarrow A + h\nu''.$$

Here $A^{*\prime}$ represents a quantum state having an energy between A and A*. The emission of the frequencies ν' and ν'' which differ from ν is called *fluorescence.* These frequencies may or may not lie in the visible spectrum.

    3. Collision. The atom A* may collide with another atom of A, or with a foreign atom or molecule. The excess electronic energy of A* may be converted by this collision into translational energy of the two atoms, in which case the excitation is *quenched*:

$$A^* + A \rightarrow A + A.$$

It is possible that fluorescence may be induced by

$$A^* + A \rightarrow A^{*\prime} + A$$

followed by

$$A^{*\prime} \rightarrow A + h\nu''.$$

If a foreign atom collides, it may simply quench the excitation or a reaction of the type

$$A^* + B \rightarrow A + B^*$$

may occur. Thereafter, B* may be quenched, or it may emit its characteristic radiation

$$B^* \rightarrow B + h\nu_B.$$

    The collision of a molecule with an excited species may also have a number of different effects. The molecule has additional degrees of freedom, rotations, and vibrations, which can accept the excess energy. The result may be

$$A^* + M \rightarrow A + M,$$

where the excess energy is simply converted to translational, vibrational, and rotational energy with no unusual effect.

The molecule may gain sufficient excess energy to be considered as an *excited* or "hot" molecule:

$$A^* + M \rightarrow A + M^*.$$

Then, M* may radiate,

$$M^* \rightarrow M + h\nu_M,$$

or may be quenched further

$$M^* + X \rightarrow M + X.$$

The M* may have an important effect on the course of a chemical reaction if the excess energy makes it more reactive than its fellows.

Collision of A* and M may result in the dissociation of M into two or more fragments, some of which may themselves be in excited states. This is represented by the process

$$A^* + M \rightarrow A + \text{Fragments}.$$

If any one of the fragments is an atom or a radical, it may initiate chains and produce a great deal of secondary chemical reaction as a result.

If the initial radiation is absorbed by a molecule, three general possibilities exist.

1. Simple excitation: $M + h\nu \rightarrow M^*.$

2. Dissociation: $M + h\nu \rightarrow \text{Fragments}.$

3. Predissociation: $M + h\nu \rightarrow M^* \rightarrow \text{Fragments}.$

In the third process, there is a time delay after the formation of M* before dissociation occurs. In the first process the excited molecule may lose the excess energy by resonance radiation, fluorescence, or by any of the collision processes described above.

Not all of the processes mentioned above are possible for every atom or molecule in any excited state. The possible transformations are limited by the energy balance and by selection rules, nor does the above list exhaust the possibilities by any means. We will now consider the chemical consequences of the various processes.

33–15 Absorption with Dissociation

A very common class of photolytic reactions consists of those in which the primary photochemical act is adsorption of the quantum by a molecule followed by dissociation of the molecule:

$$M + h\nu \rightarrow \text{Fragments}.$$

Since the fragments produced are often atoms or free radicals, this primary step frequently initiates a chain mechanism, whose occurrence may be shown up by quite large values of the quantum yield, although a chain reaction can occur with very small values of the quantum yield.

If the light quantum is to produce dissociation of a molecule, the frequency must lie in the spectral region in which continuous absorption occurs. The energy of the light quantum must be equal to or greater than the dissociation energy of the molecule. Examples of the types of dissociation which occur upon absorption are:

$$Cl_2 + hv \rightarrow Cl + Cl, \qquad \lambda < 4785 \text{ Å},$$

$$Br_2 + hv \rightarrow Br + Br^*, \qquad \lambda < 5107 \text{ Å},$$

$$I_2 + hv \rightarrow I + I^*, \qquad \lambda < 4991 \text{ Å},$$

$$NH_3 + hv \rightarrow NH_2 + H, \qquad \lambda < 2100 \text{ Å},$$

$$NO_2 + hv \rightarrow NO + O, \qquad \lambda < 3650 \text{ Å},$$

$$R-\overset{\displaystyle O}{\overset{\displaystyle \|}{C}}-R + hv \rightarrow RCO + R, \qquad \lambda < \sim 3000 \text{ Å},$$

$$R-\overset{\displaystyle O}{\overset{\displaystyle \|}{C}}-H + hv \rightarrow CHO + R, \qquad \lambda < \sim 3000 \text{ Å},$$

$$HI + hv \rightarrow H + I^*, \qquad \lambda < \sim 4000 \text{ Å}.$$

The formation of the atom in an excited state (indicated by the asterisk) is a frequent occurrence, since it is rare that the energy of the light quantum provides exactly the amount needed to dissociate the molecule into normal atoms. The excited, or "hot," atoms often influence the course of the reactions which follow the primary step.

The kinetics of two photochemical reactions will be compared with the kinetics of the thermal reactions. In the absence of light, hydrogen iodide decomposes by the elementary reaction:

$$HI + HI \rightarrow H_2 + I_2.$$

In the initial stages the rate of the reverse action can be ignored. The rate of reaction can be written

$$\frac{-d(HI)}{dt} = k(HI)^2.$$

In the photochemical reaction, at wavelengths below about 4000 Å, the mechanism is

$$HI + hv \rightarrow H + I, \qquad \text{Rate} = I_{abs},$$

$$H + HI \rightarrow H_2 + I, \qquad \text{Rate} = k_2(H)(HI),$$

$$I + I \rightarrow I_2, \qquad \text{Rate} = k_3(I)^2.$$

Other possible elementary reactions either have much higher activation energies or require three-body collisions. The rate of the reaction is

$$\frac{-d(HI)}{dt} = I_{abs} + k_2(H)(HI).$$

The steady-state requirement is

$$\frac{d(\mathrm{H})}{dt} = 0 = I_{\mathrm{abs}} - k_2(\mathrm{H})(\mathrm{HI}).$$

Combining these two equations, we obtain

$$\frac{-d(\mathrm{HI})}{dt} = 2I_{\mathrm{abs}}. \tag{33–39}$$

By definition the quantum yield is $\varphi = [-d(\mathrm{HI})/dt]/I_{\mathrm{abs}}$, so that, from Eq. (33–39), we find that $\varphi = 2$. In a variety of experimental situations, the observed value of φ is 2.

The interesting point about the photochemical reaction is that the rate, by Eq. (33–39), is simply twice the absorbed intensity and is not directly dependent on the concentration of HI. This fact implies that the reaction is very slow, since even fairly intense light sources do not produce a very large number of quanta per second. The dependence of rate on intensity can be readily verified by altering the distance between the system and the light source. The incident intensity varies inversely as the square of the distance, so for a given cell and given concentration of HI, the absorbed intensity must vary in the same way. Indirectly, the rate depends on the concentration of HI, since the absorbed intensity is dependent on concentration through Beer's law.

The photochemical reaction between H_2 and Br_2 follows a kinetic law which resembles that for the thermal reaction, in contrast to the decomposition of HI, where the kinetics are quite different. Using light of wavelength less than 5107 Å, the mechanism of the photochemical reaction between H_2 and Br_2 is

$$\mathrm{Br}_2 + h\nu \longrightarrow 2\mathrm{Br},$$

$$\mathrm{Br} + \mathrm{H}_2 \xrightarrow{k_2} \mathrm{HBr} + \mathrm{H},$$

$$\mathrm{H} + \mathrm{Br}_2 \xrightarrow{k_3} \mathrm{HBr} + \mathrm{Br},$$

$$\mathrm{H} + \mathrm{HBr} \xrightarrow{k_4} \mathrm{H}_2 + \mathrm{Br},$$

$$\mathrm{Br} + \mathrm{Br} \xrightarrow{k_5} \mathrm{Br}_2.$$

The rate of formation of HBr is the same as that given in Eq. (31–43) for the thermal reaction

$$\frac{d(\mathrm{HBr})}{dt} = k_2(\mathrm{H}_2)(\mathrm{Br}) + k_3(\mathrm{H})(\mathrm{Br}_2) - k_4(\mathrm{HBr})(\mathrm{H}). \tag{33–40}$$

The steady-state conditions for H atoms and Br atoms are

$$\frac{d(\mathrm{H})}{dt} = 0 = k_2(\mathrm{H}_2)(\mathrm{Br}) - k_3(\mathrm{H})(\mathrm{Br}_2) - k_4(\mathrm{HBr})(\mathrm{H}),$$

$$\frac{d(\mathrm{Br})}{dt} = 0 = 2I_{\mathrm{abs}} - k_2(\mathrm{H}_2)(\mathrm{Br}) + k_3(\mathrm{H})(\mathrm{Br}_2) + k_4(\mathrm{HBr})(\mathrm{H}) - k_5(\mathrm{Br})^2.$$

Addition of these two equations yields $k_5(\text{Br})^2 = 2I_{\text{abs}}$, so that

$$(\text{Br}) = \left(\frac{2I_{\text{abs}}}{k_5}\right)^{1/2}.$$

This result in either of the steady-state equations yields ultimately

$$(\text{H}) = \frac{k_2(2I_{\text{abs}}/k_5)^{1/2}(\text{H}_2)}{k_3(\text{Br}_2) + k_4(\text{HBr})}.$$

These values in Eq. (33–40) bring it to the form

$$\frac{d(\text{HBr})}{dt} = \frac{2k_2(2I_{\text{abs}}/k_5)^{1/2}(\text{H}_2)}{1 + [k_4(\text{HBr})/k_3(\text{Br}_2)]}. \qquad (33\text{–}41)$$

The expression in Eq. (33–41) is very similar to that for the thermal reaction, where the factor $(2I_{\text{abs}})^{1/2}$ is replaced by $(k_1 c_{\text{Br}_2})^{1/2}$. This means that the bromine atom concentration is maintained by the photochemical dissociation of bromine rather than the thermal dissociation. The dependence on the square root of the intensity is notable, since it has the consequence that the quantum yield is inversely proportional to the square root of the intensity:

$$\varphi = \frac{d(\text{HBr})/dt}{I_{\text{abs}}} = \frac{2k_2(2/k_5)^{1/2}(\text{H}_2)}{I_{\text{abs}}^{1/2}\{1 + [k_4(\text{HBr})/k_3(\text{Br}_2)]\}}.$$

As the intensity increases, a greater proportion of the bromine atoms formed are converted to Br_2 instead of entering the chain; most of the additional quanta are therefore wasted, and the process is less efficient. Because k_2 is very small, the quantum yield is less than unity at room temperature in spite of the fact that the HBr is formed in a chain reaction. As the temperature increases, the increase in k_2 increases the quantum yield (k_5 is nearly independent of temperature).

33–16 Photosensitized Reactions

Photosensitized reactions make up another very important class of photochemical reactions. In these reactions the reactants are mixed with a foreign gas; mercury or cadmium vapor are often used. The primary photochemical act is the absorption of the quantum by the foreign atom or molecule.

    If a mixture of hydrogen, oxygen, and mercury vapor is exposed to ultraviolet light, the mercury vapor absorbs strongly at 2537 Å with the formation of an excited mercury atom, Hg^*:

$$\text{Hg} + h\nu \rightarrow \text{Hg}^*.$$

The energy corresponding to this wavelength is 112 kcal/mole. The energy required to dissociate a molecule of hydrogen in its ground state to two hydrogen atoms in their ground state is 103.2 kcal/mole. The dissociation of oxygen requires

117.2 kcal/mole. The energy possessed by the excited mercury atom is more than enough to dissociate H_2 but not enough to dissociate O_2. The quenching reaction

$$Hg^* + H_2 \rightarrow Hg + H + H$$

introduces H atoms into the mixture, which can initiate chains to form H_2O by the usual mechanism. A reaction which is initiated by light in this way is a photosensitized reaction; the mercury is called a sensitizer.

The importance of photosensitization derives from the fact that reaction is produced in the presence of the sensitizer in circumstances where the direct photochemical dissociation is not possible. The example just cited is a case in point. Radiation of wavelength 2537 Å was absorbed by a mercury atom. The excited mercury atom dissociated a molecule of hydrogen by transferring the excitation energy in a collision. The mercury atom had 112 kcal; 103 kcal were needed for the dissociation; 9 kcal are left over and go into additional translational energy of the two hydrogen atoms and the mercury atom. If the attempt is made to dissociate H_2 directly by the process

$$H_2 + h\nu \rightarrow H + H,$$

it is found that light of $\lambda = 2537$ Å will not produce any dissociation even though it still has the 112 kcal, which is more than enough if one considers only the thermodynamics of the process. For the direct absorption to produce dissociation, the wavelength must lie in the absorption continuum; for H_2 the continuum begins at 849 Å. It follows that the absorption of light in the continuum produces at least one atom in an excited state:

$$H_2 + h\nu \rightarrow H + H^*, \qquad \lambda \leq 849 \text{ Å}.$$

The selection rules which govern the absorption of radiation forbid the direct absorption of a quantum by H_2 to yield two H atoms in their ground state; thus a quantum of light does not necessarily produce dissociation even though it may have sufficient energy. The quantum must have enough energy to produce the dissociation in the special way required by the selection rules. The transfer of energy in a collision is not limited by this requirement, and so sensitization can produce dissociation.

The decomposition of ozone sensitized by chlorine is another example of photosensitization. Ozone is stable under irradiation by visible light. The absorption continuum begins at about 2500 Å. In the presence of a little chlorine, ozone decomposes rapidly. The chlorine absorbs continuously below 4785 Å:

$$Cl_2 + h\nu \rightarrow 2Cl, \qquad \lambda \leq 4785 \text{ Å}.$$

The chlorine atoms react with ozone in a complex chain mechanism. Bromine is also effective as a sensitizer for this decomposition. The decomposition of oxalic acid sensitized by the uranyl ion was referred to earlier as an actinometric reaction.

A practical example of photosensitization is in the use of certain dyes in photographic film to render the emulsion sensitive to wavelengths which are longer than those to which it ordinarily responds; e.g., infrared photography which permits photographs of objects in the absence of visible light.

The photosynthesis of carbohydrates in plants is sensitized by chlorophyll, the principal constituent of green-plant pigment. Chlorophyll absorbs strongly in the region between 6000 to 7000 Å; this spectral region is found to be the most effective one for the photosynthetic reaction which can be represented by the equation

$$nCO_2 + nH_2O + xh\nu \rightarrow (CH_2O)_n + nO_2.$$

Several quanta of visible light are required. The mechanism of the reaction is not known, although much excellent work has been done on the problem, particularly in the last twenty years.

33-17   The Photostationary State

Absorbed light has an interesting effect on a system in chemical equilibrium. The absorption of light by a reactant can increase the rate of the forward reaction without directly influencing the rate of the reverse action; this disturbs the equilibrium. The concentration of products increases somewhat, increasing the rate of the reverse reaction. In this way the rates of the forward and reverse reaction can be brought into balance with the system having a higher concentration of products than that in the equilibrium system. This new state is not an equilibrium state but a stationary state, called a *photostationary* state.

The dimerization of anthracene offers a convenient example. The reaction

$$2A \rightarrow A_2$$

occurs upon irradiation of a solution of anthracene by ultraviolet light. A plausible mechanism is

$$A + h\nu \rightarrow A^*, \qquad \text{Rate} = I_{abs},$$

$$A^* + A \rightarrow A_2, \qquad \text{Rate} = k_2(A^*)(A),$$

(fluorescence) $\qquad A^* \rightarrow A + h\nu', \qquad \text{Rate} = k_3(A^*),$

$$A_2 \rightarrow 2A, \qquad \text{Rate} = k_4(A_2).$$

The net rate of formation of A_2 is

$$\frac{d(A_2)}{dt} = k_2(A^*)(A) - k_4(A_2). \tag{33-42}$$

In the steady state, $d(A^*)/dt = 0$, so that

$$0 = I_{abs} - k_2(A^*)(A) - k_3(A^*).$$

Hence,

$$(A^*) = \frac{I_{abs}}{k_2(A) + k_3}. \tag{33-43}$$

In the photostationary state we have the additional requirement that $d(A_2)/dt = 0$; so that Eq. (33–42) yields for the concentration of A_2,

$$(A_2) = \frac{k_2(A^*)A}{k_4}.$$

Using the value for (A^*) from Eq. (33–43), we obtain

$$(A_2) = \frac{k_2(A)I_{abs}}{k_4[k_2(A) + k_3]} = \frac{I_{abs}}{k_4\{1 + [k_3/k_2(A)]\}}.$$

If the concentration of monomer (A) is very large, then this becomes

$$(A_2) = I_{abs}/k_4. \tag{33–44}$$

The difference between Eq. (33–44) and the usual equilibrium expression $(A_2) = K(A)^2$, should be noted. In the photostationary state in the condition for which Eq. (33–44) is appropriate, the concentration of dimer is independent of the concentration of monomer.

Many other examples of the photostationary states are known. The decomposition of NO_2 occurs photochemically below 3700 Å:

$$NO_2 + h\nu \rightarrow NO + \tfrac{1}{2}O_2,$$

while the reverse reaction is a dark reaction. The maintenance of a certain amount of ozone in the upper atmosphere is probably the result of a complex photostationary state. The ozone is responsible for filtering the sun's rays so that no radiation of wavelength shorter than about 2900 to 3000 Å reaches the earth's surface. Ozone absorbs strongly at wavelengths shorter than this. This fortunate circumstance of the ozone layer makes life possible on earth. Radiation of wavelengths shorter than this produces severe damage and in many cases has a lethal effect on living cells. The effective thickness of the ozone layer is estimated at about 3 mm if the gas were at standard pressure and temperature.

33–18 Chemiluminescence

Reactions of the ordinary thermal type in which some intermediate or the product itself are formed in an electronically excited state exhibit *chemiluminescence*. The excited molecule emits a quantum of light, usually in the visible spectrum. Since the reaction may be proceeding at ordinary temperatures, the light emitted is sometimes called "cold light," presumably to contrast it with the "hot light" emitted by a flame or incandescent body. A famous example is the oxidation of 3-amino phthalic cyclic hydrazide, luminol, in alkaline solution by hydrogen peroxide. A bright green light is emitted.

The greenish glow of slowly oxidizing phosphorus is apparently due to the formation of an oxide in an excited state. The light of the firefly, the light emitted by some micro-organisms in the course of metabolism, bioluminescence, are other examples of chemiluminescence. The phosphorescence observed in marshy areas, the will o' the wisp, is apparently due to a slow oxidation of rotting organic material.

Problems

33–1. a) Using data from Table 33–1, compare the relative rates of the homogeneous decomposition of HI at 400° and 500°K.

  b) Compare the relative rates of the heterogeneous decomposition of HI on platinum at 400° and 500°K.

33–2. What conclusion can be reached about absorption on the surface from each of the following facts.

  a) The rate of decomposition of HI on platinum is proportional to the concentration of HI.

  b) On gold the rate of decomposition of HI is independent of the pressure of HI.

  c) On platinum, the rate of the reaction $SO_2 + \frac{1}{2}O_2 \rightarrow SO_3$ is inversely proportional to the pressure of SO_3.

  d) On platinum the rate of the reaction $CO_2 + H_2 \rightarrow H_2O + CO$ is proportional to the pressure of CO_2 at low CO_2 pressures and is inversely proportional to the pressure of CO_2 at high CO_2 pressures.

33–3. The diffusion coefficient of most species in aqueous solutions is of the order of 10^{-4} cm^2/sec. In a well-stirred solution, $\delta = 0.001$ cm. If the concentration of the reactant molecule is 0.01 mole/liter, what will be the rate of the reaction if the slow step is diffusion of the reactant to the surface? The concentration of the reactant at the surface may be taken as zero, since it reacts very rapidly on arrival at the surface.

33–4. The galvanometer in a potentiometer circuit can detect $\pm 10^{-6}$ A. The i_0 for hydrogen evolution is 10^{-14} A/cm^2 on mercury and 10^{-2} A/cm^2 on platinum. If the electrode area is 1 cm^2, over what range of potential will the potentiometer appear to be in balance if platinum is used as a hydrogen electrode? If mercury is used as a hydrogen electrode? (Assume the relation between i and η is linear.)

33–5. The exchange current measures the rate at which the forward and reverse reaction occur at equilibrium. The exchange current for the reaction $\frac{1}{2}H_2 \rightleftharpoons H^+ + e^-$ on platinum is 10^{-2} A/cm^2. How many hydrogen ions are formed on 1 cm^2 of a platinum surface per second? If there are 10^{15} sites for adsorption of H atoms, how many times is the surface occupied and vacated in 1 sec?

33–6. If 10% of the energy of a 100-W incandescent bulb goes into visible light having an average wavelength of 6000 Å, how many quanta of light are emitted per second?

33–7. A 0.01 molar solution of a compound transmits 20% of the sodium D line when the absorbing path is 1.50 cm. What is the molar extinction coefficient of the substance? The solvent is assumed to be completely transparent.

33–8. a) Using the mechanism for the formation of dianthracene described in Section 33–17, write the expression for the quantum yield in the initial stage of the reaction when $(A_2) = 0$.

b) The observed value of $\varphi \approx 1$. What conclusion can be reached regarding the fluorescence of A*?

33–9. A likely mechanism for the photolysis of acetaldehyde is the following:

$$CH_3CHO + hv \longrightarrow CH_3 + CHO,$$

$$CH_3 + CH_3CHO \xrightarrow{k_2} CH_4 + CH_3CO,$$

$$CH_3CO \xrightarrow{k_3} CO + CH_3,$$

$$CH_3 + CH_3 \xrightarrow{k_4} C_2H_6.$$

Derive the expressions for the rate of formation of CO and the quantum yield for CO.

Appendix One

A–1–1 Function and Derivative

The symbol $f(x)$ signifies that f is a function of x. To say that f is a function of x means that if a value of x is chosen, this value determines a corresponding value of the function. The x is called the *independent variable*; f is called the *dependent variable*.

The volume of a given mass of liquid depends on the temperature. Translating this statement into mathematics, we say that the volume is a function of temperature, or simply write the symbol $V(t)$.

Knowing that the value of f depends on the value of x, if the value of x changes, the value of f will change. Consequently, it is of interest to ask how rapidly f changes with a change in x. This information about the function is given by the derivative of the function with respect to x.

The derivative is the rate of change of the value of the function with change in the value of x. It should be noted that in general the derivative is also a function of x. To emphasize this aspect, the symbol $f'(x)$ is often used for the derivative, and we write

$$df/dx = f'(x).$$

If the derivative is positive, the value of the function increases with the value of x; if the derivative is negative, the value of the function decreases as x increases. If the derivative is zero, the curve of the function has a horizontal tangent; the function has a maximum or a minimum value.

If we ask what change in f accompanies the change in x from x_1 to x_2, the value of the derivative provides the answer. From the identity,

$$df/dx = df/dx,$$

we can write

$$df = (df/dx)\,dx.$$

This equation says that the change in the value of f, df, is equal to the rate of change with respect to x, df/dx, multiplied by dx, the change in x. If a finite change is made in x from x_1 to x_2, then the total change in f is obtained by integration:

$$\int_{f_1}^{f_2} df = \int_{x_1}^{x_2} \frac{df}{dx}\,dx, \qquad f_2 - f_1 = \int_{x_1}^{x_2} f'(x)\,dx,$$

where f_2 and f_1 are the values of f corresponding to x_2 and x_1. The fundamental definition of the derivative

$$\frac{df}{dx} = \lim_{\Delta x \to 0} \frac{\Delta f}{\Delta x}$$

leads to the geometric interpretation of the derivative as the slope of the tangent to the curve at any point.

A–1–2 The Integral

a) The integral is the limit of a sum. In the preceding paragraph, the total change in f was found by adding together (i.e., integrating) all of the small changes in the interval between x_1 and x_2.

b) The indefinite integral $\int g(x)\,dx$ always has a constant of integration associated with it. For example, evaluate the integral $\int (1/x)\,dx$. A table of integrals gives $\ln x$ as the value of the integral; the integration constant C must be added to this, so we obtain

$$\int \frac{1}{x}\,dx = \ln x + C.$$

c) The definite integral $\int_a^b g(x)\,dx$ does not have a constant of integration associated with it. If from a table of integrals we find that

$$\int g(x)\,dx = G(x) + C,$$

then

$$\int_a^b g(x)\,dx = G(b) - G(a).$$

The definite integral is a function only of the limits a and b and of any parameters other than the variable of integration which may be contained in the integrand. For example, the integral $\int_a^b g(x, \alpha)\,dx$ is a function only of a, b, and α, *and is not a function of* x.

d) The integral of a function can be represented graphically as an area. The integral $\int_a^b g(x)\,dx$ is equal to the area included between the curve of the function $g(x)$ and the x-axis and between the lines $x = a$, and $x = b$.

A–1–3 Taylor's Theorem

If we do not know the analytical form of a function, but know the values of its derivatives at some point, let us say at $x = 0$, then it is often possible to express the function as an infinite series. Assume that the function $f(x)$ can be expressed as a series:

$$f(x) = a_0 + a_1 x + \frac{a_2}{2!}x^2 + \frac{a_3}{3!}x^3 + \cdots.$$

By differentiating, we obtain

$$f'(x) = a_1 + a_2 x + \frac{a_3}{2!}x^2 + \cdots,$$

$$f''(x) = a_2 + a_3 x + \cdots,$$

$$f'''(x) = a_3 + \cdots.$$

At $x = 0$ these expressions reduce to

$$f(0) = a_0, \qquad f'(0) = a_1, \qquad f''(0) = a_2, \qquad f'''(0) = a_3, \qquad \cdots.$$

Thus, the values of the unknown coefficients in the series are expressed in terms of the values of the derivatives of the function at $x = 0$; the series becomes

$$f(x) = f(0) + f'(0)x + \frac{f''(0)}{2!}x^2 + \frac{f'''(0)}{3!}x^3 + \cdots,$$

which is Taylor's theorem. Not all functions can be expressed as a series in this way, but this expansion is often used for those functions which do behave properly. Usually, only the first two terms of the infinite series will be needed; the rest will be neglected.

A–1–4 Functions of More Than One Variable

We frequently use functions of two variables; the molar volume of a gas, for example, depends on temperature and pressure, $\overline{V} = \overline{V}(T, p)$. Written in this way, T and p are the *independent* variables, and $\overline{V}$ is the *dependent* variable. Such a function has two first derivatives: one with respect to temperature, one with respect to pressure. These are called partial derivatives and are written

$$\left(\frac{\partial \overline{V}}{\partial T}\right)_p \quad \text{and} \quad \left(\frac{\partial \overline{V}}{\partial p}\right)_T.$$

The subscript outside the parentheses in each of these symbols indicates the variable which is kept constant in the differentiation. If we ask how the molar volume changes if the temperature changes slightly at constant pressure, the answer is given by

$$d\bar{V} = \left(\frac{\partial \bar{V}}{\partial T}\right)_p dT.$$

The change in volume with change in pressure at constant temperature is given by

$$d\bar{V} = \left(\frac{\partial \bar{V}}{\partial p}\right)_T dp.$$

If *both* temperature and pressure change, then the total change in volume is the sum of the change due to temperature and the change due to pressure:

$$d\bar{V} = \left(\frac{\partial \bar{V}}{\partial T}\right)_p dT + \left(\frac{\partial \bar{V}}{\partial p}\right)_T dp.$$

This is called the *total differential* of the function. Any function of two variables $f(x, y)$ has a total differential df given by

$$df = \left(\frac{\partial f}{\partial x}\right)_y dx + \left(\frac{\partial f}{\partial y}\right)_x dy.$$

A–1–5 Solutions of Eq. (4–27)

Equation (4–27) can be written in the form

$$Af(z) = f(x)f(y),$$

where $z = x + y$. We differentiate this equation with respect to x:

$$A\frac{df(z)}{dz}\left(\frac{\partial z}{\partial x}\right) = f'(x)f(y),$$

and then with respect to y:

$$A\frac{df(z)}{dz}\left(\frac{\partial z}{\partial y}\right) = f(x)f'(y).$$

But $\partial z/\partial x = \partial(x + y)/\partial x = 1$, and also $\partial z/\partial y = 1$, so these two equations become

$$Af'(z) = f'(x)f(y), \qquad Af'(z) = f(x)f'(y).$$

The left-hand sides are the same, so $f'(x)f(y) = f'(y)f(x)$. Dividing by $f(x)f(y)$, this becomes

$$\frac{f'(x)}{f(x)} = \frac{f'(y)}{f(y)}.$$

The left-hand side of this equation is apparently a function only of x, while the right-hand side does not depend on x but only on y. This is possible only if each side of the equation is a constant, β. Thus

$$\frac{f'(x)}{f(x)} = \beta \quad \text{and therefore} \quad \frac{df(x)}{f(x)} = \beta \, dx.$$

Integrating, we obtain $\ln f(x) = \beta x + \ln A$, and therefore, $f(x) = A \exp(\beta x)$, which is the solution of the functional equation, Eq. (4–27).

Appendix Two

Table of fundamental constants

Molar volume, ideal gas, 0°C, 1 atm.	$\bar{V}^0 = 22{,}413.8$ cm^3/mole
Ice point	$T_0 = 273.15°$K
Avogadro number	$N_0 = 6.0230 \times 10^{23}$
Gas constant	$R = 0.082054$ liter·atm·deg^{-1}·mole^{-1}
	82.057 cm^3·atm·deg^{-1}·mole^{-1}
	8.3144×10^7 ergs·deg^{-1}·mole^{-1}
	8.3144 J·deg^{-1}·mole^{-1}
	1.9872 cal·deg^{-1}·mole^{-1}
Boltzmann constant	$\mathbf{k} = 1.38044 \times 10^{-16}$ erg/deg
Faraday constant	$\mathscr{F} = 96{,}490$ coulomb/equivalent
Electronic charge	$e = 4.80286 \times 10^{-10}$ esu
Electron mass	$m = 9.1083 \times 10^{-28}$ gm
Planck constant	$h = 6.6252 \times 10^{-27}$ erg·sec
Standard acceleration of gravity	$g = 980.665$ cm/sec^2
Velocity of light	$c = 2.997930 \times 10^{10}$ cm/sec

Conversion factors

1 atm $= 760$ mm $= 1.01325 \times 10^6$ dyne/cm^2

1 cal $= 4.184000$ J

1 J $= 10^7$ ergs $= 1$ V·coulomb

1 erg $= 1$ dyne·cm

1 eV $= 1.60210 \times 10^{-12}$ erg

Appendix Three

International atomic weights—1961†

Element	Symbol	Atomic number	Atomic weight	
Actinium	Ac	89		
Aluminum	Al	13	26.9815	
Americium	Am	95		
Antimony	Sb	51	121.75	
Argon	Ar	18	39.948	
Arsenic	As	33	74.9216	
Astatine	At	85		
Barium	Ba	56	137.34	
Berkelium	Bk	97		
Beryllium	Be	4	9.0122	
Bismuth	Bi	83	208.980	
Boron	B	5	10.811[a]	
Bromine	Br	35	79.909[b]	(*Continued*)

† Printed by permission of the International Union of Pure and Applied Chemistry and Butterworth Scientific Publications.
a The atomic weight varies because of natural variations in the isotopic composition of the element. The observed ranges are: boron, ± 0.003; carbon, ± 0.00005; hydrogen, ± 0.00001; oxygen, ± 0.0001; silicon, ± 0.001; sulfur, ± 0.003.
b The atomic weight is believed to have an experimental uncertainty of the following magnitude: bromine, ± 0.002; chlorine, ± 0.001; chromium, ± 0.001; iron, ± 0.003; silver, ± 0.003. For other elements the last digit given is believed to be reliable to ± 0.5.

International atomic weights—1961 (*Continued*)

Element	Symbol	Atomic number	Atomic weight
Cadmium	Cd	48	112.40
Calcium	Ca	20	40.08
Californium	Cf	98	
Carbon	C	6	12.01115[a]
Cerium	Ce	58	140.12
Cesium	Cs	55	132.905
Chlorine	Cl	17	35.453[b]
Chromium	Cr	24	51.996[b]
Cobalt	Co	27	58.9332
Copper	Cu	29	63.54
Curium	Cm	96	
Dysprosium	Dy	66	162.50
Einsteinium	Es	99	
Erbium	Er	68	167.26
Europium	Eu	63	151.96
Fermium	Fm	100	
Fluorine	F	9	18.9984
Francium	Fr	87	
Gadolinium	Gd	64	157.25
Gallium	Ga	31	69.72
Germanium	Ge	32	72.59
Gold	Au	79	196.967
Hafnium	Hf	72	178.49
Helium	He	2	4.0026
Holmium	Ho	67	164.930
Hydrogen	H	1	1.00797[a]
Indium	In	49	114.82
Iodine	I	53	126.9044
Iridium	Ir	77	192.2
Iron	Fe	26	55.847[b]
Krypton	Kr	36	83.80
Lanthanum	La	57	138.91
Lead	Pb	82	207.19
Lithium	Li	3	6.939
Lutetium	Lu	71	174.97
Magnesium	Mg	12	24.312
Manganese	Mn	25	54.9380
Mendelevium	Md	101	
Mercury	Hg	80	200.59
Molybdenum	Mo	42	95.94

International atomic weights—1961 (*Continued*)

Element	Symbol	Atomic number	Atomic weight
Neodymium	Nd	60	144.24
Neon	Ne	10	20.183
Neptunium	Np	93	
Nickel	Ni	28	58.71
Niobium	Nb	41	92.906
Nitrogen	N	7	14.0067
Nobelium	No	102	
Osmium	Os	76	190.2
Oxygen	O	8	15.9994[a]
Palladium	Pd	46	106.4
Phosphorus	P	15	30.9738
Platinum	Pt	78	195.09
Plutonium	Pu	94	
Polonium	Po	84	
Potassium	K	19	39.102
Praseodymium	Pr	59	140.907
Promethium	Pm	61	
Protactinium	Pa	91	
Radium	Ra	88	
Radon	Rn	86	
Rhenium	Re	75	186.2
Rhodium	Rh	45	102.905
Rubidium	Rb	37	85.47
Ruthenium	Ru	44	101.07
Samarium	Sm	62	150.35
Scandium	Sc	21	44.956
Selenium	Se	34	78.96
Silicon	Si	14	28.086[a]
Silver	Ag	47	107.870[b]
Sodium	Na	11	22.9898
Strontium	Sr	38	87.62
Sulfur	S	16	32.064[a]
Tantalum	Ta	73	180.948
Technetium	Tc	43	
Tellurium	Te	52	127.60
Terbium	Tb	65	158.924
Thallium	Tl	81	204.37
Thorium	Th	90	232.038
Thulium	Tm	69	168.934
Tin	Sn	50	118.69

International atomic weights—1961 (*Continued*)

Element	Symbol	Atomic number	Atomic weight
Titanium	Ti	22	47.90
Tungsten	W	74	183.85
Uranium	U	92	238.03
Vanadium	V	23	50.942
Xenon	Xe	54	131.30
Ytterbium	Yb	70	173.04
Yttrium	Y	39	88.905
Zinc	Zn	30	65.37
Zirconium	Zr	40	91.22

References

References

THERMODYNAMICS, KINETIC THEORY, STATISTICAL THERMODYNAMICS

BEATTIE, J. A. and STOCKMAYER, W. H., Chapter II in *A Treatise on Physical Chemistry*, H. S. Taylor and S. Glasstone, eds., Vol. II, 3rd ed., New York, Van Nostrand, 1951.

DENBIGH, K., *The Principles of Chemical Equilibrium*, Cambridge, 1955.

FINDLAY, A., CAMPBELL, A. N., and SMITH, N. O., *The Phase and Its Applications*, 9th ed., New York, Dover, 1951.

GOLDEN, S., *An Introduction to Theoretical Physical Chemistry*, Reading, Mass., Addison-Wesley, 1961.

GLASSTONE, S., *Thermodynamics for Chemists*, New York, Van Nostrand, 1946.

GUGGENHEIM, E. A., *Mixtures*, London, Oxford, 1952.

GUGGENHEIM, E. A., *Thermodynamics*, 3rd ed., Amsterdam, North-Holland, 1957.

GURNEY, R. W., *Introduction to Statistical Mechanics*, New York, McGraw-Hill, 1949.

HARNED, H. S. and OWEN, B. B., *The Physical Chemistry of Electrolytic Solutions*, New York, Reinhold, 1943.

HERZFELD, K. F. and SMALLWOOD, H., in *A Treatise on Physical Chemistry*, H. S. Taylor and S. Glasstone, eds., Vol. II, 3rd ed., New York, Van Nostrand, 1951.

HILDEBRAND, J. H. and SCOTT, R. L., *Solubility of Nonelectrolytes*, 3rd ed., New York, Reinhold, 1950.

HILL, T. L., *Introduction to Statistical Thermodynamics*, Reading, Mass., Addison-Wesley, 1960.

HIRSCHFELDER, J. O., CURTISS, C. F., and BIRD, R. B., *Molecular Theory of Gases and Liquids*, New York, Wiley, 1954.

LATIMER, W. M., *The Oxidation States of the Elements and Their Potentials in Aqueous Solutions*, 2nd ed., Englewood Cliffs, N. J., Prentice-Hall, 1952.

LEWIS, G. N. and RANDALL, M., *Thermodynamics*, 2nd ed., revised by K. S. Pitzer and L. Brewer, New York, McGraw-Hill, 1961.

PLANCK, M., *Treatise on Thermodynamics*, New York, Dover, 1945.

PRIGOGINE, I. and DEFAY, R., *Chemical Thermodynamics*, trans. by D. H. Everett, London, Longmans Green, 1954.

RUSHBROOKE, G. S., *Introduction to Statistical Mechanics*, New York, Oxford, 1949.

SCHRÖDINGER, E., *Statistical Thermodynamics*, Cambridge, 1946.

SEARS, F. W., *An Introduction to Thermodynamics, the Kinetic Theory of Gases, and Statistical Mechanics*, Reading, Mass., Addison-Wesley, 1950.

WAGNER, C., *Thermodynamics of Alloys*, Reading, Mass., Addison-Wesley, 1952.

WALL, F. T., *Chemical Thermodynamics*, San Francisco, W. H. Freeman, 1958.

ZEMANSKY, M., *Heat and Thermodynamics*, New York, McGraw-Hill, 1951.

STRUCTURE

BRAGG, W. L., *The Crystalline State*, Vol. I, London, G. Bell, 1949.

BOETTCHER, C. J. F., *Theory of Electric Polarization*, Amsterdam, Elsevier, 1952.

BUERGER, M. J., *X-ray Crystallography*, New York, Wiley, 1942.

BUNN, C. W., *Chemical Crystallography*, London, Oxford, 1945.

COULSON, C. A., *Valence*, London, Oxford, 1952.

CULLITY, B. D., *Elements of X-ray Diffraction*, Reading, Mass., Addison-Wesley, 1956.

DEBYE, P., *Polar Molecules*, New York, Dover, 1945.

EVANS, R. C., *Crystal Chemistry*, Cambridge, 1948.

EYRING, H., WALTER, J., and KIMBALL, G. E., *Quantum Chemistry*, New York, Wiley, 1944.

FRENKEL, J., *The Kinetic Theory of Liquids*, New York, Oxford, 1946.

GAYDON, A. G., *Dissociation Energies*, London, Chapman Hall, 1952.

GLASSTONE, S., *Theoretical Chemistry*, New York, Van Nostrand, 1944.

HEITLER, W., *Elementary Wave Mechanics*, New York, Oxford, 1945.

HERZBERG, G., *Atomic Spectra and Atomic Structure*, New York, Dover, 1944.

HERZBERG, G., *Molecular Spectra and Molecular Structure*, New York, Van Nostrand, 1950.

HERZBERG, G., *Infrared and Raman Spectra*, New York, Van Nostrand, 1945.

KIMBALL, G. E., "The Liquid State," in *A Treatise on Physical Chemistry*, 3rd ed., H. S. Taylor and S. Glasstone, eds., Vol. II, New York, Van Nostrand, 1951.

PAULING, L., *The Nature of the Chemical Bond*, 3rd ed., Ithaca, N. Y., Cornell, 1960.

PAULING, L. and WILSON, E. B., *Introduction to Quantum Mechanics*, New York, McGraw-Hill, 1935.

PIMENTEL, G. C. and McCLELLAN, A. L., *The Hydrogen Bond*, San Francisco, W. H. Freeman, 1960.

PITZER, K. S., *Quantum Chemistry*, Englewood Cliffs, N. J., Prentice-Hall, 1953.

RICE, F. O. and TELLER, E., *The Structure of Matter*, New York, Wiley, 1949.

RICHTMYER, F. K., KENNARD, E. A., and LAURITSEN, T., *Introduction to Modern Physics*, 5th ed., New York, McGraw-Hill, 1955.

SLATER, J. C., *Introduction to Chemical Physics*, New York, McGraw-Hill, 1939.

SMYTH, C. P., *Dielectric Behavior and Structure*, New York, McGraw-Hill, 1955.

SYRKIN, Y. K. and DYATKINA, M. E., *Structure of Molecules and the Chemical Bond*, Trans. by M. A. Partridge and D. O. Jordan, New York, Interscience, 1950.

WELLS, A. F., *The Third Dimension in Chemistry*, London, Oxford, 1956.

WELLS, A. F., *Structural Inorganic Chemistry*, 2nd ed., London, Oxford, 1950.

WILSON, A. H., *Semiconductors and Metals*, London, Cambridge, 1939.

KINETICS

AMIS, E. S., *Kinetics of Chemical Change in Solution*, New York, Macmillan, 1948.

BASOLO, F. and PEARSON, R. G., *Mechanisms of Inorganic Reactions*, New York, Wiley, 1958.

BOWEN, E. J., *Chemical Aspects of Light*, New York, Oxford, 1946.

FROST, A. A. and PEARSON, R. G., *Kinetics and Mechanism*, 2nd ed., New York, Wiley, 1961.

GLASSTONE, S., LAIDLER, K. J., and EYRING, H., *The Theory of Rate Processes*, New York, McGraw-Hill, 1941.

HINSCHELWOOD, C. N., *The Kinetics of Chemical Change*, New York, Oxford, 1942.

LAIDLER, K. J., *Chemical Kinetics*, New York, McGraw-Hill, 1950.

LAIDLER, K. J., *The Chemical Kinetics of Excited States*, London, Oxford, 1955.

MOELWYN-HUGHES, E. A., *The Kinetics of Reactions in Solution*, New York, Oxford, 1947.

NOYES, W. A. and LEIGHTON, P. A., *The Photochemistry of Gases*, New York, Reinhold, 1941.

ROLLEFSON, G. K. and BURTON, M., *Photochemistry and the Mechanism of Chemical Reactions*, New York, Prentice-Hall, 1939.

STEACIE, E. W. R., *Free Radical Mechanisms*, New York, Reinhold, 1946.

GENERAL

National Bureau of Standards Circular 500, *Selected Values of Thermodynamic Properties*, U. S. Government Printing Office, Washington, D. C., 1952.

American Petroleum Institute Research Project 44, *Selected Values of the Properties of Hydrocarbons*, F. D. Rossini *et al.*, Pittsburgh, Carnegie Press, 1953.

ADAM, N. K., *The Physics and Chemistry of Surfaces*, 3rd ed., London, Oxford, 1941.

BETHE, H., *Elementary Nuclear Theory*, New York, Wiley, 1947.

BUTLER, J. A. V., *Electrical Phenomena at Interfaces*, London, Methuen, 1951.

DAVIES, J. T. and RIDEAL, E. K., *Interfacial Phenomena*, New York, Academic Press, 1961.

FALKENHAGEN, H., *Electrolytes*, Trans. by R. P. Bell, London, Oxford, 1934.

GLASSTONE, S., *Introduction to Electrochemistry*, New York, Van Nostrand, 1942.

GURNEY, R. W., *Ionic Processes in Solution*, New York, McGraw-Hill, 1953.

HALLIDAY, D., *Introductory Nuclear Physics*, New York, Wiley, 1950.

HARKINS, W. D., *The Physical Chemistry of Surface Films*, New York, Reinhold, 1952.

HINSCHELWOOD, C. N., *The Structure of Physical Chemistry*, New York, Oxford, 1951.

MOELLER, T., *Inorganic Chemistry*, New York, Wiley, 1952.

POTTER, E. C., *Electrochemistry*, London, Cleaver-Hume, 1956.

STRANATHAN, J. D., *The Particles of Modern Physics*, Philadelphia, Blakiston, 1954.

Answers to Problems

Answers to Problems*

2–1. 458°C 2–2. 260 moles; 8320 gm 2–3. $8.969(10^{-4})$ gm 2–4. (a) $8.226(10^{-4})$ gm (b) 142.2 cc 2–5. $R = 0.100$ liter·atm·deg^{-1}·mole^{-1}; $N_0 = 7.34(10^{23})$; $M_H = 1.229$; $M_O = 19.50$ 2–6. $\alpha = 1/T$ 2–7. $\beta = 1/p$ 2–8. $(\partial p/\partial T)_V = \alpha/\beta$ 2–9. N_2: 0.440 atm; 46.7%; O_2:0.384 atm; 53.3%; $O_2 + N_2$: 0.824 atm 2–10. O_2:0.38 atm; 5.8%; H_2:6.15 atm; 94.2%; $O_2 + H_2$:6.53 atm 2–11. (a) 2.0% H_2O (b) 10.2 liters 2–12. 10.3 mole% H_2 2–14. 60.4% 2–15. 154.7 2–16. Denver: 633 mm; 0.833 atm; Mt. Evans; 464 mm; 0.610 atm 2–17. (p_{atm}; 50 km; 100 km; mole %; 50 km; 100 km). N_2:3.00(10^{-3}); 11.6(10^{-6}); 88.7; 87.2; O_2:0.373(10^{-3}); 0.665(10^{-6}); 11.0; 5.00; Ar:3.38(10^{-6}); 1.23(10^{-9}); 0.10; 0.0092; CO_2:5(10^{-8}); 8.33(10^{-12}); 0.0015; 6.26(10^{-5}); Ne:34.3(10^{-8}); 6.53(10^{-9}); 0.0102; 0.0491; He:2.26(10^{-6}); 1.02(10^{-6}); 0.067; 7.67; p_{total}:3.38(10^{-3}); 13.3(10^{-6}) 2–18. 175,000; polymers 2–19. (a) 5.8(10^{-18}) cm (b) Yes 2–20. 1.41 km 2–21. (a) $p_i = c_i RT$ (b) If $r_i = n_i/n_1$, then $p_1 = p/(1 + \Sigma r_i)$; $p_i = r_i p/(1 + \Sigma r_i)$ 2–23. 53 2–24. (a) If n_0' = number of molecules/cm^3 at ground level, and A = earth's area, then total number = $n_0' A(RT/Mg)$ (c) 5.26(10^{21}) gm 2–25. (a) $\bar{x}_i = (x_i^0/M_i)/\Sigma_i(x_i^0/M_i)$ (b) N_2: 0.804; O_2: 0.189; Ar: 0.007 2–26. $Z = (RT/Mg) \ln 2$.

3–1. [V/n(liter/mole), eq mix, $\alpha = 0$] 2 atm: 13.7; 12.2; 1 atm: 28.4; 24.5; $\frac{1}{2}$ atm: 60.4; 48.9 3–2. $Z = 1 + \alpha$; at $p = 0$, N_2O_4 is $2NO_2$ 3–3. 12.2 cc/mole 3–4. $a = 1.75(10^5)$ atm·cm^6·mole^{-2}; $b = 20$ cc/mole 3–5. $a = 2.10$ atm·liter^2·mole^{-2}; $b = 0.0189$ liter/mole; $R = 0.0509$ liter·atm· deg^{-1}·mole^{-1}; $a = 5.45$ atm·liter^2·mole^{-2}; $b = 0.0304$ liter/mole; $V_c = 0.0910$ liter/mole 3–6. $a = 3p_c V_c^2 T_c$; $b = \frac{1}{3}V_c$; $R = 8p_c V_c/3T_c$ 3–7. $a = e^2 p_c V_c^2$; $b = \frac{1}{2}V_c$; $R = \frac{1}{2}e^2 p_c V_c/T_c$ 3–8. (a) 0.520 liter/mole (b) 0.195 (c) 0.146 3–9. From 100 to 25°, p decreases 100-fold, $1/T$ increases by 1.2 3–10. (p_{atm}; Z); 200°: (100; 0.513); (200; 0.270); (400; 0.954); (600; 3.914); (800; 10.014); (1000; 20.12); 1000°; (100; 1.0218); (200; 1.0500); (400; 1.1288); (600; 1.2435); (800; 1.4009); (1000;

---

* Slide-rule accuracy in most instances. Solidus (/) after a unit or a number indicates "per mole"; e.g., kcal/ = kcal per mole; 21/ = 21 per mole.

1.608) 3–11. $(-dp/p) = (Mg/ZRT)\,dz$; $\ln(p/p_0) + B(p - p_0) = -Mgz/RT$ 3–14. $T = 2a/Rb$; $(\partial z/\partial p)_{max} = b^2/4a$.

CHAPTER 4

4–1. (m/sec; 300°; 500°); c_{rms}: 483; 625; $\bar{c}$: 445; 575; c_{mp}: 395; 510; speeds of H_2 are 4 times greater 4–2. (a) $c_{O_2}/c_{CCl_4} = 2.19$ (b) Same K. E. 4–3. (a) 894 cal; 1490 cal (b) $1.49(10^{-21})$ cal; $6.22(10^{-14})$ erg 4–4. (a) 96.5°K (b) 0.00925 4–5. 0.310 4–6. $\frac{1}{2}kT$ 4–7. kT: 0.572; $2kT$: 0.262; $5kT$: 0.0169; $10kT$: 1.62×10^{-4} 4–8. $3.24(10^{-8})$ cm/sec; 358 days 4–9. $(3 - 8/\pi)^{1/2}(kT/m)^{1/2}$ 4–10. $\frac{1}{2}(6)^{1/2}kT$ 4–11. (a) 1.5×10^{-22} (b) 3.0×10^{-303} (c) 0.198 (d) 3.2×10^{-14} 4–12. 0.0672; 0.198; 0.313 4–13. $(\bar{C}_v/R)_{total} = 2.696$; 3.308; 3.393 4–14. $\theta_1 = 3362$°K; $\theta_2 = 1891$°K; $\theta_3 = \theta_4 = 954.4$°K. $[\theta, (C_v/R)_{vib}]$: 3362°K, 0.001605; 1891°K, 0.07351; 954.4°K, 0.4536; 954.4°K, 0.4536 4–15. $2.58(10^{13})$ sec^{-1} 4–16. 0.00454; 0.1705; 0.7239; 0.9209; 0.9637 4–17. $4.767(10^{13})$ sec^{-1} contributes $0.03R$ to $(C_v)_{vib}$; then C_v(total) $= 3.03R = 6.02$ cal-deg^{-1}-mole^{-1}; $C_p = 4.03R = 8.01$ cal-deg^{-1}-mole^{-1}.

CHAPTER 5

5–1. 46 atm 5–3. 7690 cal 5–4. $p_\alpha = 5.44(10^5)$ atm; $p_{298} = 25.3$ mm 5–5. $1/T = (1/T_0) + (M_{air}gz/Q_{vap})(1/T_a)$; 93°C 5–9. 28.0 kcal; 1180°K 5–10. $a = \alpha_0$; $b = \frac{1}{2}(\alpha' + \alpha_0^2)$; $c = \frac{1}{6}(\alpha'' + 3\alpha_0\alpha' + \alpha_0^3)$.

CHAPTER 6

6–1. 95.6 kcal 6–2. 0.0235 cal 6–3. 2.92 kcal 6–4. $(t; t')$; (0; 0);(25; 2.52); (50; 11.6); (75; 37.6); (100; 100) 6–5. $t' = t[1 + b(t - 100)/(a + 100b)]$ 6–6. 410°.

CHAPTER 7

7–1. (a) $M = (nRT/gh)(1 - p_2/p_1)$ (b) $M' = (nRT/gh)(p_1/p_2 - 1)$ (c) $M' - M = (nRT/gh)(p_1 - p_2)^2/p_1p_2$ (d) $1.27(10^6)$ gm; $2.54(10^6)$ gm 7–2. (a) $W = RT[2 - (P'/p_1) - (p_2/P')]$ (b) $P' = (p_1p_2)^{1/2}$; $W_m = 2RT[1 - (p_2/p_1)^{1/2}]$ 7–3. $\Delta E = \Delta H = 0$; $Q = W = 40$ liter·atm 7–4. $\Delta E = \Delta H = 0$; $Q = W = 81$ liter·atm 7–5. 26.7 liter·atm 7–6. (a) -150 cal (b) $\Delta E = -225$ cal; $Q = \Delta H = -375$ cal 7–7. (a) 200 cal (b) $Q = -300$ cal; $\Delta E = -500$ cal; $\Delta H = -700$ cal 7–8. (a) $W = 0.026$ cal (b) $Q = \Delta H = 5000$ cal 7–9. I: $T_2 = 1380$°K; $Q = 0$; $\Delta E = -W = 3240$ cal; $\Delta H = 5400$ cal II: $T_2 = 1070$°K; $Q = 0$; $\Delta E = -W = 3850$ cal; $\Delta H = 5390$ cal; n moles: T_2 same; $W, \Delta E, \Delta H$ are n times larger 7–10. I: $T_2 = 756$°K; $Q = 0$; $\Delta E = -W = 1368$ cal; $\Delta H = 2280$ cal II: $T_2 = 579$°K; $Q = 0$; $\Delta E = -W = 1395$ cal; $\Delta H = 1953$ cal 7–11. I: $T_2 = 192$°K; $Q = 0$; $\Delta E = -W = -324$ cal; $\Delta H = -540$ cal II: $T_2 = 223$°K; $Q = 0$; $\Delta E = -W = -385$ cal; $\Delta H = -539$ cal 7–12. I: $T_2 = 119$°K; $Q = 0$; $\Delta E = -W = -543$ cal; $\Delta H = -905$ cal II: $T_2 = 155$°K; $Q = 0$; $\Delta E = -W = -725$ cal; $\Delta H = -1015$ cal 7–13. -850 cal 7–14. 354 atm 7–15. 100 atm 7–16. (a) $Q = \Delta H = 1480.7$ cal; $\Delta E = 1083.2$ cal; $W = 397.5$ cal (b) $Q = \Delta E = 1083.2$ cal; $\Delta H = 1480.7$ cal; $W = 0$ (In the following unless otherwise noted the unit is kcal or kcal/mole) 7–17. (a) -68.0 (b) -134.46 (c) 38.7 (d) 41.2204 (e) -30.60 (f) -202.6 (g) -42.50 (h) -30.0 (i) 42.5 7–18. (a) -68.6 (b) -133.57 (c) 39.3 (d) 40.628 (e) -28.82 (f) -202.6 (g) -41.91 (h) -30.6 (i) 41.9 7–19. 31.76 7–20. FeO: $-63.7/$; Fe_2O_3: $-196.6/$ 7–21. (a) -219.6 (b) -218.4 7–22. (a) 101.19 (b) 221.14 (c) 119.95 (d) 100.60; 219.96; 119.36 7–23. (a) 10.5195 (b) 0.592 (c) 9.928 (d) 9.772 7–24. -12.82 7–25. -52.607 7–26. (a) $-1346/$ (b) $-534/$ (c) 285.57 cal/deg 7–27. -13.71 7–28. -15.4; -16.0 7–29. -6.71; -10.02; -12.92; -16.02; -17.09; -17.68; -22.99 7–33. 133; 87.5; 102; 56; 45; 135 7–34. (a) 79 (b) 141 (c) 194 7–35. (a) CH_4: 10,400°; 5200°; C_2H_2: 15,400°; 7000° (b) CH_4: 4800°; 2000°; C_2H_2: 7100°; 2500°K.

CHAPTER 8

8–1. (a) Reverse Carnot, make $W_{comp} = 0$ (b) Forward Carnot, make $Q_{2,comp} = 0$ 8–2. 0.252
8–3. 9200 cal/min 8–4. (a) $t = 373.15(1 - 273.15/T)$ (b) $t = T - 273.15$ 8–5. (a) 80% (b) 1500°K
8–6. (a) $-R \ln 2$ (b) $-R \ln 2$ (c) $R \ln 2$ (d) Clearly, $R \ln 2 \neq 0$.

CHAPTER 9

(Units for S and ΔS are eu or eu/mole in the following.)

9–1. (a) 3.3 (b) 5.5 (c) 9.9; 16.5 9–2. 3.2 9–3. 2.43 9–4. (a) 1.39 (b) 6.95 9–5. (a) 5.65 (b) 36.96 9–6. 19.5
9–7. (a) 0.244 (b) 0.748 (c) 1.95; 5.98 9–8. 3.5462 9–9. (a) $\Delta E = 0$; $\Delta S = 1.38$ eu; $Q = W = 410$
cal; $Q = T \Delta S$ (b) $\Delta E = 0$; $\Delta S = 1.38$ eu; $Q = W = 0$; $T \Delta S > Q$ 9–10. (a) $T_2 = 228°K$; $Q = 0$;
$\Delta S = 0$; $\Delta E = -W = -216$ cal (b) $T_2 = 240°K$; $Q = 0$; $\Delta S = 0.27$; $\Delta E = -W = -180$ cal
9–11. 8.059 9–12. $(\partial S/\partial p)_T = -V\alpha$ 9–13. 11.2 gm ice + 38.8 gm water at 0°C; $\Delta S = 0.12$ eu;
$\Delta H = 0$ 9–14. (a) 0.28 gm; 0.17 eu (b) 0.77 gm; 0.005 eu (c) 34 gm; 0.14 eu (d) 123 gm; 0.40 eu
9–15. (a) -0.088 (b) -0.086 9–16. (a) -0.086 (b) $-0.086(1 - 0.0004)$ 9–18. $-1.45°$ 9–19. 3.92 eu
9–20. (a) 15 (b) 15 (c) $\frac{3}{5}$ 9–21. (a) 10 (b) 1 (c) 2; $\Delta S = \mathbf{k} \ln 2$ 9–22. 4.35/ 9–23. (x, S_{mix}): $(0; 0)$;
$(0.2; 1.0)$; $(0.4; 1.34)$; $(0.5; 1.38)$; $(0.6; 1.34)$; $(0.8; 1.0)$; $(1.0; 0)$.

CHAPTER 10

10–1. $a/\bar{V}^2$ 10–2. Set cross-derivatives equal 10–4. 0.052; 1.16 liter·atm·mole^{-1} 10–5. (a) $A = f(T) - RT \ln V$ (b) $A = f(T) - RT \ln (V - b) - a/V$ 10–6. (a) $R/(V - b)$ (b) $\Delta S = R \ln [(V_2 - b)/(V_1 - b)]$ (c) greater for vdW gas 10–8. (a) $p = T(\partial p/\partial T)_V$ (b) $p = Tf(V)$ 10–9. $\ln f = \ln p + (b - a/RT)(p/RT)$ 10–14. $C_p \mu_{JT} = (2a/RT) - b$ 10–16. $\beta p \ll \alpha T$; -0.847 cal/atm 10–19. (a)
$\bar{S} = \bar{S}°(T) - R \ln p$; $\bar{V} = RT/p$; $\bar{H} = \mu°(T) + T\bar{S}°(T) = \bar{H}°(T)$; $\bar{E} = \bar{H} - p\bar{V} = \bar{H}°(T) - RT = \bar{E}°(T)$ (b) $\bar{S} = \bar{S}°(T) - R \ln p - (ap/RT^2)$; $\bar{V} = (RT/p) + b - (a/RT)$; $\bar{H} = \bar{H}°(T) + [b - (2a/RT)]p$ where $\bar{H}°(T) = \mu°(T) + T\bar{S}°(T)$; $\bar{E} = \bar{H}°(T) - RT - (ap/RT) = \bar{E}°(T) - (ap/RT)$.

CHAPTER 11

11–2. (a) 4.52 eu (b) -1360 cal 11–3. $(y; \Delta G)$; $(0; -1332)$; $(0.2; -1811)$; $(0.4; -1863)$; $(0.6; -1667)$; $(0.8; 1169)$; $(1.0; 0)$ 11–4. (a) -8200 cal (b) $-11,300$ cal (c) -3100 cal 11–5. $(p; \mu)$; $(\frac{1}{2}; -4387$ cal); $(2; -3565)$; $(10; -2611)$; $(100; -1246)$ 11–6. 2.35(10^{-29}) 11–7. 4.2(10^{-6}); 4.2(10^{-5}) 11–8. 5.3(10^{-3}); 2.4(10^{-3}) 11–9. (a) 9.17 kcal; 22.13 kcal (b) 26.3 (c) 0.982; 0.916 11–10. (a) $G = \mu°_{H_2(g)} + \mu°_{I_2(g)} + \xi\Delta G° + 2RT \ln p + 2RT[(1 - \xi) \ln \frac{1}{2}(1 - \xi) + \xi \ln \xi]$ (b) $G = \mu°_{H_2(g)} + \mu°_{I_2(s)} + \xi\Delta G° + (1 + \xi)RT \ln p + RT[(1 - \xi) \ln (1 - \xi) + 2\xi \ln 2\xi - (1 + \xi) \ln (1 + \xi)]$
11–11. $K_p = 6.3(10^5)$; 1.6(10^{-6}) 11–12. (a) 460°K (b) $\log_{10} K_p = -(1690/T) - 0.903 \log_{10} T + 6.09$; $\Delta H° = 7740 - 1.80T$; $\Delta S° = 26.04 - 4.14 \log_{10} T$ 11–13. Mg: 694°K; Ca: 1095°K; Sr: 1370°K; Ba: 1540°K 11–14. 762 cal/mole 11–15. (a) 12.4 kcal/mole (b) 558°K (c) 0.165 mm (d) 0.0439 mm (e) 5.77 kcal/mole 11–16. 600°K: O_2: 3.88 × 10^{-33}; CO: 0.126; CO_2: 99.87; 1000°K: O_2: 5.93(10^{-20}); CO: 69.8; CO_2: 30.2. The ratio, O_2/CO_2, is constant; relatively less CO at higher pressures (b) 600°K: O_2: 3.96(10^{-33}); CO: 0.136; CO_2: 99.86; 1000°K: O_2: 1.24(10^{-19}); CO: 41.2; CO_2: 58.8 (c) O_2: 1.63(10^{-19}); CO: 22.9; CO_2: 77.1 11–19. (a) 0.166 (b) 0.344 (c) 0.166 11–21. (a) 9.84(10^{-18}); 1.50(10^{-6}); 4.12(10^{-6}); 2.49(10^{-5}); 4.60(10^{-7}); 2.03(10^{-4}) (b) No (c) A = 9.84(10^{-6}); B = 1.50(10^{-4}); C = 4.12(10^{-4}); D = 2.49(10^{-3}); E = 4.60(10^{-5}); F = 2.03(10^{-2}); G = 99.98 (d) (mole %) A = 0.0872; B = 0.414; C = 0.623; D = 1.79; E = 0.110; F = 5.28; G = 91.66 11–22. (a) 3.69(10^{-6}); (b) 1.27(10^{-6}); (c) $(K_x)_{1\,atm} = 5(K_x)_{5\,atm}$ 11–23. (c) Entropy independent of z; $\bar{H}_i = \bar{H}_i^0(T) + M_i gz$ 11–24. $\Delta G°$ increases with temperature; slope of curve changes at transition points; (298°–548°): $\Delta G° = -88,255 + 37.38(T - 298)$;

$(548°–693°)$: $\Delta G^0 = -78{,}910 + 27.4(T - 548)$; $(693°–1029°)$: $\Delta G^0 = -74{,}940 + 29.7(T - 693)$; $(1029°–1180°)$: $\Delta G^0 = -64{,}960 - 0.35(T - 1029)$; (above $1180°$): $\Delta G^0 = -65{,}010 + 22.9(T - 1180)$ 11–25. (p atm) (a) $3.71(10^{-9})$; $9.29(10^{-5})$ (b) $2.96(10^{-13})$; $1.56(10^{-8})$ (c) $7.71(10^{-5})$; 0.526 11–26. (a) 47.9 kcal; 116.5 kcal (b) $p(\text{total}) = 1$ atm: (T, P, P_2, P_4); $900°K$: $2.56(10^{-13})$; $2.91(10^{-3})$; 0.997; $1200°K$: $5.75(10^{-9})$; 0.122; 0.878.

CHAPTER 12

12–1. $168°C$; experimental value, $94.5°C$. Large discrepancy arises from assuming $\Delta C_p = 0$ 12–2. (a) $\ln p_{\text{atm}} = 10.5(1 - T_b/T)$ (b) 0.715 atm 12–3. $1158°K$; 24.4 kcal/mole; 21.1 eu/mole 12–4. (a) 11.78 kcal/mole; $487.5°K$; 24.16 eu/mole (b) 7.45 mm (c) subl: 16.76 kcal/mole; fus: 4.98 kcal/mole (d) $T < 230°K$ 12–5. $386°K$; 89 mm 12–6. S_8: 28.0 eu; P_4: 21.48 eu 12–7. $d \ln c/dT = (\Delta H_{\text{vap}} - RT)/RT^2$ 12–8. (a) $c = 1/RT_b$ (b) $\ln(T_H/273) = (\Delta H_{\text{vap}}/R)(1/T_b - 1/T_H)$ (c) (T_H, T_b); $(50°; 59°)$; $(100°; 110°)$; $(200°; 206°)$; $(300°; 297°)$; $(400°; 387°)$ (d) $(T_H; \Delta S_H; \Delta S_T)$; Ar: $(77°K; 20.2$ eu; 17.8 eu); Kr: $(110°; 19.6; 18.0)$; Xe: $(157°; 19.3; 18.3)$; O_2: $(80°; 20.4; 18.1)$; CH_4: $(101.6°; 19.2; 17.5)$; CS_2: $(324°; 19.8; 20.0)$ 12–9. $14{,}600$ atm 12–10. (a) 3400 atm (b) $-24.5°C$ 12–11. 0.017 mm 12–13. $13.8°$.

CHAPTER 13

13–1. 0.0099; 0.0050; 0.00010 13–2. (d) $M/1000$ 13–3. (a) 59.8 (b) 332 gm 13–4. (x, T); $(1.0; 273°K)$; $(0.8; 252°)$; $(0.6; 229°)$; $(0.4; 203°)$; $(0.2; 171°)$ 13–5. 3.73 deg·kgm·mole^{-1} 13–6. 242 13–7. (a) $x_2 = 0.25$ (b) $x_2 = 0.54$ 13–8. 1.74; 2.63; 3.77; 0.187; 2.40 13–9. $3.80°$; 0.018; 4.63 atm; 252 13–10. 374 cm; 0.362 atm 13–11. 2.46 atm 13–12. Above 0.77 m in water 13–13. $a = K_f$; $b = \frac{1}{2}K_f[(K_f\Delta C_p/\Delta H_0) - (2K_f/T_0) - (M/1000)]$ 13–14. (a) $x = (1 - c\overline{V}_2^0)/(1 - c\overline{V}_2^0 + c\overline{V}^0)$.

CHAPTER 14

14–3. (a) 60.44 mm (b) 0.318 14–4. $x_1 = [(p_1^0 p_2^0)^{1/2} - p_2^0]/(p_1^0 - p_2^0)$; $p = (p_1^0 p_2^0)^{1/2}$ 14–5. (a) 67.53 mm (b) $y_{C_6H_6} = 0.7626$ (c) 48.90 mm (d) $x_{C_6H_6} = 0.2374$ (e) 57.46 mm; $x_{C_6H_6} = 0.3580$; $y_{C_6H_6} = 0.6420$ 14–7. (a) $t = $ toluene; $b = $ benzene; $\exp(-\Delta S/R) = x_b\exp(-T_{0b}\Delta S/RT) + x_t\exp(-T_{0t}\Delta S/RT)$ (b) $x_b = 0.445$ 14–8. (a) $p = p_1^0 + (K_h - p_1^0)x$ (b) $p = p_1^0 K_h/[p_1^0 + (K_h - p_1^0)y_1]$ 14–19. 1.71 cc; 17.1 cc; $N_2/O_2 = 2.02$ 14–10. -3.3 kcal/mole 14–11. 380 cc.

CHAPTER 15

15–1. $x_{\text{Pb}} = 0.855$; $520°K$ 15–3. $-13°C$ 15–4. 14.8% 15–5. 62.2% 15–6. In α, β, L; $F = 2$ In $(L + \alpha)$, $(L + \beta)$, $(\alpha + \beta)$; $F = 1$ On abc; $F = 0$ 15–7. $x_{\text{Ni}} = 0.9$; $x'_{\text{Cu}} = 0.08$; $x'_{\text{Ni}} = 0.92$; $T = 1670°K$ 15–8. In $ACBe$; $F = 2$. In Aab, $bcdz$, Ade; $F = 1$. In Abd; $F = 0$ 15–10. (a) soln; $F = 2$. Soln + salt, soln + hydrate, ice + hydrate, salt + hydrate; $F = 1$. At T_{eut} and T_{trans}; $F = 0$.

CHAPTER 16

16–1. (a) 0.9982; 0.9963; 0.9905; 0.9801; 0.9685; 0.9555 (b) 1.000; 1.000; 0.999; 0.998; 0.996; 0.992 (d) 1.25; 1.25 16–2. (a, γ); $(0.061; 1.03)$; $(0.135; 1.09)$; $(0.212; 1.14)$ 16–3. (a, γ); $(0.937; 0.995)$; $(0.869; 0.990)$; $(0.799; 0.981)$ 16–4. (a) μ_i^0 is μ of pure i (b) $RT \ln \gamma_i = w(1 - x_i)^2$ (c) 1.140; 1.077; 1.035; 1.009; 1.000 16–5. (a) 0.0108; 0.0148; 0.0221; 0.0296; 0.0388; 0.0538; 0.0662 (b) 0.0145; 0.0201; 0.0306; 0.0416; 0.0557; 0.0804; 0.1039 16–6. $(a_{\pm}, a)$; (a) 0.0769; 0.00592; (b) 0.0420; $7.42(10^{-5})$ (c) 0.016; $2.56(10^{-4})$ (d) 0.0752; $3.19(10^{-5})$ (e) 0.0089; $5.72(10^{-11})$ 16–7. (a) 0.0795; 0.05; 0.05 (b) 0.15; 0.05; 0.20 16–8. 31 Å; 3.1 Å 16–9. (a) 0.736 (b) $1.68/\kappa$ 16–10. $(10^{-4}$ m; 10^{-3} m): HCl: 0.989; 0.964; $CaCl_2$: 0.964; 0.882; $ZnSO_4$: 0.912; 0.747 16–11. $\alpha = 1.37(10^{-2})$; $\alpha(\text{rough}) =$

$1.32(10^{-2})$ 16–12. (mole/liter); $1.29(10^{-5})$; $1.39(10^{-5})$; $1.56(10^{-5})$; $1.84(10^{-5})$ 16–13. (a) $1.2(10^{-3})$
(b) $7.0(10^{-5})$ (c) $4.2(10^{-4})$.

CHAPTER 17

17–1. (kcal/mole); Na^+: -62.5; Pb^{2+}: -5.8; Ag^+: 18.5 17–2. -2.52 kcal/mole 17–3. (a)
$Zn^{2+}(0.1) + 2Ag = Zn + 2Ag^+(0.01)$; -1.473 volts (b) $AgCl + Fe^{2+}(1.0) = Ag + Cl^-(0.001) +$
$Fe^{3+}(0.1)$; -0.313 volt (c) $HgO + Zn + 2OH^-(1.0) = Hg + ZnO_2^{2-}(0.1) + H_2O$; 1.343 volt (c) is
spontaneous; (a) and (b) are not 17–4. (a) 0.378 (b) $Pb + PbO_2 + 4H^+ + 2SO_4^{2-} = 2PbSO_4 +$
$2H_2O$; Yes (c) 49.6 kcal (d) $\mathscr{V} = 2.041 + 0.05916 \log_{10} a$ 17–5. (a) $Fe + Ni_2O_3 = FeO + 2NiO$
(b) Independent of a_{KOH} 17–6. (a) 0.800 volt; 0.741 volt; 0.682 volt; 0.623 volt (b) 0.328 volt
(c) -0.144 volt 17–7. (a) <0 (b) <-0.414 volt (c) Basic soln 17–8. (a) >0.401 volt (b) >1.229 volt
(c) >0.815 volt (d) Acid soln 17–9. (a) $1.6(10^{37})$ (b) $8.0(10^{16})$ (c) $1.1(10^{-3})$ (d) $8.5(10^{40})$ (e) $4.2(10^{46})$
(f) $1.7(10^{-8})$ 17–10 0.0749 volt; 0.1558 volt; 0.1900 volt 17–11. (a) $(0°; 25°)$; ΔG: -88.440;
-88.534 kcal; ΔH: -87.668; -87.025 kcal; ΔS: 2.588; 5.074 eu (b) 0.130 17–12. 0.171 17–13.
0.779 17–14. 0.299; 0.340; 0.399; 0.458; 0.511; 0.563; 0.622; 0.681; 0.722 volt 17–15. (a) $H_2(1$
atm) $= H_2(0.5 \text{ atm})$; 0.00887 volt (b) $Zn^{2+}(0.1) = Zn^{2+}(0.01)$; 0.0296 volt 17–16. 0.826; 0.01106
volt.

CHAPTER 18

18–1. 0.522 cal 18–2. 108 dyne/cm 18–3. (a) $r = 3\gamma/[\Delta H_{vap}(1 - T/T_0)]$ (b) $4.9(10^{-8})$ cm 18–4. 14.5
mm 18–5. 1.49 mm 18–6. 27.1 A 18–7. $k = 2.53$; $1/n = 0.568$ 18–8 (a) $(p; \theta)$; $(2; 0.593)$; $(5; 0.784)$;
$(10; 0.878)$; $(20; 0.935)$; $(30; 0.955)$ (b) $4.64(10^7)$ cm^2 18–10. 0.504 cal; $\Delta G_{mix} = -129$ cal.

CHAPTER 19

19–1. (erg/cm^3); $7.57(10^{-7})$; $6.13(10^{-5})$; $7.57(10^{-3})$ 19–2. (a) $5.42(10^{-6})$ (b) $7.7(10^{-3})$ 19–3. (a)
$9.66(10^{-4})$ cm (b) $5.80(10^{-4})$ cm 19–4. 4830°K 19–5. $1.21(10^{15})$ 19–6. 150 V 19–7. 0.082 V
19–8. $6.63(10^{-29})$ cm 19–9. $1.50(10^{-2})$ eV.

CHAPTER 20

20–3. $x^2(d^2/dx^2) - (d^2/dx^2)(x^2) = -2 - 4x(d/dx)$ 20–5. $P_0(x) = \sqrt{\frac{1}{2}}$; $P_1(x) = \sqrt{\frac{3}{2}}x$; $P_2(x) = \sqrt{\frac{5}{2}} \times$
$(-\frac{1}{2} + \frac{3}{2}x^2)$ 20–7. $\mathbf{M}_z\mathbf{M}_x - \mathbf{M}_x\mathbf{M}_z = ih\mathbf{M}_y$; $\mathbf{M}_y\mathbf{M}_z - \mathbf{M}_z\mathbf{M}_y = ih\mathbf{M}_x$; $\mathbf{M}^2\mathbf{M}_x = \mathbf{M}_x\mathbf{M}^2$; $\mathbf{M}^2\mathbf{M}_y =$
$\mathbf{M}_y\mathbf{M}^2$.

CHAPTER 21

21–1. $\frac{1}{2}$ if n is even; $\frac{1}{2} + (-1)^k/(n\pi)$ if n is odd $= 2k + 1$ 21–2. $\Delta E = 6.0(10^{-25})(2n + 1)$ erg; golf
balls; $n = 1.63(10^7)$ 21–3. $5.49(10^{-5})$ cm 21–4. $\pm 2.4(10^{-27})$ cm 21–6. $\langle x^2 \rangle = \beta^2(n + \frac{1}{2})$; $\Delta x =$
$\beta(n + \frac{1}{2})^{1/2}$; $\beta^2 = (hv/k)$ 21–7. $\langle p_x^2 \rangle = (h^2/\beta^2)(n + \frac{1}{2})$; $\langle p_x \rangle = 0$; $\Delta p_x = (h/\beta)(n + \frac{1}{2})^{1/2}$; $\beta^2 =$
(hv/k) 21–8. $\Delta p_x \cdot \Delta x = (h/4\pi)(2n + 1)$ 21–10. (energy, degeneracy): (3, 1); (6, 3); (9, 3); (11, 3);
(12, 1); (14, 6); (17, 3); (19, 3) 21–11. (a) $4.65(10^{-41})$ gm-cm^2; $2.39(10^{-14})$ erg (b) $1.955(10^{-39})$
gm-cm^2; $5.688(10^{-16})$ erg 21–12. (a) $\frac{1}{2}bL$ (b) $\frac{1}{2}b$ (c) $(2ab/L)[1 - (-1)^n]$ (d) $kL^2/24$ 21–13. (a) 0
(b) $\frac{3}{4}\alpha\beta^4(2n^2 + 2n + 1)$; $\beta^2 = hv/k$.

CHAPTER 22

22–1. Lyman: 1215.7Å, 1025.7Å, 972.55Å; Balmer: 6564.7Å, 4863.6Å, 4341.7Å; Paschen: 18,756Å,
12,821Å, 10,941Å 22–2. (b) $2.66a_0$ 22–3. 1s: $2.66a_0$; 2s: $9.12a_0$ 2p: $8.0a_0$; 3s: $19.5a_0$ 22–5. $6a_0$;

$5a_0$; $13.5a_0$ 22–6. ($\langle V \rangle$, $\langle K \rangle$); 1s: $-e^2/a_0$; $+e^2/2a_0$; 2s: $-e^2/4a_0$, $+e^2/8a_0$; 2p: $-e^2/4a_0$, $+e^2/8a_0$; 3s: $-e^2/9a_0$, $+e^2/18a_0$ 22–7. (a) 7 (b) 3   22–8. $21.4(10^4)$ gauss.

CHAPTER 23

23–4. Tetrahedral   23–5. Outer orbital   23–7. σ_1, σ_4, and ($\sigma_2 + \sigma_3$) are symmetric, while ($\sigma_2 - \sigma_3$) is antisymmetric.

CHAPTER 24

24–1. $\Delta\lambda = 1/21.1(J + 1)(J + 2)$; $\Delta\tilde{v} = 21.1$ cm^{-1} 24–2. $2.651(10^{-40})$ gm-cm^2; $1.28(10^{-8})$ cm   24–3. $4.272(10^{-40})$ gm-cm^2; $3.928(10^{11})$ sec^{-1} 24–4. $5.592(10^{-14})$, $16.78(10^{-14})$, $27.96(10^{-14})$ erg   24–5. (a) 1.199Å from terminal N atom; $6.676(10^{-39})$ gm-cm^2 (b) On C atom; $7.172(10^{-39})$ gm-cm^2 (c) 2.954Å from H atom; $1.846(10^{-38})$ gm-cm^2 24–6. $3.02(10^{10})$ sec^{-1}; 1.44 cal/mole   24–7. (a) $1.6288(10^{-24})$ gm; 2889 cm^{-1}; $3.1619(10^{-24})$ gm, 2073 cm^{-1}; $3.1712(10^{-24})$ gm, 2070 cm^{-1} (b) $2.6422(10^{-40})$ gm-cm^2, 21.186 cm^{-1}; $2.6462(10^{-40})$ gm-cm^2, 21.154 cm^{-1}; $5.1367(10^{-40})$ gm-cm^2, 10.898 cm^{-1}; $5.1518(10^{-40})$ gm-cm^2, 10.866 cm^{-1}.

CHAPTER 25

25–1. $\Delta E = a[(1/b) - (p/RT)] \approx a/b$ 25–2. $1.64(10^{-18})$ 25–3. 20.63 cc; 15.56 cc; 16.04 cc   25–4. $1.49(10^{-24})$ cc   25–5. 10^6 25–6. $A = n\varepsilon_m r_0^6/(6 - n)$; $B = 6\varepsilon_m r_0^n/(6 - n)$; $\varepsilon/\varepsilon_m = nr_0^6/(n - 6)r^6 - 6r_0^n/(n - 6)r^n$; $(r_0/\sigma)^{n-6} = n/6$.   25–7. $\varepsilon/\varepsilon_m = 2(r_0/r)^6 - (r_0/r)^{12}$; $\varepsilon/4\varepsilon_m = (\sigma/r)^6 - (\sigma/r)^{12}$ 25–8. (kcal/mole) (a) $-3.67(10^{-3})$ (b) $-46.3(10^{-3})$ (c) $-93.7(10^{-3})$ (d) $-216(10^{-3})$ 25–9. 0.26 kcal/mole 25–10. O^{2-}: 1.41 A; F^-: 1.00; Ne: 0.74; Na^+: 0.58; Mg^{2+}: 0.48; Al^{3+}: 0.40; Si^{4+}: 0.34.

CHAPTER 26

26–1. Co: 1.622; Mg: 1.623; Ti: 1.586; Zn: 1.861   26–2. fcc, 4; bcc, 2   26–3. 6 atoms; 2 atoms/unit cell   26–4. fcc: 26.0% empty; bcc: 32% empty   26–5. 1 Cs^+, 1 Cl^- 26–6. 4 Na^+, 4 Cl^- 26–7. fcc: 1 hole/atom; bcc: 1.5 hole/atom   26–8. $r_h/r_a = 2^{1/2} - 1 = 0.414$ 26–9. (a) 8 (b) 4 pairs (c) 2 pairs (d) 4 AB_2 units (e) 2 AB_2 units   26–10. 1 fourfold axis; 4 twofold axes; 5 mirror planes; center of symmetry   26–11. (a) (001); (100); (010)   26–12. (111) is close packed   26–13. 3.162 A   26–14. 1.544 A   26–15. $\theta_{111} = 21.6°$; $\theta_{200} = 25.2°$.

CHAPTER 27

27–1. (a) 1 (b) 1.167 (c) 1.233 (d) 1.293   27–2. 209.5 kcal; 183.2 kcal   27–3. $E(NaCl) = 178$ kcal; $E(CsCl) = 176$ kcal   27–4. (a) CsF; RbF; KF; NaF; LiF (b) KI; KBr; KCl; KF   27–5. $\sim 1:4$ 27–6. $4.1(10^{-12})$ cm^2/dyne $= 4.1(10^{-6})$ atm^{-1}.

CHAPTER 28

28–1. $S/R = -1.1653 + \frac{3}{2} \ln M + \frac{5}{2} \ln T$; for argon: $S = 36.979$ eu   28–3. (a) $C_v = Nk(\theta/T)^2 \times \exp(-\theta/T)/[1 - \exp(-\theta/T)]^2$ (b) $C_v^\infty = Nk$ (c) (θ/T; C_v/C_v^∞); (0; 1); (0.5; 0.976); (1.0; 0.920); (1.5; 0.830); (2.0; 0.720); (3.0; 0.494); (4.0; 0.303); (5.0; 0.172); (6.0; 0.0893)   28–4. (a) 935°K (b) $C_v = 0.468R$ (c) Contributes $0.936R$ (d) $C_v = 1.96(10^{-3})R$ 28–5. (a) $p = (k\theta^3/T^{1/2})(2\pi mk/h^2)^{3/2} \times \exp(W/RT)$ (b) $\Delta H_{vap} = -(W + \frac{1}{2}RT)$ 28–6. (a) Lowers value of μ (b) Diatomic.

CHAPTER 29

29–1. (a) $6.88(10^9)$ (b) $5.33(10^9)$ (c) $6.88(10^5)$ (d) $8.32(10^{28})$; $3.86(10^{28})$; $8.32(10^{20})$ 29–2. (a) $1.26(10^{-5})$ cm; $1.26(10^{-4})$ cm; $1.26(10^{-3})$ cm (b) $1.26(10^{+4})$ cm (c) $2.52(10^3)$ 29–3. $\lambda = 1/2\sigma n$; $Z_1 = 2\sigma\bar{c}n$; $Z_{11} = \sigma\bar{c}nA$; ($n =$ no./ft^2; $A =$ area). $\lambda = 31.3$ ft; $Z_1 = 3.84$; $Z_{11} = 38.4$ 29–4. $4.18(10^{-8})$ 29–5. $\kappa_{H_2}/\kappa_{O_2} = 4$ 29–6. $3.09(10^{-4})$ cal·cm^{-2}·sec^{-1} 29–7. C_2H_6 has more internal degrees of freedom, hence higher C_v. 29–8. 24.6 cc/sec 29–9. 1.86 mm 29–10. $V = (\pi/8\eta l)(p_1 - p_2)(b^2 - a^2) \times$ $[a^2 + b^2 - (b^2 - a^2)/\ln(b/a)]$; Note: If $b = a(1 + \Delta)$, where $\Delta \ll 1$, then $V = (\pi/2\eta l)(p_1 - p_2)a^4\Delta$ 29–11. 200 cal/sec 29–12. $3.276(10^{-3})$ poise 29–13. 1.6 kcal.

CHAPTER 30

30–1. $6.25(10^{18})$ 30–2. (a) 4 ohm; 0.25 ohm^{-1}; (b) 0.5 volt/cm (c) $1.275(10^4)$ amp/cm^2 (d) $3.92(10^{-5})$ ohm·cm; $2.55(10^4)$ ohm^{-1}·cm^{-1} 30–3. 188 cc 30–4. 11 hr 30–5. (a) $5.9390(10^{-3})$ cm^{-1} (b) $5.7518(10^{-3})$ ohm^{-1}·cm^{-1} 30–6. 382.1 ohm^{-1}·cm^2 30–7. $l_{H+} = 3.62(10^{-3})$; $l_{Na+} = 5.20(10^{-4})$ cm^2·volt^{-1}·sec^{-1}; $v_{H+} = 1.45(10^{-3})$; $v_{Na+} = 2.08(10^{-4})$ cm/sec 30–8. HCl: 0.179; NaCl: 0.603; KCl: 0.509; NH$_4$Cl: 0.509; CaCl$_2$: 0.562 30–9. $t_{Na+} = 0.29$; $t_{H+} = 0.20$; $t_{SO_4} = 0.51$ 30–10. $c_{HCl}/c_{NaCl} = 0.463$ 30–11. (a) 0.127; 0.0911; 0.0655 (b) $1.820(10^{-5})$ 30–12. (a) $\Delta n_c = -\Delta n_a = t_+(Q/\mathscr{F})$ (b) $-\Delta n_a = t_+(Q/\mathscr{F})$; $\Delta n_c = (t_+ - 1)(Q/\mathscr{F})$ (c) $\Delta n_a = (1 - t_+)(Q/\mathscr{F})$ 30–13. 0.82 30–14. 0.328 30–15. $1.08(10^{-10})$ 30–17. $\lambda_+ = 96.9$; $\lambda_- = 50.8$ 30–18. 410.4; 140.42; 106.36.

CHAPTER 31

31–1. (a) $c^{1/2} = c_0^{1/2} - \frac{1}{2}kt$ (b) $c^{-1/2} = c_0^{-1/2} + \frac{1}{2}kt$; $c^{1-n} = c_0^{1-n} + (n - 1)kt$ (c) $t_{1/2} = (2^{n-1} - 1)/(n - 1)kc_0^{n-1}$ 31–2. (a) $2.08(10^{-3})$ sec^{-1} (b) 139 sec 31–3. (a) $2.31(10^{-2})$ min^{-1} (b) 0.20 31–4. 72.3 hr 31–5. (a) $2.94(10^4)$ sec (b) $1.47(10^5)$ sec 31–6. $\ln[(1 - x/r_1)/(1 - x/r_2)] = 2k_1at/K^{1/2}$, where $x = c_{H_2} = c_{I_2}$; $r_1 = \frac{1}{2}a(1 + \frac{1}{2}K^{-1/2})/(1 - 1/4K)$; $r_2 = \frac{1}{2}a(1 - \frac{1}{2}K^{-1/2})/(1 - 1/4K)$; $K = k_1/k_{-1}$ 31–8. $\ln[1 - (1 + 1/K)(x/a)] = -k_1(1 + 1/K)t$ 31–9. 12.9 kcal 31–10. (a) $E^* = 35.7$ kcal; $A = 9.0(10^9)$ (b) $3.17(10^{-2})$ 31–11. $1.11(10^{-6})$ mole/liter 31–12. (a) $d(HBr)/dt = k_1(Br_2)$ (b) $d(HBr)/dt = k_1(Br_2)[k_2(H_2) - k_3(HBr)]/[k_2(H_2) + k_3(HBr)]$ 31–13. $d(CH_4)/dt = k_2(k_1/k_4)^{1/2}(CH_3CHO)^{3/2}$ 31–14. 48 kcal 31–15. $2.24(10^{-5})$; $4.48(10^{-5})$.

CHAPTER 32

32–1. $2.1(10^{-2})$ liter·mole^{-1}·sec^{-1} 32–2. 33.3 kcal 32–3. $(p; Z_3/Z_2)$; [0.1; $0.86(10^{-4})$]; [1.0; $0.86(10^{-3})$]; [10; $0.86(10^{-2})$]; [100; 0.086]; values are $\frac{1}{2}$ at 600°K 32–4. 0.296; 0.0371 32–5. ABC linear: $4(10^9)$ liter·mole^{-1}·sec^{-1}; ABC nonlinear: $4(10^{10})$ liter·mole^{-1}·sec^{-1} 32–6. (a) Decreases (b) Increases (c) No effect.

CHAPTER 33

33–1. (a) $k_{500}/k_{400} = 6(10^4)$; (b) $k_{500}/k_{400} = 33$ 33–2. (a) HI weakly absorbed (b) HI strongly absorbed (c) SO$_3$ strongly absorbed (d) CO$_2$ weakly adsorbed at low p, strongly at high p 33–3. 10^{-6} moles·cm^{-2}·sec^{-1} 33–4. Pt: $\pm 2.6(10^{-6})$ volts; Hg: $\pm 2.6(10^6)$ volts. 33–5. $6.2(10^{16})$ H$^+$ ions/sec; 62 times 33–6. $3.0(10^{19})$ quanta 33–7. 46.5 33–8. (a) $\varphi = k_2(A)/[k_2(A) + k_3]$ (b) Fluorescence is weak 33–9. $d(CO)/dt = k_2(I_{abs}/k_4)^{1/2}(CH_3CHO)$; $\varphi = k_2(CH_3CHO)/(k_4 I_{abs})^{1/2}$.

Index

Index